Advanced Fluid Mechanics and Heat Transfer for Engineers and Scientists

Meinhard T. Schobeiri

Advanced Fluid Mechanics and Heat Transfer for Engineers and Scientists

Springer

Meinhard T. Schobeiri
Department of Mechanical Engineering
Texas A&M University
College Station, TX, USA

ISBN 978-3-030-72924-0 ISBN 978-3-030-72925-7 (eBook)
https://doi.org/10.1007/978-3-030-72925-7

This Springer imprint is published by the registered company Springer Nature Switzerland AG
The registered company address is: Gewerbestrasse 11, 6330 Cham, Switzerland

Preface

The current book *Advanced Fluid Mechanics and Heat Transfer* is based on the author's four decades of industrial and academic research in the area of thermo-fluid sciences including fluid mechanics, aerothermodynamics, heat transfer, and their applications to engineering systems. It unifies the fluid mechanics and heat transfer in a unique manner. The fluid mechanics part has evolved from two of the author's graduate fluid mechanics textbooks published in 2010 and 2014. The essential elements of these textbooks are combined together and have undergone a rigorous update and enhancement resulting in the fluid mechanics part of the current book. As the content of the fluid mechanics part reveals, it includes features that are non-existent in any fluid mechanics textbook available on the market. Research facility design for creating steady and unsteady flow environment, performing steady and unsteady flow experiments using pneumatic and hot wire anemometry, explaining in detail the data acquisition and analysis and presenting the results of experiments are a few examples. The heat transfer part includes the most advanced features used for heat transfer measurement including film cooling. Here again, research facility design, performing steady and unsteady heat transfer and film cooling effectiveness experiments, detailed data acquisition, and analysis using and presenting the experimental results are a few examples. To measure the surface temperature of a test object, advanced instrumentation such as infrared thermography, liquid crystals, and pressure-/temperature-sensitive paints are used. Very detailed explanation of the above instruments along with their corresponding data acquisition and analysis is presented. The contents of this new book cover the material required in fluid mechanics and heat transfer graduate core courses in the US universities. It also covers the major parts of the PhD-level elective courses advanced fluid mechanics and heat transfer that I have been teaching at Texas A&M University for the past three decades. The unified treatment of fluid mechanics and heat transfer enables students and novice researchers to find comprehensive answers to their specific questions without consulting several books.

Fluid mechanics and heat transfer are inextricably intertwined and both are integral parts of one physical discipline. No problem from fluid mechanics that requires the calculation of the temperature can be solved using the system of Navier–Stokes and continuity equations only. Conversely, no heat transfer problem can be solved using the energy equation only without using the Navier–Stokes and continuity equations. The fact that there is no book treating this physical discipline as a unified subject in a single book and the need of the engineering and physics community, motivated me to write this book. It is primarily aimed at students of engineering, physics, and those practicing professionals who perform aero-thermo-heat transfer design tasks in the industry and would like to deepen their knowledge in this area.

The advanced character of the book requires that the reader has the basic knowledge of fluid mechanics and heat transfer. By reading through the chapters, the reader will realize how his/her level of understanding transforms from undergrad to graduate level. An example should clarify the concept of this book: The first three chapters prepare the reader for better understanding of the advanced level of fluid mechanics. Tensors in three-dimensional Euclidean space presented in Chap. 2 provide the reader with the mathematical basis that is essential for understanding the material to come. All equations are written first in coordinate invariant form that can be decomposed in any arbitrary coordinate system. Using the tensor analytical knowledge gained from Chap. 2, it is rigorously applied to the following chapters. In Chap. 3, that deals with the kinematics of flow motion, the Jacobian transformation describes in detail how a time-dependent volume integral is used to systematically derive the Reynolds transport theorem. This theorem is the essential tool to understand how it can be applied to continuity, linear momentum, angular momentum, and energy equation in integral form.

In Chaps. 4 and 5, conservation laws of fluid mechanics and thermodynamics are treated in differential and integral forms. Inviscid flow is presented in Chap. 6, where the order of magnitude of a viscosity force compared with the convective forces could be neglected. The potential flow, a special case of inviscid flow characterized by zero vorticity, exhibited a major topic in fluid mechanics in pre-CFD era. In recent years, however, its relevance has been diminished. Despite this fact, I presented it in this book for two reasons. First, despite its major shortcomings to describe the flow pattern directly close to the surface, because it does not satisfy the no-slip condition, it reflects a reasonably good picture of the flow outside the boundary layer. Second, combined with the boundary layer calculation procedure, it helps acquiring a reasonably accurate picture of the flow field outside and inside the boundary layer. This, of course, is valid as long as the boundary layer is not separated. It is also worth noting that today the combined method mentioned above is still used by many engine designers at the early stage of design and development before using CFD packages for final design. Also, in the context of the potential flow, two topics, namely the conformal transformation and the vorticity theorems, are presented and discussed.

Following the inviscid flow discussion, Chap. 7 deals with the particular issues of viscous laminar flow through curved channels at negative, zero, and positive pressure gradients. Exact solution of Navier–Stokes equation delivers the velocity distribution in those channels. The combination of the Navier–Stokes equations with energy equation delivers the temperature distribution within these channels. These examples demonstrate the impact of the wall curvature and pressure gradient on fluid mechanics and heat transfer behavior of these curved channels. Thus, the content of Chap. 7 seamlessly merges into Chap. 8 that starts with the stability of laminar flow, followed by laminar–turbulent transition, its intermittent behavior, and its implementation into Navier–Stokes equations. Averaging the Navier–Stokes equation that includes the intermittency function leading to the Reynolds-averaged Navier–Stokes equation (RANS) concludes Chap. 8.

In discussing the RANS-equations, two quantities have to be accurately modeled. One is the intermittency function, and the other is the Reynolds stress tensor with its nine components. Inaccurate modeling of these two quantities leads to a multiplicative error of their product. The transition was already discussed in Chap. 8 but the Reynolds stress tensor remains to be modeled. This, however, requires the knowledge and understanding of turbulence before attempts are made to model it and this topic is treated in Chap. 9.

In Chap. 9, I tried to present the quintessence of the subject turbulence that is required for a graduate-level engineering and physics courses and to critically discuss several different turbulence models. The two-equation models, $k - \mathbf{g}$, $k - \mathbf{T}$ and the combination of both, that are widely used, render satisfactory qualitative results when they are applied to relatively simple flow cases. However, simulating a complex flow situation, where different parameters are involved, leads to results, whose proper interpretations require the understanding of the model shortcomings. Efforts have been made to illustrate these shortcomings by simulating more complex cases and compare the numerical results with the experiment.

While Chap. 9 predominantly deals with the wall turbulence, Chap. 10 treats different aspects of free turbulent flows and their general relevance in engineering. Among different free turbulent flows, the process of development and decay of wakes under positive, zero, and negative pressure gradients is of particular engineering relevance. With the aid of the characteristics developed in Chap. 10, this process of wake development and decay can be described accurately. Following Chap. 10, two chapters deal with the: (a) *boundary layer aerodynamics* and (b) *boundary layer heat transfer*. These two chapters are of fundamental importance for fluid mechanics and heat transfer: On the one hand, the velocity distribution and its gradient within the boundary are responsible for the wall friction distribution and thus the total pressure loss within an engineering system. On the other hand, the velocity fluctuation distribution within the boundary layer determines the transfer of mass, momentum, and energy to the boundary layer and thus specifies the temperature distribution on the wall surface.

Chapter 11 is entirely dedicated to boundary layer aerodynamics. Without the knowledge of boundary layer physics, no aircraft wings, no turbine and compressor blades, no gas turbines, and jet engines could be designed in the form we know and use them today. The boundary layer is a thin viscous layer very close to the surface. L. Prandtl was the first to recognize the crucial role that this viscous layer plays in aerodynamics. Based on his experimental observations, Prandtl found that the effect of viscosity is confined to a thin viscous layer that he called, *the boundary layer*. The distribution of the velocity and its fluctuation within the boundary determines the wall friction, the heat transfer coefficient, and the film cooling effectiveness of the surface exposed to a fluid flow. Based on his thorough experiments and sound dimensional analysis, he developed his *boundary layer theory* that up to today is used in the areas mentioned above.

Conducting boundary layer experiments on today's test objects are totally different from those by Prantl. The plate length that Prantl used for his boundary layer measurement was more than 10 m long generating a boundary layer thickness of 50 cm and more. He used pneumatic probes with fish mouth geometry with a lateral dimension of approximately 2 mm for traversing his 50 cm boundary layer. The today's turbine or compressor blades generate maximum boundary layer thicknesses of 3–5 mm or less. To be able to traverse such boundary layer thicknesses, hot wire sensors are used with the sensing wire diameters of 0.25–5 μm. These sensors are attached to an anemometer with data acquisition frequency of 100 kHz and above detailed in Chap. 11.

Measurement of boundary layer flow occupies a full section of this chapter. Two subsections detail design facilities for boundary layer research in stationary and rotating frame of references. For generation of periodic unsteady wake flows that emulate the interaction between the stationary and rotating frame, two facilities are introduced that generate periodic unsteady wake flows. Instrumentation, data acquisition of steady and periodic unsteady data, sampling, and ensemble averaging exhibit another subsection in Chap. 11. More detailed data acquisition and analysis are given in Chap. 13. Several case studies presented that show the steady and unsteady boundary layer experiments. For periodic unsteady flow, distribution of the mean and the fluctuation velocities as functions or time demonstrate the important role of the fluctuation velocity in transferring mass, momentum, and energy to the boundary layer.

Chapter 12 is fully dedicated to boundary layer heat transfer. After a brief introduction, equations for heat transfer calculation are presented. The introduction is followed by the section that describes the instrumentation for temperature measurement. To measure the surface temperature of a test object, four different instrumentations are introduced, namely thermocouples, infrared thermography, liquid crystals, and pressure-/temperature-sensitive paints. This section is followed by the calculation procedure of individual contributors to obtain the heat transfer coefficient. This section is followed by presenting experimental examples.

In selecting the material for experimental examples, I have been influenced by the needs of mechanical engineers. The first example starts with measuring the heat transfer coefficient distribution along the concave and convex sides of a curved plate. Both sides of the curved plate are covered by spacial liquid crystal sheets. To show the effect of impinging the periodic unsteady wakes on the concave and convex surfaces, the reduced frequency of the wake generator was varied from 0.0 (steady state) to 5.166. While the above example dealt with the zero-longitudinal pressure gradient, the following case study includes the variation of pressure gradient, re-number, turbulence intensity, and reduced frequency.

A major section treats the heat transfer problems using pressure-/temperature-sensitive paints (SP/TSP). After explaining the working principle of this method, its calibration, data acquisition, and analysis, it is applied to a rotating frame for measuring film cooling effectiveness. Two representative cases are presented: Case

(1) measures the film cooling effectiveness on tip of turbine blades with different geometries. Case (2) presents the film cooling effectiveness of a turbine blade that rotates at up to 3000 rpm. The major portion of the material presented in this chapter is quite new and is in accord with the structure of this book. The discourse differs substantially from the conventional heat transfer book and maybe even new to experienced reader. In addition to explaining the facility design, instrumentation, data acquisition, and analysis presented in Chaps. 11 and 12, I found necessary to add two chapters that deal with the above subjects in a more detailed fashion. Chapter 14 deals with fluid mechanics measurements and Chap. 15 with all aspects of heat transfer measurement at advanced level.

Chapter 13 deals with the compressible flow. At first glance, this topic seems to be in dissonance with the rest of the book. Despite this, I decided to integrate it into this book for two reasons: (1) Due to a complete change of the flow pattern from subsonic to supersonic, associated with a system of oblique shocks, makes it imperative to present this topic in an advanced engineering fluid text; (2) Unsteady compressible flow with moving shockwaves occurs frequently in many engines such as transonic turbines and compressors, operating in off-design, and even design conditions. A simple example is the shock tube, where the shock front hits one end of the tube to be reflected to the other end. A set of steady-state conservation laws does not describe this unsteady phenomenon. An entire set of unsteady differential equations must be called upon which is presented in Chap. 13. Arriving at this point, the students need to know the basics of gas dynamics. I had two options, either refer the reader to existing gas dynamics textbooks, or present a concise account of what is most essential in following this chapter. I decided for the second option.

At the end of each chapter, there is a section that entails problems and projects. In selecting the problems, I carefully selected those from the book *Fluid Mechanics Problems and Solutions* by Prof. Spurk of Technische Universität Darmstadt which I translated in 1997. This book contains a number of highly advanced problems followed by very detailed solutions. I strongly recommend this book to those instructors who are in charge of teaching graduate fluid mechanics as a source of advanced problems. My sincere thanks go to Prof. Spurk, my former Co-advisor, for giving me the permission. Besides the problems, a number of demanding projects are presented that are aimed at getting the readers involved in solving CFD-type of problems. In typing several thousand equations, errors may occur. I tried hard to eliminate typing, spelling, and other errors, but I have no doubt that some remain to be found by readers. In this case, I sincerely appreciate the reader notifying me of any mistakes found; the electronic address is given below. I also welcome any comments or suggestions regarding the improvement of future editions of the book.

My sincere thanks are due to many fine individuals and institutions. First and foremost, I would like to thank the faculty of the Technische Universität Darmstadt from whom I received my entire engineering education. In particular, I owe my thanks and appreciation to my deceased thesis adviser Prof. Dr.-Ing. H. Pfeil and my co-adviser Prof. Dr.-Ing. J. Spruk for their continuous support. I finalized major chapters of the manuscript of the first book I mentioned previously during my sabbatical in Germany where I received the Alexander von Humboldt Prize. I am indebted to the

Alexander von Humboldt Foundation for this Prize and the material support for my research sabbatical in Germany. My thanks are extended to Prof. Bernd Stoffel, Prof. Ditmar Hennecke, and Dipl. Ing. Bernd Matyschok for providing me with a very congenial working environment.

I am also indebted to TAMU administration for partially supporting my sabbatical which helped me in finalizing the book. Special thanks are due to Mr. Kelly Minnis who converted the WordPerfect manuscript into the typesetting version with latex.

Last, but not least, my special thanks go to my family, Susan, and Wilfried for their support throughout this endeavor.

College Station, TX, USA
October 2020

Meinhard T. Schobeiri
tschobeiri@tamu.edu

Contents

Nomenclature

Symbols

$\boldsymbol{A}$	Acceleration vector
b	Wake width
c	Complex eigenfunction, $c = cr + ici$
c	Speed of sound
c_p, c_v	Specific heat capacities
c_w	Specific heat capacities of a hot wire
C	Von Kármán constant
C_D	Drag coefficient
C_f	Friction coefficient
C_p	Pressure coefficient
D	Deformation tensor
D	Total differential operator in absolute frame of reference
D	Van Driest's damping function
D_R	Total differential operator in relative frame of reference
e	Specific total energy
e_i	Orthonormal unit vector
E	Source (+), sink (-) strength
E	Total energy
$E(k)$	Energy spectrum
f_S	Sampling frequency
$\boldsymbol{F}$	Force
$F(z)$	Complex function
g_i, g^i	Co-, contravariant base vectors in orthogonal coordinate system
g_{ij}, g^{ij}	Co-, contravariant metric coefficients
$\boldsymbol{G}_i$	Transformation vector
h, H	Specific static, total enthalpy
H_{12}	Boundary layer momentum form factor, $H_{12} = \delta_1/\delta_2$
H_{13}	Boundary layer energy form factor, $H_{32} = \delta_3/\delta_2$
$\dot{q}$	Heat flux

$I(\boldsymbol{x}, t)$	Intermittency function
I_1, I_2, I_3	Principle invariants of deformation tensor
J	Jacobian transformation function
k	Thermal conductivity
$\boldsymbol{k}$	Wave number vector
K	Specific kinetic energy
l_m	Prandtl mixing length
$\boldsymbol{L}_{ij}(\boldsymbol{x}, \boldsymbol{t})$	Turbulence length scale
m	Mass
$\dot{m}$	Mass flow
M	Mach number
$\boldsymbol{M}$	Vector of moment of momentum
$\boldsymbol{M}_a$	Axial vector of moment of momentum
$\boldsymbol{n}$	Normal unit vector
N	Navier–Stokes operator
Nu	Nusselt number
p	Static pressure
$\tilde{\boldsymbol{p}}$	Deterministic pressure fluctuation
p^+	Dimensionless pressure gradient
$\boldsymbol{p}'$	Random pressure fluctuation
P, p_0	Total (stagnation) pressure, $P = p + DV^2/2$
Pr	Prandtl number
Pr_e	Effective Prandtl number
Pr_t	Turbulent Prandtl number
q	Specific thermal energy
Q	Thermal energy
$\dot{\boldsymbol{q}}$	Heat flux vector
R	Radius in conformal transformation
Re	Reynolds number
Re_{crit}	Critical Re
$\boldsymbol{R}(\boldsymbol{x}, \boldsymbol{t}, \boldsymbol{r}, \boldsymbol{\tau})$	Correlation second-order tensor
s	Specific entropy
St	Stanton number
Str	Strouhal number
$S, S(t)$	Fixed, time-dependent surface
t	Time
t	Tangential unit vector
$\boldsymbol{T}_{ij}(\boldsymbol{x}, \boldsymbol{t})$	Turbulence timescale
T	Static temperature
T	Stress tensor, $\boldsymbol{T} = e_i e_j \tau - ij$
T_0	Stagnation or total temperature
Tr	Trace of second-order tensor
$T_n(y)$	Chebyshev polynomial of the first kind
u	Specific internal energy
u	Velocity

u_τ	Wall friction velocity
u^+	Dimensionless wall velocity, $u^+ = u/u_\tau$
U	Undisturbed potential velocity
$\underline{U}$	Rotational velocity vector
$\overline{\boldsymbol{U}_1}$	Time-averaged wake velocity defect
$\overline{\boldsymbol{U}_\mathrm{I}}$	Time-averaged wake momentum defect
$\overline{\boldsymbol{U}_{1m}}$	Maximum velocity defect
v	Specific volume
V	Volume
V_0	Fixed volume
$V(t)$	Time-dependent volume
$\boldsymbol{V}$	Absolute velocity vector
$\boldsymbol{V}_L$	Velocity vector, laminar solution
$\boldsymbol{V}_T$	Velocity vector, turbulent solution
$\tilde{\boldsymbol{V}}$	Deterministic velocity fluctuation vector
$\bar{\boldsymbol{V}}$	Mean velocity vector
$\boldsymbol{V}'$	Random velocity fluctuating vector
V_t, V^j	Co- and contravariant component of a velocity vector
$\langle \boldsymbol{V} \rangle$	Ensemble-averaged velocity vector
w_m	Specific shaft power
W	Mechanical energy
$\dot{\boldsymbol{W}}$	Mechanical energy flow (power)
$\dot{\boldsymbol{W}}_{sh}$	Shaft power
$\boldsymbol{W}$	Relative velocity vectors
x_i	Coordinates
y^+	Dimensionless wall distance, $y^+ = u_\tau y/\nu$
z	Complex variable

Greek Symbols, Operators

α	Heat transfer coefficient
α	Real quantity in disturbance stream function
β_i	Disturbance amplification factor
β_r	Circular disturbance frequency
$\gamma(\boldsymbol{x})$	Time-averaged intermittency factor, $\gamma(\boldsymbol{x}) = \bar{I}$
$\langle \gamma(\boldsymbol{t}) \rangle$	Ensemble-averaged intermittency at a fixed position
$\langle \gamma(\boldsymbol{t}) \rangle_{\max}$	Ensemble-averaged maximum intermittency at a fixed position
$\langle \gamma(\boldsymbol{t}) \rangle_{\min}$	Ensemble-averaged minimum intermittency at a fixed position
Γ	Circulation strength
Γ	Relative intermittency
Γ	Circulation vector
Γ^i_{jk}	Christoffel symbol
$\gamma_{\min}$, $\gamma_{\max}$	Minimum, maximum intermittency

δ	Kronecker delta
$\delta_1, \delta_2, \delta_3$	Boundary layer displacement, momentum, energy thickness
ε	Turbulence dissipation
ε_h	Eddy diffusivity
ε_m	Eddy viscosity
ε_{ijk}	Permutation symbol
ζ	Dimensionless periodic parameter
ζ	Kolmogorov's length scale
ζ	Total pressure loss coefficient
Θ	Shock expansion angle
$\Theta_{\mathbf{ij}}(\boldsymbol{k}_1, \boldsymbol{t})$	One-dimensional spectral function
κ	Isentropic exponent
κ	Ratio of specific heats
κ	Von Kármán constant
λ	Disturbance wavelength
λ	Eigenvalue
λ	Taylor microlength scale
λ	Tangent unit vector
μ	Absolute viscosity
μ	Mach angle
ν	Expansion angle
ν	Kinematic viscosity
ξ	Dimensionless coordinate, $\xi = x/L$
ξ	Position vector in material coordinate system
ν	Dimensionless coordinate, $\nu = y/L$
ν	Kolmogorov's length scale
π	Pressure ratio
Π	Stress tensor, $\Pi = e_i e_j \pi$
ρ	Density
$\boldsymbol{\rho}_{ij}(\boldsymbol{x}, \boldsymbol{t}, \boldsymbol{r}, \boldsymbol{\tau})$	Dimensionless correlation coefficient
τ	Kolmogorov's timescale
τ_0, τ_W	Wall sear stress
υ	Kolmogorov's velocity scale
$\boldsymbol{\varphi}_1$	Dimensionless wake velocity defect
Φ	Dissipation function
Φ, ψ	Potential, stream function
$\Phi(\boldsymbol{k}, \boldsymbol{t})$	Spectral tensor
Ψ	Mass flow function
ξ	Complex function
ω	Angular velocity
$\boldsymbol{\omega}$	Vorticity vector
Ω	Rotation tensor

Subscripts, Superscripts

∞	Freestream
a, t	Axial, tangential
ex	Exit
in	Inlet
max	Maximum
min	Minimum
s	Isentropic
t	Turbulent
w	Wall
-	Time averaged
’	Random fluctuation
~	Deterministic fluctuation
*	Dimensionless
+	Wall functions

Chapter 1
Introduction

The structure of thermo-fluid sciences rests on three pillars, namely fluid mechanics, thermodynamics, and heat transfer. While fluid mechanics' principles are involved in open system thermodynamics processes, they play a primary role in every convective heat transfer problem. Fluid mechanics deals with the motion of *fluid particles* and describe their behavior under any dynamic condition where the particle velocity may range from low subsonic to hypersonic. It also includes the special case termed fluid statics, where the fluid velocity approaches zero. Fluids are encountered in various forms including homogeneous liquids, unsaturated, saturated, and superheated vapors, polymers and inhomogeneous liquids and gases. As we will see in the following chapters, only a few equations govern the motion of a fluid that consists of molecules. At microscopic level, the molecules continuously interact with each other moving with random velocities. The degree of interaction and the mutual exchange of momentum between the molecules increases with increasing temperature, thus, contributing to an intensive and random molecular motion.

1.1 Continuum Hypothesis

The random motion mentioned above, however, does not allow to define a molecular velocity at a fixed spatial position. To circumvent this dilemma, particularly for gases, we consider the mass contained in a volume element δV_G which has the same order of magnitude as the volume spanned by the mean free path of the gas molecules. The volume δV_G has a comparable order of magnitude for a molecule of a liquid δV_L. Thus, a fluid can be treated as a continuum if the volume δV_G occupied by the mass δm does not experience excessive changes. This implies that the ratio

M. T. Schobeiri, *Advanced Fluid Mechanics and Heat Transfer for Engineers and Scientists*, https://doi.org/10.1007/978-3-030-72925-7_1

$$\rho = \lim_{\delta V_G \to 0} \left(\frac{\delta m}{\delta V_G} \right) \tag{1.1}$$

does not depend upon the volume δV_G. This is known as the continuum hypothesis that holds for systems, whose dimensions are much larger than the mean free path of the molecules. Accepting this hypothesis, one may think of a *fluid particle* as a collection of molecules that moves with a velocity that is equal to the average velocity of all molecules that are contained in the fluid particle. With this assumption, the density defined in Eq. (1.1) is considered as a point function that can be dealt with as a thermodynamic property of the system. If the p-v-T behavior of a fluid is given, the density at any position vector **x** and time t can immediately be determined by providing an information about two other thermodynamic properties. For fluids that are frequently used in technical applications, the p-v-T behavior is available from experiments in the form of p-v, h-s, or T-s tables or diagrams. For computational purposes, the experimental points are fitted with a series of algebraic equations that allow a quick determination of density by using two arbitrary thermodynamic properties.

1.2 Molecular Viscosity

Molecular viscosity is the fluid property that causes friction. Figure 1.1 gives a clear physical picture of the friction in a viscous fluid. A flat plate placed at the top of a particular viscous fluid is moving with a uniform velocity $V_1 = U$ relative to the stationary bottom wall.

The following observations were made during experimentation:

1. In order to move the plate, a certain force F_1 must be exerted in x_1-direction.
2. The fluid sticks to the plate surface that moves with the velocity **U**.
3. The velocity difference between the stationary bottom wall and the moving top wall causes a velocity change which is, in this particular case, linear.
4. The force F_1 is directly proportional to the velocity change and the area of the plate.

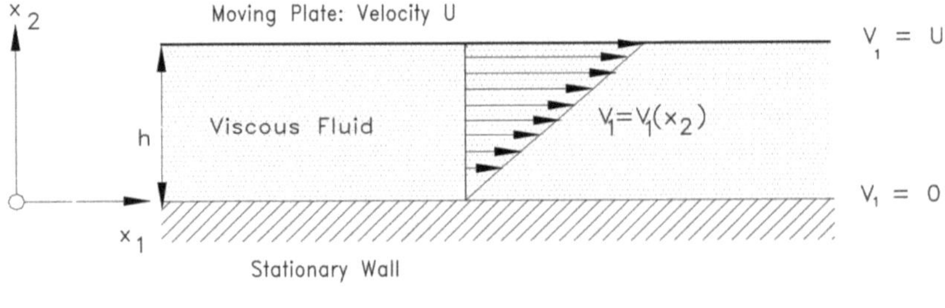

Fig. 1.1 Viscous fluid between a moving and a stationary flat surface

These observations lead to the conclusion that one may set:

$$F_1 \propto A\frac{dV_1}{dx_2}. \tag{1.2}$$

Multiplying the proportionality (1.2) by a factor μ which is the substance property *viscosity*, results in an equation for the friction force in x_1-direction:

$$F_1 = \mu A\frac{dV_1}{dx_2}. \tag{1.3}$$

The subsequent division of Eq. (1.3) by the plate area A gives the shear stress component τ_{21}:

$$\tau_{21} = \mu\frac{dV_1}{dx_2}. \tag{1.4}$$

Equation (1.4) is the Newton's equation of viscosity for this particular case. The first subscript refers to the plane perpendicular to the x_2-coordinate; the second refers to the direction of shear stress. Equation (1.4) is valid for a two-dimensional flow of a particular class of fluids, *the Newtonian Fluids*, whose shear stress is linearly proportional to the velocity change. The general three-dimensional version derived and discussed in Chap. 4 is:

$$\mathbf{T} = \lambda(\nabla\cdot\mathbf{V})\mathbf{I} + 2\mu\mathbf{D} \tag{1.5}$$

with $\mathbf{D}$ as the deformation tensor. The coefficient λ is given by $\lambda = \bar{\mu} - 2/3\mu$, with μ as the absolute viscosity and $\bar{\mu}$ the bulk viscosity. Inserting Eq. (1.5) into the equation of motion (see Chap. 4), the resulting equation independently developed by Navier [1] and Stokes [2] completely describes the motion of a viscous fluid. In a coordinate invariant form the Navier-Stokes equation reads:

$$\frac{D\mathbf{V}}{Dt} = \frac{1}{\rho}\nabla\cdot[(-p + \lambda\nabla\cdot\mathbf{V})\mathbf{I} + 2\mu\mathbf{D}] + \mathbf{g}. \tag{1.6}$$

Although Eq. (1.6) has been known since the publication of the famous paper by Navier in 1823, with the exception of few special cases, it was not possible to find solutions for cases of practical interests. Neglecting the viscosity term significantly reduces the degree of difficulty in finding a solution for Eq. (1.6). This simplification, however, leads to results that do not account for the viscous nature of the fluid, therefore they do not reflect the real flow situations. This is particularly true for the flow regions that are close to the surface. Consider the suction surface of a wing subjected to an air flow as shown in Fig. 1.2.

Two flow layers are distinguished: (1) a very thin layer close to the surface, called the *boundary layer*, where the viscosity effect is predominant and (2) an external layer where the viscosity may be neglected. As a result, the fluid outside the bound-

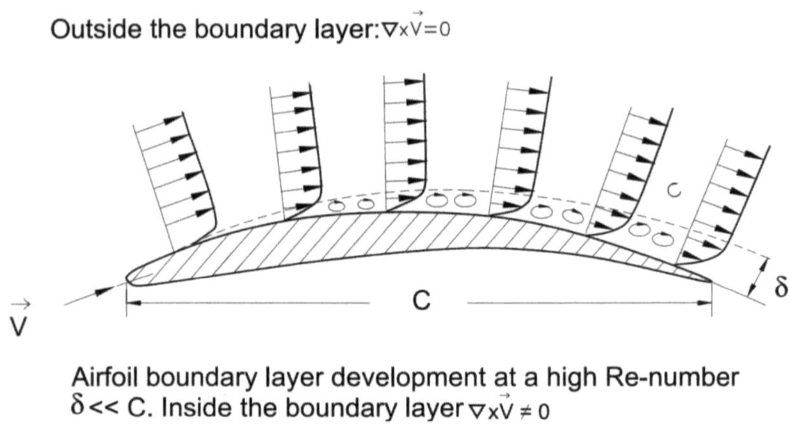

Fig. 1.2 Boundary layer development along the suction surface of a wing, the effect of viscosity diminishes outside the boundary layer

ary layer may be considered *inviscid*. In this case, the Navier-Stokes equation is reduced to the Euler equation of motion that can be solved. Prandtl [3] and von Kármán [4] significantly simplified the governing system of partial differential equations and derived an integral method to solve for boundary layer momentum deficiency thickness for incompressible steady flow. Although the integral method is capable of providing useful information about the boundary layer integral parameters such as momentum thickness or wall friction, it is not able to provide detail information about the velocity distribution within the boundary layer. Likewise, cases with flow separation cannot be treated. Furthermore, it contains several empirical correlations that have to be adjusted from case to case. To partially circumvent the above deficiencies, the integral method can be replaced by a differential method.

Although the introduction of boundary layer theory was a major breakthrough in fluid mechanics, its field of applications is limited. With the introduction of powerful numerical methods and high speed computers, it is now possible to solve the Navier-Stokes equations for laminar (see Sect. 1.3.1) flows. To find solutions for turbulent (see Sect. 1.3.1) flows, the equations are averaged leading to *Reynolds averaged* Navier-Stokes equations (RANS). The averaging process creates a new second order tensor called the *Reynolds stress tensor*, with nine unknowns. The numerical solution of RANS, however, requires modeling the Reynolds stress tensor. In the last three decades, a variety of turbulence models have been developed including single algebraic and multi-equation models. The trend in computation fluid dynamics goes toward a direct numerical simulation (DNS) of Navier-Stokes equations, avoiding time averaging and turbulence modeling altogether.

1.3 Flow Classification

1.3.1 Velocity Pattern: Laminar, Intermittent, Turbulent Flow

Laminar flow is characterized by the smooth motion of fluid particles with no random fluctuations present. This characteristic is illustrated in Fig. 1.3a by measuring the velocity distribution $\mathbf{V} = V(\mathbf{x})$ of a statistically steady flow at an arbitrary position vector **x.** As Fig. 1.3 reveals, the velocity distribution for laminar flow does not have any time-dependent random fluctuations. In contrast, random fluctuations are inherent characteristics of a turbulent flow. Figure 1.3b shows the velocity distribution for a turbulent flow with random fluctuations. For a statistically steady flow, the velocity distribution is time dependent, given by $\mathbf{V} = V(\mathbf{x}, t)$.

It can be decomposed as a constant mean velocity $\bar{V}(\mathbf{x})$ and random fluctuations $\mathbf{V}'(\mathbf{x}, \mathbf{t})$:

$$V(\mathbf{x}, t) = \bar{\mathbf{V}}(x) + \mathbf{V}'(\mathbf{x}, t). \tag{1.7}$$

At this point, the question may arise under which condition the flow pattern may change from laminar to turbulent. To answer this question, consider the experiment by Reynolds [5] late nineteenth century, who injected dye streak into a pipe flow as shown in Fig. 1.4.

At a lower velocity, Fig. 1.4a, no fluctuation was observed and the dye filament followed the flow direction. At certain distances, the diffusion process that was gradually taking place caused a complete mixing of the dye with the main fluid. Increasing the velocity, Fig. 1.4b however, changed the flow picture completely. The orderly motion of the dye with a short laminar length, shown in Fig. 1.4b, changed into a transitional mode that started with a *sinus-like wave*, which we discuss in detail in Chap. 8. The transitional mode was followed by a strong fluctuating turbulent motion. This resulted in a rapid mixing of the dye with the main fluid. To explain this phenomenon, Reynolds introduced a dimensionless parameter, named after him later as the Reynolds number:

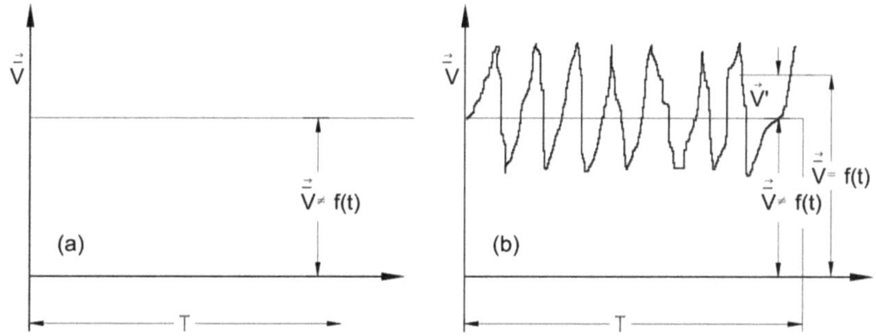

Fig. 1.3 **a** Laminar flow velocity, **b** turbulent flow velocity at an arbitrary position vector

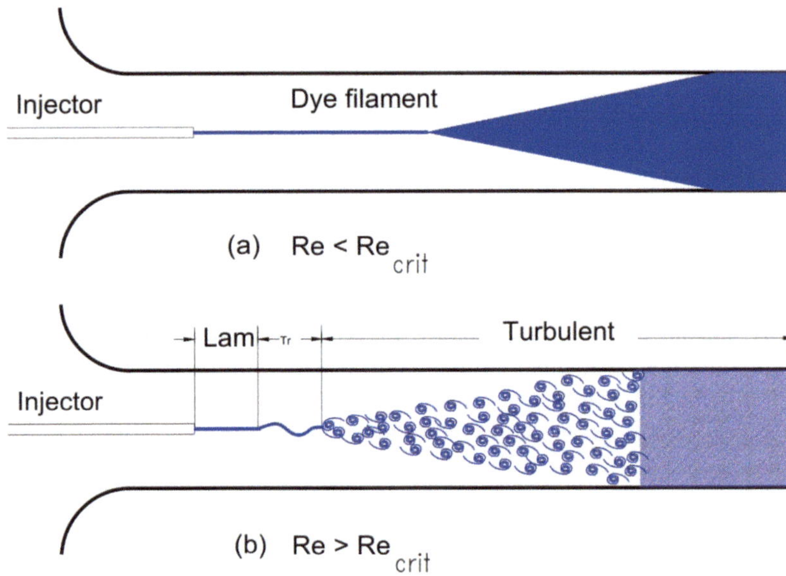

Fig. 1.4 Dye experiment by Reynolds, a subcritical, b super critical

$$\mathrm{Re} = \frac{\rho V D}{\mu} \tag{1.8}$$

with μ as the absolute viscosity, ρ the density, D the pipe diameter and V the flow velocity. For a *critical Reynolds number*, $\mathrm{Re} < \mathrm{Re}_{criit} \approx 2300$, a laminar flow pattern was observed. Keeping the pipe geometry, as well as the flow substance the same, an increase in velocity resulting in a Reynold number $\mathrm{Re} > \mathrm{Re}_{criit}$ changed the flow pattern. As Fig. 1.4b shows, the initially laminar flow underwent a transition followed by random turbulent fluctuations causing a strong mixing of the dye with the main fluid.

Similar flow behavior is observed in boundary layer flow along bodies. As an example, Fig. 1.5 shows the changes of the flow pattern within the boundary layer along a flat plate at zero pressure gradient.

Following an arbitrary streamline within the boundary layer within the flow region ①, a stable laminar flow is established that starts from the leading edge and extends to the point of inception of the unstable two-dimensional *Tollmien-Schlichting waves*. Region ② includes the following subsets: (a) the onset of the unstable two-dimensional Tollmien-Schlichting waves, (b) the bursts of turbulence in places with high vorticity, (c) the intermittent formation of turbulent spots with high vortical core at intense fluctuation. Region ③ indicates the coalescence of turbulent spots into a fully developed turbulent boundary layer. The issue is discussed in detail in Chap. 4.

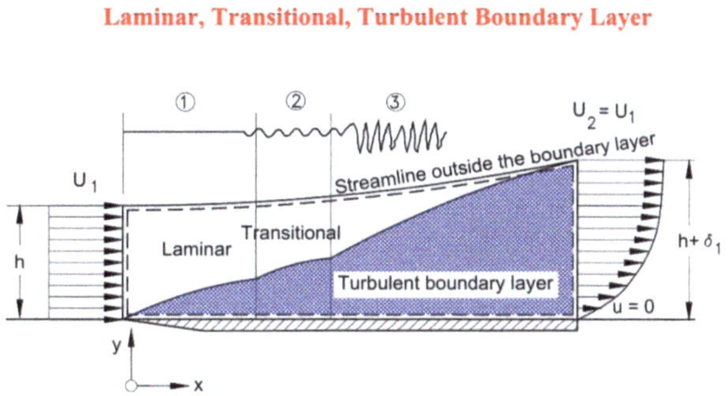

Fig. 1.5 Boundary layer development along a flat plate

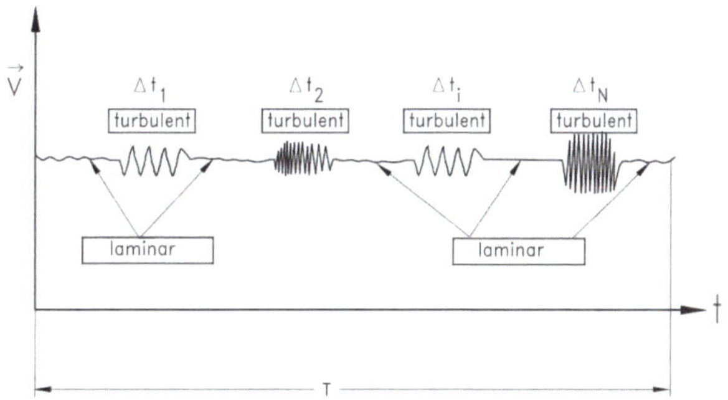

Fig. 1.6 Intermittent behavior of a transitional flow

The transitional region is characterized by an intermittently laminar-turbulent pattern described by the *intermittency factor* γ. For a statistically steady flow, this factor is defined as the ratio of the sum of all time intervals, within which the flow is turbulent divided by the period of the observation time T as shown in Fig. 1.6 and Eq. (1.9):

$$\gamma = \frac{\sum_{i=1}^{n} \Delta t_i}{T}. \tag{1.9}$$

In Eq. (1.9) n is the number of Δt_i-intervals. The result of an experimental study along a curved plate at zero pressure gradient is plotted in Fig. 1.6.

The intermittency value $\gamma = 0$ indicates a fully laminar flow, for $0 < \gamma < 1$ the flow is transitional, and for $\gamma = 1$ the flow is fully turbulent. Figure 1.7 illustrates the intermittency distribution within the boundary layer along a flat plate with $\mathrm{Re}_x = V_1 x_1/\nu$ as the Reynolds number with V_1 as the velocity component in x_1-direction. Up to Re =15,000, the flow is sub-critically stable laminar with $\gamma = 0$. The onset of transition starts at Re = 15,000 and continues until $\gamma = 1$ has been reached. This point indicates the beginning of a fully turbulent boundary layer flow.

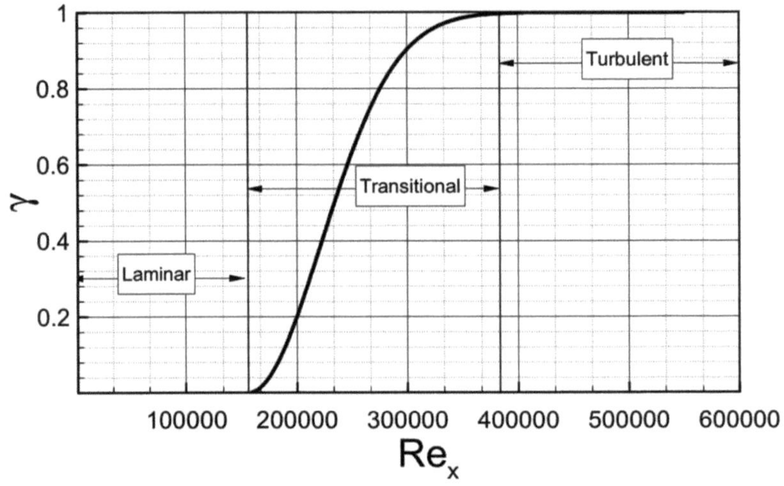

Fig. 1.7 Distribution of intermittency factor along a flat plate

1.3.2 Change of Density, Incompressible, Compressible Flow

Fluid density generally changes with pressure and temperature. As the Mollier diagram for steam shows, the density of water in the liquid state changes insignificantly with pressure. In contrast, significant changes are observed when water changes the state from liquid to vapor. A similar situation is observed for other gases.

Considering a statistically steady liquid flow with negligibly small changes in density, the flow is termed *incompressible*. For gas flows, however, the density change is a function of the flow Mach number.

Figure 1.8 depicts relative changes of different flow properties as functions of the flow Mach number. Up to $M = 0.3$, the relative changes of density may be considered negligibly small meaning that the flow may be considered incompressible. For Mach numbers $M > 0.3$, density changes cannot be neglected. In case the flow velocity approaches the speed of sound, $M = 1.0$, the flow pattern undergoes a drastic change associated with shock waves.

The density classification based on flow Mach number gives a practical idea about the density change. A more adequate definition whether the flow can be considered compressible or incompressible is given by the condition $D\rho/Dt = 0$, which in conjunction with the continuity equation results in $\nabla \cdot \mathbf{V} = 0$. This is the condition for a flow to be considered incompressible. This issue is discussed in more detail in Chap. 4.

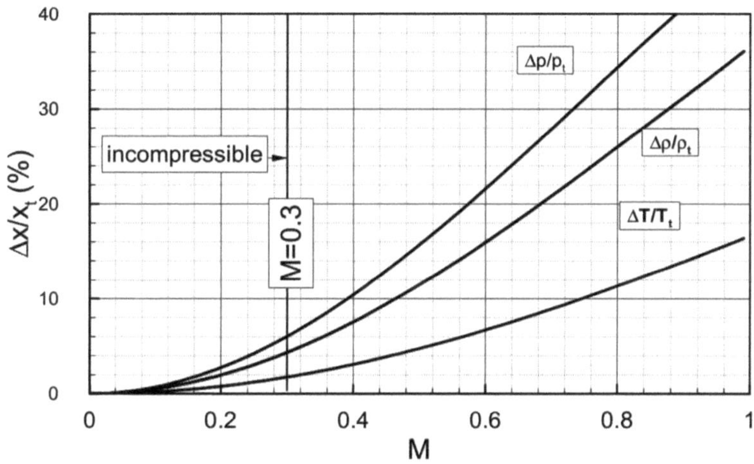

Fig. 1.8 Density, pressure, and temperature changes as a function of the flow Mach number

1.3.3 *Statistically Steady Flow, Unsteady Flow*

Figure 1.9 illustrates the nature of the statistically steady and unsteady flow types. As an example, Fig. 1.9a shows the velocity distribution of a statistically steady turbulent pipe flow with a constant mean. Figure 1.9b represents the turbulent velocity of a statistically unsteady flow discharging from a container under pressure. As seen, the mean velocity is a function of time. A periodic unsteady turbulent flow through a reciprocating engine is represented by Fig. 1.9c. In both unsteady cases, the unsteady mean is the result of an *ensemble averaging* process that we discuss in Chap. 10.

In Fig. 1.9, random fluctuations typical of a turbulent flow are superimposed on the mean flow. For steady or unsteady laminar flows where the Reynolds number is below the critical one, the velocity distributions do not have random component as shown in Fig. 1.10.

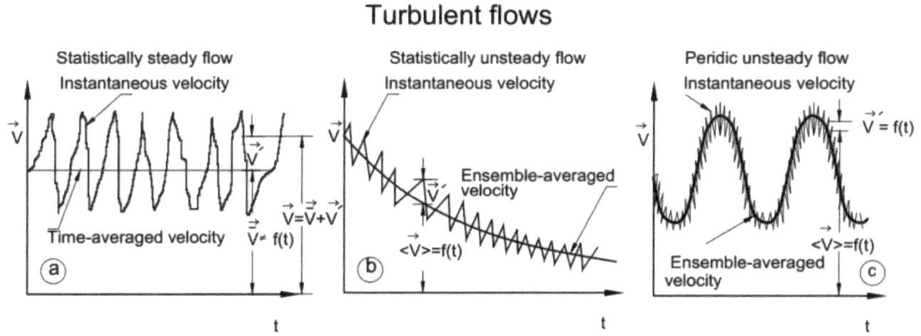

Fig. 1.9 Statistically steady and unsteady turbulent flows

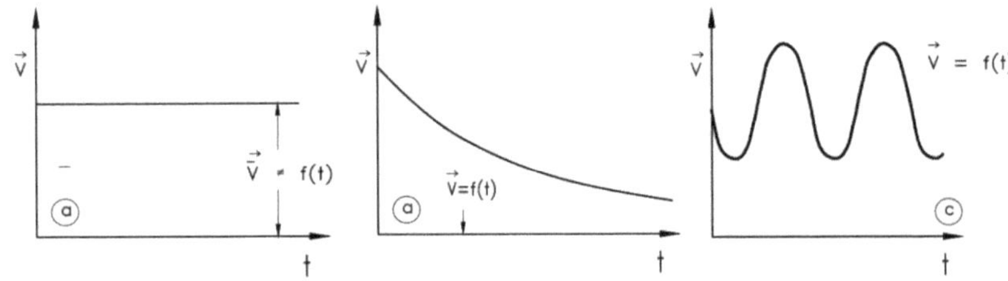

Fig. 1.10 Steady and unsteady laminar flows

1.4 Shear-Deformation Behavior of Fluids

As briefly discussed in Sect. 1.2, there is a relationship between the shear stress τ_{21} and the deformation rate dV_1/dx_2. Fluids which exhibits a linear shear-deformation behavior are called *Newtonian Fluids*. There are, however, many fluids which exhibit a nonlinear shear- deformation behavior. Figure 1.11 shows qualitatively the behavior of few of these fluids. More details are found among others in [6].

While the pseudoplastic fluids are characterized by a degressive slope, dilatant fluids exhibit progressive slops. For these type of fluids the shear stress tensor can be described as a polynomial function of deformation tensor, where the degree of polynomials and the coefficients are determined from experiments.

Those fluids with linear behavior which will not deform unless certain initial stress $(\tau_{21})_0$ is exceeded are called Bingham fluids. It should be noted that most of the fluid used in engineering applications belong to the Newtonian Class.

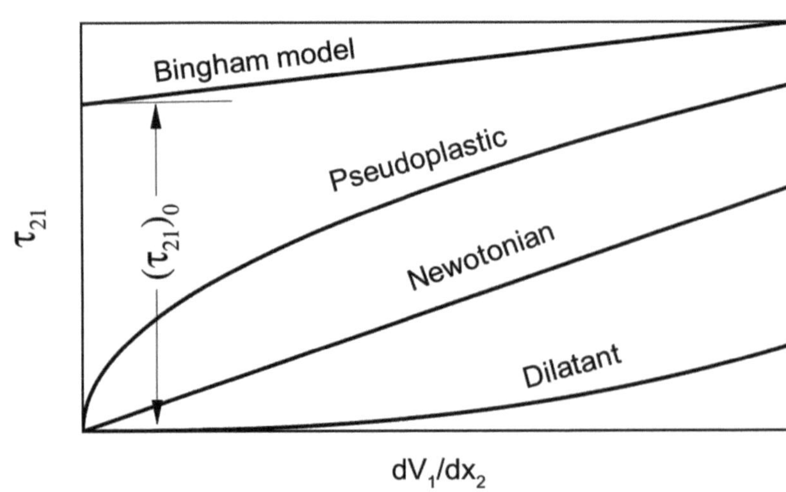

Fig. 1.11 Shear-deformation behavior of different fluids

Chapter 2
Vector and Tensor Analysis, Applications to Fluid Mechanics

2.1 Tensors in Three-Dimensional Euclidean Space

In this section, we briefly introduce tensors, their significance to fluid dynamics and their applications. The tensor analysis is a powerful tool that enables the reader to study and to understand more effectively the fundamentals of fluid mechanics. Once the basics of tensor analysis are understood, the reader will be able to derive all conservation laws of fluid mechanics without memorizing any single equation. In this section, we focus on the tensor analytical application rather than mathematical details and proofs that are not primarily relevant to engineering students. To avoid unnecessary repetition, we present the definition of tensors from a unified point of view and use exclusively the three-dimensional Euclidean space, with $N = 3$ as the number of dimensions. The material presented in this chapter has drawn from classical tensor and vector analysis texts, among others those mentioned in References. It is tailored to specific needs of fluid mechanics and is considered to be helpful for readers with limited knowledge of tensor analysis.

The quantities encountered in fluid dynamics are *tensors.* A physical quantity which has a *definite magnitude* but not a *definite direction*exhibits a *zeroth-order tensor*, which is a special category of tensors. In a N-dimensional Euclidean space, a zeroth-order tensor has $N^0 = 1$ component, which is basically its magnitude. In physical sciences, this category of tensors is well known as a *scalar* quantity, which has a definite magnitude but not a definite direction. Examples are: mass m, volume v, thermal energy Q (heat), mechanical energy W (work) and the entire thermo-fluid dynamic properties such as density ρ, temperature T , enthalpy h, entropy s , etc.

In contrast to the zeroth-order tensor, a *first-order* tensor encompasses physical quantities with a *definite magnitude* with $N^1(N^1 = 3^1 = 3)$ components and a *definite direction* that can be decomposed in $N^1 = 3$ directions. This special category of tensors is known as *vector*. Distance $\mathbf{X}$, velocity $\mathbf{V}$, acceleration A, force F and moment of momentum M are few examples. A vector quantity is *invariant* with respect to a given category of coordinate systems. Changing the coordinate system

M. T. Schobeiri, *Advanced Fluid Mechanics and Heat Transfer for Engineers and Scientists*, https://doi.org/10.1007/978-3-030-72925-7_2

by applying certain transformation rules, the vector components undergo certain changes resulting in a new set of components that are related, in a definite way, to the old ones. As we will see later, the order of the above tensors can be reduced if they are multiplied with each other in a *scalar* manner. The mechanical energy $W = \mathbf{F}.\mathbf{X}$ is a representative example, that shows how a tensor order can be reduced. The reduction of order of tensors is called *contraction.*

A *second order tensor* is a quantity, which has N^2 definite components and N^2 definite directions (in three-dimensional Euclidean space: $N^2 = 9$). General stress tensor Π, normal stress tensor Σ, shear stress tensor T, deformation tensor $\mathbf{D}$ and rotation tensor Ω are few examples. Unlike the zeroth and first order tensors (scalars and vectors), the second and higher order tensors cannot be directly geometrically interpreted. However, they can easily be interpreted by looking at their pertinent force components, as seen later in Sect. 2.5.4.

2.1.1 Index Notation

In a three-dimensional Euclidean space, any arbitrary first order tensor or vector can be decomposed into three components. In a Cartesian coordinate system shown in Fig. 2.1, the *base vectors* in x_1, x_2, x_3 directions $\mathbf{e_1}$, $\mathbf{e_2}$, $\mathbf{e_3}$ are perpendicular to each other and have the magnitude of unity, therefore, they are called *orthonormal unit vectors*. Furthermore, these base vectors are not dependent upon the coordinates, therefore, their derivatives with respect to any coordinates are identically zero. In contrast, in a general curvilinear coordinate system (discussed in Appendix A) the base vectors do not have the magnitude of unity. They depend on the curvilinear coordinates, thus, their derivatives with respect to the coordinates do not vanish.

As an example, vector A with its components A_1, A_2 and A_3 in a Cartesian coordinate system shown in Fig. 2.1 is written as:

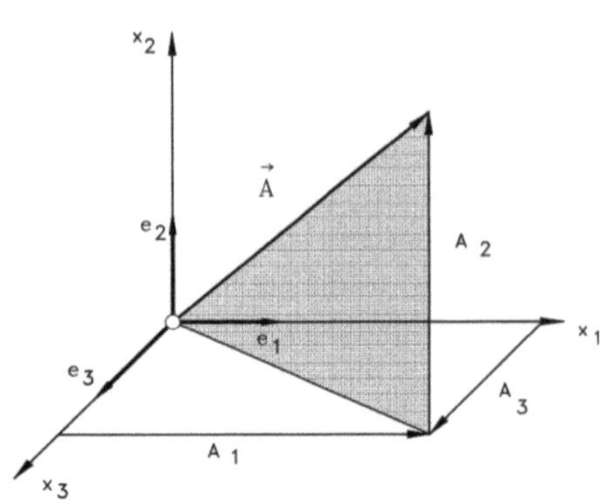

Fig. 2.1 Vector decomposition in a Cartesian coordinate system

$$\mathbf{A} = e_1 A_1 + e_2 A_2 + e_3 A_3 = \sum_{i=1}^{N=3} e_i A_i. \tag{2.1}$$

According to Einstein's summation convention, it can be written as:

$$\mathbf{A} = e_i A_i. \tag{2.2}$$

The above form is called the *index notation.* Whenever the same index (in the above equation i) appears twice, the summation is carried out from 1 to N ($N = 3$ for Euclidean space).

2.2 Vector Operations: Scalar, Vector and Tensor Products

2.2.1 *Scalar Product*

Scalar or dot product of two vectors results in a scalar quantity $\mathbf{A} \cdot \mathbf{B} = \mathbf{C}$. We apply the Einstein's summation convention defined in Eq. (2.2) to the above vectors:

$$(e_i A_i) \cdot (e_j B_j) = C. \tag{2.3}$$

We rearrange the unit vectors and the components separately:

$$(e_i \cdot e_j) A_i B_j = C. \tag{2.4}$$

In Cartesian coordinate system, the scalar product of two unit vectors is called *Kronecker delta*, which is:

$$\delta_{ij} = e_i \cdot e_j = 1 \text{ for } i = j, \delta_{ij} = e_i \cdot e_j = 0 \text{ for } i \neq j \tag{2.5}$$

with δ_{ij} as Kronecker delta. Using the Kronecker delta, we get:

$$(e_i \cdot e_j) A_i B_j = \delta_{ij} A_i B_j. \tag{2.6}$$

The non-zero components are found only for $i = j$, or $\delta_{ij} = 1$, which means in the above equation the index $j$must be replaced by $i$resulting in:

$$\mathbf{A} \cdot \mathbf{B} = \mathbf{A_i B_i} = \mathbf{A_1 B_1} + \mathbf{A_2 B_2} + \mathbf{A_3 B_3} = \mathbf{C} \tag{2.7}$$

with scalar C as the result of scalar multiplication.

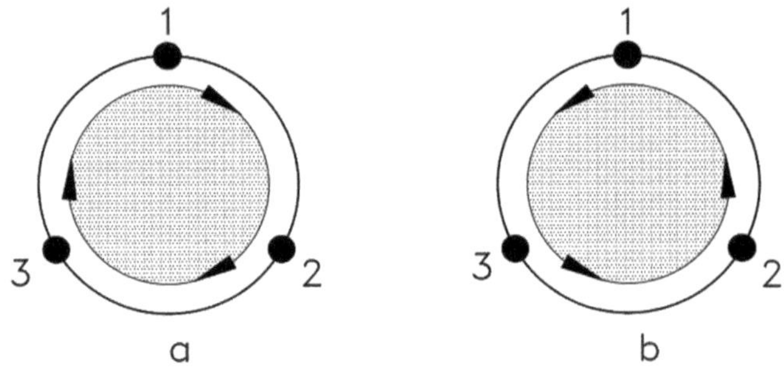

Fig. 2.2 Permutation symbol, **a** positive , **b** negative permutation

2.2.2 *Vector or Cross Product*

The vector product of two vectors is a vector that is perpendicular to the plane described by those two vectors. Example:

$$\mathbf{F} \times \mathbf{R} = \mathbf{M} \text{ or } \mathbf{A} \times \mathbf{B} = \mathbf{C} \tag{2.8}$$

with **C** as the resulting vector. We apply the index notation to Eq. (2.8):

$$\mathbf{A} \times \mathbf{B} = (\mathbf{e_i A_i}) \times (\mathbf{e_j B_j}) = \varepsilon_{ijk} \boldsymbol{e}_k \boldsymbol{A}_i \boldsymbol{B}_j \tag{2.9}$$

with ε_{ijk} as the permutation symbol with the following definition illustrated in Fig. 2.2:

$\varepsilon_{ijk} = 0$ for $i = j$, $j = k$ or $i = k$: (e.g. 122)
$\varepsilon_{ijk} = 1$ for cyclic permutation: (e.g. 123)
$\varepsilon_{ijk} = -1$ for anticyclic permutation: (e.g.132).

Using the above definition, the vector product is given by:

$$\mathbf{C} = \mathbf{e_k C_k} = \varepsilon_{ijk} \boldsymbol{e}_k \boldsymbol{A}_i \boldsymbol{B}_j. \tag{2.10}$$

2.2.3 *Tensor Product*

The tensor product is a product of two or more vectors, where the unit vectors are not subject to scalar or vector operation. Consider the following *tensor operation*:

$$\Phi = \mathbf{AB} = (\mathbf{e_i A_i})(\mathbf{e_j B_j}) = \mathbf{e_i e_j A_i B_j}. \tag{2.11}$$

The result of this purely mathematical operation is a second order tensor with nine components:

$$\Phi = \begin{matrix} e_1(e_1A_1B_1 + e_2A_1B_2 + e_3A_1B_3)+ \\ e_2(e_1A_2B_1 + e_2A_2B_2 + e_3A_2B_3)+ \\ e_3(e_1A_3B_1 + e_2A_3B_2 + e_3A_3B_3) \end{matrix} \quad (2.12)$$

The operation with any tensor such as the above second order one acquires a physical meaning if it is multiplied with a vector (or another tensor) in scalar manner. Consider the scalar product of the vector **C** and the second order tensor Φ. The result of this operation is a *first order tensor* or a vector. The following example should clarify this:

$$\mathbf{D} = \mathbf{C} \cdot \Phi = \mathbf{C} \cdot (\mathbf{AB}) = \mathbf{e_k C_k} \cdot (\mathbf{e_i e_j})\mathbf{A_i B_j}. \quad (2.13)$$

Rearranging the unit vectors and the components separately:

$$\mathbf{D} = \mathbf{C} \cdot \Phi = \mathbf{C} \cdot (\mathbf{AB}) = \mathbf{e_k} \cdot (\mathbf{e_i e_j})\mathbf{C_k A_i B_j}. \quad (2.14)$$

It should be pointed out that in the above equation, the unit vector $\mathbf{e_k}$ must be multiplied with the closest unit vector namely $\mathbf{e_i}$

$$\mathbf{D} = \mathbf{C} \cdot \Phi = \mathbf{C} \cdot (\mathbf{AB}) = \delta_{\mathbf{ki}}(\mathbf{e_j})\mathbf{C_k A_i B_j} = \mathbf{e_j C_i A_i B_j}. \quad (2.15)$$

The result of this tensor operation is a vector with the same direction as vector **B**. Different results are obtained if the positions of the terms in a dot product of a vector with a tensor are reversed as shown in the following operation:

$$\mathbf{E} = \Phi \cdot \mathbf{C} = (\mathbf{AB}) \cdot \mathbf{C} = \mathbf{e_i A_i B_j} \delta_{\mathbf{jk}} \mathbf{C_k} = \mathbf{e_i A_i B_j C_j}. \quad (2.16)$$

The result of this operation is a vector in direction of **A**. Thus, the products $\mathbf{E} = \Phi \cdot \mathbf{C}$ is different from $\mathbf{D} = \mathbf{C} \cdot \Phi$.

2.3 Contraction of Tensors

As shown above, the scalar product of a second order tensor with a first order one is a first order tensor or a vector. This operation is called contraction. The *trace of a second order tensor* is a tensor of zeroth order, which is a result of a contraction and is a scalar quantity.

$$\text{Tr}\,(\Phi) = e_i \cdot e_j \Phi_{ij} = \delta_{ij}\Phi_{ij} = \Phi_{ii} = \Phi_{11} + \Phi_{22} + \Phi_{33}. \quad (2.17)$$

As can be shown easily, the trace of a second order tensor is the sum of the diagonal element of the *matrix* Φ_{ij}. If the tensor $\boldsymbol{\Phi}$ itself is the result of a contraction of two second order tensors $\boldsymbol{\Pi}$ and **D**:

$$\boldsymbol{\Phi} = \boldsymbol{\Pi} \cdot \mathbf{D} = \mathbf{e_i e_j} \Pi_{\mathbf{ij}} \cdot \mathbf{e_k e_l D_{kl}} = \mathbf{e_i e_l} \delta_{\mathbf{jk}} \Pi_{\mathbf{ij}} \mathbf{D_{kl}} = \mathbf{e_i e_l} \Pi_{\mathbf{ik}} \mathbf{D_{kl}} \quad (2.18)$$

then the *Tr(Φ)* is:

$$\mathrm{Tr}(\mathbf{\Phi}) = e_i \cdot e_l \Pi_{ik} \mathbf{D_{kl}} = \delta_{\mathbf{il}} \Pi_{\mathbf{ik}} \mathbf{D_{kl}} = \Pi_{\mathbf{lk}} \mathbf{D_{kl}}. \tag{2.19}$$

2.4 Differential Operators in Fluid Mechanics

In fluid mechanics, the particles of the working medium undergo a time dependent or unsteady motion. The flow quantities such as the velocity **V** and the thermodynamic properties of the working substance such as pressure p, temperature T, density ρ or any arbitrary flow quantity Q are generally functions of space and time: $\mathbf{V} = \mathbf{V}(\mathbf{xt}), \mathbf{p} = \mathbf{p}(\mathbf{x}, \mathbf{t}), \mathbf{T} = \mathbf{T}(\mathbf{x}, \mathbf{t}), \rho = \rho(\mathbf{x}, \mathbf{t})$. During the flow process, these quantities generally change with respect to time and space. The following operators account for the *substantial*, *spatial*, and *temporal* changes of the flow quantities.

2.4.1 *Substantial Derivatives*

The *temporal* and *spatial change* of the above quantities is described most appropriately by the *substantial* or *material derivative*. Generally, the substantial derivative of a flow quantity Q, which may be a scalar, a vector or a tensor valued function, is given by:

$$DQ = \frac{\partial Q}{\partial t} dt + dQ. \tag{2.20}$$

The operator D represents the *substantial or material* change of the quantity Q, the first term on the right hand side of Eq. (2.20) represents the *local* or *temporal change* of the quantity Q with respect to a fixed position vector **x**. The operator d symbolizes the *spatial or convective change* of the same quantity with respect to a fixed instant of time. The convective change of Q may be expressed as:

$$dQ = \frac{\partial Q}{\partial x_1} dx_1 + \frac{\partial Q}{\partial x_2} dx_2 + \frac{\partial Q}{\partial x_3} dx_3. \tag{2.21}$$

A simple rearrangement of the above equation results in:

$$dQ = (e_1 dx_1 + e_2 dx_2 + e_3 dx_3) \cdot \left(e_1 \frac{\partial}{\partial x_1} + e_2 \frac{\partial}{\partial x_2} + e_3 \frac{\partial}{\partial x_3} \right) Q. \tag{2.22}$$

Scalar multiplication of the expressions in the two parentheses of Eq. (2.22) results in Eq. (2.21).

2.4.2 *Differential Operator* ∇

The expression in the second parenthesis of Eq. (2.22) is the *spatial differential operator* ∇ (*nabla, del*) which has a vector character. In Cartesian coordinate system,the operator nabla ∇ is defined as:

$$\nabla = \left(e_1 \frac{\partial}{\partial x_1} + e_2 \frac{\partial}{\partial x_2} + e_3 \frac{\partial}{\partial x_3} \right) = e_i \frac{\partial}{\partial x_i}. \tag{2.23}$$

Using the above differential operator, the change of the quantity Q is written as:

$$dQ = d\mathbf{x} \cdot \nabla \mathbf{Q}. \tag{2.24}$$

The increment dQ of Eq. (2.24) is obtained either by applying the product $d\mathbf{x} \cdot \nabla$, or by taking the dot product of the vector $d\mathbf{x}$ and ∇Q. If Q is a scalar quantity, then ∇Q is a vector or a *first order tensor* with definite components. In this case ,∇Q is called the *gradient* of the scalar field. Equation (2.24) indicates that the spatial change of the quantity Q assumes a maximum if the vector ∇Q (*gradient* of Q) is parallel to the vector $d\mathbf{x}$. If the vector ∇Q is perpendicular to the vector $d\mathbf{x}$, their product will be zero. This is only possible, if the spatial change $d\mathbf{x}$ occurs on a surface with Q=*const.* Consequently, the quantity Q does not experience any changes. The physical interpretation of this statement is found in Fig. 2.3. The scalar field is

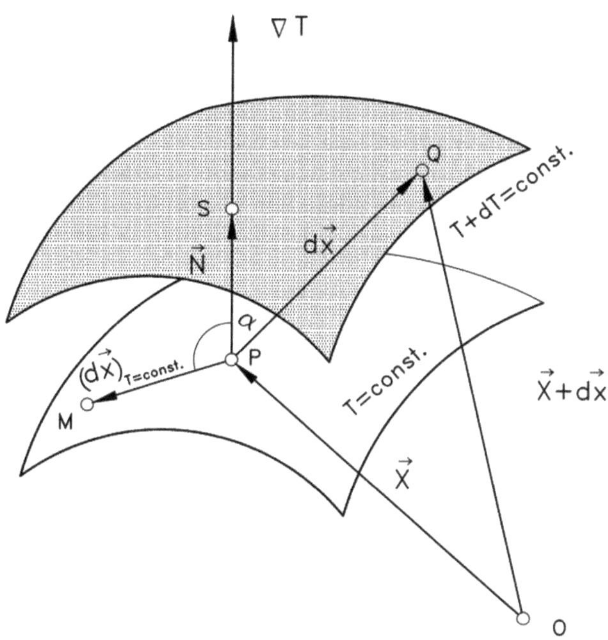

Fig. 2.3 Physical explanation of the gradient of scalar field

represented by the point function temperature that changes from the surfaces T to the surface $T+dT$. In Fig. 2.3, the gradient of the temperature field is shown as ∇T, which is perpendicular to the surface $T = const.$ at point P. The temperature probe located at P moves on the surface $T=const.$ to the point M, thus measuring no changes in temperature ($\alpha = \pi/2$, $\cos\alpha = 0$). However, the same probe experiences a certain change in temperature by moving to the point Q, which is characterized by a higher temperature $T + dT$ ($0 < \alpha < \pi/2$). The change dT can immediately be measured, if the probe is moved parallel to the vector ∇T. In this case, the displacement dx (see Fig. 2.3) is the shortest ($\alpha = 0$, $\cos\alpha = 1$). Performing the similar operation for a vector quantity as seen in Eq. (2.21) yields:

$$d\mathbf{V} = \frac{\partial \mathbf{V}}{\partial \mathbf{x}_1}\mathbf{dx}_1 + \frac{\partial \mathbf{V}}{\partial \mathbf{x}_2}\mathbf{dx}_2 + \frac{\partial \mathbf{V}}{\partial \mathbf{x}_3}\mathbf{dx}_3. \tag{2.25}$$

The right-hand side of Eq. (2.25) is identical with:

$$d\mathbf{V} = (\mathbf{dx} \cdot \nabla)\mathbf{V}. \tag{2.26}$$

In Eq. (2.26) the product $d\mathbf{x} \cdot \nabla$ can be considered as an operator that is applied to the vector $\mathbf{V}$ resulting in an increment of the velocity vector. Performing the scalar multiplication between dx and ∇ gives:

$$d\mathbf{V} = (\mathbf{dx} \cdot \nabla)\mathbf{V} = \mathbf{dx_i}\frac{\partial \mathbf{V}}{\partial \mathbf{x_i}} = \mathbf{e_j}\mathbf{dx_i}\frac{\partial V_j}{\partial \mathbf{x_i}} \equiv \mathbf{dx} \cdot \nabla\mathbf{V} \tag{2.27}$$

with $\nabla\mathbf{V}$ as the gradient of the vector field which is a second order tensor. To perform the differential operation, first the ∇ operator is applied to the vector V, resulting in a second order tensor. This tensor is then multiplied with the vector $d\mathbf{x}$ in a scalar manner that results in a first order tensor or a vector. From this operation, it follows that spatial change of the velocity component can be expressed as the scalar product of the vector $d\mathbf{x}$ and the second order tensor $\nabla\mathbf{V}$, which represents the spatial gradient of the velocity vector. Using the spatial derivative from Eq. (2.27), the substantial change of the velocity is obtained by:

$$D\mathbf{V} = \frac{\partial \mathbf{V}}{\partial \mathbf{t}}\mathbf{dt} + \mathbf{dV} \tag{2.28}$$

where the spatial change of the velocity is expressed as:

$$d\mathbf{V} = \mathbf{dx} \cdot \nabla\mathbf{V}. \tag{2.29}$$

Dividing Eq. (2.29) by dt yields the *convective* part of the acceleration vector:

$$\frac{d\mathbf{V}}{dt} = \left(\frac{d\mathbf{x}}{dt}\right) \cdot (\nabla\mathbf{V}) = \mathbf{V} \cdot \nabla\mathbf{V}. \tag{2.30}$$

The substantial acceleration is then:

$$\frac{D\mathbf{V}}{dt} \equiv \frac{D\mathbf{V}}{Dt} = \frac{\partial \mathbf{V}}{\partial t} + \mathbf{V} \cdot \nabla \mathbf{V}. \tag{2.31}$$

The differential dt may symbolically be replaced by Dt indicating the material character of the derivatives. Applying the index notation to velocity vector and Nabla operator, performing the vector operation , and using the Kronecker delta, the index notation of the *material acceleration A* is:

$$\mathbf{A} = \mathbf{e_i A_i} = \mathbf{e_i}\frac{\partial \mathbf{V_i}}{\partial \mathbf{t}} + \mathbf{e_i V_j}\frac{\partial \mathbf{V_i}}{\partial \mathbf{x_j}}. \tag{2.32}$$

Equation (2.32) is valid only for Cartesian coordinate system, where the unit vectors do not depend upon the coordinates and are constant. Thus, their derivatives with respect to the coordinates disappear identically. To arrive at Eq. (2.32) with a unified index i, we renamed the indices. To decompose the above acceleration vector into three components, we cancel the unit vector from both side in Eq. (2.32) and get:

$$A_i = \frac{\partial V_i}{\partial t} + V_j \frac{\partial V_i}{\partial x_j}. \tag{2.33}$$

To find the components in x_i-direction, the index i assumes subsequently the values from 1 to 3, while the summation convention is applied to the free index j. As a result we obtain the three components:

$$\begin{aligned} A_1 &= \tfrac{\partial V_1}{\partial t} + V_1 \tfrac{\partial V_1}{\partial x_1} + V_2 \tfrac{\partial V_1}{\partial x_2} + V_3 \tfrac{\partial V_1}{\partial x_3} \\ A_2 &= \tfrac{\partial V_2}{\partial t} + V_1 \tfrac{\partial V_2}{\partial x_1} + V_2 \tfrac{\partial V_2}{\partial x_2} + V_3 \tfrac{\partial V_2}{\partial x_3} \\ A_3 &= \tfrac{\partial V_3}{\partial t} + V_1 \tfrac{\partial V_3}{\partial x_1} + V_2 \tfrac{\partial V_3}{\partial x_2} + V_3 \tfrac{\partial V_3}{\partial x_3}. \end{aligned} \tag{2.34}$$

2.5 Operator ∇ Applied to Different Functions

This section summarizes the applications of nabla operator to different functions. As mentioned previously, the spatial differential operator ∇ has a vector character. If it acts on a scalar function, such as temperature, pressure, enthalpy etc., the result is a vector and is called the *gradient* of the corresponding scalar field, such as gradient of temperature, pressure, etc. (see also previous discussion of the physical interpretation of ∇Q). If, on the other hand, ∇ acts on a vector, three different cases are distinguished.

2.5.1 Scalar Product of ∇ and V

This operation is called the *divergence of the vector* **V**. The result is a zeroth-order tensor or a scalar quantity. Using the index notation, the divergence of **V** is written as:

$$\nabla \cdot \mathbf{V} = (\mathbf{e_i}\frac{\partial}{\partial \mathbf{x_i}}).(\mathbf{e_j}\mathbf{V_j}) = \delta_{\mathbf{ij}}\frac{\partial}{\partial \mathbf{x_i}}\mathbf{V_j} = \frac{\partial \mathbf{V_i}}{\partial \mathbf{x_i}}. \tag{2.35}$$

The physical interpretation of this purely mathematical operation is shown in Fig. 2.4. The mass flow balance for a steady incompressible flow through an infinitesimal volume $dv = dx_1dx_2dx_3$ is shown in Fig. 2.4. We first establish the entering and exiting mass flows through the cube side areas perpendicular to x_1-direction given by $dA_1 = dx_2dx_3$:

$$\dot{m}_{x_{1en}} = \rho(dx_2dx_3)V_1. \tag{2.36}$$

$$\dot{m}_{x_{1ex}} = \rho(dx_2dx_3)\left(V_1 + \frac{\partial \mathbf{V_1}}{\partial x_1}dx_1\right). \tag{2.37}$$

Repeating the same procedure for the cube side areas perpendicular to x_2 and x_3 directions given by $dA_2 = dx_3dx_1$ and $dA_3 = dx_1dx_2$ and subtracting the entering mass flows from the exiting ones, we obtain the net mass net flow balances through the infinitesimal differential volume as:

$$\rho(dx_1dx_2dx_3)\left(\frac{\partial V_1}{\partial x_1} + \frac{\partial V_2}{\partial x_2} + \frac{\partial V_3}{\partial x_3}\right) = \rho dv\nabla \cdot \mathbf{V}) = \mathbf{0}. \tag{2.38}$$

The right hand side of Eq. (2.38) is a product of three terms, the density ρ, the differential volume dV and the divergence of the vector **V**. Since the first two terms are not zero, the divergence of the vector must disappear. As result, we find:

$$\nabla \cdot \mathbf{V} = \mathbf{0}. \tag{2.39}$$

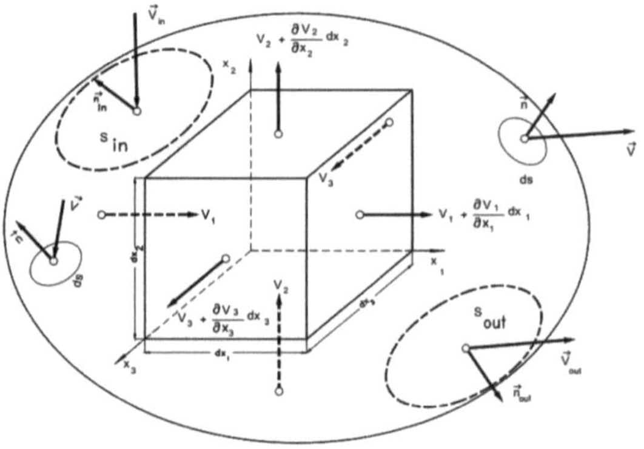

Fig. 2.4 Physical interpretation of $\nabla.V$

Equation (2.39) expresses the continuity equation for an incompressible flow, as we will see in the following chapters.

2.5.2 *Vector Product*

This operation is called the *rotation or curl* of the velocity vector **V**. Its result is a first-order tensor or a vector quantity. Using the index notation, the curl of **V** is written as:

$$\nabla \times \mathbf{V} = (e_i \frac{\partial}{\partial \mathbf{x_i}}) \times (\mathbf{e_j V_j}) = \boldsymbol{\epsilon}_{\mathbf{ijk}} \mathbf{e_k} \frac{\partial \mathbf{V_j}}{\partial \mathbf{x_i}} \equiv \boldsymbol{\omega}. \tag{2.40}$$

The curl of the velocity vector is known as *vorticity*, $\omega = \nabla \times V$. As we will see later, the vorticity plays a crucial role in fluid mechanics. It is a characteristic of a *rotational* flow. For viscous flows encountered in engineering applications, the curl $\omega = \nabla \times V$ is always different from zero. To simplify the flow situation and to solve the equation of motion, as we will discuss later, the vorticity vector $\omega = \nabla \times V$, can under certain conditions, be set equal to zero. This special case is called *the irrotational flow*.

2.5.3 *Tensor Product of ∇ and V*

This operation is called the *gradient* of the velocity vector **V**. Its result is a second tensor. Using the index notation, the gradient of the vector **V** is written as:

$$\nabla \mathbf{V} = (\mathbf{e_i} \frac{\partial}{\partial \mathbf{x_i}})(\mathbf{e_j V_j}) = \mathbf{e_i e_j} \frac{\partial \mathbf{V_j}}{\partial \mathbf{x_i}}. \tag{2.41}$$

Equation (2.41) is a second order tensor with nine components and describes the deformation and the rotation kinematics of the fluid particle. As we saw previously, the scalar multiplication of this tensor with the velocity vector, $\mathbf{V} \cdot \nabla \mathbf{V}$ resulted in the convective part of the acceleration vector, Eq. (2.32). In addition to the applications we discussed, ∇ can be applied to a product of two or more vectors by using the Leibnitz's chain rule of differentiation:

$$\nabla(\mathbf{U} \cdot \mathbf{V}) = \mathbf{U} \cdot \nabla \mathbf{V} + \mathbf{V} \cdot \nabla \mathbf{U} + \mathbf{U} \times (\nabla \times \mathbf{V}) + \mathbf{V} \times (\nabla \times \mathbf{U}). \tag{2.42}$$

For $\mathbf{U} = \mathbf{V}$, Equation (2.42) becomes $\nabla(\mathbf{V} \cdot \mathbf{V}) = \mathbf{2V} \cdot \nabla \mathbf{V} + \mathbf{2V} \times (\nabla \times \mathbf{V})$ or

$$\mathbf{V} \cdot \nabla \mathbf{V} = \frac{\mathbf{1}}{\mathbf{2}} \nabla(\mathbf{V} \cdot \mathbf{V}) - \mathbf{V} \times (\nabla \times \mathbf{V}) = \frac{\mathbf{1}}{\mathbf{2}} \nabla(\mathbf{V^2}) - \mathbf{V} \times (\nabla \times \mathbf{V}). \tag{2.43}$$

Equation (2.43) is used to express the convective part of the acceleration in terms of the gradient of kinetic energy of the flow.

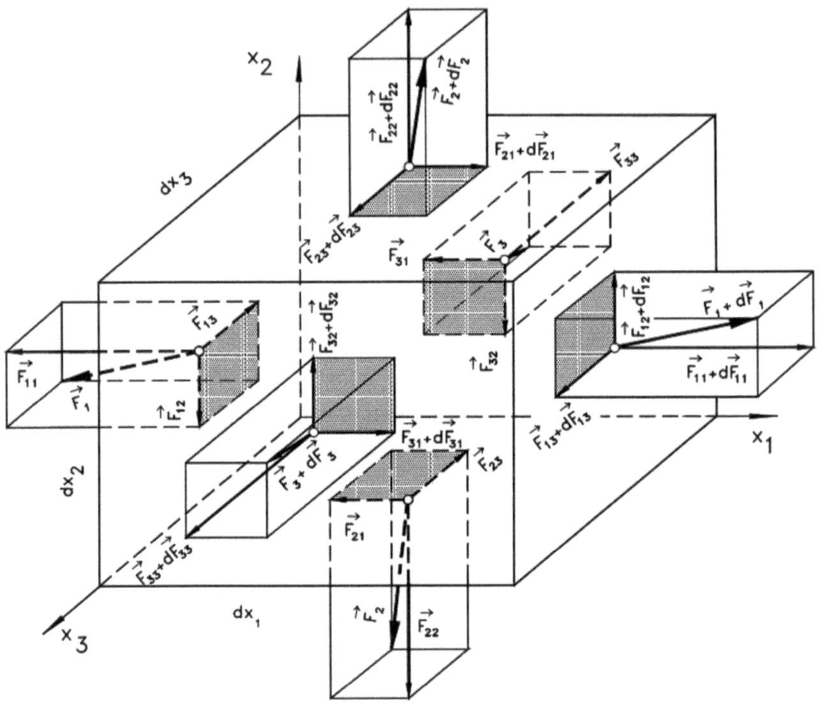

Fig. 2.5 Fluid element under a general three-dimensional stress condition

2.5.4 Scalar Product of ∇ and a Second Order Tensor

Consider a fluid element with sides dx_1, dx_2, dx_3 parallel to the axis of a Cartesian coordinate system, Fig. 2.5. The fluid element is under a general three-dimensional stress condition. The force vectors acting on the surfaces, which are perpendicular to the coordinates x_1, x_2, and x_3 are denoted by $\mathbf{F_1}$, $\mathbf{F_2}$ and $\mathbf{F_3}$. The opposite surfaces are subject to forces that have experienced infinitesimal changes $\mathbf{F_1} + \mathbf{dF_1}\mathbf{F_2} + \mathbf{dF_2}$ and $\mathbf{F_3} + \mathbf{dF_3}$. Each of these force vectors is decomposed into three components F_{ij} according to the coordinate system defined in Fig. 2.5.

The first index *i* refers to the axis, to which the fluid element surface is perpendicular, whereas the second index *j* indicates the direction of the force component. We divide the individual components of the above force vectors by their corresponding area of the fluid element side. The results of these divisions exhibit the components of a second order stress tensor represented by Π as shown in Fig. 2.6. As an example, we take the force component F_{11} and divide it by the corresponding area dx_2dx_3 results in $\frac{F_{11}}{dx_2dx_3} = \pi_{11}$. Correspondingly, we divide the force component on the opposite surface $F_{11} + dF_{11}$ by the same area dx_2dx_3 and obtain $\frac{F_{11}+dF_{11}}{dx_2dx_3} = \pi_{11} + \frac{\partial \pi_{11}}{\partial x_1}dx_1$. In a similar way we find the remaining stress components, which are shown in Fig. 2.6.

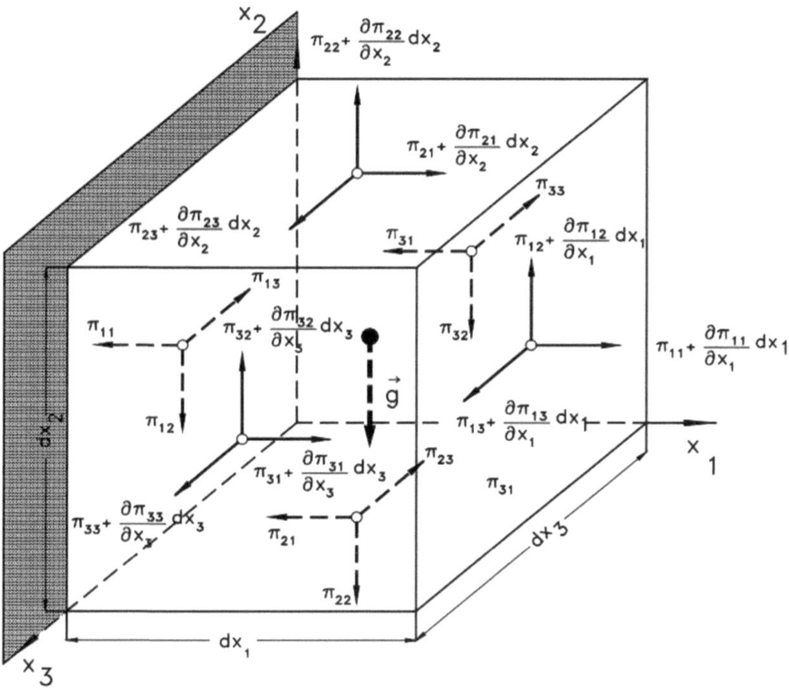

Fig. 2.6 General stress condition

The tensor $\mathbf{\Pi} = e_i e_j \pi_{ij}$ has nine components π_{ij} as the result of forces that are acting on surfaces. Similar to the force components, the first index i refers to the axis, to which the fluid element surface is perpendicular, whereas the second index j indicates the direction of the stress component.

Considering the stress situation in Fig. 2.6, we are now interested in finding the resultant force acting on the fluid particle that occupies the volume element $dv = dx_1 dx_2 dx_3$.

For this purpose, we look at the two opposite surfaces that are perpendicular to the axis x_1 as shown in Fig. 2.7. As this figure shows, we are dealing with three stress components on each surface, from which one on each side is the *normal stress* component such as π_{11} and $\pi_{11} + \frac{\partial \pi_{11}}{\partial x_1} dx_1$. The remaining components are the shear stress components such as π_{12} and $\pi_{12} + \frac{\partial \pi_{12}}{\partial x_1} dx_1$. According to Fig. 2.7 the force balance in x_1-directions is:

$$e_1 \left(\pi_{11} + \frac{\partial \pi_{11}}{\partial x_1} dx_1 - \pi_{11} \right) dx_2 dx_3 = e_1 \frac{\partial \pi_{11}}{\partial x_1} dx_1 dx_2 dx_3$$

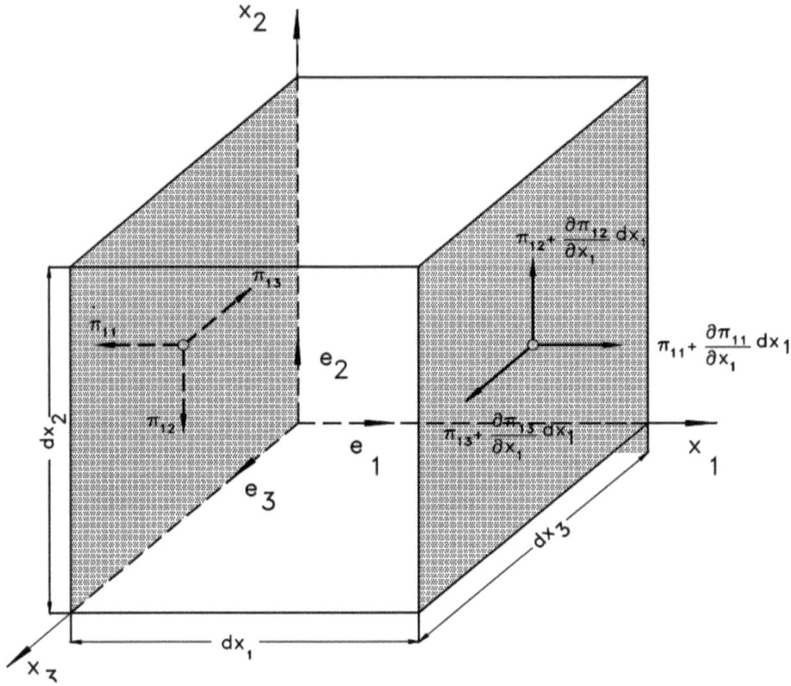

Fig. 2.7 Stresses on two opposite walls

and in x_2-direction, we find

$$e_2\left(\pi_{12}+\frac{\partial \pi_{12}}{\partial x_1}dx_1-\pi_{12}\right)dx_2dx_3=e_2\frac{\partial \pi_{12}}{\partial x_1}dx_1dx_2dx_3.$$

Similarly, in x_3, we obtain

$$e_3\left(\pi_{13}+\frac{\partial \pi_{13}}{\partial x_1}dx_1-\pi_{13}\right)dx_2dx_3=e_3\frac{\partial \pi_{13}}{\partial x_1}dx_1dx_2dx_3.$$

Thus, the resultant force acting on these two opposite surfaces is:

$$d\mathbf{F}_1=\left(\mathbf{e_1}\frac{\boldsymbol{\partial \pi_{11}}}{\boldsymbol{\partial x_1}}+\mathbf{e_2}\frac{\boldsymbol{\partial \pi_{12}}}{\boldsymbol{\partial x_1}}+\mathbf{e_3}\frac{\boldsymbol{\partial \pi_{13}}}{\boldsymbol{\partial x_1}}\right)\mathbf{dx_1dx_2dx_3}.$$

In a similar way, we find the forces acting on the other four surfaces. The total resulting forces acting on the entire surface of the element are obtained by adding the nine components. Defining the volume element $dv = dx_1dx_2dx_3$, we divide the results by dv and obtain the resulting force vector that is acting on the volume element.

$$\frac{d\mathbf{F}}{dv} = e_1\left[\frac{\partial\pi_{11}}{\partial x_1} + \frac{\partial\pi_{21}}{\partial x_2} + \frac{\partial\pi_{31}}{\partial x_3}\right] + \\ + e_2\left[\frac{\partial\pi_{12}}{\partial x_1} + \frac{\partial\pi_{22}}{\partial x_2} + \frac{\partial\pi_{32}}{\partial x_3}\right] + \\ + e_3\left[\frac{\partial\pi_{13}}{\partial x_1} + \frac{\partial\pi_{23}}{\partial x_2} + \frac{\partial\pi_{33}}{\partial x_3}\right]. \tag{2.44}$$

Since the stress tensor Π is written as:

$$\mathbf{\Pi} = e_i e_j \pi_{ij} \tag{2.45}$$

it can be easily shown that:

$$d\mathbf{F} = \nabla \cdot \mathbf{\Pi dv}. \tag{2.46}$$

The expression $\nabla \cdot \Pi$ is a scalar differentiation of the second order stress tensor and is called the divergence of the tensor field $\mathbf{\Pi}$. We conclude that the force acting on the surface of a fluid element is the divergence of its stress tensor. The stress tensor is usually divided into its normal and shear stress parts. For an incompressible fluid it can be written as

$$\mathbf{\Pi} = -\mathbf{Ip} + \mathbf{T} \tag{2.47}$$

with $\mathbf{I}p$ as the normal and $\mathbf{T}$ as the shear stress tensor. The normal stress tensor is a product of the unit tensor $\mathbf{I} = \mathbf{e_i e_j}\delta_{\mathbf{ij}}$ and the pressure p. Inserting Equation (2.47) into Eq. (2.46) leads to

$$\frac{d\mathbf{F}}{dv} = \nabla \cdot \mathbf{\Pi} = -\nabla p + \nabla \cdot \mathbf{T}. \tag{2.48}$$

Its components are

$$\frac{dF_i}{dv} = -\frac{\partial p}{\partial x_i} + \frac{\partial \tau_{ji}}{\partial x_j}. \tag{2.49}$$

2.5.5 *Eigenvalue and Eigenvector of a Second Order Tensor*

The velocity gradient expressed in Eq. (2.41) can be decomposed into a symmetric deformation tensor $\mathbf{D}$ and an antisymmetric rotation tensor Ω:

$$\nabla\mathbf{V} = \frac{\mathbf{1}}{\mathbf{2}}\left(\nabla\mathbf{V} + \nabla\mathbf{V}^{\mathbf{T}}\right) + \frac{\mathbf{1}}{\mathbf{2}}\left(\nabla\mathbf{V} - \nabla\mathbf{V}^{\mathbf{T}}\right) = \mathbf{D} + \mathbf{\Omega}. \tag{2.50}$$

A scalar multiplication of $\mathbf{D}$ with any arbitrary vector $\mathbf{A}$ may result in a vector, which has an arbitrary direction. However, there exists a particular vector $\mathbf{V}$ such that its scalar multiplication with $\mathbf{D}$ results in a vector, which is parallel to $\mathbf{V}$ but has different magnitude:

$$\mathbf{D} \cdot \mathbf{V} = \boldsymbol{\lambda}\mathbf{V} \tag{2.51}$$

with **V** as the eigenvector and λ the eigenvalue of the second order tensor **D**. Since any vector can be expressed as a scalar product of the unit tensor and the vector itself $I \cdot \mathbf{V} = \mathbf{V}$, we may write:

$$\mathbf{D} \cdot \mathbf{V} = \boldsymbol{\lambda}\mathbf{V} = \boldsymbol{\lambda}(\mathbf{I} \cdot \mathbf{V}) \tag{2.52}$$

that can be rearranged as:

$$(\mathbf{D} - \boldsymbol{\lambda}\mathbf{I}) \cdot \mathbf{V} = 0. \tag{2.53}$$

The index notation gives:

$$e_j(D_{ij}V_i - \delta_{ij}\lambda V_i) = 0 \tag{2.54}$$

or

$$D_{ij}V_i = \delta_{ij}\lambda V_i. \tag{2.55}$$

Expanding Eq. (2.54) gives a system of linear equations,

$$\begin{aligned} D_{11}V_1 + D_{12}V_2 + D_{13}V_3 &= \lambda V_1 \\ D_{21}V_1 + D_{22}V_2 + D_{23}V_3 &= \lambda V_2 \\ D_{31}V_1 + D_{32}V_2 + D_{33}V_3 &= \lambda V_3. \end{aligned} \tag{2.56}$$

A nontrivial solution of these Eq. (2.55) is possible if and only if the following determinant vanishes:

$$\det(\mathbf{D} - \boldsymbol{\lambda}\mathbf{I}) = \mathbf{0} \tag{2.57}$$

or in index notation:

$$\det(D_{ij} - \delta_{ij}\lambda) = 0. \tag{2.58}$$

Expanding Eq. (2.57) results in

$$\det \begin{pmatrix} D_{11} - \lambda & D_{12} & D_{13} \\ D_{21} & D_{22} - \lambda & D_{23} \\ D_{31} & D_{32} & D_{33} - \lambda \end{pmatrix} = 0. \tag{2.59}$$

After expanding the above determinant, we obtain an algebraic equation in λ in the following form

$$\lambda^3 - I_{1D}\lambda^2 + I_{2D}\lambda - I_{3D} = 0 \tag{2.60}$$

where I_{1D}, I_{2D} and I_{3D} are *invariants* of the tensor D defined as:

$$I_{1D} = \text{ Tr } D = D_{ii} = D_{11} + D_{22} + D_{33} \tag{2.61}$$

$$I_{2D} = \frac{1}{2}(I_{1D}^2 - D : D) = \frac{1}{2}(D_{ii}D_{jj} - D_{ij}D_{ij}) \tag{2.62}$$

$$I_{3D} = \det(D_{ij}). \tag{2.63}$$

The roots of Eq. (2.59) $\lambda^1, \lambda^2, \lambda^3$ are known as the eigenvalues of the tensor D. The superscript $n = 1, 2, 3$ refers to the roots of Eq. (2.59)—not to be confused with the component of a vector.

2.6 Problems

Problem 2.1 Show that $\nabla \times (\nabla\phi) = 0$ with ϕ as a scalar function.

Problem 2.2 Show that $\nabla \cdot (\nabla \times \mathbf{V}) = \mathbf{0}$ with $\mathbf{V}$ as a vector function.

Problem 2.3 Show that $\nabla \times (\Phi\mathbf{A}) = \Phi(\nabla \times \mathbf{A}) + (\nabla\phi) \times \mathbf{A}$ with ϕ as a scalar and $\mathbf{A}$ a vector function.

Problem 2.4 Show that $\nabla \times (\nabla \times \mathbf{A}) = -\nabla^2\mathbf{A} + \nabla(\nabla \cdot \mathbf{A})$

Problem 2.5 A scalar function is given as $r = \sqrt{x_1^2 + x_2^2 + x_3^2}$, find ∇r.

Problem 2.6 Show that $\nabla \times (\Phi\mathbf{A}) = \Phi(\nabla \times \mathbf{A}) + (\nabla\Phi) \times \mathbf{A}$ with ϕ and $\mathbf{A}$ as a scalar, vector function, respectively.

Problem 2.7 A scalar function is given as $f(x_1, x_2, x_3) = 2x_1^3x_2^2x_3^4$ find ∇f and $\nabla \cdot \nabla f$.

Problem 2.8 An incompressible flow field with water as the working fluid is given by the following vector function, where the coordinates are measured in meters.

$$\mathbf{V} = \mathbf{e_1(x_1 + 2x_2)e^{-t}} - \mathbf{e_2 2x_2 e^{-t}}$$

(a) Find the substantial acceleration of a fluid particle in vector form.
(b) Decompose the acceleration into the components, specify the nature of the flow.
(c) Using the Euler equation of motion:

$$\frac{D\mathbf{V}}{Dt} = -\frac{1}{\varrho}\nabla p + \mathbf{g}, \text{ where } \mathbf{g} = -\mathbf{e_3 g} = -\nabla(\mathbf{gz})$$

d) Find the pressure gradient at the $p(x_1, x_2) = (1, 2)$.

Problem 2.9 Starting from the above Euler equation of motion for inviscid incompressible flow obtain: a) the energy equation by multiplying the equation of motion with a differential displacement using the vector identity $\mathbf{V} \cdot \nabla\mathbf{V} = \nabla(\mathbf{V} \cdot \mathbf{V})/\mathbf{2} - \mathbf{V} \times (\nabla \times \mathbf{V})$.

Problem 2.10 The velocity field is given by:

$$u_1 = 0$$
$$u_2 = a(x_1x_2 - x_3^2)e^{-B(t-t_0)}$$
$$u_3 = a(x_2^2 - x_1x_3)e^{-B(t-t_0)}$$

(a) with B as a constant. Determine the components of the velocity gradient tensor. Start with the coordinate invariant form of the tensor, use index notation, write components and then plug functions in.
(b) Determine the components of the deformation tensor. Start with the coordinate invariant form of the tensor, use index notation, decompose into components and then plug values in.
(c) Determine the components of the rotation tensor. Start with the coordinate invariant form of the tensor, use index notation, decompose into components and then plug the values in.

Problem 2.11 The velocity field is given by:

$$u_1 = -\frac{\omega}{h}x_2x_3$$
$$u_2 = +\frac{\omega}{h}x_1x_3$$
$$u_3 = 0$$

(a) Determine the components of the velocity gradient tensor.
(b) Determine the components of the deformation tensor.
(c) Determine the components of the rotation tensor.

Problem 2.12 The Navier- Stokes equation is given by :

$$\frac{D\mathbf{V}}{Dt} = -\frac{1}{\varrho}\nabla p + \nu\Delta\mathbf{V} + \mathbf{g}, \text{ where } \mathbf{g} = -\mathbf{e}_3 g = -\nabla(\mathbf{gz})$$

(a) Give the index notation.
(b) Give the three components.

Chapter 3
Kinematics of Fluid Motion

3.1 Material and Spatial Description of the Flow Field

3.1.1 Material Description

Engineering fluid dynamic design process has been experiencing a continuous progress using the Computational Fluid Dynamics (CFD) tools. The use of CFD-tools opens a new perspective in simulating the complex three-dimensional (3-D) engineering flow fields. Understanding the details of the flow motion and the interpretation of the numerical results require a thorough comprehension of fluid mechanics laws and the kinematics of fluid motion. Kinematics is treated in many fluid mechanics texts. Aris [13] and Spurk [14] give an excellent account of the subject. In the following sections, a compact and illustrative treatment is given to cover the needs of engineers.

The kinematics is the description of the fluid motion and the particles without taking into account how the motion is brought about. It disregards the forces that cause the fluid motion. As a result, in the context of kinematics, no conservation laws of motion will be dealt with. Consequently, the results of kinematic studies can be applied to all types of fluids and exhibit the ground work that is necessary for describing the dynamics of the fluid. The motion of a fluid particle with respect to a reference coordinate system is in general given by a time dependent position vector $\mathbf{x}(\mathbf{t})$, Fig. 3.1

To identify the motion of a particle or a *material point* labeled as ξ^1 at a certain instance of time $t = t_0 = 0$, we introduce the position vector $\boldsymbol{\xi} = \boldsymbol{x}(t_0)$. Thus, the motion of the fluid is described by the vector:

$$\boldsymbol{x} = \boldsymbol{x}(\boldsymbol{\xi}, t, x_i = x_i(\boldsymbol{\xi}, t) \tag{3.1}$$

with x_i as the components of vector $\boldsymbol{x}$, as explained in Chap. 2. Equation (3.1) describes the path of a material point that has an initial position vector $\boldsymbol{\xi}$ that characterizes or better labels the material point at $t = t_0$. We refer to this description as the

M. T. Schobeiri, *Advanced Fluid Mechanics and Heat Transfer for Engineers and Scientists*, https://doi.org/10.1007/978-3-030-72925-7_3

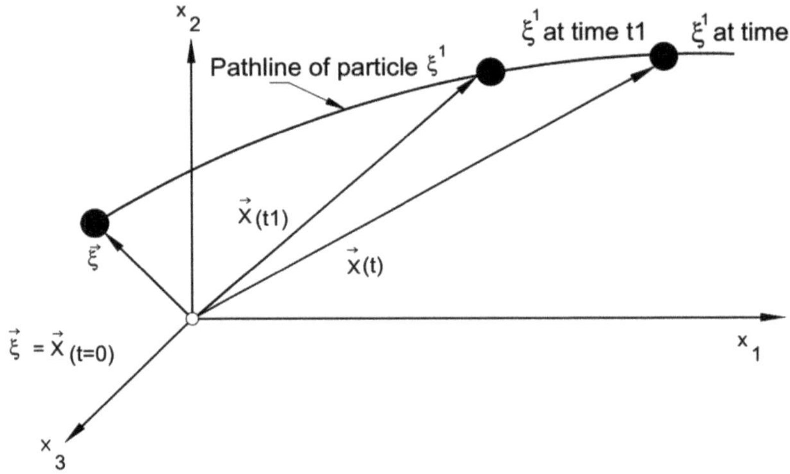

Fig. 3.1 Material description of a fluid particle motion

material description also called *Lagrangian* description. Considering another material point labeled ξ^i with different $\boldsymbol{\xi}$-coordinates, their paths are similarly described by Eq. (3.1). If we assume that the motion is continuous and single valued, then the inversion of Eq. (3.1) must give the initial position $\boldsymbol{\xi}$ or material coordinate of each fluid particle which may be at any position $\boldsymbol{x}$ and any instant of time t; that is,

$$\boldsymbol{\xi} = \boldsymbol{\xi}(\boldsymbol{x}, t), \xi_i = \xi_i(\boldsymbol{x}, t). \tag{3.2}$$

The necessary and sufficient condition for an inverse function to exist is that the *Jacobian transformation function*

$$J = \left(\varepsilon_{kmn} \frac{\partial x_k}{\partial \xi_1} \frac{\partial x_m}{\partial \xi_2}, \frac{\partial x_n}{\partial \xi_3} \right) \tag{3.3}$$

does not vanish. Because of the significance of the Jacobian transformation function to fluid mechanics, we derive this function in the following section.

3.1.2 Jacobian Transformation Function and its Material Derivative

We consider a differential *material volume* at the time $t = 0$, to which we attach the reference coordinate system $\boldsymbol{\xi}_1, \boldsymbol{\xi}_2, \boldsymbol{\xi}_3$, as shown in Fig. 3.2.

At the time $t = 0$, the reference coordinate system is fixed so that the *undeformed* differential material volume dV_0 (Figs. 3.2 and 3.3) can be described as:

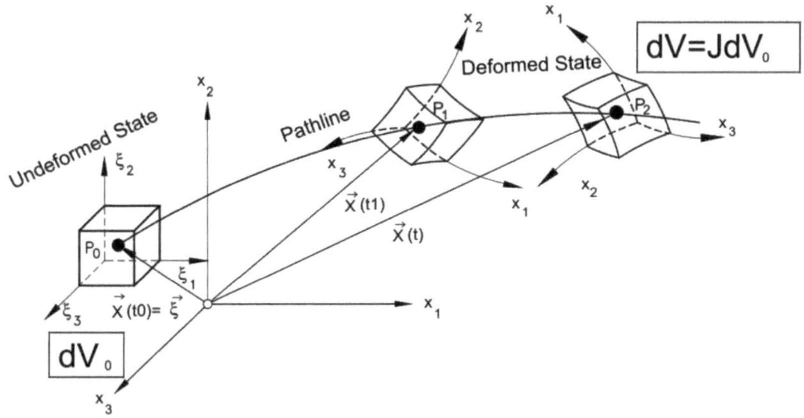

Fig. 3.2 Deformation of a differential volume at different instant of time

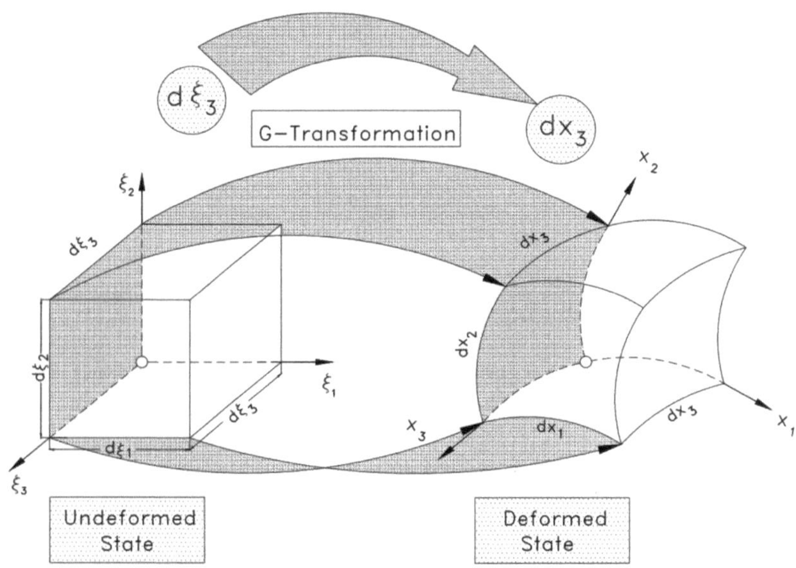

Fig. 3.3 Transformation of $d\xi_1$, $d\xi_2$, $d\xi_3$ into dx_1, dx_2, and dx_3 using G-transformation

$$dV_0 = (e_1 d\xi_1 \times e_2 d\xi_2) \cdot e_3 d\xi_3 = d\xi_1 d\xi_2 d\xi_3. \tag{3.4}$$

Moving through the space, the differential material volume may undergo certain deformation and rotation. As deformation takes place, the sides of the material volume initially given as $d\xi_i$ would be convected into a non-rectangular, or curvilinear form. The changes of the deformed coordinates are then:

$$dx = \frac{\partial x}{\partial \xi_1} d\xi_1 + \frac{\partial x}{\partial \xi_2} d\xi_2 + \frac{\partial x}{\partial \xi_3} d\xi_3 = \frac{\partial x}{\partial \xi_i} d\xi_i. \tag{3.5}$$

Using the index notation for the position vector $\boldsymbol{x} = \boldsymbol{e}_k x_k$, Eq. (3.5) may be rearranged in the following way:

$$d\boldsymbol{x} = e_k \frac{\partial x_k}{\partial \xi_i} d\xi_i \equiv \boldsymbol{G}_i d\xi_i \tag{3.6}$$

with the vector $\boldsymbol{G}_i$ as where

$$\boldsymbol{G}_i \equiv e_k \frac{\partial x_k}{\partial \xi_i}. \tag{3.7}$$

G_i is a *transformation vector function* that transforms the differential changes $d\xi_i$ into dx_i. Figure 3.3 illustrates the deformation of the material volume and the transformation mechanism. The new deformed differential volume is obtained by:

$$dV = (d\boldsymbol{x}_1 \times d\mathbf{x_2}) \cdot \mathbf{d}\boldsymbol{x}_3. \tag{3.8}$$

Introducing Eq. (3.6) into Eq. (3.8) leads to:

$$dV = (\boldsymbol{G}_1 d\xi_1 \times \mathbf{G_2 d_{,2}}) \cdot \boldsymbol{G}_3\mathbf{d}_{,3}. \tag{3.9}$$

Inserting $\boldsymbol{G}_i$ from Eq. (3.7) into Eq. (3.9) and considering Sect. 2.2.2, we arrive at:

$$dV = \left(e_k \frac{\partial x_k}{\partial \xi_1} \times e_m \frac{\partial x_m}{\partial \xi_2} \right) \cdot e_n \frac{\partial x_n}{\partial \xi_3} d\xi_1 d\xi_2 d\xi_3. \tag{3.10}$$

Now we replace the vector product and the following scalar product of the two unit vectors in Eq. (3.10) with the permutation symbol and the Kronecker delta:

$$dV = \varepsilon_{kml} \delta_{\mathrm{ln}} \frac{\partial x_k}{\partial \xi_1} \frac{\partial x_m}{\partial \xi_2} \frac{\partial x_n}{\partial \xi_3} d\xi_1 d\xi_2 d\xi_3. \tag{3.11}$$

Applying the Kronecker delta to the terms with the indices l and n, we arrive at:

$$dV = \left(\varepsilon_{kmn} \frac{\partial x_k}{\partial \xi_1} \frac{\partial x_m}{\partial \xi_2} \frac{\partial x_n}{\partial \xi_3} \right) (d\xi_1 \xi_2 d\xi_3). \tag{3.12}$$

The expression in first parenthesis in Eq. (3.12) represents the *Jacobian function* J.

$$J = \left(\varepsilon_{kmn} \frac{\partial x_k}{\partial \xi_1} \frac{\partial x_m}{\partial \xi_2} \frac{\partial x_n}{\partial \xi_3} \right). \tag{3.13}$$

The second parenthesis in Eq. (3.12) represents the initial infinitesimal material volume in the *undeformed state* at $t = 0$, described by Eq. (3.3). Using these terms, Eq. (3.12) is rewritten as:

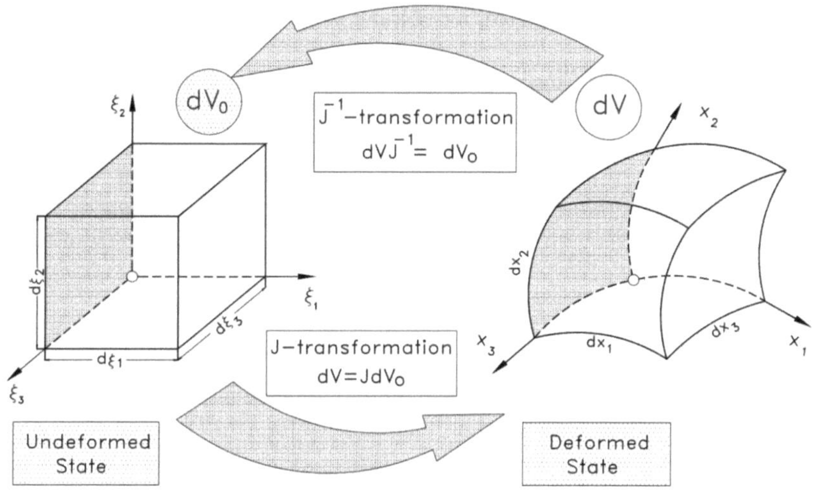

Fig. 3.4 Jacobian transformation of a material volume, change of states

$$dV = JdV_0 \tag{3.14}$$

where dV represents the differential volume in the *deformed state*, dV_0 has the same differential volume in the undeformed state at the time $t = 0$. The transformation function J is also called the *Jacobian functional determinant.* Performing the permutation in Eq. (3.13), the determinant is given as:

$$J = \det \begin{pmatrix} \frac{\partial x_1}{\partial \xi_1} & \frac{\partial x_2}{\partial \xi_1} & \frac{\partial x_3}{\partial \xi_1} \\ \frac{\partial x_1}{\partial \xi_2} & \frac{\partial x_2}{\partial \xi_2} & \frac{\partial x_3}{\partial \xi_2} \\ \frac{\partial x_1}{\partial \xi_3} & \frac{\partial x_2}{\partial \xi_3} & \frac{\partial x_3}{\partial \xi_3} \end{pmatrix}. \tag{3.15}$$

With the Jacobian functional determinant, we now have a necessary tool to directly relate any time dependent differential volume $dV = dV(t)$ to its fixed reference volume dV_0 at the reference time $t = 0$ as shown in Fig. 3.4.

The Jacobian transformation function and its material derivative are the fundamental tools to understand the conservation laws using the integral analysis in conjunction with control volume method. To complete this section, we briefly discuss the material derivative of the Jacobian function.

As the volume element dV follows the motion from $\boldsymbol{x} = \boldsymbol{x}(\boldsymbol{\xi}, t)$ to $\boldsymbol{x} = \boldsymbol{x}(\boldsymbol{\xi}, t + dt)$ it changes and, as a result, the Jacobian transformation function undergoes a time change. To calculate this change, we determine the material derivative of J:

$$\frac{DJ}{Dt} = \frac{\partial J}{\partial t} + \boldsymbol{V} \cdot \nabla J. \tag{3.16}$$

Inserting Eq. (3.13) into Eq. (3.16), we obtain:

$$\frac{DJ}{Dt} = \frac{\partial}{\partial t}\left(\varepsilon_{kmn}\frac{\partial x_k}{\partial \xi_1}\frac{\partial x_m}{\partial \xi_2}\frac{\partial x_n}{\partial \xi_3}\right) + V_j\frac{\partial}{\partial x_j}\left(\varepsilon_{kmn}\frac{\partial x_k}{\partial \xi_1}\frac{\partial x_m}{\partial \xi_2}\frac{\partial x_n}{\partial \xi_3}\right). \tag{3.17}$$

Let us consider an arbitrary element of the Jacobian determinant, for example $\partial x_1/\partial \xi_2$. Since the reference coordinate $\xi \neq f(t)$ is not a function of time and is fixed, the differentials with respect to t and ξ_2, can be interchanged resulting in:

$$\frac{\partial}{\partial t}\left(\frac{\partial x_1}{\partial \xi_2}\right) = \frac{\partial}{\partial \xi_2}\left(\frac{\partial x_1}{\partial t}\right) = \frac{\partial V_1}{\partial \xi_2}. \tag{3.18}$$

Performing similar operations for all elements of the Jacobian determinant and noting that the second expression on the right-hand side of Eq. (3.17) identically vanishes, we arrive at:

$$\frac{DJ}{Dt} = \left(\frac{\partial V_1}{\partial x_1} + \frac{\partial V_2}{\partial x_2} + \frac{\partial V_3}{\partial x_3}\right) J. \tag{3.19}$$

The expression in the parenthesis of Eq. 3.19 is the well known divergence of the velocity vector. Using vector notation, Eq. (3.19) becomes:

$$\frac{DJ}{Dt} = (\nabla \cdot \boldsymbol{V})J. \tag{3.20}$$

3.1.3 Velocity, Acceleration of Material Points

The *velocity* and the *acceleration* of a material point are given by:

$$\boldsymbol{V} = \frac{d\mathbf{x}}{dt}, \boldsymbol{A} = \frac{d^2\boldsymbol{x}}{dt^2}. \tag{3.21}$$

The velocity of the material point is written as:

$$\boldsymbol{V}(\boldsymbol{\xi}, t) = \left[\frac{\partial \boldsymbol{x}}{\partial t}\right]_{\boldsymbol{\xi}}, V_i(\xi_j, t) = \left[\frac{\partial x_i}{\partial t}\right]_{\xi_j} \tag{3.22}$$

where the subscript $\boldsymbol{\xi}$ refers to a fixed material point. The acceleration can be obtained form:

$$\boldsymbol{A}(\boldsymbol{\xi}, t) = \left[\frac{\partial \boldsymbol{V}}{\partial t}\right]_{\boldsymbol{\xi}}, A_i(\xi_j, t) = \left[\frac{\partial V_i}{\partial t}\right]_{\xi_j}. \tag{3.23}$$

As seen from Eqs. (3.22) and (3.23), the derivatives were taken at a fixed $\boldsymbol{\xi}$; it is the time derivative for the $\boldsymbol{\xi}$th material point, such as $\boldsymbol{\xi}^1$, $\boldsymbol{\xi}^2$, $\boldsymbol{\xi}^3$, $\boldsymbol{\xi}^i$,.... and $\boldsymbol{\xi}^n$. Regarding the differentiation, confusion is highly unlikely to arise, since $\boldsymbol{\xi}$ is not a function of time. The introduction of the term *material description* we used on several occasions is obviously descriptive, because the variable $\boldsymbol{\xi}$ directly labels the material point at the reference time $t = 0$.

3.1.4 Spatial Description

The material description we discussed in the previous section deals with the motion of the individual particles of a continuum, and is used in *continuum mechanics*. In fluid dynamics, we are primarily interested in determining the flow quantities such as velocity, acceleration, density, temperature, pressure, and etc., at fixed points in space. For example, determining the three-dimensional distribution of temperature, pressure and shear stress helps engineers design turbines, compressors, combustion engines, etc. with higher efficiencies. For this purpose, we introduce the *spatial description*, which is also called the *Euler description*. The independent variables for the spatial descriptions are the space characterized by the position vector $\boldsymbol{x}$ and the time t. Consider the transformation of Eq. (3.1), where $\boldsymbol{\xi}$ is solved in terms of $\boldsymbol{x}$:

$$\boldsymbol{\xi} = \boldsymbol{\xi}(\boldsymbol{x}, t), \xi_i = \xi_i(x_j, t). \tag{3.24}$$

The position vector $\boldsymbol{\xi}$ in the velocity of the material element $\boldsymbol{V}(\boldsymbol{\xi}, t)$ is replaced by Eq. (3.24):

$$\boldsymbol{V}(\boldsymbol{\xi}, t) = \boldsymbol{V}\,[\boldsymbol{\xi}(\boldsymbol{x}, t), t] = \boldsymbol{V}(\boldsymbol{x}, t). \tag{3.25}$$

For a fixed $\boldsymbol{x}$, Eq. (3.25) exhibits the velocity at the spatially fixed position $\boldsymbol{x}$ as a function of time. On the other hand, for a fixed t, Eq. (3.25) describes the velocity at the time t. With Eq. (3.25), any quantity described in spatial coordinates can be transformed into material coordinates provided the Jacobian transformation function J, which we discussed in the previous section, does not vanish. If the velocity is known in a spatial coordinate system, the path of the particle can be determined as the integral solution of the differential equation with the initial condition $\boldsymbol{x}(t_0)$ along the path $\boldsymbol{x} = \boldsymbol{x}(\boldsymbol{\xi}, t)$ from the following relation:

$$\frac{d\boldsymbol{x}}{dt} = \boldsymbol{V}(\boldsymbol{x}, t), \frac{dx_i}{dt} = V_i(x_j, t). \tag{3.26}$$

3.2 Translation, Deformation, Rotation

During a general three-dimensional motion, a fluid particle undergoes a translational and rotational motion which may be associated with deformation. The velocity of a particle at a given spatial, temporal position $(\boldsymbol{x} + d\boldsymbol{x}, t)$ can be related to the velocity at $(\boldsymbol{x}, t)$ by using the following Taylor expansion:

$$\boldsymbol{V}(\mathbf{x}+\mathbf{dx},\mathbf{t}) = \boldsymbol{V}(\mathbf{x},\mathbf{t}) + \mathbf{d}\boldsymbol{V}. \tag{3.27}$$

Inserting in Eq. (3.27) for the differential velocity change $d\boldsymbol{V} = d\boldsymbol{x} \cdot \nabla \boldsymbol{V}$, Eq. (3.27) is re-written as:

$$\boldsymbol{V}(\boldsymbol{x}+d\boldsymbol{x},t) = \boldsymbol{V}(\boldsymbol{x},t) + d\boldsymbol{x} \cdot \nabla \boldsymbol{V}. \tag{3.28}$$

The first term on the right-hand side of Eq. (3.28) represents the translational motion of the fluid particle. The second expression is a scalar product of the differential displacement $d\boldsymbol{x}$ and the *velocity gradient* $\nabla \boldsymbol{V}$. We decompose the velocity gradient, which is a second order tensor, into two parts resulting in the following *identity*:

$$\nabla \boldsymbol{V} = \frac{1}{2}\left(\nabla \boldsymbol{V} + \nabla \boldsymbol{V}^T\right) + \frac{1}{2}\left(\nabla \boldsymbol{V} - \nabla \boldsymbol{V}^T\right). \tag{3.29}$$

The superscript T indicates that the matrix elements of the second order tensor $\nabla \boldsymbol{V}^T$ are the transpositions of the matrix elements that pertain to the second order tensor $\nabla \boldsymbol{V}$. The first term in the right-hand side represents the *deformation tensor*, which is a symmetric second order tensor:

$$\boldsymbol{D} = \frac{1}{2}\left(\nabla \boldsymbol{V} + \nabla \boldsymbol{V}^T\right) = e_i e_j D_{ij} = \frac{1}{2} e_i e_j \left(\frac{\partial V_i}{\partial x_j} + \frac{\partial V_j}{\partial x_i}\right) \tag{3.30}$$

with components:

$$D_{ij} = \frac{1}{2}\left(\frac{\partial V_i}{\partial x_j} + \frac{\partial V_j}{\partial x_i}\right). \tag{3.31}$$

The second term of Eq. (3.28) is called the rotation or vorticity tensor, which is antisymmetric and is given by:

$$\boldsymbol{\Omega} = \frac{1}{2}\left(\nabla \boldsymbol{V} - \nabla \boldsymbol{V}^T\right) = e_i e_j \Omega_{ij} = \frac{1}{2} e_i e_j \left(\frac{\partial V_j}{\partial x_i} - \frac{\partial V_i}{\partial x_j}\right). \tag{3.32}$$

The components are:

$$\Omega_{ij} = \frac{1}{2}\left(\frac{\partial V_j}{\partial x_i} - \frac{\partial V_i}{\partial x_j}\right). \tag{3.33}$$

Inserting Eqs. (3.30) and (3.32) into Eq. (3.28), we arrive at:

$$\boldsymbol{V}(\boldsymbol{x}+d\boldsymbol{x},t) = \boldsymbol{V}(\boldsymbol{x},t) + d\boldsymbol{x} \cdot \boldsymbol{D} + d\boldsymbol{x} \cdot \boldsymbol{\Omega}. \tag{3.34}$$

Equation (3.34) describes the kinematics of the fluid particle, which has a combined translational and rotational motion and undergoes a deformation.

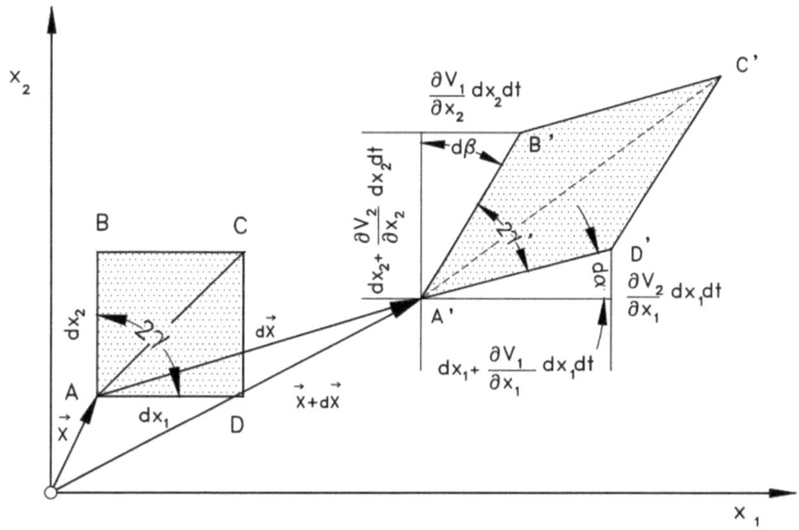

Fig. 3.5 Translation, rotation and deformation details of a fluid particle

Figure 3.5 illustrates the geometric representation of the rotation and deformation, [15]. Consider the fluid particle with a square-shaped cross section in the $x_1 - x_2$ plane at the time t. The position of this particle is given by the position vector $\boldsymbol{x} = \boldsymbol{x}(t)$. By moving through the flow field, the particle experiences translational motion to a new position $\boldsymbol{x} + \boldsymbol{dx}$. This motion may be associated with a rotational motion and a deformation. The deformation is illustrated by the initial and final state of diagonal $A - C$, which is stretched to $A' - C'$ and the change of the angle 2γ to $2\gamma'$. The rotational motion can be appropriately illustrated by the rotation of the diagonal by the angle $d\phi_3 = \gamma' + d\alpha - \gamma$, where γ' can be eliminated using the relation $2\gamma' + d\alpha + d\beta = 2\gamma$. As a result, we obtain the infinitesimal rotation angle:

$$d\phi_3 = \frac{1}{2}\left(d\alpha - d\beta\right) \tag{3.35}$$

where the subscript 3 denotes the direction of the rotation axis, which is parallel to x_3, Fig. 3.5. Referring to Fig. 3.5, direct relationships between $d\alpha$, $d\beta$ and the velocity gradients can be established by:

$$d\alpha \approx \tan(d\alpha) = \frac{\frac{\partial V_2}{\partial x_1} dx_1 dt}{dx_1 + \frac{\partial V_1}{\partial x_1} dx_1 dt} \approx \frac{\partial V_2}{\partial x_1} dt. \tag{3.36}$$

A similar relationship is given for the angle change $d\beta$:

$$d\beta \approx \tan(d\beta) = \frac{\frac{\partial V_1}{\partial x_2} dx_2 dt}{dx_2 + \frac{\partial V_2}{\partial x_2} dx_2 dt} \approx \frac{\partial V_1}{\partial x_2} dt. \tag{3.37}$$

Substituting Eqs. (3.36) and (3.37) into Eq. (3.35), the *rotation rate* in the x_3-direction is found:

$$\frac{d\phi_3}{dt} = \frac{1}{2}\left(\frac{\partial V_2}{\partial x_1} - \frac{\partial V_1}{\partial x_2}\right). \tag{3.38}$$

Executing the same procedure, the other two components are:

$$\frac{d\phi_1}{dt} = \frac{1}{2}\left(\frac{\partial V_3}{\partial x_2} - \frac{\partial V_2}{\partial x_3}\right), \ \frac{d\phi_2}{dt} = \frac{1}{2}\left(\frac{\partial V_1}{\partial x_3} - \frac{\partial V_3}{\partial x_1}\right). \tag{3.39}$$

The above three terms in Eqs. (3.38) and (3.39) may be recognized as one-half of the three components of the *vorticity vector* $\boldsymbol{\omega}$, which is:

$$\boldsymbol{\omega} \equiv \nabla \times \boldsymbol{V} = \varepsilon_{ijk} e_k \frac{\partial V_j}{\partial x_i}$$
$$\boldsymbol{\omega} = e_1 \left(\frac{\partial V_3}{\partial x_2} - \frac{\partial V_2}{\partial x_3}\right) + e_2 \left(\frac{\partial V_1}{\partial x_3} - \frac{\partial V_3}{\partial x_1}\right) + e_3 \left(\frac{\partial V_2}{\partial x_1} - \frac{\partial V_1}{\partial x_2}\right). \tag{3.40}$$

Examining the elements of the rotation matrix,

$$\Omega_{ij} = \begin{pmatrix} 0 & \frac{1}{2}\left(\frac{\partial V_2}{\partial x_1} - \frac{\partial V_1}{\partial x_2}\right) & \frac{1}{2}\left(\frac{\partial V_3}{\partial x_1} - \frac{\partial V_1}{\partial x_3}\right) \\ \frac{1}{2}\left(\frac{\partial V_1}{\partial x_2} - \frac{\partial V_2}{\partial x_1}\right) & 0 & \frac{1}{2}\left(\frac{\partial V_3}{\partial x_2} - \frac{\partial V_2}{\partial x_3}\right) \\ \frac{1}{2}\left(\frac{\partial V_1}{\partial x_3} - \frac{\partial V_3}{\partial x_1}\right) & \frac{1}{2}\left(\frac{\partial V_2}{\partial x_3} - \frac{\partial V_3}{\partial x_2}\right) & 0 \end{pmatrix} \tag{3.41}$$

we notice that the diagonal elements of the above antisymmetric tensor are zero and only three of the six non-zero elements are distinct. Except for the factor of one-half, these three distinct components are the same as those making up the vorticity vector. Comparing the components of the vorticity vector $\boldsymbol{\omega}$ given by Eq. (3.40) and the three distinct terms of Eq. (3.41), we conclude that the components of the vorticity vector, except for the factor of one-half, are identical with the *axial vector* of the antisymmetric tensor, Eq. (3.41). The axial vector of the second order tensor $\boldsymbol{\Omega}$ is the double scalar product of the third order permutation tensor $\boldsymbol{\varepsilon} = e_i e_j e_k \varepsilon_{ijk}$ with $\boldsymbol{\Omega}$:

$$\boldsymbol{\varepsilon} : \boldsymbol{\Omega} = \varepsilon_{ijk} e_i e_j e_k : e_m e_n \Omega_{mn} = e_i \varepsilon_{ijk} \frac{1}{2}\left(\frac{\partial V_j}{\partial x_k} - \frac{\partial V_k}{\partial x_j}\right). \tag{3.42}$$

Expanding Eq. (3.42) results in:

$$\varepsilon : \boldsymbol{\Omega} = e_1\left(\frac{\partial V_2}{\partial x_3} - \frac{\partial V_3}{\partial x_2}\right) + e_2\left(\frac{\partial V_3}{\partial x_1} - \frac{\partial V_1}{\partial x_3}\right) + e_3\left(\frac{\partial V_1}{\partial x_2} - \frac{\partial V_2}{\partial x_1}\right). \quad (3.43)$$

Comparing Eq. (3.43) to Eq. (3.40) shows that the right-hand side of Eq. (3.43) multiplied with a negative sign is exactly equal the right-hand side of Eq. (3.40). This indicates that the axial vector of the rotation tensor is equal to the negative rotation vector and can be expressed as:

$$\nabla \times \boldsymbol{V} = -\varepsilon : \boldsymbol{\Omega} = -e_i\varepsilon_{ijk}\frac{1}{2}\left(\frac{\partial V_j}{\partial x_k} - \frac{\partial V_k}{\partial x_j}\right). \quad (3.44)$$

The existence of the vorticity vector $\boldsymbol{\omega}$ and therefore, the rotation tensor $\boldsymbol{\Omega}$, is a characteristic of viscous flows that in general undergoes a *rotational motion*. This is particularly true for boundary layer flows, where the fluid particles move very close to the solid boundaries. In this region, the wall shear stress forces (friction forces) cause a combined deformation and rotation of the fluid particle. In contrast, for *inviscid flows*, or the flow regions, where the viscosity effect may be neglected, the rotation vector $\boldsymbol{\omega}$ may vanish if the flow can be considered isentropic. This ideal case is called *potential flow*, where the rotation vector $\nabla \times \boldsymbol{V} = 0$ in the entire potential flow field.

3.3 Reynolds Transport Theorem

The conservation laws in integral form are, strictly speaking, valid for *closed systems*, where the *mass* does not cross the *system boundary*. In fluid mechanics, however, we are dealing with *open systems*, where the *mass flow* continuously crosses the system boundary. To apply the conservation laws to open systems, we briefly provide the necessary mathematical tools. In this section, we treat the volume integral of an arbitrary field quantity $f(\boldsymbol{X}, t)$ by deriving the *Reynolds transport theorem*. This is an important kinematic relation that we will use in Chap. 4.

The field quantity $f(\boldsymbol{X}, t)$ may be a zeroth, first or second order tensor valued function, such as mass, velocity vector, and stress tensor. The time dependent volume under consideration with a given time dependent surface moves through the flow field and may experience dilatation, compression and deformation. It is assumed to contain the same fluid particles at any time and therefore, it is called the material volume. The volume integral of the quantity $f(\boldsymbol{X}, t)$:

$$F(t) = \int_{v(t)} f(\boldsymbol{X}, t)dv \quad (3.45)$$

is a function of time only. The integration must be carried out over the varying volume $v(t)$. The material change of the quantity $F(t)$ is expressed as:

$$\frac{DF(t)}{Dt} = \frac{D}{Dt} \int\limits_{v(t)} f(\boldsymbol{X}, t) dv. \tag{3.46}$$

Since the shape of the volume $v(t)$ changes with time, the differentiation and integration cannot be interchanged. However, Eq. (3.46) permits the transformation of the time dependent volume $v(t)$ into the fixed volume v_0 at time $t = 0$ by using the Jacobian transformation function:

$$\frac{DF(t)}{Dt} = \frac{D}{Dt} \int\limits_{v_0} f(\boldsymbol{X}, t) J dv_0. \tag{3.47}$$

With this operation in Eq. (3.47), it is possible to interchange the sequence of differentiation and integration:

$$\frac{DF(t)}{Dt} = \int\limits_{v_0} \frac{D}{Dt} f(\boldsymbol{X}, t) J) dv_0. \tag{3.48}$$

The chain differentiation of the expression within the parenthesis results in

$$\frac{DF(t)}{Dt} = \int\limits_{v_0} \left(J \frac{D}{Dt} f(\boldsymbol{X}, t) + f(\boldsymbol{X}, t) \frac{DJ}{Dt} \right) dv_0. \tag{3.49}$$

Introducing the material derivative of the Jacobian function, Eq. (3.19) into Eq. (3.49) yields:

$$\frac{DF(t)}{Dt} = \int\limits_{v_0} \left(\frac{D}{Dt} f(\boldsymbol{X}, t) + f(\boldsymbol{X}, t) \nabla \cdot \boldsymbol{V} \right) J dv_0. \tag{3.50}$$

Eq. (3.50) permits the back transformation of the fixed volume integral into the time dependent volume integral:

$$\frac{DF(t)}{Dt} = \int\limits_{v(t)} \left(\frac{D}{Dt} f(\boldsymbol{X}, t) + f(\boldsymbol{X}, t) \nabla \cdot \boldsymbol{V} \right) dv. \tag{3.51}$$

According to A4.1, the first term in the parenthesis can be written as:

$$\frac{Df}{Dt} = \frac{\partial f}{\partial t} + \boldsymbol{V} \cdot \nabla f. \tag{3.52}$$

Introducing Eq. (3.52) into Eq. (3.51) results in:

$$\frac{DF(t)}{Dt} = \int\limits_{v(t)} \left(\frac{\partial}{\partial t} f(\boldsymbol{X}, t) + \mathbf{V} \cdot \nabla \mathbf{f}(\boldsymbol{X}, \mathbf{t}) + \mathbf{f}(\boldsymbol{X}, \mathbf{t}) \nabla \cdot \boldsymbol{V} \right) dv. \tag{3.53}$$

The chain rule applied to the second and third term in Eq. (3.53) yields:

$$\frac{DF(t)}{Dt} = \int\limits_{v(t)} \left\{ \frac{\partial}{\partial t} f(\boldsymbol{X}, t) + \nabla \cdot (f(\boldsymbol{X}, t)\boldsymbol{V}) \right\} dv. \tag{3.54}$$

The second volume integral in Eq. (3.54) can be converted into a surface integral by applying the Gauss' divergence theorem:

$$\int\limits_{v(t)} \nabla \cdot (f(\boldsymbol{X}, t)\boldsymbol{V})\, dv = \int\limits_{S(t)} f(\boldsymbol{X}, t)\boldsymbol{V} \cdot \boldsymbol{n} dS \tag{3.55}$$

where $\boldsymbol{V}$ represents the *flux velocity* and $\boldsymbol{n}$ the unit vector normal to the surface. Inserting Eq. (3.55) into Eq. (3.54) results in the following final equation, which is called the *Reynolds transport theorem*

$$\frac{DF(t)}{Dt} = \int\limits_{v(t)} \frac{\partial}{\partial(t)} f(\boldsymbol{X}, t) dv + \int\limits_{S(t)} f(\boldsymbol{X}, t)\mathbf{V} \cdot \boldsymbol{n}\mathbf{dS}. \tag{3.56}$$

Equation (3.56) is valid for any system boundary with time the dependent volume $V(t)$ and surface $S(t)$ at any time, including the time $t = t_0$, where the volume $V = V_C$ and the surface $S = S_C$ assume fixed values. We call V_C and S_C the *control volume* and *control surface*.

3.4 Pathline, Streamline, Streakline

Equation (3.23) indicates that the path of a material point is tangential to its velocity. Consequently, the pathline can be defined as the trajectory of a material point, in this case, a fluid particle over a period of time. Pathline is inherent in material description. Figure3.6 exhibits the pathlines of material points labeled as ξ^k. In spatial description of the flow, we deal with the *streamlines* rather than pathlines. Consider a time dependent (*unsteady flow*) velocity field at time t, where each position $\boldsymbol{x}$ is associated with a velocity vector. The streamlines are curves whose tangent directions have the same directions as the velocity vectors, Fig. 3.7. To find an analytical expression for the description of a streamline, we define a unit tangent vector to the streamline curve S (Fig. 3.7). The direction of this unit tangent vector is then identical with the direction of the velocity vector at the vector position $\boldsymbol{X}$. As illustrated in Fig. 3.7, we define the tangent unit vector as:

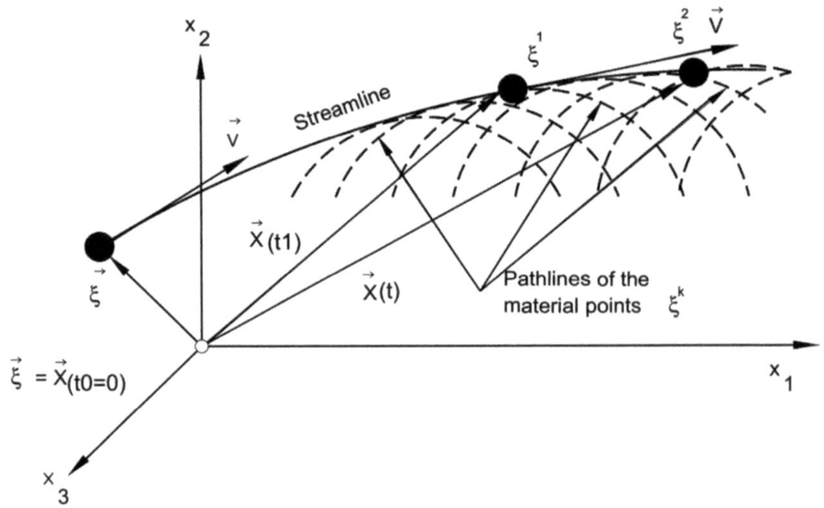

Fig. 3.6 Representation of pathlines and streamline

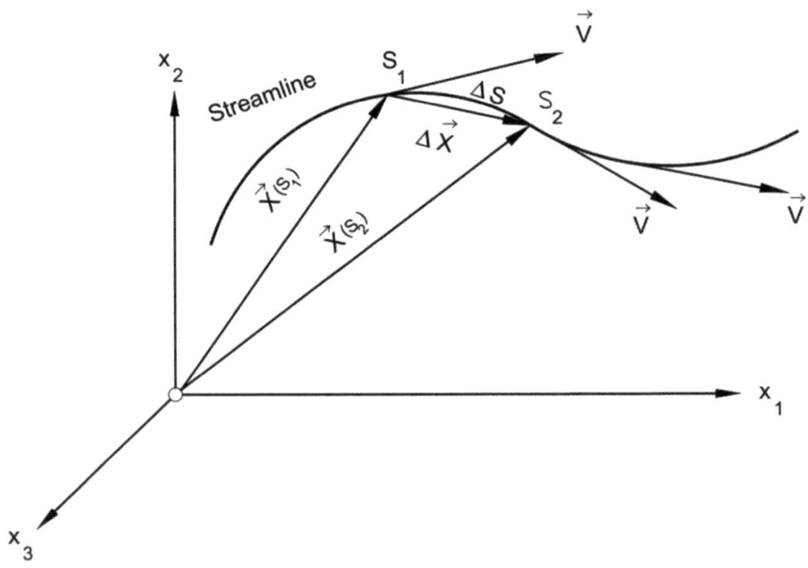

Fig. 3.7 Construction of a streamline

$$\boldsymbol{\lambda} = \lim_{\Delta S \to 0} \left(\frac{\boldsymbol{X}(S_2) - \boldsymbol{X}(S_1)}{S_2 - S_1} \right) = \frac{d\boldsymbol{X}}{dS}, (t = \text{const.}) \tag{3.57}$$

Considering the spatial description of the velocity vector, the unit vector tangent to the velocity vector is constructed by:

$$\boldsymbol{\lambda} = \frac{d\boldsymbol{X}}{dS} = \frac{\boldsymbol{V}}{|\boldsymbol{V}|}; \ \frac{dX_i}{dS} = \frac{V_i}{|\boldsymbol{V}|} = \frac{V_i}{\sqrt{(V_j V_j)}}. \tag{3.58}$$

Applying the summation convention, the streamline is fully described by the following three differential equations:

$$\frac{dX_1}{dS} = \frac{V_1}{|\boldsymbol{V}|}; \ \frac{dX_2}{dS} = \frac{V_2}{|\boldsymbol{V}|}; \ \frac{dX_3}{dS} = \frac{V_3}{|\boldsymbol{V}|}. \tag{3.59}$$

The infinitesimal arc length dS can easily be eliminated by rearranging Eq. (3.59)

$$\frac{dX_1}{dX_2} = \frac{V_1}{V_2}; \ \frac{dX_2}{dX_3} = \frac{V_2}{V_3}; \ \frac{dX_1}{dX_3} = \frac{V_1}{V_3}. \tag{3.60}$$

A *streakline* represents the fluid motion in a way that an observer can easily see. It is a curve traced out by all particles passing through some fixed point. The plum of smoke from a cigarette represents a streakline (we neglect the lateral diffusion of the smoke particle). A streakline at a fixed time t is the connecting line or the locus of different fluid particles passing through a fixed location $\boldsymbol{y}$ at a different time τ. The $\boldsymbol{x} = \boldsymbol{x}(\boldsymbol{y}, \tau)$ pathlines of the particles are given by the equation $\boldsymbol{x} = \boldsymbol{x}(\boldsymbol{\xi}, t)$. Solving this equation for $\boldsymbol{x} = \boldsymbol{x}(\boldsymbol{\xi}, t)$ and replacing $\boldsymbol{x}$ by the coordinates of the fixed location

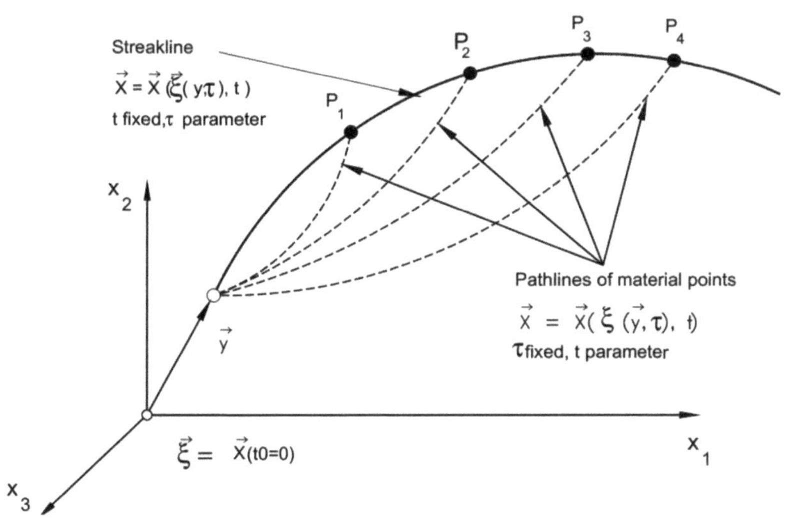

Fig. 3.8 Construction of a streamline

$\boldsymbol{y}$, and setting $t = \tau$, we locate the fluid particles $\boldsymbol{\xi}$, that were passing through the fixed location $\boldsymbol{y}$ at time τ. The pathline coordinates are obtained from $\boldsymbol{x} = \boldsymbol{x}(\boldsymbol{\xi}(\boldsymbol{y}, \tau), t)$. Thus, at the fixed time t we obtain the streakline as a curve, which connects the different fluid particles passing though a fixed spatial location $\boldsymbol{y}$ at different time τ. Figure 3.8 explains this statement graphically.

3.5 Problems

Problem 3.1 The material description of a fluid motion is given by the pathline equations

$$x_1 = \xi_1,$$
$$x_2 = k\xi_1^2 t^2 + \xi_2,$$
$$x_3 = \xi_3$$

with k as a constant having a dimension, such that the dimensional integrity of both sides of the above equation systems is preserved. Show that the Jacobian determinant $J = \det(\partial x_i / \partial \xi_j)$ does not vanish and obtain the transformation $\boldsymbol{\xi} = \boldsymbol{\xi}(x, t)$.

Problem 3.2 The fluid motion is described by:

$$x_1 = \xi_1,$$
$$x_2 = \frac{1}{2}(\xi_2 + \xi_3)\, e^{at} + \frac{1}{2}(\xi_2 - \xi_3)\, e^{-at},$$
$$x_3 = \frac{1}{2}(\xi_2 + \xi_3)\, e^{at} - \frac{1}{2}(\xi_2 - \xi_3)\, e^{-at}$$

(a) Show that the Jacobian determinant does not vanish.
(b) Determine the velocity and acceleration components

  (1) in material coordinates $V_i(\xi_j, t)$, $A_i(\xi_j, t)$,
  (2) in spatial coordinates $V_i(x_j, t)$, $A_i(x_j, t)$.

Problem 3.3 The motion of a fluid is given by the material description

$$x_1 = \left(\xi_1^2 + \xi_2^2\right)^{1/2} \cos\left[\frac{\Omega t}{\xi_1^2 + \xi_2^2} + \arctan\left(\frac{\xi_2}{\xi_1}\right)\right],$$
$$x_2 = \left(\xi_1^2 + \xi_2^2\right)^{1/2} \sin\left[\frac{\Omega t}{\xi_1^2 + \xi_2^2} + \arctan\left(\frac{\xi_2}{\xi_1}\right)\right],$$
$$x_3 = \xi_3.$$

(a) Find the equation of pathlines in an implicit form and show that for the position vector $\mathbf{x}$ at time $t = 0$ the identities: $x_1 = \pm\xi_1$ and $x_2 = \pm\xi_2$ are valid.
(b) Calculate the components of the velocity $V_i(\xi_j, t)$.
(c) Determine the velocity field $V_i(x_k, t)$ and the acceleration field $A_i(x_k, t$.
(d) Explain the equation of streamline through the point (x_{10}, x_2).

Problem 3.4 The motion of a fluid is described in the material coordinate by:

$$x_1 = \xi_1 e^{at},$$
$$x_2 = \xi_2 e^{at},$$
$$x_3 = \xi_3 e^{-2at}$$

with $a =$ const. and $\boldsymbol{\xi} = \boldsymbol{\xi}(x, t = 0)$.

a) Calculate the velocity and acceleration components $V_i(\xi_j, t$ and $A_i(\xi_j, t)$ in material coordinates.
b) Determine the spatial description of the velocity and acceleration components $V_i(x_k, t)$ and $A_i(x_k, t)$ by eliminating the material coordinate $\xi_j = \xi_j(x_k, t)$ in the results obtained in (a).
c) Find the acceleration components using the substantial derivatives of $V_i(x_k, t)$.
d) Is this a potential flow? If yes, find the potential function.

Problem 3.5 Given is the following unsteady velocity field:

$$V_1 = \frac{1}{t_0 + t} x_1,$$
$$V_2 = U,$$
$$V_3 = 0\,(t_0 = \text{ const.}, U = \text{const.})\,.$$

(a) Find the equation of streamlines through the point $(x_{10}, x_{20}, x_{30}$ at time t.
(b) Express the pathline equation of a fluid particle with a material coordinate $\mathbf{x(t = 0)} = \boldsymbol{\xi}$.
(c) Determine the particle velocity along its pathline.
(d) Find the equation for streaklines.

Problem 3.6 The nozzle of a water hose is vertically located at $\mathbf{y} = \mathbf{e_2 H}$ and oscillates with the angle $\alpha = \alpha(t)$. Water leaves the nozzle with a constant exit velocity U. Neglecting the air forces exerted on the water jet, determine:

(a) the velocity components $V_i(t)$ of a fluid particle which was at the nozzle exit at the time τ,
(b) its pathline for $\mathbf{x(t = 0)} = \boldsymbol{\xi}$,
(c) the equation of streaklines.
(d) Has this type of flow streamlines?

Problem 3.7 The components of a velocity flow field $V_i(\xi_j)$ are given by

$$V_1 = a\,(x_1 + x_2)$$
$$V_2 = a\,(x_1 - x_2)$$
$$V_3 = W$$

with the constants a and W. Determine

(a) the divergence $\nabla \cdot \mathbf{V}$ of the flow field,
(b) the rotation $\nabla \times \mathbf{V}$,
(c) the parametric representation of the pathlines $x_i = x_i(\xi_j, t)$ with $\xi_j = x_j(t = 0)$,
(d) nonparametric representation of the projection of the pathlines in x_1, x_2-plane by eliminating the curve parameter t,
(e) the projection of the streamlines in x_1, x_2-plane by integrating the differential equations for the streamlines.

Problem 3.8 The velocity components of an unsteady two-dimensional flow field are given by

$$V_1 = (a + b \sin \omega t) x_1$$
$$V_2 = -(a + b \sin \omega t) x_2,$$

(a) Find the equation of streamline through the point (x_{10}, x_{20})?
(b) Find the equation of pathline for a fluid particle at the time $t = 0$ and the location $\mathbf{x}(t = 0) =$ ¸
(c) Find the equation of the streaklines through the origin ($\mathbf{y} = 0$).
(d) What is the velocity change that a probe would measure if it moves along the streamline $x_{1^B} = x_{2^B} = c_0 t$?

Problem 3.9 The velocity vector of a plane, unsteady flow field is given in cylindrical coordinates (r, φ) by $\mathbf{V} = \frac{\mathbf{1}}{\mathbf{r}}\,(\mathbf{A_0 e_r} + \mathbf{B_0 (1 + at) e_{\cdot}})$ with the dimensional constants A_0, B_0, a. Using cylindrical coordinates, calculate

(a) the equation of streamline through the location $P(r = r_0, \varphi = 0)$ and
(b) the pathline equation of a fluid particle, which was at time $t = 0$ at location P.

Problem 3.10 The velocity components of an unsteady flow field are given as

$$u_1 = 0,$$
$$u_2 = A\left(x_1 x_2 - x_3^2\right) e^{-B(t-t_0)}$$
$$u_3 = A\left(x_2^2 - x_1 x_3\right) e^{-B(t-t_0)}.$$

Determine the components of

(a) the velocity gradient tensor,
(b) the deformation tensor $\mathbf{D}$ and the spin tensor Ω, as well as
(c) curl of $\mathbf{u}$ at point $P = (1, 0, 3)$ and time $t = t_0$.

Chapter 4
Differential Balances in Fluid Mechanics

In this and the following chapter, we present the conservation laws of fluid mechanics that are necessary to understand the basics of flow physics from a unified point of view. The main subject of this chapter is the differential treatment of the conservation laws of fluid mechanics, namely conservation law of mass, linear momentum, angular momentum, and energy. In many engineering applications, such as in turbomachinery, the fluid particles change the frame of reference from a *stationary frame* followed by a rotating *one*. The absolute frame of reference is rigidly connected with the stationary parts, such as casings, inlets, and exits of a turbine, a compressor, a stationary gas turbine or a jet engine, whereas the relative frame is attached to the rotating shaft, thereby turning with certain angular velocity about the machine axis. By changing the frame of reference from an absolute frame to a relative one, certain flow quantities remain unchanged, such as normal stress tensor, shear stress tensor, and deformation tensor. These quantities are indifferent with regard to a change of frame of reference. However, there are other quantities that undergo changes when moving from a stationary frame to a rotating one. Velocity, acceleration, and rotation tensor are a few. We first apply these laws to the stationary or absolute frame of reference, then to the rotating one.

The differential analysis is of primary significance to all engineering applications such as compressor, turbine, combustion chamber, inlet, and exit diffuser, where a detailed knowledge of flow quantities, such as velocity, pressure, temperature, entropy, and force distributions, are required. A complete set of independent conservation laws exhibits a system of partial differential equations that describes the motion of a fluid particle. Once this differential equation system is defined, its solution delivers the detailed information about the flow quantities within the computational domain with given initial and boundary conditions.

M. T. Schobeiri, *Advanced Fluid Mechanics and Heat Transfer for Engineers and Scientists*, https://doi.org/10.1007/978-3-030-72925-7_4

4.1 Mass Flow Balance in Stationary Frame of Reference

The conservation law of mass requires that the mass contained in a material volume $v = v(t)$, must be constant:

$$m = \int_{v(t)} \rho dv. \tag{4.1}$$

Consequently, Eq. (4.1) requires that the substantial changes of the above mass must disappear:

$$\frac{Dm}{Dt} = \frac{D}{Dt}\int_{v(t)} \rho dv = 0. \tag{4.2}$$

Using the Reynolds transport theorem (see Chap. 3), the conservation of mass, Eq. (4.2), results in:

$$\frac{D}{Dt}\int_{v(t)} \rho dv = \int_{v(t)} \left(\frac{\partial \rho}{\partial t} + \nabla \cdot (\rho \mathbf{V})\right) dv = 0. \tag{4.3}$$

Since this integral in Eq. (4.3) is zero, the integrand in the bracket must vanish identically. A s a result, we may write the continuity equation for unsteady and compressible flow as:

$$\frac{\partial \rho}{\partial t} + \nabla.(\rho \mathbf{V}) = 0. \tag{4.4}$$

Equation (4.4) is a coordinate invariant equation. Its index notation in the Cartesian coordinate system given is:

$$\frac{\partial \rho}{\partial t} + \frac{\partial (\rho V_i)}{\partial x_i} = 0. \tag{4.5}$$

Expanding Eq. (4.5), we get:

$$\frac{\partial \rho}{\partial t} + \frac{\partial (\rho V_1)}{\partial x_1} + \frac{\partial (\rho V_2)}{\partial x_2} + \frac{\partial (\rho V_3)}{\partial x_3} = 0. \tag{4.6}$$

For an orthogonal curvilinear coordinate system, the continuity Eq. (4.6) for an incompressible fluid is written as (see Appendix Eq. (A.34)a):

$$\frac{\partial \rho}{\partial t} + \nabla \cdot (\rho \mathbf{V}) = \frac{\partial \rho}{\partial \mathbf{t}} + (\rho \mathbf{V}^{\mathrm{i}})_{,\mathrm{i}} + (\rho \mathbf{V}^{\mathrm{j}})\Gamma^{\mathrm{i}}_{\mathrm{ij}} = \mathbf{0}. \tag{4.7}$$

Applying Eq. (4.7) to a cylindrical coordinate system with the Christoffel symbols, Eq. (4.8), from Appendix Eq. (A.56), we have

$$\left(\Gamma^1_{lm}\right) = \begin{pmatrix} 0 & -0 & 0 \\ 0 & -r & 0 \\ 0 & -0 & 0 \end{pmatrix}, \left(\Gamma^2_{lm}\right) = \begin{pmatrix} 0 & 1/r & 0 \\ 1/r & 0 & 0 \\ 0 & 0 & 0 \end{pmatrix}, \left(\Gamma^3_{lm}\right) = \begin{pmatrix} 0 & 0 & 0 \\ 0 & 0 & 0 \\ 0 & 0 & 0 \end{pmatrix} \quad (4.8)$$

and introducing the physical components, Eq. (4.7) becomes:

$$\frac{\partial \rho}{\partial t} + \frac{\partial(\rho r V_r)}{r \partial r} + \frac{1}{r}\frac{\partial(\rho V_\theta)}{\partial \theta} + \frac{\partial(\rho V_z)}{\partial z} = 0. \quad (4.9)$$

Equation (4.9) is valid only for cylindrical coordinate systems. To apply the continuity balance to any arbitrary orthogonal coordinate system, one has to determine first the Christoffel symbols as outlined in Appendix A and then find the continuity equation.

4.1.1 Incompressibility Condition

The condition for a working medium to be considered as incompressible is that the substantial change of its density along the flow path vanishes. This means that:

$$\frac{D\rho}{Dt} = \frac{\partial \rho}{\partial t} + \mathbf{V} \cdot \nabla \rho = 0. \quad (4.10)$$

Inserting Eq. (4.10) into Eq. (4.4) and performing the chain differentiation of the second term in Eq. (4.4) namely, $\nabla \cdot (\rho \mathbf{V}) = \rho \nabla \cdot \mathbf{V} + \mathbf{V} \cdot \nabla \rho$, the continuity equation for an incompressible flow reduces to:

$$\nabla \cdot \mathbf{V} = 0. \quad (4.11)$$

In a Cartesian coordinate system, Eq. (4.11) can be expanded as written in Eq. (4.12):

$$\frac{\partial V_1}{\partial x_1} + \frac{\partial V_2}{\partial x_2} + \frac{\partial V_3}{\partial x_3} = 0. \quad (4.12)$$

In an orthogonal, curvilinear coordinate system, the continuity balance for an incompressible fluid is:

$$\nabla \cdot \mathbf{V} = V^{\mathrm{i}}_{,\mathrm{i}} + V^{\mathrm{j}} \Gamma^{\mathrm{i}}_{\mathrm{ij}} = 0. \quad (4.13)$$

Inserting the Christoffel symbols into Eq. (4.13) and the physical components for cylindrical coordinate systems, we obtain the continuity equation in terms of its physical components (4.14):

$$\frac{\partial(r V_r)}{r \partial r} + \frac{1}{r}\frac{\partial(V_\theta)}{\partial \theta} + \frac{\partial(V_z)}{\partial z} = 0. \quad (4.14)$$

4.2 Differential Momentum Balance in Stationary Frame of Reference

In addition to the continuity equation we treated above, the detailed calculation of the entire flow field through different engineering devices and components requires the equation of motion in differential form. In the following, we provide the equation of motion in differential form in a four-dimensional time-space coordinate. We start from Newton's second law of motion and apply it to an infinitesimal fluid element shown in Fig. 4.1, with mass dm for which the equilibrium condition is written as:

$$dm\mathbf{A} = \mathbf{dF}. \tag{4.15}$$

The acceleration vector **A** is the well known material derivative (see Chap. 2):

$$\mathbf{A} = \frac{\mathbf{DV}}{\mathrm{Dt}} = \frac{\partial \mathbf{V}}{\partial \mathrm{t}} + \mathbf{V} \cdot \nabla \mathbf{V}. \tag{4.16}$$

In Eq. (4.16), **A** is the acceleration vector and $d\mathbf{F}$ the vector sum of all forces exerting on the fluid element. In the absence of magnetic, electric or other extraneous effects, the force $d\mathbf{F}$ is equal to the vector sum of the surface force $d\mathbf{F_s}$ acting on the particle

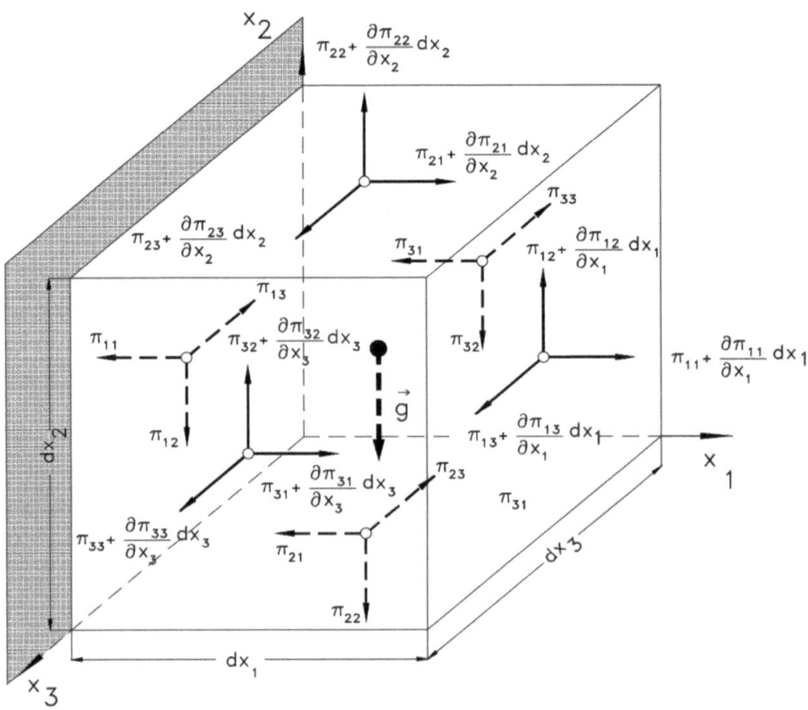

Fig. 4.1 Surface and gravitational forces acting on a volume element

surface and the gravity force $dm/\mathbf{g}$ as shown in Fig. 4.1. Inserting Eq. (4.16) into Eq. (4.15), we arrive at:

$$dm\left(\frac{\partial \mathbf{V}}{\partial t} + \mathbf{V}\cdot\nabla\mathbf{V}\right) = d\mathbf{F}_s + \mathbf{dmg}. \tag{4.17}$$

Consider the fluid element shown in Fig. 4.1 with sides dx_1, dx_2, dx_3 parallel to the axis of a Cartesian coordinate system. The stresses acting on the surfaces of this element are represented by the stress tensor $\mathbf{\Pi}$ which has the components π_{ij} that produce surface forces. The first index i refer to the axis, on which the fluid element surface is perpendicular, whereas the second index j indicates the direction of the stress component. Considering the stress situation in Fig. 4.1, the following resultant forces are acting on the surface dx_2dx_3 perpendicular to the x_1 axis:

$$e_1\frac{\partial \pi_{11}}{\partial x_1}dx_1dx_2dx_3,\ e_2\frac{\partial \pi_{12}}{\partial x_1}dx_1dx_2dx_3,\ e_3\frac{\partial \pi_{13}}{\partial x_1}dx_1dx_2dx_3. \tag{4.18}$$

The total resulting forces acting on the entire surface of the element are obtained by adding the nine components that result in Eq. (4.19):

$$\begin{aligned}\frac{d\mathbf{F_S}}{dv} = e_1\left(\frac{\partial \pi_{11}}{\partial x_1} + \frac{\partial \pi_{21}}{\partial x_2} + \frac{\partial \pi_{31}}{\partial x_3}\right) + e_2\left(\frac{\partial \pi_{12}}{\partial x_1} + \frac{\partial \pi_{22}}{\partial x_2} + \frac{\partial \pi_{32}}{\partial x_3}\right)\\ + e_3\left(\frac{\partial \pi_{13}}{\partial x_1} + \frac{\partial \pi_{23}}{\partial x_2} + \frac{\partial \pi_{33}}{\partial x_3}\right).\end{aligned} \tag{4.19}$$

Since the stress tensor $\mathbf{\Pi}$ and the volume of the fluid element are written as:

$$\mathbf{\Pi} = \mathrm{e_i e_j}\pi_{\mathrm{ij}},\ \mathrm{dv} = \mathrm{dx_1dx_2dx_3}. \tag{4.20}$$

It can be easily shown that Eq. (4.19) is the divergence of the stress tensor expressed in Eq. (4.20)

$$\frac{d\mathbf{F_S}}{dv} = \nabla\cdot\mathbf{\Pi}. \tag{4.21}$$

The expression $\nabla\cdot\Pi$ is the scalar differentiation of the second order tensor $\mathbf{\Pi}$ and is called divergence of the tensor field $\mathbf{\Pi}$ which is a first order tensor or a vector. Inserting Eq. (4.21) into Eq. (4.17) and divide both sides by dm, results in the following **Cauchy** equation of motion:

$$\frac{\partial \mathbf{V}}{\partial t} + \mathbf{V}\cdot\nabla\mathbf{V}\frac{1}{\rho}\nabla\cdot\mathbf{\Pi} + \mathbf{g}. \tag{4.22}$$

The stress tensor in Eq. (4.22) can be expressed in terms of deformation tensor, as we will see in the following section.

4.2.1 Relationship Between Stress Tensor and Deformation Tensor

Since the surface forces resulting from the stress tensor cause a deformation of fluid particles, it is obvious that one might attempt to find a functional relationship between the stress tensor and the velocity gradient:

$$\boldsymbol{\Pi} = f(\nabla \mathbf{V}). \tag{4.23}$$

As we saw in Chap. 2, the velocity gradient in Eq. (4.23) can be decomposed into an symmetric part called deformation tensor and an anti-symmetric part, called rotation or vorticity tensor:

$$\nabla \mathbf{V} = \frac{1}{2}\left(\nabla \mathbf{V} + \nabla \mathbf{V}^{T}\right) + \frac{1}{2}\left(\nabla \mathbf{V} - \nabla \mathbf{V}^{T}\right) = \mathbf{D} + \boldsymbol{\Omega}. \tag{4.24}$$

Consequently, the stress tensor may be set:

$$\boldsymbol{\Pi} = f(\nabla \mathbf{V}) = f(\mathbf{D}, \boldsymbol{\Omega}) \tag{4.25}$$

with the deformation tensor as:

$$\mathbf{e}_i \mathbf{e}_j \mathbf{D}_{ij} = \frac{1}{2} e_i e_j \left(\frac{\partial V_i}{\partial x_j} + \frac{\partial V_j}{\partial x_i}\right) \tag{4.26}$$

and the rotation tensor, which is antisymmetric, and is given by Eq. (2.27):

$$\boldsymbol{\Omega} = e_i e_j \Omega_{ij} = \frac{1}{2} e_i e_j \left(\frac{\partial V_j}{\partial x_i} - \frac{\partial V_i}{\partial x_j}\right). \tag{4.27}$$

Since the stress tensor $\boldsymbol{\Pi}$ in Eq. (4.25) is a *frame indifferent quantity*, it remains unchanged or invariant under any changes of frame of reference. Moving from an absolute frame into a relative one exhibits such a change in frame of reference. Thus, the stress tensor $\boldsymbol{\Pi}$ satisfies the principle of *frame indifference*, also called the *principle of material objectivity*. To satisfy the objectivity principle, the arguments of the functional f must also be frame indifferent quantities. This is true only for the first argument $\mathbf{D}$ in Eq. (4.25). The second argument $\boldsymbol{\Omega}$ in Eq. (4.25) is not a frame indifferent quantity. As a consequence, the stress tensor is a function of deformation tensor $\mathbf{D}$ only.

$$\boldsymbol{\Pi} = f(\mathbf{D}). \tag{4.28}$$

A general form of Eq. (4.28) may be a polynomial in $\mathbf{D}$ as suggested in [16]

$$\boldsymbol{\Pi} = \mathbf{f}_1 \mathbf{I} + \mathbf{f}_2 \mathbf{D} + \mathbf{f}_3 (\mathbf{D} \cdot \mathbf{D}) \tag{4.29}$$

with $\mathbf{I} = \mathrm{e_i e_j} \delta_{\mathrm{ij}}$ as the unit Kronecker tensor. To fulfill the frame indifference requirement, the functions f_i must be invariant. This means they depend on either the thermodynamic quantities, such as pressure, or the following three-principal invariant of the deformation tensor:

$$I_{1D} = \mathrm{Tr}\mathbf{D} = \nabla \cdot \mathbf{V} = \mathbf{D}_{\mathrm{ii}} \tag{4.30}$$

$$I_{3D} = \det D_{ij} \tag{4.31}$$

$$I_{2D} = \frac{1}{2}\left(I_{1D}^2 - \mathbf{D} : \mathbf{D}\right) = \frac{1}{2}\left(D_{ii} D_{jj} - D_{ij} D_{ij}\right). \tag{4.32}$$

Of particular interest is the category of those fluids for which there is a linear relationship between the stress tensor and the deformation tensor. Many working fluids used in engineering applications, such as air, steam, and combustion gases, belong to this category. They are called the **Newtonian fluids** for which Eq. (4.29) is reduced to:

$$\mathbf{\Pi} = \mathbf{f}_1 \mathbf{I} + \mathbf{f}_2 \mathbf{D} \tag{4.33}$$

where the functions f_1 and f_2 are given by:

$$f_1 = (-p + \lambda \nabla \cdot \mathbf{V},\ \mathrm{f}_2 = 2\mu,\ \bar{\mu} = \lambda + \frac{2}{3}\mu \tag{4.34}$$

with μ as the absolute viscosity and $\bar{\mu}$ as the bulk viscosity. Introducing Eq. (4.34) into Eq. (4.29) results in the **Cauchy-Poisson law**:

$$\mathbf{\Pi} = (-\mathrm{p} + \lambda \nabla \cdot \mathbf{V},\ \mathbf{I} + 2\mu \mathbf{D} = -\mathrm{p}\mathbf{I} + \lambda \nabla \cdot \mathbf{V},\ \mathbf{I} + 2\mu \mathbf{D} = -\mathrm{p}\mathbf{I} + \mathbf{T}. \tag{4.35}$$

In Eq. (4.35), the terms with the coefficients involving viscosity are grouped together leading to a pressure tensor $-p\mathbf{I}$ and a *friction stress tensor* $\mathbf{T}$ that reads:

$$\mathbf{T} = \lambda(\nabla \cdot \mathbf{V})\mathbf{I} + 2\mu \mathbf{D}. \tag{4.36}$$

The first term on the right-hand side of Eq. (4.35) associated with the unit Kronecker tensor, $p\mathbf{I}$, represents the contribution of the thermodynamic pressure to the normal stress. The second term, $(\lambda \nabla \cdot \mathbf{V})\mathbf{I}$, exhibits a normal stress contribution caused by a volume dilatation or compression due to the compressibility of the working medium. For an incompressible medium, this term identically vanishes. The coefficient λ related to the coefficient of shear viscosity μ and the bulk viscosity $\bar{\mu}$ is given in Eq. (4.34) as $\bar{\mu} = \lambda + 2/3\mu$. For most of the fluids used in engineering applications, the bulk viscosity may be approximated as $\bar{\mu} = \lambda + 2/3\mu = 0$ leading to $\lambda = -2/3\mu$. This relation, frequently called the Stokes' relation, is valid for monoatomic gases [17]. Finally, the last term expresses a direct relationship between the shear stress tensor and the deformation tensor. For an incompressible fluid, Eq. (4.35) reduces to:

$$\mathbf{\Pi} = -\mathrm{p}\mathbf{I} + 2\mu\mathbf{D}. \tag{4.37}$$

4.2.2 Navier-Stokes Equation of Motion

Inserting Eq. (4.35) into Eq. (4.22) we obtain:

$$\rho\frac{\partial V}{\partial t} + \rho\mathbf{V}\cdot\nabla\mathbf{V} = \nabla\cdot[(-\mathbf{p} + \lambda\nabla\cdot\mathbf{V})\mathbf{I} + 2\mu\mathbf{D}] + \rho\mathbf{g}. \tag{4.38}$$

This is referred to as the **Navier-Stokes equation for compressible fluids**. In Eq. (4.38), the coefficient λ can be expressed in terms of shear viscosity μ. This, however, requires rearranging the second and third term in the bracket by using the index notation. Assuming that the viscosities μ, and thus λ are not varying spatially, for $\nabla\cdot(\lambda\nabla\cdot\mathbf{VI})$ we may write:

$$\nabla\cdot(\lambda\nabla\cdot\mathbf{VI}) = \left(\mathrm{e_i}\frac{\partial}{\partial \mathrm{x_i}}\right)\cdot\left(\lambda\frac{\partial \mathrm{V_j}}{\partial \mathrm{x_j}}\mathrm{e_k e_l}\delta_{\mathrm{kl}}\right) = \lambda\delta_{\mathrm{ik}}\delta_{\mathrm{kl}}\mathrm{e_l}\frac{\partial^2 \mathrm{V_j}}{\partial \mathrm{x_i}\partial \mathrm{x_j}}$$

$$\nabla\cdot(\lambda\nabla\cdot\mathbf{VI}) = \lambda\mathrm{e_l}\frac{\partial^2 \mathrm{V_j}}{\partial \mathrm{x_i}\partial \mathrm{x_J}} = \lambda\mathrm{e_i}\frac{\partial}{\partial \mathrm{x_i}}\left(\frac{\partial \mathrm{V_j}}{\partial \mathrm{x_j}}\right) = \lambda\nabla(\nabla\cdot\mathbf{V}) \tag{4.39}$$

We apply the same procedure to $\nabla\cdot(2\mu\mathbf{D})$:

$$\nabla\cdot(2\mu\mathbf{D}) = 2\mu\left(\mathrm{e_i}\frac{\partial}{\partial \mathrm{x_i}}\right)\cdot\left[\frac{1}{2}\mathrm{e_j e_k}\left(\frac{\partial \mathrm{V_j}}{\partial \mathrm{x_k}} + \frac{\partial \mathrm{V_k}}{\partial \mathrm{x_j}}\right)\right]$$

$$\nabla\cdot(2\mu\mathbf{D}) = \mu\left[\mathrm{e_k}\frac{\partial}{\partial \mathrm{x_k}}\left(\frac{\partial \mathrm{V_i}}{\partial \mathrm{x_i}}\right) + \mathrm{e_k}\frac{\partial^2 \mathrm{V_k}}{\partial \mathrm{x_i}\partial \mathrm{x_i}}\right] = \mu\left[\nabla\left(\nabla\cdot\mathbf{V}\right) + \Delta\mathbf{V}\right]. \tag{4.40}$$

Introducing Eqs. (4.39) and (4.40) into Eq. (4.38), we arrive at

$$\rho\left(\frac{\partial\mathbf{V}}{\partial t} + \mathbf{V}\cdot\nabla\mathbf{V}\right) = -\nabla p + (\lambda + \mu)\nabla(\nabla\cdot\mathbf{V}) + \mu\Delta\mathbf{V} + \rho\mathbf{g}. \tag{4.41}$$

For $\lambda = -2/3\mu$, Eq. (4.41) results in:

$$\rho\left(\frac{\partial\mathbf{V}}{\partial t} + \mathbf{V}\cdot\nabla\mathbf{V}\right) = -\nabla p + \frac{\mu}{3}\nabla(\nabla\cdot\mathbf{V}) + \mu\Delta\mathbf{V} + \rho\mathbf{g}. \tag{4.42}$$

For incompressible flows with constant shear viscosity, Eq. (4.38) reduces to:

$$\rho\frac{\partial\mathbf{V}}{\partial t} + \rho\mathbf{V}\cdot\nabla\mathbf{V} = \nabla\cdot(-\mathbf{pI} + 2\mu\mathbf{D}) + \rho\mathbf{g}. \tag{4.43}$$

Performing the differentiation on the right-hand side and dividing by ρ leads to:

$$\frac{\partial \mathbf{V}}{\partial t} + \mathbf{V} \cdot \nabla \mathbf{V} = -\frac{1}{\rho} \nabla \mathbf{p} + \nu \Delta \mathbf{V} + \mathbf{g} \tag{4.44}$$

with $\nu = \mu/\rho$ as the kinematic viscosity and $\Delta = \nabla \cdot \nabla = \nabla^2$ as the *Laplace operator.* Equation (4.38) or its special case, Eq. (4.44) with the equation of continuity and energy, exhibits a system of partial differential equations. This system describes the flow field completely. Its solution yields the detailed distribution of flow quantities. In many engineering applications, with the exception of hydro power generation, the contribution of the gravitational term $\mathbf{g} = \mathbf{e}_i \mathbf{g}_i = \mathbf{e}_3 \mathbf{g}_3$ compared to the other terms is negligibly small. Equation (4.44) in Cartesian index notation is written as:

$$\frac{\partial V_i}{\partial t} + V_j \frac{\partial V_i}{\partial x_j} = -\frac{1}{\rho} \frac{\partial p}{\partial x_i} + \nu \frac{\partial^2 V_i}{\partial x_j \partial x_j} + g_i. \tag{4.45}$$

Using the Einstein summation convention, the three components of Eq. (4.45) are:

$$\begin{aligned}
\frac{\partial V_1}{\partial t} + V_1 \frac{\partial V_1}{\partial x_1} + V_2 \frac{\partial V_1}{\partial x_2} + V_3 \frac{\partial V_1}{\partial x_3} &= -\frac{1}{\rho} \frac{\partial p}{\partial x_1} + \nu \left(\frac{\partial^2 V_1}{\partial x_1^2} + \frac{\partial^2 V_1}{\partial x_2^2} + \frac{\partial^2 V_1}{\partial x_3^2} \right) \\
\frac{\partial V_2}{\partial t} + V_1 \frac{\partial V_2}{\partial x_1} + V_2 \frac{\partial V_2}{\partial x_2} + V_3 \frac{\partial V_2}{\partial x_3} &= -\frac{1}{\rho} \frac{\partial p}{\partial x_2} + \nu \left(\frac{\partial^2 V_2}{\partial x_1^2} + \frac{\partial^2 V_2}{\partial x_2^2} + \frac{\partial^2 V_2}{\partial x_3^2} \right) \\
\frac{\partial V_3}{\partial t} + V_1 \frac{\partial V_3}{\partial x_1} + V_2 \frac{\partial V_3}{\partial x_2} + V_3 \frac{\partial V_3}{\partial x_3} &= -\frac{1}{\rho} \frac{\partial p}{\partial x_3} + \nu \left(\frac{\partial^2 V_3}{\partial x_1^2} + \frac{\partial^2 V_3}{\partial x_2^2} + \frac{\partial^2 V_3}{\partial x_3^2} \right) + g_3.
\end{aligned} \tag{4.46}$$

To obtain the components of the Navier-Stokes equation in an orthogonal curvilinear coordinate system, we use metric coefficients, Christoffel symbols, and the index notation outlined in Appendix A:

$$\begin{aligned}
\mathbf{g}_i \left(\frac{\partial \mathrm{V}^{\mathrm{i}}}{\partial \mathrm{t}} + \mathrm{V}^{\mathrm{j}} \mathrm{V}^{\mathrm{i}}_{,\mathrm{j}} + \mathrm{V}^{\mathrm{j}} \mathrm{V}^{\mathrm{k}} \Gamma^{\mathrm{i}}_{\mathrm{kj}} \right) = -\frac{1}{\rho} \mathbf{g}_{\mathrm{i}} g^{\mathrm{ji}} \mathrm{p}_{,\mathrm{j}} + \nu \mathbf{g}_{\mathrm{m}} \Big[& \mathrm{V}^{\mathrm{m}}_{,\mathrm{ik}} + \\
& V^n_{,i} \Gamma^m_{nk} + V^n_{,k} \Gamma^m_{ni} - V^m_{,j} \Gamma^j_{ik} + \\
& V^p \left(\Gamma^n_{pi} \Gamma^m_{nk} - \Gamma^j_{ik} \Gamma^m_{pj} + \Gamma^m_{pi,k} \right) \Big] g^{ik}.
\end{aligned} \tag{4.47}$$

Using the Christoffel symbols and the physical components for a cylindrical coordinate system as specified in Appendix A, we arrive at the component of Navier-Stokes equation in r-direction:

$$\frac{\partial V_r}{\partial t} + V_r \frac{\partial V_r}{\partial r} + \frac{V_\Theta}{r} \frac{\partial V_r}{\partial \theta} + V_z \frac{\partial V_r}{\partial z} - V_{\frac{\Theta^2}{r}} = -\frac{1}{\rho} \frac{\partial p}{\partial r} +$$
$$\nu \left(\frac{\partial^2 V_r}{\partial r^2} + \frac{1}{r^2} \frac{\partial^2 V_r}{\partial \Theta^2} + \frac{\partial^2 V_r}{\partial z^2} - 2 \frac{\partial V_\Theta}{r^2 \partial \Theta} + \frac{\partial V_r}{r \partial r} - \frac{V_r}{r^2} \right) \tag{4.48}$$

in θ-direction,

$$\frac{\partial V_\Theta}{\partial t} + V_r \frac{\partial V_\Theta}{\partial r} + V_{\frac{\Theta}{r}} \frac{\partial V_\Theta}{\partial \Theta} V_z \frac{\partial V_\Theta}{\partial z} + \frac{V_r V_\Theta}{r} = -\frac{1}{\rho} \frac{\partial p}{r \partial \Theta}$$
$$+ \nu \left(\frac{\partial^2 V_\Theta}{\partial r^2} + \frac{1}{r^2} \frac{\partial^2 V_\Theta}{\partial \Theta^2} + \frac{\partial^2 V_\Theta}{\partial z^2} + \frac{2}{r^2} \frac{\partial V_r}{\partial \Theta} + \frac{1}{r} \frac{\partial V_\Theta}{\partial r} - \frac{V_\theta}{r^2} \right) \tag{4.49}$$

and in z-direction:

$$\frac{\partial V_z}{\partial t} + V_r \frac{\partial V_z}{\partial r} + V_{\frac{\Theta}{r}} \frac{\partial V_z}{\partial z} + V_z \frac{\partial V_z}{\partial z} = -\frac{1}{\rho} \frac{\partial p}{\partial z} +$$
$$\nu \left(\frac{\partial^2 V_z}{\partial r^2} + \frac{\partial^2 V_z}{r^2 \partial \Theta^2} + \frac{\partial^2 V_z}{\partial z^2} + \frac{1}{r} \frac{\partial V_z}{\partial r} \right) \tag{4.50}$$

4.2.3 Special Case: Euler Equation of Motion

For the special case of steady inviscid flow (no viscosity), Eq. (4.44) is reduced to:

$$\frac{\partial \mathbf{V}}{\partial t} + \mathbf{V} \cdot (\nabla \mathbf{V}) = -\frac{1}{\rho} \nabla \mathrm{p} + \mathbf{g}. \tag{4.51}$$

This equation is called *Euler equation of motion.* Its index notation is:

$$\frac{\partial V_i}{\partial t} + V_j \frac{\partial V_i}{\partial x_j} = -\frac{1}{\rho} \frac{\partial p}{\partial x_i} + g_i. \tag{4.52}$$

Replacing the convective term in Eq. (4.51) by the following vector identity:

$$\mathbf{V} \cdot \nabla \mathbf{V} = \nabla (\mathbf{V} \cdot \mathbf{V})/2 - \mathbf{V} \times (\nabla \times V) \tag{4.53}$$

we find that the convective acceleration is expressed in terms of the gradient of the kinetic energy $\nabla(\mathbf{V} \cdot \mathbf{V})/2 = \nabla(\mathbf{V}^2/2)$, and a second term which is a vector product of the velocity and the vorticity vector $\nabla \times \mathbf{V}$. If the flow field under investigation allows us to assume a zero vorticity within certain flow regions, then we may assign a *potential* to the velocity field that significantly simplifies the equation system. This assumption is permissible for the flow region outside the boundary layer and is discussed more in detail in Chap. 6.

Before proceeding with the conservation of energy, in context of the Euler equation that describes the motion of inviscid flows, it is appropriate to present the *Bernoulli* equation, which exhibits a special integral form of Euler equations. For this purpose, we first rearrange the gravitational acceleration vector by introducing a scalar surface potential z whose gradient ∇z has the same direction as the unit vector in x_3-direction. Furthermore, it has only one component that points in the negative x_3-direction. As a result, we may write $\mathbf{g} = -\mathbf{e}_3\mathbf{g} = -\mathbf{g}\nabla(\mathbf{z})$. Thus, the Euler equation of motion assumes the following form:

$$\frac{\partial \mathbf{V}}{\partial t} + \nabla(\frac{V^2}{2} + gz) + \frac{1}{\rho}\nabla p = \mathbf{V} \times (\nabla \times V). \quad (4.54)$$

Equation (4.54) shows that despite the inviscid flow assumption, it contains vorticities that are inherent in viscous flows and cause additional entropy production. This can be expressed in terms of the second law of thermodynamics, $Tds = dh - dp/\rho$, where the changes of entropy, enthalpy, and static pressure, ds, dh, dp, or other thermodynamic properties are expressed in terms of the product of their gradients and a differential displacement as shown by Eq. (2.24) $dQ = d\mathbf{X} \cdot \nabla \mathrm{Q}$. Replacing the quantity Q by the following properties, we obtain:

$$ds = d\mathbf{X} \cdot \nabla \mathrm{s},\ \mathrm{dh} = \mathrm{d}\mathbf{X} \cdot \nabla \mathrm{h},\ \mathrm{dp} = \mathrm{d}\mathbf{X} \cdot \nabla \mathrm{p} \quad (4.55)$$

with s as the specific entropy, h as the specific static enthalpy and p the static pressure. Inserting the above property changes into the first law of thermodynamics, $Tds = dh - dp/\rho$, we find:

$$\frac{\partial \mathbf{V}}{\partial t} + \nabla(h + \frac{V^2}{2} + gz) = \mathbf{V} \times (\nabla \times V) + \mathbf{T}\nabla\mathbf{s}. \quad (4.56)$$

As we comprehensively discuss in Chap. 5, the expression in the parentheses on the left-hand side of Eq. (4.56) is the total enthalpy $H = (h + V^2/2 + gz)$. In the absence of mechanical or thermal energy addition or rejection, H remains constant meaning that its gradient ∇H vanishes. Furthermore, for steady flow cases, Eq. (4.56) reduces to:

$$\mathbf{V} \times (\nabla \times V) = -\mathbf{T}\nabla\mathbf{s}. \quad (4.57)$$

Equation (4.57) is an important result that establishes a direct relation between the vorticity and the entropy production in inviscid flows. In a flow field with discontinuities as a result of the presence of shock waves, there are always jumps in velocities across the shock front causing vorticity production and therefore, changes in entropy. The Bernoulli equation can be obtained as a scalar product of the Euler differential Eq. (4.51) and a differential displacement vector. Figure 4.2 shows different displacement vectors that, in principle, may be used. The vector $d\mathbf{X}^*$ shows the differential distance between two neighboring fluid particles located at positions A and B at the same time t with the thermodynamic states shown in Fig. 4.2. The

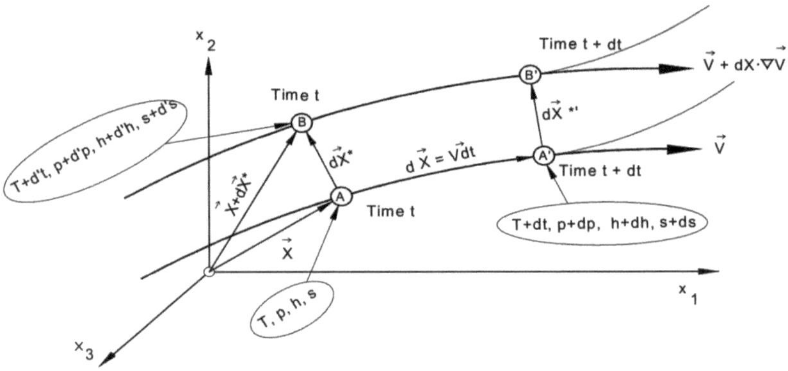

Fig. 4.2 Fluid particles at different thermodynamic state conditions

particles move along their flow paths and at $t + dt$, they occupy the positions A' and B'. The distance AA' is denoted by $d\mathbf{X}$. The thermodynamic conditions at A', denoted by $T + dT,\ p + dp,\ h + dh,\ s + ds$, indicate that the changes this particle has undergone are different from those of particle B. Thus, the vector $d\mathbf{X} = \mathbf{V}dt$ is the appropriate vector which we choose to multiply with the Euler equation of motion. For steady flow cases, the differential distance $d\mathbf{X}$ along the particle path is identical with a distance along a streamline. Thus, the multiplication of Euler Eq. (4.56) with the differential displacement $d\mathbf{X} = \mathbf{Vdt}$, gives:

$$\mathbf{V} \cdot \left(\frac{\partial \mathbf{V}}{\partial \mathrm{t}}\right) \mathbf{dt} + \mathrm{d}\mathbf{X} \cdot \nabla(\frac{\mathrm{V}^2}{2} + \mathrm{gz}) + \mathrm{d}\mathbf{X} \cdot \frac{\nabla \mathrm{p}}{\rho} = \mathbf{V} \cdot (\mathbf{V} \times \nabla \times \mathrm{V})\, \mathrm{dt}. \quad (4.58)$$

The terms in Eq. (4.58) must be rearranged as follows. The first term starts with a scalar product of two vectors, eliminating the Kronecker delta and utilizing the Einstein summation convention results in:

$$\mathbf{V} \cdot \left(\frac{\partial \mathbf{V}}{\partial \mathrm{t}}\right) \mathbf{dt} = \mathrm{V_i}\frac{\partial \mathrm{V_i}}{\partial \mathrm{t}}\mathrm{dt} = \frac{1}{2}\frac{\partial(\mathrm{V_i V_i})}{\partial \mathrm{t}}\mathrm{dt} = \frac{1}{2}\frac{\partial \mathrm{V}^2}{\partial \mathrm{t}}\mathrm{dt} = \mathrm{V}\frac{\partial \mathrm{V}}{\partial \mathrm{t}}\mathrm{dt} = \frac{\partial \mathrm{V}}{\partial \mathrm{t}}\mathrm{dX}. \quad (4.59)$$

For rearrangement of the second and third term, we use Eq. (2.24) $dQ = d\mathbf{X} \cdot \nabla \mathrm{Q}$. The fourth term in Eq. (4.58) identically vanishes because the vector $\mathbf{V}$ is perpendicular to the vector $\mathbf{V} \times \nabla \times \mathbf{V}$. As a result, we obtain:

$$\left(\frac{\partial V}{\partial t}\right) dX + d(\frac{V^2}{2} + gz) + \frac{dp}{\rho} = 0. \quad (4.60)$$

Integrating Eq. (4.60) results in:

$$\int \frac{\partial V}{\partial t} dX + \frac{V^2}{2} + gz + \int \frac{dp}{\rho} = C. \quad (4.61)$$

Integrating Eq. (4.60) from the initial point B to the final point E, we arrive at:

$$\int_B^E \frac{\partial V}{\partial t} dX + \frac{V_B^2}{2} + g z_B + B \int^{E \frac{dp}{\rho}} = \frac{V_E^2}{2} + g z_E. \tag{4.62}$$

For an unsteady, incompressible flow, the integration of Eq. (4.60) delivers:

$$\rho \int \frac{\partial V}{\partial t} dX + \rho \frac{V^2}{2} + \rho g z + p = C. \tag{4.63}$$

And finally, for a steady, incompressible flow, Eq. (4.63) is reduced to:

$$p + \rho \frac{V^2}{2} + \rho g z = C \tag{4.64}$$

which is the Bernoulli equation.

4.3 Some Discussions on Navier-Stokes Equations

The flow in engineering applications, such as in a turbine, a compressor or a combustion engine is characterized by a three-dimensional, highly unsteady motion with random fluctuations due to the interactions between the stator and rotor rows. Considering the flows within the blade boundary layer, based on the blade geometry and pressure gradient, three distinctly different flow patterns are be identified: 1) laminar flow (or non-turbulent flow) characterized by the absence of stochastic motions, 2) turbulent flow, where flow pattern is determined by a fully stochastic motion of fluid particles, and 3) transitional flow characterized by intermittently switching from laminar to turbulent at the same spatial position. Of the three patterns, the predominant one is the transitional flow pattern. The Navier-Stokes equations presented in this section generally describe the steady and unsteady flows through a variety of engineering components. Using a direct numerical simulation (DNS) approach delivers the most accurate results [18]. However, the computational domain must be at least as large as the physical domain. As extensively discussed in [16], the application of DNS, for the time being, is restricted to simple flows at low Reynolds numbers. For calculating the complex flow field with a reasonable time frame, the Reynolds averaged Navier-Stokes (RANS) can be used. This issue is discussed in Sect. 4.6.

4.4 Energy Balance in Stationary Frame of Reference to Fluid Mechanics

For the complete description of flow process, the total energy equation is presented. This equation includes mechanical and thermal energy balances.

4.4.1 Mechanical Energy

The mechanical energy balance is obtained by the scalar multiplication of the equation of motion, Eq. (4.22), with the local velocity vector:

$$\rho \mathbf{V} \cdot \frac{\mathrm{D}\mathbf{V}}{\mathrm{Dt}} = \mathbf{V} \cdot (\nabla \cdot \mathbf{\Pi}) + \rho \mathbf{g} \cdot \mathbf{V}. \tag{4.65}$$

The first expression on the right hand side of Eq. (4.65) is modified using the following vector identity:

$$\mathbf{V} \cdot (\nabla \cdot \mathbf{\Pi}) = \nabla \cdot (\mathbf{\Pi} \cdot \mathbf{V}) - \mathrm{Tr}\,(\mathbf{\Pi} \cdot \nabla \mathbf{V}). \tag{4.66}$$

The velocity gradient $\nabla \mathbf{V}$ can be decomposed into deformation $\mathbf{D}$ and rotation $\mathbf{\Omega}$ part as shown in Eq. (4.24):

$$\nabla \mathbf{V} = \frac{1}{2}\left(\nabla \mathbf{V} + \nabla \mathbf{V}^{\mathbf{T}}\right) + \frac{1}{2}\left(\nabla \mathbf{V} - \nabla \mathbf{V}^{T}\right) = \mathbf{D} + \mathbf{\Omega}. \tag{4.67}$$

With this operation, the trace of the second order tensor in Eq. (4.66) is calculated from:

$$\mathrm{Tr}\,(\mathbf{\Pi} \cdot \nabla \mathbf{V}) = \mathbf{\Pi} : \mathbf{D} + \mathbf{\Pi} : \mathbf{\Omega}. \tag{4.68}$$

Since the second term on the right-hand side of Eq. (4.68) vanishes identically, Eq. (4.66) reduces to:

$$\mathbf{V} \cdot (\nabla \cdot \mathbf{\Pi}) = \nabla \cdot (\mathbf{\Pi} \cdot \mathbf{V}) - \mathbf{\Pi} : \mathbf{D}. \tag{4.69}$$

As a result, the mechanical energy balance, Eq. (4.65), becomes:

$$\rho \frac{D}{Dt}\left(\frac{V^2}{2}\right) = \nabla \cdot (\mathbf{V} \cdot \mathbf{\Pi}) - \mathbf{\Pi} : \mathbf{D} + \rho \mathbf{g} \cdot \mathbf{V}. \tag{4.70}$$

Incorporating Eq. (4.35) for Newtonian fluids into Eq. (4.70) leads to:

$$\begin{aligned} \rho \frac{D}{Dt}\left(\frac{V^2}{2}\right) &= \nabla \cdot [\mathbf{V} \cdot (-p + \lambda \nabla \cdot \mathbf{V}) I + 2\mu \mathbf{V} \cdot \mathbf{D}] \\ &\quad - [(-p + \lambda \nabla \cdot \mathbf{V}) \nabla \cdot \mathbf{V} + 2\mu \mathbf{D} : \mathbf{D}] + \rho \mathbf{V} \cdot \mathbf{g} \end{aligned} \tag{4.71}$$

with $\mathbf{I} : \mathbf{D} = \nabla \cdot \mathbf{V}$. For incompressible flow, Eq. (4.71) is reduced to

$$\rho \frac{D}{Dt}\left(\frac{V^2}{V}\right) = \nabla \cdot [V \cdot (-pI) + 2\mu \mathbf{V} \cdot \mathbf{D}] - 2\mu \mathbf{D} : \mathbf{D} + \rho \mathbf{V} \cdot \mathbf{g}. \tag{4.72}$$

The index notation of Eq. (4.72) is

$$\frac{\partial\left(V_jV_j/2\right)}{\partial t}+V_k\frac{\partial\left(V_jV_j/2\right)}{\partial x_k}=-\frac{\partial}{\partial x_i}\left(\frac{p}{\rho}V_i\right)+\nu\frac{\partial}{\partial x_i}V_j\left(\frac{\partial V_i}{\partial x_j}+\frac{\partial V_j}{\partial x_i}\right)$$

$$-\nu\left(\frac{\partial V_i}{\partial x_j}+\frac{\partial V_j}{\partial x_i}\right)\frac{\partial V_j}{\partial x_i}+\varrho V_i g_i. \tag{4.73}$$

The sum of the last two terms in the second bracket of Eq. (4.71) is called the*dissipation function*:

$$\Phi=\lambda(\nabla\cdot\mathbf{V})(\nabla\cdot\mathbf{V})+2\mu\mathbf{D}:\mathbf{D}. \tag{4.74}$$

The dissipation function Eq. (4.74) is identical with the double scalar product between the friction stress tensor and the deformation tensor:

$$\Phi=\mathbf{T}:\mathbf{D}. \tag{4.75}$$

The index notation of Eq. (4.74) is

$$\Phi=\lambda\left(\frac{\partial V_i}{\partial x_i}\right)^2+\frac{2}{4}\mu\left(\frac{\partial V_i}{\partial x_j}+\frac{\partial V_j}{\partial x_i}\right)\left(\frac{\partial V_i}{\partial x_j}+\frac{\partial V_j}{\partial x_i}\right). \tag{4.76}$$

Expanding Eq. (4.76) results in:

$$\Phi=\lambda\left(\frac{\partial V_1}{\partial x_1}+\frac{\partial V_2}{\partial x_2}+\frac{\partial V_3}{\partial x_3}\right)^2+2\mu\left[\left(\frac{\partial V_1}{\partial x_1}\right)^2+\left(\frac{\partial V_2}{\partial x_2}\right)^2+\left(\frac{\partial V_3}{\partial x_3}\right)^2\right]+$$
$$\mu\left[\left(\frac{\partial V_1}{\partial x_2}+\frac{\partial V_2}{\partial x_1}\right)^2+\left(\frac{\partial V_1}{\partial x_3}+\frac{\partial V_3}{\partial x_1}\right)^2+\left(\frac{\partial V_2}{\partial x_3}+\frac{\partial V_3}{\partial x_2}\right)^2\right]. \tag{4.77}$$

In Eq. (4.77) the coefficient λ can be replaced by $\lambda=\bar{\mu}-2/3\mu$ from Eq. (4.34). For an incompressible flow, Eq. (4.74) reduces to:

$$\Phi=2\mu\mathbf{D}:\mathbf{D} \tag{4.78}$$

and as a result, Eq. (4.77) is written as:

$$\Phi=2\mu\left[\left(\frac{\partial V_1}{\partial x_1}\right)^2+\left(\frac{\partial V_2}{\partial x_2}\right)^2+\left(\frac{\partial V_3}{\partial x_3}\right)^2\right]+$$
$$\mu\left[\left(\frac{\partial V_1}{\partial x_2}+\frac{\partial V_2}{\partial x_1}\right)^2+\left(\frac{\partial V_1}{\partial x_3}+\frac{\partial V_3}{\partial x_1}\right)^2+\left(\frac{\partial V_2}{\partial x_3}+\frac{\partial V_3}{\partial x_2}\right)^2\right] \tag{4.79}$$

The dissipation function indicates the amount of mechanical energy dissipated as heat, which is due to the deformation caused by viscosity. Consider a viscous flow along a flat plate, an aircraft wing, a compressor stator or turbine rotor blade or any other engineering surfaces exposed to a flow. Close to the wall in the *boundary layer region,* the velocity experiences a high deformation because of the no-slip condition. By moving outside the boundary layer, the rate of deformation decreases leading to lower dissipation. To analyze the individual terms in the equation of energy and to demonstrate the role of shear stress and its effect on the dissipation of mechanical energy, we introduced the friction stress tensor, Eq. (4.36)

$$\mathbf{T} = \lambda(\nabla \cdot \mathbf{V})\mathbf{I} + 2\mu\mathbf{D}. \tag{4.80}$$

The off-diagonal elements of this tensor represent the shear stress components and characterize the *shear-deformative behavior* of this tensor. The diagonal elements of this tensor

$$T_{ii} = \lambda \frac{\partial V_i}{\partial x_i} + 2\mu D_{ii} \tag{4.81}$$

exhibit additional contributions caused by the volume dilatation or compression due to the compressibility of the working medium as mentioned before. This contribution is added to the normal stress components of the pressure tensor $p\mathbf{I} = \mathrm{e_i e_j} \delta_{\mathrm{ij}} \mathrm{p}$. For an incompressible flow with $\nabla \cdot \mathbf{V} = \mathbf{D}_{\mathrm{ii}} = \mathbf{0}$, these terms identically disappear. Inserting Eq. (4.80) into Eq. (4.71) we arrive at:

$$\rho \frac{D}{Dt}\left(\frac{V^2}{2}\right) = -\mathbf{V} \cdot \nabla p + \nabla \cdot (\mathbf{T} \cdot \mathbf{V}) - \mathbf{T} : \mathbf{D} + \rho \mathbf{V} \cdot \mathbf{g} \tag{4.82}$$

$$\mathbf{V} \cdot \nabla \left(\mathrm{p} + \frac{1}{2}\rho \mathrm{V}^2 + \rho \mathrm{gz}\right) \equiv \mathrm{d}\left(\mathrm{p} + \frac{1}{2}\rho \mathrm{V}^2 + \rho \mathrm{gz}\right) = 0. \tag{4.83}$$

Equation (4.82) exhibits the mechanical energy balance in differential form. The first term on the right-hand side represents the rate of mechanical energy due to the change of pressure acting on the volume element. The second term is the rate of work done by viscous forces on the fluid particle. The third term represents the rate of irreversible mechanical energy due to the friction stress. It dissipates as heat and increases the internal energy of the system. This term corresponds to the dissipation function defined by Eq. (4.78). Finally, the forth term represents the mechanical energy necessary to overcome the gravity force acted on the fluid particle. Equation (4.82) exhibits the general differential form of mechanical energy balance for a viscous flow. For a steady, incompressible, inviscid flow, Eq. (4.82) is simplified as: where the vector $\mathbf{g}$ is replaced by $\mathbf{g} = -\mathrm{g}\nabla \mathrm{z}$. Integration of the above equation leads to the Bernoulli equation of energy:

$$p + \frac{1}{2}\rho V^2 + \rho g z = \text{ Constant.} \tag{4.84}$$

This equation is easily derived by multiplying the Euler equation of motion with a differential displacement.

4.4.2 Thermal Energy Balance

The thermal energy balance is described by the first law of thermodynamics which is postulated for a closed "thermostatic system". For this system, properties, such as temperature, pressure, entropy, internal energy, etc., have no spatial gradients. Since in an open system the thermodynamic properties undergo temporal and spatial changes, the classical first law must be formulated under open system conditions. To do so, we start from an open system within which a steady flow process takes place and replaces the differential operator, d, from classical thermodynamics by the substantial differential operator D. This operation implies the requirement that the thermodynamic system under consideration be at least in a locally stable equilibrium state. Starting from the first law for an internally irreversible system:

$$du = \delta Q - pdv + |\delta w_f| \tag{4.85}$$

where u is the internal energy, and Q[1] the specific thermal energy added to or removed from the system, p the thermodynamic pressure, v the specific volume, and δw_f the part of mechanical energy dissipated as heat by the internal friction. The subscript f refers to the irreversible nature of the process caused by internal friction. Applying the differential operator D:

$$\frac{Du}{Dt} = \delta \dot{Q} - p\frac{Dv}{Dt} + \delta \dot{w}_f \tag{4.86}$$

where $\delta \dot{Q}$ is the rate of thermal energy added (or removed) to or from the open system per unit mass and time. It can be expressed as the divergence of the thermal energy flux vector $\delta \dot{Q} = -\nabla \cdot \dot{\mathbf{q}}/\rho$. The rate of the mechanical energy dissipated as heat $\delta \dot{w}_f$ is identical to the third term $\mathbf{T} : \mathbf{D}/\rho$ in Eq. (4.82):

$$\frac{Du}{Dt} = -\frac{\nabla \cdot \dot{\mathbf{q}}}{\rho} - p\frac{Dv}{Dt} + \frac{\mathbf{T} : \mathbf{D}}{\rho}. \tag{4.87}$$

The negative sign of $-\nabla \cdot \dot{\mathbf{q}}$ is introduced to account for a positive heat transfer to the system. Furthermore, since the first term on the left-hand side is per unit mass and time, the divergence of the heat flux vector $\nabla \cdot \dot{\mathbf{q}}$ as well as the dissipation term $\mathbf{T} : \mathbf{D}$ had to be divided by the density to preserve the dimensional integrity. Multiplying both sides of Eq. (4.87) with ρ, we obtain

$$\rho\frac{Du}{Dt} = -\nabla \cdot \dot{\mathbf{q}} - \rho p\frac{Dv}{Dt} + \mathbf{T} : \mathbf{D}. \tag{4.88}$$

---

[1] Usually Q (kJ) represents the thermal energy and q (kJ/kg) the specific thermal energy. However, in order to avoid confusion that may arise using the same symbol for specific thermal energy and the heat flux vector $\dot{\mathbf{q}}$, we use Q for the specific thermal energy. We will use q, wherever there is no reason for confusion.

In Eq. (4.88), first we replace the specific volume v by $1/\rho$ and consider the continuity equation

$$\frac{D\rho}{Dt} = -\rho\nabla\cdot\mathbf{V} \tag{4.89}$$

then we insert Eq. (4.89) into Eq. (4.88) and arrive at:

$$\rho\frac{Du}{Dt} = -\nabla\cdot\dot{\mathbf{q}} - p\nabla\cdot\mathbf{V} + \mathbf{T}:\mathbf{D}. \tag{4.90}$$

In Eq. (4.90) the internal energy can be related to the temperature by the thermodynamic relation $u = c_v T$ with c_v as the specific heat at constant volume. The heat flux vector $\dot{\mathbf{q}}$ can also be expressed in terms of temperature using the *Fourier heat conduction law*. For an *isotropic medium*, the Fourier law of heat conduction is written as:

$$\dot{\mathbf{q}} = -k\nabla T \tag{4.91}$$

with k (kJ/msec K) as the thermal conductivity. Introducing Eq. (4.91) into Eq. (4.90), for an incompressible fluid we get:

$$\rho C_v\frac{DT}{Dt} = k\nabla^2 T + 2\mu\mathbf{D}:\mathbf{D}. \tag{4.92}$$

For a steady flow, Eq. (4.92) can be simplified as:

$$C_v\mathbf{V}\cdot\nabla\mathbf{T} = \frac{\mathbf{k}}{\rho}\nabla^2\mathbf{T} + \mathbf{2}\nu\mathbf{D}:\mathbf{D}. \tag{4.93}$$

The thermal energy equation can equally be expressed in terms of enthalpy

$$dh = \delta Q + vdp + |\delta w_f|. \tag{4.94}$$

Following exactly the same procedure that has lead to Eq. (4.90), we find

$$\rho\frac{Dh}{Dt} = -\nabla\cdot\dot{\mathbf{q}} + \frac{Dp}{Dt} + \mathbf{T}:\mathbf{D} = \mathbf{k}\nabla^2\mathbf{T} + \frac{\mathbf{Dp}}{\mathbf{Dt}} + \mathbf{T}:\mathbf{D}. \tag{4.95}$$

Introducing the temperature in terms of $h = c_p T$:

$$c_p\frac{DT}{Dt} = \frac{k}{\rho}\nabla^2 T + \frac{1}{\rho},\frac{Dp}{Dt} + \mathbf{T}:\mathbf{D}. \tag{4.96}$$

The index notation of Eq. (4.96) reads:

$$c_p\left(\frac{\partial T}{\partial t} + V_i\frac{\partial T}{\partial x_i}\right) = \frac{\kappa}{\rho}\left(\frac{\partial^2 T}{\partial x_i\partial x_i}\right) + \frac{1}{\rho}\left(\frac{\partial p}{\partial t} + V_i\frac{\partial p}{\partial x_i}\right) + \frac{1}{\rho}\Phi \tag{4.97}$$

and taking Φ from Eq. (4.79), we can expand Eq. (4.97) to arrive at:

$$
\begin{aligned}
c_p\left(\frac{\partial T}{\partial t}+V_1\frac{\partial T}{\partial x_1}+V_2\frac{\partial T}{\partial x_2}+V_3\frac{\partial T}{\partial x_3}\right)&=\frac{\kappa}{\rho}\left(\frac{\partial^2 T}{\partial x_1^2}+\frac{\partial^2 T^2}{\partial x_2}+\frac{\partial^2 T}{\partial x_3^2}\right)\\
+\frac{1}{\rho}\left(\frac{\partial p}{\partial t}+V_1\frac{\partial p}{\partial x_1}+V_2\frac{\partial p}{\partial x_2}+V_3\frac{\partial p}{\partial x_3}\right)&+\frac{1}{\rho}\Phi.
\end{aligned}
\tag{4.98}
$$

4.4.3 *Total Energy*

The combination of the mechanical and thermal energy balances, Eqs. (4.82) and (4.90), results in the following *total energy equation:*

$$\rho\frac{D}{Dt}(u+\frac{V^2}{2})=-\nabla\cdot\dot{\mathbf{q}}-\nabla\cdot(p\mathbf{V})+\nabla\cdot(\mathbf{T}\cdot\mathbf{V})+\rho\mathbf{V}\cdot\mathbf{g}. \tag{4.99}$$

We may rearrange the second and third term on the right-hand side of Eq. (4.99)

$$\rho\frac{D}{Dt}(u+\frac{V^2}{2})=-\nabla\cdot\dot{\mathbf{q}}+\nabla\cdot\left[\mathbf{V}\cdot(-\mathbf{pI}+\mathbf{T})\right]+\rho\mathbf{V}\cdot\mathbf{g}. \tag{4.100}$$

The argument inside the parenthesis within the bracket exhibits the stress tensor

$$\rho\frac{D}{Dt}(u+\frac{V^2}{2})=-\nabla\cdot\dot{\mathbf{q}}+\nabla\cdot(\mathbf{V}\cdot\mathbf{\Pi})+\rho\mathbf{V}\cdot\mathbf{g} \tag{4.101}$$

and the second term on the right-hand side of Eq. (4.101) constitutes the mechanical energy necessary to overcome the surface forces. The heat flux vector in Eq. (4.101) can be replaced by the Fourier Eq. (4.91) that gives

$$\rho\frac{D}{Dt}(u+\frac{V^2}{2})=k\nabla^2 T+\nabla\cdot(\mathbf{V}\cdot\mathbf{\Pi})+\rho\mathbf{V}\cdot\mathbf{g}. \tag{4.102}$$

Equation (4.101) may be written in different forms using different thermodynamic properties. Since in an open system enthalpy is used, we replace the internal energy by the enthalpy $h=u+pv$ and find

$$\rho\frac{D}{Dt}(h+\frac{V^2}{2})=\frac{\partial p}{\partial t}-\nabla\cdot\dot{\mathbf{q}}+\nabla\cdot(\mathbf{T}\cdot\mathbf{V})+\rho\mathbf{V}\cdot\mathbf{g} \tag{4.103}$$

and with the Fourier Eq. (4.91)

$$\rho\frac{D}{Dt}(h+\frac{V^2}{2})=\kappa\nabla^2 T+\frac{\partial p}{\partial t}+\nabla\cdot(\mathbf{T}\cdot\mathbf{V})+\rho\mathbf{V}\cdot\mathbf{g}. \tag{4.104}$$

The expression in the parenthesis on the left-hand side is called the *total enthalpy* which is defined as $H = h + V^2/2$. With this definition, the re-arrangement of Eq. (4.104) gives

$$\rho \frac{DH}{Dt} = \rho \left(\frac{\partial H}{\partial t} + \mathbf{V} \cdot \nabla H \right) = k\nabla^2 T + \frac{\partial p}{\partial t} + \nabla \cdot (\mathbf{T} \cdot \mathbf{V}) + \rho \mathbf{V} \cdot \mathbf{g} \quad (4.105)$$

and its index notation reads

$$\rho \left(\frac{\partial H}{\partial t} + V_i \frac{\partial H}{\partial x_i} \right) = k \frac{\partial^2 T}{\partial x_i \partial x_i} + \frac{\partial p}{\partial t} + \frac{\partial}{\partial x_i} \left(T_{ij} V_j \right) + \rho V_i g_i. \quad (4.106)$$

The gravitational term is $\mathbf{V} \cdot \mathbf{g}$ can be expressed as $\mathbf{V} \cdot \mathbf{g} = -\mathbf{V} \cdot \nabla(gz)$ which can be written as $\mathbf{V} \cdot \nabla(\mathbf{gz}) = d\mathbf{X}/\mathbf{dt} \cdot \nabla(\mathbf{gz}) = \mathbf{d}(\mathbf{gz})/\mathbf{dt}$.

4.4.4 Entropy Balance

The second law of thermodynamics expressed in terms of internal energy as

$$ds = \frac{\delta Q}{T} = \frac{du + pdv}{T}. \quad (4.107)$$

The infinitesimal heat δQ added to or rejected from the system may include the heat generated by the irreversible dissipation process. Replacing the differential d by the material differential operators, we arrive at:

$$T \frac{Ds}{Dt} = \frac{Du}{Dt} + p \frac{Dv}{Dt}. \quad (4.108)$$

The right-hand side of Eq. (4.108) is expressed by Eq. (4.90) as:

$$\frac{Du}{Dt} + p \frac{Dv}{Dt} = -\frac{1}{\rho} \nabla \cdot \dot{\mathbf{q}} + \frac{1}{\rho} \mathbf{T} : \mathbf{D} \quad (4.109)$$

replacing the left-hand side of Eq. (4.109) by Eq. (4.108) results in

$$\rho \frac{Ds}{Dt} = -\frac{1}{T} \nabla \cdot \dot{\mathbf{q}} + \frac{1}{T} \mathbf{T} : \mathbf{D}. \quad (4.110)$$

The second term on the right-hand side, which includes the second order friction tensor $\mathbf{T}$ is the dissipation function Eq. (4.74)

$$\rho \frac{Ds}{Dt} = -\frac{1}{T} \nabla \cdot \dot{\mathbf{q}} + \frac{1}{T} \Phi. \quad (4.111)$$

This equation shows clearly that the total entropy change Ds/Dt generally consists of two parts. The first part is the entropy change due to a reversible heat supply to the system (addition or rejection) and may assume positive, zero, or negative values. The second term exhibits the entropy production due to the irreversible dissipation and is always positive. Thus, Eq. (4.111) may be modified as:

$$\rho\frac{Ds}{Dt} = \rho\left(\frac{Ds}{Dt}\right)_{\text{v}} + \rho\left(\frac{Ds}{Dt}\right)_{\text{irr}} \tag{4.112}$$

with $\rho\left(\frac{Ds}{Dt}\right)_{\text{rev}} = -\frac{1}{T}\nabla\cdot\dot{\mathbf{q}}$ and $\rho\left(\frac{Ds}{Dt}\right)_{\text{irr}} = \frac{\Phi}{T}$. The reversible part exhibits the heat added/rejected reversibly to/from the system, thus the entropy change can assume positive or negative values, whereas, for the irreversible, the entropy change is always positive.

4.5 Differential Balances in Rotating Frame of Reference

4.5.1 Velocity and Acceleration in Rotating Frame

We consider now a rotating frame of reference that is attached to the rotor, thus, turns with an angular velocity $\boldsymbol{\omega}$ about the machine axis. From a stationary observer point of view, a fluid particle that travels through a rotation frame has at an arbitrary time t, the position vector $\mathbf{r}$, and a *relative velocity* $\mathbf{W}$. In addition, it is subjected to the inherent rotation of the frame, causing the fluid particle to rotate with the velocity $\boldsymbol{\omega}\times\mathbf{r}$. Thus, the observer located outside the rotating frame observes the velocity

$$\mathbf{V} = \mathbf{W} + \boldsymbol{\omega}\times\mathbf{r}. \tag{4.113}$$

Inserting Eq. (4.113) into Eq. (4.16), the substantial acceleration is found

$$\frac{D\mathbf{V}}{Dt} = \frac{\partial(\mathbf{W}+\boldsymbol{\omega}\times\mathbf{r})}{\partial t} + (\mathbf{W}+\boldsymbol{\omega}\times\mathbf{r})\cdot\nabla(\mathbf{W}+\boldsymbol{\omega}\times\mathbf{r}). \tag{4.114}$$

We multiply Eq. (4.114) out and find

$$\frac{D\mathbf{V}}{Dt} = \frac{\partial\mathbf{W}}{\partial t} + \frac{\partial(\boldsymbol{\omega}\times\mathbf{r})}{\partial t} + \mathbf{W}\cdot\nabla\mathbf{W}$$

$$+\,\mathbf{W}\cdot\nabla(\boldsymbol{\omega}\times\mathbf{r}) + (\boldsymbol{\omega}\times\mathbf{r})\cdot\nabla\mathbf{W} + (\boldsymbol{\omega}\times\mathbf{r})\cdot\nabla(\boldsymbol{\omega}\times\mathbf{r}). \tag{4.115}$$

Investigating the terms in Eq. (4.115), we begin with the second term on the right-hand side

$$\frac{\partial(\boldsymbol{\omega}\times\mathbf{r})}{\partial t} = \boldsymbol{\omega}\times\frac{\partial\mathbf{r}}{\partial t} + \frac{\partial\boldsymbol{\omega}}{\partial t}\times\mathbf{r} = \frac{\partial\boldsymbol{\omega}}{\partial t}\times\mathbf{r} \tag{4.116}$$

since in the first term on the right-hand side of Eq. (4.116) for a fixed radius vector $\partial \mathbf{r}/\partial \mathbf{t} = \mathbf{0}$. Furthermore, the last three terms of Eq. (4.115) are:

$$(\boldsymbol{\omega} \times \mathbf{r})\cdot\nabla\mathbf{W} = \boldsymbol{\omega} \times \mathbf{W},\ \mathbf{W}\cdot\nabla(\boldsymbol{\omega} \times \mathbf{r}) = \boldsymbol{\omega} \times \mathbf{W},$$

$$\text{and } (\boldsymbol{\omega} \times \mathbf{r})\cdot\nabla(\boldsymbol{\omega} \times \mathbf{r}) = \boldsymbol{\omega} \times \boldsymbol{\omega} \times \mathbf{r}. \tag{4.117}$$

Detailed derivations of Eq. (4.117) are given in Vavra [19]. Considering Eqs. (4.116) and (4.117), Eq. (4.115) becomes

$$\frac{D\mathbf{V}}{Dt} = \frac{\partial \mathbf{W}}{\partial t} + \frac{\partial \omega}{\partial t} \times \mathbf{r} + \mathbf{W}\cdot\nabla\mathbf{W} + \boldsymbol{\omega} \times (\boldsymbol{\omega} \times \mathbf{r}) + 2\boldsymbol{\omega} \times \mathbf{W}. \tag{4.118}$$

The first term on the right-hand side $\partial\mathbf{W}/\partial t$ expresses the local acceleration of the velocity field within the relative frame of reference. In the second term, $\partial\omega/\partial t$ is the angular velocity acceleration. It is non-zero during any transient operation of components with rotating shaft, such as a turbomachine, where the shaft speed experience changes. The third term, $\mathbf{W}\cdot\nabla\mathbf{W}$, constitutes the convective term within the relative frame of reference. The forth term is the centrifugal force. Finally, the last term in Eq. (4.118), $2\boldsymbol{\omega} \times \mathbf{W}$, is called the Coriolis acceleration. It can be equal zero only if the relative velocity vector $\mathbf{W}$ and the angular velocity vector $\boldsymbol{\omega}$ are parallel. As shown in Eq. (4.117), two terms contributed to producing the Coriolis acceleration. The first term originates from the spatial changes of $\mathbf{W}$ because of the rotation. The second one from the changes in circumferential velocity, $\nabla(\boldsymbol{\omega} \times \mathbf{r})$, in the direction of $\mathbf{W}$. For the case that $\boldsymbol{\omega}$ and $\mathbf{W}$ are parallel, both terms become zero. The centrifugal acceleration and Coriolis accelerations are fictitious forces that are produced as a result of transformation from absolute into a relative frame of reference. Figure 4.2 shows the direction of the Coriolis force which is perpendicular to the plane described by the two vectors $\boldsymbol{\omega}$ and $\mathbf{W}$. The force vector, $\boldsymbol{\omega} \times (\boldsymbol{\omega} \times \mathbf{r})$, is perpendicular and pointing toward the axis of rotation. The direction of the radius vector $\mathbf{e}_r$ is expressed in terms of the radius gradient ∇R.

4.5.2 Continuity Equation in Rotating Frame of Reference

Inserting the velocity vector from Eq. (4.113) into the continuity equation for absolute frame of reference, Eq. (4.4), we obtain:

$$\frac{\partial \rho}{\partial t} + \nabla\cdot[\rho(\mathbf{W} + \boldsymbol{\omega} \times \mathbf{r})] = 0. \tag{4.119}$$

When we expand the second term in Eq. (4.119), we find:

$$\frac{\partial \rho}{\partial t} + (\boldsymbol{\omega} \times \mathbf{r})\cdot\nabla\rho + \mathbf{W}\cdot\nabla\rho + \rho\nabla\cdot\mathbf{W} + \rho\nabla\cdot(\boldsymbol{\omega} \times \mathbf{r}) = 0. \tag{4.120}$$

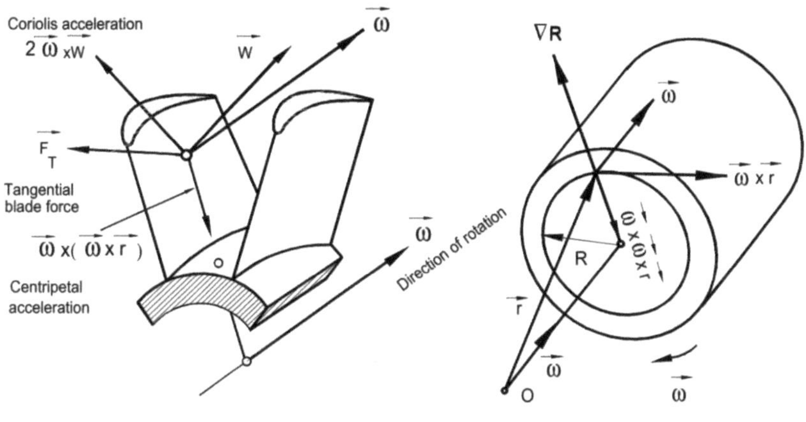

Fig. 4.3 Coriolis and centripetal forces created by the rotating frame of reference

After a simple rearrangement, Eq. (4.121) leads to:

$$\frac{\partial \rho}{\partial t} + (\mathbf{W} + \boldsymbol{\omega} \times \mathbf{r}) \cdot \nabla \rho + \rho \nabla \cdot \mathbf{W} + \rho \nabla \cdot (\boldsymbol{\omega} \times \mathbf{r}) = 0. \tag{4.121}$$

It is necessary to discuss the individual terms in Eq. (4.120) before rearranging them. The first term indicates the time rate of change of density at a fixed station in an absolute (stationary) frame of reference. The second term involves the spatial change of density registered by a stationary observer. Combining the first and second terms expresses the time rate of change of the density within the rotating frame of reference:

$$\frac{\partial_R \rho}{\partial t} \equiv \frac{\partial \rho}{\partial t} + (\boldsymbol{\omega} \times \mathbf{r}) \cdot \nabla \rho. \tag{4.122}$$

From Eq. (4.122), it becomes clear that in cases where the local change of the density in an absolute frame might be zero, $\partial \rho / \partial t = 0$, in a rotating frame of reference, it will become a function of time $\partial \rho_R / \partial t \not\equiv 0$. Since the product $(\boldsymbol{\omega} \times \mathbf{r}) \cdot \nabla \rho$ exhibits the circumferential change of the density in the rotating frame, it can vanish only if the flow within the rotating frame is considered axisymmetric. Since the last term in Eq. (4.120), $\nabla \cdot (\boldsymbol{\omega} \times \mathbf{r}) = 0$, identically vanishes, the equation of continuity in a rotating frame reduces to:

$$\frac{\partial_R \rho}{\partial t} + \mathbf{W} \cdot \nabla \rho + \rho \nabla \cdot \mathbf{W} = \frac{\partial_R \rho}{\partial t} + \nabla \cdot (\rho \mathbf{W}) = 0. \tag{4.123}$$

Equation (4.123) has the same form as Eq. (4.4), however, the spatial operator ∇ as well as the time derivative $\partial_R / \partial t = 0$ refer to the relative frame of reference. Since the flow in the rotor is understood exclusively with reference to a relative frame of reference, from now on it is unnecessary to differentiate between the operators and the time derivatives.

4.5.3 Equation of Motion in Rotating Frame of Reference

Replacing the acceleration in Eq. (4.22) by the expression obtained in Eq. (4.118):

$$\frac{\partial \mathbf{W}}{\partial t} + \frac{\partial \boldsymbol{\omega}}{\partial t} \times \mathbf{r} + \mathbf{W} \cdot \nabla \mathbf{W} + \boldsymbol{\omega} \times (\boldsymbol{\omega} \times \mathbf{r}) 2\boldsymbol{\omega} \times \mathbf{W} = \frac{1}{\rho} \nabla \cdot \Pi + \mathbf{g} \tag{4.124}$$

and replacing stress tensor Π by Eq. (4.35), $\Pi = -p\mathbf{I} + \lambda(\nabla \cdot \mathbf{V})I + 2\mu\mathbf{D}$, Eq. (4.124) becomes:

$$\frac{\partial \mathbf{W}}{\partial t} + \frac{\partial \boldsymbol{\omega}}{\partial t} \times \mathbf{r} + \mathbf{W} \cdot \nabla \mathbf{W} + \boldsymbol{\omega} \times (\boldsymbol{\omega} \times \mathbf{r}) + 2\boldsymbol{\omega} \times \mathbf{W} = \frac{1}{\rho} \nabla \cdot [-p\mathbf{I} + \lambda(\nabla \cdot \mathbf{V})\mathbf{I} + 2\mu\mathbf{D}] + \mathbf{g}. \tag{4.125}$$

Combining the last two terms in the bracket as $\nabla \cdot [\lambda(\nabla \cdot \mathbf{V})\mathbf{I} + 2\mu\mathbf{D}]/\rho \equiv -\mathbf{f}$, and setting for $\mathbf{g} = -\nabla(\mathrm{gz})$, we re-arrange Equation (4.125) as:

$$\frac{\partial \mathbf{W}}{\partial t} + \frac{\partial \boldsymbol{\omega}}{\partial t} \times \mathbf{r} + \mathbf{W} \cdot \nabla \mathbf{W} + \boldsymbol{\omega} \times (\boldsymbol{\omega} \times \mathbf{r}) + 2\boldsymbol{\omega} \times \mathbf{W} = -\frac{1}{\rho} \nabla p - \mathbf{f} - \nabla(\mathrm{gz}). \tag{4.126}$$

The friction force $\mathbf{f}$ was given a negative sign since it opposes the flow motion and causes energy dissipation. Using the Clausius entropy relation, the pressure gradient can be expressed in terms of enthalpy and entropy gradients:

$$\delta q = T ds = dh - v dp. \tag{4.127}$$

The thermodynamic properties s, h, and p are uniform continuous scalar point functions whose changes are expressed as:

$$ds = d\mathbf{X} \cdot \nabla \mathrm{s},\ \mathrm{dh} = \mathrm{d}\mathbf{X} \cdot \nabla \mathrm{h},\ \mathrm{dp} = \mathrm{d}\mathbf{X} \cdot \nabla \mathrm{p},\ \mathrm{ds} = \mathrm{d}\mathbf{X} \cdot \nabla \mathrm{s}, \tag{4.128}$$

with $d\mathbf{X}$ as the differential displacement along the path of the fluid particle. We replace the quantities in Eq. (4.127) by those in Eq. (4.128) and arrive at:

$$d\mathbf{X} \cdot \left(\mathbf{T}\nabla\mathbf{s} - \nabla \mathrm{h} + \frac{\nabla \mathbf{p}}{\rho} \right) = 0. \tag{4.129}$$

Since the differential displacement in Eq. (4.129), $d\mathbf{X} \not\equiv 0$, the vector sum in the bracket must vanish

$$T\nabla s - \nabla h + \frac{\nabla p}{\rho} = 0. \tag{4.130}$$

Replacing the pressure gradient term in Eq. (4.126) by Eq. (4.130), we find

$$\frac{\partial \mathbf{W}}{\partial t} + \frac{\partial \boldsymbol{\omega}}{\partial t} \times \mathbf{r} + \mathbf{W} \cdot \nabla \mathbf{W} + \boldsymbol{\omega} \times (\boldsymbol{\omega} \times \mathbf{r}) + 2\boldsymbol{\omega} \times \mathbf{W} = -\nabla(\mathrm{h+gz}) + \mathbf{T}\nabla\mathrm{s} - \mathbf{f}. \tag{4.131}$$

Further treatment of Eq. (4.131) requires a re-arrangement of few terms. As Fig. 4.3 shows, the centrifugal acceleration points in the negative direction of the gradient of the radius vector and can be written as:

$$\boldsymbol{\omega} \times (\boldsymbol{\omega} \times \mathbf{r}) = -\omega^2 \mathrm{R} \nabla \mathrm{R} = -\nabla \left(\frac{\omega^2 \mathrm{R}^2}{2} \right). \tag{4.132}$$

With Eq. (4.132), the equation of motion in a relative frame of reference becomes:

$$\frac{\partial \mathbf{W}}{\partial t} + \frac{\partial \boldsymbol{\omega}}{\partial t} \times \mathbf{r} + \mathbf{W} \cdot \nabla \mathbf{W} + \nabla(\mathrm{h} - \frac{\omega^2 \mathrm{R}^2}{2} + \mathrm{gz}) = -2\boldsymbol{\omega} \times \mathbf{W} + \mathrm{T}\nabla\mathrm{s} - \mathbf{f}. \tag{4.133}$$

Using the vector identity,

$$\mathbf{W} \cdot \nabla \mathbf{W} = \nabla \left(\frac{W^2}{2} \right) - \mathbf{W} \times (\nabla \times \mathbf{W}). \tag{4.134}$$

Equation (4.133) is modified as:

$$\frac{\partial \mathbf{W}}{\partial t} + \frac{\partial \boldsymbol{\omega}}{\partial t} \times \mathbf{r} + \nabla \left(h + \frac{W^2}{2} - \frac{\omega^2 R^2}{2} + \mathbf{gz} \right) = -2\boldsymbol{\omega} \times \mathbf{W} + \mathbf{W} \times \nabla \times \mathbf{W}) + \mathrm{T}\nabla\mathrm{s} - \mathbf{f}. \tag{4.135}$$

For a constant rotational speed and with $\boldsymbol{\omega} \times \mathbf{W} = -\mathbf{W} \times \boldsymbol{\omega}$, we find,

$$\frac{\partial \mathbf{W}}{\partial t} + \nabla \left(h + \frac{W^2}{2} - \frac{\omega^2 R^2}{2} + gz \right) = 2\mathbf{W} \times \boldsymbol{\omega} + \mathbf{W} \times (\nabla \times \mathbf{W}) + \mathrm{T}\nabla\mathrm{s} - \mathbf{f}. \tag{4.136}$$

We introduce now the concept of the *relative total enthalpy*:

$$H_R = h + \frac{W^2}{2} - \frac{\omega^2 R^2}{2} + gz. \tag{4.137}$$

4.5.4 Energy Equation in Rotating Frame of Reference

The energy equation for rotating frame of reference is simply obtained by multiplying the equation of motion with a differential displacement $d\mathbf{r}^*_\mathbf{R} = \mathbf{Wdt}$ along the path of a particle that moves within a rotating frame of reference. It is given by,

$$\mathbf{W}dt \cdot \left[\frac{\partial \mathbf{W}}{\partial t} + \nabla\left(h + \frac{W^2}{2} - \frac{\omega^2 R^2}{2} + gz\right)\right] =$$

$$\mathbf{W}dt \cdot [2\mathbf{W} \times \boldsymbol{\omega} + \mathbf{W} \times (\nabla \times \mathbf{W}) + \mathrm{T}\nabla \mathrm{s} - \mathbf{f}]\,. \tag{4.138}$$

Multiplying out and re-arranging the terms, we find:

$$\frac{\partial}{\partial t}\left(\frac{W^2}{2}\right) + d_R\left(h + \frac{W^2}{2} - \frac{\omega^2 R^2}{2} + gz\right) =$$

$$\mathbf{W}dt \cdot [2\mathbf{W} \times \boldsymbol{\omega} + \mathbf{W} \times (\nabla \times \mathbf{W}) + \mathrm{T}\nabla \mathrm{s} - \mathbf{f}]\,. \tag{4.139}$$

In Eq. (4.139), d_R denotes the changes in a relative frame of reference. Since the vectors $\mathbf{W} \times \boldsymbol{\omega}$ and $\mathbf{W} \times (\nabla \times \mathbf{W})$ are perpendicular to $\mathbf{W}$ their scalar products with $\mathbf{W}$ are zero. As a result, Eq. (4.139) reduces to:

$$\frac{\partial}{\partial t}\left(\frac{W^2}{2}\right) + d_R\left(h + \frac{W^2}{2} - \frac{\omega^2 R^2}{2} + gz\right) = d\mathbf{r}_R^* \cdot (T\nabla s - \mathbf{f})\,. \tag{4.140}$$

Multiplying out the right-hand side and considering the identity $d_R s = d\mathbf{r}_\mathbf{R}^* \cdot (\nabla \mathbf{s})$, Eq. (4.140) is modified as:

$$\frac{\partial}{\partial t}\left(\frac{W^2}{2}\right) + d_R\left(h + \frac{W^2}{2} - \frac{\omega^2 R^2}{2} + gz\right) = T d_R s - dt\mathbf{W} \cdot \mathbf{f}. \tag{4.141}$$

The term $T d_R s \equiv \delta_R q$ is identified as heat that consists of two contributions. The first contribution comes from heat supplied or removed from a fluid particle that moves along its path within the relative frame of reference. We call this contribution the reversible part, $\delta_R q_{\mathrm{rev}}$. The second contribution is the irreversible part due to the internal friction and dissipation of mechanical energy into heat, which is identical with the friction work, $\delta_R q_{\mathrm{irr}} = dt\mathbf{W} \cdot \mathbf{f}$. We summarize the above statement in the following relation:

$$T d_R s \equiv \delta_R q = \delta_R q_{\mathrm{rev}} + \delta_R q_{\mathrm{irr}} = \delta_R q_{\mathrm{rev}} + dt\mathbf{W} \cdot \mathbf{f}. \tag{4.142}$$

A simple re-arrangement of Eq. (4.142) yields:

$$T d_R s - dt\mathbf{W} \cdot \mathbf{f} = \delta_\mathrm{R} \mathrm{q}_{\mathrm{rev}}. \tag{4.143}$$

We insert Eq. (4.143) into Eq. (4.141) and obtain:

$$\frac{\partial}{\partial t}\left(\frac{W^2}{2}\right) + d_R\left(h + \frac{W^2}{2} - \frac{\omega^2 R^2}{2} + gz\right) = \delta_\mathrm{R} \mathrm{q}_{\mathrm{rev}}. \tag{4.144}$$

With Eq. (4.144), the changes of relative total enthalpy in a relative frame of reference along the path of a fluid particle is expressed as:

$$d_R\left(h + \frac{w^2}{2} - \frac{\omega^2 R^2}{2} + gz\right) = \delta_R q_{\text{rev}} - \frac{\partial}{\partial t}\left(\frac{W^2}{2}\right). \tag{4.145}$$

Only for adiabatic steady flow inside the rotating frame of reference the enthalpy change is zero resulting in:

$$h + \frac{W^2}{2} - \frac{\omega^2 R^2}{2} + gz = \text{const.} \tag{4.146}$$

It should be pointed out that Eq. (4.146) is strictly valid along the path of a fluid particle. If the flow within the relative frame can be approximated as steady, then Eq. (4.146) is also valid along the streamline. Its value changes however, by moving from one streamline to the next. For a rotating frame such as turbine or a compressor rotor row, under the above assumption, Eq. (4.146) is written as:

$$\left(h + \frac{W^2}{2} - \frac{\omega^2 R^2}{2} + gz\right)_2 = \left(h + \frac{W^2}{2} - \frac{\omega^2 R^2}{2} + gz\right)_3 \tag{4.147}$$

where the subscripts 2 and 3 in Eq. (4.147) refer to the inlet and exit station of the rotor row.

4.6 Problems

Problem 4.1 Incompressible Newtonian fluid with constant density and viscosity flows between two parallel plates with infinite width. Body forces are neglected. Given are the plate height h, the components of the pressure gradient, $\frac{\partial p}{\partial x_1} = -K\ \frac{\partial p}{\partial x_2} \equiv 0,\ \frac{\partial p}{\partial x_3} \equiv 0$, the velocity field between the plates $u_1(x_2) = \frac{K}{2\mu}\left(\frac{h^2}{4} - x_2^2\right),\ u_2 \equiv 0,\ u_3 \equiv 0$, the density ϱ and the absolute viscosity μ.

(a) Show that the given velocity field satisfies the continuity and the Navier-Stokes equation.
(b) Determine the components of the stress tensor.
(c) Calculate the dissipation function Φ.
(d) Find the energy per unit depth, length, and time dissipated in heat within the gap.
(e) Calculate the principal stresses and their directions.

Problem 4.2 Newtonian fluid flows through the sketched channel, Fig. 4.4, with infinite extensions in x_1- and x_3-direction and the height h. The plane flow is steady, the density ϱ and the viscosity μ are assumed to be constant, and body forces are

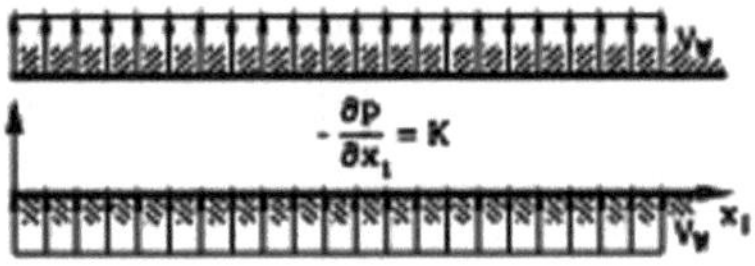

Fig. 4.4 Viscous flow through a two porous channel

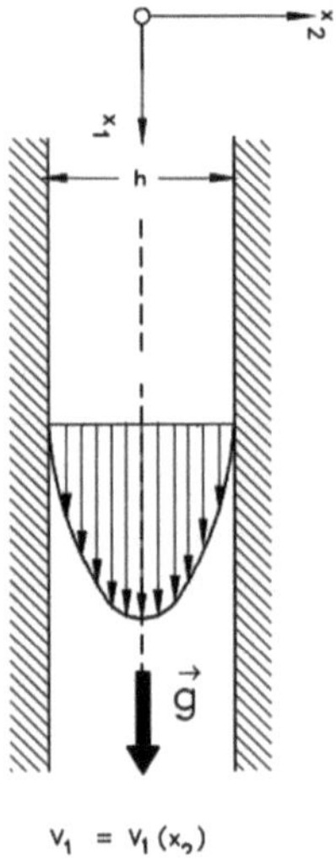

Fig. 4.5 Viscous flow through a two dimensional channel positioned vertically

neglected. The top and bottom wall are porous such that a constant normal velocity component V_W can be established at the walls. The pressure gradient in x_1-direction is constant ($\partial p/\partial x_1 = -K$). Because of the infinite extension of the channel, the velocity distribution does not depend upon x_1. The variables ϱ, μ, K, h, V_W are given.

(a) Using the continuity equation, calculate the distribution of the velocity component in x_2-direction $u_2(x_2)$.
(b) Simplify the x_1-component of the Navier-Stokes equation for this problem.
(c) Give the boundary condition for the velocity component u_1.
(d) Calculate the velocity distribution $u_1(x_2)$. (Hint: After solving the homogeneous differential equation, the particular solution of the inhomogeneous differential equation can be found setting $u_{1_p} =$ const. x_2).

Problem 4.3 A *Newtonian* fluid with constant density and viscosity flows *steadily* through a two dimensional vertically positioned channel with the width h shown in Fig. 4.5. The motion of the fluid is described by the Navier Stokes equations. The flow is subjected to the gravitational acceleration $\mathbf{g} = e_1 g$ and a constant pressure gradient in flow direction x_1. Assume that $V_2 = V_3 = 0$

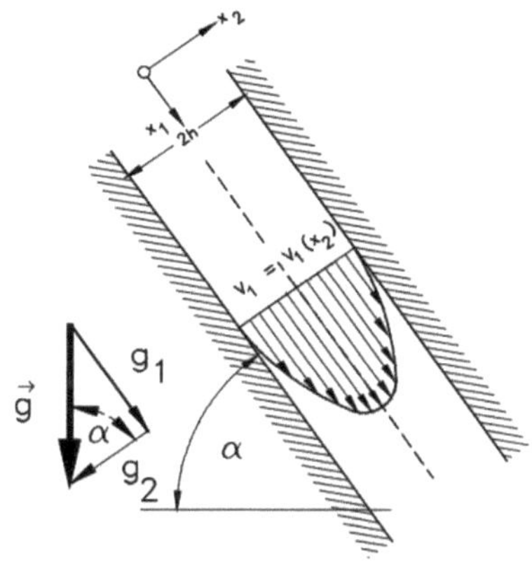

Fig. 4.6 Viscous flow through a two dimensional channel positioned at an angle

(a) Determine the solution of the Navier-Stokes equations.
(b) Write a computer program; show the velocity distributions for the following cases: (a) For $K = 0$, (b) $K > 0$, and (c) $K < 0$.
(c) For which K there is no flow?

Problem 4.4 A Newtonian fluid with constant density and viscosity flows *steadily* through a two dimensional channel positioned at an angle α shown in Fig. 4.6 with the width 2h. The motion of the fluid is described by the Navier Stokes equations. The flow is subjected to the gravitational acceleration $\mathbf{g} = \mathbf{e}_1 \mathbf{g}_1 + \mathbf{e}_2 \mathbf{g}_2$ and a constant pressure gradient in flow direction x_1. Assume that $V_2 = V_3 = 0$.

(a) Determine the solution of the Navier-Stokes Equations. Write a computer program and plot the velocity distributions for: (a) For $K = 0$, (b) $K > 0$, and (c) $K < 0$.
(b) For which K there is no flow?

Problem 4.5 River water considered a Newtonian fluid with constant viscosity and density steadily flows down an inclined river bed at a constant height h as shown in the Fig. 4.7. The motion of the fluid is described by the Navier-Stokes equation. Along the sloped river bed, the flow is driven by the gravitational acceleration and its free surface is subjected to the constant atmospheric pressure p_{atm}. The air viscosity at the free surface is negligible compared to the water viscosity. Furthermore, we assume that the flow is unidirectional in x_1 direction.

(a) Decompose the Navier-Stokes equation into its components.
(b) Show that the $\partial V_1/\partial x_1 = 0$.
(c) Solve the Navier-Stokes equations and find the velocity distribution in x_2-direction.
(d) Determine the velocity ratio $V_1/V_{1\,max}$.
(e) Determine the river mass flow $\dot{m}$.

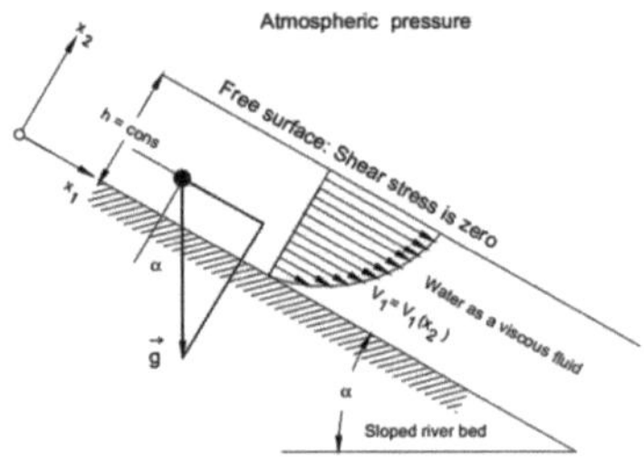

Fig. 4.7 Viscous flow through a river at ab angle

Problem 4.6 Give the index notation of the friction stress tensor Eq. (4.36) and its matrix and show that the diagonal elements of the matrix is identical with Eq. (4.81).

Problem 4.7 Give the index notation of the energy Eq. (4.71) and expand the result.

Problem 4.8 Insert the equation of continuity into the equation of energy (4.86) to arrive at Eq. (4.87).

Problem 4.9 Give the index notation of Eq. (4.93) and expand the result.

Problem 4.10 Give the index notation of Eq. (4.111) and expand the result.

Problem 4.11 We reconsider the flow calculated in Problem 4.1 and assume a calorically perfect fluid with a constant heat conductivity κ. We further assume a constant temperature at the top wall T_0 and a full heat insulation at the bottom wall.

(a) Calculate the temperature distribution $T(x_2)$ in the gap.
(b) Find the temperature at the bottom wall.
(c) Determine the heat flux per unit area through the top wall.
(d) Calculate the entropy increase Ds/Dt of the fluid inside the gap.

Chapter 5
Integral Balances in Fluid Mechanics

In the following sections, we summarize the conservation laws in an integral form essential for applying to fluid mechanics. Using the Reynolds transport theorem explained in Chap. 2, we start with the continuity equation which will be followed by the equation of linear momentum, angular momentum, and the energy.

5.1 Mass Flow Balance

We apply the Reynolds transport theorem by substituting the function $f(\boldsymbol{X}, t)$ in Chap. 2 by the density of the flow field:

$$m = \int_{v(t)} \rho(\boldsymbol{X}, t) dv \tag{5.1}$$

where the density generally changes with space and time. To obtain the integral formulation, the Reynolds transport theorem from Chap. 2 is applied. The requirement that the mass be constant leads to:

$$\frac{Dm}{Dt} = \int_{v(t)} \frac{\partial}{\partial t} \rho(\boldsymbol{X}, t) dv + \int_{S(t)} \rho(\boldsymbol{X}, t) \boldsymbol{V} \cdot \boldsymbol{n} dS = 0 \tag{5.2}$$

with $v(t)$ and $S(t)$ as the time dependent volume and surface of the integral boundaries. If the density does not undergo a time change (steady flow), the above equation is reduced to:

$$\int_{S(t)} \rho(\boldsymbol{X}, t) \boldsymbol{V} \cdot \boldsymbol{n} dS = 0. \tag{5.3}$$

M. T. Schobeiri, *Advanced Fluid Mechanics and Heat Transfer for Engineers and Scientists*, https://doi.org/10.1007/978-3-030-72925-7_5

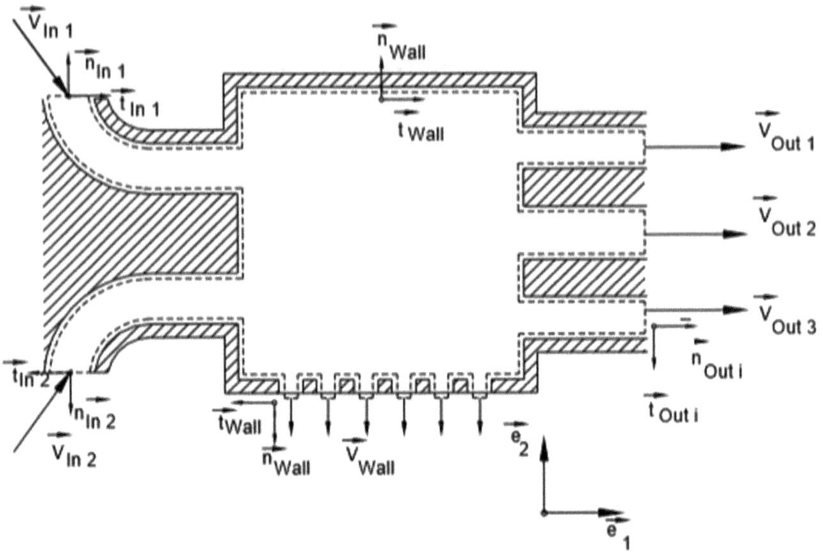

Fig. 5.1 Control volume, unit normal and tangential vectors

For practical purposes, a fixed control volume is considered where the integration must be carried out over the entire control surface:

$$\int_{S_C} \rho \boldsymbol{V} \cdot \boldsymbol{n} dS = \int_{S_{C\text{in}}} \rho \boldsymbol{V} \cdot \boldsymbol{n} dS + \int_{S_{C\text{out}}} \rho \boldsymbol{V} \cdot \boldsymbol{n} dS + \int_{S_{C\text{wall}}} \rho \boldsymbol{V} \cdot \boldsymbol{n} dS = 0. \tag{5.4}$$

The control surface may consist of one or more inlets, one or more exits, and may include porous walls, as shown in Fig. 5.1. For such a case, Eq. (5.4) is expanded as:

$$\begin{aligned} &\int_{S_{C\text{in}_1}} \rho \boldsymbol{V} \cdot \boldsymbol{n} dS + \int_{S_{C\text{in}_2}} \rho \boldsymbol{V} \cdot \boldsymbol{n} dS + \int_{S_{C\text{out}_1}} \rho \boldsymbol{V} \cdot \boldsymbol{n} dS \\ &+ \int_{S_{C\text{out}_2}} \rho \boldsymbol{V} \cdot \boldsymbol{n} dS + \int_{S_{C\text{out}_3}} \rho \boldsymbol{V} \cdot \boldsymbol{n} dS + \int_{S_{C\text{wall}}} \rho \boldsymbol{V} \cdot \boldsymbol{n} dS = 0 \end{aligned} \tag{5.5}$$

As shown in Fig. 5.1 and by convention, the normal unit vectors, $\boldsymbol{n}_{\text{in}}, \boldsymbol{n}_{\text{out}}, \boldsymbol{n}_{\text{Wall}}$, point away from the region bounded by the control surface. Similarly, the tangential unit vectors, $\boldsymbol{t}_{\text{in}}, \boldsymbol{t}_{\text{out}}, \boldsymbol{t}_{\text{Wall}}$, point in the direction of shear stresses. A representative example where the integral over the wall surface does not vanish is a film cooled turbine blade with discrete film cooling hole distribution along the blade suction and pressure surfaces, as shown in Fig. 5.2. To establish the mass flow balance through a turbine or cascade blade channel, the control volume should be placed in such a way that it includes quantities that we consider as known as well as those we seek

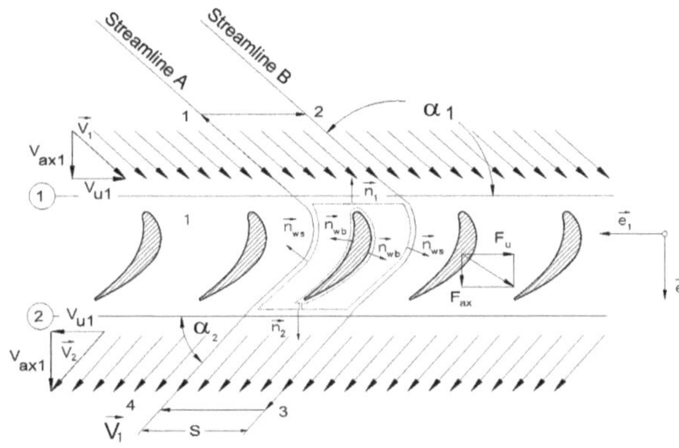

Fig. 5.2 Flow through a rectilinear turbine cascade with discrete film cooling holes

to find. For the turbine cascade in Fig. 5.2, the appropriate control surface consists of the surfaces AB, BC, CD, and DA. The two surfaces, BC and DA, are portions of two neighboring streamlines. Because of the periodicity of the flow through the cascade, the surface integrals along these streamlines will cancel each other. As a result, the mass flow balance reads:

$$\int_{S_{C\mathrm{in}}} \rho \boldsymbol{V} \cdot \boldsymbol{n} dS + \int_{S_{C\mathrm{out}}} \rho \boldsymbol{V} \cdot \boldsymbol{n} dS + \int_{S_{C\mathrm{wall}}} \rho \boldsymbol{V} \cdot \boldsymbol{n} dS = 0. \tag{5.6}$$

The last surface integral accounts for the mass flow injection through the film cooling holes. If there is no mass diffusion through the wall surfaces, the last integral in Eq. (5.6) will vanish, leading to:

$$\int_{S_C} \rho \boldsymbol{V} \cdot \boldsymbol{n} dS = \int_{S_{C\mathrm{in}}} \rho \boldsymbol{V} \cdot \boldsymbol{n} dS + \int_{S_{C\mathrm{out}}} \rho \boldsymbol{V} \cdot \boldsymbol{n} dS = 0. \tag{5.7}$$

5.2 Balance of Linear Momentum

The momentum equation in integral form applied to a control volume determines the integral flow quantities such as lift force, drag forces, average pressure, averaged temperature, and entropy. The motion of a material volume is described by the Newton's second law of motion which states that mass times acceleration is the sum of all external forces acting on the system. In the absence of electrodynamic, electrostatic, and magnetic forces, the external forces can be summarized as the surface forces and the gravitational forces:

$$m\frac{D\boldsymbol{V}}{Dt} = \boldsymbol{F}_S + \boldsymbol{F}_G. \tag{5.8}$$

Equation (5.8) is valid for a closed system with a system boundary that may undergo deformation, rotation, expansion or compression. In an engineering component subjected to flow, however, there is no closed system with a defined system boundary. The mass is continuously flowing from one point within a component to another point. Thus, in general, we deal with *mass flow* rather than mass. Consequently, Eq. (5.8) must be modified in such a way that it is applicable to a *predefined control volume* with mass flow passing through it. This requires applying the Reynolds transport theorem to a control volume, as we already discussed in the previous section. For this purpose, we prepare Eq. (5.8) before proceeding with the Reynolds transport theorem. In the following steps, we add a zero-term to Eq. (5.8):

$$\frac{Dm}{Dt} = 0, \quad \boldsymbol{V}\frac{Dm}{Dt} = 0. \tag{5.9}$$

Adding this term to Eq. (5.8) leads to:

$$m\frac{D\boldsymbol{V}}{Dt} + \boldsymbol{V}\frac{Dm}{Dt} = \boldsymbol{F}_S + \boldsymbol{F}_G. \tag{5.10}$$

Using the Leibnitz's chain rule of differentiation, Eq. (5.10) can be rearranged as:

$$\frac{D}{Dt}(m\boldsymbol{V}) = \boldsymbol{F}_S + \boldsymbol{F}_G. \tag{5.11}$$

Applying the Reynolds transport theorem to the left-hand side of Eq. (5.11), we arrive at:

$$\frac{D}{Dt}(m\boldsymbol{V}) = \int\limits_{v(t)} \left(\frac{\partial(\rho\boldsymbol{V})}{\partial t} + \nabla\cdot(\rho\boldsymbol{V}\boldsymbol{V})\right)dv$$
$$\frac{D}{Dt}(m\boldsymbol{V}) = \int\limits_{v(t)} \left(\frac{\partial(\rho\boldsymbol{V})}{\partial t}\right)dv + \int\limits_{v(t)} \nabla\cdot(\rho\boldsymbol{V}\boldsymbol{V})dv \tag{5.12}$$

and replace the second volume integral by a surface integral using the Gauss conversion theorem (see Chap. 3):

$$\frac{D}{Dt}(m\boldsymbol{V}) = \int\limits_{v(t)} \left(\frac{\partial(\rho\boldsymbol{V})}{\partial t}\right)dv + \int\limits_{S(t)} \boldsymbol{n}\cdot(\rho\boldsymbol{V}\boldsymbol{V})dS. \tag{5.13}$$

We consider now the surface and gravitational forces acting on the moving material volume under investigation. The first term on the right-hand side of Eq. (5.10) represents the resultant surface force acting on the entire control surface. It can be written as the integral of a scalar product of the normal unit vector with the total stress tensor acting on the surface element ds:

$$\boldsymbol{F}_S = \int_{S_C} d\boldsymbol{F}_S = \int_{S_C} \boldsymbol{n} \cdot \boldsymbol{\Pi} dS. \tag{5.14}$$

The product of the normal vector and the stress tensor gives a stress vector which can be decomposed into a normal and a shear stress force

$$\boldsymbol{n} \cdot \boldsymbol{\Pi} = -\boldsymbol{n}p - \boldsymbol{t}\tau \tag{5.15}$$

with $\boldsymbol{n}$ as the normal unit vector that points away from the surface and $\boldsymbol{t}$ as the tangential unit vector. The negative signs of $\boldsymbol{n}$ and $\boldsymbol{t}$ have been chosen to indicate that the pressure p and the shear stress τ are exerted by the surroundings *on the surface* S. Thus, the surface force acting on a differential surface is:

$$d\boldsymbol{F}_s = -\boldsymbol{n}pdS - \boldsymbol{t}\tau dS. \tag{5.16}$$

Inserting Eq. (5.16) into Eq. (5.11) and considering Eq. (5.12), we arrive at:

$$\int_{v_C} \left(\frac{\partial(\rho \boldsymbol{V})}{\partial t}\right) dv + \int_{S_C} \boldsymbol{n} \cdot (\rho \boldsymbol{V}\boldsymbol{V}) dS = \int_{S_C} (-\boldsymbol{n}p - \boldsymbol{t}\tau) dS + \boldsymbol{G}. \tag{5.17}$$

Since the control volume does not change with time (fixed), Eq. (5.17) becomes with $\rho dv = dm$:

$$\frac{\partial}{\partial t} \int_{V_C} \boldsymbol{V} dm + \int_{S_C} \boldsymbol{n} \cdot (\rho \boldsymbol{V}\boldsymbol{V}) dS = \int_{S_C} (-\boldsymbol{n}p - \boldsymbol{t}\tau) dS + \boldsymbol{G}. \tag{5.18}$$

In Eq. (5.18) the integration must be carried out over the entire control surface. For a control surface consisting of inlet, exit, and wall surfaces, the second integral on the left-hand side gives:

$$\int_{S_C} \boldsymbol{n} \cdot (\rho \boldsymbol{V}\boldsymbol{V}) dS = \int_{S_{C\text{in}}} \boldsymbol{n} \cdot (\rho \boldsymbol{V}\boldsymbol{V}) dS + \int_{S_{C\text{out}}} \boldsymbol{n} \cdot (\rho \boldsymbol{V}\boldsymbol{V}) dS + \int_{S_{C\text{wall}}} \boldsymbol{n} \cdot (\rho \boldsymbol{V}\boldsymbol{V}) dS. \tag{5.19}$$

Evaluating the integrands on the right-hand side of Eq. (5.19) by considering the directions of the unit vectors shown in Fig. 5.3, we find for the single inlet cross section:

$$\int_{S_C} \boldsymbol{n} \cdot (\rho \boldsymbol{V}\boldsymbol{V}) dS = \int_{S_{C\text{in}}} \boldsymbol{V}(\rho \boldsymbol{n} \cdot \boldsymbol{V} dS) = \int_{S_{C\text{in}}} \boldsymbol{V}(-\rho \boldsymbol{e}_1 \cdot \boldsymbol{e}_1 V_{\text{in}} dS)$$
$$= -\int_{S_{C\text{in}}} \boldsymbol{V} d\dot{m}. \tag{5.20}$$

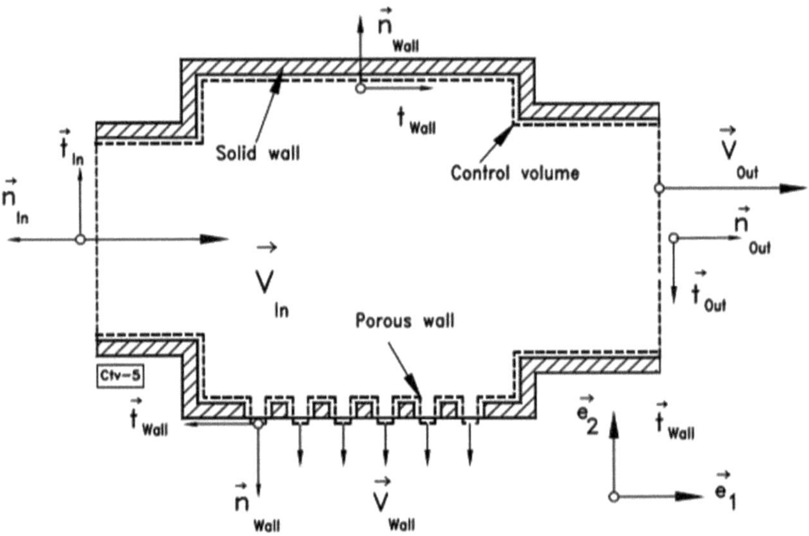

Fig. 5.3 Control volume, single inlet, single and outlet and porous wall

In the case of a control volume with multiple inlets as Fig. 5.1 shows, we need to integrate over the entire inlet cross sections. For the exit cross section we obtain:

$$\int_{S_{C\text{out}}} \boldsymbol{n}\cdot(\rho \boldsymbol{V}\boldsymbol{V})dS = \int_{S_{C\text{out}}} \boldsymbol{V}(\rho\boldsymbol{n}\cdot\boldsymbol{V}dS) = \int_{S_{C\text{out}}} \boldsymbol{V}(\rho\boldsymbol{e}_1\cdot\boldsymbol{e}_1 V_{\text{out}}dS)$$

$$\int_{S_{C\text{out}}} \boldsymbol{n}\cdot(\rho \boldsymbol{V}\boldsymbol{V})dS = \int_{Sout} \boldsymbol{V}d\dot{m}. \tag{5.21}$$

And, finally, for the wall:

$$\int_{S_{C\text{wall}}} \boldsymbol{n}\cdot(\rho \boldsymbol{V}\boldsymbol{V})dS = \int_{S_{C\text{wall}}} \rho(\boldsymbol{n}\cdot\boldsymbol{V})\boldsymbol{V}dS = \int_{S_{C\text{wall}}} \rho(-\boldsymbol{e}_2)\cdot(-\boldsymbol{e}_2 V_{\text{wall}})\boldsymbol{V}dS$$

$$= \int_{S_{C\text{wall}}} \boldsymbol{V}d\dot{m} \int_{S_{C\text{wall}}} \boldsymbol{n}\cdot(\rho \boldsymbol{V}\boldsymbol{V})dS = \int_{S_{C\text{wall}}} \boldsymbol{V}d\dot{m}. \tag{5.22}$$

Inserting Eqs. (5.20) through (5.22) into Eq. (5.19) and the results into Eq. (5.18), we obtain a relation that includes the mass flow through the control volume:

$$\frac{\partial}{\partial t}\int_{V_C} \boldsymbol{V}dm + \int_{S_{C\text{out}}} \boldsymbol{V}d\dot{m} - \int_{S_{C\text{in}}} \boldsymbol{V}d\dot{m} + \int_{S_{C\text{wall}}} \boldsymbol{V}d\dot{m} = \int_{S_C} (-\boldsymbol{n}p - \boldsymbol{t}\tau)dS + \boldsymbol{G}. \tag{5.23}$$

The first term expresses the total momentum exchange of all particles contained in the region (control volume) under consideration, at the time t, because of velocity changes produced by a non-steady flow. For a steady flow, it vanishes. The second and third integral are *leaving* and *entering velocity momenta*. The fourth term exhibits the velocity momentum through the wall. This term is different from zero if the wall is porous (permeable) or has perforations or slots that may be used for different purposes such as cooled turbine blades, Fig. 5.2, boundary layer suctions, etc. For a solid wall, this term, of course, vanishes identically. The first and the second integral on the right-hand side of Eq. (5.23) are momentum contributions due to the action of static pressure and the shear stresses. These integrals must be taken over the entire bounding surface that includes inlet, exit, and wall surfaces, Fig. 5.3:

$$\int_{S_C}(-\boldsymbol{n}p)dS + \int_{S_C}(-\boldsymbol{t}\tau)dS = \int_{S_{C\text{out}}}(-\boldsymbol{n}p)dS + \int_{S_{C\text{out}}}(-\boldsymbol{t}\tau)dS + \int_{S_{C\text{in}}}(-\boldsymbol{n}p)dS + \int_{S_{C\text{in}}}(-\boldsymbol{t}\tau)dS + \int_{S_{C\text{wall}}}(-\boldsymbol{n}p)dS + \int_{S_{C\text{wall}}}(-\boldsymbol{t}\tau)dS. \quad (5.24)$$

According to the convention in Fig. 5.1, the direction of unit normal vectors $\boldsymbol{n}_{\text{in}}$, $\boldsymbol{n}_{\text{out}}$, and $\boldsymbol{n}_{\text{wall}}$ point away from the region bounded by the control surface S_C. The last two integrals in Eq. (5.24) determine the reaction forces. To demonstrate the physical significance of the reaction force, we consider a rectilinear turbine cascade, Fig. 5.4.
The reaction force $\boldsymbol{R}$ which is exerted by the flow on the surface S_{Cw}, that is, on the turbine blade wall between the stations (1) and (2) and the body, is therefore:

$$\boldsymbol{F}_R = -\boldsymbol{F}_{\text{Flow}} = -\int_{S_{C\text{wall}}}(-\boldsymbol{n}p)dS - \int_{S_{C\text{wall}}}(-\boldsymbol{t}\tau)dS = \int_{S_{C\text{wall}}}(\boldsymbol{n}p)dS + \int_{S_{C\text{wall}}}(\boldsymbol{t}\tau)dS \quad (5.25)$$

As Eq. (5.25) indicates, the flow force equals the negative value of the two last integrals. Considering a steady flow and implementing Eq. (5.25) into Eq. (5.23), the reaction forces can be determined using the relationship:

$$\boldsymbol{F}_R = \int_{S_{C\text{in}}}\boldsymbol{V}d\dot{m} - \int_{S_{C\text{out}}}\boldsymbol{V}d\dot{m} - \int_{S_{C\text{wall}}}\boldsymbol{V}d\dot{m} + \int_{S_{C\text{in}}}(-\boldsymbol{n}p)dS + \int_{S_{C\text{in}}}(-\boldsymbol{t}\tau)dS + \int_{S_{C\text{out}}}(-\boldsymbol{n}p)dS + \int_{S_{C\text{out}}}(-\boldsymbol{t}\tau)dS + \boldsymbol{G}. \quad (5.26)$$

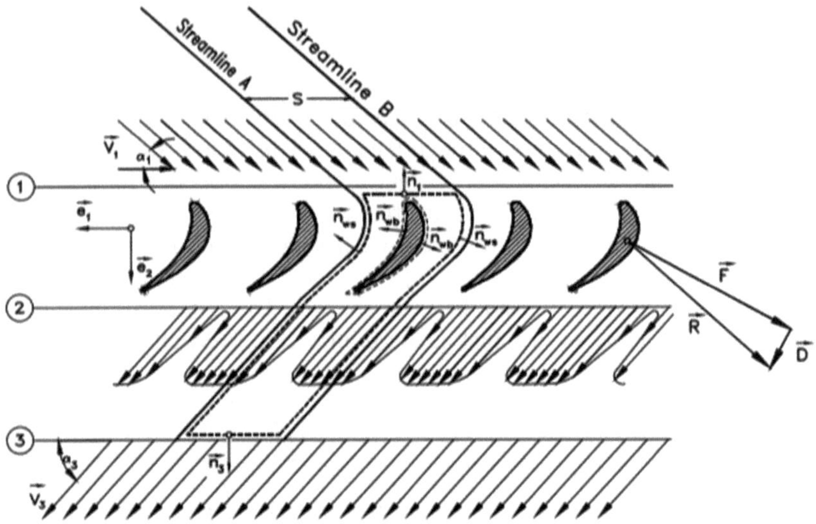

Fig. 5.4 Reaction force on a turbine blade with ***F*** and ***D*** as the lift, drag forces, and ***R*** the resultant force

The vector equation (5.26) can be decomposed into three components. An order of magnitude estimation suggests that the shear stress terms at the inlet and outlet are, in general, very small compared to the other terms. It should be pointed out that, the wall shear stress is already included in the resultant force $\boldsymbol{F}_R$.

5.3 Balance of Moment of Momentum

One of the application field of conservation of moment of momentum is turbomachinery. This allows a better understanding of the physics in an illustrative manner. To establish the conservation law of moment of momentum for a time dependent material volume, we start from the second law of Newton, Eq. (5.18):

$$m\frac{D\boldsymbol{V}}{Dt} = \sum \boldsymbol{F} = \boldsymbol{F}_S + \boldsymbol{G} = \int_{V(t)} \nabla \cdot \boldsymbol{\Pi} dv + \boldsymbol{G}. \tag{5.27}$$

The moment of the force given by Eq. (5.27) is then

$$m\boldsymbol{X} \times \frac{D\boldsymbol{V}}{Dt} = \sum \boldsymbol{X} \times F \tag{5.28}$$

with ***X*** as the position vector originating from a fixed point. To rearrange Eq. (5.28) for further analysis, its left-hand side is extended by adding the following zero-term identities:

$$\boldsymbol{V} \times \boldsymbol{V} = \frac{D\boldsymbol{X}}{Dt} \times \boldsymbol{V} = m\frac{D\boldsymbol{X}}{Dt} \times \boldsymbol{V} = 0 \tag{5.29}$$

and:

$$\frac{Dm}{Dt} = 0 = \boldsymbol{X} \times \boldsymbol{V}\frac{Dm}{Dt} = 0. \tag{5.30}$$

Introducing the identities (5.29) and (5.30) into Eq. (5.28), we arrive at:

$$m\boldsymbol{X} \times \frac{D\boldsymbol{V}}{Dt} + m\frac{D\boldsymbol{X}}{Dt} \times \boldsymbol{V} + \boldsymbol{X} \times \boldsymbol{V}\frac{Dm}{Dt} = \sum \boldsymbol{X} \times \boldsymbol{F}. \tag{5.31}$$

Using the Leibnitz's chain differential rule, a simple rearrangement of Eq. (5.31) allows the application of the Reynolds transport theorem as follows:

$$\frac{D(m\boldsymbol{X} \times \boldsymbol{V})}{Dt} = \sum \boldsymbol{X} \times \boldsymbol{F}. \tag{5.32}$$

Since $m = \int_{v(t)} \rho dv$, Eq. (5.32) can be written as:

$$\frac{D}{Dt}\int_{v(t)} (\rho\boldsymbol{X} \times \boldsymbol{V})dv = \sum \boldsymbol{X} \times \boldsymbol{F} \tag{5.33}$$

with $v(t)$ as the time dependent volume of the integral boundary. We apply the Reynolds transport theorem and the Gauss conversion theorem (Chap. 2) to the left-hand side of Eq. (5.33) and arrive at:

$$\frac{D}{Dt}\int_{V_C} (\rho\boldsymbol{X} \times \boldsymbol{V})dv = \int_{V_C} \left(\frac{\partial(\rho\boldsymbol{X} \times \boldsymbol{V})}{\partial t}dv\right) - \int_{S_C} \boldsymbol{n} \cdot (\rho\boldsymbol{V}\boldsymbol{V} \times \boldsymbol{X})dS. \tag{5.34}$$

We now interchange the sequence of multiplication for the vector product inside the parenthesis of the second integral in Eq. (5.34), $\rho\boldsymbol{V}\boldsymbol{V} \times \boldsymbol{X} = -\rho\boldsymbol{V} \times \boldsymbol{X}\boldsymbol{V}$, and obtain:

$$\frac{D}{Dt}\int_{V_C} (\rho\boldsymbol{X} \times \boldsymbol{V})dv = \int_{V_C} \left(\frac{\partial(\rho\boldsymbol{X} \times \boldsymbol{V})}{\partial t}dv\right) + \int_{S_C} \boldsymbol{n} \cdot (\rho\boldsymbol{V}\boldsymbol{X} \times \boldsymbol{V})\, dS. \tag{5.35}$$

Introducing the mass flow $\boldsymbol{n} \cdot \boldsymbol{V}ds = d\dot{m}$, Eq. (5.35) results in:

$$\frac{D}{Dt}\int_{V_C} (\rho\boldsymbol{X} \times \boldsymbol{V})dv = \int_{V_C} \left(\frac{\partial(\rho\boldsymbol{X} \times \boldsymbol{V})}{\partial t}dv\right) + \int_{S_C} (\boldsymbol{X} \times \boldsymbol{V})\, d\dot{m}. \tag{5.36}$$

The surface integral has to be carried out over the entire control surface S_C.

$$\frac{D}{Dt}\int_{V_C}(\rho \boldsymbol{X}\times \boldsymbol{V})dv = \int_{V_C}\left(\frac{\partial(\rho \boldsymbol{X}\times \boldsymbol{V})}{\partial t}dv\right) + \int_{S_{C2}}(\boldsymbol{X}\times \boldsymbol{V})d\dot{m} - \int_{S_{C1}}(\boldsymbol{X}\times \boldsymbol{V})d\dot{m}. \quad (5.37)$$

Now we consider the moment of momentum of all other forces on the right-hand side of Eq. (5.33):

$$\sum \boldsymbol{X}\times \boldsymbol{F} = \int_{S_C}\boldsymbol{X}\times(-\boldsymbol{n}p)dS + \int_{S_C}\boldsymbol{X}\times(-\boldsymbol{t}\tau)dS + \int_{V_C}\boldsymbol{X}\times \boldsymbol{g}dm. \quad (5.38)$$

Since the right side of Eq. (5.33) is equal to the right side of Eq. (5.38), the equation of moment of momentum can be presented in a more compact form that contains the contributions of velocity, pressure and shear stress momenta:

$$\int_{V_{(t)}}\left(\frac{\partial(\rho \boldsymbol{X}\times \boldsymbol{V})}{\partial t}dv\right) + \int_{S_{C\text{out}}}(\boldsymbol{X}\times \boldsymbol{V})d\dot{m} - \int_{S_{C\text{in}}}(\boldsymbol{X}\times \boldsymbol{V})d\dot{m} = \int_{S_C}\boldsymbol{X}\times(-\boldsymbol{n}p)dS + \int_{S_C}\boldsymbol{X}\times(-\boldsymbol{t}\tau)dS + \int_{V_C}\boldsymbol{X}\times \boldsymbol{g}dm. \quad (5.39)$$

The integration of the first two integrals on the right-hand side have to be performed; over $S_{C\text{in}}$, $S_{C\text{out}}$ and $S_{C\text{wall}}$.

$$\int_{S_C}\boldsymbol{X}\times(-\boldsymbol{n}p)dS + \int_{S_C}\boldsymbol{X}\times(-\boldsymbol{t}\tau)dS = \int_{S_{C\text{out}}}\boldsymbol{X}\times(-\boldsymbol{n}p)dS + \int_{S_{C\text{out}}}\boldsymbol{X}\times(-\boldsymbol{t}\tau)dS + \int_{S_{C\text{in}}}\boldsymbol{X}\times(-\boldsymbol{n}p)dS + \int_{S_{C\text{in}}}\boldsymbol{X}\times(-\boldsymbol{t}\tau)dS + \int_{S_{C\text{wall}}}\boldsymbol{X}\times(-\boldsymbol{n}p)dS + \int_{S_{C\text{wall}}}\boldsymbol{X}\times(-\boldsymbol{t}\tau)dS. \quad (5.40)$$

Similar to the expression for the reaction force, the last two integrals on the right-hand side of Eq. (5.40) determine the reaction moment, $\boldsymbol{M}_0$. This reaction moment is exerted by the flow on the solid boundary S_W of the system with respect to a fixed point such as the coordinate origin shown in Fig. 5.5. This figure exhibits the flow through a mixed axial-radial compressor stage where the flow undergoes a change in the radial direction associated with certain deflection from the inlet at station S_1

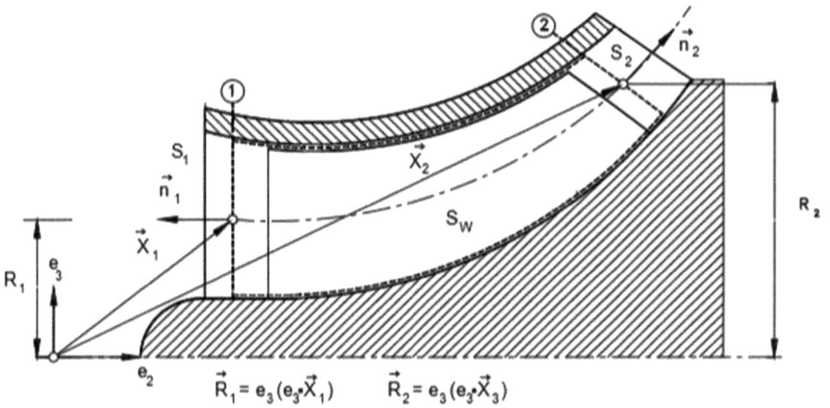

Fig. 5.5 A mixed flow compressor with control surfaces

to the exit at station S_2. A fixed control volume is placed on the rotor that includes a compressor blade. The normal unit vectors at the inlet and exit are used to establish the mass flow balances at stations 1 and 2. The *wall surface* S_W represents one blade surface (pressure or suction surface) that is projected on the drawing plane. The reaction moment consists of the moment by the surface shear and pressure forces:

$$\boldsymbol{M}_0 = \int\limits_{S_{C\text{wall}}} \boldsymbol{X} \times (\boldsymbol{t}\tau) dS + \int\limits_{S_{C\text{wall}}} \boldsymbol{X} \times (\boldsymbol{n}p) dS, \tag{5.41}$$

From Eq. (5.41) it is seen that the last two integrals of Eq. (5.40) are equal to—$\boldsymbol{M}_0$. Therefore, from Eqs. (5.40) and (5.41), the moment $\boldsymbol{M}_0$ exerted by the flow on the solid boundary S_W with respect to a fixed point using the station numbers in Fig. 5.5 is:

$$\begin{aligned}
\boldsymbol{M}_0 = &-\frac{\partial}{\partial t}\left(\int\limits_{V_C} \boldsymbol{X} \times \boldsymbol{V} dm\right) + \int\limits_{S_1} (\boldsymbol{X} \times \boldsymbol{V}) d\dot{m} - \int\limits_{S_2} (\boldsymbol{X} \times \boldsymbol{V}) d\dot{m} \\
&\int\limits_{S_1} (\boldsymbol{X} \times (-\boldsymbol{n}p) dS)_1 + \int\limits_{S_1} (\boldsymbol{X} \times (-\boldsymbol{t}\tau) dS)_1 \\
&+ \int\limits_{S_2} (\boldsymbol{X} \times (-\boldsymbol{n}p) dS)_2 + \int\limits_{S_2} (\boldsymbol{X} \times (-\boldsymbol{t}\tau) dS)_2 \\
&+ \int\limits_{V_C} \boldsymbol{X} \times \boldsymbol{g} dm.
\end{aligned} \tag{5.42}$$

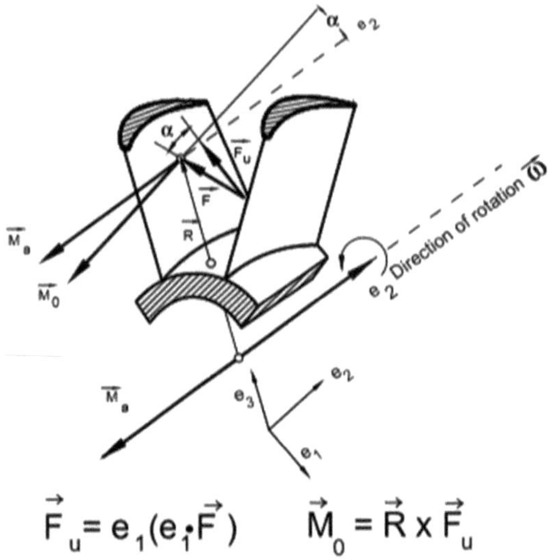

Fig. 5.6 Illustration of the axial moment by projecting the reaction moment M_0 on axial direction e_2

Equation (5.42) describes the moment of momentum in general form. The first integral on the right-hand side expresses the angular momentum contribution due to the unsteadiness. The second and third term represents the contribution due to the velocity momenta at the inlet and exit. The forth and sixth terms are formally the contributions of pressure momenta at the inlet and exit. The shear stress integrals and the fifth and seventh terms, representing the moment due to shear stresses at the inlet and exit, are usually ignored in practical cases. For applications to turbomachines, Eq. (5.42) can be used to determine the moment that the flow exerts on a turbine or compressor cascade. Of practical interest is the *axial moment* $M = M_a$ which acts on the cascade with respect to the axis of rotation. The moment $M = M_a$ is equal to the component of the moment vector parallel to the axis of revolution. It is also identical with the moment vector built by the product of the local radius vector $\boldsymbol{R}$ and the circumferential component of force vector $\boldsymbol{F}_u$. Details are shown in Fig. 5.6.

The axial moment is

$$\boldsymbol{M} = \boldsymbol{M}_a = e_2(e_2 \cdot \boldsymbol{M}_0). \tag{5.43}$$

Neglecting the contribution of the shear stress terms at the inlet and performing the above scalar multiplication, we further evaluate the contribution of the velocity terms in Eq. (5.42). To illustrate the details we utilize Fig. 5.7. The first velocity integral term in Eq. (5.42) is

$$e_2 \left[\int_{S_1} e_2 \cdot (\boldsymbol{X} \times \boldsymbol{V}) d\dot{m} \right] = e_2 \left[\int_{S_1} e_2 \cdot (e_3 R_1 \times \boldsymbol{V}) d\dot{m} \right]. \tag{5.44}$$

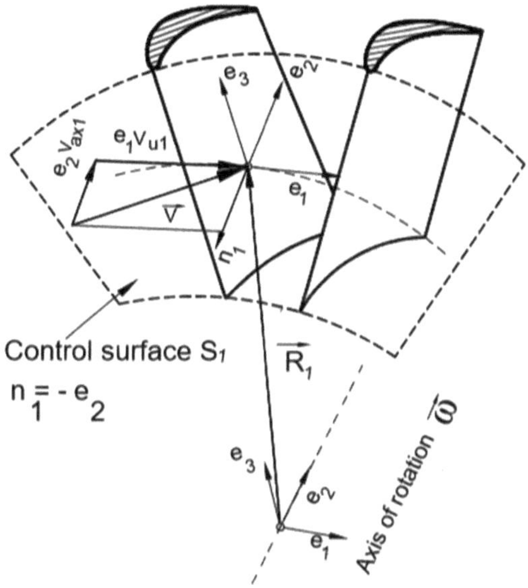

Fig. 5.7 Explaining the velocity integral at the inlet of a turbine blade

Without loss of generality we assume a two dimensional velocity vector with one component in circumferential and one in axial direction. Thus, Eq. (5.44) reads:

$$e_2 \left[\int\limits_{S_1} e_2 \cdot (\boldsymbol{X} \times \boldsymbol{V}) d\dot{m} \right] = e_2 \left\{ \int\limits_{S_1} e_2 \cdot [e_3 R_1 \times (e_1 V_{u1} + e_2 V_{ax1})] d\dot{m} \right\}. \quad (5.45)$$

In Eq. (5.44), the expression inside the integral is

$$\begin{aligned} & e_2 \cdot [e_3 R_1 \times (e_1 V_u + e_2 V_{ax})] d\dot{m} \\ & = e_2 \cdot (e_3 \times e_1 R_1 V_u + e_3 \times e_2 R_1 V_{ax}) d\dot{m} \end{aligned} \quad (5.46)$$

Now we consider the following vector operations

$$e_3 \times e_1 = e_2, \text{ and } e_3 \times e_2 = -e_1 \quad (5.47)$$

which we implement into Eq. (5.46) and arrive at

$$e_2 \cdot [e_3 R_1 \times (e_1 V_u + e_2 V_{ax})] d\dot{m} = e_2 \cdot (e_2 R_1 V_u - e_1 R_1 V_{ax}) d\dot{m} = R_1 V_u d\dot{m}. \quad (5.48)$$

With Eq. (5.48), Eq. (5.45) becomes

$$e_2 [\int\limits_{S_1} e_2 \cdot (\boldsymbol{X} \times \boldsymbol{V}) d\dot{m}] = e_2 \left(\int\limits_{S_1} R_1 V_{u1} d\dot{m} \right). \quad (5.49)$$

In an analogous way we have the velocity integral at the exit

$$e_2\left[\int_{S_2} e_2 \cdot (\boldsymbol{X} \times \boldsymbol{V})d\dot{m}\right] = e_2\left(\int_{S_2} R_2 V_{u2} d\dot{m}\right). \tag{5.50}$$

And finally the pressure integrals in Eq. (5.42)

$$e_2 \int_{S_1} e_2 \cdot (e_3 R \times (-\boldsymbol{n}p))dS \tag{5.51}$$

with

$$\boldsymbol{n} = -e_2, \; e_3 \times e_2 = -e_1 \tag{5.52}$$

we arrive at

$$e_2 \int_{S_1} e_2 \cdot e_1 RpdS = 0. \tag{5.53}$$

In an analogous manner the pressure integrals at the exit vanishes identically. Furthermore, the moment contribution of gravitational force will vanish. With this premise, Eq. (5.43) reduces to:

$$\boldsymbol{M}_a = -\frac{\partial}{\partial t}\left(\int_{V_C} \boldsymbol{X} \times \boldsymbol{V} dm\right) + \int_{S_1} e_2(R_1 V_{u1})\, d\dot{m} - \int_{S_2} e_2(R_2 V_{u2}) d\dot{m} \tag{5.54}$$

with V_u as the absolute velocity component in circumferential direction. For steady flow, Eq. (5.54) reduces to:

$$\boldsymbol{M}_a = \int_{S_1} e_2(R_1 V_{u1}) d\dot{m} - \int_{S_2} e_2(R_2 V_{u2}) d\dot{m}. \tag{5.55}$$

As shown in Fig. 5.6, the direction of the axial moment is identical with the direction of the shaft axis. For the case where the velocity distributions at the inlet and exit of the channel are fully uniform and the turbomachine is rotating with the angular velocity $\boldsymbol{\omega}$, the power consumed by a compressor stage or produced by a turbine stage is calculated from:

$$P = \boldsymbol{\omega} \cdot \boldsymbol{M}_a = \boldsymbol{\omega} \cdot e_2 \dot{m}(R_1 V_{u1} - R_2 V_{u2}) = \dot{m}(U_1 V_{u1} - U_2 V_{u2}). \tag{5.56}$$

Although the application of the conservation laws are extensively discussed in the following chapters, it is found necessary to present a simple example of how the moment of momentum is obtained by utilizing the *velocity diagram* of a single-

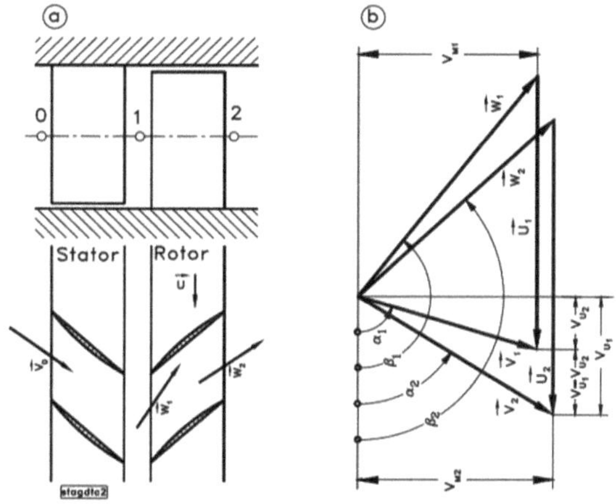

Fig. 5.8 A single-stage axial compressor (**a**), velocity diagram (**b**). The circumferential velocity difference $(V_{u1} - V_{u2})$ is responsible for the power consumption

stage axial compressor. Figure 5.8a represents a single stage axial compressor with the constant hub and tip diameters. We consider the flow situation at the mid-section. The flow is first deflected by the *stator row*, Fig. 5.8a (bottom). Entering the rotor row, the fluid particle moves through a rotating frame where the rotational velocity $\boldsymbol{U}$ is superimposed on the relative velocity $\boldsymbol{W}$.

The constant radii at the inlet and exit of the mid-section with $\omega R_1 = \omega R_2$ result in a constant circumferential velocity, $U_1 = U_2 = U$. As a consequence, Eq. (5.56) simplifies as $P = \dot{m}U(V_{u1} - V_{u2})$. The expression in the parenthesis, $(V_{u1} - V_{u2})$, is shown in the velocity diagram, Fig. 5.8b. It states that the compressor power consumption is related to the flow deflection expressed in terms of the circumferential velocity difference. The larger the difference $(V_{u1} - V_{u2})$ is, the higher the pressure ratio that the compressor produces. However, for each type of compressor design (axial, radial, subsonic, super sonic) there is always a limit to this difference, which is dictated by the flow separation.

For the case where no blades are installed inside the channel and the axial velocity distributions at the inlet and exit of the channel are fully uniform, Eq. (5.55) is reduced to:

$$R_1 V_{u1} = R_2 V_{u2} = \text{const.} \tag{5.57}$$

This is the so called free vortex flow.

5.4 Balance of Energy

The conservation law of energy in integral form, which we discuss in the following sections, is based on the thermodynamic principals, primarily the first law of thermodynamics for open systems and time independent control volumes. It is fully independent of the conservation law of energy derived for fluid mechanics. How-

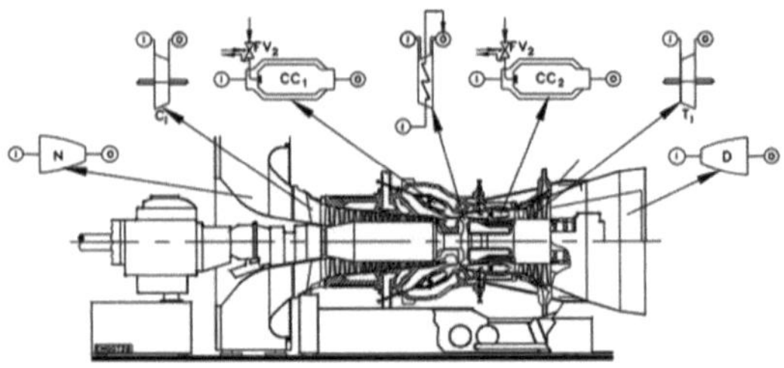

Fig. 5.9 A modern power generation gas turbine engine with a single shaft, two combustion chambers, a multi-stage compressor, a single-stage reheat turbine and a multi-stage turbine

ever, it *implicitly* contains the irreversibility aspects described by the dissipation process in the previous chapter. The contribution of the irreversibility is *explicitly* expressed by using the *Clausius-Gibbs* entropy equation, known as the second law of thermodynamics. The energy equation is applied to a variety of engineering devices such as internal combustion engines, jet engines, steam and gas turbine engines and their components in which a *chain of energy conversion processes* takes place. As an example, Fig. 5.9 shows a high performance gas turbine engine with several components to which we apply the results of our derivations.

In this chapter, we apply the conservation law of energy to a material volume with a system boundary that moves through the space where it may undergo deformation, rotation, and translation. The first law of thermodynamics in integral form states that if we add thermal energy (heat) Q and mechanical energy (work) W to a closed system, the total energy of the system E experiences a change from initial state E_1 to the final state E_2. Expressing in terms of energy balance, we have:

$$Q + W = E_2 - E_1. \tag{5.58}$$

The total energy E is the sum of internal, kinetic and potential energies,

$$E = U + \frac{1}{2}mV^2 + mgz. \tag{5.59}$$

In order to apply the conservation law of energy to a control volume, we divide Eq. (5.59) by the mass m to arrive at the *specific total energy*,

$$E = me = m\left(u + \frac{1}{2}V^2 + gz\right) \tag{5.60}$$

with e as the specific total energy. Similar to the conservation laws of mass, momentum, and moment of momentum, we ask for substantial change of the total energy, i.e.:

$$\frac{D(Q+W)}{Dt} = \dot{Q} + \dot{W} = \frac{DE}{Dt} = \frac{D(me)}{Dt} \tag{5.61}$$

with $\dot{Q}$ and $\dot{W}$ as the thermal and mechanical energy flow, respectively. Since $m = \int_{v(t)} \rho dv$, Eq. (5.61) yields:

$$\dot{Q} + \dot{W} = \frac{D}{Dt} \int\limits_{v(t)} \rho e dv. \tag{5.62}$$

To apply the the conservation of energy to a control volume, we use the Reynolds transport theorem. Using the Jacobi-transformation function $dv = J dv_0$, and introducing the time fixed volume v_0, we arrive at:

$$\dot{Q} + \dot{W} = \int\limits_{v_0} \left(J \frac{D(\rho e)}{Dt} + \frac{DJ}{Dt} \rho e \right) dv_0. \tag{5.63}$$

We now introduce $DJ/Dt = J\nabla \cdot \boldsymbol{V}$ for the substantial derivative of the Jacobian

$$\dot{Q} + \dot{W} = \int\limits_{v(t)} \left(\frac{D(\rho e)}{Dt} + \rho e \nabla \cdot \boldsymbol{V} \right) dv. \tag{5.64}$$

Developing the first integral term:

$$\dot{Q} + \dot{W} = \int\limits_{v(t)} \left(\frac{\partial(\varrho e)}{\partial t} + \boldsymbol{V} \cdot \nabla(\varrho e) + \varrho e \nabla \cdot \boldsymbol{V} \right) dv. \tag{5.65}$$

Application of the chain rule to the second and third term yields:

$$\dot{Q} + \dot{W} = \int\limits_{v(t)} \frac{\partial(\varrho e)}{\partial t} dv + \int\limits_{v(t)} \nabla \cdot (\rho e \boldsymbol{V}) dv. \tag{5.66}$$

With Gauss-Divergence Theorem:

$$\dot{Q} + \dot{W} = \int\limits_{v(t)} \frac{\partial(\varrho e)}{\partial t} dv + \int\limits_{S(t)} (\varrho e \boldsymbol{n} \cdot \boldsymbol{V}) dS. \tag{5.67}$$

The above equation is valid for any volume $v(t)$ including $v(t = 0)$ which might be a fixed control volume. In Eq. (5.67), the integration must be carried out over the entire control surface. For a control surface consisting of inlet, exit, and wall surfaces (Fig. 5.4), the second integral on the left-hand side gives:

$$\int\limits_{S_C} (\varrho e \boldsymbol{n} \cdot \boldsymbol{V}) dS = \int\limits_{S_{C\mathrm{in}}} e(\varrho \boldsymbol{n} \cdot \boldsymbol{V}) dS + \int\limits_{S_{C\mathrm{out}}} e(\varrho \boldsymbol{n} \cdot \boldsymbol{V}) dS. \tag{5.68}$$

Evaluating the integrands on the right-hand side of Eq. (5.68) by considering the directions of the unit vectors shown in Fig. 5.4, $\boldsymbol{n}_{\text{in}} = -\boldsymbol{e}_1$, $\boldsymbol{n}_{\text{out}} = +\boldsymbol{e}_1$, we find for the inlet cross-section:

$$\int_{S_{C\text{in}}} e(\varrho \boldsymbol{n} \cdot \boldsymbol{V}) dS = \int_{S_{C\text{in}}} e(-\varrho \boldsymbol{e}_1 \cdot \boldsymbol{e}_1 V_{\text{in}} dS) = -\int_{S_{C\text{in}}} e d\dot{m}. \tag{5.69}$$

For the exit cross-section we obtain:

$$\int_{S_{C\text{out}}} e(\varrho \boldsymbol{n} \cdot \boldsymbol{V}) dS = \int_{S_{C\text{out}}} e(\varrho \boldsymbol{e}_1 \cdot \boldsymbol{e}_1 V_{\text{in}} dS) = \int_{S_{C\text{out}}} e d\dot{m}. \tag{5.70}$$

Inserting the Eqs. (5.69) and (5.70) into Eq. (5.67), we obtain the energy equation for a control volume:

$$\dot{Q} + \dot{W} = \int_{V_C} \frac{\partial(\varrho e)}{\partial t} dv + \int_{S_{C\text{out}}} e d\dot{m} - \int_{S_{C\text{in}}} e d\dot{m} \tag{5.71}$$

with the specific total energy, $e = u + \frac{1}{2}V^2 + gz$,

$$\dot{Q} + \dot{W} = \int_{V_C} \frac{\partial\left[\varrho\left(u + \frac{1}{2}V^2 + gz\right)\right]}{\partial t} dv + \int_{S_{C\text{out}}} \left(u + \frac{1}{2V^2} + gz\right) d\dot{m} - \int_{S_{C\text{in}}} \left(u + \frac{1}{2V^2} + gz\right) d\dot{m}. \tag{5.72}$$

For uniform velocity distributions, Eq. (5.72) is reduced to:

$$\dot{Q} + \dot{W} = \int_{V_C} \frac{\partial\left[\varrho\left(u + \frac{1}{2}V^2 + gz\right)\right]}{\partial t} dv + \dot{m}_{\text{out}} \left(u + \frac{1}{2V^2} + gz\right) - \dot{m}_{\text{in}} \left(u + \frac{1}{2V^2} + gz\right). \tag{5.73}$$

The mechanical energy flow $\dot{W}$ consists of the shaft power $\dot{W}_{\text{shaft}}$ and the mechanical energy flow $\dot{W}_{\text{flow}}$ which is needed to overcome the shear and normal stresses at the system or control volume boundaries:

$$\dot{W} = \dot{W}_{\text{Shaft}} + \dot{W}_{\text{Flow}}. \tag{5.74}$$

The shaft power is the sum of the net shaft power and the power dissipated by the bearings $\dot{W}_{\text{shaft}} = \dot{W}_{\text{net}} + \dot{W}_{\text{bearings}}$. The second term in Eq. (5.74) is the power

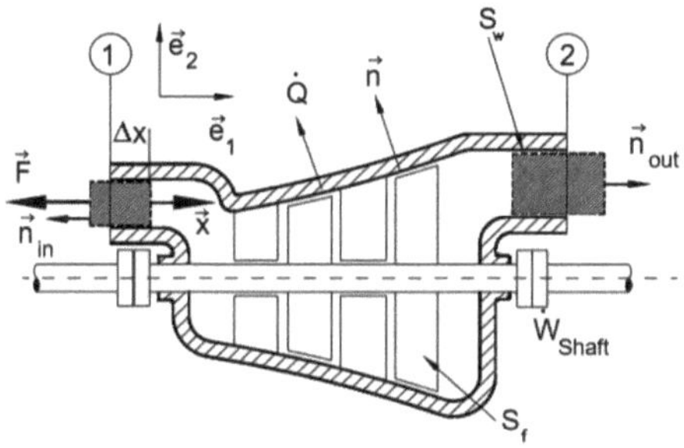

Fig. 5.10 Explanation of the flow forces, sketch of a turbine component

required to overcome the normal and shear stress forces at the inlet and exit of the system. It is the product of the flow force vector $\boldsymbol{F}$ and the displacement vector $d\boldsymbol{X}$, Fig. 5.10.

$$\dot{W}_{\text{Flow}} = \int_{\text{in}}^{\text{out}} d\dot{W}_F = \int_{\text{in}}^{\text{out}} d\left(\frac{\boldsymbol{F} \cdot d\boldsymbol{X}}{dt}\right). \tag{5.75}$$

Consider a turbine component, Fig. 5.10, where the working fluid (gas or steam) enters the inlet station. To force the differential mass dm into the turbine, which is under high pressure, a force is required that must compensate the pressure and the shear stress forces at the inlet. Figure 5.10 exhibits a simplified schematic of one of the turbine components in Fig. 5.9. It shows the directions of the forces and the displacements. At the inlet, the force vector $\boldsymbol{F}$ is expressed in terms of pressure and the inlet area and is oriented toward negative $\boldsymbol{e}_1$-direction. The displacement vector $d\boldsymbol{X}$ has the positive direction. As a result, the product:

$$\boldsymbol{F} \cdot d\boldsymbol{X} = -e_1 \cdot e_1 psdx = -pdV \tag{5.76}$$

is negative. The differential volume can be expressed as the product of the specific volume and the differential mass.
Replacing in Eq. (5.76) the differential volume dV with vdm $(dV = vdm)$ and dividing the result by dt, we arrive at:

$$\frac{\boldsymbol{F} \cdot d\boldsymbol{x}}{dt} = -pv\frac{dm}{dt} = -pv\dot{m} \tag{5.77}$$

with v as the specific volume of the fluid. Inserting Eq. (5.77) into Eq. (5.75) and assuming a constant mass flow, the integration from inlet to outlet

$$\dot{W}_{\text{Flow}} = -\int_{\text{in}}^{\text{out}} d(\dot{m}pv) = -[(\dot{m}pv)_{\text{Out}} - (\dot{m}pv)_{\text{In}}]. \tag{5.78}$$

To eliminate the internal energy from the equation of energy for open systems, we introduce the enthalpy $h = u + pv$ and introduce Eq. (5.78) into Eq. (5.73) and obtain:

$$\dot{Q} + \dot{W}_{\text{Shaft}} = \int_{V_C} \frac{\partial}{\partial t}\left[\rho\left(u + \frac{1}{2}V^2 + gz\right)\right] dv$$
$$+ \dot{m}_{\text{out}}\left(h + \frac{1}{2V^2} + gz\right)_{\text{out}}$$
$$- \dot{m}_{\text{in}}\left(h + \frac{1}{2V^2} + gz\right)_{\text{in}}. \quad (5.79)$$

For a fixed control volume, the volume integral can be rearranged as:

$$\int_{V_C} \frac{\partial(\varrho e)}{\partial t} dv = \frac{\partial}{\partial t} \int_{V_C} (\varrho e) dv. \quad (5.80)$$

We set $\int_{CV} \rho e dv = E_{CV}$, and since E_{CV} can only change with time, the partial derivative is replaced by the ordinary one, $\partial/\partial t \equiv d/dt$. As a result, we obtain:

$$\dot{Q} + \dot{W}_{\text{Shaft}} = \frac{dE}{dt} + \dot{m}_{\text{out}}\left(h + \frac{1}{2V^2} + gz\right)_{\text{out}} - \dot{m}_{\text{in}}\left(h + \frac{1}{2V^2} + gz\right)_{\text{in}}. \quad (5.81)$$

Equation (5.81) exhibits the general form of energy equation for an open system with a fixed control volume. For technical applications, several special cases are applied which we will discuss in the following.

5.4.1 Energy Balance Special Case 1: Steady Flow

If a power generating or consuming machine such as a turbine or a compressor operates in a steady design point, the first term on the right-hand side of Eq. (5.81) disappears, $dE/dt = 0$, which leads to:

$$\dot{Q} + \dot{W}_{\text{Shaft}} = \dot{m}_{\text{out}}\left(h + \frac{1}{2}V^2 + gz\right)_{\text{out}} - \dot{m}_{\text{in}}\left(h + \frac{1}{2}V^2 + gz\right)_{\text{in}}. \quad (5.82)$$

Equation (5.82) is the energy balance for a machine with heat addition or rejection $\dot{Q}$ and the shaft power supplied or consumed $\dot{W}_{\text{shaft}}$.

5.4.2 Energy Balance Special Case 2: Steady Flow, Constant Mass Flow

In many applications, the mass flow remains constant from the inlet to the exit of the machine. Examples are uncooled turbines and compressors where no mass flow is added during the compression or expansion process. In this case, Eq. (5.82) reduces to:

$$\dot{Q} + \dot{W}_{\text{Shaft}} = \dot{m}\left[\left(h + \frac{1}{2}V^2 + gz\right)_{\text{out}} - \left(h + \frac{1}{2}V^2 + gz\right)_{\text{in}}\right]. \tag{5.83}$$

Now, we define the *specific total enthalpy*

$$H \equiv h + \frac{1}{2}V^2 + gz \tag{5.84}$$

and insert it into Eq. (5.83), from which we get:

$$\dot{Q} + \dot{W}_{\text{Shaft}} = \dot{m}(H_{\text{out}} - H_{\text{in}}). \tag{5.85}$$

In Eqs. (5.83) or (5.85), the contribution of Δgz compared to Δh and $\Delta V^2/2$ is negligibly small. Using the above equation, the energy balance for the major components of the gas turbine engine shown in Fig. 5.9 can be established as detailed in the following section.

5.5 Application of Energy Balance to Engineering Components and Systems

For the following reasons, we selected a gas turbine as an appropriate system. For better understanding, we apply the energy balance to the gas turbine components. The gas turbine engine shown in Fig. 5.9 consists of a variety of components to which the energy balance in different form can be applied. These components can be categorized in three groups. The first group entails all those components that serve either the mass flow transport from one point of the engine to another or the conversion of kinetic energy into the potential energy and vice versa. Pipes, diffusers, nozzles, and throttle valves are representative examples of the first group. Within this group no thermal or mechanical energy (shaft work) is exchanged with the surroundings. In thermodynamic sense these components are assumed adiabatic. The second group contains those components within which thermal energy is generated or exchanged with the surroundings. Combustion chambers and heat exchangers are typical examples of these components. Thermodynamically speaking, in these cases we are dealing with *diabatic systems*. Finally, the third group includes components within which thermal and mechanical energy is exchanged. In the following sections, each group is treated individually.

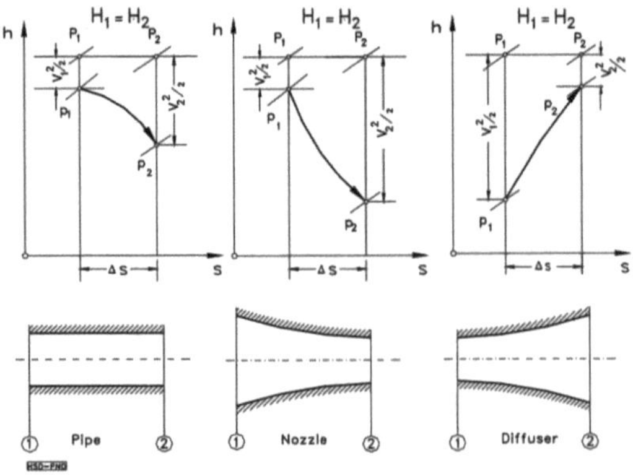

Fig. 5.11 Energy transfer in pipes, nozzles, and diffusers

5.5.1 Application: Pipe, Diffuser, Nozzle

Pipes, nozzles, diffusers, compressor and turbine stator cascades, as well as throttle devices are a few examples. For this group, Eq. (5.85) reduces to:

$$H_{out} - H_{In} = 0, \text{ or } H_{out} = H_{in} = \text{Const.} \tag{5.86}$$

Replacing the subscripts *in* and *out* by 1 and 2, the h-s-diagrams of the pipe, nozzle, and diffuser are shown in Fig. 5.11.

As this figure shows, the viscous flow causes entropy increase which results in a reduction of the total pressure from P_1 to P_2. The total pressure is the sum of static pressure, dynamic pressure, and the pressure due to the change of height:

$$P = p + \frac{1}{2}\rho V^2 + \rho g z \tag{5.87}$$

neglecting the contribution of Δgz results in the following relation for total pressure loss:

$$\Delta P = P_1 - P_{12} = \left(p + \frac{1}{2}\varrho V^2\right)_1 - \left(p + \frac{1}{2}\varrho V^2\right)_2. \tag{5.88}$$

The area under the process-line reflects the irreversibility due to the internal friction which results in total pressure drop.

5.5.2 Application: Combustion Chamber

As indicated, combustion chambers or heat exchangers are typical examples of the components belonging to the group within which heat transfer or conversion of chemical into thermal energy takes place. The energy balance is:

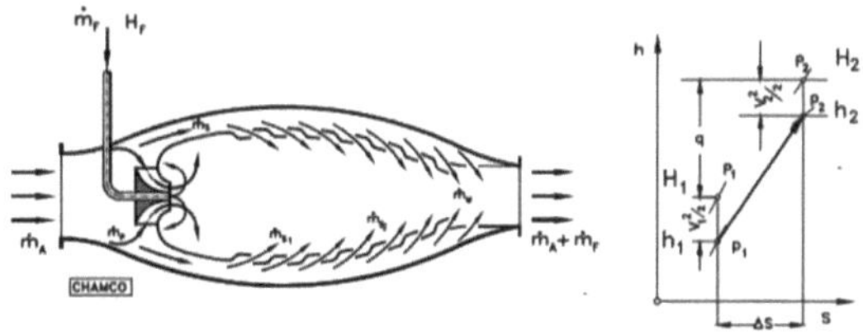

Fig. 5.12 Schematic of a gas turbine combustion chamber, h-s diagram. Fuel mass flow $\dot{m}_F$, primary air mass flow $\dot{m}_p$, secondary air mass flow $\dot{m}_S$, and the mixing air mass flow $\dot{m}_M$

$$\dot{Q} = \dot{m}(H_2 - H_1). \tag{5.89}$$

As a result, the total enthalpy at the exit is the sum of the inlet total enthalpy plus the heat added to the system. Introducing the specific thermal energy, $q = \dot{Q}/\dot{m}$, we find

$$H_2 = H_1 + q. \tag{5.90}$$

Figure 5.12 shows a schematic of a typical gas turbine combustion chamber where the combustion air and fuel are mixed and burned leading to a combustion gas with an increased exit temperature and enthalpy. The combustion process is shown in Fig. 5.12 where a simplified model of a combustion chamber is presented.

The flow and combustion process within the combustion chamber is associated with entropy increases due to the heat addition and internal friction inside the chamber. The internal friction, the wall friction, and particularly the mixing process of the primary and secondary air mass flows $\dot{m}_p, \dot{m}_s$ causes pressure decreases of up to 5%. The thermal energy per unit mass is shown in an h-s-diagram, Fig. 5.12 as q. It corresponds to the total enthalpy difference.

5.5.3 *Application: Turbo-shafts, Energy Extraction, Consumption*

Within this group, mechanical and thermal energy transfers to/from surroundings take place. Turbines and compressors are two representative examples. The energy balance in general form is:

$$\dot{Q} + \dot{W}_{\text{Shaft}} = \frac{dE}{dt} + \dot{m}_{\text{Out}}\left(h + \frac{1}{2V^2} + gz\right)_{\text{out}} - \dot{m}_{\text{In}}\left(h + \frac{1}{2V^2} + gz\right)_{\text{In}}. \tag{5.91}$$

We distinguish in the following cases where we consider steady flow only. Thus, the first term on the right-hand side, $dE/dt = 0$, disappears.

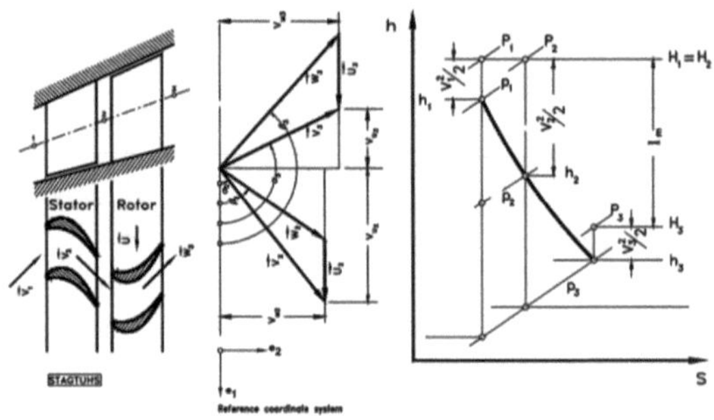

Fig. 5.13 Turbine stage consisting of a stator and a rotor row (left), velocity diagram (middle), h-s-diagram (right)

5.5.3.1 Uncooled Turbine

We start with and adiabatic (uncooled) turbine component where no heat exchange between the turbine blades and the turbine working medium takes place i.e., $\dot{Q} = 0$. The mass flows at the inlet and exit are the same. Figure 5.13 shows a turbine stage which consists of a stator and a rotor row. The stator row, with several blades, deflects the flow to the following rotor row which turns with angular velocity ω. The process of conversion of total energy into mechanical energy takes place within the rotor.

Following the nomenclature in Fig. 5.13, we introduce the specific stage mechanical energy $l_m = \dot{W}_{\text{Shaft}}/\dot{m}$. Considering the h-s-diagram in Fig. 5.13 for adiabatic turbine, $\dot{Q} = 0$, Eq. (5.91) reduces to:

$$-\,l_m = H_3 - H_1 = (h_3 - h_1) + \frac{1}{2}(V_3^2 - V_1^2). \tag{5.92}$$

The negative sign of l_m indicates that energy is rejected from the system (to the surroundings). The h-s-diagram in Fig. 5.13 shows the expansion process within the stator, where the total enthalpy within the stator $H_1 = H_2$ remains constant.

Changes of the total enthalpy occur within the rotor, where the total energy of the working medium is partially converted into mechanical energy. In addition, the *stage velocity diagram* is also shown in Fig. 5.13. This diagram shows the flow angle deflection within both stator and rotor blades. As we saw from Eq. (5.56), the stage power is given by:

$$P = \omega \cdot \boldsymbol{M}_a = \omega \cdot e_2 \dot{m}(R_1 V_{u1} - R_2 V_{u2}) = \dot{m}(U_1 V_{u1} - U_2 V_{u2}). \tag{5.93}$$

Dividing the above equation by the mass flow, we find:

$$\frac{\dot{P}}{m} = l_m = U_1 V_{u1} - U_2 V_{u2}. \tag{5.94}$$

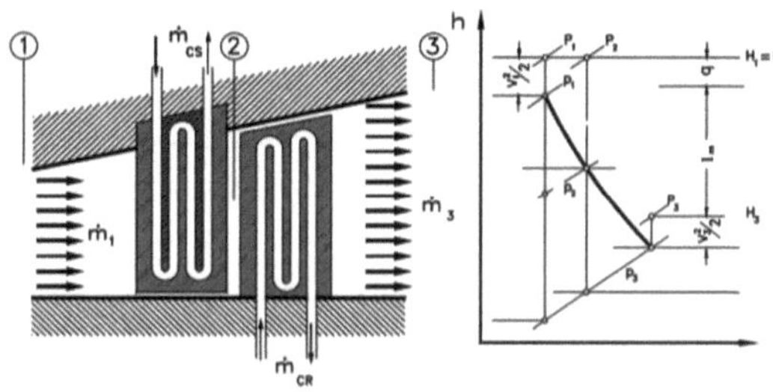

Fig. 5.14 A simplified schematic of a cooled turbine blade with internal cooling channels, hs-diagram

This equation can be found using the energy Eq. (5.92) by replacing the static enthalpies in Eq. (5.92) by the kinetic energies using trigonometric relations and the angle definition given by the velocity diagram in Fig. 5.13.

5.5.3.2 Cooled Turbine

As the second case, we consider a cooled (diabatic) gas turbine blade where a heat exchange between the turbine material and the cooling medium takes place. The schematic of such a gas turbine blade is shown in Fig. 5.14. In a high performance gas turbine engine, the front stages of the turbine component are exposed to temperatures that are close to the melting point of the turbine blade material. In order to protect the blades, a substantial amount of heat must be removed from the blades. One of the cooling methods currently used introduces cooling air into the turbine cooling channels. Inside these channels, intensive heat transfer from the blade material to the cooling medium takes place, resulting in a substantial reduction of the blade surface temperature.

The process of expansion and heat transfer is depicted in the h-s diagram shown in Fig. 5.14. Assuming a steady flow through the turbine and neglecting the potential energy by the gravitational force, the energy equation reads:

$$\dot{Q} + \dot{W}_{\text{Shaft}} = \dot{m}_{\text{Out}}\left(h + \frac{1}{2}V^2\right) - \dot{m}_{\text{In}}\left(h + \frac{1}{2}V^2\right). \tag{5.95}$$

Since in this particular type of cooling scheme the cooling mass flows through the stator and rotor, $\dot{m}_{CS}$ and $\dot{m}_{CR}$, do not join the main turbine mass flow, the inlet and exit mass flows are the same, $\dot{m}_{\text{In}} = \dot{m}_{\text{Out}} = \dot{m}$. We introduce the specific heat $q \equiv \dot{Q}/\dot{m}_{\text{In}}$ which is transferred from the stator and rotor blades to the cooling mass flows, $\dot{m}_{CS}$ and $\dot{m}_{CR}$. Considering the negative signs of the specific mechanical energy l_m and the heat q, we obtain from Eq. (5.95):

$$q + l_m = \left(h + \frac{1}{2}V^2\right)_{\text{In}} - \left(h + \frac{1}{2}V^2\right)_{\text{Out}}. \tag{5.96}$$

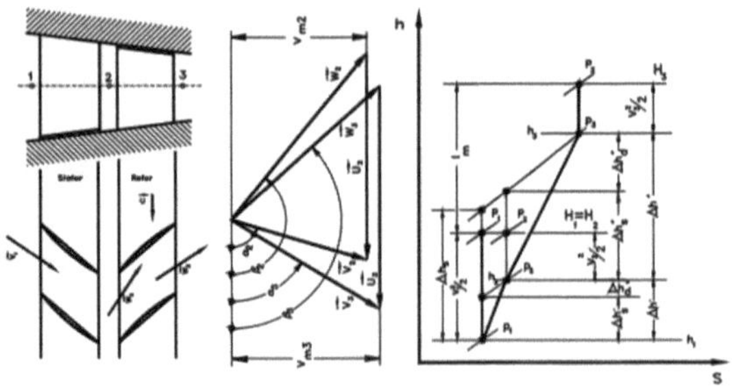

Fig. 5.15 A compressor stage (left) with velocity diagram (middle) and the compression h-s-diagram (right). P: total pressure, p static pressure

The h-s diagram in Fig. 5.14 shows the specific stage mechanical energy l_m and the heat transferred from the turbine stage blade material q. From this diagram, we can see that the turbine specific stage mechanical energy has been reduced by the amount of the heat rejected from the blades.

5.5.3.3 Uncooled Compressor

Figure 5.15 shows an uncooled compressor stage which consists of a stator and a rotor row.

Similar to the turbine stage, the stator row with several blades deflects the flow to the following rotor row which is turning with an angular velocity ω. The process of conversion of mechanical energy into total energy takes place within the rotor. As in the case of a turbine component, we follow the nomenclature of Fig. 5.15 for mechanical energy transfer and introduce the specific stage mechanical energy $l_m = \dot{W}_{\text{Shaft}}/\dot{m}$. Considering the h-s-diagram in Fig. 5.15, Eq. (5.91) modifies as:

$$l_m = H_3 - H_1. \tag{5.97}$$

The positive sign of l_m indicates that the energy is consumed by the system (from the surroundings). In the h-s-diagram of Fig. 5.15, the enthalpy differences $\Delta h'$, $\Delta h''$ refer to the polytropic enthalpy differences in stator and rotor, respectively. The corresponding isentropic enthalpy differences for stator and rotor are shown as $\Delta h'_s$, $\Delta h''_s$. In addition, the h-s-diagram shows the enthalpy differences due to the dissipation process for stator and rotor as $\Delta h'_d$ and $\Delta h''_d$.

5.6 Irreversibility, Entropy Increase, Total Pressure Loss

The total pressure losses within a component can be calculated using the second law of thermodynamics:

$$ds = \frac{\delta q}{T} = \frac{du + pdv}{T} = \frac{dh - vdp}{T}. \tag{5.98}$$

Using the generalized thermodynamic relations, we find:

$$ds = \frac{c_v}{T}dT + \left(\frac{\partial p}{\partial T}\right)_v dv \tag{5.99}$$

or in terms of c_p:

$$ds = \frac{c_P}{T}dT - \left(\frac{\partial v}{\partial T}\right)_p dp. \tag{5.100}$$

With:

$$dh = c_p dT + \left[v - T\left(\frac{\partial v}{\partial T}\right)_p\right]dp. \tag{5.101}$$

For the working media used in thermal turbomachines such as steam, air, and combustion gases, the thermodynamic properties can be taken from appropriate gas and steam tables. In general, the specific heats at constant pressure c_p and constant specific volume c_v are functions of temperature.

Figure 5.16 shows the specific heat at constant pressure as a function of temperature with the *fuel/air ratio* μ as parameter. The dry air is characterized by $\mu = 0$ and no moisture. As seen at lower temperatures, changes of c_p are not significant. However, increasing the temperature results in higher specific heat. In the case of combustion gases, the addition of fuel in a combustion chamber causes a change in the gas constant R and additional increase in c_p. At moderate pressures, the ideal gas relation can be applied

$$pv = RT; \frac{\partial v}{\partial T} = \frac{R}{p} \tag{5.102}$$

and the entropy change can be obtained using Eq. (5.99) in terms of enthalpy or internal energy:

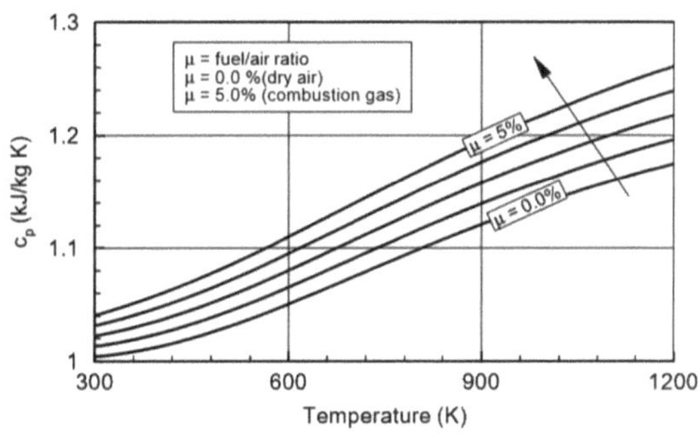

Fig. 5.16 Temperature dependency of the specific heat c_p for air and combustion gases at different fuel/air ratios

$$ds = \frac{c_p}{T}dT - R\frac{dp}{p},\ ds = \frac{c_v}{T}dT + R\frac{dv}{v}. \tag{5.103}$$

Assuming lower temperatures where c_p and c_v can be approximated as constant, the entropy change is calculated by integrating Eq. (5.103):

$$\Delta s = c_p \ln\left(\frac{T_2}{T_1}\right) - R\ln\left(\frac{p_2}{p_1}\right) = c_p \ln\left[\left(\frac{T_2}{T_1}\right)\left(\frac{p_1}{p_2}\right)^{\frac{\kappa-1}{\kappa}}\right]. \tag{5.104}$$

$$\Delta s = c_v \ln\left(\frac{T_2}{T_1}\right) + R\ln\left(\frac{v_2}{v_1}\right) = c_v \ln\left[\left(\frac{T_2}{T_1}\right)\left(\frac{v_2}{v_1}\right)^{\kappa-1}\right]. \tag{5.105}$$

Equations (5.104) and (5.105) are valid under perfect gas assumption, c_p and $c_v \neq f(T)$, for estimating the entropy changes. For dry or moist air as working media in compressors, and combustion gases as the working media in turbines and combustion chambers, appropriate gas tables must be used in order to avoid significant errors.

5.6.1 Application of Second Law to Engineering Components

To calculate the entropy increase as a result of an irreversible process, a flow through a simple nozzle or a turbine is considered. The expansion process for both devices are shown in Fig. 5.16. The entropy change is obtained using the second law of thermodynamics:

$$\Delta s = c_p \ln\left(\frac{T_2}{T_1}\right) - R\ln\left(\frac{p_2}{p_1}\right) \tag{5.106}$$

where p_1, p_2 and T_1, T_2 are static pressures and temperatures, respectively. These quantities can be related to the total pressure p_{01}, p_{02} and total temperature T_0, T_{02} by the following simple modification:

$$\frac{p_2}{p_1} = \frac{\left(\frac{p_2}{p_{02}}\right)}{\left(\frac{p_1}{p_{01}}\right)}\left(\frac{p_{02}}{p_{01}}\right). \tag{5.107}$$

Introducing the temperature relation by applying the isentropic relation with pv^κ =const.,

$$\left(\frac{p_2}{p_{02}}\right) = \left(\frac{T_2}{T_{01}}\right)^{\frac{\kappa}{\kappa-1}},\ \text{and}\ \left(\frac{p_1}{p_{01}}\right) = \left(\frac{T_1}{T_{01}}\right)^{\frac{\kappa}{\kappa-1}} \tag{5.108}$$

and inserting Eq. (5.108) into Eq. (5.107) leads to:

$$\frac{p_2}{p_1} = \frac{\left(\frac{T_2}{T_{02}}\right)^{\frac{\kappa}{\kappa-1}}}{\left(\frac{T_1}{T_{01}}\right)^{\frac{\kappa}{\kappa-1}}} \left(\frac{p_{02}}{p_{01}}\right). \tag{5.109}$$

If we assume that the fluid is a perfect gas with $c_p \not\equiv f(T)$ =const., then we may set $T_{01} = T_{02}$. With this assumption Eq. (5.109) simplifies as:

$$\frac{p_2}{p_1} = \left(\frac{T_2}{T_1}\right)^{\frac{\kappa}{\kappa-1}} \left(\frac{p_{02}}{p_{01}}\right). \tag{5.110}$$

The entropy change obtained form Eq. (5.106) becomes:

$$\Delta s = c_p \ln\left(\frac{T_2}{T_1}\right) - R\frac{\kappa}{\kappa - 1} \ln\left(\frac{T_2}{T_1}\right) - R\ln\left(\frac{p_{02}}{p_{01}}\right). \tag{5.111}$$

With:

$$c_p = R\frac{\kappa}{\kappa - 1} \tag{5.112}$$

the first two terms on the right-hand side of Eq. (5.111) cancel each other leading to:

$$\Delta s = R\ln\left(\frac{p_{01}}{p_{02}}\right) = -R\ln\left(\frac{p_{02}}{p_{01}}\right) = -R\ln\frac{p_{01} - \Delta p_0}{p_{01}}. \tag{5.113}$$

Thus, the entropy change is directly related to the total pressure loss. We introduce the total pressure loss coefficient ζ:

$$\zeta = \frac{\Delta p_0}{p_{01}} \tag{5.114}$$

then we have:

$$\Delta s = -R\ln(1 - \zeta) \tag{5.115}$$

or

$$\zeta = 1 - e^{\frac{-\Delta s}{R}}. \tag{5.116}$$

If the total pressure loss coefficient is known, the entropy change can be calculated using Eq. (5.116). The loss correlations are developed empirically based on experimental data.

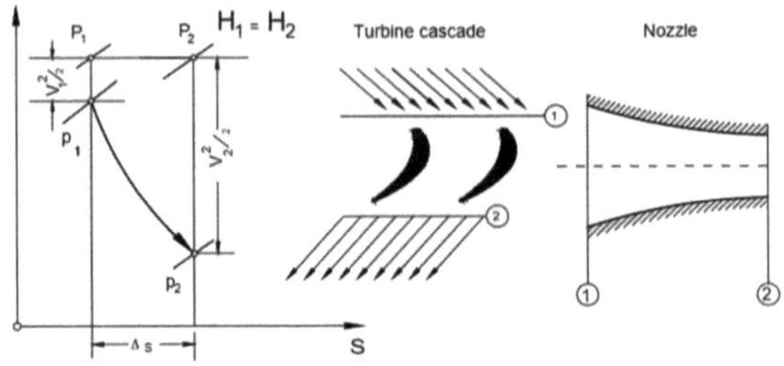

Fig. 5.17 Total pressure loss and entropy increase in a nozzle and a turbine cascade

5.7 Theory of Thermal Turbomachinery Stages

Turbomachines are devices within which conversion of total energy of a working medium into mechanical energy and vice-versa takes place. Turbomachines are generally divided into two main categories. The first category is used primarily to produce power. It includes, among others, steam turbines, gas turbines, and hydraulic turbines. The main function of the second category is to increase the total pressure of the working fluid by consuming power. It includes compressors, pumps, and fans. Gas turbines are also used for thrust generation and utilized in small aircrafts as propeller gas turbines and as high performance jet engines in medium and large size civil and military aircrafts. While the power generation gas turbines have a *single spool* with a multi-stage compressor, a combustion chamber and a multi-stage turbine, the aircraft engines may have up to three-spools that rotate at different frequency. The turbine component of gas and steam turbines are of axial type design, the compressor may be of axial or radial design, depending on their required mass flow. For small scale gas turbines, the application of radial compressors is more common than the axial ones. This is also true for the turbine component of turbochargers. The subject of turbomachinery aero-thermodynamic design is treated, among others, in [20–22].

5.7.1 *Energy Transfer in Turbomachinery Stages*

The energy transfer in turbomachinery is established by means of the stages. A *turbomachinery stage* comprises a row of fixed guide vanes called *stator blades*, and a row of rotating blades termed *rotor*. To elevate the total pressure of a working fluid, *compressor stages* are used that partially convert the mechanical energy into potential energy. According to the conservation law of energy, this energy increase requires an external energy input which must be added to the system in the form of mechanical energy. Figure 5.17a, b schematically represent two single stages within a multi-stage high pressure turbine and a high pressure compressor environment with constant mean diameter. These stages consist of one stator row followed by one rotor row. To define a unified station nomenclature for compressor and turbine

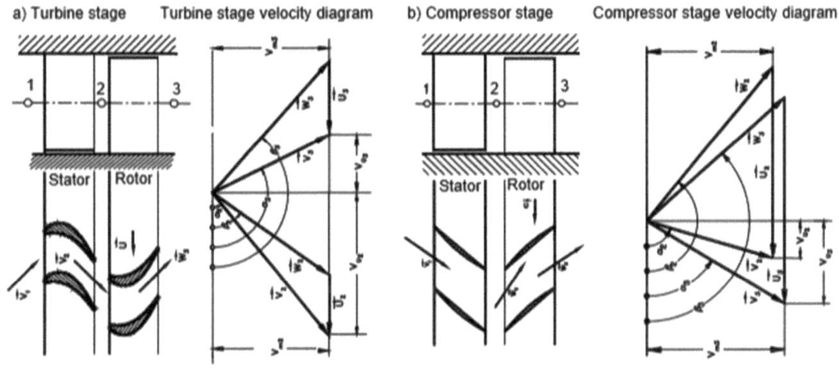

Fig. 5.18 **a** Turbine and **b** compressor stage configurations with the stator rotor arrangements and velocity diagrams

stages, we identify with station number 1 as the inlet of the stator, followed by station 2 as the rotor inlet and 3, rotor exit. The absolute and relative flow angles are counted counterclockwise from a horizontal line. This convention allows an easier calculation of the off-design behavior of compressor and turbine stages during a transient operation.

The working fluid enters the stator row with an *absolute velocity* vector $\boldsymbol{V}_1$ and an absolute inlet flow angle α_1. It is deflected and exits the stator row at an exit flow angle α_2 in direction of the rotor's leading edge. The expansion process through the turbine stage, Fig. 5.18a, in connection with the rotational motion of the rotor causes a major portion of the total energy of the working medium to convert into the shaft power. Conversely, in the compressor stage shown in Fig. 5.18b, the compression process converts a major part of the mechanical energy input into the potential energy causing the total pressure to rise. In general, the compression process resulting in a decrease of specific volume requires a decrease in flow path cross sectional area. In contrast, the expansion process in a multi-stage turbine causes a continuous increase in specific volume which requires an increase in flow path cross section, Fig. 5.18.

5.7.2 Energy Transfer in Relative Systems

Since the rotor operates in a relative frame of reference (relative system), the energy conversion mechanism is quite different from that of a stator (absolute system). A fluid particle that moves with a relative velocity $\boldsymbol{W}$ within the relative system that rotates with the angular velocity $\boldsymbol{\omega}$ has an absolute velocity:

$$\boldsymbol{V} = \boldsymbol{W} + \boldsymbol{\omega} \times \boldsymbol{R} = \boldsymbol{W} + \boldsymbol{U}, \boldsymbol{\omega} \times \boldsymbol{R} = \boldsymbol{U} \tag{5.117}$$

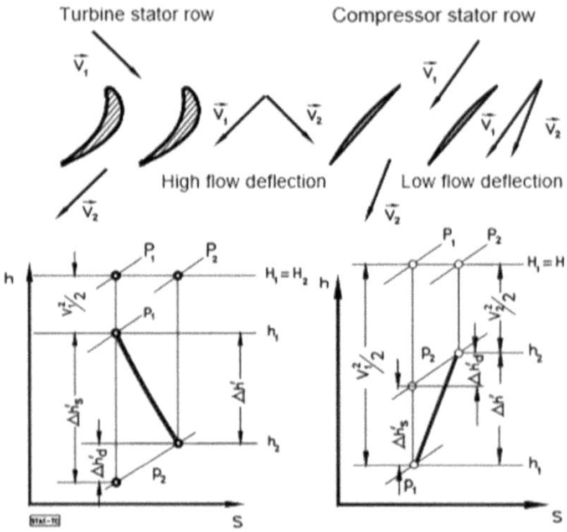

Fig. 5.19 Expansion and compression process through a turbine and a compressor stator row

with $\boldsymbol{R}$ in Eq. (5.117) as the radius vector of the particle in the relative system. Introducing the absolute velocity vector $\boldsymbol{V}$ in the equation of motion and multiplying the results with a relative differential displacement dR, we obtain the energy equation for an adiabatic steady flow within a rotating relative system:

$$d\left(h + \frac{1}{2}W^2 - \frac{\omega^2 R^2}{2} + gz\right) = 0 \tag{5.118}$$

or the relative total enthalpy:

$$H_r = h + \frac{1}{2}W^2 - \frac{\omega^2 R^2}{2} + gz = \text{const.} \tag{5.119}$$

Neglecting the gravitational term, $gz \approx 0$, Eq. (5.122) can be written as:

$$h_1 + \frac{1}{2}W_1^2 - \frac{1}{2}U_1^2 = h_2 + \frac{1}{2}W_2^2 - \frac{1}{2}U_2^2. \tag{5.120}$$

Equation (5.120) is the energy equation transformed into a relative system. As can be seen, the transformation of kinetic energy undergoes a change while the transformation of static enthalpy is frame indifferent. With these equations in connection with the energy balance, we can analyze the energy transfer within an arbitrary turbine or compressor stage.

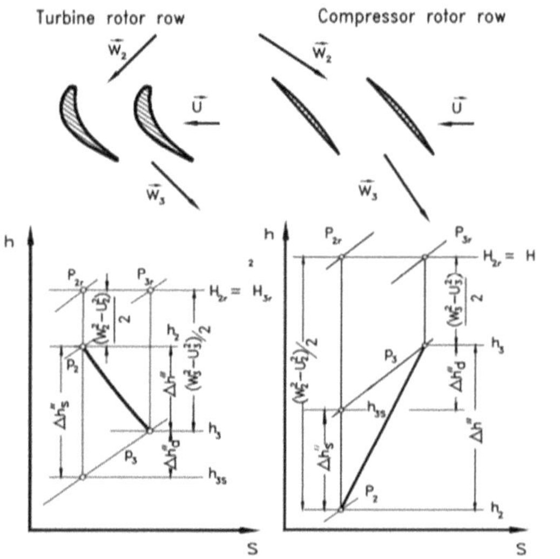

Fig. 5.20 Expansion and compression process through a turbine and a compressor rotor row

5.7.3 *Unified Treatment of Turbine and Compressor Stages*

In this chapter, compressor and turbine stages are treated from a unified physical point of view. Figures 5.19 and 5.20 show the decomposition of a turbine and a compressor stage into their stator and rotor rows. The primes "′" and "″" refer to stator and rotor rows, respectively. As seen, the difference between the isentropic and the polytropic enthalpy difference is expressed in terms of dissipation $\Delta h'_d = \Delta h'_s - \Delta h'$ for turbines and $\Delta h'_d = \Delta h' - \Delta h'_s$ for compressors. For the stator, the energy balance requires that $H_2 = H_1$. This leads to:

$$h_1 - h_1 = \Delta h' = \frac{1}{2}(V_2^2 - V_1^2). \tag{5.121}$$

Moving to the relative frame of reference, the relative total enthalpy $H_{r2} = H_{r3}$ remains constant. Thus, the energy equation for the rotor is according to Fig. 5.20:

$$h_2 - h_3 = \Delta h'' = \frac{1}{2}(W_3^2 - W_2^2 + U_2^2 - U_3^2). \tag{5.122}$$

The stage specific shaft power balance requires:

$$1_m = H_1 - H_3 = (h_1 - h_2) - (h_3 - h_2) + \frac{1}{2}(V_1^2 - V_3^2). \tag{5.123}$$

Inserting Eqs. (5.121) and (5.122) into Eq. (5.123) yields :

$$1_m = \frac{1}{2}[(V_2^2 - V_3^2) + (W_3^2 - W_2^2) + (U_2^2 - U_3^2)]. \tag{5.124}$$

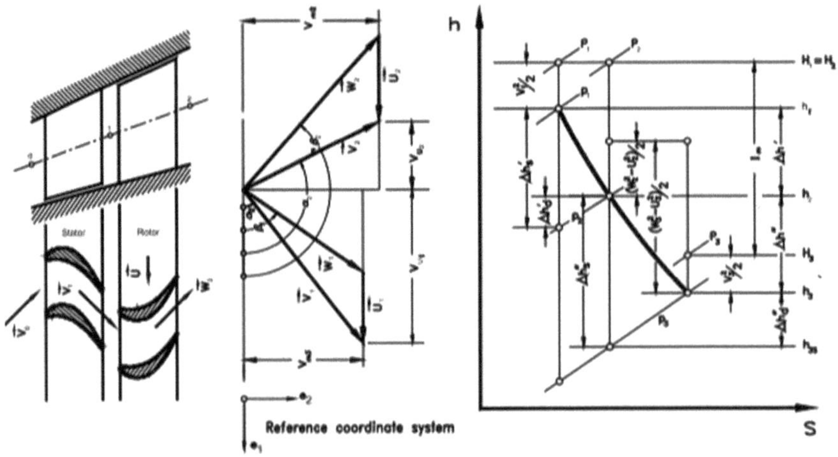

Fig. 5.21 A turbine stage (left) with the velocity diagram (middle) and the expansion process (right). The direction of the unit vector e_1 is identical with the rotational direction

Equation (5.124), known as the *Euler Turbine Equation,* indicates that the stage work can be expressed simply in terms of absolute, relative, and rotational kinetic energies. This equation is equally applicable to turbine stages that *generate* shaft power and to compressor stages that *consume* one. In the case of a turbine stage, the sign of the specific mechanical energy l_m is negative, which indicates that energy is removed from the system (power generation). In compressor cases, it is positive because energy is added to the system (power consumption). Figures 5.21 and 5.22 show the stage configuration, the velocity diagram and the expansion, compression process within a single stage turbine and compressor.

Before proceeding with velocity diagrams, it is of interest to evaluate the individual kinetic energy differences in Eq. (5.124). If we wish to design a turbine or a compressor stage with a high specific shaft power l_m for a particular rotational speed, then we have two options: (1) we increase the *flow deflection* that leads to an increase in $(V_2^2 - V_3^2)$ or, (2) we increase the radial difference between the inlet and the exit that leads to a larger $(U_2^2 - U_3^2)$. While option (1) is used in axial stages, option (2) is primarily applied to radial stages. Radial turbine design is used for small size turbines such as turbocharger or as power generation component of an open-cycle ocean thermal energy conversion plant as reported detailed in [24, 25]. Using the trigonometric relation with the *angle convention* from the velocity diagram in Figs. 5.21 and 5.22, we express the velocity vectors in terms of their components which are incorporated in Eq. (5.124) leading to the *stage specific shaft power*:

$$l_m = U_2 V_{u2} + U_3 V_{u3}. \tag{5.125}$$

Equation (5.125) is valid for axial, radial, and mixed flow turbines and compressors. The stage shaft power is then calculated by

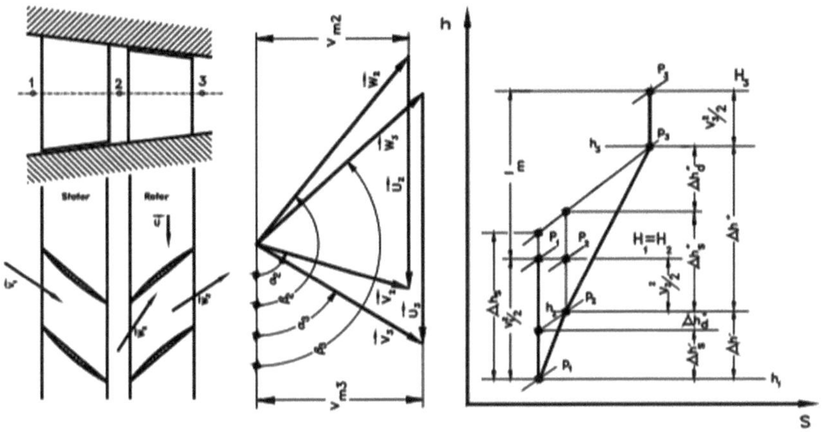

Fig. 5.22 A compressor stage (left) with the velocity diagram (middle) and the compression process (right)

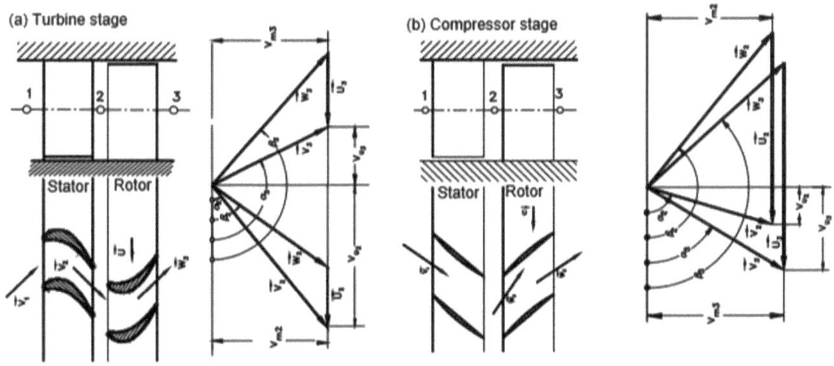

Fig. 5.23 Turbine and compressor stages with the velocity diagrams

$$P = \dot{m} l_m = \dot{m}(U_2 V_{u2} + U_3 V_{u3}). \quad (5.126)$$

A similar relation was obtained in Sect. 5.3, Eq. (5.56), from the scalar product of moment of momentum and the angular velocity. There we found the power as $P = \dot{m}U(V_{u1} - V_{u2})$. In order to avoid confusions that may arise from different signs, it should be pointed out that in Sect. 5.3, no angle convention was introduced and the negative sign in Eq. (5.56) was the result of the formal derivation of the conservation law of moment of momentum. This negative sign implies that V_{u1} and V_{u2} point in the same direction. The unified angle convention introduced in Figs. 5.21 and 5.22, however, takes the actual direction of the velocity components with regard to a predefined coordinate system.

5.8 Dimensionless Stage Parameters

Equation (5.124) exhibits a direct relation between the specific stage shaft power l_m and the kinetic energies. The velocities from which these kinetic energies are built can be taken from the corresponding stage velocity diagram. The objective of this chapter is to introduce dimensionless stage parameters that completely determine the stage velocity diagram. These stage parameters exhibit a set of unified relations for compressor and turbine stages respectively.

Starting from a turbine or compressor stage with constant mean diameter and axial components, shown in Fig. 5.23, we define the dimensionless stage parameters that describe the stage velocity diagram of a *normal stage*.
A normal stage is encountered within the high pressure (HP) part of multi-stage turbines or compressors and is characterized by $\boldsymbol{U}_3 = \boldsymbol{U}_2$, $\boldsymbol{V}_3 = \boldsymbol{V}_1$, $V_{m1} = V_{m3}$, and $\alpha_1 = \alpha_3$. The similarity of the velocity diagrams allows using the same blade profile throughout the HP-turbine or compressor, thus, significantly reducing manufacturing costs.

We define the stage flow coefficient ϕ as the ratio of the meridional velocity component and the circumferential component. For this particular case, the meridional component is identical with the axial component:

$$\phi = \frac{V_{m3}}{U_3}. \tag{5.127}$$

The stage flow coefficient ϕ in Eq. (5.127) is a characteristic for the mass flow behavior through the stage. The *stage load coefficient* λ is defined as the ratio of the specific stage mechanical energy l_m and the exit circumferential kinetic energy U_3^2. This coefficient directly relates the flow deflection given by the velocity diagram with the specific stage mechanical energy:

$$\lambda = \frac{l_m}{U_3^2}. \tag{5.128}$$

The stage load coefficient λ in Eq. (5.128) describes the work capability of the stage. It is also a measure for the stage loading. The *stage enthalpy coefficient* ψ represents the ratio of the isentropic stage mechanical energy and the exit circumferential kinetic energy U_3^2.

$$\psi = \frac{l_s}{U_3^2}. \tag{5.129}$$

The stage enthalpy coefficient represents the stage isentropic enthalpy difference within the stage. Furthermore, we define the *stage degree of reaction* r, which is the ratio of the static enthalpy difference used in the rotor row divided by the static enthalpy difference used in the entire stage:

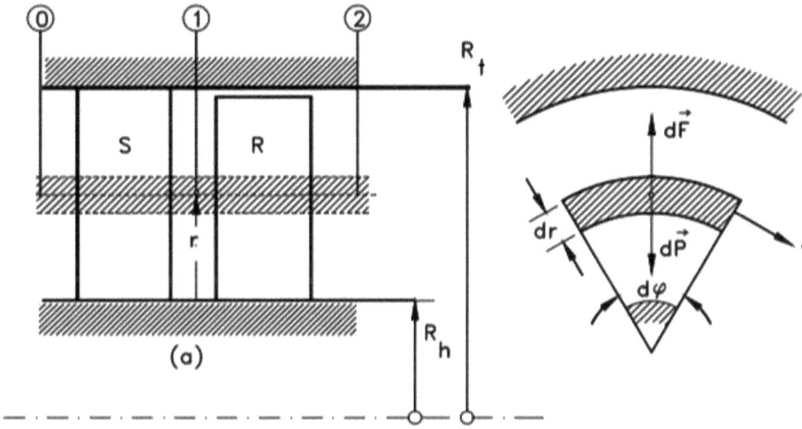

Fig. 5.24 Explanation for simple radial equilibrium within the axial gap between the stator and rotor blade

$$r = \frac{\Delta h''}{\Delta h'' + \Delta h'}. \tag{5.130}$$

The degree of reaction r indicates the portion of energy transferred in the rotor blading. Using Eqs. (5.124) and (5.125), expressing the velocity vectors by their corresponding components and inserting the results into Eq. (5.130), we arrive at:

$$r = \frac{1}{2}\frac{W_{u3} - W_{u2}}{U}. \tag{5.131}$$

5.8.1 Simple Radial Equilibrium to Determine r

Expressing the relationship between the degree of reaction and the blade height requires the knowledge of the *radial equilibrium* condition within the axial gaps between the stator and rotor blades. In a fully three-dimensional turbomachinery flow, describing the radial equilibrium condition is a complicated issue. Attempts to numerically analyze the issue of the radial equilibrium have encountered divergence problems. The streamline curvature method based on an axisymmetric assumption exhibits a reasonable and practical solution [23]. For the simple cases we discuss in this Chapter, we further simplify the radial equilibrium condition to arrive at simple relationships between the degree of reaction and the blade height.

The fluid particles in compressors and turbines experience a rotational and translational motion. For the simple turbine and compressor cases case we discussed in this Chapter the rotating fluid is subjected to centrifugal forces that must be balanced by the pressure gradient in order to maintain the radial equilibrium. Consider an infinitesimal sector of an annulus with unit depth containing the fluid element which is rotating with tangential velocity V_u in an absolute frame of reference. The centrifugal force acting on the element is shown in Fig. 5.24. Since the fluid element is in radial equilibrium, the centrifugal force per unit width is obtained from:

$$dF = dm\frac{V_u^2}{R} \tag{5.132}$$

with $dm = \rho R dR d\phi$. The centrifugal force is kept in balance by the pressure forces:

$$\frac{dp}{dR} = \rho\frac{V_u^2}{R}. \tag{5.133}$$

This result can also be obtained by decomposing the Euler equation of motion (4.51) for inviscid flows in its three components in a cylindrical coordinate system. The assumptions needed to arrive at Eq. (5.133) are:

$$\frac{\partial V_r}{\partial R} \simeq 0, \text{ Axial symmetric: } \frac{\partial V_r}{\partial \phi} = 0, \frac{\partial V_r}{\partial z} \simeq 0. \tag{5.134}$$

With these assumptions, Eq. (5.133) yields:

$$\frac{1}{\rho}\frac{\partial p}{\partial R} = \frac{V_u^2}{R}. \tag{5.135}$$

Equation (5.135) is identical with Eq. (5.133), where the partial differential ∂ is replaced by d because of the assumptions in Eq. (5.134). Calculation of a static pressure gradient requires additional information from the total pressure relation. For this purpose, we apply the Bernoulli equation, neglecting the gravitational term and differentiating the results in radial direction:

$$\frac{dP}{dR} = \frac{dp}{dR} + \rho V_u \frac{dV_u}{dR} + \rho V_{ax}\frac{dV_{ax}}{dR} + \rho V_r \frac{dV_r}{dR}. \tag{5.136}$$

The assumption of a constant total pressure P =const. and a constant axial component V_{ax} =const. simplifies Eq. (5.136) to:

$$\frac{dp}{dR} + \rho V_u \frac{dV_u}{dR} = 0, \text{ or } \frac{dp}{dR} = -\rho V_u \frac{dV_u}{dR}. \tag{5.137}$$

Equating (5.135) and (5.137) and separating the variables results in:

$$\frac{dV_u}{V_u} + \frac{dR}{R} = 0. \tag{5.138}$$

The integration of Eq. (5.138) leads to $V_u R$ =const. This type of flow is called free vortex flow and fulfills the requirement to be potential flow, $\nabla \times V = 0$. We use this relation to rearrange the specific stage mechanical energy:

$$l_m = U_2 V_{u2} + U_3 V_{u3} = \omega(R_2 V_{u2} + R_3 V_{u3}). \tag{5.139}$$

Going back to Fig. 5.18, at station (2) the swirl is $R_2 V_{u2}$ =const.= K_2; likewise at station 3 the swirl is $R_3 V_{u3} = K_3$. Since ω =const., the specific stage power is constant:

$$l_m = (K_2 + K_3)\omega = \text{const.} \tag{5.140}$$

Equation (5.140) implies that for a stage with constant spanwise meridional components and constant total pressure from hub to tip, the specific stage power is constant over the entire blade height. To express the degree of reaction in the spanwise direction, we replace the enthalpy differences in Eq. (5.130) by pressure differences. For this purpose we apply the first law for an adiabatic process through stator and rotor blades expressed in terms of $\Delta h'' = \bar{v}''\Delta p''$, $\Delta h' = \bar{v}'\Delta p'$ with $\bar{v}$ as the averaged specific volume. It leads to:

$$r = \frac{\bar{v}''\Delta p''}{\bar{v}''\Delta p'' + \bar{v}'\Delta p'} = \frac{\Delta p''}{\Delta p'' + v\frac{\bar{v}'}{\bar{v}''}\Delta p'} \cong \frac{p_2 - p_3}{p_1 - p_3}. \tag{5.141}$$

In the above equation, the ratio of specific volumes was approximated as $\bar{v}', /\bar{v}'' \cong 1$. This approximation is admissible for low Mach number ranges.

Considering $R_2 V_{u2}$ =const., the integration of Eq. (5.137) for station 1 from an arbitrary radius R to the mean radius R_m yields,

$$(p_1 - p_{m1}) = \frac{\rho}{2}(V_{um})_1^2 \left(1 - \frac{R_m^2}{R^2}\right)_1. \tag{5.142}$$

Similarly, at station (2) we have,

$$(p_2 - p_{m2}) = \frac{\rho}{2}(V_{um})_2^2 \left(1 - \frac{R_m^2}{R^2}\right)_2 \tag{5.143}$$

and finally, at station (3) we arrive at:

$$(p_3 - p_{m3}) = \frac{\rho}{2}(V_{um})_3^2 \left(1 - \frac{R_m^2}{R^2}\right)_3 \tag{5.144}$$

with $(R_m)_1 = (R_m)_2 = (R_m)_3$, and $V_{um3} = V_{um1}$. Introducing Eqs. (5.142), (5.143) and (5.144) into Eq. (5.141), we finally arrive at a simple relationship for the degree of reaction:

$$\frac{1 - r}{1 - r_m} = \frac{R_m^2}{R^2}. \tag{5.145}$$

From a turbomachinery design point of view, it is of interest to estimate the degree of reaction at the hub radius by inserting the corresponding radii into Eq. (5.145). As a result, we find:

$$\frac{1-r}{1-r_h} = \left(\frac{R_h}{R}\right)^2. \tag{5.146}$$

If, for example, the degree of reaction at the mean diameter is set at $r_m = 50\%$, Eq. (5.145) immediately calculates r at any radius R. It should be mentioned that, for a turbine, a negative degree of reaction at the hub may lead to flow separation and is not desired. Likewise, for the compressor, r should not exceed the value of 100%.

Equation (5.146) represents a simple radial equilibrium condition which allows the calculation of the inlet flow angle in a radial direction using the free vortex relation $V_u R$ =const. from Eq. (5.138):

$$V_u R = \text{const.};\ R = \frac{\text{const.}}{V_u}. \tag{5.147}$$

This leads to determination of the inlet flow angle in a spanwise direction,

$$\frac{R_m}{R} = \frac{\cot \alpha_1}{\cot \alpha_{1m}}. \tag{5.148}$$

5.8.2 *Effect of Degree of Reaction on the Stage Configuration*

The value of r has major design consequences. For turbine blades with $r = 0$, as shown in Fig. 5.25a, the flow is deflected in the rotor blades without any enthalpy changes. As a consequence, the magnitude of the inlet and exit velocity vectors are the same and the entire stage static enthalpy difference is partially converted within the stator row. Note that the flow channel cross section remains constant. For $r = 0.5$, shown in Fig. 5.25b, a fully symmetric blade configuration is established. Figure 5.25c shows a turbine stage with $r > 0.5$. In this case, the flow deflection inside the rotor row is much greater than the one inside the stator row. In the past, mainly two types of stages were common designs in steam turbines.

The stage with a constant pressure across the rotor blading ($p_2 = p_3$) called *action stage*, was used frequently. This turbine stage was designed such that the exit absolute velocity vector $\boldsymbol{V}_3$ was swirl free. It is most appropriate for the design of single stage turbines and as the last stage of a multi-stage turbine. The *exit loss*, which corresponds to the kinetic energy of the exiting mass flow, becomes a minimum by using a swirl free absolute velocity. The stage with $r = 0.5$ is called the *reaction stage*.

5.8.3 *Effect of Stage Load Coefficient on Stage Power*

The stage load coefficient λ defined in Eq. (5.139) is an important parameter which describes the stage capability to generate/consume shaft power. A turbine stage with low flow deflection, thus, low specific stage load coefficient λ, generates lower specific stage power l_m. To increase l_m, blades with higher flow deflection are used that produce higher stage load coefficient λ. The effect of an increased λ is shown in Fig. 5.25 where three different bladings are plotted. The top blading with the stage load coefficient $\lambda = 1$ has lower deflection. The middle blading has a moderate flow deflection and moderate $\lambda = 2$ which delivers the stage power twice as high as the top blading. Finally, the bottom blading with $\lambda = 3$, delivers three times the stage power as the first one. In the practice of turbine design, among other things, two major parameters must be considered. These are the specific load coefficients and the stage polytropic efficiencies. Lower deflection generally yields higher stage polytropic efficiency, but many stages are needed to produce the required turbine power. However, the same turbine power may be established by a higher stage flow deflection and, thus, a higher λ at the expense of the stage efficiency. Increasing the stage load coefficient has the advantage of significantly reducing the stage number, thus, lowering the engine weight and manufacturing cost. In aircraft engine design practice, one of the most critical issues besides the thermal efficiency of the engine, is the thrust/weight ratio. Reducing the stage numbers may lead to a desired thrust/weight ratio. While a high turbine stage efficiency has top priority in power generation steam and gas turbine design, the thrust/weight ratio is the major parameter for aircraft engine designers.

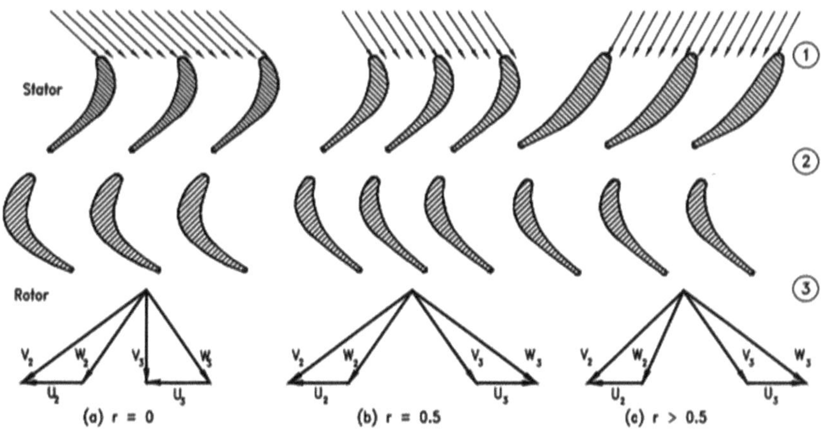

Fig. 5.25 Effect of degree of reaction on the stage configuration

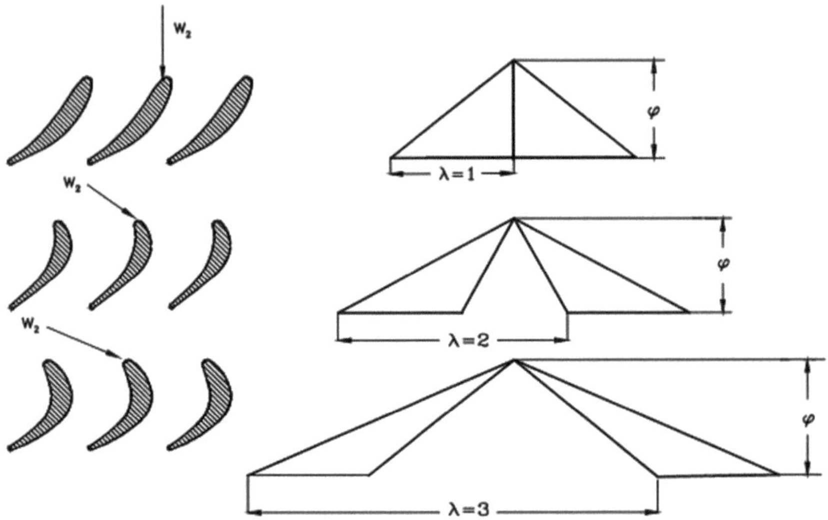

Fig. 5.26 Dimensionless stage velocity diagram to explain the effect of stage load coefficient λ on flow deflection and blade geometry, $r = 0.5$

5.9 Unified Description of a Turbomachinery Stage

The following sections treat turbine and compressor stages from a unified standpoint. Axial, mixed flow, and radial flow turbines and compressors follow the same thermodynamic conservation principles. Special treatments are indicated when dealing with aerodynamic behavior and loss mechanisms. While the turbine aerodynamics is characterized by a negative pressure gradient environment, the compression process operates in a positive (adverse) pressure gradient environment. As a consequence, partial or total flow separation may occur on compressor blade surfaces leading to partial stall or surge. On the other hand, with the exception of some minor local separation bubbles on the suction surface of highly loaded low pressure turbine blades, the turbine operates normally without major flow separation or breakdown. These two distinctively different aerodynamic behaviors are due to different pressure gradient environments. In this section, we present a set of algebraic equations that describes the turbine and compressor stages with constant mean diameter.

5.9.1 Unified Description of Stage with Constant Mean Diameter

For a turbine or compressor stage with constant mean diameter (Fig. 5.27), we present a set of equations that describe the stage by means of the dimensionless parameters such as stage flow coefficient ϕ, stage load coefficient λ, degree of reaction r, and the flow angles. From the velocity diagram with the angle definition in Fig. 5.27, we obtain the flow angles:

$$\cot\alpha_2 = \frac{U_2 + W_{u2}}{V_{ax}} = \frac{1}{\phi}\left(1 + \frac{W_{u2}}{U}\right) = \frac{1}{\phi}\left(1 - r + \frac{\lambda}{2}\right)$$
$$\cot\alpha_3 = -\frac{W_{u2} - U_2}{V_{ax}} = -\frac{1}{\phi}\left(\frac{W_{u3} - U}{U}\right) = \frac{1}{\phi}\left(1 - r - \frac{\lambda}{2}\right). \tag{5.149}$$

Similarly, we find the other flow angles, thus, we summarize:

$$\cot\alpha_2 = \frac{1}{\phi}\left(1 - r + \frac{\lambda}{2}\right), \cot\beta_2 = \frac{1}{\phi}\left(\frac{\lambda}{2} - r\right)$$
$$\cot\alpha_3 = \frac{1}{\phi}\left(1 - r - \frac{\lambda}{2}\right), \cot\beta_3 = -\frac{1}{\phi}\left(\frac{\lambda}{2} + r\right). \tag{5.150}$$

The stage load coefficient can be calculated from:

$$\lambda = \phi(\cot\alpha_2 - \cot\beta_3) - 1. \tag{5.151}$$

As seen from Eq. (5.150), one is dealing with seven unknowns and only four equations. To obtain a solution, assumptions need to be made relative to the remaining three unknowns. These may include any of the following parameters: α_2, β_3, ϕ, λ, or r. The criteria for selecting these parameters are discussed in details in [23].

The preceding discussions that have led to Eqs. (5.150) and (5.151) deal with compressor and turbine stages with constant hub and tip diameters. These equations cannot be applied to cases where the diameter, circumferential, and meridional velocities are not constant. Examples are axial flow turbine and compressor types shown in Figs. 5.21 and 5.22, radial inflow (centripetal) turbines, and centrifugal compressors. In these cases, the meridional velocity ratio and the diameter are no longer constant. The dimensionless parameters for these cases are summarized below:

$$\mu = \frac{V_{m2}}{V_{m3}}, \nu = \frac{R_2}{R_3} = \frac{U_2}{U_3}, \phi = \frac{V_{m3}}{U_3}, \lambda = \frac{1_m}{U_3^2}, r = \frac{\Delta h''}{\Delta h' + \Delta h''}. \tag{5.152}$$

As seen, two more parameters, namely the meridional velocity ratio μ and the diameter ratio ν, are added to the list of unknowns resulting in four equations and nine unknowns. The set of four equations and the discussions how to select the five remaining parameters to solve these equations are given in [23].

5.10 Turbine and Compressor Cascade Flow Forces

The preceding section was dedicated to the energy transfer within turbomachinery stages. The stage shaft power production or consumption in turbines and compressors were treated from a unifying point of view by introducing a set of dimensionless parameters. As shown, the stage power is the result of the scalar product between the moment of momentum acting on the rotor and the angular velocity. The moment of momentum in turn was brought about by the forces acting on rotor blades. The blade forces are obtained by applying the conservation equation of linear momentum to the turbine or compressor cascade under investigation. In this section, we first assume an inviscid flow for which we establish the relationship between the *lift force* and *circulation*. Then we consider the viscosity effect that causes friction or drag forces on the blading.

5.10.1 Blade Force in an Inviscid Flow Field

Starting from a given turbine cascade with the inlet and exit flow angles shown in Fig. 5.27, the blade force can be obtained by applying the linear momentum principles to the control volume with the unit normal vectors and the coordinate system shown in Fig. 5.27. Applying Eq. (5.26), the blade inviscid force is obtained from:

$$\boldsymbol{F}_i = \dot{m}\boldsymbol{V}_1 - \dot{m}\boldsymbol{V}_2 - \boldsymbol{n}_1 p_1 sh - \boldsymbol{n}_2 p_2 sh \tag{5.153}$$

with the subscript i that refers to inviscid flow, s as the spacing and h as the blade height that can be assumed unity. The relationship between the control volume normal unit vectors and the unit vectors pertaining to the coordinate system is given by $\boldsymbol{n}_1 = -\boldsymbol{e}_2$ and $\boldsymbol{n}_2 = \boldsymbol{e}_2$. The velocities in Eq. (5.153) can be expressed in terms of circumferential as well as axial components:

$$\boldsymbol{F}_i = -\boldsymbol{e}_1\dot{m}[(V_{u1} + V_{u2})] + \boldsymbol{e}_2[\dot{m}(V_{ax1} - V_{ax2}) + (p_1 - p_2)sh] \tag{5.154}$$

with $V_{ax1} = V_{ax2}$ as a result of incompressible flow assumption and $V_{u1} \not\equiv V_{u2}$ from Fig. 5.22. Equation (5.154) rearranged as:

$$\boldsymbol{F}_i = -\boldsymbol{e}_1\dot{m}(V_{u1} + V_{u2}) + \boldsymbol{e}_2(p_1 - p_2)sh = \boldsymbol{e}_1 F_u + \boldsymbol{e}_2 F_{ax} \tag{5.155}$$

with the circumferential and axial components

$$F_u = -\dot{m}(V_{u1} + V_{u2}) \text{ and } F_{ax} = (p_1 - p_2)sh. \tag{5.156}$$

The static pressure difference in Eq. (5.156) is obtained from the following Bernoulli equation:

$$p_{01} = p_{02}$$
$$p_1 - p_2 = \frac{1}{2}\rho(V_2^2 - V_1^2) = \frac{1}{2}\rho(V_{u2}^2 - V_{u1}^2). \quad (5.157)$$

Inserting the pressure difference along with the mass flow $\dot{m} = \rho V_{ax} sh$ into Eq. (5.156) and the blade height $h = 1$, we obtain the axial as well as the circumferential components of the lift force:

$$\left.\begin{aligned} F_{ax} &= \tfrac{1}{2}\varrho(V_{u2} + V_{u1})(V_{u2} - V_{u1})s \\ F_u &= -\varrho V_{ax}(V_{u2} + V_{u1})s \end{aligned}\right\} \quad (5.158)$$

From Eq. (5.157) and considering Eq. (5.158), the *lift force* vector for the inviscid flow is:

$$\boldsymbol{F}_i = \rho s(V_{u2} + V_{u1})\left[-\boldsymbol{e}_1 V_{ax} + \boldsymbol{e}_2 \frac{V_{u2} - V_{u1}}{2}\right]. \quad (5.159)$$

This means that the direction of the blade force is identical with the direction of the vector within the brackets. To further evaluate the inviscid force $\boldsymbol{F}_i$, we calculate the mean velocity vector V_∞:

$$\boldsymbol{V}_\infty = \frac{\boldsymbol{V}_1 + \boldsymbol{V}_2}{2} = \frac{1}{2}\boldsymbol{e}_1(V_{u2} - V_{u1}) + \boldsymbol{e}_2 V_{ax} \quad (5.160)$$

and the circulation around the profile shown in Fig. 5.27 is:

$$\Gamma = \oint_C \boldsymbol{V} \cdot d\boldsymbol{c} \quad (5.161)$$

with the closed curve $C \equiv (12341)$ as the boundary of the contour integral (5.161) and $d\boldsymbol{c}$ a differential element along C and $\boldsymbol{V}$, the velocity vector. The closed curve is placed around the blade profile so that it consists of two streamlines that are apart by the spacing s. Performing the contour integral around the closed curve c, we find:

$$\Gamma = V_{u1}s + \int_2^3 \boldsymbol{V} \cdot d\boldsymbol{c} + V_{u2}s + \int_4^1 \boldsymbol{V} \cdot d\boldsymbol{c}. \quad (5.162)$$

Since the following integrals cancel each other:

$$\int_2^3 \boldsymbol{V} \cdot d\boldsymbol{c} = -\int_4^1 \boldsymbol{V} \cdot d\boldsymbol{c}. \quad (5.163)$$

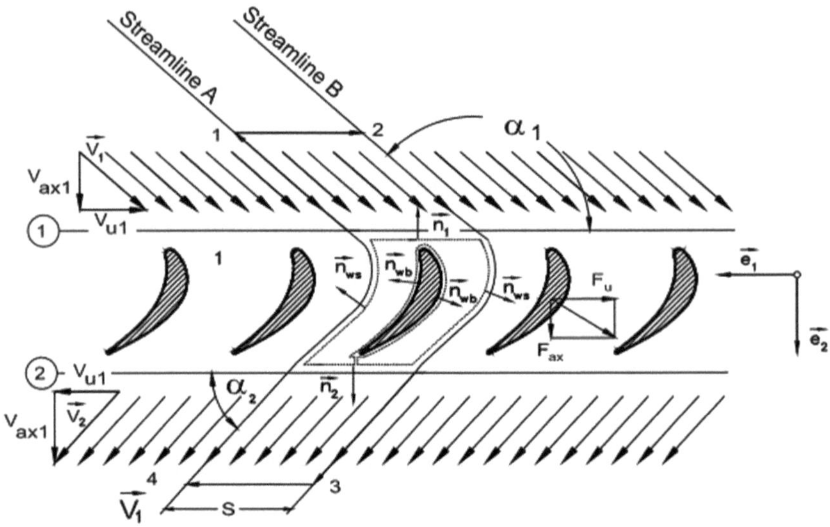

Fig. 5.27 Inviscid incompressible flow through a turbine cascade, calculation of blade forces

We obtain the circulation and, thus, the circulation vector:

$$\Gamma = (V_{u2} + V_{u1})s \text{ with the direction } \boldsymbol{e}_3 = -\boldsymbol{e}_2 \times \boldsymbol{e}_1$$
$$\boldsymbol{\Gamma} = (\boldsymbol{e}_2 \times \boldsymbol{e}_1)s(V_{u2} + V_{u1}) = (-\boldsymbol{e}_3)s(V_{u1} + V_{u2}). \tag{5.164}$$

The vector product of the circulation vector and the mean velocity vector gives

$$\boldsymbol{V}_\infty \times \Gamma = \left[\frac{1}{2}\boldsymbol{e}_2(V_{u2} - V_{u1}) - \boldsymbol{e}_1 V_{ax}\right](V_{u2} + V_{u1})s. \tag{5.165}$$

Comparing Eq. (5.165) with Eq. (5.159), we arrive at the inviscid flow force:

$$\boldsymbol{F}_i = \varrho \boldsymbol{V}_\infty \times \boldsymbol{\Gamma}. \tag{5.166}$$

This is the well-known Kutta-Joukowsky lift-equation for inviscid flow. Expressing F_i in terms of V_∞, the inviscid lift force for a turbine cascade is:

$$F_i = \rho V_\infty (V_{u2} + V_{u1})s. \tag{5.167}$$

Figure 5.28 exhibits a single blade taken from a turbine cascade with the velocities $\boldsymbol{V}_1$, $\boldsymbol{V}_2$, $\boldsymbol{V}_\infty$, as well as the circulation vector $\boldsymbol{\Gamma}$, and the force vector $\boldsymbol{F}_i$. As seen, the inviscid flow force vector $\boldsymbol{F}_i$ is perpendicular to the plane spanned by the mean velocity vector $\boldsymbol{V}_\infty$ and the circulation vector $\boldsymbol{\Gamma}$. Equation (5.166) holds for any arbitrary body that might have a circulation around it regardless of the body shape.

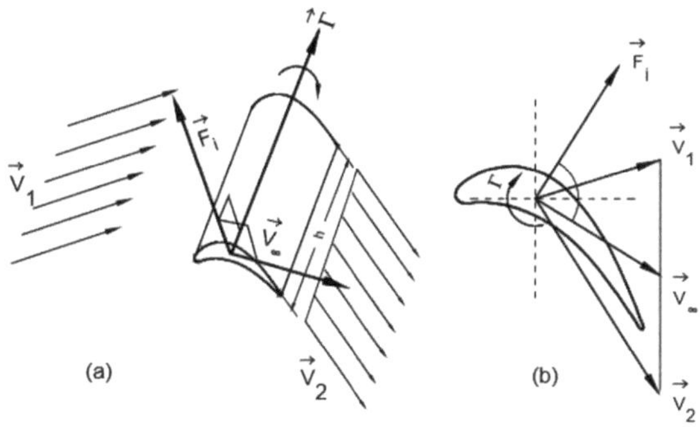

Fig. 5.28 A turbine blade in an inviscid flow field with velocity, circulation and force vector, **a** Schematic 3-D-view, **b** top view

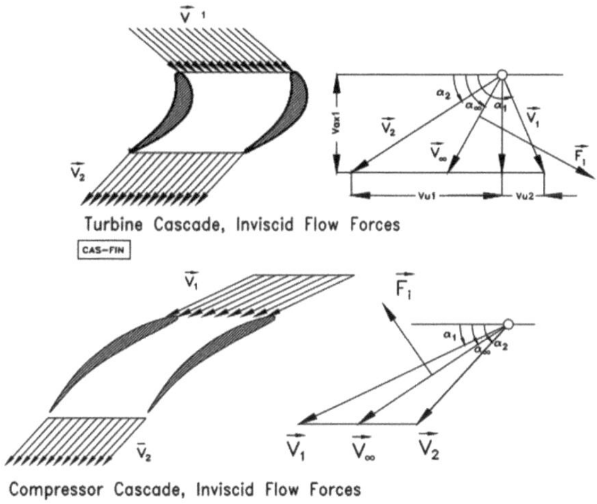

Fig. 5.29 Turbine (top) and Compressor cascade (bottom) with velocity diagram and inviscid flow forces.

Thus, it is valid for turbine and compressor cascades and exhibits the fundamental relation in inviscid flow aerodynamics. As shown in Fig. 5.28, the inviscid flow force (inviscid lift) is perpendicular to the plane spanned by the mean velocity vector $\boldsymbol{V}_\infty$ and the circulation vector $\boldsymbol{\Gamma}$.

Figure 5.29 exhibits the inviscid flow forces acting on a turbine and a compressor cascade. The flow deflection through the cascades is shown using the velocity diagram for an accelerated flow (turbine) and decelerated flow (compressor). The lift force can be non-dimensionalized by dividing Eq. (5.167) by a product of the exit dynamic pressure $\rho V^2/2$ and the projected area $A = ch$ with the height $h = 1$. Thus, the *lift coefficient* is obtained from:

$$C_L = \frac{F_i}{\frac{\rho}{2}V_2^2 c} = \left[\frac{2V_\infty(V_{u2} + V_{u1})}{V_2^2}\right]\frac{s}{c}. \tag{5.168}$$

As shown in the following section, the above relationship can be expressed in terms of the cascade flow angles and the geometry.

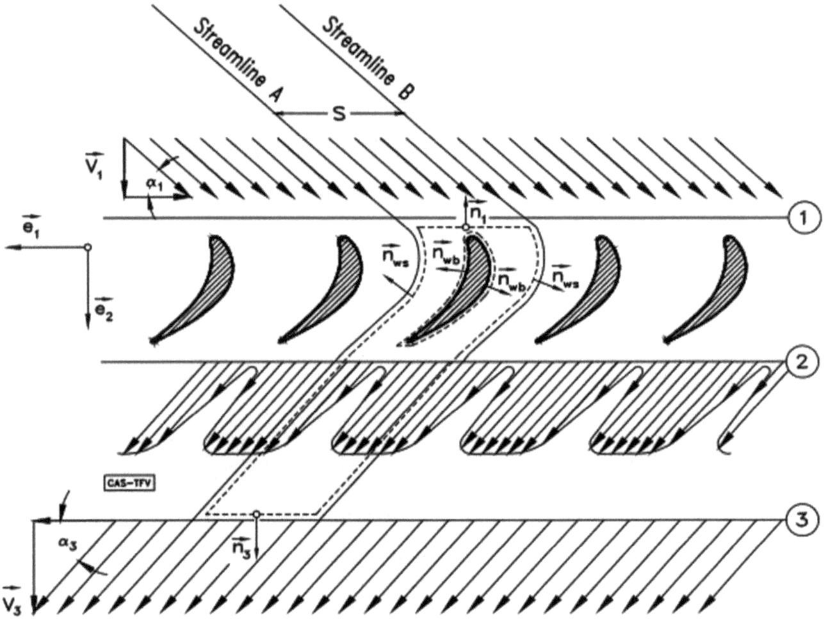

Fig. 5.30 Viscous flow through a turbine cascade. Station ① has a uniform velocity distribution. At station ② wakes are generated by the trailing edge- and boundary layer thickness and are mixed out at ③.

5.10.2 Blade Forces in a Viscous Flow Field

The working fluids in turbomachinery, whether air, combustion gas, steam or other substances, are always viscous. The blades are subjected to the viscous flow and undergo shear stresses with *no-slip condition* on blades, casing and hub surfaces, resulting in boundary layer developments. Furthermore, the blades have certain definite trailing edge thicknesses. These thicknesses together with the boundary layer thickness, generate a spatially periodic wake flow downstream of each cascade as shown in Fig. 5.30.

The presence of the shear stresses cause drag forces that reduce the total pressure. In order to calculate the blade forces, the momentum Eq. (5.153) can be applied to the viscous flows. As seen from Eq. (5.156), the circumferential component remains unchanged. The axial component, however, changes in accordance with the pressure difference as shown in the following relations:

$$F_u = -\rho V_{ax}(V_{u2} + V_{u1})sh$$
$$F_{ax} = (p_1 - p_2)sh. \tag{5.169}$$

The blade height h in Eq. (5.169) may be assumed as unity. For a viscous flow, the static pressure difference cannot be calculated by the Bernoulli equation. In this case, the total pressure drop must be taken into consideration. We define the total pressure loss coefficient:

$$\zeta \equiv \frac{P_1 - P_2}{\frac{1}{2}\varrho V_2^2} \tag{5.170}$$

with P_1 and P_2 as the averaged total pressure at stations 1 and 2. Inserting for the total pressure the sum of static and dynamic pressures, we get the static pressure difference as:

$$p_1 - p_2 = \frac{\rho}{2}(V_2^2 - V_1^2) + \zeta\frac{\rho}{2}V_2^2. \tag{5.171}$$

Incorporating Eq. (5.171) into the axial component of the blade force in Eq. (5.169) yields:

$$F_{ax} = \frac{\rho}{2}(V_2^2 - V_1^2)s + \zeta\frac{\rho}{2}V_2^2 s. \tag{5.172}$$

We introduce the velocity components into Eq. (5.172) and assume that for an incompressible flow the axial components of the inlet and exit flows are the same. As a result, Eq. (5.172) reduces to:

$$F_{ax} = \frac{\rho}{2}(V_{u2}^2 - V_{u1}^2)s + \zeta\frac{\rho}{2}V_2^2 s. \tag{5.173}$$

The second term on the right-hand side exhibits the axial component of drag forces accounting for the viscous nature of a frictional flow shown in Fig. 5.31. Thus, the axial projection of the drag force is obtained from:

$$D_{ax} = \zeta\frac{\rho}{2}V_2^2 s. \tag{5.174}$$

Figure 5.31 exhibits the turbine and compressor cascade flow forces, including the lift and drag forces on each cascade for viscous flow where the periodic exit velocity distribution caused by the wakes, and shown in Fig. 5.30, is completely mixed out resulting in an averaged uniform velocity distribution, Fig. 5.30.
With Eq. (5.174), the loss coefficient is directly related to the drag force. Since the drag force D is in the direction of V_∞, its axial projection D_{ax} can be written as:

$$D_{ax} = \frac{D}{\sin\alpha_\infty}. \tag{5.175}$$

Assuming the blade height $h = 1$, we define the drag and lift coefficients as:

$$C_D = \frac{D}{\frac{\varrho}{2}V_2^2 c} C_L = \frac{F}{\frac{\varrho}{2}V_2^2 c}. \tag{5.176}$$

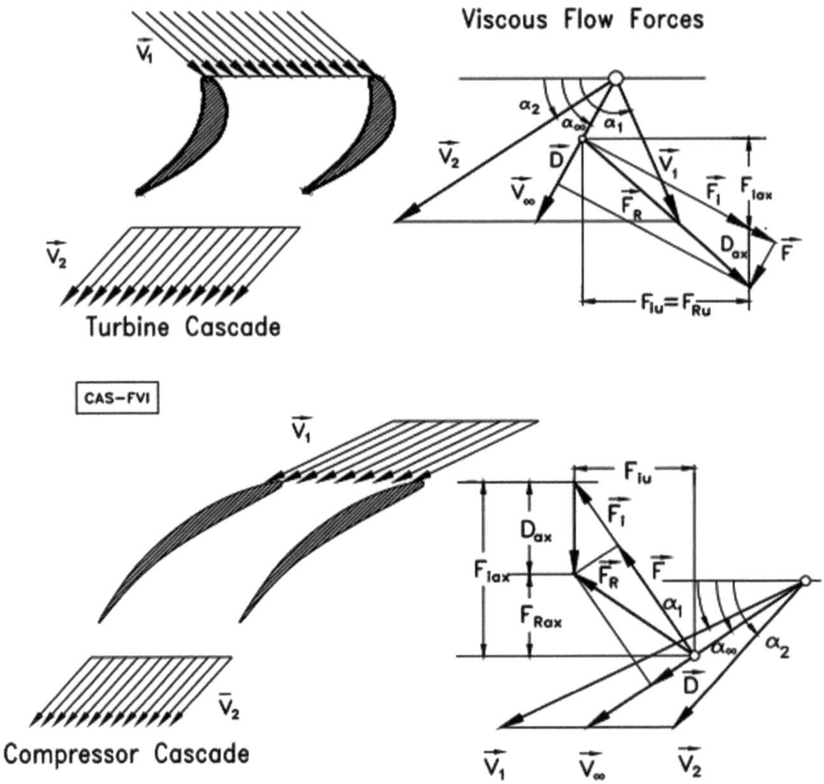

Fig. 5.31 Viscous flow forces on a turbine blade (top) and a compressor blade (bottom). The resultant force is decomposed into a drag and a lift component

Introducing the drag coefficient C_D, we obtain a direct relationship between the loss and drag coefficient:

$$\zeta = C_D \frac{c}{s} \frac{1}{\sin \alpha_\infty}. \tag{5.177}$$

The magnitude of the viscous lift force is the projection of the resultant force $\boldsymbol{F}_R$ on the plane perpendicular to $\boldsymbol{V}_\infty$:

$$F = F_i + D_{ax} \cos \alpha_\infty. \tag{5.178}$$

Using the lift coefficient defined previously and inserting the lift force, we find

$$C_L = \frac{2V_\infty (V_{u2} + V_{u1})}{V_2^2} \frac{s}{c} + \zeta \frac{s}{c} \cos \alpha_\infty. \tag{5.179}$$

Introducing the cascade solidity $\sigma = c/s$ into Eq. (5.179) results in:

$$C_L \frac{c}{s} \equiv C_L \sigma = \frac{2V_\infty (V_{u2} + V_{u1})}{V_2^2} + \zeta \cos \alpha_\infty. \tag{5.180}$$

The lift-solidity coefficient is a characteristic quantity for the cascade aerodynamic loading. Using the flow angles defined in Fig. 5.32, the relationship for the lift-solidity coefficient becomes:

$$C_L \sigma = 2 \frac{\sin^2 \alpha_2}{\sin \alpha_\infty} (\cot \alpha_2 - \cot \alpha_1) + \zeta \cos \alpha_\infty \tag{5.181}$$

with:

$$\cot \alpha_\infty = \frac{1}{2} (\cot \alpha_2 + \cot \alpha_1). \tag{5.182}$$

For a preliminary design, the contribution of the second term in Eq. (5.181) compared with the first term can be neglected. However, for the final design, the loss coefficient ζ needs to be calculated as detailed in [23] and inserted into Eq. (5.181).

Figure 5.32 shows the results as a function of the inlet flow angle with the exit flow angle as the parameter for turbine and compressor cascades. As an example, a turbine cascade with an inlet flow angle of $\alpha_1 = 132°$, and an exit flow of $\alpha_2 = 30°$ resulting in a total flow deflection of $\Theta = 102°$, has a positive lift-solidity coefficient of $C_L \sigma = 2.0$. This relatively high lift coefficient is responsible for generating large blade forces and, thus, a high blade specific power for the rotor. In contrast, a compressor cascade with an inlet flow angle of $\alpha_1 = 60°$ and an exit flow of $\alpha_2 = 80°$ which result in a total compressor cascade flow deflection of only $\Theta = 20°$, has a lift-solidity coefficient of $C_L \sigma = -0.8$. This leads to a much lower blade force and, thus, lower specific mechanical energy input for the compressor rotor. The numbers in the above example are typical for compressor and turbine blades. The high lift-solidity coefficient for a turbine cascade is representative of the physical process within a highly accelerated flow around a turbine blade where, despite a high flow deflection, no flow separation occurs. On the other hand, in case of a compressor cascade, a moderate flow deflection, such as the one mentioned above, may result in flow separation. The difference between the turbine and compressor cascade flow behavior is explained by the nature of boundary layer flow around the turbine and compressor cascades. In a compressor cascade, the boundary layer flow is subjected to two co-acting decelerating effects, the wall shear stress dictated by the viscous nature of the fluid and the positive pressure gradient imposed by the cascade geometry. A fluid particle within the boundary layer that has inherently lower kinetic energy compared to a particle outside the boundary layer has to overcome the pressure forces due to the governing positive pressure gradient. As a result, this particle continuously decelerates, comes to a rest, and separates. In the case of a turbine cascade, the decelerating effect of the shear stress forces is counteracted by the accelerating effect of the negative pressure gradient that predominates a turbine cascade flow.

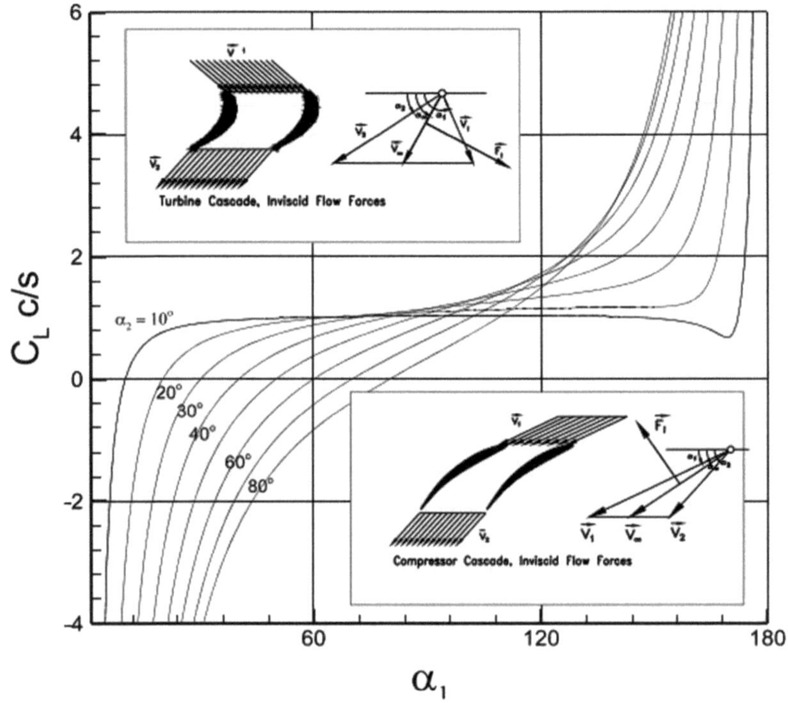

Fig. 5.32 Lift-solidity coefficient as a function of inlet flow angle ″1 with the exit flow angle α_2 as parameter for turbine and compressor cascades

5.10.3 Effect of Solidity on Blade Profile Losses

Equation (5.181) exhibits a fundamental relationship between the lift coefficient, the solidity, the inlet and exit flow angle, and the loss coefficient ζ. The question is, how the profile loss ζ will change if the solidity σ changes. The solidity has the major influence on the flow behavior within the blading. If the spacing is too small, the number of blades is large and the friction losses dominate. Increasing the spacing, which is identical to reducing the number of blades, at first causes a reduction of friction losses. Further increasing the spacing decreases the friction losses and also reduces the guidance of the fluid that results in flow separation leading to additional losses. With definite spacing, there is an equilibrium between the separation and friction losses. At this point, the profile loss $\zeta = \zeta_{\text{friction}} + \zeta_{\text{separation}}$ is at a minimum. The corresponding spacing/chord ratio has an optimum, which is shown in Fig. 5.33. To find the optimum solidity for a variety of turbine and compressor cascades, a series of comprehensive experimental studies have been performed by several researchers. A detailed discussion of the results of these studies is presented in [23].

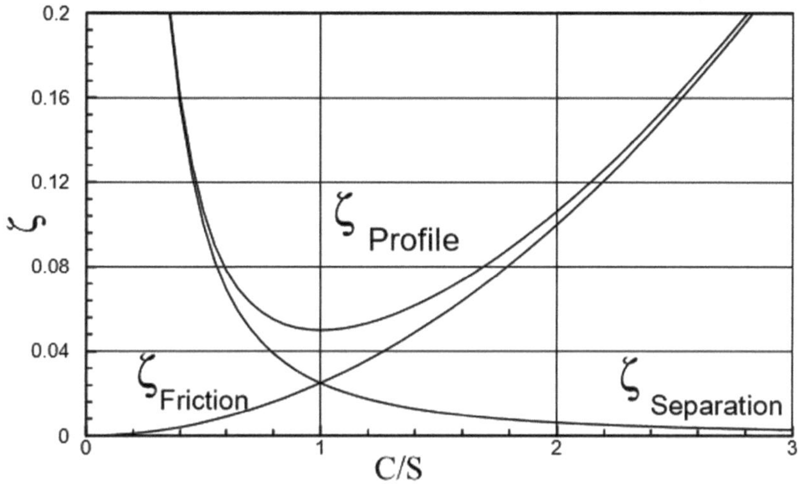

Fig. 5.33 Profile loss coefficient as a function of chord/spacing ratio

The relationship for the lift-solidity coefficient derived in the preceding sections is restricted to turbine and compressor stages with constant inner and outer diameters. This geometry is encountered in high pressure turbines or compressor components, where the streamlines are almost parallel to the machine axis. In this special case, the stream surfaces are cylindrical with almost constant diameter. In a general case such as the intermediate and low pressure turbine and compressor stages, however, the stream surfaces have different radii. The meridional velocity component may also change from station to station. In order to calculate the blade lift-solidity coefficient correctly, the radius and the meridional velocity changes must be taken into account. Detailed discussions on this and turbomachinery aero-thermodynamic topics are found in [23].

5.11 Problems, Project

Problem 5.1 A one-dimensional unsteady flow is given by the following velocity $u = \frac{2}{y+1}\left(\frac{x}{t} - a_0\right)$ and density field $\frac{\varrho}{\varrho_0} = \left(\frac{\gamma-1}{\gamma+1}\frac{x}{t}\frac{1}{a_0} + \frac{2}{\gamma+1}\right)^{\frac{2}{\gamma-1}}$.

(a) Calculate the substantial change of the density.
(b) Check the validity of the continuity equation
$$\frac{D\varrho}{Dt} + \varrho\frac{\partial u}{\partial x} = 0$$

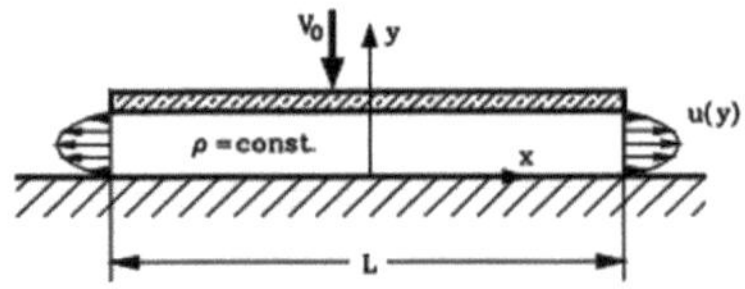

Fig. 5.34 Flow in a gap with top wall moving downward

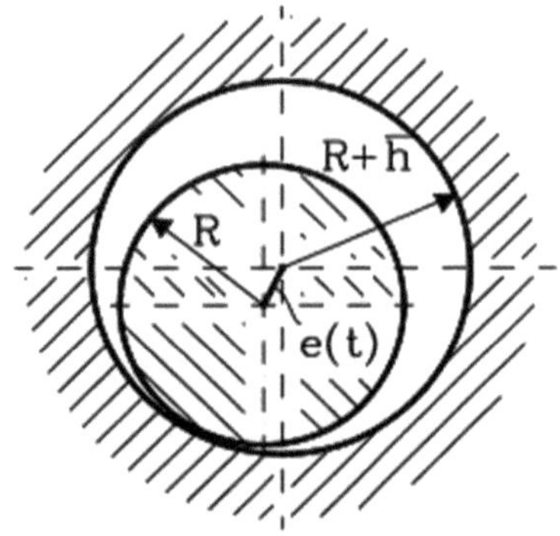

Fig. 5.35 Shaft oscillating in a bearing casing

Problem 5.2 The gap shown in Fig. 5.34 has the length L, the height $h(t)$, and is filled with a fluid of constant density. The top wall of the gap moves downward with the velocity V_0. The velocity distribution at the exit is

(a) Given the values: U_0, h_0, L, ϱ, determine the function of the gap height for $h(t = 0) = h_0$.
(b) Calculate the maximum velocity U_0 at the exit.

Problem 5.3 Figure 5.35 shows an oscillating journal bearing with the eccentricity $e = e(t)$ and the shaft radius R. The shaft rotates with a constant rotational speed ω. We assume that the bearing has an infinite width in the direction perpendicular to the shaft axis. For $\bar{h}R \ll 1$ the clearance distribution $h(x_1, t)$ in x_1-direction can be unwrapped and the following assumption can be made:

$$\frac{h(x_1, t)}{\bar{h}} = 1 + \varepsilon \cos \omega t \cos \frac{x_1}{R}.$$

The density of the fluid ϱ is constant and the volume flux per unit width $\dot{V}(0, t)$ at location $x_1 = 0$ is known. Given the following quantities $\dot{V}(0, t)$, ϵ, ω, R, $\bar{h}$, calculate the volume flux per unit width $\dot{V}(x_{10}, t)$ as a function of time at x_{10}.

Problem 5.4 Incompressible fluid flows over a flat plate, Fig. 5.36, (width b, length L) with constant velocity U_0. The viscosity effect causes a boundary layer with the thickness $\delta(x_1)$. Outside the boundary layer, the velocity is $u_1 = U_0$ =const. We assume that the velocity distribution within the boundary layer follows a sine function with no-slip condition at the wall. Given $\delta = \delta(x_1)$ and $\delta_L = \delta(x_1 = L)$, and $\frac{u_1}{U0} = \sin\left(\frac{\pi}{2}\frac{x_2}{\delta}\right)$ for $0 \leq \frac{x_2}{\delta(x_1)} \leq 1$ and $\frac{u_1}{U0} = 1$ for $\frac{x_2}{\delta(x_1)} > 1$

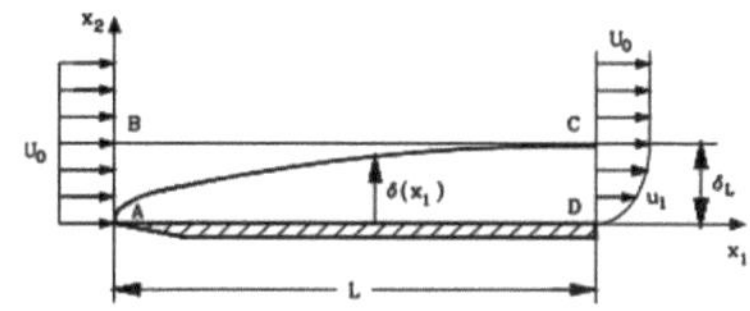

Fig. 5.36 Flow over a flat plate

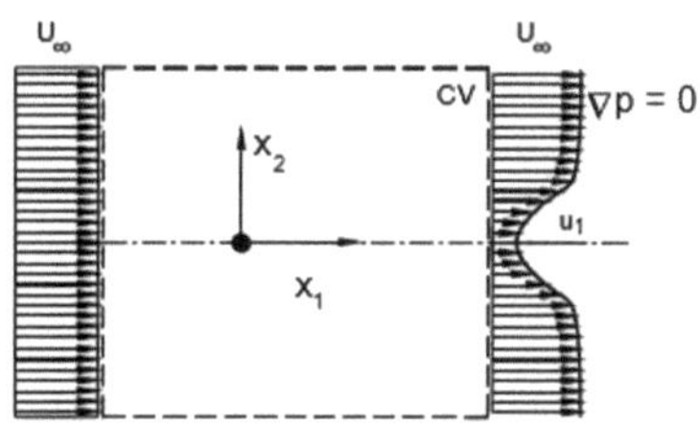

Fig. 5.37 Viscous flow past a circular cylinder

(a) Determine the mass flow through the surface BC of the sketched control volume.
(b) Calculate the velocity field within the boundary layer $u_i(x_j)$.
(c) Calculate the mass flow through BC using $u_2(x_1, x_2 = \delta)$.

Problem 5.5 Fluid with constant velocity U_∞ and density ϱ flows past an infinitely long cylinder, Fig. 5.37. The flow direction coincides with the symmetry axis and the only force on the body is then the drag force F_D. Downstream of the body a wake flow is generated where the velocity u_1 is less than U_∞. With a given u_1/U_∞ calculate the drag force F_D per unit depth acting on the body.

Chapter 6
Inviscid Potential Flows

As discussed in Chap. 4, generally the motion of fluids encountered in engineering applications is described by the Navier-Stokes equations. Considering today's computational fluid dynamics capabilities, it is possible to numerically solve the Navier-Stokes equations for laminar flows (no turbulent fluctuations), transitional flows (using appropriate intermittency models), and turbulent flow (utilizing appropriate turbulence models). Given today's computational capabilities, one may argue at this juncture that there is no need to artificially subdivide the flow regime into different categories such as incompressible, compressible, viscid or inviscid ones. However, based on the degree of complexity of the flow under investigation, a computational simulation may take up to several days, weeks, and even months for direct Navier-stokes simulations (DNS). The difficulties associated with solving the Navier-Stokes equations are caused by the existence of the viscosity terms in the Navier-Stokes equations.

Measuring the velocity distributions encountered in engineering applications such as in a pipe flow, flow around a compressor or turbine blade, or along the wing of an aircraft, we find that the effect of viscosity is confined to a very thin layer called the boundary layer with a local thickness δ. As we discuss in Chap. 11, comprehensive experimental investigations performed earlier by Prandtl [26, 27] show that the boundary layer thickness δ compared to the length L of the subject under investigation is very small. In the vicinity of the wall, because of the no-slip condition, the velocity is $\boldsymbol{V}_{\text{wall}} = 0$. Moving away from the wall towards the edge of the boundary layer, the velocity continuously increases until it reaches the velocity at the edge of the boundary layer $\boldsymbol{V} = V_\delta$. Within the boundary layer, the flow is characterized by non-zero vorticity $\nabla \times \boldsymbol{V} \neq 0$. No major changes in velocity magnitude is expected outside the boundary layer, provided that the surface of the subject under investigation does not have a curvature. In case of surfaces with convex or concave curvatures, the velocity outside the boundary layer changes in lateral direction.

M. T. Schobeiri, *Advanced Fluid Mechanics and Heat Transfer for Engineers and Scientists*, https://doi.org/10.1007/978-3-030-72925-7_6

Outside the boundary layer, the effect of the viscosity can be neglected as long as the Reynolds number is high enough (Re = 100,000 and above) indicating that the convective flow forces are much larger than the shear stress forces. Theoretically, the boundary layer thickness approaches zero as the Reynolds number tends to infinity. In this case, the flow can be assumed as irrotational, which is then characterized by zero vorticity $\nabla \times \boldsymbol{V} = 0$. Thus, as Prandtl suggested, the flow may be decomposed into two distinct regions, the vortical inner region, called the boundary layer, where the viscosity effect is predominant, and the non-vortical region outside the boundary layer.

The flow in the outer region can be calculated using the Euler equation of motion, while the boundary layer method can be applied for calculating the viscous flow within the inner region. Combining these two methods allows calculation of the flow field in a sufficiently accurate manner as long as the boundary layer is not separated. Figure 6.1 exhibits the velocity distributions along the suction surface of an airfoil. While in case (a) the viscosity is accounted for, in case (b) it is neglected. Thus, the flow is assumed irrotational, which is characterized by $\nabla \times \boldsymbol{V} = 0$. As a consequence of this assumption, the velocity on the surface has a non-zero tangential component, which is in contrast to the reality. These type of flows are called *potential flows* which is the subject of the following sections.

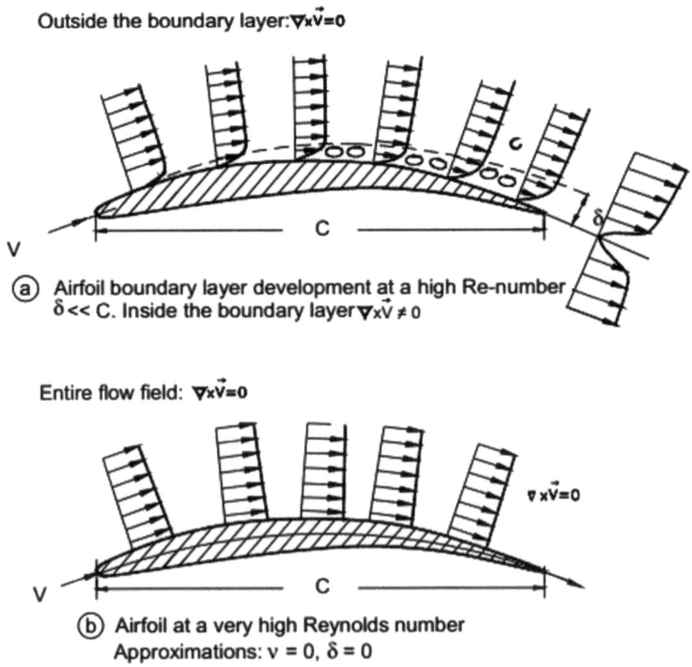

Fig. 6.1 Velocity distribution inside and outside the boundary layer along the suction surface of a subsonic compressor, (b) velocity at zero viscosity

6.1 Incompressible Potential Flows

As seen in Chap. 3, an incompressible flow satisfies the condition $D\rho/Dt = 0$ which, in conjunction with the continuity equation, leads to $\nabla \cdot \boldsymbol{V} = 0$. Furthermore, we assume that the flow is irrotational with $\nabla \times \boldsymbol{V} = 0$ everywhere in the flow field. This assumption, which significantly simplifies the mathematical treatment of the flow field, allows introduction of a scalar function called the *velocity potential* Φ, from which the velocity vector and its components are derived as the gradient of the potential Φ:

$$\boldsymbol{V} = \nabla\Phi,\, e_i V_i = e_i \frac{\partial \Phi}{\partial x_i}. \tag{6.1}$$

Expanding the index notation results in:

$$\boldsymbol{V} = e_i \frac{\partial \Phi}{\partial x_i} = e_1 \frac{\partial \Phi}{\partial x_1} + e_2 \frac{\partial \Phi}{\partial x_2} + e_3 \frac{\partial \Phi}{\partial x_3}. \tag{6.2}$$

Inserting Eq. (6.1) into the continuity equation for incompressible flow $\nabla \cdot \boldsymbol{V} = 0$, we arrive at:

$$\nabla \cdot \boldsymbol{V} = \nabla \cdot \nabla\Phi = \Delta\Phi = 0. \tag{6.3}$$

Equation (6.3) is the Laplace equation decomposed as:

$$\frac{\partial^2 \Phi}{\partial x_i \partial x_i} = 0. \tag{6.4}$$

The Laplace equation (6.4) is an *elliptic, linear* partial differential equation encountered in many branches of engineering and physics such as electromagnetism, heat conduction, and theory of elasticity. It can be solved using appropriate boundary conditions. The introduction of the velocity potential Φ in conjunction with the Bernoulli equation having a constant that has the same value everywhere in the flow field significantly reduces the solution efforts. The solution of the Laplace equation simultaneously satisfies the continuity condition $\nabla \cdot \boldsymbol{V} = 0$ (no divergence) as well as the irrotationality condition $\nabla \times \boldsymbol{V} = 0$. In addition, the solution has to satisfy the boundary conditions dictated by the solid surfaces that the potential flow is exposed to. As a simple example, we will consider a potential flow past a flat surface. On the surface and at infinity, the solution has to satisfy the following two boundary conditions:

$$BC1 : (V_2)_{\text{surface}} = \frac{\partial \Phi}{\partial x_2} = 0,\; BC2 : (V_1)_{\text{at}\, x_1 = \infty} = \frac{\partial \Phi}{\partial x_1} = V_\infty \tag{6.5}$$

with x_1, x_2 as the coordinates in longitudinal and lateral directions, respectively. The boundary condition $BC1$ requires that on the surface, the normal component of the velocity must vanish, whereas the boundary condition $BC2$ necessitates that the velocity has to approach V_∞ as x_1 approaches infinity. There are not many potential flow functions with practical significance that can deliver analytic solutions satisfying the boundary conditions (6.5). A function that satisfies the Laplace equation and possesses continuous second derivatives is called analytic. The linear nature of the Laplace equation allows superposition of analytical functions to build a new harmonic function that satisfies the above boundary conditions. This unique characteristic of the Laplace equation allows an *indirect* approach by composing harmonic functions that consist of individual functions with the known solutions. Prandtl and his co-workers, among others [28, 29], were the first to provide a list of those individual functions, based on complex analysis. The complex analysis exhibits a powerful tool to deal with the potential theory in general and the potential flow in particular. It is found in almost every fluid mechanics textbook that has a chapter dealing with potential flow. While they all share the same underlying mathematics, the style of describing the subject to engineering students differ. A very compact and precise description of this subject matter is found in an excellent textbook by Spurk [30].

6.2 Complex Potential for Plane Flows

Plane potential flows that satisfy the Laplace equation are treated most effectively using complex variables. These flows differ from other two-dimensional flows (with two independent variables) because two independent variables, x and y, can be combined into one complex variable:

$$z = x + iy = r\cos\theta + ir\sin\theta = r(\cos\theta + i\sin\theta) = re^{i\theta} \tag{6.6}$$

with $i = \sqrt{-1}$. The complex variable z and its conjugate complex $\bar{z}$ are shown in Fig. 6.2. The z-components on x and y-axis are real ($\Re$) and imaginary ($\Im$) which are parts of the variable z.

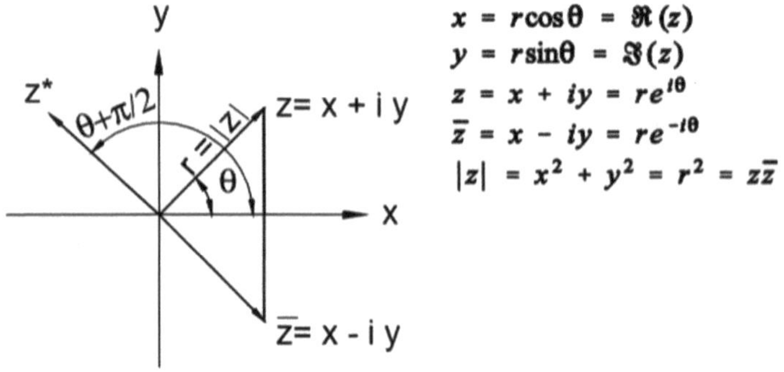

Fig. 6.2 Complex variables

Since every analytic function of the complex coordinate z satisfies Laplace's equation, the computation of both the direct and indirect problems becomes considerably easier. If we know the flow past a cylindrical body whose cross-sectional surface is simply connected (e.g. circular cylinder), then according to the *Riemann mapping theorem*, we can obtain the flow past any other cylinder using a *conformal transformation*. By this theorem, every simple connected region in the complex plane can be mapped into the inside of the unit circle. By doing this, in principle, we have solved the problem of flow past a body, and we only need to find a suitable mapping function.

The complex function $F(z)$ is called *analytic (holomorphic)*, if it is complex differentiable at every point z, where the limit

$$\lim_{\Delta z \to 0} \frac{F(z+\Delta z) - F(z)}{\Delta z} = \frac{dF}{dz} \tag{6.7}$$

exists and is independent of the path from z to $z + \Delta z$. If this requirement is not satisfied, the point is a *singular point*. Along a path parallel to the x axis, the relation

$$\frac{dF}{dz} = \frac{\partial F}{\partial x} \tag{6.8}$$

holds and the same holds for the path parallel to the y axis

$$\frac{dF}{dz} = \frac{\partial F}{\partial (iy)}. \tag{6.9}$$

Since every complex function $F(z)$ is of the form

$$F(z) = \Phi(x, y) + i\psi(x, y);\ dF = d\Phi + id\Psi \tag{6.10}$$

we then have

$$\frac{\partial F}{\partial x} = \frac{\partial \Phi}{\partial x} + i\frac{\partial \psi}{\partial x} = \frac{1}{i}\frac{\partial \Phi}{\partial y} + \frac{\partial \psi}{\partial y} = \frac{1}{i}\frac{\partial F}{\partial y}. \tag{6.11}$$

Clearly for the derivative to exist, it is necessary that

$$\frac{\partial \Phi}{\partial x} = \frac{\partial \psi}{\partial y} \text{ and } \frac{\partial \Phi}{\partial y} = -\frac{\partial \psi}{\partial x} \tag{6.12}$$

holds true. Equation (6.12) called the *Cauchy-Riemann differential equations* are sufficient for the existence of the derivative of $F(z)$. We can also show easily that both the real part $\Re(F) = \Phi(x, y)$, and the imaginary part, $\Im(F) = \psi(x, y)$, satisfy the Laplace's equation. To do this, we differentiate the first differential equation in Eq. (6.12) by x and the second by y and add the results. We then see that Φ satisfies Laplace's equation. If we differentiate the first by y and the second by x and subtract the results, we see that the same also holds for ψ. Both functions can, therefore, serve as the velocity potential of a plane flow. We choose Φ as the velocity potential and shall now consider the physical meaning of ψ. The velocity vector, as the gradient of scalar potential Φ, is obtained from:

$$\boldsymbol{V} = \nabla\Phi = e_1\frac{\partial\Phi}{\partial x_1} + e_2\frac{\partial\Phi}{\partial x_2} = e_1V_1 + e_2V_2 \equiv e_1u + e_2v. \tag{6.13}$$

To comply with the nomenclature generally used in two-dimensional complex analysis, we replaced in Eq. (6.13) the components V_1 and V_2 by u and v, respectively. Because of Cauchy-Riemann condition (6.12) we also have

$$\nabla\psi = e_1\frac{\partial\psi}{\partial x_1} + e_2\frac{\partial\psi}{\partial x_2} = -e_1v + e_2u \tag{6.14}$$

with $x_1 \equiv x$ and $x_2 \equiv y$, respectively. From $\nabla\Phi \cdot \nabla\psi = 0$ we conclude that $\nabla\psi$ is perpendicular to the velocity vector $\boldsymbol{V}$, and therefore ψ =const. are streamlines. Thus, we have identified ψ as a stream function and note that introducing a stream function is not restricted to potential flows. Constructing an array of streamlines, we define a particular streamline that is identical with the body contour, which is exposed to a potential flow by assigning a constant to ψ. In this case,

$$\psi = 0 \tag{6.15}$$

represents the equation of the body contour. With ψ known, we obtain the velocity vector directly from the following relationship

$$\boldsymbol{V} = \nabla\psi \times e_3 \text{ or } V_i = \varepsilon_{ij3}\frac{\partial\psi}{\partial x_j} \tag{6.16}$$

therefore

$$V_1 = u = \frac{\partial\psi}{\partial y},\ V_2 = v = -\frac{\partial\psi}{\partial x}, \tag{6.17}$$

so that the continuity equation $\frac{\partial u}{\partial x} + \frac{\partial v}{\partial y} = 0$ is identically satisfied. The velocity components can be most easily calculated using

$$\frac{dF}{dz} = \frac{\partial F}{\partial x} = \frac{\partial\Phi}{\partial x} + i\frac{\partial\psi}{\partial x} = u - iv, \tag{6.18}$$

as the *complex conjugate velocity*

$$\frac{dF}{dz} = \bar{w} = u - iv, \text{ similarly, we find: } \overline{\frac{dF}{dz}} = w = u + iv \tag{6.19}$$

as the mirror image of the *complex velocity* $w = u + iv$ at the real axis.

6.2.1 Elements of Potential Flow

As mentioned previously, the Laplace equation allows any linear combination of complex functions that satisfy the Laplace requirement. In the following, first we discuss the basic elements of complex potentials that are used for superposition purposes.

6.2.1.1 Translational Flows

Translational flows in x-direction, y-direction and at an angle are shown in Fig. 6.3. The complex function of the translational flow is defined as

$$F(z) = (U_\infty - iV_\infty)z \tag{6.20}$$

or

$$F = (U_\infty x + V_\infty y) + i(U_\infty y - V_\infty x). \tag{6.21}$$

For a horizontal flow from left to right and vertical stream upward Eq. (6.21) is reduced to:

$$F(z) = U_\infty z \tag{6.22}$$

and

$$F(z) = -iV_\infty z. \tag{6.23}$$

Because of Eq. (6.10), we find

$$\Phi = U_\infty x + V_\infty y \tag{6.24}$$

and

$$\psi = U_\infty y - V_\infty x. \tag{6.25}$$

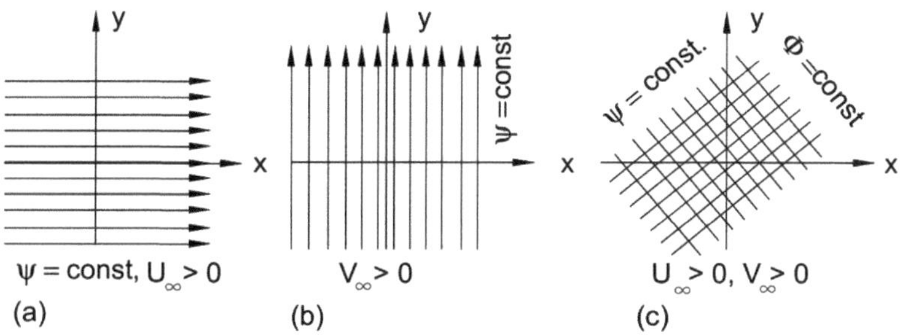

Fig. 6.3 Uniform Flows, **a** parallel to x-axis, **b** parallel to y-axis **c** flow velocity at an angle

For a streamline with Ψ =const., we obtain $y = xV_\infty/U_\infty + C$, where the constant C can be varied to construct the desired streamlines. The complex conjugate velocity is found as

$$\frac{dF}{dz} = U_\infty - iV_\infty. \tag{6.26}$$

6.2.1.2 Sources and Sinks

Sources and sinks shown in Fig. 6.4 are represented by the complex potential

$$F(z) = \frac{E}{2\pi}\ln z \tag{6.27}$$

that is located at the origin with positive E as the source strength and negative E as the strength of the sink. Replacing $z = re^{i\theta}$ leads to

$$F = \frac{E}{2\pi}(1nr + i\theta) = \Phi + i\Psi \tag{6.28}$$

from which the velocity potential and the stream function are determined as

$$\Phi = \frac{E}{2\pi}\ln r \tag{6.29}$$

and

$$\psi = \frac{E}{2\pi}\theta. \tag{6.30}$$

As shown in Fig. 6.4, the streamlines ψ =const. and the potentials Φ =const. are straight lines through the origin and concentric circles (r =const.), respectively.

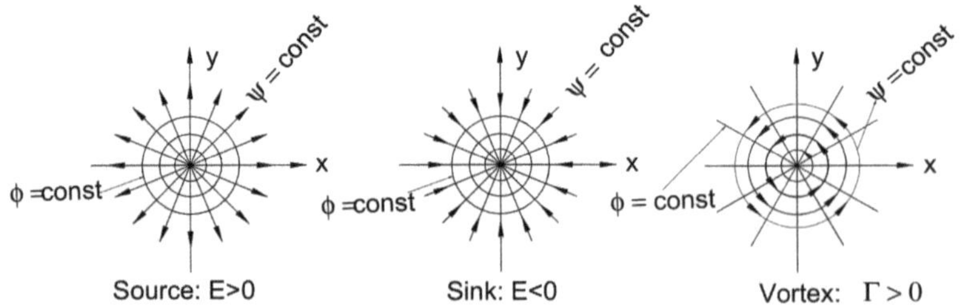

Fig. 6.4 Plane source, sink and vortex located at the origin

6.2.1.3 Potential Vortex

The potential vortex shown in Fig. 6.4 is represented by the complex potential:

$$F = \pm i \frac{\Gamma}{2\pi} 1nz. \tag{6.31}$$

with Γ as the vortex strength. The positive sign refers to the counter clockwise circulation direction, Fig. 6.4, whereas the negative sign indicates the clockwise direction. In polar coordinates, Eq. (6.31), a vortex with a clockwise circulation direction can be written as

$$F = -i \frac{\Gamma}{2\pi} (1nr + i\theta), \tag{6.32}$$

therefore

$$\Phi = +\frac{\Gamma}{2\pi}\theta \tag{6.33}$$

and

$$\psi = -\frac{\Gamma}{2\pi} 1nr. \tag{6.34}$$

As seen in Fig. 6.4, the streamlines ψ =const. and the potentials Φ =const. are concentric circles (r =const.) and straight lines through the origin, respectively.

6.2.1.4 Dipole Flow

This element also called doublet is actually a superposition of a source and a sink that are arranged on the real axis at a distance $\pm c$ from the origin. Taking advantage of the superposition principle applied to a source-sink pair described by Eq. (6.27), we find

$$F(z) = \frac{E}{2\pi} 1n(z + c) - \frac{E}{2\pi} 1n(z - c) \tag{6.35}$$

which is rearranged as

$$F(z) = \frac{E}{2\pi} \ln \left(\frac{z + c}{z - c} \right). \tag{6.36}$$

Using the Taylor expansion of the expression in the parentheses results in:

$$E \ln \left(\frac{z + c}{z - c} \right) = 2\frac{Ec}{z} + \frac{2}{3}\frac{Ec^3}{z^3} + \frac{2}{5}\frac{Ec^5}{z^5} + \cdots \tag{6.37}$$

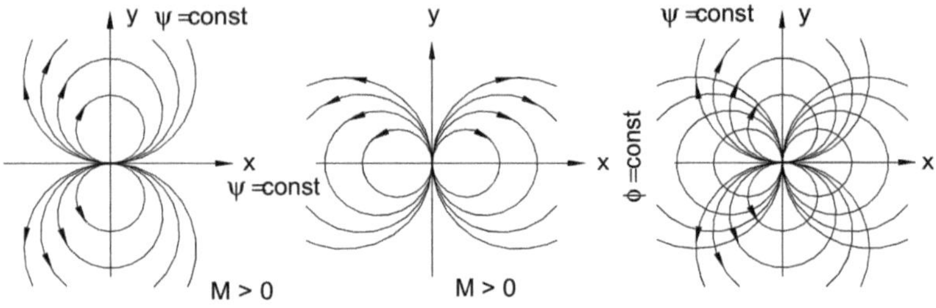

Fig. 6.5 Streamlines and equipotenial lines of a dipole

we set now $M = 2Ec$, and as c approaches zero we find:

$$E \ln\left(\frac{z+c}{z-c}\right) = \frac{M}{z} \tag{6.38}$$

resulting in a simple relationship for $F(z)$

$$F(z) = \frac{M}{2\pi}\frac{1}{z} = \frac{M}{2\pi}\frac{1}{r(\cos\theta + i\sin\theta)} = \frac{M}{2\pi r}(\cos\theta - i\sin\theta), \tag{6.39}$$

or

$$F = \frac{M}{2\pi}\frac{1}{r}(\cos\theta - i\sin\theta) = \frac{M}{2\pi}\frac{1}{r^2}(x - iy), \tag{6.40}$$

from which we read off directly:

$$\phi = +\frac{M}{2\pi}\frac{\cos\theta}{r} = \frac{M}{2\pi}\frac{x}{r^2} \tag{6.41}$$

and

$$\psi = -\frac{M}{2\pi}\frac{\sin\theta}{r} = -\frac{M}{2\pi}\frac{y}{r^2}. \tag{6.42}$$

For ψ =const. we obtain with $\sin\theta = y/r$

$$r^2 = x^2 + y^2 = -\frac{M}{C}y \tag{6.43}$$

that is, the streamlines and potential lines are circles which are tangent to the x- and y-axis at the origin shown in Fig. 6.5.

6.2.1.5 Corner Flow

Figure 6.6 shows the streamline plot for different corner flow configurations described in the following section.

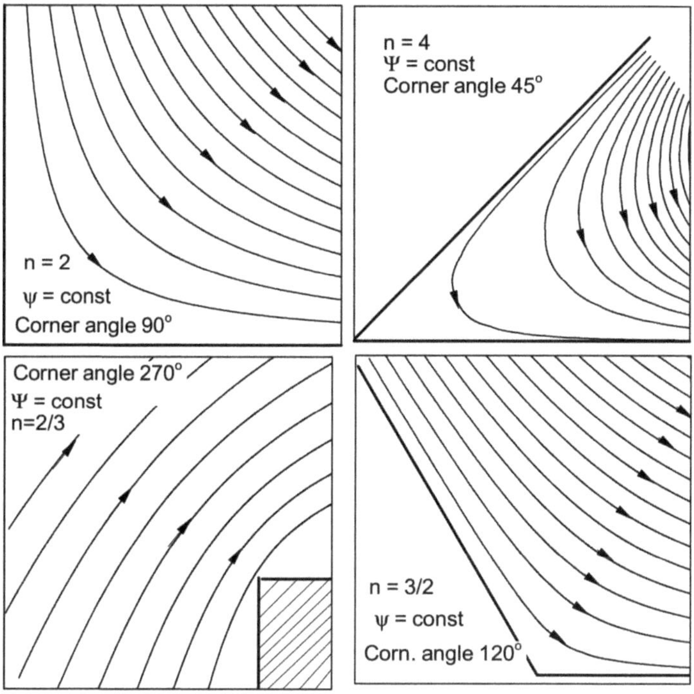

Fig. 6.6 Corner flow with the exponent n as a parameter

The complex potential of this element is described by

$$F(z) = \frac{a}{n}z^n \tag{6.44}$$

with $z = re^{i\theta}$ it follows that

$$F = \frac{a}{n}r^n(\cos n\theta + i\sin n\theta) \tag{6.45}$$

and therefore

$$\Phi = \frac{a}{n}r^n\cos n\theta, \tag{6.46}$$

and

$$\psi = \frac{a}{n}r^n\sin n\theta. \tag{6.47}$$

For the magnitude of the velocity, we obtain

$$|\boldsymbol{V}| = \left|\frac{dF}{dz}\right| = |az^{n-1}| = |a|r^{n-1}. \tag{6.48}$$

The streamlines are constructed by setting ψ =const. The corner walls are represented by the streamlines $\psi = 0$, thus, $\sin n\theta = 0$ or $\theta = k\pi/n(k = 0, 1, 2, \ldots)$.

6.3 Superposition of Potential Flow Elements

This section presents the superposition of the few basic elements we discussed above to arrive at a more complicated flow picture. It contains the superposition of source and vortex, uniform flow, source and dipole flow, uniform flow and combined dipole and vortex flow.

6.3.1 Superposition of a Uniform Flow and a Source

Combining the uniform complex potential, Eq. (6.22) with the source potential Eq. (6.27), leads to a new complex potential that satisfies the Laplace equation:

$$F(z) = U_\infty z + \frac{E}{2\pi} 1nz. \tag{6.49}$$

Decomposing Eq. (6.49) into its real and imaginary parts, we arrive at stream function

$$\Psi = U_\infty r \sin\theta + \frac{E\theta}{2\pi} \tag{6.50}$$

and the potential function

$$\Phi = U_\infty r \cos\theta + \frac{E}{2\pi} \ln r. \tag{6.51}$$

Figure 6.7 exhibits the streamlines resulting from superposition Eq. (6.50).
The velocity components are obtained by differentiating Eq. (6.49) with respect to the complex variable z and decomposing the result into its real and imaginary parts, we find the velocities in x- and y-direction as:

$$u = U_\infty + \frac{E}{2\pi}\frac{x}{x^2+y^2},\ v = \frac{E}{2\pi}\frac{y}{x^2+y^2}. \tag{6.52}$$

Figure 6.7 exhibits the streamlines resulting from superposition Eq. (6.50). As seen, the location of the stagnation point is on the x-axis at an angle $\Theta = \pi$ and x, y positions, found by setting in Eq. (6.52) $u = v = 0$ respectively. This results in $x_{St} = -E/(2\pi U_\infty)$, $y = 0$, which, in conjunction with Eq. (6.50), determines the stagnation streamline $\Psi = \Psi_{St} E/2$.

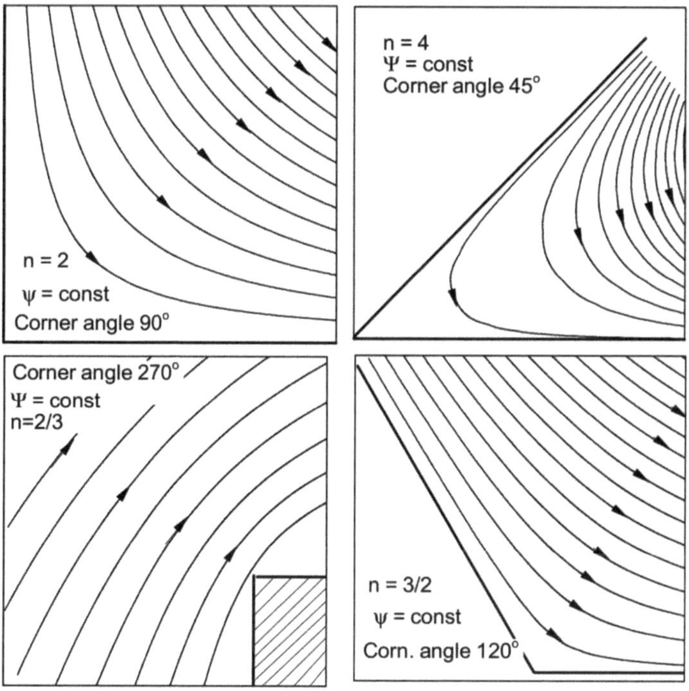

Fig. 6.7 Superposition of a uniform flow and a source to construct a flow past a plane semi-infinite body

6.3.2 *Superposition of a Translational Flow and a Dipole*

The potential flow around a circular cylinder can be simulated by a combination of a translational flow described by Eq. (6.22) and a dipole as defined by Eq. (6.39).

$$F(z) = U_\infty z + \frac{M}{2\pi}\frac{1}{z} = V_\infty\left(z + \frac{R^2}{z}\right). \tag{6.53}$$

In Eq. (6.53) we introduced $R^2 = M/(2\pi V_\infty)$. Inserting $z = re^{i\theta}$, Eq. (6.53) results in

$$F(r\theta) = U_\infty\left(r + \frac{R^2}{r}\right)\cos\theta + iU_\infty\left(r - \frac{R^2}{r}\right)\sin\theta \tag{6.54}$$

and therefore

$$\phi = U_\infty\left(r + \frac{R^2}{r}\right)\cos\theta \tag{6.55}$$

and

$$\psi = U_\infty \left(r - \frac{R^2}{r} \right) \sin\theta. \tag{6.56}$$

The stagnation streamline $\Psi = \Psi_S = 0$ is found by setting $r = R$ and $\theta = 0, \pi$, respectively. From the complex conjugate velocity

$$\frac{dF}{dz} = U_\infty \left(1 - \frac{R^2}{z^2} \right) \tag{6.57}$$

and by setting $dF/dz = 0$ we find the location of the stagnation points at $z = \pm R$. The velocity components in r and θ directions are determined from:

$$V_r = \frac{\partial \Phi}{\partial r} = U_\infty \left(1 - \frac{R^2}{r^2} \right) \cos\theta \tag{6.58}$$

and

$$V_\theta = \frac{1}{r}\frac{\partial \Phi}{\partial \theta} = -U_\infty \left(1 + \frac{R^2}{r^2} \right) \sin\theta. \tag{6.59}$$

Using the Bernoulli equation for inviscid flows

$$p + \frac{1}{2}\rho V^2 = p_\infty + \frac{1}{2}\rho U_\infty^2 \tag{6.60}$$

and setting $V^2 = V_r^2 + V_\theta^2$, we find the pressure coefficient as

$$C_p = \frac{p - p_\infty}{\frac{1}{2}\rho U_\infty^2} = 1 - \frac{V^2}{U_\infty^2} = 1 - 4\sin^2\theta. \tag{6.61}$$

Figure 6.8 shows the results of the superposition of a translational potential flow with a dipole flow simulating the inviscid potential flow past a cylinder. The cylinder with a radius $R = 1$ separates the dipole streamlines that constitute the interior of the cylinder from the exterior streamlines pertinent to the translational potential flow.

Figure 6.9 shows a symmetric C_p-distribution around the cylinder. Maximum C_p is obtained at the stagnation points A and C with $\theta = 0°$ and $180°$. Strong suction with $C_p = -3$ occurs at points B and D with $\theta = 90°$ and $270°$. Figure 6.9 also shows the distribution of the pressure gradient. Changes of pressure gradient from negative to positive values are observed at the top and the bottom of the cylinder marked with B and D. Integrating the pressure distribution around the cylinder surface results in a zero net reaction force in the streamwise direction. This means that the cylinder exposed to a potential flow experiences no drag force. As a consequence of the boundary conditions and the assumption of irrotationality, the flow is fully attached

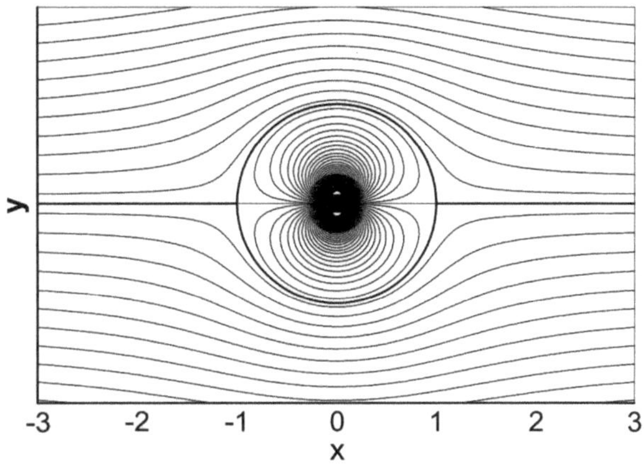

Fig. 6.8 Superposition of a translational flow and a dipole simulating a potential flow past a circular cylinder without circulation

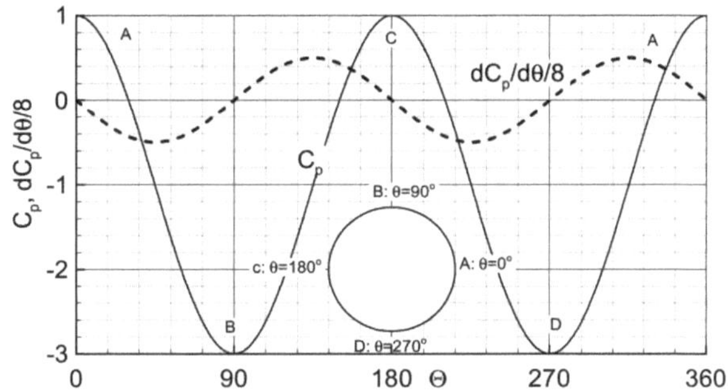

Fig. 6.9 Pressure distribution and the distribution of the pressure gradient around a cylinder exposed to a potential flow

to the cylinder surface with non-zero tangential components. In reality, there is always a drag force acting on the cylinder surface. The existence of the shear stress caused by the flow viscosity in conjunction with boundary layer instability and separation due to the change of pressure gradient, causes the pressure distribution to significantly deviate from the potential flow solution. The maximum velocity $V_{max} = 2V_\infty$ is reached for $r = R$, i.e. on the body at points B and D with $\theta = 90°$ and $270°$. At the stagnation points A and C, the velocity diminishes ($V_{St} = 0$).

6.3.3 *Superposition of a Translational Flow, a Dipole and a Vortex*

Adding to the case discussed in Sect. 6.3.2, a potential vortex in the clockwise direction (negative) and the potential flow past a rotating circular cylinder is simulated. This superposition is possible since a potential vortex satisfies the kinematic boundary condition. The complex potential of this flow is

$$F(z) = U_\infty \left(z + \frac{R^2}{z}\right) - i\frac{\Gamma}{2\pi}\ln\left(\frac{z}{R}\right). \tag{6.62}$$

Extracting its real and imaginary parts, we find the velocity potential and the stream function as

$$\phi = U_\infty \left(r + \frac{R^2}{r}\right)\cos\theta + \frac{\Gamma}{2\pi}\theta \tag{6.63}$$

and

$$\psi = U_\infty \left(r - \frac{R^2}{r}\right)\sin\theta - \frac{\Gamma}{2\pi}\ln(r/R). \tag{6.64}$$

The function $F(z)$ represents the flow past a circular cylinder with the circulation strength Γ as a parameter. When $\Gamma > 0$, the vortex has a counterclockwise direction, whereas a $\Gamma < 0$ refers to a clockwise direction. Equation (6.64) simulates the flow around a cylinder with a radius R that rotates with an angular velocity Ω. The circulation around the cylinder in Eq. (6.64) is obtained using the relationship:

$$\Gamma = \oint_{(C)} \boldsymbol{V} \cdot d\boldsymbol{C} = \int_0^{2\pi} V_\theta R d\theta \tag{6.65}$$

with $V_\theta = R\Omega$ and Ω as the angular velocity of the rotating cylinder. The stagnation points on the cylinder contour are computed from:

$$V_\theta = \frac{1}{r}\frac{\partial\phi}{\partial\theta}\bigg|_{r=R} = -2U_\infty \sin\theta + \frac{\Gamma}{2\pi}\frac{1}{R}. \tag{6.66}$$

Setting Eq. (6.66) equal to zero, we obtain the angular positions of the stagnation points:

$$\sin\theta = \frac{\Gamma}{4\pi}\frac{1}{U_\infty R}, \theta = \text{arc sin}\left(\frac{\Gamma}{4\pi}\frac{1}{U_\infty R}\right). \tag{6.67}$$

Note that within a cycle of 2π there are two solutions for each $4\pi U_\infty R > |\Gamma| > 0$. For $|\Gamma| = 4\pi U_\infty R$, there is only one solution for Θ meaning that there is one stagnation point. Figure 6.10 shows the flow pictures for different negative circulation values Γ. As seen in Fig. 6.9, for $\Gamma = 0$, the front and rear stagnation points were located at $\theta_{S1} = 180°$ and $\theta_{S2} = 0°$. Imposing a negative circulation (clockwise) causes the two stagnation points to move closer together, Fig. 6.10a. Consequently, the streamlines on top of the cylinder are crowded together, while the bottom streamlines are spaced farther apart leading to larger flow velocities above the cylinder than below. This results in higher pressure below the cylinder than above. Increasing the vortex strength moves the streamlines closer together, Fig. 6.10b–d. Further increase of circulation strength to reach $|\Gamma| = 4\pi RU_\infty$ causes the two stagnation points to merge together leading to a single point at an angle of $\theta = 3\pi/4$, Fig. 6.10e. For

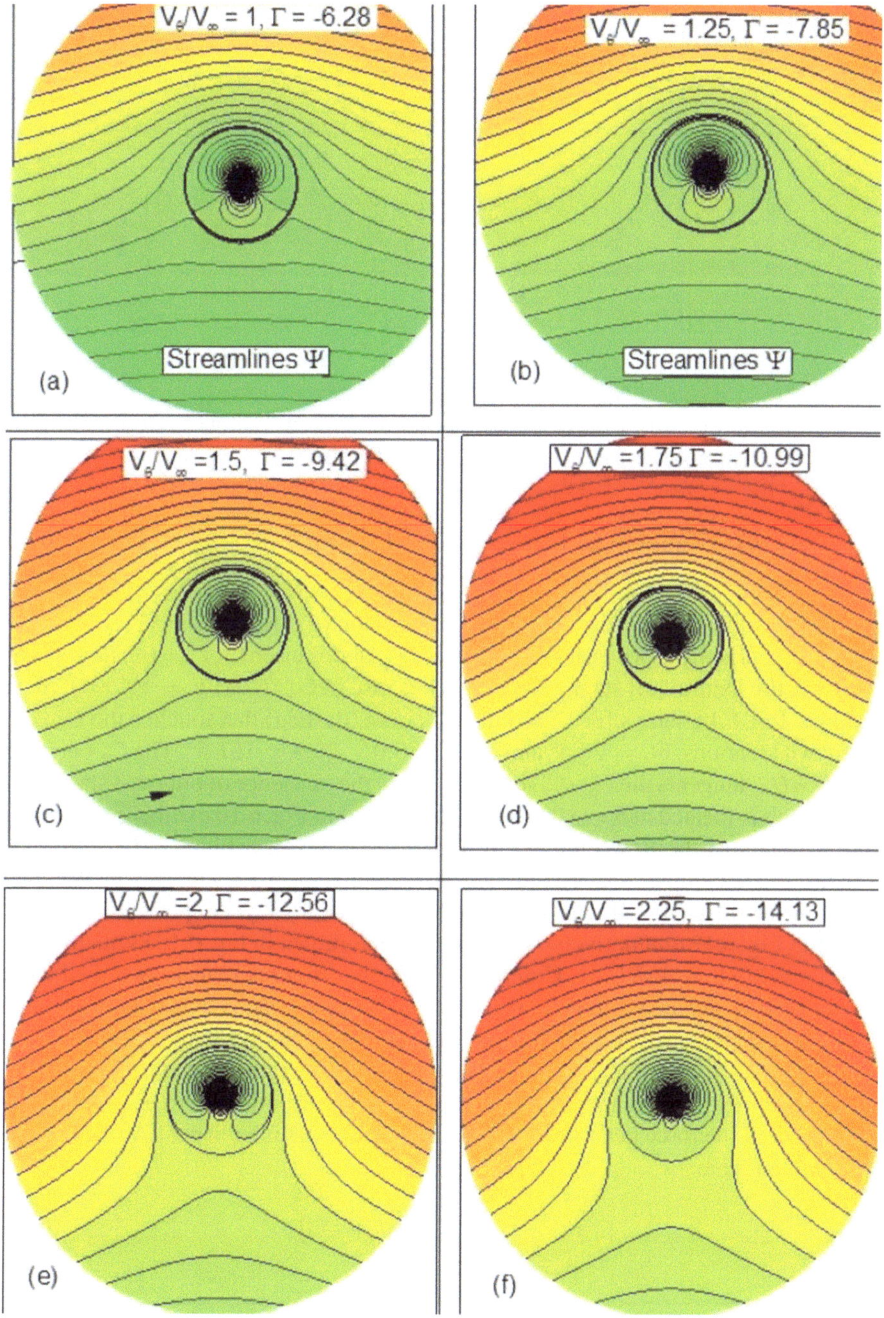

Fig. 6.10 Potential flow past a circular cylinder with clockwise circulation Γ as parameter

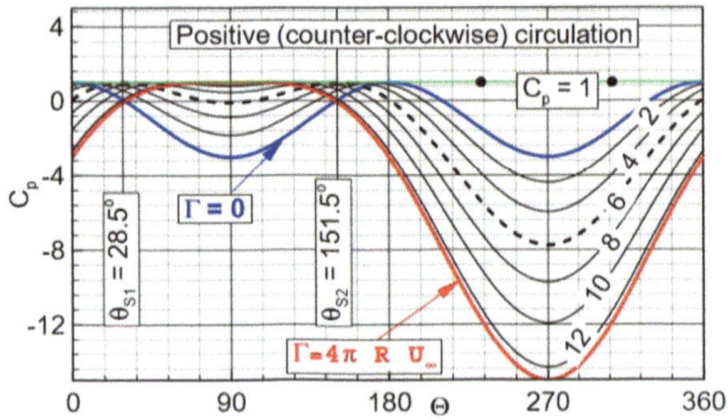

Fig. 6.11 Distribution of pressure coefficient C_p as a function of θ with the Γ as parameter

$|\Gamma| > 4\pi R U_\infty$, Fig. 6.10f, the stagnation point moves out. Using the Bernoulli equation (6.60), the pressure coefficient is calculated from:

$$Cp = \frac{p - p_\infty}{\frac{1}{2}\rho U_\infty^2} = 1 - \frac{V^2}{U_\infty^2} = 1 - \left(-2\sin\Theta + \frac{\Gamma}{2\pi R U_\infty}\right)^2 \quad (6.68)$$

Figure 6.11 shows the C_p-distributions as a function of θ for different $\Gamma > 0$ values. For comparison purposes, it also entails the C_p distribution for $\Gamma = 0$. The constant $C_p = 1$ is the locus of all stagnation points because in Eq. (6.68) $V = V_\Theta = 0$. As an example, Fig. 6.11 shows for $\Gamma = 6$ the C_p-curve (dashed) that touches the $C_p = 1$-line at two locations: $\Theta_s = 28.5°$ and $\Theta_s = 180.0 - 28.5 = 151.5°$. The C_p-curve for $\Gamma = 4\pi R U_\infty$ tangents the $C_p = 1$-line indicating the existence of only one stagnation point which is located at $\theta = 90$. The results plotted in Fig. 6.10 exhibit qualitatively similar trends as the flow pictures by Prandtl [31] display. Figure 6.12 shows the flow around a cylinder that rotates with a velocity ratio $V_\theta / U_\infty = 4.0$. With the exception of different velocity ratios and the boundary layer separation, the flow pattern in Fig. 6.12 qualitatively resembles the one shown in Fig. 6.10e with the ratio $V_\theta / U_\infty = 2.0 (\Gamma = 4\pi R U_\infty)$. Comparing these two figures reveals two distinctive characteristics: (1) the viscous flow exposed to a major positive pressure gradient as we discussed previously leads to flow separation that is clearly visible in Fig. 6.12, (2) in order for the two stagnation points to merge, a velocity ratio far above $\Gamma = 4\pi R V_\infty$ must be applied. Calculating the static pressure distribution from Eq. (6.68), the component of the force per unit of depth acting on the cylinder in positive x-direction is calculated from:

$$F_x = -\int_{o}^{2\pi} pR\cos\theta d\theta = 0 \quad (6.69)$$

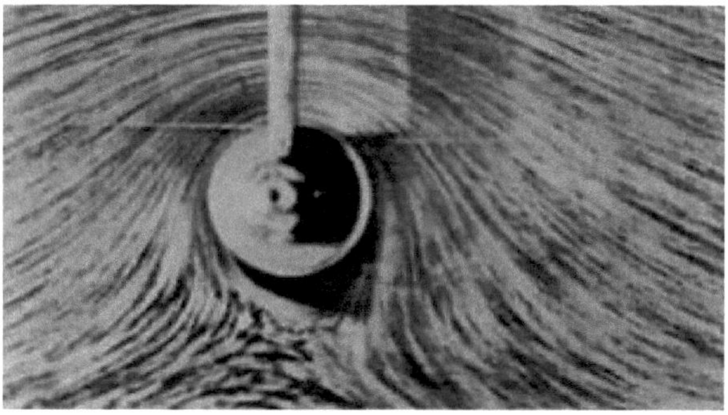

Fig. 6.12 Viscous flow around a rotating cylinder from [6]. Unlike the potential flow the viscous flow has caused a separation zone

which vanishes for symmetry reasons. The component in positive y-direction is:

$$F_y = -\int_{\circ}^{2\pi} pR\sin\theta d\theta. \tag{6.70}$$

Inserting the static pressure from Eq. (6.68) into Eq. (6.70) results in:

$$F_y = \varrho V_\infty \Gamma. \tag{6.71}$$

Note that F_y is positive for counter clockwise direction $\Gamma < 0$. In the following sections dealing with the Kutta-Joukowsky theorem, it is shown that Eq. (6.71) is generally valid for any two-dimensional body regardless of its shape. This equation was independently developed by W. Kutta (1867–1944) and N. Joukowsky (1847–1921). The generalized relationship in vector form is:

$$\boldsymbol{L} = \varrho \boldsymbol{V}_\infty \times \boldsymbol{\Gamma}. \tag{6.72}$$

Equation (6.72) states that a body of any arbitrary shape exposed to a potential flow with the velocity $\boldsymbol{V}_\infty$ and the circulation vector $\boldsymbol{\Gamma}$ generates a lift vector $\boldsymbol{L}$ (per unit depth) which is perpendicular to the plane spanned by $\boldsymbol{V}_\infty$ and $\boldsymbol{\Gamma}$.

Figure 6.13 exhibits an airfoil that is subjected to an inviscid flow with the velocity $\boldsymbol{V_1}$. The airfoil shown in Fig. 6.13a has a certain camber which causes the velocity direction to deflect by an amount of δ, Fig. 6.13b, whose magnitude determines the magnitude of the circulation Γ. Immediately downstream of the training edge, the velocity assumes the value $\boldsymbol{V_2}$. These two velocities form the mean velocity vector $\boldsymbol{V}_\infty$ which determines the magnitude of the lift force as well. The direction of the velocity vector $\boldsymbol{V}_\infty$, the circulation vector $\boldsymbol{\Gamma}$ vector and the lift vector $\boldsymbol{L}$ are shown in Fig. 6.13b. In this case, the circulation is generated by the flow velocity deflection δ between $\boldsymbol{V}_1$ and $\boldsymbol{V}_2$ caused by passing over the airfoil, resulting in a mean

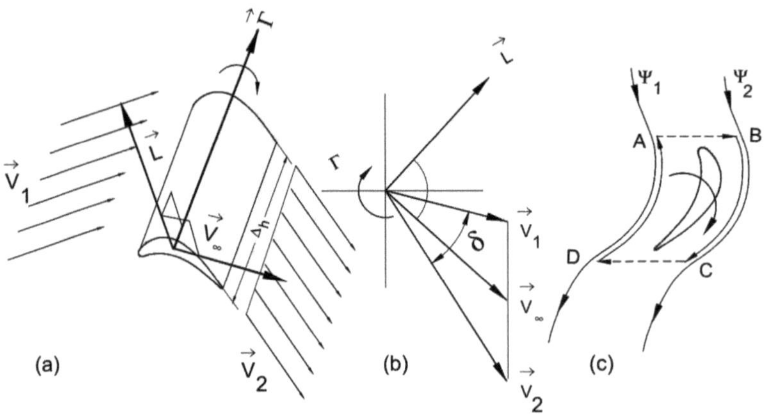

Fig. 6.13 Lift force acting on an airfoil of height $\Delta h = 1$

velocity vector $\boldsymbol{V}_\infty = (\boldsymbol{V}_1 + \boldsymbol{V}_2)2$. Using Eq. (6.65) and considering Fig. 6.13c, the expansion of circulation integral leads to:

$$\Gamma = \int_A^B \boldsymbol{V} \cdot d\boldsymbol{c} + \int_B^C \boldsymbol{V} \cdot d\boldsymbol{c} \int_C^D \boldsymbol{V} \cdot d\boldsymbol{c} + \int_D^A \boldsymbol{V} \cdot d\boldsymbol{c} = \int_A^B \boldsymbol{V} \cdot d\boldsymbol{c} + \int_C^D \boldsymbol{V} \cdot d\boldsymbol{c}. \quad (6.73)$$

The integrals in Eq. (6.73) from B to C and D to A are performed along two adjacent streamlines Ψ_1 and Ψ_2. These two integrals are equal and opposite, therefore, they cancel each other. For a positive value of Γ obtained from Eq. (6.73) along a closed curve surrounding the blade (Fig. 6.13c), its vector $\boldsymbol{\Gamma}$ must form a right-handed screw with the chosen direction. It should be pointed out that the circulation integral can be carried out around any curve that surrounds the body. In carrying out the integration only for practical reasons, we chose the closed curve ABCDA.

As shown in Fig. 6.13, the specific lift force $\boldsymbol{L}$ (force per unit of depth) forms a right-handed screw with the direction obtained by rotating $\boldsymbol{V}_\infty$ toward $\boldsymbol{\Gamma}$, such that $\boldsymbol{V}_\infty$ forms the angle of 90° between these two vectors. As seen, the circulation does not necessarily need to be generated by rotation. Any flow deflection caused by passing over a body that generates certain flow deflection produces circulation and, therefore, a lift. The phenomenon of a rotating body in a cross flow that experiences a lift is called *Magnus effect*.

6.3.4 Superposition of a Uniform Flow, Source, and Sink

Figure 6.14 exhibits the superposition of a uniform flow, a source and a sink discussed in the following section.

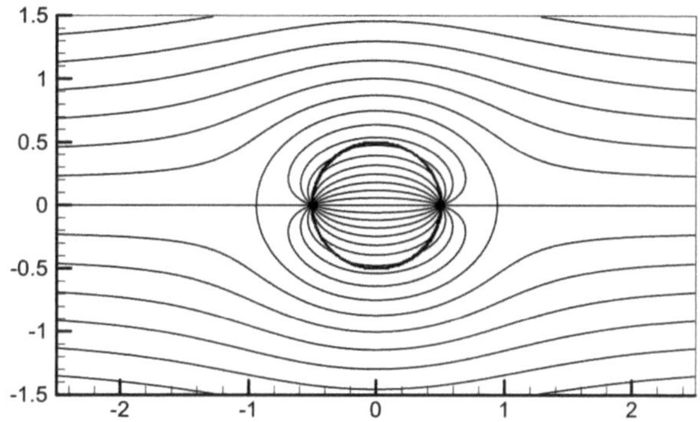

Fig. 6.14 Superposition of a uniform flow, a source and a sink

Combining the uniform complex potential, Eq. (6.22) with a source located at $z = -a$ and a sink at $z = +a$ with the complex potentials described by Eq. (6.27), we obtain a new complex potential that satisfies the Laplace equation:

$$F(z) = V_\infty z + \frac{E}{2\pi} \ln(z+a) - \frac{E}{2\pi} \ln(z-a) \text{ or}$$
$$F(z) = V_\infty z + \frac{E}{2\pi} \ln\left(\frac{z+a}{z-a}\right). \tag{6.74}$$

Decomposing Eq. (6.74) into its real and imaginary parts, we arrive at the potential function

$$\Phi = V_\infty x + \frac{E}{4\pi} \ln\left(\frac{(x+a)^2 + y^2}{(x-a)^2 + y^2}\right) \tag{6.75}$$

and the stream function

$$\Psi = -V_\infty y - \frac{E}{2\pi}\left[\arctan\left(\frac{y}{x+a}\right) - \arctan\left(\frac{y}{x-a}\right)\right]. \tag{6.76}$$

Figure 6.14 exhibits the streamlines resulting from superposition Eq. (6.76). The velocity components are obtained by differentiating Eq. (6.76) with respect to the complex variable z and decomposing the result into its real and imaginary parts. The process of differentiation and decomposition is the same as shown in Sect. 6.2.

6.3.5 Superposition of a Source and a Vortex

The superposition of a source and a vortex is plotted in Fig. 6.15 and discussed below. As in the last example, we combine the complex potential of a source, Eq. (6.27), and a vortex with the complex potential described by Eq. (6.31). The source and the vortex are located at the origin.

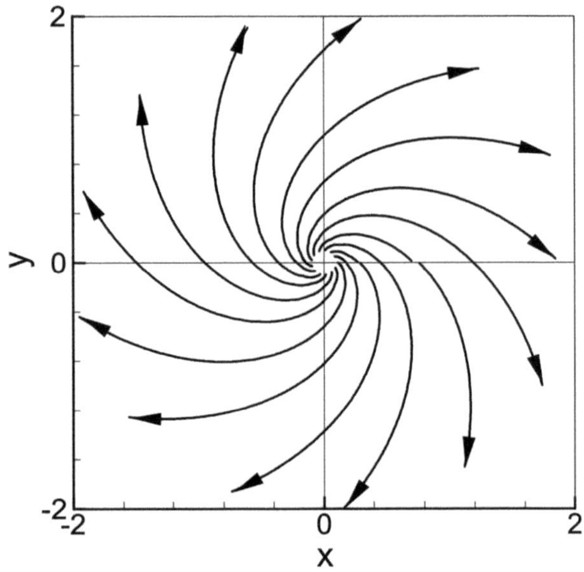

Fig. 6.15 Superposition of a source and vortex

$$F(z) = \frac{E}{2\pi} \ln z - i\frac{\Gamma}{2\pi} \ln z. \tag{6.77}$$

Decomposing Eq. (6.77) into its real and imaginary parts, we arrive at the potential function

$$\Phi = \frac{E}{2\pi} \ln r + \frac{\Gamma}{2\pi}\theta \tag{6.78}$$

and the stream function

$$\Psi = \frac{E}{2\pi}\theta - \frac{\Gamma}{2\pi} \ln r. \tag{6.79}$$

The combination of source and a vortex is frequently referred to as a logarithmic spiral. The streamlines are plotted in Fig. 6.15.

6.3.6 Blasius Theorem

In this section, we utilize the complex analysis to provide the equation structure that is needed to derive the Kutta-Joukowsky lift equation from a potential theoretical point of view. As we saw in Chap. 5, any force exerted on any body of any shape that is subjected to a viscous or inviscid flow can be calculated using the integral balance of linear momentum. Thus, the following procedure is an alternative that can be applied only to potential flow. Equation (5.25) applied to a two-dimensional body results in:

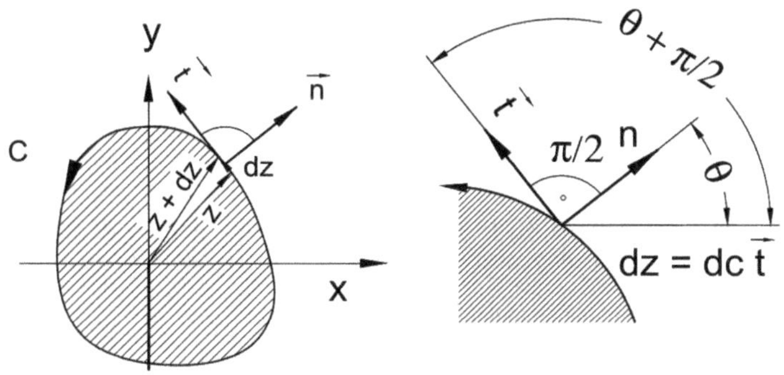

Fig. 6.16 Relation between the normal and tangential unit vectors ***n*** and ***t***

$$\boldsymbol{R} = \int_{S_B} \boldsymbol{n} p dS + \int_{S_B} \boldsymbol{t} \tau dS \tag{6.80}$$

with ***n*** and ***t*** as the normal and tangential unit vectors shown in Fig. 6.16. For an an inviscid irrotational flow, Eq. (6.80) is reduced to:

$$\boldsymbol{R} = \int_{S_B} \boldsymbol{n} p dS. \tag{6.81}$$

The integration has to be performed over the entire body surface. For a two-dimensional body, the differential surface element is $dS = wdC$ with w as the width of the body and dC as a differential element of the body contour C, Fig. 6.16.

By moving along the contour, the normal unit vector changes the direction. However, it can be related to the velocity direction along the body contour. For the normal and tangential unit vectors, we may write:

$$\boldsymbol{n} = \cos\theta + i \sin\theta; \; \boldsymbol{t} = \cos(\theta + \pi/2) + i \sin(\theta + \pi/2). \tag{6.82}$$

Expanding ***t*** in Eq. (6.82) and multiplying the results with $-i$, we find:

$$-i\boldsymbol{t} = \boldsymbol{n}. \tag{6.83}$$

Inserting Eq. (6.83) into Eq. (6.81), we get:

$$\boldsymbol{R} = -i \int_{S_B} \boldsymbol{t} p dS. \tag{6.84}$$

With $dS = wdC, tdC = dz$ and assuming $w = 1$, we find:

$$\boldsymbol{R} = -i \oint_C p dz. \tag{6.85}$$

Thus, the surface integral in Eq. (6.84) is converted into a *contour integral* (closed line integral). The static pressure p is calculated from Bernoulli equation:

$$p = P - \frac{\rho}{2} V^2 = p_0 - \frac{\rho}{2} \frac{dF}{dz} \overline{\frac{dF}{dz}} \tag{6.86}$$

with $P = p_0$ as the constant total pressure. We substitute the static pressure in Eq. (6.85) by Eq. (6.86) and obtain the force vector:

$$\boldsymbol{R} = i\frac{\rho}{2} \oint_C \frac{dF}{dz} \overline{\frac{dF}{dz}} dz. \tag{6.87}$$

In Eq. (6.87), the constant total pressure term does not appear because the contour integral over the constant total pressure vanishes. Since the contour of the body is a closed curve with Ψ =const. and $d\Psi = 0$ from Eq. (6.10), it follows that:

$$dF = d\Phi = \overline{dF} = dF \tag{6.88}$$

and, thus, Eq. (6.87) reduces to:

$$\boldsymbol{R} = i\frac{\rho}{2} \oint_C \left(\overline{\frac{dF}{dz}} \right)^2 dz. \tag{6.89}$$

Equation (6.89) is called the *Blasius Theorem*.

6.4 Kutta-Joukowski Theorem

With Eq. (6.89) we are now able to calculate the force acting on a cylinder of arbitrary contour using the Kutta-Joukowsky lift equation. We assume the cylinder is exposed to a flow with the velocity vector $\boldsymbol{V}_\infty$ and the components $V_{\infty x} + iV_{\infty y}$ at infinity. We further assume that there are no singularities outside the body, although there will be inside in order to represent the body and to produce the lift. The velocity field can be represented by a *Laurent series* of the form

$$\frac{dF}{dz} = \bar{V} = u - iv = A_0 + A_1\left(\frac{1}{z}\right) + A_2\left(\frac{1}{z^2}\right) + A_3\left(\frac{1}{z^3}\right) + \ldots = \sum_{n=0}^{\infty} A_n\left(\frac{1}{z^n}\right) \tag{6.90}$$

with $\bar{V}$ as the conjugate velocity vector. The integration of Eq. (6.90) yields the complex potential

$$F(z) = A_0 z + A_1 \ln z - \sum_{n=2}^{\infty} \frac{A_n}{n-1}\left(\frac{1}{z^{n-1}}\right) + \text{const.} \tag{6.91}$$

The boundary condition at infinity requires

$$\frac{dF}{dz}|_\infty = V_{\infty x} - iV_{\infty y} \tag{6.92}$$

which determines the coefficient A_0

$$A_0 = U_{\infty x} - iV_{\infty y} = \bar{V}_\infty. \tag{6.93}$$

To calculate the coefficient A_1 we integrate $(u - iv)$ around the contour of the body:

$$\oint_{(C)} \frac{dF}{dz} dz = \oint_{(C)} (u - iv) dz = \oint_{(C)} (u - iv)(dx + idy). \tag{6.94}$$

Performing the multiplication of the right-hand side integrand of Eq. (6.94) leads to:

$$\oint_{(C)} \frac{dF}{dz} dz = \oint_{(C)} (udx + vdy) + i \oint_{(C)} (udy - vdx). \tag{6.95}$$

The first integral on the right-hand side is the circulation defined in Eq. (6.65). The second term is the closed integral of derivative of the stream function ψ:

$$\oint_{(C)} (u - iv) dz = \oint_{(C)} \boldsymbol{V} \cdot d\boldsymbol{C} + i \oint_{(C)} d\psi \tag{6.96}$$

since Ψ is constant along the contour, its derivative $d\psi$ vanishes. As a result, with the definition of the circulation Eq. (6.65), we have:

$$\oint_{(C)} (u - iv) dz = \oint_{(C)} \boldsymbol{V} \cdot d\boldsymbol{C} = \Gamma. \tag{6.97}$$

According to the *residue theorem* of complex analysis, if C is a closed curve, and if $f(z)$ is analytic within and on C except at a finite number of singular points in the interior of C, then

$$\oint_{(C)} f(z)dz = 2\pi i(r_1 + r_2 + r_3 + \cdots + r_n). \tag{6.98}$$

Since the Laurent series has only one essential singularity ($z = 0$), then from Eq. (6.97) we have

$$\oint_{(C)} \frac{1}{z^n}dz = 2\pi i, \text{ for } n = 1, \text{ and } \oint_{(C)} \frac{1}{z^n}dz = 0, \text{ for } n \geq 2. \tag{6.99}$$

Implementing the results from Eq. (6.99) into Eq. (6.97), we find the coefficient A_1 from:

$$\oint_{(C)} (u - iv)dz = 2\pi i A_1 = \Gamma. \tag{6.100}$$

To calculate the force vector acting on the body, we utilize the Laurent series, Eq. (6.90), that describes the velocity as an analytic function and construct the following equation:

$$\left(\frac{dF}{dz}\right)^2 = A_0^2 + 2A_0A_1\left(\frac{1}{z}\right) + (A_1^2 + 2A_0A_2)\left(\frac{1}{z}\right)^2 + \cdots \tag{6.101}$$

Taking the contour integral of Eq. (6.101),

$$\oint_{(C)} \left(\frac{dF}{dz}\right)^2 dz = \oint_{(C)} \left[A_0^2 + 2A_0A_1\left(\frac{1}{z}\right) + (A_1^2 + 2A_0A_2)\left(\frac{1}{z}\right)^2 + \cdots\right]dz. \tag{6.102}$$

The first term in Eq. (6.102) is a contour integral over the constant $A_0 = \bar{V}_\infty$ taken from Eq. (6.93). Its contour integral vanishes. The contour integral of the second term is $2\pi i$ because of Eq. (6.99) for $n = 1$. The contour integral of the third and all higher order terms vanishes due to Eq. (6.99) for $n \geq 2$. As a result we get:

$$\oint_{(C)} \left(\frac{dF}{dz}\right)^2 dz = \oint_{(C)} 2A_0A_1\left(\frac{1}{z}\right) dz = 4A_0A_1\pi i. \tag{6.103}$$

With Eqs. (6.93) and (6.100), Eq. (6.103) becomes:

$$\oint_{(C)} \left(\frac{dF}{dz}\right)^2 dz = 2\Gamma \bar{V}_\infty \tag{6.104}$$

and then from Eq. (6.89), we find the *Kutta-Joukowski* lift equation.

$$\boldsymbol{R} = i\varrho\Gamma\bar{V}_\infty. \tag{6.105}$$

Equation (6.105) shows that the lift force is proportional to the flow velocity, the circulation and the density, regardless of the shape of the body. The result of Eq. (6.83) suggests that the lift force $\boldsymbol{R}$ is perpendicular to the velocity vector. To prove this statement, we re-write Eq. (6.105) using the following relations from complex analysis:

$$\begin{aligned}
e^{\pm i\theta} &= \cos\theta \pm i\sin\theta \\
e^{\pm i\pi/2} &= \pm i \\
e^{-i\pi} &= -1 \\
e^{2i\pi} &+ 1
\end{aligned} \tag{6.106}$$

the force vector equation is modified as:

$$\boldsymbol{R} = i\varrho\Gamma|V_\infty|e^{-i\theta} = \varrho\Gamma|V_\infty|e^{-i\theta}e^{i\pi/2} = \varrho\Gamma|V_\infty|e^{i(\pi/2-\theta)} \tag{6.107}$$

where exponential expression determines the direction of the force vector. Is the direction of the conjugate velocity vector $e^{-i\theta}$, so is the direction of the force vector $e^{i(\pi/2-\theta)}$, meaning that the force vector and the conjugate velocity vector are perpendicular to each other. Figure 6.17 shows the effect of a circulation sign. As seen, the sign of the circulation determines if the force vector points up- or downward. Figure 6.17a shows a clockwise (negative) circulation that generated an upward lift, whereas, the counterclockwise (positive) circulation created a downward lift force.

6.5 Conformal Transformation

The method of conformal transformation was used extensively in the pre-CFD era to reduce a more complicated flow configuration to a simpler one amenable to mathematical treatment. We already learned to compose a more complicated flow by using the superposition principle. In Sect. 6.3.5, we treated the flow past a rotating circular cylinder. With conformal transformation treated in this section, it is possible to transform the flow past a circular cylinder to a flow pasta cylinder of arbitrary contour such as an airfoil. As long as no separation of the boundary layer occurs in the real

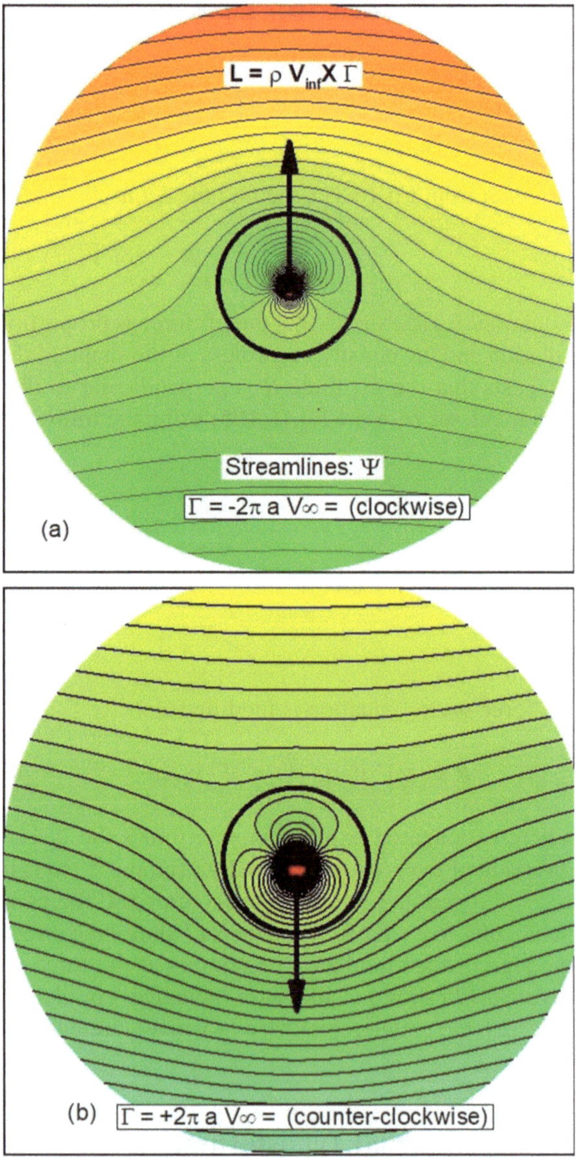

Fig. 6.17 Circulation sign and lift direction, **a** negative circulation (clockwise) causes a positive lift, **b** positive circulation (counter clockwise) causes a negative lift

flow, potential theory describes the actual flow behavior reasonably well. For this reason the potential flow pasta circular cylinder still has some technical importance. Conformal transformation is a major subject of many complex analysis textbooks and advanced engineering mathematics, among others [29, 32, 33]. In explaining the basics of conformal transformation, the following introductory section is presented.

6.5.1 Conformal Transformation, Basic Principles

Consider two planes, one is the z-plane, in which the point $z = x + iy$ is located and the other is the ζ-plane in which the point $\zeta = \xi + i\eta$ is to be plotted. Let there be a function $\zeta = f(z)$ that facilitates the transformation of point z in z-plane into the ζ-plane. The function $\zeta = f(z)$, thus, defines a mapping or transformation of z-plane onto ζ-plane. Figure 6.18 is a simple example of transformation of points that constitute the straight lines in z-plane onto corresponding points in ζ-plane. Consider the transformation function

$$\zeta = \xi + i\eta = z^2 = (x + iy)^2 = x^2 - y^2 + 2ixy. \tag{6.108}$$

Comparing the real and imaginary parts, it follows that

$$\xi = x^2 - y^2 \eta = 2xy. \tag{6.109}$$

As seen in Fig. 6.18, constant lines $x = C_x$ in the z-plane are mapped onto parabolas open to the left. Furthermore, Fig. 6.18 suggests that the magnitudes of angles between the x =const. and y =const. in z-plane are preserved, when transforming into ζ-plane. Eliminating y from Eq. (6.109) leads to

$$\xi = C_x^2 - \frac{\eta^2}{4C_x^2}. \tag{6.110}$$

For $C_x = 0$ (y axis) the parabolae coincide with the negative ξ axis. Lines $y = C_y$ are mapped onto parabolae open to the right:

$$\xi = \frac{\eta^2}{4C_y^2} - C_y^2 \tag{6.111}$$

where for $C_y = 0$ (x axis) the parabolas lie along the positive ξ axis. Before getting into transformation details, it is important to know when the transformation equation can be solved for x and y as *single-valued* functions of ξ and η, that is, when the transformation has a single-valued inverse. As we saw in Chap. 3, Eq. (3.15) the condition for this is that the Jacobian determinant of the transformation

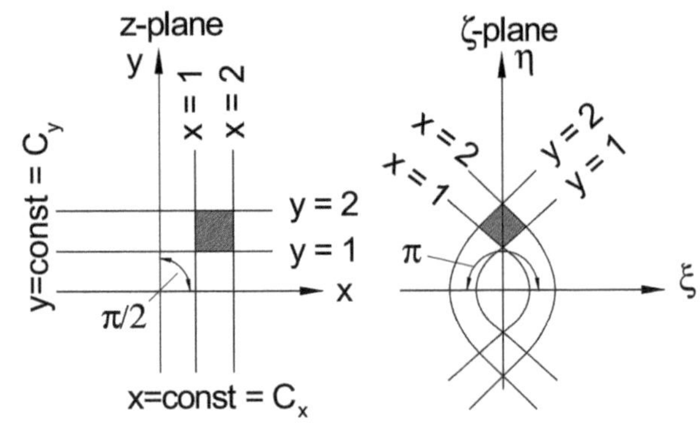

Fig. 6.18 Conformal transformation from z-plane onto ζ-plane

$$J\left(\frac{\xi,\eta}{x,y}\right) = \det\begin{pmatrix} \frac{\partial \xi}{\partial x} & \frac{\partial \xi}{\partial y} \\ \frac{\partial \eta}{\partial x} & \frac{\partial \eta}{\partial y} \end{pmatrix} \not\equiv 0 \tag{6.112}$$

does not vanish. Since $\zeta = f(z)$ is assumed to be analytic, ξ and η must satisfy the Chauchy-Riemann equations. Substituting them into the Jacobian determinant, we have

$$\begin{aligned} J\left(\frac{\xi,\eta}{x,y}\right) &= \det\begin{pmatrix} \frac{\partial \xi}{\partial x} & \frac{-\partial \eta}{\partial x} \\ \frac{\partial \eta}{\partial x} & \frac{\partial \xi}{\partial x} \end{pmatrix} = \left(\frac{\partial \xi}{\partial x}\right)^2 + \left(\frac{\partial \eta}{\partial x}\right)^2 \\ &= |\frac{\partial \xi}{\partial x} + i\frac{\partial \eta}{\partial x}|^2 = |f'(z)|^2. \end{aligned} \tag{6.113}$$

A transformation which has the following properties, is called conformal:

(a) If the function $\zeta = f(z)$ has a single-valued inverse in the neighborhood of any point where the derivative of the transformation function is non-zero,
(b) If in the mapping the lengths of infintesimal segments, regardless of their direction, are altered by a factor $|f'(z)|$ which depend only on the point from which the segments are drawn,
(c) If the angles are preserved in magnitude and sense; the case where the vertex of the angle is an n-fold zero of $f'(z)$ is the only exception,
(d) If the velocity potential $\Phi(x, y)$ is the solution of the Laplace equation and when $\Phi(x, y)$ is transformed into $\Phi(\xi, \eta)$, then the transformation will satisfy the Laplace equation also.

A transformation function $\zeta = f(z)$ with the properties defined above, is analytic. Conversely, it can be shown that if the mapping $\xi = \xi(x, y)$, $\eta = \eta(x, y)$ is conformal, and if their first partial derivatives are continuous, then $\zeta = \xi + i, \eta = f(z)$ is an analytic function.

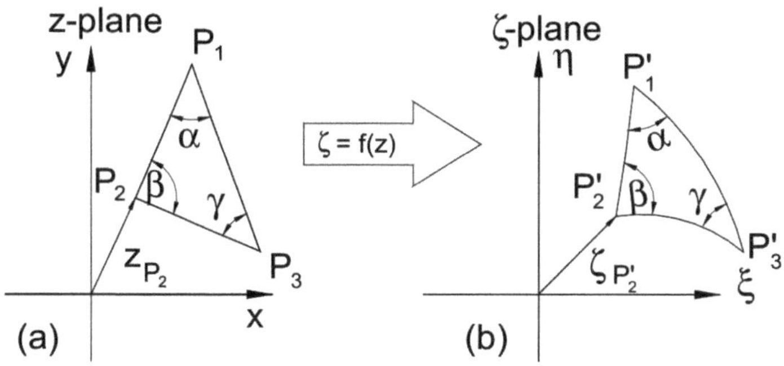

Fig. 6.19 Transformation of straight segments and angles from z-plane into ζ-plane

From (c), it follows that at the origin in Fig. 6.18, as a singular point of the transformation, the derivative $f' = d\zeta/dz$ has a simple zero, and the mapping is no longer conformal at this point. At a simple zero, the angle between two line elements, such as the x and y axes $(\pi/2)$, is doubled in the ζ plane (π). In general we have: At a zero of order n of $f'(z)$, the angle is altered by a factor $(n+1)$ (branch point of order n).

The graphical representation of the conformal transformation principle is shown in Fig. 6.19. The sides of the triangle in the z-plane shown in Fig. 6.19 are transformed onto curves in the ζ-plane. As seen, straight segments in the z-plane transformed into curved segments in the ζ-plane, while the angles between intersecting segments remain the same.

6.5.2 *Kutta-Joukowsky Transformation*

Before treating the Kutta-Joukowsky transformation, a brief description of the transformation process is given below. We consider the mapping of a circular cylinder from the z-plane onto the ζ-plane. Using a mapping function, the region outside the cylinder in the z-plane is mapped onto the region outside another cylinder in the ζ-plane. Let P and Q be the corresponding points in the z- and ζ-planes respectively. The potential at the point P is

$$F(z) = \Phi + i\Psi. \tag{6.114}$$

The point Q has the same potential, and we obtain it by insertion of the mapping function

$$F(z) = F(z(\zeta)) = F(\zeta). \tag{6.115}$$

Taking the first derivative of Eq. (6.115) with respect to ζ, we obtain the complex conjugate velocity $\bar{V}_\zeta$ in the ζ plane from

$$\bar{V}_\zeta(\zeta) = \frac{dF}{d\zeta}. \tag{6.116}$$

Considering z to be a parameter, we calculate the value of the potential at the point z. Using the transformation function $\zeta = f(z)$ we determine the value of ζ which corresponds to z. At this point ζ, the potential then has the same value as at the point z. To determine the velocity in the ζ plane, we form

$$\frac{dF}{d\zeta} = \frac{dF}{dz}\frac{dz}{d\zeta} = \frac{dF}{dz}\left(\frac{d\zeta}{dz}\right)^{-1} \tag{6.117}$$

after introducing Eq. (6.116) into Eq. (6.117) and considering $\bar{V}_z(z) = dF/dz$, Eq. (6.117) is rearranged as

$$\bar{V}_\zeta(\zeta) = \bar{V}_z(z)\left(\frac{d\zeta}{dz}\right)^{-1}. \tag{6.118}$$

Equation (6.118) expresses the relationship between the velocity in ζ-plane and the one in z-plane. Thus, to compute the velocity at a point in the ζ plane we divide the velocity at the corresponding point in the z plane by $d\zeta/dz$. The derivative $dF\ d\zeta$ exists at all points where $d\zeta/dz \neq 0$. At singular points with $d\zeta/dz = 0$, the complex conjugate velocity in the ζ plane $\bar{V}_\zeta(\zeta) = dF/d\zeta$ becomes infinite, if it is not equal to zero at the corresponding point in the z plane.

6.5.3 Joukowsky Transformation

The conformal transformation method introduced by Joukowski allows mapping an unknown flow past a cylindrical airfoil to a known flow past a circular cylinder. Using the method of conformal transformation, we can obtain the direct solution of the flow past a cylinder of an arbitrary cross section. Although numerical methods of solution of the direct problem have now superseded the method of conformal mapping, it has still retained its fundamental importance [30]. In what follows, we shall examine several flow cases using the *Joukowsky* transformation function.

$$\zeta = f(z) = z + \frac{a^2}{z}, \text{ with } z = re^i\theta, \zeta = \xi + i\eta. \tag{6.119}$$

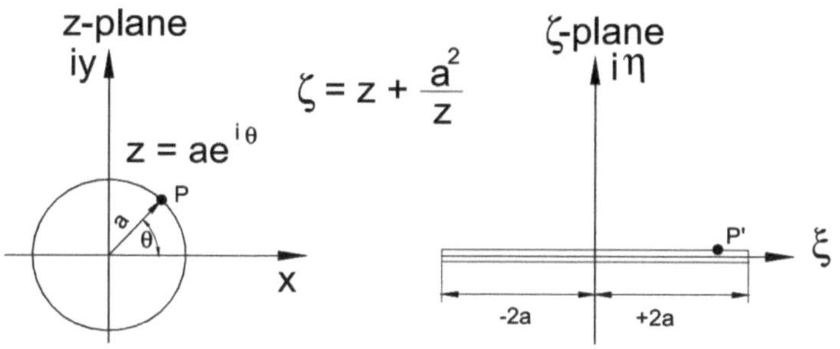

Fig. 6.20 Transformation of a circle onto a slit (straight line section)

6.5.3.1 Circle-Flat Plate Transformation

Decomposing Eq. (6.119) into its real and imaginary parts, we obtain:

$$\xi = \left(r + \frac{a^2}{r})\right) \cos\Theta, \eta = \left(r - \frac{a^2}{r}\right) \sin\Theta. \tag{6.120}$$

The function $f(z)$ maps a circle with radius $r = a$ in the z plane onto a "slit" in the ζ plane. Equation (6.120) delivers the coordinates:

$$\xi = 2a\cos\theta, \eta = 0 \tag{6.121}$$

with ξ as a real independent variable in the ζ-plane. As the point P with the angle θ moves in z-plane from 0 to 2π, (Fig. 6.20), its image p' moves from $+2a$ to $-2a$ in the ζ-plane. With the complex potential Eq. (6.53)

$$F(z) = V_\infty\left(z + \frac{R^2}{z}\right) \tag{6.122}$$

and setting $R = a$, the Joukowski transformation function directly provides the potential in the ζ plane as

$$F(\zeta) = V_\infty\zeta. \tag{6.123}$$

6.5.3.2 Circle-Ellipse Transformation

For this transformation, the circle center is still at the origin of the z-plane. Now if we map a circle with radius b which is smaller or larger than the mapping constant a, we obtain an ellipse. Replacing r by $b(b \neq a)$, Eq. (6.120) becomes

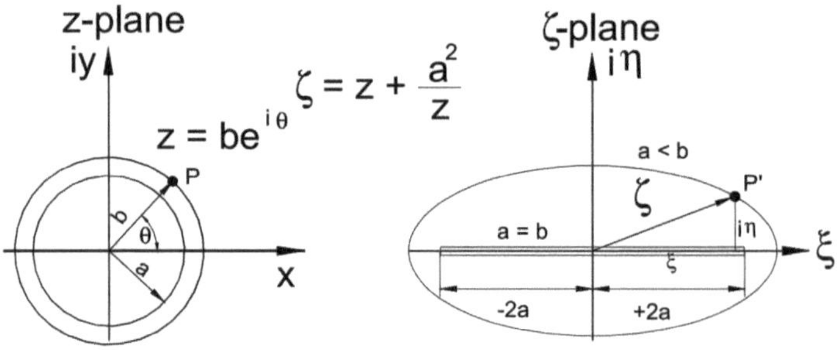

Fig. 6.21 Conformal transformation of a circle to an ellipse

$$\xi = \left(b + \frac{a^2}{b}\right)\cos\Theta, \eta = \left(b - \frac{a^2}{b}\right)\sin\Theta. \tag{6.124}$$

Eliminating θ from Eq. (6.124), we find:

$$\cos^2\theta = \left(\frac{\xi}{b + a^2/b}\right)^2, \sin^2\eta = \left(\frac{\eta}{b - a^2/b}\right)^2 \tag{6.125}$$

and with $\sin^2\theta + \cos^2\theta = 1$, we obtain the equation of ellipse as:

$$\left(\frac{\xi}{b + a^2/b}\right)^2 + \left(\frac{\eta}{b - a^2/b}\right)^2 = 1. \tag{6.126}$$

Equation (6.126) describes an ellipse, plotted in Fig. 6.21, with the major and minor axes which are given as the denominators in Eq. (6.126). In Fig. 6.21, $b > a$, however any ellipse may be constructed by varying the ratio b/a.

6.5.3.3 Circle-Symmetric Airfoil Transformation

A set of symmetrical airfoils can be constructed by shifting the center of the circle with the radius b by Δx along the x-axis on the z-plane as shown in Fig. 6.22. An eccentricity $\varepsilon = e/a$ with $e = \Delta x$ is defined that determines the thickness of the airfoil. The radius of the circle is determined by:

$$b = (1 + \varepsilon)a. \tag{6.127}$$

Thus, the magnitude of the eccentricity defines the slenderness of the airfoil. For $\varepsilon = 0$, the circle is mapped into a slit, as seen in Fig. 6.20. Due to zero flow deflection, the symmetrical airfoils at zero-angle of attack do not generate circulation and, therefore, no lift. Similar profiles are used in turbomachinery design practices such as *base profiles* to be superimposed on *camberlines*.

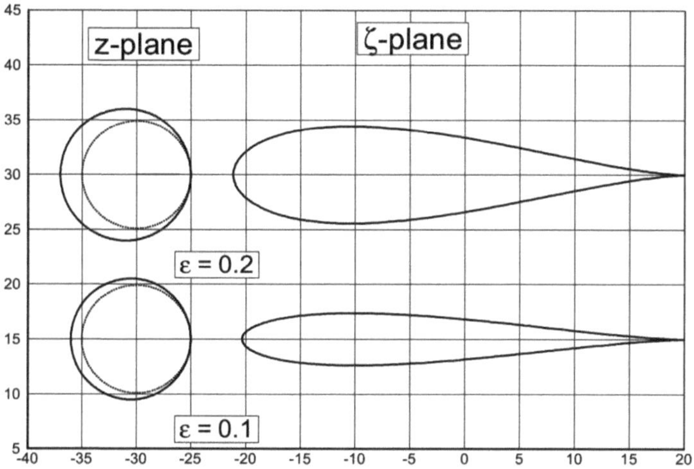

Fig. 6.22 Transformation of a circle into a symmetric airfoil

6.5.3.4 Circle-Cambered Airfoil Transformation

To generate airfoils that produce circulation and, therefore, lift, the profile must be cambered. In this case, the circle with the radius b is displaced horizontally as well as vertically relative to the origin of the circle with the radius a. To generate a systematic set of profiles, we need to know how the circle b is to be displaced relative to the origin of circle a. Only three parameters define the shape of the cambered profiles. These are: (a) Eccentricity e, angle α, and the intersection angle β. With these three parameters, the displacements in x- and y-directions as well as the radius of the circle b to be mapped onto the ζ-plane are calculated using the following relations from Fig. 6.23. Figure 6.24 shows a family of profiles generated by varying the above parameters.

$$\varepsilon = e/a$$
$$\overline{OC} = a\sin\gamma = e\sin(\beta/2)$$
$$\gamma = \arcsin\left(\frac{e}{a}\sin\beta/2\right) = \arcsin(\varepsilon\sin\beta/2)$$
$$\overline{AB} = b,\ \overline{OB} = a,\ b = a\cos\gamma + a\varepsilon\cos(\beta/2)$$
$$b/a = \cos\gamma + \varepsilon\cos(\beta/2)$$
$$\Delta x = \overline{OD} = e\cos(\alpha + \beta/2)$$
$$\Delta y = \overline{DA} = e\sin(\alpha + \beta/2)$$

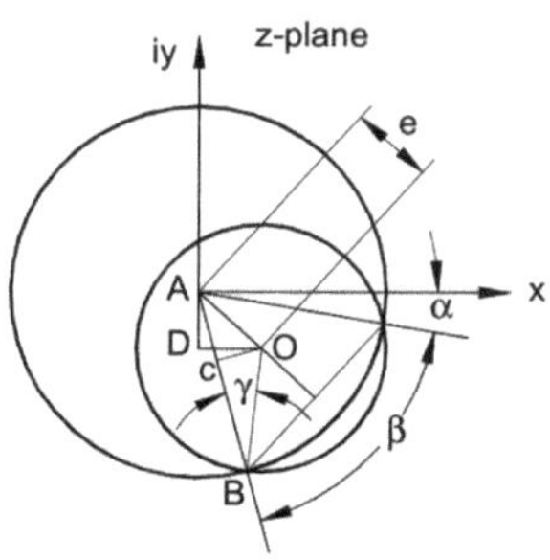

Fig. 6.23 Construction of cambered airfoils

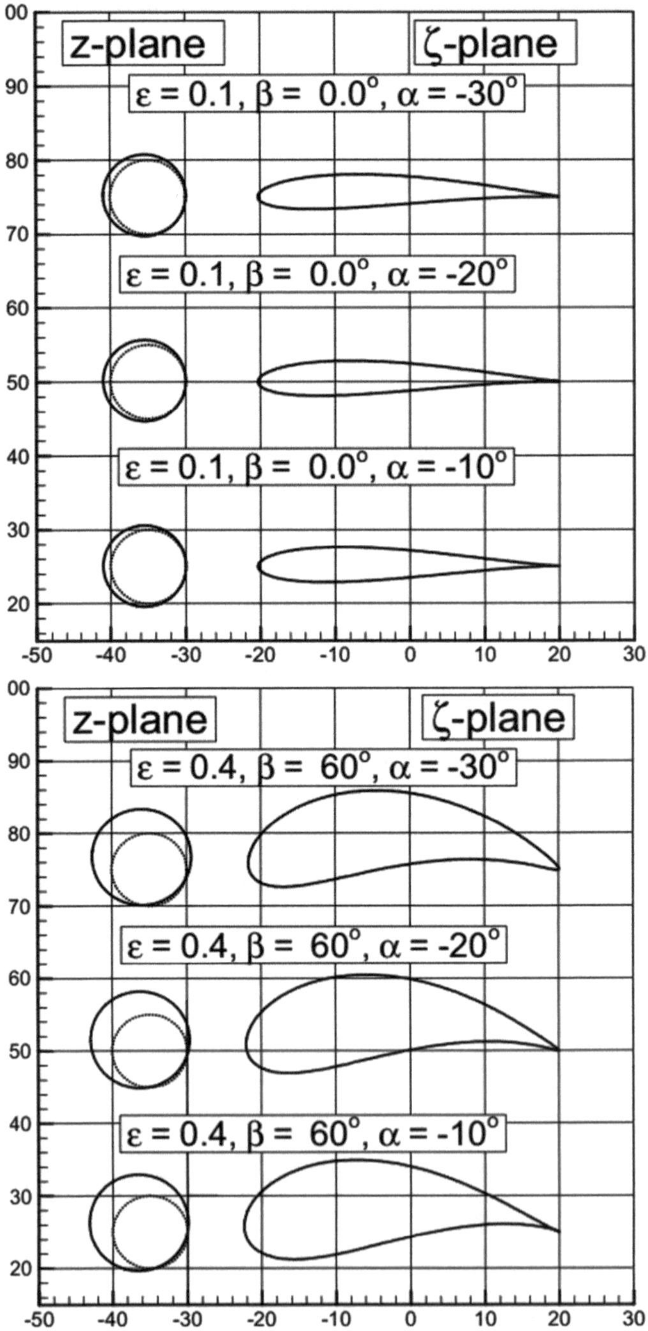

Fig. 6.24 Cambered airfoils constructed by conformal transformation

Starting with a small eccentricity of $\varepsilon = 10\%$, we set $\beta = 0°$ and vary the angle α from $-10°$ to $30°$. This configuration indicates that the two circles have tangents at the angle α. At this small eccentricity, slender profiles are generated that resemble low subsonic compressor blade profiles. Increasing the magnitude of α results in an increase of the profile cambers. If the angle β is different from zero, then the two circles intersect each other, as shown in Fig. 6.23. This is also shown in Fig. 6.24 with $\varepsilon = 0.4$, $\beta = 60°$ and α varied from $-10°$ to $-30°$. The resulting profiles resemble the turbine profiles.

6.5.3.5 Circulation, Lift, Kutta Condition

The conformal transformation we discussed previously allows, among others, the generation of asymmetric airfoils with prescribed cambers. These airfoils resemble profiles that are utilized as aircrafts wings, compressors and turbine blade profiles. The significance of the cambered profiles is to generate the necessary force to lift the aircraft, to generate higher total pressure (compressors), and to produce power (turbine). Generation of lift, however, requires the existence of circulation as we briefly discussed in Sect. 6.4. In the context of the potential flow analysis, certain conditions must be fulfilled to bring about a circulation which is a prerequisite for lift generation. Figure 6.25 exhibits the potential flow around one of those cambered airfoils we designed in the previous section.

The corresponding configuration in the z-plane is the flow around a circle with the circulation Γ and an angle of attack α, Fig. 6.25a. The complex potential of this configuration is almost the same as in Eq. (6.62) with the exception being that the axis of the dipole flow is turned by the angle α. Performing a simple coordinate transformation by substituting in the dipole part of Eq. (6.62) $z = r^{i\theta}$ by $z = r^{i(\theta-\alpha)}$ results in:

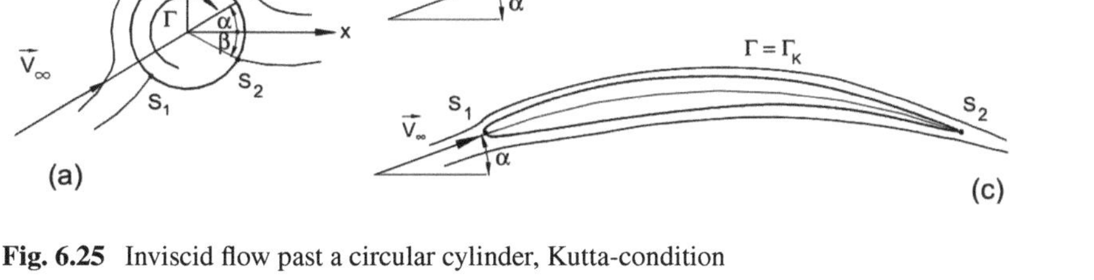

Fig. 6.25 Inviscid flow past a circular cylinder, Kutta-condition

$$F(z) = V_\infty \left(ze^{-i\alpha} + \frac{R^2}{z} e^{i\alpha} \right) - i \frac{\Gamma}{2\pi} \ln \left(\frac{ze^{-i\alpha}}{R} \right). \tag{6.128}$$

Assuming a circulation in the clockwise direction, Fig. 6.25a, two stagnation points S_1 and S_2 are present in the z-plane. In the ζ-plane, the transformation of the front stagnation point S_1 may be located on the pressure surface (concave side) of the blade, while the rear S_2 may be located on the suction side (convex side), Fig. 6.25b. Considering the flow situation at the sharp trailing edge, the fluid particles move from the pressure surface (concave side) of the blade to the suction surface (convex side) with an infinitely large velocity. Increasing the circulation causes both stagnation points to move. For a particular $\Gamma = \Gamma_K$, the Kutta-circulation, the rear stagnation point S_2 coincides with the trailing edge. At this point the velocity is zero. Known as the *Kutta condition*, it specifies that for an airfoil under inviscid flow conditions, to generate enough circulation, the rear stagnation point must coincide with the trailing edge. To satisfy this condition we resort to the complex potential Eq. (6.10) with $F(z) = \Phi + i\Psi$ with the derivative

$$\frac{df(z)}{d\zeta} = \frac{df(z)}{dz} \frac{dz}{d\zeta} = \frac{df(z)/dz}{d\zeta/dz} = u - iv. \tag{6.129}$$

Using the Joukowsky transformation function, we find for

$$\frac{df(z)}{d\zeta} = \frac{df(z)}{dz} \left(\frac{z^2}{z^2 - a^2} \right). \tag{6.130}$$

For $z \to \infty$, the expression in the parentheses approaches unity resulting in

$$\frac{df(z)}{d\zeta} = \frac{df(z)}{dz}. \tag{6.131}$$

As a consequence, we will have

$$u(z) = u(\zeta), v(z) = v(\zeta). \tag{6.132}$$

For the rear stagnation point to satisfy the Kutta-condition, both components must identically disappear leading to

$$u(z) = u(\zeta) = v(z) = v(\zeta) = 0. \tag{6.133}$$

Among an infinite number of circulation values that generate an infinite number of stagnation points distributed all over the profile surfaces, there is only one circulation that places the rear stagnation point at the trailing edge. To find this particular circulation we differentiate Eq. (6.128) with respect to z

$$\frac{dF(z)}{dz} = V_\infty\left(e^{-i\alpha} - \frac{R^2}{z^2}e^{i\alpha}\right) - i\frac{\Gamma}{2\pi}\frac{1}{z}. \tag{6.134}$$

Substituting $z == Re^{i\theta}$, we find

$$\frac{dF(z)}{dz} = V_\infty(e^{-i\alpha} - e^{i\alpha}e^{-2i\theta}) - i\frac{\Gamma}{2\pi R}e^{-i\theta} \tag{6.135}$$

or

$$\frac{dF(z)}{dz} = V_\infty e^{-i\theta}(e^{i(\theta-\alpha)} - e^{-i(\theta-\alpha)}) - i\frac{\Gamma}{2\pi R}e^{-i\theta}. \tag{6.136}$$

Now we replace the exponential expressions by the trigonometric one to get

$$\frac{dF(z)}{dz} = i\left(2V_\infty \sin(\theta-\alpha) - \frac{\Gamma}{2\pi R}\right)e^{-i\theta} \tag{6.137}$$

and with $i = e^{i\pi/2}$ we obtain

$$\frac{dF(z)}{dz} = \left(2V_\infty \sin(\theta-\alpha) - \frac{\Gamma}{2\pi R}\right)e^{i(\pi/2-\theta)}. \tag{6.138}$$

From Eq. (6.131), we conclude that the velocity in both z- and ζ-plane must vanish. As a consequence, the magnitude of the velocity vector included in parentheses of Eq. (6.138) must disappear

$$2V_\infty \sin(\theta-\alpha) + \frac{\Gamma}{2\pi R} = 0. \tag{6.139}$$

Now we set in Eq. (6.139) $\theta = \theta_{S_2} = -\beta$ and find

$$\frac{\Gamma}{2\pi R} = -2V_\infty \sin(\alpha+\beta). \tag{6.140}$$

Which gives the circulation that satisfies the Kutta-condition

$$\Gamma = \Gamma_{\text{Kutta}} = -4\pi R V_\infty \sin(\alpha+\beta) \tag{6.141}$$

meaning that the image of the rear stagnation point S_2 in ζ-plane lies at the trailing edge. The value of Γ_{Kutta} depends on the parameters R, β, the angle of attack α, and on the undisturbed velocity V_∞. In calculating the Kutta-circulation, we used the value of the circulation around the circular cylinder in z-plane. The circulation around the profile in ζ-plane is calculated from:

$$\Gamma = \oint_{C_\zeta} \bar{w}_\zeta(\zeta)d\zeta = \oint_{C_\zeta} \bar{w}_z(z)\frac{dz}{d\zeta}d\zeta = \oint_{C_z} \bar{w}_z(z)dz. \tag{6.142}$$

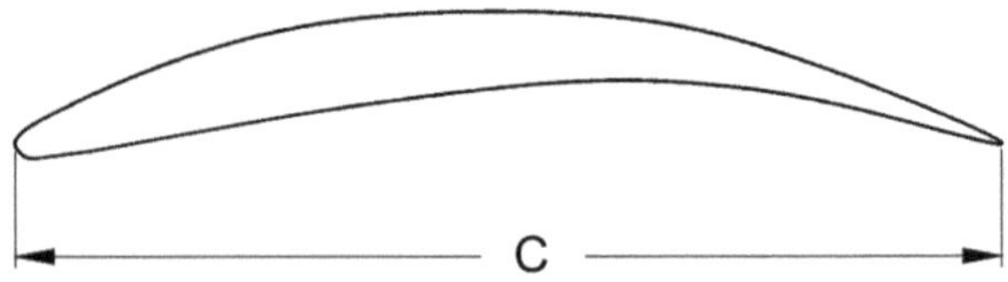

Fig. 6.26 Airfoil with chord length C

As seen from Eq. (6.142), the circulation in the ζ-plane is exactly the same as in the z-plane.

The force vector per unit depth on the airfoil is calculated from the Kutta-Joukowski theorem Eq. (6.105), where we note that the conjugate velocity $\bar{V}_\infty$ is now to be replaced by $\bar{V}_\infty e^{-i\alpha}$. Inserting Eq. (6.141) into Eq. (6.105), we obtain

$$\boldsymbol{R} = -i4\pi R\varrho V_\infty^2 e^{-i\alpha} \sin(\alpha + \beta) = 4\pi R\varrho V_\infty^2 e^{-i(\alpha+\pi/2)} \sin(\alpha + \beta) \tag{6.143}$$

where we replaced $-i = e^{-i\pi/2}$. The magnitude of the force is therefore

$$|R| = 4\pi R\varrho V_\infty^2 \sin(\alpha + \beta). \tag{6.144}$$

We introduce the dimensionless *lift coefficient* by dividing Eq. (6.144) by the *convective force* per unit dept:

$$c_L = \frac{|F|}{(\varrho/2)V_\infty^2 c} = 8\pi \frac{R}{C} \sin(\alpha + \beta) \tag{6.145}$$

where c is the chord length of the airfoil (Fig. 6.26) which can be calculated from the mapping function.

For $\beta = 0$ and $R = a$ the circle in the z-plane again is transformed onto a flat plate of length $c = 4a$. Its lift coefficient is

$$c_l = 2\pi \sin \alpha. \tag{6.146}$$

Figure 6.27 qualitatively reflects the differences between the lift coefficient as a function of the angle of attack α obtained from inviscid flow theory and the one from experiments without specifying any particular airfoil geometry.

While the inviscid flow theory does not account for boundary layer development and separation (solid line, no symbols), the experimental results show the pre- and post- stall differences. Differences in pre-stall C_L, is due to the boundary layer momentum deficiency caused by the wall shear stress as a result of the viscosity that has caused drag forces, Fig. 6.27a. At the point of maximum C_L, the boundary layer is still attached but is close to separation ($\tau = \mu\partial u/\partial y \approx 0$). The separation point depends on profile geometry, angle of attack α (or more specific: angle of incidence i) and the flow condition (Reynolds, and Mach number). Once the incidence angle exceeds the separation limit, partial or full stall will follow, Fig. 6.27b. The inviscid

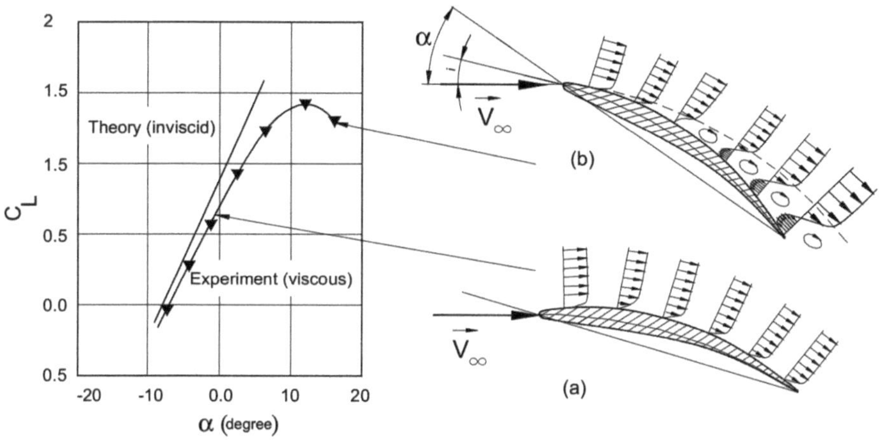

Fig. 6.27 Theoretical and experimental lift coefficient

flow theory does not account for any of the effects mentioned above. The Kutta-condition which had to be satisfied for an inviscid flow to generate circulation and, therefore, lift, has no relevance in a real viscous environment. The boundary layer development on suction and pressure surfaces that mix at the trailing edge plane puts the rear stagnation point where it belongs regardless of what the Kutta-condition dictates. This statement is also valid for the case that some minor separation may occur. In this case, the trailing edge is somehow submerged into a wake region and the profile generates a higher drag and a lower lift based on the severity of the separation.

6.6 Vortex Theorems

The previous section discussed the role of circulation and its significance for lift generation. The following sections deal with different aspects of circulation that are integral parts of inviscid flow analysis. We briefly present the two related theorems by *Thomson* and *Helmholz*.

6.6.1 *Thomson Theorem*

The circulation defined in Eq. (6.65) as the line integral of the velocity $\boldsymbol{V}$ along the closed curve C shown in Fig. 6.28 can be converted into a surface integral by means of the Stoke's theorem. The proof of this theorem is an integral part of engineering mathematics textbooks. It is also found in great detail in Vavra [34].

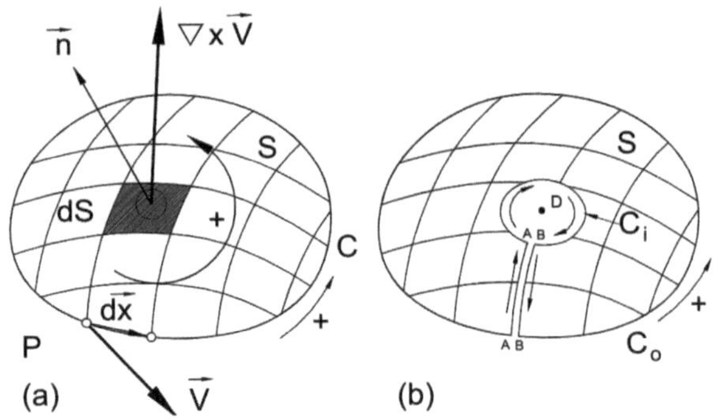

Fig. 6.28 Relationship between the circulation closed integral and the rotation surface integral using Stoke's theorem

The Stoke's theorem gives the relationship between the circulation which is a closed integral and the surface integral of the rotation vector $\nabla \times bfV$. It is summarized in the following equation:

$$\Gamma = \oint_{(c)} \boldsymbol{V} \cdot d\boldsymbol{x} = \int_{(S)} (\nabla \times \boldsymbol{V}) \cdot \boldsymbol{n} dS \tag{6.147}$$

with $\boldsymbol{n}$ as the unit vector normal to the differential surface element dS. In an unsteady flow, the circulation along a fixed curve is a function of time. The stoke's theorem is valid for a curve that bounds a *simply connected surface*, as exhibited in Fig. 6.28a. If the surface is *doubly-connected* as shown in Fig. 6.28b, infinitely thin cuts such as *AA* and *BB* can be placed such that a simply connected surface is established and the Stoke's equation can be applied.
Since the line integrals along *AA* and *BB* cancel each other out, Eq. (6.147) can be written as:

$$\oint_{(c_o)} \boldsymbol{V} \cdot d\boldsymbol{x} + \oint_{(c_i)} \boldsymbol{V} \cdot d\boldsymbol{x} = \int_{(S)} (\nabla \times \boldsymbol{V}) \cdot \boldsymbol{n} dS \tag{6.148}$$

or

$$\Gamma_0 = \Gamma_i + \int_{(S)} (\nabla \times \boldsymbol{V}) \cdot \boldsymbol{n} dS. \tag{6.149}$$

Equation (6.149) is valid only if the curve C_i encloses all discontinuities and singularities that lie inside C_0. Equation (6.147) and its modified version (6.148) describe the circulation at the time t. As the fluid particles move at time $t + dt$, another curve is formed and correspondingly the circulation undergoes a temporal change.

The substantial derivative of the circulation determines the rate of change of Γ

$$\frac{D\Gamma}{Dt} = \frac{\partial \Gamma}{\partial t} + \boldsymbol{V} \cdot \nabla \Gamma = \frac{\partial \Gamma}{\partial t} + \frac{d\Gamma}{dt} \tag{6.150}$$

with $\partial\Gamma/\partial t$ as the local change of the circulation and $d\Gamma/dt = \boldsymbol{V}\cdot\nabla\Gamma$ as the convective change. Introducing the circulation Γ into the local term in Eq. (6.150), we find

$$\frac{D\Gamma}{Dt} = \frac{\partial}{\partial t}\oint_{(c)} \boldsymbol{V}\cdot d\boldsymbol{x} + \frac{d\Gamma}{dt}. \tag{6.151}$$

Since the infinitesimal curve element dx will not change by unsteady velocity change, we may write

$$\frac{D\Gamma}{Dt} = \oint_{(c)} \frac{\partial}{\partial t}(\boldsymbol{V}\cdot d\boldsymbol{x}) + \frac{d\Gamma}{dt} = \oint_{(c)} \frac{\partial \boldsymbol{V}}{\partial t}\cdot d\boldsymbol{x} + \oint_{(c)} \boldsymbol{V}\cdot d\boldsymbol{V} + \frac{d\Gamma}{dt}$$

$$\frac{D\Gamma}{Dt} = \oint_{(c)} \frac{\partial \boldsymbol{V}}{\partial t}\cdot d\boldsymbol{x} + \oint_{(c)} d\left(\frac{V^2}{2}\right) + \frac{d\Gamma}{dt}. \tag{6.152}$$

Since the integrand $V^2/2$ in Eq. (6.152) is an exact differential, hence not a path function, its integral along a closed curve vanishes. The last term in Eq. (6.152) is determined by applying Stoke's theorem to the surface S between curves $C1$ and $C2$ sketched in Fig. 6.29.

The cuts AA' and BB' convert the surface S into a simply connected surface bound by a closed curve. The counter clockwise direction is indicated by (+) whereas the clockwise direction is (−), thus, the integration along $C2$ is $-(\Gamma + d\Gamma)$. The line integrals along AA' and BB' cancel each other out. We now obtain the convective change $d\Gamma$ by implementing the Stokes theorem and integrating over the surface S that leads to

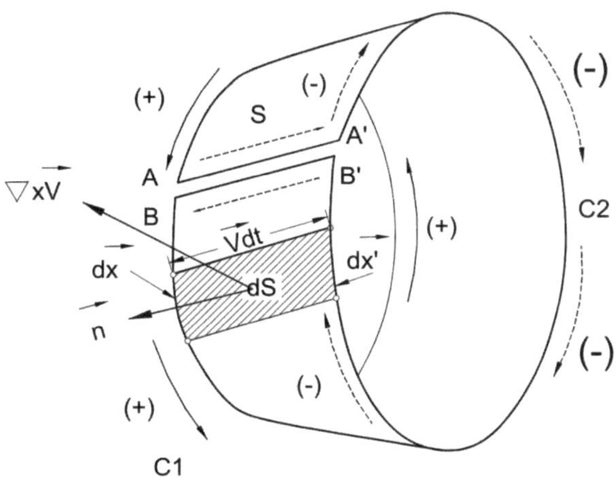

Fig. 6.29 Geometric interpretation of the relationship between the differential circulation and the surface integral of rotation

$$\Gamma - (\Gamma + d\Gamma) = \int_{(S)} (\nabla \times \boldsymbol{V}) \cdot \boldsymbol{n} dS. \tag{6.153}$$

From Fig. 6.29 $\boldsymbol{n}dS$ is expressed in terms of

$$\boldsymbol{n}dS = d\boldsymbol{x} \times \boldsymbol{V}dt. \tag{6.154}$$

With Eq. (6.154), Eq. (6.153) is modified as

$$\Gamma - (\Gamma + d\Gamma) = \int_{(S)} (\nabla \times \boldsymbol{V}) \cdot (d\boldsymbol{x} \times \boldsymbol{V})dt. \tag{6.155}$$

Re-arranging the right-hand side of Eq. (6.155) using the triple scalar product, we arrive at

$$\frac{d\Gamma}{dt} = -\int_{(S)} \boldsymbol{V} \times (\nabla \times \boldsymbol{V}) \cdot d\boldsymbol{x}. \tag{6.156}$$

The negative sign in Eq. (6.156) stems from re-arranging the sequence of vector multiplication. Introducing Eq. (6.156) into Eq. (6.152), we find

$$\frac{D\Gamma}{Dt} = \oint_{(c)} \frac{\partial \boldsymbol{V}}{\partial t} \cdot d\boldsymbol{x} - \int_{(S)} \boldsymbol{V} \times (\nabla \times \boldsymbol{V}) \cdot d\boldsymbol{x} \tag{6.157}$$

or

$$\frac{D\Gamma}{Dt} = \oint_{(c)} \left(\frac{\partial \boldsymbol{V}}{\partial t} - \boldsymbol{V} \times (\nabla \times \boldsymbol{V}) \right) \cdot d\boldsymbol{x}. \tag{6.158}$$

As seen in Eq. (6.158), the surface integral in Eq. (6.157) is replaced by a curve integral. This is admissible because the changes of $\boldsymbol{V}$ and $\nabla x \boldsymbol{V}$ between the C_1 and C_2 are so small that the surface integral reduces to an integration along the curve C. The expression within the parentheses in Eq. (6.158) can be replaced by Eq. (4.55) resulting in:

$$\frac{D\Gamma}{Dt} = \oint_{(C)} (-\nabla H + T\nabla s) \cdot d\boldsymbol{x}. \tag{6.159}$$

With H as the total enthalpy $H = (h + V^2/2 + gz)$. Introducing $\nabla H \cdot d\boldsymbol{x} = dh$ and $\nabla s \cdot d\boldsymbol{x} = ds$ into Eq. (6.159), we arrive at:

$$\frac{D\Gamma}{Dt} = -\oint_{(C)} dH + \oint_{(C)} Tds. \tag{6.160}$$

The integrand of the first integral represents an exact differential, whose integral around a closed curve C vanishes reducing Eq. (6.160) to:

$$\frac{D\Gamma}{Dt} = \oint_{(C)} T ds. \tag{6.161}$$

If we take into consideration the frictional force per unit mass given in Eq. (4.43) caused by the viscosity, then Eq. (6.159) is enhanced as

$$\frac{D\Gamma}{Dt} = \oint_{(C)} (-\nabla H + T\nabla s + \boldsymbol{f}) \cdot d\boldsymbol{x} \tag{6.162}$$

with $\boldsymbol{f}$ as the friction force that is defined for a flow with constant viscosity as:

$$\boldsymbol{f} = \nu \left[\Delta \boldsymbol{V} + \frac{1}{3} \nabla (\nabla \cdot \boldsymbol{V}) \right]. \tag{6.163}$$

Assuming incompressible flow and the following vector identity

$$\Delta \boldsymbol{V} = \nabla^2 \boldsymbol{V} \equiv \nabla(\nabla \cdot \boldsymbol{V}) - \nabla \times (\nabla \times \boldsymbol{V}). \tag{6.164}$$

Equation (6.163) reduces to

$$\boldsymbol{f} = -\nu \nabla \times (\nabla \times \boldsymbol{V}). \tag{6.165}$$

Thus, for an inviscid flow, the substantial change of the circulation reduces to

$$\frac{D\Gamma}{Dt} = 0, \text{ or } \Gamma = \text{const.} \tag{6.166}$$

Equation (6.166) states that for an inviscid irrotational, flow the substantial change of the circulation is zero, meaning that the circulation remains constant. This is the Kelvin's theorem (see Thompson [35]). This theorem was first discovered and proved by Helmholtz [36]. It implies that the line integral along a closed curve in a homogeneous non-viscous fluid is constant for all times if the flow is under the influence of an irrotational field force. The use of material derivatives emphasizes the circulation around a closed material curve.

In deriving the Kelvin's theorem, we used Eq. (4.55), which incorporates the total enthalpy. Equally, we can use Eq. (4.53) which includes the pressure-density term $\nabla p/\rho$. If we assume that the density is a function of pressure only $\rho = \rho(p)$ (the fluid is called *barotropic*, a perfect gas is such a fluid with $p\rho^{-\kappa}$ =const.), then we may set $\nabla p/\rho \equiv \nabla P$ with $P = p/\rho$, thus, $\nabla P \cdot d\boldsymbol{x} = dP$, whose curve integral vanishes.

6.6.2 Generation of Circulation

In the preceding sections we derived the relationship for lift as a function of circulation (Eqs. 6.105 and 6.143) assuming that a circulation is superposed on the translational flow past the body, without explaining how this circulation has been brought about. The question that needs to be answered is: how can the existence of such a circulation flow be explained? To answer this question we revert to the flow visualization experiments by Prandtl [31] taken from an airfoil subjected to different flow modes. Figure 6.30 reflect the physical contents of images presented in [31].

We assume that at first the fluid is at rest, Fig. 6.30a, so that the line integral of the velocity along a curve completely surrounding the airfoil is zero, because all velocities are zero. This would correspond to a potential flow situation without circulation immediately after starting Fig. 6.30b. According to Thomson's theorem, Eq. (6.166), the circulation in a frictionless fluid must remain constant (in this case equal to zero) at all times including the moment when the fluid is suddenly put into a uniform translatory motion with respect to the airfoil. This is apparently in contradiction to the experimental fact that there is a circulation around the airfoil. Considering the infinitely large velocity around the sharp trailing edge in Fig. 6.30b of the airfoil, one could suggest that the flow, at the first moment after starting, might be a potential flow without circulation. The presence of the viscosity in the boundary layer, however, causes this large velocity to develop into a surface of discontinuity, Fig. 6.30c. At the sharp trailing edge, the viscosity of the real fluid causes an equalization of the velocity jump,

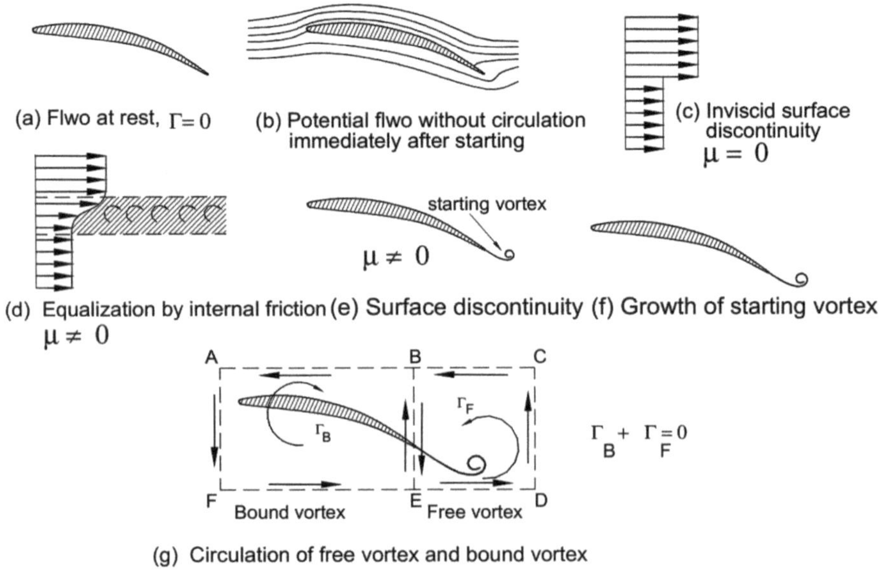

Fig. 6.30 Circulation, free vortex Γ_F, bound vortex Γ_B

leading to a layer of finite thickness which is occupied by vortices, Fig. 6.30d. This vortical layer, then, is rolled up to a vortex, the so-called *starting vortex*, Fig. 6.30e, f. This vortex, according to the theorems of Helmholtz (treated in the following section), is always associated with the same particles of fluid, is washed away with the fluid, and is convected downstream as a free vortex. Since this *free vortex* has a non-zero magnitude, its existence clearly contradicts the Thomson's theorem. Assuming the validity of the Thomson's theorem, the process of starting must have generated another vortex with the same magnitude but in the opposite direction so that the sum of their strengths vanishes. In fact, the existence of the free vortex is always associated with the existence of another vortex called *bound vortex*, Fig. 6.30g. Calculating the circulation around the closed curve $C \equiv ABCDFA$, $C_B \equiv ABEFA$, and $C_F \equiv BCDFB$, we find $\Gamma = \oint_{(C)} \boldsymbol{V} \cdot d\boldsymbol{C} = \oint_{(C_B)} \boldsymbol{V} \cdot d\boldsymbol{C} + \oint_{(C_F)} \boldsymbol{V} \cdot d\boldsymbol{C} = \Gamma_B + \Gamma_F = 0$ from which we conclude that $\Gamma_B = -\Gamma_F$. This result is confirmed experimentally verifying the validity of the Thomson's vortex theorem. The most important feature essential for upholding the Thompson's theorem is the viscosity effect, without which no vortices can be produced.

In generating the vortex images presented in [6] that we summarized in Fig. 6.30, Prandtl first kept the airfoil in a fixed position that was exposed to a moving fluid. In a second set of experiments, he moved the airfoil relative to undisturbed fluid. The same phenomenon was observed in both cases.

6.6.3 Helmholtz Theorems

Research on vortex flow has been initiated by the fundamental paper [36] of H.L.F. Helmholtz (1821–1894), a physicist and a professor of physiology and anatomy at the University of Königsberg, Bonn, Heidelberg and Berlin. In his paper, Helmholtz established his three theorems of vortex motion. Assuming incompressible frictionless fluids subjected to flow forces defined by a potential, Helmholtz [36] published a paper about the vortex motions in which he stated his vortex theorems. These theorems are translated from German and appear in an excellent textbook by Prandtl and Tietiens [31]. They reflect the quintessence of the vortex flow motion treated in Helmholtz original work. Before starting with the discussion of Helmholtz theorems, it is helpful to become familiar with the nomenclature sketched in Fig. 6.31.

A *vortex line*, Fig. 6.31a is a line tangent to the rotation vector $\nabla \times \boldsymbol{V}$. The vortex lines may form a *vortex tube*, Fig. 6.31b. Reducing the cross sections of a vortex tube to an infinitely small size, we obtain a *vortex filament*. Thus, a vortex filament is essentially a vortex tube with an infinitely small cross section but a finite value of circulation. This particular configuration allows one to apply the Stoke's theorem Eq. (6.147) without integrating the rotation vector

$$\Gamma = (\nabla \times \boldsymbol{V}) \cdot \boldsymbol{n} dS. \tag{6.167}$$

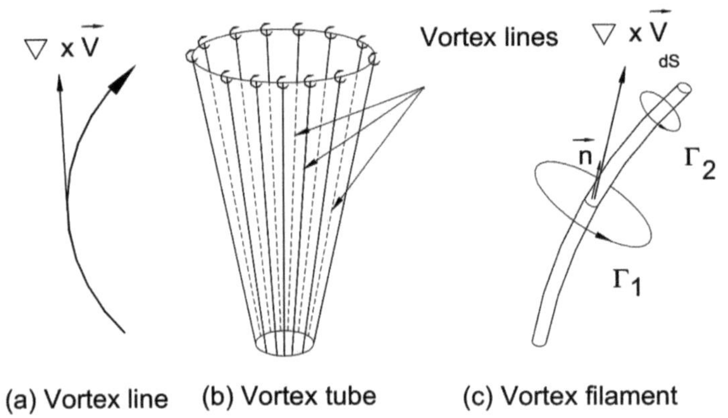

Fig. 6.31 Illustration of different vortex types

Since the unit vector $\boldsymbol{n}$ is parallel to the rotation vector $\nabla \times \boldsymbol{V}$, we may re-arrange Eq. (6.167)

$$\nabla \times \boldsymbol{V} = \boldsymbol{n}\frac{\Gamma}{dS}. \tag{6.168}$$

Since ds is, per definition, infinitely small and Γ has a finite value, the rotation vector $\nabla \times \boldsymbol{V}$ must be infinitely large, indicating that the vortex filaments represent a singularity. This and many other types of singularities are used for dealing with more complicated issues particularly in aerodynamics.

Ignoring the friction forces and assuming that there exists a potential acting on fluid particles, Helmholtz formulated in his original paper [36], three theorems: The first theorem states that no fluid particle can have a rotation if it did not originally rotate. This theorem reflects the physical content of the differential part of Eq. (6.166). The second theorem states that the fluid particles, which at any time are part of a vortex line, always belong to that same vortex line. This theorem is the consequence of the integral part of Eq. (6.166), stating that the circulation remains constant. The third theorem states that the product of the cross section area and angular velocity of an infinitely thin vortex filament remains constant over the whole length of the filament and keeps the same value even when the vortex moves. It further states that the vortex filament must, therefore, be either closed curves or end on the boundaries of the fluid.

In what follows, the mathematical structure of the Helmholtz theorems is presented. It should be pointed out that, all three theorems deal with kinematic conditions. The Helmholtz theorems are also treated by Prandtl and Titiens [31], Spurk [30], Vavra [34] and Kotschin et al. [37].

Figure 6.32 may help better understand the physical content of the following derivation. At time t, a differential element of a vortex filament with a cross section ds and the height dx contains a certain number of fluid particles with the density ρ. The mass of this element is then $dm = \rho dx dS$. The vorticity vector $\omega = \nabla \times \boldsymbol{V}$

Fig. 6.32 Explaining the Helmholtz vortex theorems

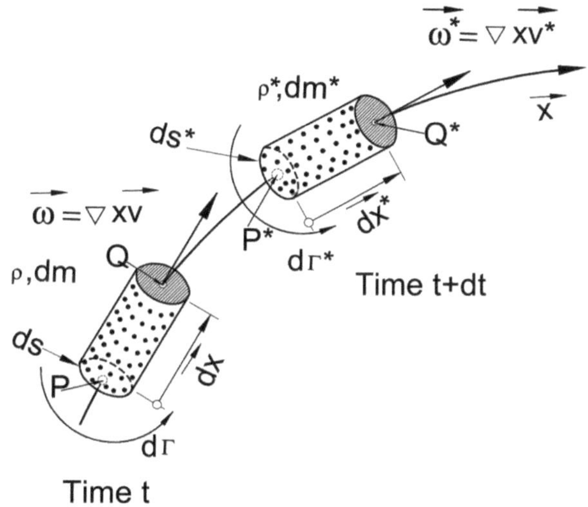

(Sect. 4.2.3) is parallel to the vector $\boldsymbol{dx}$ and perpendicular to the cross section ds. Using the Stoke's theorem, the circulation of the vortex element is:

$$d\Gamma = (\nabla \times \boldsymbol{V}) \cdot \boldsymbol{n} dS = \boldsymbol{\omega} \cdot \boldsymbol{n} dS = \omega dS. \tag{6.169}$$

The element moves through the space, where its new kinematic conditions at the time. The second theorem requires that the fluid particles contained in the differential volume element $dv = dxdS$ at the time t must be the same that are contained the differential volume element $dv^* = dx^*dS^*$ at the time $t + dt$ which means that

$$dm = dm^* = \rho^* dx^* dS^* = \rho dx dS. \tag{6.170}$$

From the third theorem, we infer that

$$d\Gamma = d\Gamma^* = dS^* \omega^* = dS\omega. \tag{6.171}$$

As a consequence of the third theorem and integrating Eq. (6.171), it follows that for a vortex tube with varying cross sections, the product of $\Delta S^* \omega^* = \Delta S \omega$ remains constant, leading to

$$\frac{\Delta S_1}{\Delta S} = \frac{\omega}{\omega_1}, \omega = \frac{\Delta S_1}{\Delta S} \omega_1. \tag{6.172}$$

From Eq. (6.172) we conclude that, decreasing the cross section of the vortex tube leads to an increase of vorticity vectors (angular velocity). Dividing Eq. (6.170) by Eq. (6.171), we obtain

$$\frac{\rho dx}{\omega} = \frac{\rho^* dx^*}{\omega^*}. \tag{6.173}$$

The second theorem also implies that the directions of $d\boldsymbol{x}$ and $d\boldsymbol{x}^*$ must be the same as the directions of the vectors $\boldsymbol{\omega}$ and ω^* respectively, or

$$\frac{d\boldsymbol{x}}{dx} = \frac{\boldsymbol{\omega}}{\omega}, \frac{d\boldsymbol{x}^*}{dx^*} = \frac{\boldsymbol{\omega}^*}{\omega^*}. \tag{6.174}$$

With Eqs. (6.173) and (6.174) we find

$$d\boldsymbol{x}^* = \frac{\rho}{\rho^*}, \frac{\boldsymbol{\omega}^*}{\omega} dx. \tag{6.175}$$

Considering the kinematics in Fig. 6.32, when the element moves from point P to P^* or Q to Q^*, respectively, the vector $d\boldsymbol{x}$ experiences the following change

$$d\boldsymbol{x}^* - d\boldsymbol{x} = d\boldsymbol{V} dt = d\boldsymbol{x} \cdot \nabla \boldsymbol{V} dt. \tag{6.176}$$

Considering Eq. (6.174), we find

$$d\boldsymbol{x}^* = d\boldsymbol{x} + d\boldsymbol{V} dt = d\boldsymbol{x} \cdot \nabla \boldsymbol{V} = dx\frac{\boldsymbol{\omega}}{\omega} + dx\frac{\boldsymbol{\omega}}{\omega} \cdot \nabla \boldsymbol{V} dt. \tag{6.177}$$

The vorticity vector, as well as the density, experiences the following *material* changes

$$\boldsymbol{\omega}^* - \boldsymbol{\omega} = \frac{D\boldsymbol{\omega}}{Dt} dt$$

$$\rho^* - \rho = \frac{D\rho}{Dt} dt. \tag{6.178}$$

Introducing Eqs. (6.178) and (6.177) into Eq. (6.175) and neglecting higher order terms, we obtain

$$\frac{D\boldsymbol{\omega}}{Dt} - \boldsymbol{\omega} \cdot \nabla \boldsymbol{V} - \frac{1}{\rho\boldsymbol{\omega}} \frac{D\rho}{Dt} = 0. \tag{6.179}$$

The combination of the substantial change of density given by Eq. (4.10) and the equation of continuity (4.4) gives

$$\frac{D\rho}{Dt} = -\rho \nabla \cdot \boldsymbol{V} \tag{6.180}$$

Equation (6.180) with $\boldsymbol{\omega} = \nabla \times \boldsymbol{V}$ inserted into Eq. (6.17) results in

$$\frac{D(\nabla \times \boldsymbol{V})}{Dt} - (\nabla \times \boldsymbol{V}) \cdot \nabla \boldsymbol{V} + (\nabla \times \boldsymbol{V}) \nabla \cdot \boldsymbol{V} = 0. \tag{6.181}$$

Equation (6.181) is called Helmholtz derivative of the vorticity vector $\boldsymbol{\omega} = \nabla \times \boldsymbol{V}$. It satisfies all three theorems of Helmholtz. It clearly indicates that if the flow is irrotational along the path of its particles, the material change of $\nabla \times \boldsymbol{V}$ is zero. Thus, if the flow was initially irrotational, it must remain irrotational in the entire flow field. Using the following vector identity: $\nabla \times (\boldsymbol{U} \times \boldsymbol{V}) = \boldsymbol{V} \cdot \nabla \boldsymbol{U} - \boldsymbol{U} \cdot \nabla \boldsymbol{V} + \boldsymbol{U} \nabla \cdot \boldsymbol{V} - \boldsymbol{V} \nabla \cdot \boldsymbol{U}$ and replacing the vector $\boldsymbol{U}$ with the vector $\nabla \times \boldsymbol{V}$, the second and third term in Eq. (6.181) becomes $-(\nabla \times \boldsymbol{V}) \cdot \nabla \boldsymbol{V} + (\nabla \times \boldsymbol{V}) \nabla \cdot \boldsymbol{V} = \nabla \times (\nabla \times \boldsymbol{V}) \times \boldsymbol{V} + \boldsymbol{V} \nabla \cdot (\nabla \times \boldsymbol{V}) - \boldsymbol{V} \cdot \nabla (\nabla \times \boldsymbol{V})$ which allows rewriting Eq. (6.181) as

$$\frac{D(\nabla \times \boldsymbol{V})}{Dt} + \nabla \times (\nabla \times \boldsymbol{V}) \times \boldsymbol{V} + \boldsymbol{V} \nabla \cdot (\nabla \times \boldsymbol{V}) - \boldsymbol{V} \cdot \nabla (\nabla \times \boldsymbol{V}) = 0. \quad (6.182)$$

Replacing the substantial differential by the sum of its local and convective parts, we arrive at

$$\frac{\partial (\nabla \times \boldsymbol{V})}{\partial t} - \nabla \times [\boldsymbol{V} \times (\nabla \times \boldsymbol{V})] + \boldsymbol{V} \nabla \cdot (\nabla \times \boldsymbol{V}) = 0. \quad (6.183)$$

Expanding the last term in Eq. (6.183) shows that it is zero. Furthermore, since the operator ∇ is a time independent spatial operator, it can be moved out of the differential, causing Eq. (6.183) to further reduce to

$$\nabla \times \left[\frac{\partial \boldsymbol{V}}{\partial t} - \boldsymbol{V} \times (\nabla \times \boldsymbol{V}) \right] = 0. \quad (6.184)$$

Equation (6.184) is the result of the three Helmholtz theorems, which are purely kinematic conditions, without applying the conservation law of motion. It is valid for inviscid fluids where the forces can be expressed in terms of gradient of a potential. The expression in the brackets is obtained if we rearrange the equation of motion (4.55) by considering an isentropic flow $T \nabla s = 0$ with constant total enthalpy $\nabla H = 0$:

$$\frac{\partial \boldsymbol{V}}{\partial t} - \boldsymbol{V} \times (\nabla \times V) = 0. \quad (6.185)$$

As seen, taking the rotation (curl) of Eq. (6.185) leads to Eq. (6.184). It should be pointed out that, applying rotation does not generate another independent conservation law. It has, rather, produced another version of the same physical principle, which in this case, confirms the three theorems of Helmholtz.

6.6.4 *Induced Velocity Field, Law of Bio-Savart*

We consider now an isolated vortex filament with the strength Γ imbedded in an inviscid irrotational flow environment, as shown in Fig. 6.33.

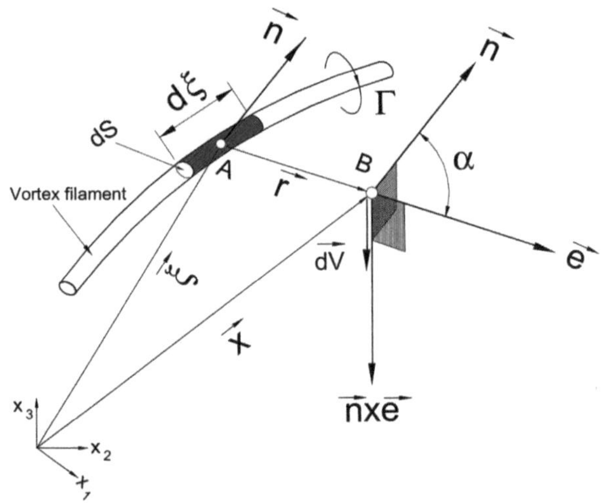

Fig. 6.33 Velocity field induced by an isolated vortex filament

In a distance $\boldsymbol{r}$ from the point A, a differential element $d\xi$ of the vortex filament at the point B, *induces* a differential velocity vector field $d\boldsymbol{V}$. The velocity vector is perpendicular to the plane spanned by the normal unit vector $\boldsymbol{n}$ and the unit vector $\boldsymbol{e}$ in the $\boldsymbol{r}$-direction. The unit vector $\boldsymbol{n}$ is perpendicular to the infinitesimal cross section dS, whereas the unit vector $\boldsymbol{e}$ points from the center point of the element A to the position B, where the velocity $d\boldsymbol{V}$ is being induced. The relationship describing the velocity field is analogous to the one discovered by Bio and Savart through electrodynamic experiments. It describes the magnetic field induced by a current through a conducting wire. In an aerodynamic context, the conducting wire corresponds to the vortex filament, its current corresponds to the vortex strength Γ and the induced magnetic field corresponds to the induced velocity field.

To present the derivation, first we provide the mathematical tool essential to arriving at the Bio-Savart law. Let us decompose an arbitrary vector point function $\boldsymbol{V}$ into an irrotational part that can be expressed in terms of the gradient of a potential and a rotational or *solenoidal* part

$$\boldsymbol{V} = \nabla\Phi + \nabla \times \boldsymbol{U} \tag{6.186}$$

with Φ as a scalar potential and $\nabla \times \boldsymbol{U}$ as the solenoidal part of Eq. (6.186). Taking the curl of Eq. (6.186) gives

$$\nabla \times \boldsymbol{V} = \nabla \times (\nabla\Phi) + \nabla \times (\nabla \times \boldsymbol{U}) = \nabla \times (\nabla \times \boldsymbol{U}). \tag{6.187}$$

The first term on the right-hand side of Eq. (6.187) is the curl of the gradient of the scalar field Φ that identically vanishes. The divergence of Eq. (6.186) delivers

$$\nabla \cdot \boldsymbol{V} = \Delta\Phi + \nabla \cdot (\nabla \times \boldsymbol{U}) = \nabla^2\Phi = \Delta\Phi \tag{6.188}$$

with $\nabla \cdot \nabla \Phi = \Delta \Phi$ and $\nabla \cdot (\nabla \times \boldsymbol{U}) \equiv 0$. Equation (6.188) is an inhomogeneous partial differential equation called *Poisson's equation*. What makes the Poisson's equation (6.188) a special case where $\Delta \Phi = 0$, is the Laplace equation we treated in the preceding sections. Using the vector identity for $\nabla \times (\nabla \times \boldsymbol{U}) = \nabla(\nabla \cdot \boldsymbol{U}) - \Delta \boldsymbol{U}$, Eq. (6.187) reads

$$\nabla \times \boldsymbol{V} = \nabla(\nabla \cdot \boldsymbol{U}) - \Delta \boldsymbol{U}. \tag{6.189}$$

It can be shown that, it is possible to set $\nabla \cdot \boldsymbol{U} = 0$ without loss of generality. This step reduces Eq. (6.189) to

$$\nabla \times \boldsymbol{V} = -\Delta \boldsymbol{U} = -\nabla^2 \boldsymbol{U}. \tag{6.190}$$

After this preparation, we turn our attention to Eq. (6.188) with the solution

$$\Phi = -\frac{1}{4\pi} \int_{(\infty)} \frac{1}{r} (\nabla \cdot \boldsymbol{V}) dv. \tag{6.191}$$

The integral boundary (∞) indicates that the integration has to be carried out over the entire space (volume integral). Similarly, the solution for differential equation (6.190) is

$$\boldsymbol{U} = \frac{1}{4\pi} \int_{(\infty)} \frac{1}{r} (\nabla \times \boldsymbol{V}) dv. \tag{6.192}$$

Introducing in Eq. (6.186) Φ from Eq. (6.191) and U from Eq. (6.192) gives

$$\boldsymbol{V} = -\frac{1}{4\pi} \nabla \left[\int_{(\infty)} \frac{1}{r} (\nabla \cdot \boldsymbol{V}) dv \right] + \frac{1}{4\pi} \nabla \times \left[\int_{(\infty)} \frac{1}{r} (\nabla \times \boldsymbol{V}) dv \right]. \tag{6.193}$$

Equation (6.193) indicates that the irrotational part of a decomposed velocity field is determined by the divergence of the vector field, whereas the solenoidal (rotational) part is obtained if the rotation (curl) of the vector field is known. For an incompressible flow, the condition $\nabla \cdot \boldsymbol{V} = 0$ must be satisfied, causing Eq. (6.193) to reduce to

$$\boldsymbol{V} = \frac{1}{4\pi} \nabla \times \left[\int_{(\infty)} \frac{1}{r} (\nabla \times \boldsymbol{V}) dv \right]. \tag{6.194}$$

The flow field with the velocity described by Eq. (6.194) is irrotational everywhere except in the space occupied by the isolated vortex filament with the known constant strength Γ. To evaluate the integral, first we set for the differential volume element $dv = dSd\xi$ with dS as the vortex cross section and $d\xi$ as a differential length element, shown in Fig. 6.32. Further, we replace the curl vector in Eq. (6.194) by $\nabla \times \boldsymbol{V} dS =$

$\boldsymbol{n}\Gamma$ from Eq. (6.168). Since the entire flow field is irrotational with the exception of the space occupied by the isolated vortex filament, the integration of Eq. (6.194) needs to be carried out over the length of the filament only. Hence, Eq. (6.194) is reduced to

$$\boldsymbol{V}_B = \frac{1}{4\pi}\nabla \times \left(\int_{(L)} \frac{1}{r}\boldsymbol{n}\Gamma d\xi\right). \tag{6.195}$$

Since the vortex strength of the filament Γ is constant, it can be moved out of the integral sign leading to

$$\boldsymbol{V}_B = \frac{\Gamma}{4\pi}\nabla \times \left(\int_{(L)} \frac{\boldsymbol{n}}{r} d\xi\right). \tag{6.196}$$

Of particular interest is the velocity induced by the differential element $d\xi$ of the filament at a given fixed point B

$$d\boldsymbol{V}_B = \frac{\Gamma}{4\pi}\nabla \times \left(\frac{\boldsymbol{n}}{r}\right) d\xi. \tag{6.197}$$

To evaluate the curl of the ratio $\boldsymbol{n}r$, we apply the spatial differential operator to the argument in parentheses, perform chain differentiation accounting for the direction of the vector $\boldsymbol{r}$ which is $\boldsymbol{e}$, and obtain

$$\nabla \times \left(\frac{\boldsymbol{n}}{r}\right) = -\frac{\boldsymbol{e} \times \boldsymbol{n}}{r^2} = \frac{\boldsymbol{n} \times \boldsymbol{e}}{r^2} \tag{6.198}$$

which we then insert into Eq. (6.198). As a result, we find

$$d\boldsymbol{V}_B = \frac{\Gamma}{4\pi r^2}\boldsymbol{n} \times \boldsymbol{e} d\xi. \tag{6.199}$$

Integrating Eq. (6.199) leads to

$$\boldsymbol{V}_B = \frac{\Gamma}{4\pi}\int_{(L)} \frac{\boldsymbol{n} \times \boldsymbol{e}}{r^2} d\xi. \tag{6.200}$$

Equation (6.199) is the Bio-Savart law for inviscid flow. As seen from the preceding derivation, only kinematic conditions were applied leading to arrival at the Bio-Savart law, which is a kinematic relation. It was originally discovered by calculating the induced electromagnetic field strength $\boldsymbol{dB}$ at point B by a differential element $d\xi$ of a wire with the current I that moves in direction of $\boldsymbol{d\xi}$. The version used in electrodynamics is

$$d\boldsymbol{B} = \frac{\mu I}{4\pi r^2}\boldsymbol{n} \times \boldsymbol{e} d\xi \tag{6.201}$$

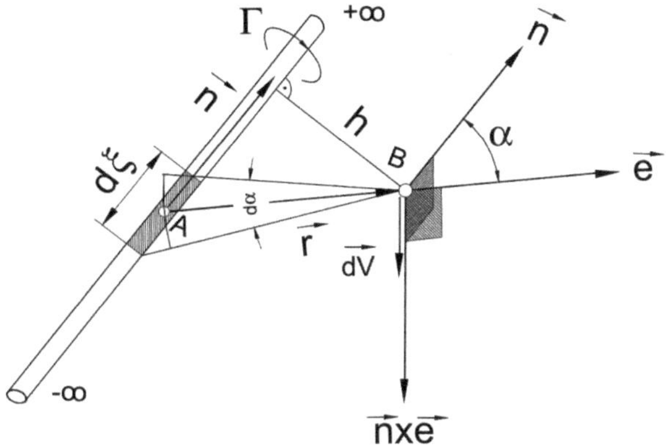

Fig. 6.34 Induced velocity by a straight vortex filament with infinite length and strength Γ

with μ as the permeability of the medium surrounding the wire and I the electric current.

Applying Eq. (6.200) to a straight vortex filament of infinite length and the strength Γ, Fig. 6.33, the magnitude of the induced velocity is

$$\boldsymbol{V}_B = \frac{\Gamma}{4\pi} \int_{-\infty}^{\infty} \frac{\sin\alpha}{r^2} d\xi. \tag{6.202}$$

From Fig. 6.34, we find the following relationships are obtained. Introducing Eq. (6.203) into Eq. (6.202) results in

$$d\xi = \frac{r d\alpha}{\sin\alpha}, \; r = \frac{h}{\sin\alpha} \tag{6.203}$$

$$\boldsymbol{V}_B = \frac{\Gamma}{4\pi h} \int_0^{\pi} \sin\alpha d\alpha = -\frac{\Gamma}{4\pi h} \cos\alpha|_0^{\pi} = \frac{\Gamma}{2\pi h}. \tag{6.204}$$

To calculate the velocity induced by a vortex filament of finite length L, the integration boundaries of Eq. (6.204) need to be replaced by α_1 and α_2, shown in Fig. 6.35. The integration results in:

$$\boldsymbol{V}_B = \frac{\Gamma}{4\pi h} \int_{\alpha_1}^{\alpha_2} \sin\alpha d\alpha = \frac{\Gamma}{4\pi h}(\cos\alpha_1 - \cos\alpha_2). \tag{6.205}$$

Setting $\alpha_1 = 0$ and $\alpha_2 = \pi/2$, for a semi-infinite vortex filament, we obtain the induced velocity

$$\boldsymbol{V}_B = \frac{\Gamma}{4\pi h}. \tag{6.206}$$

Fig. 6.35 Calculation of induced velocity

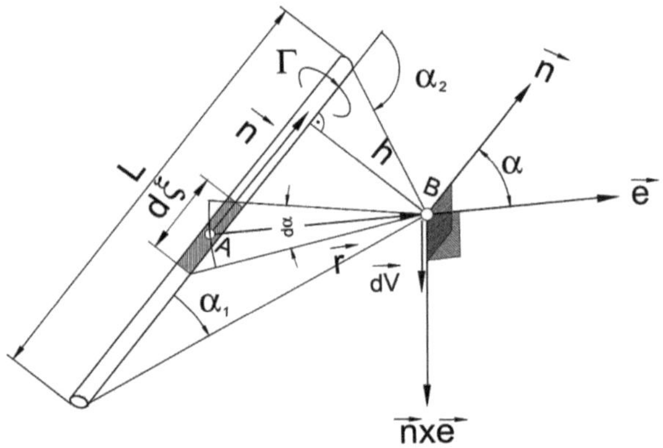

The experimental findings by Prandtl, and the subsequent discussion in Sect. 6.6.3 relative to the circulation generation around a two-dimensional airfoil, has led to the assumption that a two-dimensional airfoil of an infinite span with a bound vortex can simply be represented by a vortex filament of infinite length and the same vortex strength. In this case, the lift force acting on a wing was determined by the Kutta-Joukowsky equation (6.72) with Γ as a constant circulation. A constant circulation assumption does not hold for wings of finite length because the circulation around a finite wing changes from the center of the wing to both ends. However, it can be used as a useful tool for estimating the lift force.

6.6.5 Induced Drag Force

So far, we have been dealing with two-dimensional airfoils of infinite span with a bound vortex of constant strength. The superposition of a circulation with a parallel flow generated a lift force, which is the result of the pressure difference between the suction surface (convex surface) and the pressure surface (concave surface). In the case of an airfoil with a finite span, the pressure difference at both tips of the airfoil causes a secondary flow motion.

Figure 6.36a shows the inception of the secondary flow on both tips of a wing. This secondary flow creates tip vortices which induce downward velocities that change the flow pattern of a two-dimensional flow to a three-dimensional one. At the tips, the pressure difference and, thus, the circulation, disappears leading to a circulation distribution that varies from the mid-section of the wing towards both tips, Fig. 6.36b. Immediately behind the trailing edge, a surface separates the flow which has passed over the suction surface from that which passed over the pressure surface. A surface of discontinuity is formed which is occupied by free vortices, Fig. 6.36c and d, as detailed in Sect. 6.6.3. This vortical layer is unstable and rolls itself up to form two discrete vortices with opposite circulation directions, Fig. 6.36e, f. These vortices

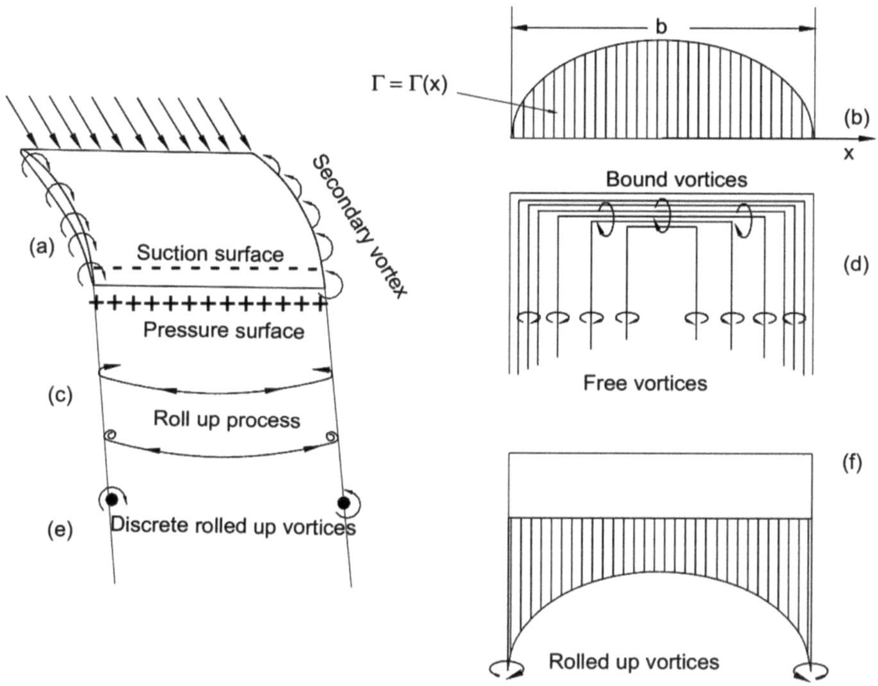

Fig. 6.36 Inception of secondary flow on tips of a wing

are responsible for inducing a downward velocity $\boldsymbol{w}_{\text{ind}}$ which is superimposed on the undisturbed velocity $\boldsymbol{V}_\infty$, changing the effective angle of attack from α_∞ to $\alpha = \alpha_\infty - \varepsilon$ and the resultant velocity to $\boldsymbol{V}_R$, as shown in Fig. 6.37. According to the Kutta-Joukowsky theorem, in an inviscid flow field, the lift force is perpendicular to the plane spanned by the velocity and the circulation vectors.

Considering an infinitesimal lift force brought about by an infinitesimal wing span dx as $dL = \rho V_\infty \Gamma dx$, the infinitesimal induced drag is calculated from $d\boldsymbol{D} = d\boldsymbol{L} \tan \varepsilon$ with ε as the induced angle. Since $V_\infty >> w_{\text{ind}}$, we may approximate $\tan \varepsilon = \varepsilon$, which leads to

$$dD = dL \frac{w_{\text{ind}}}{V_\infty}. \tag{6.207}$$

Integrating Eq. (6.207) gives

$$D_{\text{ind}} = \int_{-b/2}^{b/2} \frac{w_{\text{ind}}}{V_\infty} dL = \int_{-b/2}^{b/2} \frac{w_{\text{ind}}}{V_\infty} V_\infty \Gamma(x) dx = \int_{-b/2}^{b/2} w_{\text{ind}} \Gamma(x) dx. \tag{6.208}$$

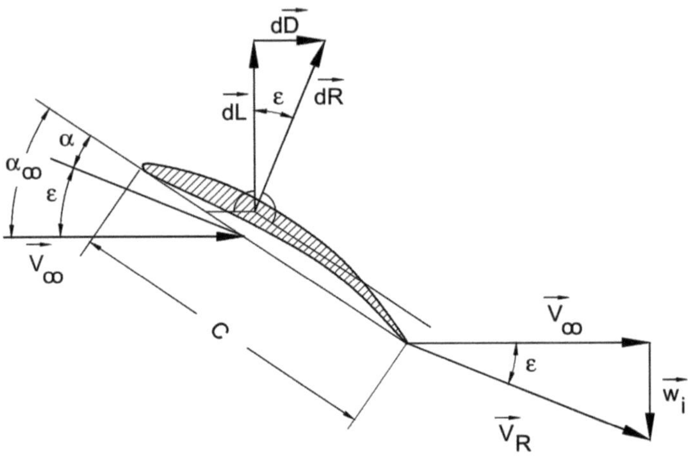

Fig. 6.37 Induced velocity and drag by secondary vortices

As Eq. (6.208) indicates, the main parameter determining the induced drag is the circulation function Γ and its distribution. Thus, the induced drag can be calculated if the Γ-distribution is known.

For an elliptic distribution of $\Gamma(x) = \Gamma_0\sqrt{1-(x/s)^2}$ and $s = b/2$, Eq. (6.208) can be integrated analytically as given below

$$D_{\text{ind}} = \frac{2L^2}{\pi\rho V_\infty^2 b^2} \tag{6.209}$$

with the lift force $L = (\pi/4)\rho b V_\infty \Gamma_0$, the induced drag $D_i = (\pi/8)\rho\Gamma_0^2$ and the induced velocity $w_i(x) = \Gamma_0/2b$. The induced drag coefficient is found by dividing Eq. (6.209) by $\rho/2V_\infty^2 bC$

$$C_{D_{\text{ind}}} = \frac{C_L^2}{\pi}\frac{c}{b}. \tag{6.210}$$

The total drag force acting on a wing of finite span with an arbitrary geometry is the sum of the viscous drag force and the induced drag force. To overcome the induced drag, additional mechanical energy must be provided externally. In contrast to the drag force caused by boundary layer momentum deficit along the surface of the wing, the induced (inviscid) drag arises from a change of the flow direction due to the downward velocity field.

The problematic of calculation of lift, drag, and other aerodynamic quantities is treated comprehensively in the book by Schlichting and Truckenbrodt [150].

6.7 Problems

Problem 6.1 Superimposing a point source and a parallel flow results in a flow around an infinitely long body.

(a) Find the velocity potential of the flow.
(b) By expanding the velocity components, show that the flow in the neighborhood of the stagnation point corresponds to a stagnation point flow with z as the symmetry axis.
(c) Write computer program to calculate the streamlines and the potential lines.
(d) Plot the stagnation flow.

Problem 6.2 In an incompressible, plane potential flow with the potential $F_1(z)$, a circular cylinder with the radius a is inserted at the origin. As a result, the resulting complex potential $F_2(z)$ of the new flow is $F_2(z) = F_1(z) + \bar{F}_1\left(\frac{a^2}{z}\right)$, where $\bar{F}_1$ is the conjugate complex potential.

(a) Calculate the complex potential of a circular cylinder (radius a) at $z = 0$ in a source flow (strength m, source at $z = b$).
(b) Show here that the circle $z = ae^{i\varphi}$ is a streamline.
(c) Plot the streamlines.
(d) Calculate the velocity potential. Where is the stagnation point located?
(e) Calculate the force on the cylinder with Blasius' theorem.

Problem 6.3 Using Kutta-Joukowsky transformation function:

(a) Map a flow parallel to the x-axis around a circle with a radius a in z-plane onto the surface of an ellipse in ζ-plane, decompose the transformation function into its real and imaginary parts and plot the ellipse and the circle.
(b) Plots on the ellipse the streamlines and potential lines.
(c) Find the velocity components in ζ-plane.
(d) Plot the c_p-distribution.
(e) Consider the velocity vector $\boldsymbol{V}_\infty$ with an of attack α impinging on the circular cylinder in z-plane and repeat case (a) to (d). Hint: You may start with this version and make some angle variation with $\alpha = 0$ as one of the cases.

Problem 6.4 Using Kutta-Joukowsky transformation function:

(a) Map a flow parallel to the x-axis around a circle with a radius a in z-plane onto the surface of an airfoil in ζ-plane, decompose the transformation function into its real and imaginary parts and plot the ellipse and the circle.
(b) Plots on the ellipse the streamlines and potential lines.

(c) Find the velocity components in ζ-plane.
(d) Plot the c_p-distribution.
(e) Consider the velocity vector $\boldsymbol{V}_\infty$ with an of attack α impinging on the circular cylinder in z-plane and repeat case (a) to (d). Hint: You may start with this version and make some angle variation with $\alpha_\infty = 0$ as one of the cases.

Problem 6.5 A circular cylinder with a radius a is located in a plane, inviscid, potential flow. The angle of attack of the undisturbed, translational flow is α.

(a) Find the complex potential of the flow.
(b) Calculate the position of the stagnation points, plot the streamlines.
(c) Which body contour do we obtain, if we map the circular cylinder by the mapping function $\zeta = (z + a^2/z)e^{-i\alpha}$ onto the ζ-plane?
(d) Find the position of the stagnation points in the ζ-plane.
(e) What type of flow do we get for $|\zeta| \rightarrow \infty$?
(f) Calculate the pressure distribution along the body contour in the z-plane.

Given: a, α, U_∞, ϱ, p_∞.

Chapter 7
Viscous Laminar Flow

As briefly discussed in Chap. 4, the motion of Newtonian fluids is described by the Navier-Stokes equations. Due to the non-linear nature of these equations and the general complexity of the flow geometry, analytical solutions of Navier-Stoke's equations has been exhibiting a major problem in fluid mechanics. The continuous development in the area of computer technology and the introduction of powerful numerical methods in the last two decades have brought a breakthrough in the area of *Computational Fluid mechanics* (CFD). Using CFD-methods, viscous flow problems within arbitrary channel geometries can be solved numerically regardless the complexity of the geometry. This requires significant computational efforts. An adequate treatment of CFD-methods is beyond the scope of this book. However, in the context of this course, in Chap. 9, we present the essential features of the computational fluid mechanics that are necessary for the basic understanding of the physics behind CFD. This includes a rather detailed introduction into turbulence and its modeling.

In this chapter, we introduce a class of exact solutions of the Navier-Stokes equations for the two-dimensional *laminar* flow, a special case of viscus flows, where the velocity does not exhibit a *random* characteristic. Exact analytical solutions are found only for few cases, where the flow can be assumed *unidirectional*. This implies that the velocity vector has a component in longitudinal direction only that may change in lateral direction. A general overview of a class of exact solutions for viscous laminar flows through two-dimensional channels is found in Schlichting [39]. In a few curved channels, where the velocity vector of a two-dimensional flow has generally two components, the coordinate system can be transformed such that the velocity vector has only one direction in a curvilinear coordinate system. In the following sections, several cases are presented that are of fundamental significance for understanding the motion of viscous flows.

M. T. Schobeiri, *Advanced Fluid Mechanics and Heat Transfer for Engineers and Scientists*, https://doi.org/10.1007/978-3-030-72925-7_7

7.1 Steady Viscous Flow Through a Curved Channel

Solving the Navier-Stokes equation, we investigate the influence of curvature and pressure gradient on the flow temperature and velocity distribution. The flows within curved channels under adverse, zero, and favorable pressure gradients are encountered in numerous practical devices such as compressor and turbine blades, diffusers and nozzles. Within these devices the distribution of flow quantities such as the temperature and velocity and consequently the flow behavior are affected primarily by the curvature and pressure gradient. To calculate the above quantities, conservation laws of fluid mechanics and thermodynamics are applied. For an incompressible Newtonian fluid, the Navier-Stokes equation describes the flow motion completely. This equation has exact solutions for only a few special cases. For the major part of practical problems encountered in applied fluid mechanics, however, it is hardly possible to find any exact solutions. This deficiency is in part due to the complexity of the individual flow field and its geometry under consideration. Despite this fact, the existence of exact solutions of fluid mechanics problems including the velocity and temperature distribution within viscous flows are of particular interest to the computational fluid dynamics (CFD) community dealing with development of CFD-codes. A comprehensive code assessment and validation requires both the experimental verification and theoretical confirmation. For the latter case, a comparison with existing exact solutions exhibits an appropriate procedure to demonstrate the code capability. For symmetric flows through channels with positive and negative pressure gradients exact solutions are found by Jeffery [40]. For asymmetric curved channels with convex and concave walls, exact solutions of the Navier-Stokes equation are found by Schobeiri [42, 43], where the influence of the wall curvature on the velocity distribution is discussed. Furthermore, a class of approximate solutions of Navier-Stokes is presented in [44].

This section treats the influence of curvature and pressure gradient on temperature and velocity distributions by solving the energy and momentum equations. Under the assumption that the flow is two dimensional, steady, incompressible, and has constant viscosity, the conservation laws of fluid mechanics and thermodynamics are transformed into a curvilinear coordinate system. The system describes the two-dimensional, asymmetrically curved channels with convex and concave walls. As a result, exact solutions for the equation of energy as well as the Navier-Stokes equation are found.

7.1.1 Case I: Conservation Laws

To determine the influence of curvature and pressure gradient on temperature distribution, the velocity distribution must be known. This requires the solution of continuity and the Navier-Stokes equations. As the first conservation law, the continuity equation in coordinate invariant form is:

$$\nabla \cdot \mathbf{V} = \mathbf{0}. \tag{7.1}$$

For a curvilinear coordinate system, Equation (7.1) can be written as [see Eqs. (4.7) and (A.36)]:

$$V^i_{;i} + V^k \Gamma^i_{ki} = 0 \tag{7.2}$$

with $\mathbf{V}$ as the velocity vector that is decomposed in its contravariant components V^i in ξ_i_direction. For a two-dimensional flow, we prescribe that the velocity component normal to the flow direction must vanish. As a result, the integration of Eq. (7.2) must fulfill both the continuity and the Navier-Stokes equations. This is possible only if the Christoffel symbols Γ^i_{ki} are not functions of the coordinates themselves. The corresponding channel with the curvilinear coordinate is then obtained from the transformation:

$$w = -\frac{2}{a+ib} \ln z \text{ with } z = x + iy \text{ and } w = \xi_1 + i\xi_2 \tag{7.3}$$

with ξ_i as the orthogonal curvilinear coordinate system.

$$x = e^{-\frac{1}{2}(a\xi_1 - b\xi_2)} \cos\left(\frac{a\xi_1 + b\xi_2}{2}\right)$$

$$y = -e^{-\frac{1}{2}(a\xi_1 - b\xi_2)} \sin\left(\frac{a\xi_2 + b\xi_1}{2}\right) \tag{7.4}$$

with a and b as real constants that define the configuration of the channel and ξ_1 and ξ_2 as the orthogonal curvilinear coordinates. The corresponding metric coefficients and Christoffel symbols are:

$$g^{11} = g^{22} = \frac{4}{a^2+b^2} e^{a\xi_1 - b\xi_2},\ g^{12} = g^{21} = 0 \tag{7.5}$$

$$\Gamma^1_{kl} = \frac{1}{2}\begin{pmatrix} -a & b \\ b & +a \end{pmatrix} \quad \Gamma^2_{kl} = -\frac{1}{2}\begin{pmatrix} b & a \\ a & -b \end{pmatrix}. \tag{7.6}$$

With Eqs. (7.5) and (7.6) and the requirement that the velocity component in ξ_2 must vanish, the integration of the continuity Eq. (7.2) leads to:

$$V^1 = \frac{4\nu}{a^2+b^2} F(\xi_2) e^{a\xi_1 - b\xi_2} \tag{7.7}$$

where V^1 is the contravariant component of the velocity in the ξ_1-direction, ν is the kinematic viscosity, and $F = F(\xi_{2m})$ is a function to be determined. Thus, the only physical component of the velocity vector is the one in the ξ_1-direction, for which we may omit the superscript 1 and set:

$$U \equiv V^{*1} = \frac{V^1}{\sqrt{g}^{11}} = \frac{2\nu}{\sqrt{a^2 + b^2}} F(\xi_2) e^{\frac{1}{2}(a\xi_1 - b\xi_2)}. \tag{7.8}$$

Equation (7.8) must strictly satisfy the Navier-Stokes equation in order to be an exact solution. As discussed in Chap. 4, the conservation law of motion for the steady Newtonian fluids is represented by the Navier-Stokes Equation (4.37) that describes the flow motion completely. Neglecting the body forces, its coordinate invariant form for incompressible flow with constant viscosity reads:

$$\mathbf{V} \cdot \nabla \mathbf{V} = -\frac{1}{\rho} \nabla \mathbf{p} + \mathbf{v} \nabla^2 \mathbf{V} \tag{7.9}$$

with ρ as the density. The Navier-Stokes Equation (7.9) decomposed in its contravariant components is written as [see Eq. (A.73)]

$$\begin{aligned} V^i V^j_i + V^i V^k \Gamma^j_{ki} = & -\frac{1}{\rho} p_i g^{ij} + \upsilon [V^j_{ik} + V^m_i \Gamma^j_{mk} + V^m_k \Gamma^j_{mi} \\ & - V^j_m \Gamma^m_{ik} + V^m \Gamma^n_{mi} \Gamma^j_{nk} - V^m \Gamma^q_{ik} \Gamma^j_{mq} + V^m \Gamma^j_{mi,k}] g^{ik}. \end{aligned} \tag{7.10}$$

For the two-dimensional flow with $V^3 = 0$, Eq. (7.10) leads in ξ_1-direction to:

$$\begin{aligned} & V^1 V^1_1 + (V^1)^2 \Gamma^1_{11} = -\frac{1}{\rho} p_1 g^{11} + \\ & \upsilon [V^1_{11} + V^1_{22} + V^1_1 (\Gamma^1_{11} - \Gamma^1_{22}) + V^1_2 (2\Gamma^1_{12} - \Gamma^2_{11} - \Gamma^2_{22})] g^{11} \end{aligned} \tag{7.11}$$

and in the ξ_2-direction:

$$(V^1)^2 \Gamma^2_{11} = -\frac{1}{\rho} p_2 g^{22} + \upsilon [2V^1_{11} \Gamma^2_{11} + 2V^1_2 \Gamma^2_{12}] g^{22}. \tag{7.12}$$

Introducing the integration results of the continuity Eq. (7.7) into the system of differential Eqs. (7.11) and (7.12) and eliminating the pressure terms, the result of the first integration is:

$$F'' - 2bF' + (a^2 + b^2)F - \frac{a}{2} F^2 + K_1 = 0 \tag{7.13}$$

with K_1 as the integration constant. Dividing Eq. (7.13) by its maximum value $F_{\max}$, the dimensionless velocity function is obtained from:

$$\Phi'' - 2b\Phi' + (a^2 + b^2)\Phi - \frac{a}{2} F_{\max} \Phi^2 + C_1 = 0 \tag{7.14}$$

where

$$\Phi = \frac{F}{F_{\max}} \text{ and } C_1 = \frac{K_1}{F_{\max}}. \tag{7.15}$$

The significant parameter affecting the flow within the curved channel is the Reynolds number, which is defined as:

$$\mathrm{Re} = \frac{\Delta s \cdot U_m}{v} \tag{7.16}$$

where Δs and U_m are the distance and the maximum velocity in the ξ_1-direction. The latter is obtained by setting in Eq. (7.8), the coordinate ξ_2 equal to $\xi_2 = \xi_{2_{\max}}$:

$$U_m = \frac{2\nu}{\sqrt{a^2b^2}} F_{\max} e^{\frac{1}{2}(a\xi_1 - b\xi_{2\max})}. \tag{7.17}$$

With the distance Δs:

$$\Delta s = -\sqrt{\frac{a^2+b^2}{a}} e^{-\frac{1}{2}(a\xi_1 - b\xi_{2\max})} \tag{7.18}$$

the Reynolds number is:

$$\mathrm{Re} = -\frac{2}{a} F_{\max}. \tag{7.19}$$

Introducing Eqs. (7.19) into (7.14) leads to:

$$\Phi'' - 2b\Phi' + (a^2+b^2)\Phi + \frac{a^2}{4}\,\mathrm{Re}\,\Phi^2 + C_1 = 0. \tag{7.20}$$

7.1.2 Case I: Solution of the Navier-Stokes Equation

Equation (7.20) describes the motion of viscous flows through curved channels pertaining to the coordinate transformation discussed in Sect. 7.1.1. It includes both the Navier-Stokes and continuity equations that are reduced to a single, ordinary, nonlinear, second-order differential equation. The solutions of Eq. (7.20), $\Phi = \Phi(\xi_2)$ are functions of the coordinate ξ_2 and incorporate the Reynolds number as parameter. Special cases of Eq. (7.20) are the purely radial flow, where $a = -2$ and $b = 0$, and the flow through concentric cylinders with $a = 0$ and $b = 1$. For those cases analytical and numerical solutions were found in [40, 41]. Based on Jeffery-Hammel's solutions, Milsaps and Pohlhausen [45] calculated the temperature distribution within the straight wall diffuser and nozzle. Extensive discussions by Schlichting [39] underscore the importance of those flows from a general theoretical point of view. To show the effect of the curvature and pressure gradient on the temperature and velocity distribution, an asymmetrically curved channel with convex and concave walls is generated by choosing $a = -1$ and $b = 1$, Schobeiri [42, 43].

For the solution of Eq. (7.20), a numerical integration procedure is applied. Starting from the initial conditions specified below and the determination of constant C_1, an iteration method is developed that reduces the boundary-value problem to an initial one. The solution of differential Eq. (7.20) must fulfill the governing initial and boundary conditions. The boundary conditions are given by the non-slip conditions at the channel walls:

$$\begin{aligned}\xi_2 = \xi_{2_{B1}} \equiv 0.1,\ \Phi = \Phi_{B1} \equiv 0\\ \xi_2 = \xi_{2_{B2}} \equiv 0.5,\ \Phi = \Phi_{B2} \equiv 0\end{aligned} \tag{7.21}$$

where the indices $B1$ and $B2$ refer to the convex and concave channel walls. The initial condition is described by the maximum value of the velocity distribution and its position $\xi_2 = \xi_{2_{\max}}$, which is unknown for the time being:

$$\xi_2 = \xi_{2_{\max}},\ \Phi = \Phi_{\max} = \pm 1,\ \Phi' = \Phi'_{\max} = 0. \tag{7.22}$$

The positive sign of Φ indicates an increase of the cross-section area in direction of decreasing ξ_1, which is associated with the positive pressure gradient. The negative sign characterizes the accelerated flow in direction of increasing ξ_1, where negative pressure gradient prevails. The constant C_1 in Eq. (7.20) specifies the solution of Eq. (7.20) and significantly affects the convergence speed. It must be determined so that the above boundary and initial conditions are identically fulfilled. The following iteration method enables precise calculation of C_1. Starting from Eq. (7.20),

$$\Phi'' = \Psi'' + C_1 \tag{7.23}$$

where

$$\Psi'' \equiv -2b\Phi' + (a^2 + b^2)\Phi + \frac{a^2}{4}\,\mathrm{Re}\,\Phi^2. \tag{7.24}$$

Integration of Eq. (7.24) between $\xi_{2\max}$ and ξ_{2B1} leads to:

$$\Phi_{\max} - \Phi_{B1(i)} = \Delta\Psi_{(i)} + \frac{C_{1(i)}}{2}(\xi_{2\max}^2 - \xi_{2B1}^2) \tag{7.25}$$

where $C_{1(I)m}$ is the constant calculated at the i-th iteration step that can result in boundary value $\Phi_{B1(I)m} \neq 0$. Similarly, we obtain a relation for the constant $C_{1(Im+1)}$ that corresponds to $\Phi_{B1(Im+1)} = 0$:

$$\Phi_{\max} = \Delta\Psi_{(i+1)} + \frac{C_{1(i+1)}}{2}(\xi_{2\max}^2 - \xi_{2B1}^2). \tag{7.26}$$

By subtracting Eq. (7.25) from Eq. (7.26) and introducing:

$$\Delta\Psi_{(i+1)} - \Delta\Psi_{(i)} \equiv (1-\eta)\Phi_{B1(i)} \tag{7.27}$$

where $\eta \leq 1$, the constant C_1 is calculated from the following iteration function:

$$C_{1(i+1)} = \frac{2\eta\Phi_{B1(i)}}{\xi_{2m}^2 - \xi_{2R1}^2} + C_{1(i)}. \tag{7.28}$$

For Reynolds number range Re< 3500, the precise value of the constant C_1 is obtained within a few iteration steps by setting $\eta = 1$. For higher Reynolds numbers the factor $\eta \sim 0.5$ has proved to be effective. To start the iteration process, the constant $C_{1(1)}$ should have the same order of magnitude as the Reynolds number. The initial value for $\xi_{2_{\max(1)}}$ can be estimated from:

$$\xi_{2\max(i)} = \frac{\xi_{2_{B1}} + \xi_{2_{B2}}}{2}. \tag{7.29}$$

With $\xi_{2\max(Im)}$, the constant C_1 from Equation (7.28), and the initial and boundary conditions from Eqs. (7.21) and (7.22), the zero at $\xi_{2B2(I)}$ is found by Newton's iteration method. The improved zero is obtained from:

$$\xi_{2\max(i+1)} = \xi_{2\max(i)} - (\xi_{2B2(i+1)} - \xi_{2B2(i)}). \tag{7.30}$$

The new value from Eq. (7.30) leads to improved $\xi_{2B2(Im+1)}$. If the absolute difference $| \xi_{2B2(im+1)} - \xi_{2B2(i)} | = \varepsilon \leq 10^{-6}$, the required accuracy has been obtained; otherwise the iteration procedure is repeated until ε is reached.

7.1.3 *Case I: Curved Channel, Negative Pressure Gradient*

Once the solution of Eq. (7.20) is found, the dimensionless velocity distribution is obtained from Eq. (7.17):

$$\frac{U}{U_m} \equiv \frac{V^*}{V_m^*} = \Phi e^{\frac{1}{2}b\xi_{2\max} - \xi_2)}. \tag{7.31}$$

As seen earlier, the solution $\Phi = \Phi(\xi_2)$ is a function of the coordinate ξ_2 only and incorporates the Reynolds number as a parameter. Thus, the velocity distributions represented by Eq. (7.31) exhibit similar solutions. An asymmetrically curved channel with convex and concave walls is generated by choosing $a = -1$ and $b = 1$. As shown in Fig. 7.1, the negative pressure gradient is established by an asymmetrically convergent channel with convex and concave walls. For Reynolds number Re = 500 the velocity distributions at the coordinate $\xi_1 = 3.8$ exhibit an almost parabolic shape with the maximum close to $\xi_2 = 0.3$. For the similarity reasons explained above, similar velocity distribution is found and plotted at $u = 0.38$ for the same Reynolds number. Increasing the Reynolds number to Re = 750, 1000 respectively results in steeper velocity slopes at both walls (Fig. 7.1). As a consequence, the velocity

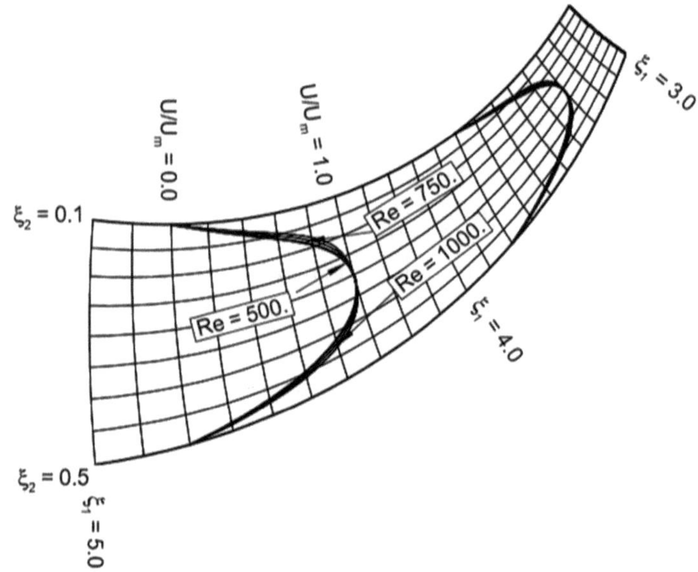

Fig. 7.1 Accelerated laminar flow through a two-dimensional curved channel at different Reynolds numbers

profile tends to become fuller, particularly for higher Reynolds numbers. As shown, the viscosity effect is restricted predominantly to the wall regions and continuously reduces by increasing the Reynolds number. This behavior again justifies the Prandtl assumption for higher Reynolds number to divide the flow field into a viscous and an inviscid flow zone. For Reynolds numbers up to Re = 5000, velocity distributions can be calculated without convergence problems. Thus for an accelerated flow, the stability of the laminar flow and the transition from laminar into turbulent flow are apparently extended to higher Reynolds numbers as expected.

7.1.4 Case I: Curved Channel, Positive Pressure Gradient

The positive pressure gradient within the asymmetrically curved channel discussed above is created by reversing the flow direction. Figure 7.2 shows the flow at different Reynolds numbers. As shown in Fig. 7.2, for Re = 500, the velocity distribution on the concave wall is fully attached. The fluid particles moving in streamwise direction are exposed to three different type of forces: (1) the wall shear stress force acting in opposite direction decelerates the fluid particle; (2) the decelerating effect of the wall shear stress is intensified by the pressure forces which also act in opposite direction causing the flow to further decelerate; and (3) the centrifugal force caused by the channel curvature pushes the fluid particle away from the convex wall towards

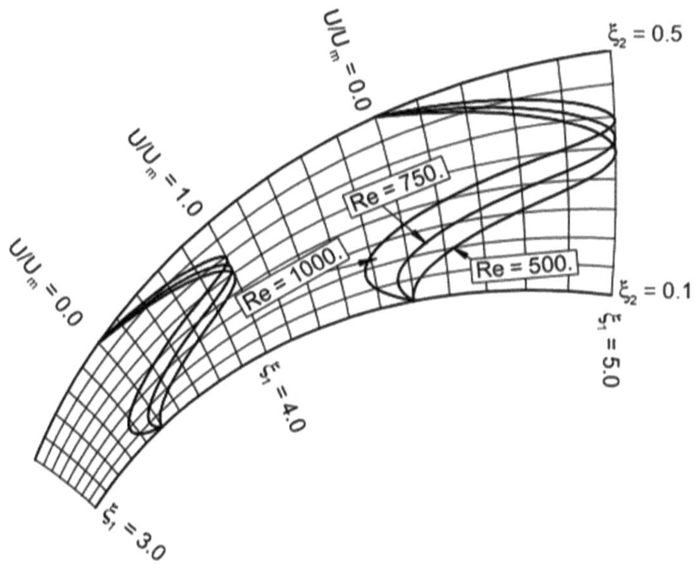

Fig. 7.2 Decelerated laminar flow through a two-dimensional channel at different Reynolds number

the concave one increasing the susceptibility of flow to separation. The interaction of these three forces increase the tendency for separating along the convex wall. Increasing the Reynolds number to Re = 1500 causes the flow separation on the convex channel wall. In this case the laminar low along the convex surface is, while the non-separated portion appears as a laminar jet attaching to the concave wall.

7.1.5 Case II: Radial Flow, Positive Pressure Gradient

As shown in Fig. 7.2, the combination of the channel curvature and the positive pressure gradient has caused a flow separation on the convex wall, whereas no separation occurred on the concave wall. From fluid mechanical point of view, we are interested in determining the effect of pressure gradient on the velocity distribution in the absence of curvature. To investigate this, we generate a channel with straight wall geometry by setting $a = -2$, and $b = 0$. With these new constants, Eq. (7.20) reduces to:

$$\Phi'' + 4\Phi + \text{ Re } \Phi^2 + C_1 = 0. \tag{7.32}$$

This special case constitutes a purely radial laminar flow through a channel with straight walls and is known as the Hamel-flow [41]. The results are shown in Fig. 7.3, where the velocity distributions are plotted for three different Reynolds numbers. Close to the wall at Re = 500, the flow exhibits a tendency for separation on both walls. Increasing the Reynolds number to Re = 750 and 1500 respectively causes

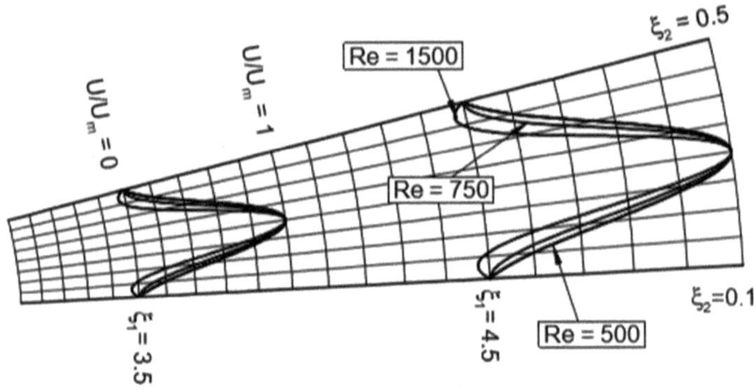

Fig. 7.3 Decelerated laminar flow through a two-dimensional channel with straight walls at different Reynolds numbers

the flow separation on both walls. A comparison with the results in Fig. 7.2 clearly indicates that the difference in velocity distributions is attributed to the nature of wall curvature.

7.2 Temperature Distribution

To determine the temperature distribution within the curved channel we combine the mechanical and thermal energy balances as we discussed in Sects. 4.4.1 and 4.4.2:

$$\rho \frac{Du}{Dt} = -\nabla \cdot \dot{\mathbf{q}} - p(\nabla \cdot \mathbf{V}) + \mathbf{T} : \nabla \mathbf{V} \tag{7.33}$$

with U as the internal energy, $\dot{\mathbf{q}}$ the heat flux vector and $\mathbf{T}$ the shear stress tensor. Considering the thermodynamic relationship, for the steady incompressible flow, Eq. (7.33) reduces to:

$$\rho c_v \frac{dT}{dt} = -\nabla \cdot \dot{\mathbf{q}} + \mathbf{T} : \nabla \mathbf{V} \tag{7.34}$$

where c_v is the specific heat capacity at constant volume and T is the temperature of the working medium. Using the identity for the velocity gradient, the Fourier equation of conduction, and the Stokes relation:

$$\nabla \mathbf{V} = \mathbf{D} + \Omega, \dot{\mathbf{q}} = -\kappa \nabla \mathbf{T}, \mathbf{T} = 2\mu \mathbf{D} \tag{7.35}$$

with κ as the thermal conductivity, μ the absolute viscosity, and D and Ω as second-order tensors of the deformation and rotation, respectively (see Chap. 4). Introducing Eqs. (7.35) into (7.34) and considering the identity $\mathbf{D} : \Omega = \mathbf{0}$ the result of this operation leads to the differential equation of the energy in terms of temperature:

$$c_v(\nabla T)\cdot \mathbf{V} = \frac{\kappa}{\rho}\nabla^2 T + 2\upsilon \mathbf{D}:\mathbf{D}. \tag{7.36}$$

The above differential equation is invariant with respect to coordinate system transformation. We first write Eq. (7.36) in contravariant form:

$$\begin{aligned} c_v T_i V^i =& \frac{k}{\rho} g^{ik}(T_{ik} - T_j \Gamma^j_{ik}) + \frac{1}{2}\nu(V_{k,1} + V_{1,k} - V_m\Gamma^m_{kl} - V_m\Gamma^m_{lk}) \\ & \times (g^{1i}V^k_i + g^{ki}V^1_i + g^{1i}V^m\Gamma^k_{mi} + g^{ki}V^m\Gamma^1_{mi}) \end{aligned} \tag{7.37}$$

and insert Γ^k_{mn} from Eqs. (7.6) into (7.37) and considering the assumptions made at the beginning, Eq. (7.37) is reduced to:

$$c_v T_1 V^1 = \frac{\kappa}{\rho}(T_{,11} + T_{,22}) + \upsilon(aV_1 V^1_{,1} - bV_1 V^1_{,2} + V_{1,2}V^1_{,2})g^{11}. \tag{7.38}$$

7.2.1 *Case I: Solution of Energy Equation*

Equation (7.38) is a second order, nonlinear, partial differential equation, in which the temperature $T = T(\xi_1, \xi_2)$. It can be reduced to an ordinary differential equation by making the following ansatz:

$$T = T(\xi_1, \xi_2) = \frac{4}{a^2 + b^2} G e^{(a\xi_1 - b\xi_2)} + T_w \tag{7.39}$$

with $G = G(\xi_2)$ and T_w as the wall temperature assumed to be constant. Incorporating Eqs. (7.39) and (7.7) into (7.38) results in:

$$ac_v \upsilon G F = \frac{\kappa}{\rho}\left[(a^2 + b^2)G - 2bG' + G''\right] + \upsilon^3\left[(a^2 + b^2)F^2 - 2bFF' + F'^2\right]. \tag{7.40}$$

Dividing Equation (7.40) by $F^2_{\max}$ and introducing the Reynolds number from Eq. (7.19) leads to:

$$\begin{aligned} &\Theta'' - 2b\Theta' + (a^2 + b^2)\Theta + \frac{a^2}{2}\,\mathrm{Pr}\,\mathrm{Re}\,\Phi\Theta \\ &+ \frac{a^2}{4}\,\mathrm{Pr}\,\mathrm{Re}\,[(a^2 + b^2)\Phi^2 - 2b\Phi\Phi' + \Phi'^2] = 0. \end{aligned} \tag{7.41}$$

In Equation (7.41) the function Θ is defined as $\Theta = c_v G/\upsilon^2\,\mathrm{Re}$, with Pr$= \rho\upsilon c_p/\kappa$, as the Prandtl number. For gases the Prandtl number is around 0.7 and for water around 7. Detailed distributions of the values for the absolute viscosity μ, the thermal conductivity κ and the Prandtl number for dry air can be taken from Fig. 7.4. These values change slightly if the humidity ratio $\omega = m_{\mathrm{water}}/m_{\mathrm{air}}$ increases from 0% to 10%.

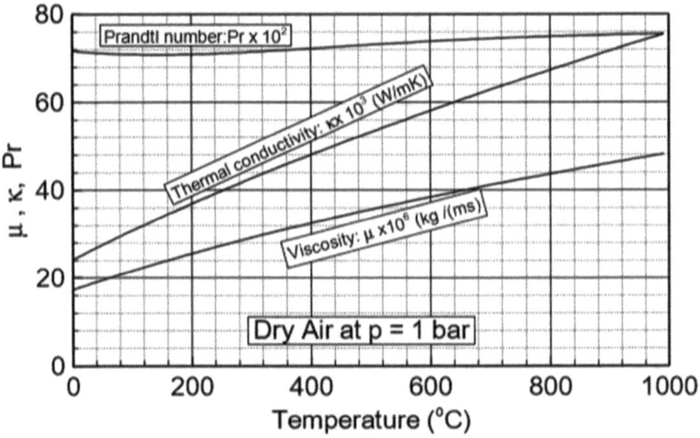

Fig. 7.4 Absolute viscosity, thermal conductivity and Pr-number as a function of temperature for dry air at $p = 1$ bar

The terms Φ and Φ' are given as the solution of Eq. (7.14). The solution of the ordinary, nonlinear, second-order differential Eq. (7.41) must satisfy the following boundary conditions:

$$\begin{aligned} &\xi_2 = \xi_{2_{B1}} \equiv 0.1;\ \Phi = \Phi_{B1} \equiv 0;\ \Theta = \Theta_{B1} \equiv 0 \\ &\xi_2 = \xi_{2_{B2}} \equiv 0.5;\ \Phi = \Phi_{B2} \equiv 0;\ \Theta = \Theta_{B2} \equiv 0. \end{aligned} \tag{7.42}$$

To find the solution of Eq. (7.41) it must first be combined with the equation of motion (7.14). For the solution of the resulting system of two nonlinear, second-order differential equations, a numerical procedure based on the Predictor-Corrector method is applied. Starting from $\xi_2 = \xi_{2B1}$, and already known Φ'_{B1} from Sect. 7.1.2, Θ'_{B1} is first estimated that may lead to $\Theta_{B2} \neq 0$. The correct value can be obtained quickly with the iteration function:

$$\Theta'_{B1(i+1)} = \Theta'_{B1(i)} - \Theta_{B2(i-1)} \frac{\Theta'_{B1(i)} - \Theta'_{B1(i-1)}}{\Theta_{B2(i)} - \Theta_{B2(i-1)}}. \tag{7.43}$$

The iteration process is repeated until the accuracy ε is reached:

$$|\Theta_{B2(i+1)} - \Theta_{B2(i)}| \leq \varepsilon = 10^{-6}. \tag{7.44}$$

7.2.2 *Case I: Curved Channel, Negative Pressure Gradient*

The effect of the different wall curvatures on temperature distributions is shown in Fig. 7.5 by asymmetrical temperature slopes at the convex and concave walls. For

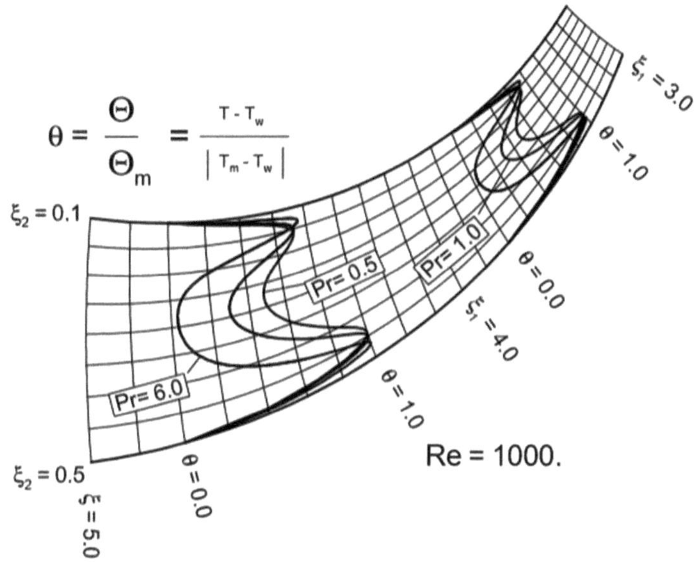

Fig. 7.5 Dimensionless temperature distribution for an accelerated laminar flow through a two-dimensional curved channel with Re- and Pr numbers as parameters, T_m = maximum temperature, T_W = wall temperature

the accelerated flow with Re = 500, Fig. 7.5 shows the dimensionless temperature distribution for different Prandtl numbers as parameter. As a consequence of energy dissipation, the temperature distribution near the channel walls experiences a steep gradient, with the maxima located close to the concave wall.
By approaching the channel middle, the temperature gradient gradually decreases for small Prandtl numbers and sharply for large ones. Increasing the Reynolds number to Re = 3500 causes pronounced temperature boundary layers, particularly for higher Prandtl numbers. Moving towards the channel middle, the temperature distribution exhibits almost a constant value slightly above the wall temperature.

7.2.3 Case I: Curved Channel, Positive Pressure Gradient

For a positive pressure gradient and Re = 500, temperature distributions are shown in Fig. 7.6 with Prandtl number as the parameter. As with the accelerated flow, high energy dissipation occurs near the channel walls. When approaching the middle of the channel, the temperature gradient changes sign. This effect might contribute to the instability of the flow field under a positive pressure gradient.

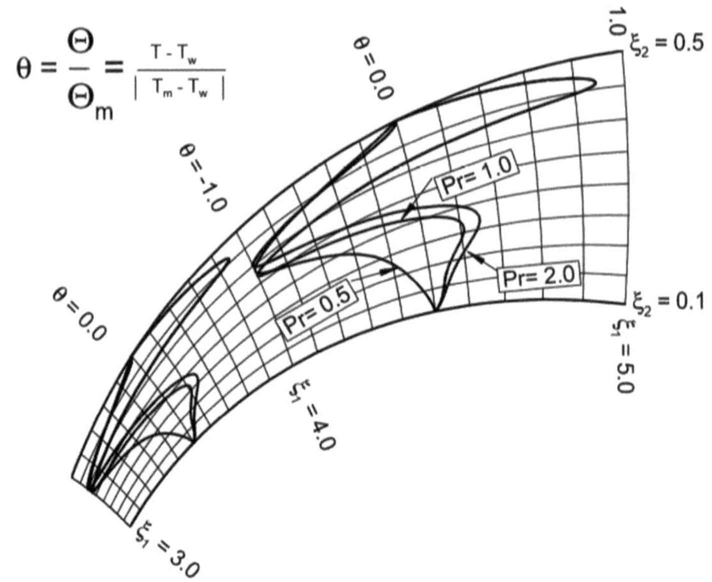

Fig. 7.6 Dimensionless temperature distribution for laminar decelerated flow through a two-dimensional curved channel with Re- and Pr- numbers as parameters, T_m = minimum temperature, T_W = wall temperature

For a positive pressure gradient and Re = 500, temperature distributions are shown in Fig. 7.6 with Prandtl number as the parameter. As with the accelerated flow, high energy dissipation occurs near the channel walls. When approaching the middle of the channel, the temperature gradient changes sign. This effect might contribute to the instability of the flow field under a positive pressure gradient.

7.2.4 Case II: Radial Flow, Positive Pressure Gradient

The effect of the different wall curvatures on temperature distributions in Fig. 7.7 is illustrated by asymmetric temperature slopes at the convex and concave walls. As we discussed in Sect. 7.1.5, the pressure gradient and the wall curvature were responsible for flow separation. In this section we investigate the effect of pressure gradient in the absence of wall curvature. Similar to the Case I in Sect. 7.1.5 we construct a straight walled channel by setting $a = -2$, and $b = 0$. With these new constants, Eq. (7.41) reduces to:

$$\Theta'' + 4\Theta + 2\,\mathrm{Pr}\,\mathrm{Re}\,\Phi\Theta + \mathrm{Pr}\,\mathrm{Re}\,[4\Phi^2 + \Phi'^2] = 0. \tag{7.45}$$

To obtain the temperature distribution, Eq. (7.45) must be combined with Eq. (7.32), which is the exact solution of the Navier-Stokes equation. The solution is presented in Fig. 7.7.

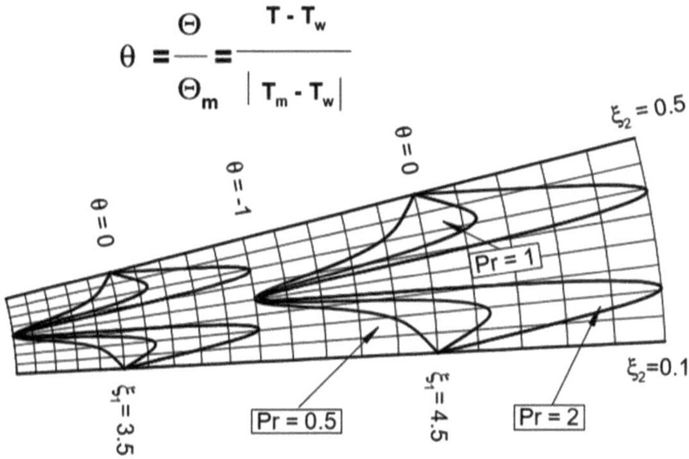

Fig. 7.7 Dimensionless temperature distribution within a straight walled channel with positive pressure gradient for Re = 1500 and different Pr-numbers, T_m = minimum temperature, T_W = wall temperature

As expected, the corresponding temperature distributions have symmetric profiles. A comparison with the results in Fig. 7.6 clearly indicates that the difference in temperature distributions is attributed to the wall curvature. As we saw, pronounced temperature boundary layer characteristics are exhibited for accelerated flow with higher Prandtl numbers. The separation tendency in the case of decelerated flow is apparent in the temperature distribution.

Another interesting case, namely the flow through concentric cylinders can be constructed by setting $a = 0$ and $b = 1$. In this case the Navier-Stokes and energy equation are:

$$\Phi'' - 2\Phi' + \Phi + C_1 = 0 \tag{7.46}$$

$$\Theta'' - 2\Theta' + 2\Theta + \text{ Pr } (\Phi^2 - 2\Phi\Phi' + \Phi'^2) = 0. \tag{7.47}$$

As seen from the above equation, all terms with Re- number disappeared leading to the results that the temperature and velocity distribution do not dependent on Re-number.

7.3 Steady Parallel Flows

As we saw in Sect. 7.1, the velocity distribution within the curved channel has in coordinate system only one physical component $U \equiv V^{*1}$ (in ξ_1-direction). In Cartesian coordinate system, however the velocity component $U \equiv V^{*1}$ has effectively two components, one in x_1-and x_2-direction. In this section we present exact solutions of

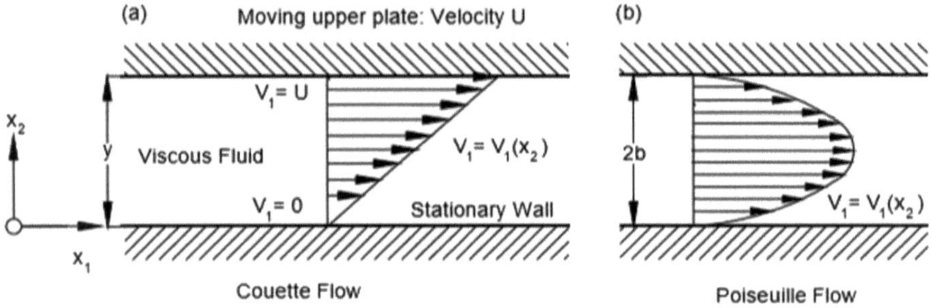

Fig. 7.8 Velocity distributions in Couette flow (**a**) and Poiseulle parallel flow (**b**)

the Navier-Stokes equations for parallel flows with only one component in a Cartesian coordinate system. These type of flows constitute a simple class of viscous fluid motions. Couette flow, Couette-Poiseuille flow, and Hagen Poiseuille flow are the classical examples of these flows.

7.3.1 Couette Flow Between Two Parallel Walls

A flow between two parallel flat plates, from which one is moving with the translational velocity U and the other is at rest as shown as shown in Fig. 7.8a is called *Couette flow*.
Since $V_2 = V_3 = 0$, the continuity Eq. (4.11) is reduced to:

$$\nabla \cdot \mathbf{V} = \frac{\partial \mathbf{V_1}}{\partial \mathbf{x_1}} = \mathbf{0}. \tag{7.48}$$

Similarly, the x_1- component of the Navier-Stokes equations reads

$$0 = -\frac{1}{\varrho}\frac{\partial p}{\partial x_1} + \upsilon\frac{\partial^2 V_1}{\partial x_2^2}. \tag{7.49}$$

Implementing the above assumption into the second component of Navier-Stokes equation leads to:

$$\frac{\partial p}{\partial x_2} = 0. \tag{7.50}$$

On the one hand Eq. (7.50) states that pressure may change in x_1-direction. On the other hand, the second term on the right hand side of Eq. (7.49) requires that $\partial p/\partial x_1$ be either a constant or a function of x_2. Since Eq. (7.50) excludes the latter, it follows that $\partial p/\partial x_1$ must be a constant that may assume positive, zero, and negative values. For further analysis, we set in Eq. (7.49) $\partial p/\partial x_1 = -K$ and obtain the solution of the resulting ordinary second order differential equation:

$$\mu\frac{d^2V_1}{dx_2^2} = -K. \tag{7.51}$$

Integrating Eq. (7.51) twice leads us to the general solution

$$V_1(x_2) = -\frac{K}{2\mu}x_2^2 + C_1x_2 + C_2. \tag{7.52}$$

Among many solutions of Eq. (7.51) we seek a specific solution that satisfies the following boundary conditions:

$$BC1 : V_{1_{(x_2=0)}} = 0 \text{ and } BC2\text{: } V_{1_{(x_2=h)}} = U. \tag{7.53}$$

As a result we find:

$$C_1 = \frac{U}{h} + \frac{K}{2\mu}hC_2 = 0. \tag{7.54}$$

Thus the solution of the boundary value problem (7.51) is

$$\frac{V_1(x_2)}{U} = \frac{x_2}{h} + \frac{Kh^2}{2\mu U}\left(1 - \frac{x_2}{h}\right)\frac{x_2}{h}. \tag{7.55}$$

For $K = 0$ we find the simple shearing *Couette flow* solution

$$\frac{V_1(x_2)}{U} = \frac{x_2}{h}. \tag{7.56}$$

For $K \neq 0$ we find the *Couette-Poiseuille flow* (Fig. 7.9), which is a superposition of Couette flow and *Poiseuille flow* expressed in terms of Eq. (7.55) The application of the superposition principle is permissible to this and similar cases, where the nonlinear convective terms disappear leading to linear differential equations such as Eq. (7.55).

The above Couette flow bounded by two parallel walls may be thought of as a flow through the gap between two concentric cylinders with radii approaching infinity. In case that radii are finite, the Navier-Stoke's equation can be substantially simplified by using cylindrical coordinate system.

7.3.2 Couette Flow Between Two Concentric Cylinders

Exact solution of Navier-Stoke's equations can also be found for this case. In contrast to the parallel flat walls discussed above, we use two concentric cylinders as the bounding walls that may rotate with different rotational velocities. In this case it is most convenient to use the cylindrical coordinate system for decomposing the Navier-Stoke's equation into its components. We assume that the flow moves in

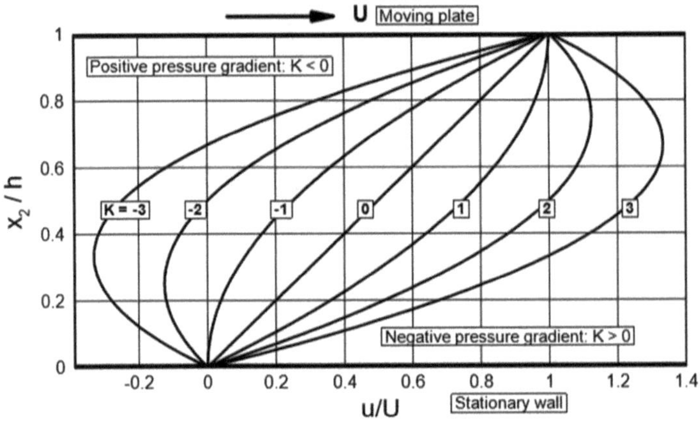

Fig. 7.9 Velocity distribution in Couette flow with pressure gradient

circumferential direction only meaning that the components in radial and axial components are zero everywhere. Furthermore, we assume that the flow is axisymmetric which implies that the pressure in circumferential direction is constant. Implementing these assumptions into the Navier-Stoke's equations (A.74–A.76), the radial component is simplified to

$$\frac{V_\Theta^2}{r} = \frac{1}{\rho}\frac{\partial p}{\partial r} \tag{7.57}$$

and the circumferential component simplifies as

$$\frac{\partial^2 V_\Theta}{\partial r^2} + \frac{1}{r}\frac{\partial V_\Theta}{\partial r} - \frac{V_\Theta}{r^2} = 0. \tag{7.58}$$

Since the velocity is in circumferential direction and changes in radial direction only, we set $V_\theta \equiv u(r)$ and replace the partial derivatives by ordinary ones. As a result we find

$$\frac{u^2}{r} = \frac{1}{\rho}\frac{dp}{dr} \tag{7.59}$$

and

$$\frac{d^2u}{dr^2} + \frac{1}{r}\frac{du}{dr} - \frac{u}{r^2} = 0. \tag{7.60}$$

The solution of Eqs. (7.59) and (7.60) must satisfy the following boundary conditions at the inner and outer cylinder

$$\text{for } r = R_I : u_I = R_I\Omega_I, \text{ and } r = R_0\colon u_0 = R_0\Omega_0 \tag{7.61}$$

where the angular velocity of the outer cylinder may assume negative, zero, or positive values. Using the above boundary conditions, the solution of Eq. (7.60) is

$$u(r) = \frac{1}{R_0^2 - R_I^2}\left[r(\Omega_0 R_0^2 - \Omega_I R_I^2) - \frac{R_I^2 R_0^2}{r}(\Omega_0 - \Omega_I)\right]. \tag{7.62}$$

Introducing dimensionless parameters $\omega = \Omega_0/\Omega_I$, $\rho_I = R_i$, R_0 and $\rho = r/R_0$ Eq. (7.57) is re-arranged as

$$\frac{u(r)}{u_I} = \frac{1}{1-\rho_I^2}\left[\rho\left(\frac{\omega - \rho_I^2}{\rho_I}\right) - \frac{\rho_I}{\rho}(\omega - 1)\right]. \tag{7.63}$$

For the outer cylinder at rest, $\omega = 0$, Eq. (7.63) is reduced to

$$\frac{u(r)}{u_I} = \frac{\rho_I}{1-\rho_I^2}\,\frac{1-\rho^2}{\rho}. \tag{7.64}$$

In a similar approach utilizing Eq. (7.62), a dimensionless expression can be derived that relates $u(r)$ to the surface velocity of the rotating outer cylinder u_0. Assuming the inner cylinder is at rest, while outer cylinder is rotating, we find

$$\frac{u(r)}{u_0} = \frac{\rho_I}{1-\rho_I^2}\left(\frac{\rho}{\rho_I} - \frac{\rho_I}{\rho}\right). \tag{7.65}$$

Figure 7.10 represents the dimensionless velocity distribution in radial direction with $\rho_I = R_I/R_0$ as a parameter for (a) inner cylinder rotating and outer cylinder at rest and (b) inner cylinder at rest and outer cylinder rotating. As the figures show, when $\rho_I = R_I/R_0$ approaches unity the velocity distributions look very similar to flat wall Couette flow for zero pressure gradient-curve plotted in Fig. 7.9. Eqs. (7.64) and (7.65) allow calculating the wall shear stress on the inner and outer cylinder walls using the shear stress relation $\tau_W = \mu(\partial u/\partial r)_W$ with $(\partial u/\partial r)_W = (du/r)_W$ as the velocity slope at the wall for this particular case. The resulting shear stress force and the moment of momentum acting on the surface per unit cylinder depth is calculated from

$$F_S = 2\pi R\mu\left(\frac{du}{dr}\right)_W \tag{7.66}$$

$$M_S = 2\pi R^2\mu\left(\frac{du}{dr}\right)_W. \tag{7.67}$$

Equation (7.67) may be used to experimentally determine the viscosity of the working fluid. With the measured moment of momentum, the angular velocity and the given geometry, the viscosity can be obtained.

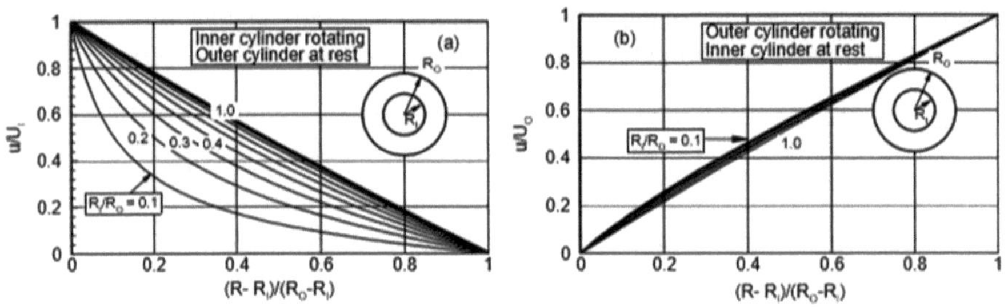

Fig. 7.10 Velocity distribution between two concentric cylinders with rotation, **a** inner cylinder while outer cylinder at rest, **b** inner cylinder at rest, while outer cylinder rotating

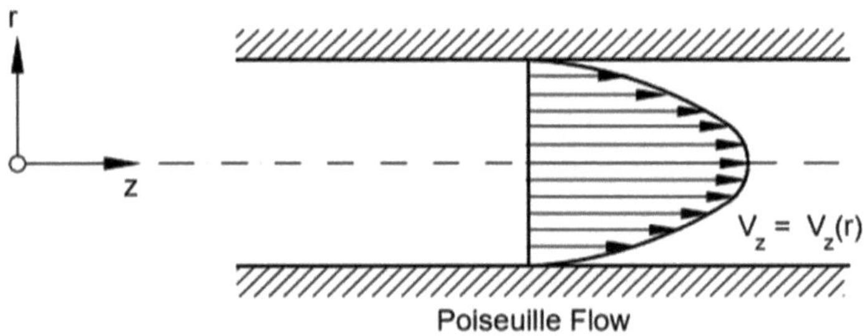

Fig. 7.11 Parabolic velocity distribution in a channel with circular cross section

7.3.3 Hagen-Poiseuille Flow

Axisymmetric laminar flow through a straight circular pipe called Hagen-Poiseuille flow is shown in Fig. 7.11. The velociy distribution in radial direction is obtained as an exact solution of the Navier-Stoke's equations.

Similar to the case discussed preciously, we use the cylindrical coordinate system to decompose the Navier-Stoke's equations in circumferential, radial and axial directions. The no-slip condition at the wall requires that
$V_r = V_\theta = V_z = 0$. We assume that $V_r = V_\theta = 0$ everywhere and require that the flow be axisymmetric ($\partial\partial_\theta = 0$). The continuity equation in cylindrical coordinates (see appendix A) gives

$$\frac{\partial V_z}{\partial z} = 0 \text{ or } V_z = V_z(r). \tag{7.68}$$

Because of the above assumptions, the r-component of the Navier-Stokes equations is reduced to

$$0 = \frac{\partial p}{\partial r}. \tag{7.69}$$

Likewise, all terms of the Navier-Stokes equation in the θ direction vanish identically leaving the z-component as the only non-zero component.

$$0 = -\frac{\partial p}{\partial z} + \mu \left[\frac{\partial^2 V_z}{\partial r^2} + \frac{1}{r}\frac{\partial V_z}{\partial r} \right]. \tag{7.70}$$

Since the expression in the bracket of Eq. (7.70) is only a function of r and considering the axisymmetric assumption, $\partial p/\partial\theta = 0$, and Eq. (7.69), the pressure gradient $\partial p/\partial z \equiv dp/dz$ must be a constant implying that the pressure p is a linear function of z. As before we set $dp/dz = -K$ and re-arrange Eq. (7.70)

$$-\frac{K}{\mu} = \frac{1}{r}\frac{d}{dr}\left[r\frac{dV_z}{dr} \right] \tag{7.71}$$

which, integrated twice, gives

$$V_z(r) = -\frac{Kr^2}{4\mu} + C_1 \ln r + C_2. \tag{7.72}$$

With the maximum velocity located $r = 0$ and the no-slip condition at $r = R$, the solution of Eq. (7.72) is found as

$$V_z(r) = \frac{K}{4\mu}(R^2 - r^2). \tag{7.73}$$

The pressure gradient can be expressed in terms of the maximum velocity by setting in Eq. (7.73) $r = 0$ which results in $V_{\max} = KR^2/4\mu$. Thus, the dimensionless velocity distribution is

$$\frac{V_z(r)}{V_{\max}} = 1 - \left(\frac{r}{R}\right)^2. \tag{7.74}$$

As Eq. (7.73) indicates, the pressure gradient $dp/dz = -K$ is a parameter determining the velocity distribution. Since the pressure drop in a pipe may be set proportional to the averaged dynamic pressure $\rho/2\bar{V}^2$ with $\bar{V}$ obtained from continuity equation:

$$\bar{V}_z = \frac{1}{\pi R^2}\int_0^R \pi V_z(r) r dr = \frac{V_{\max}}{2} = \frac{KR^2}{8\mu}. \tag{7.75}$$

The pressure gradient can be approximated as:

$$\frac{dp}{dz} \approx \frac{\Delta p}{\Delta z} = \frac{p_1 - p_2}{l} \tag{7.76}$$

with Δp as the pressure drop across the pipe length l. Introducing a dimensionless pressure loss coefficient ζ,

$$\zeta = \frac{\Delta p}{\frac{1}{2}\rho \bar{V}_z^2} \tag{7.77}$$

considering Eqs. (7.76) and (7.75), we find:

$$\zeta = 64\frac{l}{D}\frac{\mu}{\rho D\bar{V}_z} = \frac{l}{D}\frac{64}{\text{Re}} = \frac{l}{D}\lambda \tag{7.78}$$

with Re= $\rho \bar{V}_z D/\mu$ and the friction coefficient $\lambda = 64/\text{Re}$.

7.4 Unsteady Laminar Flows

So far, we have treated steady laminar flows through channels with curved walls, straight walls and pipes, for which exact solutions were found. There are also few unsteady flow cases for which exact solutions of Navier-stoke's equations still exist. To describe the solution procedure, in the following two different cases will be presented. More examples are found in Schlichting [39].

7.4.1 Flow Near Oscillating Flat Plate, Stokes-Rayleigh Problem

We consider laminar flow between two plane infinitely extending plates with a distance h from each other, where the lower plate oscillates in its plane. A very detailed discussion of this case is found in an excellent textbook by Spurk [46] which is reflected here. Similar to the cases presented previously, the unsteady flow under investigation is unidirectional, where the corresponding assumptions are applicable. This implies that there exist only non-zero velocity component, which we set $V_1 = V_1(x_2)$ and simply as $V_1 = V_1(x_2, t) \equiv u = u(y, t)$. The wall oscillation velocity is given by

$$u_w = u_{y=0} = U(t) = \hat{U}\cos(\omega t). \tag{7.79}$$

Using the complex notation, the wall velocity reads

$$u_w = U(t) = \hat{U}e^{i\omega t} \tag{7.80}$$

where only the real part $\Re(e^{iwt})$ has physical meaning. Utilizing the velocity distribution

$$u = f(yt) \tag{7.81}$$

the u-component of the Navier-Stokes equations is written as:

$$\frac{\partial u}{\partial t} = -\frac{1}{\varrho}\frac{\partial p}{\partial x} + v\frac{\partial^2 u}{\partial y^2}. \tag{7.82}$$

Since the flow motion is caused by oscillation of the lower wall with the no-slip condition, pressure changes in x-direction can be excluded leading to $\partial p \partial x = 0$, thus the boundary conditions at the lower and upper wall are given as:

$$BC1 : u(0, t) = u_w = \hat{U}e^{i\omega t},\ BC2: u(h, t) = 0. \tag{7.83}$$

Since we are interested in the oscillation state after the initial transients have died away, we do not need to include time t in boundary conditions. Considering the boundary conditions (7.83), we may set

$$u(y, t) = \hat{U}e^{i\omega t} f(y) \tag{7.84}$$

where the $f(y)$, which is to be determined, has to satisfy the boundary conditions (7.83).

$$BC1 : f(0) = 1,\ BC2: f(h) = 0. \tag{7.85}$$

Inserting Eqs. (7.84) into (7.82), the partial differential Eq. (7.82) is reduced to an ordinary differential equation with constant (complex) coefficients

$$f'' - \frac{i\omega}{v} f = 0, \tag{7.86}$$

where $f'' = d^2 f/dy^2$. From the solution $f(y) = e^{\lambda y}$ we obtain the characteristic polynomial

$$\lambda^2 - \frac{i\omega}{\nu} = 0, \tag{7.87}$$

with the roots

$$\lambda = \pm(1 + i)\sqrt{\frac{\omega}{2\nu}}. \tag{7.88}$$

With (7.88), the general solution of Eq. (7.86) can be written in the form

$$f(y) = A \sin h(1 + i)\sqrt{\omega/vy} + B \cos h(1 + i)\sqrt{\omega/2vy}, \tag{7.89}$$

from which, using the boundary conditions (7.83), we find the special solution

$$f(y) = \frac{\sin h\{(1 + i)(h - y)\sqrt{\omega/2v}\}}{\sin h\{(1 + i)\sqrt{\omega/2vh}\}} \tag{7.90}$$

which inserted into Eq. (7.84) gives the velocity distribution

$$u(y,t) = \hat{U}\Re\left\{e^{i\omega t}\frac{\sin h\{(1+i)(1-y/h)\sqrt{\omega h^2/2v}\}}{\sin h\{(1+i)\sqrt{\omega h^2/2v}\}}\right\}. \tag{7.91}$$

In Eq. (7.91) the dimensionless argument h^2/v represents a time scale for diffusion of oscillating motion across the channel height. The following two limiting cases discussed in [46] are presented in this section:

$$\omega h^2/\nu \ll 1 \tag{7.92}$$

$$\omega h^2/\nu \gg 1. \tag{7.93}$$

In the first case this time is much smaller than the typical oscillation time $1/\omega$, i.e. the diffusion process adjusts at every instant the velocity field to the steady shearing flow with the instantaneous wall velocity $u_w(t)$. This is what is called *quasi-steady* flow. Using the first term of the expansion of the hyperbolic sine function for small arguments we have

$$u = \hat{U}\Re\left\{e^{i\omega t}\frac{(1+i)(1-y/h)\sqrt{\omega h^2/2\nu}}{(1+i)\sqrt{\omega h^2/2\nu}}\right\} \tag{7.94}$$

and deduce that

$$u = \hat{U}\cos(\omega t)(1-y/h) = U(1-y/h). \tag{7.95}$$

Equation (7.95) corresponds to the simple Couette flow (7.56) where the upper plate represents the moving wall. We also obtain this limiting case if the kinematic viscosity v tends to infinity. In the limit $\omega h^2/v \gg 1$ we use the asymptotic form of the hyperbolic sine function and write Eq. (7.91) in the form

$$u = \hat{U}\Re(e^{-\sqrt{\omega/2\nu}y}e^{i(\omega t-\sqrt{\omega/2\nu}y)}) \tag{7.96}$$

or

$$u = \hat{U}e^{-\sqrt{\omega/2\nu}y}\cos(\omega t - \sqrt{\omega/2\nu}y). \tag{7.97}$$

The distance h no longer appears in Eq. (7.97). Measured in units $\lambda = \sqrt{2v/\omega}$ the upper wall is at infinity. Relative to the variable y the solutions also have a wave form; we call these *shearing waves* of wavelength λ. To obtain the velocity at the wall, we set in Eq. (7.97) $y = 0$ and arrive at:

$$u_W = \hat{U}\cos(\omega t). \tag{7.98}$$

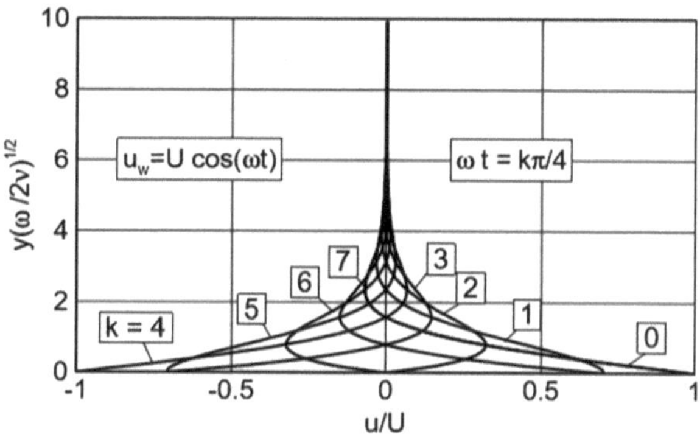

Fig. 7.12 Unsteady velocity distribution caused by oscillating the bottom wall

The velocity distribution described by Eq. (7.97) is plotted in Fig. 7.12 for different k-values in the parameter $\omega t = k\pi/4$.

7.4.2 Influence of Viscosity on Vortex Decay

Reconsider the case of two concentric rotating cylinders we treated in Sect. 7.3.2 with the velocity distribution described by Eq. (7.62). Setting $\Omega_2 = 0$ and assuming that the outer radius goes to infinity, while the inner radius approaches an infinitesimally small size similar to the one of a vortex filament (see Sect. 6.6.3), Eq. (7.62) reduces to:

$$u(r) = \frac{R^2\Omega}{r} \tag{7.99}$$

with R as the radius of the inner cylinder (filament) and Ω its angular velocity. Equations (7.99) and (7.100) describe the velocity around a vortex filament with the strength $\Gamma = 2\pi R^2\Omega = 2\pi Ru$ (see Sect. 6.6.3). For a constant circulation within an inviscid flow field the velocity at an arbitrary radius r is

$$u(r) = \frac{\Gamma}{2\pi r}. \tag{7.100}$$

Equation (7.100) implies that the flow velocity at the center of the vortex $r = 0$ becomes infinity indicating a discontinuity at the center of the vortex. We now suppose that cylinder which is rotating with an angular velocity Ω and is embedded in a viscous environment suddenly stops rotating at time $t = 0$. This triggers a transient event, where the flow velocity continuously decreases as a result of viscous diffusion. This transient event is described by the Navier-Stokes equations (4.47). From Eqs. (7.99) and (7.100) it follows that the streamlines are concentric circles (see also Sect. 6.2.2.3). Thus, the flow may be assumed to be unidirectional in circumferential direction with $V_\theta = V_\theta(r, t)$, implying that $V_r = V_z = 0$. This requires that the pressure gradient in circumferential direction must vanish. As a consequence, Eq. (4.47) reduces to:

$$\frac{\partial V_\Theta}{\partial t} = \nu\left(\frac{\partial^2 V_\Theta}{\partial r^2} + \frac{1}{r}\frac{\partial V_\Theta}{\partial r} - V\frac{\Theta}{r^2}\right). \tag{7.101}$$

The solution of Eq. (7.101) must satisfy the following boundary conditions:

$$\begin{aligned} &BC1\text{: at } t = 0,\ V_\theta(r, 0) = \Gamma/2\pi r \\ &BC2\text{: at } r = 0,\ V_\theta(0, t) = 0. \end{aligned} \tag{7.102}$$

To find the solution for Eq. (7.101), we introduce a dimensionless parameter $\eta = r^2/4\nu t$ such that Eq. (7.101) is transformed into an ordinary differential equation in terms of $V_\theta = f(\eta)$ with η as an the independent variable leading to:

$$f'' + f' = 0 \tag{7.103}$$

with the solution;

$$f = 1 - e^\eta \tag{7.104}$$

that results in the solution for the circumferential velocity:

$$u \equiv V_\theta = \frac{\Gamma}{2\pi r}(1 - e^{-r^2/4\nu t}). \tag{7.105}$$

Setting in Eq. (7.105) $t = 0$, we obtain the reference velocity:

$$U_0 \equiv V_{\theta_0} = \frac{\Gamma}{2\pi r_0} \tag{7.106}$$

which represents the velocity of the vortex in an inviscid flow field. Using Eq. (7.106), the nondimensionalized version of Eq. (7.105) is

$$\frac{u}{U_0} = \frac{r_0}{r}(1 - e^{-r^2/4\nu t}). \tag{7.107}$$

Equation (7.107) represents an exact solution of the Navier-Stokes equation that describes the distribution of the circumferential velocity component of a decaying vortex as a function of radial distance and time. It was derived by Oseen [47]. The velocity distributions described by Eq. (7.107) are plotted in Fig. 7.13a, b.

Figure 7.13a shows the velocity distribution in radial direction with dimensionless time as a parameter. The dashed curve with $\nu t/r_0^2 = 0$ represents the irrotational solution with the origin as the singularity. For $\nu t/r_0^2 > 0$ the damping effect of the viscosity is clearly visible. However, at $r/r_0 = 1$ all viscous (rotational) solutions approach the inviscid (irrotational) solution. Figure 7.13b exhibits the velocity decay

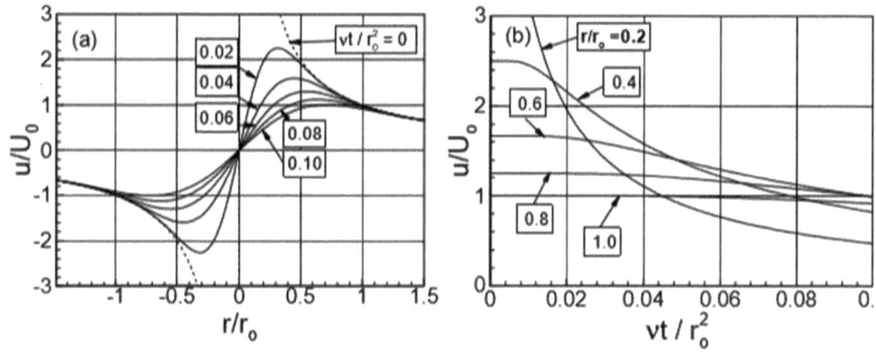

Fig. 7.13 Velocity distribution caused by a decaying vortex, **a** dimensionless velocity in radial direction with dimensionless time as parameter, **b** temporal change of dimensionless velocity with dimensionless radius as parameter

for each r/r_0-ratio. The rotational behavior of the unsteady vortex decay described by Eq. (7.105) can be shown explicitly by calculating the vorticity $\boldsymbol{\omega} = \nabla \times \mathbf{V}$ which has, in this particulare case, only one non-zero component:

$$\omega_z = \frac{1}{r}\frac{\partial (r V_\theta)}{\partial r}. \tag{7.108}$$

Substituting in Eq. (7.108) V_θ by Eq. (7.105), we find:

$$\omega_z = \frac{\Gamma}{4\pi \nu t} e^{\left(-\frac{r^2}{4\nu t}\right)}. \tag{7.109}$$

Equation (7.109) shows that for $t = 0$, the solution is irrotational, while for $t > 0$ it becomes rotational.

7.5 Problems

Problem 7.1 A Newtonian fluid with constant density and viscosity flows steadily through a two dimensional vertically positioned channel with the width $2h$ shown in Fig. 7.14. The motion of the fluid is described by the Navier Stokes equations. The flow is subjected to the gravitational acceleration $\mathbf{g} = \mathbf{e}_1 \mathbf{g}$ and a constant pressure gradient in flow x_1 direction. Assume that $V_2 = V_3 = 0$.

(a) Determine the solution of the Navier-Stokes equations.
(b) Write a computer program, show the velocity distributions for the following cases: a) For $K = 0$,(b) $K > 0$, and (c) $K < 0$. c) For which K there is no flow?

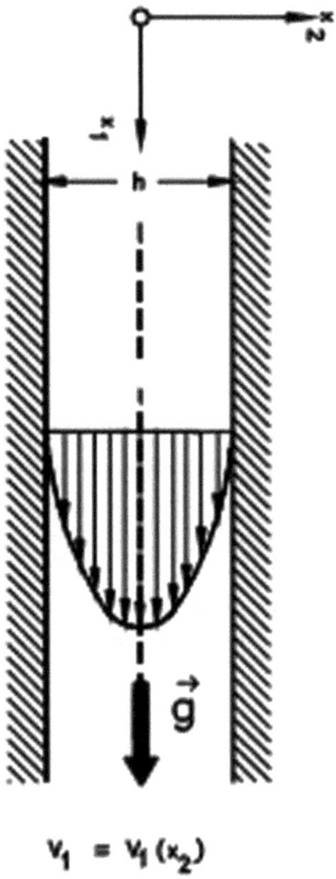

Fig. 7.14 Viscous flow through a vertical channel [275]

Problem 7.2 Newtonian fluid with constant density and viscosity flows *steadily* through a two dimensional channel positioned at an angle α shown in Fig. 7.15 with the width $2h$. The motion of the fluid is described by the Navier-Stokes equations. The flow is subjected to the gravitational acceleration $\mathbf{g} = \mathbf{e}_1 g_1 + \mathbf{e}_2 g_2$ and a constant pressure gradient in flow direction x_1. Assume that

(a) Determine the solution of the Navier-Stokes Equations.
(b) Write a computer program and plot the velocity distributions for: (a) For $K = 0$,(b) $K > 0$, and (c) $K < 0$.
(c) For which K there is no flow?

Problem 7.3 River water considered a Newtonian fluid with constant viscosity and density steadily flows down an inclined river bed at a constant height h as shown in the Fig. 7.16. The motion of the fluid is described by the Navier-Stokes equation. Along the sloped river bed, the flow is driven by the gravitational acceleration and its free surface is subjected to the constant atmospheric pressure p_{atm}. The air viscosity at the free surface is negligible compared to the water viscosity. Furthermore, we assume that the flow is unidirectional in x_1 direction.

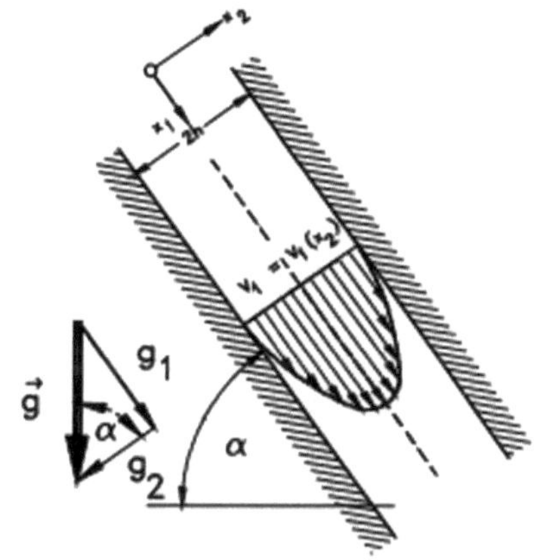

Fig. 7.15 Viscous flow through a channel positioned at an angle [275]

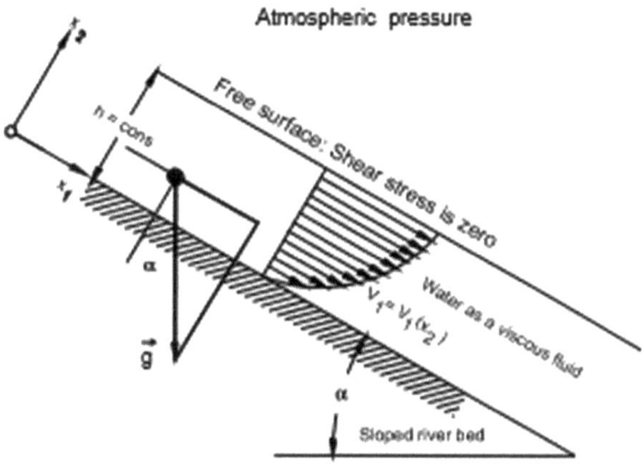

Fig. 7.16 River flow with open surface exposed to atmospheric pressure [275]

(a) Decompose the Navier-Stokes equation into its components.
(b) Show that the $\partial V_1/\partial x_1 = 0$.
(c) Solve the Navier-Stokes equations and find the velocity distribution in x_2 direction.
(d) Determine the velocity ratio $V_1 = V_{1\,\max}$.
(e) Determine the river mass flow $\dot{m}$.

Problem 7.4 Incompressible Newtonian fluid with constant density and viscosity flows between two parallel plates with infinite width. Body forces are neglected. Given are the plate height h, the components of the pressure gradient $\frac{\partial p}{partialx_1} = -K$, $\frac{partialp}{\partial x_2} \equiv 0$, $\frac{\partial p}{\partial x_3} \equiv 0$. and the velocity field between the plates is given by: $V_1(x_2) = \frac{K}{2}\mu\left(\frac{h^2}{4} - x_2^2\right)$, $V_2 \equiv 0$, $V_3 \equiv 0$.

(a) Show that the given velocity field satisfies the continuity and the Navier-Stokes equation.
(b) Determine the components of the stress tensor.

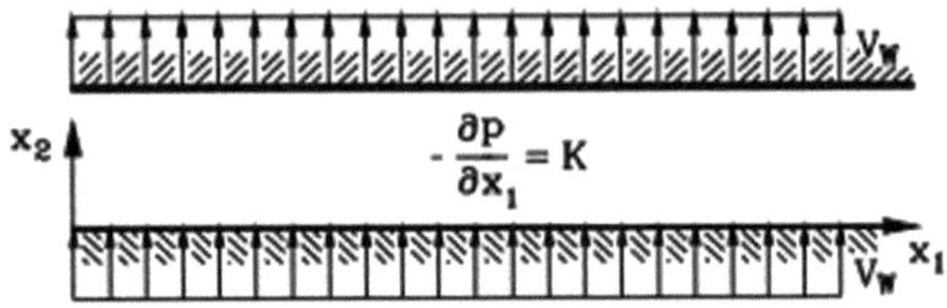

Fig. 7.17 Viscous flow through a porous channel at constant pressure gradient [275]

(c) Calculate the dissipation function Φ.
(d) Find the energy per unit depth, length, and time dissipated in heat within the gap.
(e) Calculate the principal stresses and their directions.

Problem 7.5 Reconsider the flow calculated in Problem 7.4 and assume a calorically perfect fluid with a constant heat conductivity λ. Further assume a constant temperature at the top wall T_0 and a full heat insulation at the bottom wall.

(a) Determine the temperature distribution $T(x_2)$ in the gap.
(b) Find the temperature at the bottom wall.
(c) Determine the heat flux per unit area through the top wall.
(d) Calculate the entropy increase Ds/Dt of the fluid inside the gap.

Problem 7.6 Newtonian fluid flows through the channel shown in Fig. 7.17 with infinite extensions in x_1- and x_3- direction and the height h. The plane flow is steady, the density ϱ and the viscosity μ: are assumed to be constant, and body forces are neglected. The top and bottom wall are porous such that a constant normal velocity component V_{2W} can be established at the walls. The pressure gradient in x_1-direction is constant ($\partial p/\partial x_1 = -K$). Because of the infinite extension of the channel, the velocity distribution does not depend upon x_1.

(a) Using the continuity equation calculate the distribution of the velocity component in x_2-direction $V_2(x_2)$.
(b) Simplify the x1- component of the Navier-Stokes equation for this problem.
(c) Give the boundary condition for the velocity component u_1.
(d) Calculate the velocity distribution $V_1(x_2)$. Hint: After solving the homogeneous differential equation, the particular solution of the inhomogeneous differential equation can be found setting V_{1_p} =const.x_2.

Given: ϱ, μ, K, h, V_W

Problem 7.7 Newtonian fluid (ϱ, μ =const.) flows steadily through the channel, Fig. 7.18, (height $2h$). In the middle of the channel, an infinitely thin splitter plate is mounted. The channel walls move with a constant velocity U in positive x_1-direction. The two fluid streams separated by the plate are mixed at the end of the plate. At station [2], a new velocity profile $u_1 = u_1(x_2)$ is developed that does not change anymore with x_1. The body forces can be neglected.

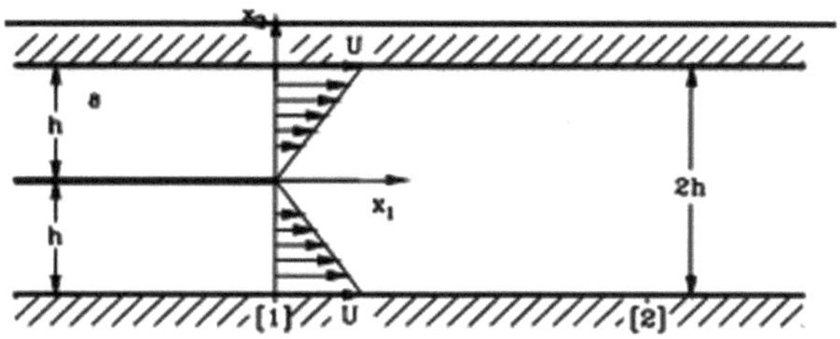

Fig. 7.18 Channel with geometry described in problem 7.7 [275]

(a) Using the equation of motion, show that the pressure gradient $\partial p/\partial x_1$ downstream of [2] does not change.
(b) Calculate the volume flux per unit depth $\dot{V}$ at station [1].
(c) Obtain the velocity profile $u_1 = u_1(x_2)$ at station [2] using the no-slip condition at $x_2 = \pm h$ and the requirement that the volume flux at stations [2] must be the same as at [1]. Show that the pressure gradient must be different from zero, resulting in a pressure driven Couette flow.
(d) Calculate the pressure gradient.

Given: h, U, ϱ, η

Problem 7.8 A curved channel is described by the following orthogonal curvilinear coordinate system: $z = e^{-\frac{1}{2}(a+ib)w}$ with $z = x + iy$, and $w = \xi_1 i \xi_2$

(a) Find the base vectors, metric coefficients, and Christoffel symbols;
(b) Generate a grid for ξ_1 from 3.0 to 5.0 and ξ_2 from 0.1 to 0.5.
(c) Transform the continuity and Navier-Stokes equation into this curvilinear coordinate system.
(d) Solve the Navier Stokes equations and plot the velocity distribution for the logarithmic spiral with $a = -1$ and $b = 1$.
(e) Find the Navier-Stokes solutions for purely radial flow by setting $a = -2$ and $b = 0$ and the flow through concentric cylinders with $a = 0$ and $b = 1$.

Problem 7.9 A two-dimensional symmetric curved channel is described by the following orthogonal curvilinear coordinate system: $z = \frac{1}{2}w^2$, with $z = x + iy$, and $w = \xi_1 + i\xi_2$.

(a) Find the base vectors, metric coefficients, and Christoffel symbols;
(b) Generate a grid for ξ_1 from 10 to 15 and ξ_2 from 0 to $\pm$ 0.8.
(c) Transform the continuity and Navier-Stokes equation into this curvilinear coordinate system.
(d) Assume that the velocity component in ξ_2-direction compared to the component in $\xi_1$1-direction can be neglected. Give an approximate solution of the Navier Stokes equations and plot the velocity distribution.

Problem 7.10 A curved channel is described by the following orthogonal curvilinear coordinate system: $\bar{z} = \frac{1}{w}$ with $\bar{z} = x + iy$, and $w = \xi_1 + i\xi_2$.

(a) Find the base vectors, metric coefficients, and Christoffel symbols; Generate a grid for >1 from 0.2 to 0.1 and >2 from 0 to $\pm$ 0.008.
(b) Transform the continuity and Navier-Stokes equation into this curvilinear coordinate system.
(c) Assume that the velocity component in ξ_2-direction compared to the component in ξ_1-direction can be neglected. Solve the Navier Stokes equations and plot the velocity distribution.

Chapter 8
Laminar-Turbulent Transition

8.1 Stability of Laminar Flow

This Chapter is devoted to the complex problematic of laminar flow stability, intermittency, steady and unsteady boundary transition. The phenomena of stability of laminar flows, transition, and turbulence were systematically studied first by Reynolds [48] in the eighties of the eighteenth century. Schlichting [49, 50] and in his classical textbook *Boundary Layer Theory* [51] gives an excellent treatment of these complex flow phenomena and critically reviews the contributions up to 1979, where the seventh and last edition of his book appeared. In this chapter, we first treat the fundamental issues pertaining to the subject matter followed by original contributions recently made in the area of steady and unsteady boundary layer transition.

In Chap. 7, we have presented several exacts solutions of the Navier-Stokes equations, where at given Reynolds numbers, the effect of curvature and pressure gradient on the velocity and temperature distributions were discussed. To perform the integration process without encountering numerical instabilities, we have utilized Reynolds numbers ranging from 500 to 5000. For the particular geometry pertaining to the positive pressure gradient (decelerated flow), the highest Reynolds number we could apply without numerical instability was about Re = 1500. For a negative pressure gradient (accelerated flow) and the same geometry, but with reversed flow direction, a Reynolds number as high as Re = 5000 could be used. For higher Reynolds numbers, numerical instabilities occurred, indicating the sensitivity of the laminar flow at positive pressure gradient with respect to higher Reynolds numbers. In fact, for a given geometry, there is always a definite Reynolds number, the *critical Reynolds number,* Re_{crit}, above which the flow pattern changes drastically. The numerical value of this critical Reynolds number, however, depends, strongly, among other things, on pressure gradient, inlet flow conditions, and surface roughness. For a steady flow through a pipe with very smooth surface and no inlet disturbance, the critical Reynolds number is approximately

M. T. Schobeiri, *Advanced Fluid Mechanics and Heat Transfer for Engineers and Scientists*, https://doi.org/10.1007/978-3-030-72925-7_8

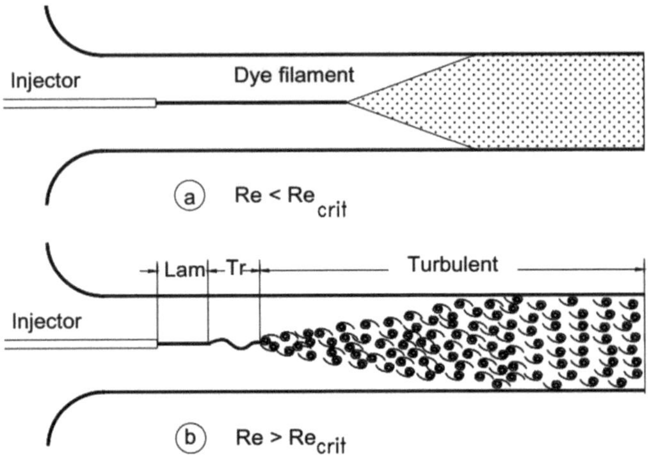

Fig. 8.1 On the stability of laminar flow, historic dye filament experiment by Reynolds [48]; **a** laminar flow for $\mathrm{Re}<\mathrm{Re}_{\mathrm{crit}}$; **b** turbulent flow for $\mathrm{Re}>\mathrm{Re}_{\mathrm{crit}}$

$$\mathrm{Re}_{\mathrm{crit}} = \left(\frac{\bar{U}d}{\nu}\right)_{\mathrm{crit}} \approx 2300. \tag{8.1}$$

Keeping the working medium and the geometry the same, a change in the flow velocity results in a change of the Reynolds number. For the above flow, the laminar flow pattern is sustained as long as $\mathrm{Re}<\mathrm{Re}_{\mathrm{crit}}$. At this Reynolds number, the flow exhibits a well orderly pattern and the fluid particles travel along neighboring layers. Approaching the $\mathrm{Re}_{\mathrm{crit}}$ and eventually increasing it beyond the critical one $\mathrm{Re}>\mathrm{Re}_{\mathrm{crit}}$ causes the flow pattern to change drastically. The orderly pattern ceases to exist. This drastic change of the flow pattern is demonstrated by the classic dye filament experiment conducted by Reynolds [48] and reconstructed in Fig. 8.1. At a low Reynolds number $\mathrm{Re} < 2000$, the filament remained laminar with sharply defined boundaries in th center of the stream that spread slowly due to molecular diffusion, Fig. 8.1a. Increasing the Reynolds number above the critical one changed the flow pattern completely leading to a strong mixing of the dye filament particles with the main flow. At a Reynolds number $\mathrm{Re}>\mathrm{Re}_{\mathrm{crit}}$ the particles of the dye filament were subjected not only to a longitudinal motion but also to a lateral motion with a high frequency random fluctuation superimposed on the main (longitudinal) motion. This high frequency random fluctuation which is inherently three dimensional characterizes the new flow pattern that is termed *turbulence*. As soon as the flow becomes turbulent the filament diffuses into the stream and the fluid becomes uniformly colored in a short distance downstream of the dye injector as seen in Fig. 8.1b.

8.2 Laminar-Turbulent Transition

Increasing the Reynolds number from a subcritical to a supercritical range, the flow undergoes a *laminar- turbulent transition process*. This transition process relates the stable, subcritical laminar state to the stable, supercritical turbulent state and is of fundamental importance for the entire engineering fluid mechanics. As indicated above, the complex process of transition is affected by several parameters, the most significant ones are the Reynolds number, pressure gradient, surface roughness, and the external disturbance (turbulence intensity $Tu = \sqrt{V'2}/\bar{V}$) in the free stream. To understand the fundamentals of the laminar turbulent transition, we try first to reduce the number of parameters affecting the transition process. This is done effectively by investigating the transition within the boundary layer along a flat plate with a smooth surface at zero pressure gradient. This is particularly important for the development of boundary layer and its onset which is primarily responsible for the inception and magnitude of the drag forces that exert on any surface exposed to a flow field. Figure 8.2 schematically explains the transition process that takes place within the boundary layer along a flat plate at zero pressure gradient.

Starting from the leading edge, the viscous flow along the plate generates two distinctly different flow regimes. Close to the wall, where the viscosity effect is predominant, a thin *boundary layer* is developed, within which the velocity grows from zero at the wall (no-slip condition) to a definite magnitude at the edge of the boundary layer (the boundary layer and its theory is extensively discussed in Chap. 11). Inside this thin viscous layer the flow initially constitutes a *stable laminar boundary layer* flow that starts from the leading edge and extends over a certain range ①. By further passing over the plate surface, the first indications of the laminar flow instability appear in form of infinitesimal unstable two-dimensional disturbance waves that are referred to as Tollmien-Schlichting waves ②. Further downstream,

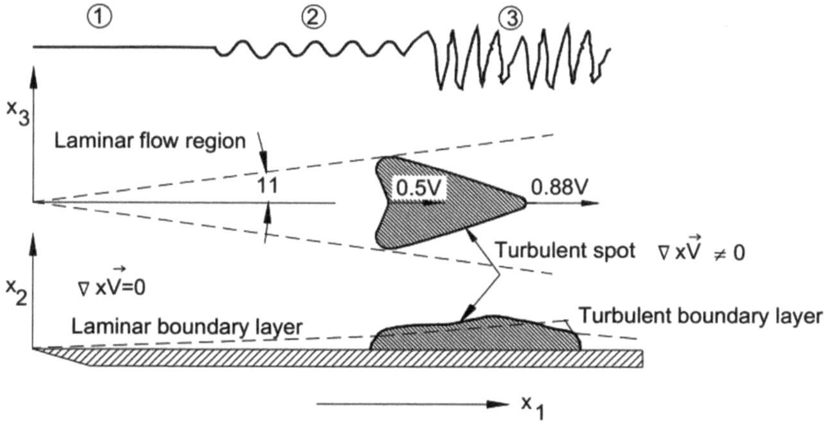

Fig. 8.2 Transition process along a flat plate at zero-pressure gradient sketched by Schubauer and Klebanoff [6]

discrete turbulent spots with highly vortical core appear *intermittently* and randomly ③. Inside these wedge-like spots the flow is predominantly turbulent with $\nabla \times \mathbf{V} \neq \mathbf{0}$, whereas outside the spots it is laminar. According to the experiments by Schubauer and Klebanoff [53], the leading edge of a turbulent spot moves with a velocity of $V_{le} = 0.88U$, whereas its trailing edge moves with a lower velocity of $V_{te} = 0.5U$. As a consequence, the spot continuously undergoing deformation decomposes and builds new sets of turbulence spots with increasingly random fluctuations characteristic of a turbulent flow. Schubauer and Klebanoff [53] also noted the existence of a *calmed* region trails behind the turbulent spot. This region was named calmed because the flow is not receptive to disturbances.

Analytical investigations by McCormick [53] indicate that artificially created turbulent spot does not persist if the Reynolds number satisfies the condition $Re_{\delta 1} < 500$ which results from linear stability theory. Schlichting [51] summarized the transition process as follows:

① A stable laminar flow is established that starts from the leading edge and extends to the point of inception of the unstable two-dimensional *Tollmien-Schlichting waves.*
② Onset of the unstable two-dimensional Tollmien-Schlichting waves.
③ Development of unstable, three-dimensional waves and the formation of vortex cascades.

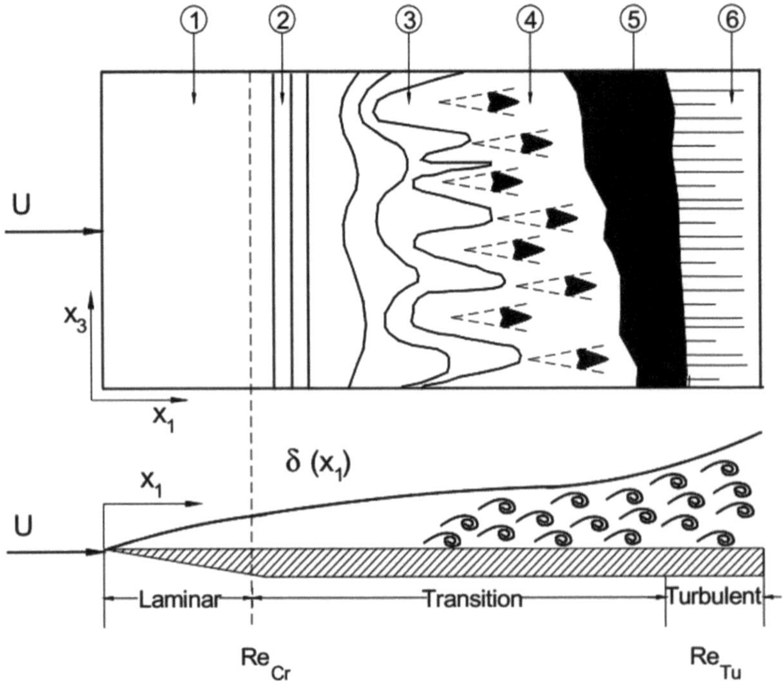

Fig. 8.3 Sketch of transition process in the boundary layer along a flat plate at zero pressure gradient, a composite picture of features in [6] after [7]

④ Bursts of turbulence in places with high vorticity.
⑤ Intermittent formation of turbulent spots with high vortical core at intense fluctuation.
⑥ Coalescence of turbulent spots into a fully developed turbulent boundary layer.

White [54] presented the a simplifying sketch, Fig. 8.3, of transition process of a *disturbance free* flow along a smooth flat plate at zero pressure gradient by assembling the essential elements of transition measured by Schubauer and Klebanoff [52].

The process of flow transition from laminar to turbulent in the sequence discussed above takes place at low level of freestream turbulence intensity of 0.1% or less. In this case, the presence of Tollmien-Schlichting waves are clearly present leading to a process of *natural transition*. In many engineering applications, particularly in turbomachinery flows, where the main stream is periodic unsteady associated with highly turbulent fluctuations. The boundary layer transition mainly occurs bypassing the amplification of Tollmien-Schlichting waves. This type of transition is called *bypass transition* [55].

8.3 Stability of Laminar Flows

The transition process described briefly above have been the subject of ongoing theoretical and experimental investigations for more than half of a century. A-priori predicting the transition process flows is based on the assumption that laminar flow stability is affected by small external disturbances. In case of internal flows though pipes, nozzles, diffusers, turbine or compressor blades channels, these disturbances may originate, for example, in the inlet, whereas in the case of a boundary layer on a solid body that is exposed to a flow may be due to wall roughness or disturbance in the external flow. In this connection, we exclude external disturbances that accelerate the transition start. We also exclude the effect of pressure gradient on the transition process, assuming a flow at zero-pressure gradient. Thus, we restrict our self to investigating the effect of small disturbances on the stability of laminar flows. A stable laminar flow continues to remain stable as long as the small disturbances die out with time. On the other hand the laminar flow becomes unstable if the disturbances increase with time and there is possibility of transition into turbulent.

8.3.1 Stability of Small Disturbances

We consider a *statistically steady* flow motion, on which a *small disturbance* is superimposed. This particular flow is characterized by a constant mean velocity vector field $\bar{\mathbf{V}}(\mathbf{x})$ and its corresponding pressure $\bar{p}(\mathbf{x})$. We assume that the small disturbances we superimpose on the main flow is inherently unsteady, three-dimensional and is described by its vector filed $\tilde{\mathbf{V}}(\mathbf{x}, t)$ and its pressure disturbance $\tilde{p}(\mathbf{x}, \mathbf{t})$. In contrast to the random fluctuations which characterize turbulent flows, the disturbance field is of *deterministic* nature that is why we denote the disturbances with a tilde (~) as opposed to a prime (′), which we use for random fluctuations. Thus, the resulting motion has the velocity vector field:

$$\mathbf{V}(\mathbf{X}, t) = \bar{\mathbf{V}}(\mathbf{X}) + \tilde{\mathbf{V}}(\mathbf{X}, t) \tag{8.2}$$

and the pressure field:

$$\mathbf{p}(\mathbf{X}, t) = \bar{\mathbf{p}}(\mathbf{X}) + \tilde{\mathbf{p}}(\mathbf{X}, t). \tag{8.3}$$

Assuming that $|\tilde{\mathbf{V}}(\mathbf{X}, t)|LL|\bar{\mathbf{V}}(\mathbf{X})|$ and $\tilde{p}(\mathbf{X}, t)LL\bar{\mathbf{p}}(\mathbf{X})$, we introduce Eqs. (8.2) and (8.3) into the Navier Stokes equation (4.43):

$$\frac{\partial(\bar{\mathbf{V}} + \tilde{\bar{V}})}{\partial t} + (\bar{\mathbf{V}} + \tilde{\mathbf{V}}) \cdot \nabla(\bar{\mathbf{V}} + \tilde{\mathbf{V}}) = -\frac{1}{\rho}\nabla(\bar{p} + \tilde{p}) + \nu\Delta(\bar{\mathbf{V}} + \tilde{\mathbf{V}}). \tag{8.4}$$

Performing the differentiation and multiplication, we arrive at:

$$\frac{\partial \tilde{\mathbf{V}}}{\partial t} + \bar{\mathbf{V}} \cdot \nabla\bar{\mathbf{V}} + \bar{\mathbf{V}} \cdot \nabla\tilde{\mathbf{V}} + \tilde{\mathbf{V}} \cdot \nabla\bar{\mathbf{V}} + \tilde{\mathbf{V}} \cdot \nabla\tilde{\mathbf{V}} = -\frac{1}{\rho}(\nabla\bar{p} + \nabla\tilde{p}) + \nu(\Delta\bar{\mathbf{V}} + \Delta\tilde{\mathbf{V}}). \tag{8.5}$$

The *small disturbance* leading to linear stability theory requires that the nonlinear disturbance terms be neglected. This results in

$$\frac{\partial \tilde{\mathbf{V}}}{\partial t} + \bar{\mathbf{V}}.\nabla\bar{\mathbf{V}} + \bar{\mathbf{V}}.\nabla\tilde{\mathbf{V}} + \tilde{\mathbf{V}}.\nabla\bar{\mathbf{V}} = -\frac{1}{\rho}\nabla\bar{p} + \nu\Delta\bar{\mathbf{V}} - \frac{1}{\rho}\nabla\tilde{p} + \nu\Delta\tilde{\mathbf{V}}. \tag{8.6}$$

Equation (8.6) is the composition of the main motion flow superimposed by a disturbance. The velocity vector $\bar{\mathbf{V}}(\mathbf{x})$ constitutes the Navier-Stokes solution of the main laminar flow. Since the solution of the main laminar flow satisfies the Navier-Stokes equation (8.6) must also fulfill the Navier-Stokes equation. As a consequence, we have:

$$\frac{\partial \tilde{\mathbf{V}}}{\partial t} + \bar{\mathbf{V}}.\nabla\tilde{\mathbf{V}} + \tilde{\mathbf{V}}.\nabla\bar{\mathbf{V}} = -\frac{1}{\rho}\nabla\tilde{p} + \nu\Delta\tilde{\mathbf{V}}. \tag{8.7}$$

Equation (8.7) in Cartesian index notation is written as

$$\frac{\partial \tilde{V}_i}{\partial t} + \bar{V}_j\frac{\partial \tilde{V}_i}{\partial x_j} + \tilde{V}_j\frac{\partial \bar{V}_i}{\partial x_j} = -\frac{1}{\rho}\frac{\partial \tilde{p}}{\partial x_i} + \nu\frac{\partial^2 \tilde{V}_i}{\partial x_j \partial x_j}. \tag{8.8}$$

Equation (8.8) describes the motion of a three-dimensional disturbance field superimposed on a three-dimensional laminar main flow field. In order to find an analytic solution that determines the stability of the main flow, we have to make two further simplifying assumptions. The first assumption implies that the main flow is unidirectional in the sense defined in Chap. 7. Thus, the main flow is assumed to be two-dimensional, where the velocity vector in streamwise direction changes only in lateral direction. The second assumption concerns the disturbance field. In this case, we also assume the disturbance field to be two-dimensional too. The first assumption is considered less controversial, since the experimental verification shows that in an unidirectional flow, the lateral component can be neglected compared with the longitudinal one. As an example, the boundary layer flow along a flat plate at zero pressure gradient can be regarded as a good approximation. The second assumption concerning the spatial two dimensionality of the disturbance flow is not quite obvious and may raise objections that the disturbances need not be two dimensional at all. Squire [56] performed a stability analysis using disturbances which were periodic also in z-direction and found that a two dimensional laminar flow becomes unstable at higher Reynolds number if the disturbance is assumed to be three-dimensional than when it is supposed to be two-dimensional. This means that a two-dimensional disturbance causes an earlier instability leading to lower critical Reynolds numbers. Furthermore, the use of two-dimensional disturbance leads faster to the linear stability equation, which may also be achieved using a three-dimensional disturbance assumption. With these assumption, the decomposition of Eq. (8.8) in its components yields:

$$\frac{\partial \tilde{V}_1}{\partial t} + \bar{V}_1 \frac{\partial \tilde{V}_1}{\partial x_1} + \tilde{V}_2 \frac{\partial \bar{V}_1}{\partial x_2} = -\frac{1}{\rho}\frac{\partial \tilde{p}}{\partial x_1} + \nu \left(\frac{\partial^2 \tilde{V}_1}{\partial x_1^2} + \frac{\partial^2 \tilde{V}_1}{\partial x_2^2}\right)$$

$$\frac{\partial \tilde{V}_2}{\partial t} + \bar{V}_1 \frac{\partial \tilde{V}_2}{\partial x_1} = -\frac{1}{\rho}\frac{\partial \tilde{p}}{\partial x_2} + \nu \left(\frac{\partial^2 \tilde{V}_2}{\partial x_2^2} + \frac{\partial^2 \tilde{V}_2}{\partial x_2^2}\right). \tag{8.9}$$

The continuity equation for incompressible flow (4.11) yields:

$$\nabla \cdot (\bar{\mathbf{V}} + \tilde{\mathbf{V}}) = 0, \nabla.\tilde{\mathbf{V}} = 0 \tag{8.10}$$

with $\nabla \cdot \bar{\mathbf{V}}(\mathbf{x}) = 0$, Eq. (8.10) decomposed as

$$\frac{\partial \tilde{V}_1}{\partial x_1} + \frac{\partial \tilde{V}_2}{\partial x_2} = 0. \tag{8.11}$$

With Eqs. (8.9) and (8.11) we have three-equations to solve three unknowns, namely $\tilde{V}_1 \tilde{V}_2$ and $\tilde{p}$. The solution is presented in the following section.

8.3.2 The Orr-Sommerfeld Stability Equation

Before proceeding with the stability analysis, for the sake of simplicity, we set in Eq. (8.9) $\bar{V}_1 \equiv U$, $\tilde{V}_1 \equiv \tilde{u}$, $\tilde{V}_2 \equiv \tilde{v}$, $x_1 = x$, $x_2 = y$ and find

$$\frac{\partial \tilde{u}}{\partial t} + U\frac{\partial \tilde{u}}{\partial x} + \tilde{v}\frac{\partial U}{\partial y} = -\frac{1}{\rho}\frac{\partial \tilde{p}}{\partial x} + \nu\left(\frac{\partial^2 \tilde{u}}{\partial x^2} + \frac{\partial^2 \tilde{u}}{\partial y^2}\right)$$

$$\frac{\partial \tilde{v}}{\partial t} + U\frac{\partial \tilde{v}}{\partial x} = -\frac{1}{\rho}\frac{\partial \tilde{p}}{\partial y} + \nu\left(\frac{\partial^2 \tilde{v}}{\partial x^2} + \frac{\partial^2 \tilde{v}_2}{\partial y^2}\right). \tag{8.12}$$

For the disturbance field superimposed on the main laminar flow we introduce the following complex stream function:

$$\psi(x, y, t) = \phi(y)e^{i(\alpha x - \beta t)}. \tag{8.13}$$

In Eq. (8.13) ϕ is the complex function of disturbance amplitude which is assumed to be a function of y only. The stream function can be decomposed into a real and an imaginary part:

$$\psi(x, y, t) = \psi_{\Re}(x, y, t) + i\psi(x, y, t)_{\Im} \tag{8.14}$$

from which only the real part

$$\Re(\psi) = e^{i\beta t}[\phi_{\Re}\cos(\alpha x - \beta_{\Re} t) - \phi_{\Im}\sin(\alpha x - \beta_{\Re} t)] \tag{8.15}$$

has a physical meaning. Similarly the complex amplitude is decomposed into a real and an imaginary part:

$$\phi(y, t) = \phi_{\Re}(x, y, t) + i\phi(y, t)_{\Im}. \tag{8.16}$$

While α is a real quantity and is related to the wavelength $\lambda = 2\pi/\alpha$, the quantity β is complex and consists of a real and an imaginary part

$$\beta = \beta_r + i\beta_i \tag{8.17}$$

with β_r as the oscillation frequency of the perturbation field and β_i as the amplification/damping factor of the disturbance. For $\beta_i \prec 0$, disturbances are damped and stable laminar flow persists. On the other hand, disturbances are amplified if $\beta_i \succ 0$. In this case instability may drastically change the flow pattern from laminar to turbulent. We now introduce the following ratio:

$$c = \frac{\beta}{\alpha} = \frac{\beta_r}{\alpha} + \frac{\beta_i}{\alpha} = c_r + ic_i \tag{8.18}$$

with c_r as the wave propagation velocity and c_i the damping factor. The components of the perturbation velocity are obtained from the stream function as:

$$\tilde{u} = \frac{\partial \psi}{\partial y} = \phi'(y)e^{i(\alpha x - \beta t)}; \tilde{v} = -\frac{\partial \psi}{\partial x} = -i\alpha\phi(y)e^{i(\alpha x - \beta t)}. \quad (8.19)$$

Introducing Eqs. (8.19) into (8.12) and eliminating the pressure terms by differentiating the first component of the Navier-Stokes equation with respect to y and the second with respect to x respectively and subtracting the results from each other, we obtain

$$(U - c)(\phi'' - \alpha^2\phi) - U''\phi = -\frac{i}{\alpha \mathrm{Re}}(\phi'''' - 2\alpha^2\phi'' + \alpha^4\phi). \quad (8.20)$$

Equation (8.20) referred to as the *Orr-Sommerfeld*-equation was derived by Orr [57] and independently Sommerfeld [58]. It constitutes the fundamental differential equation for stability of laminar flows in dimensionless form. The velocities are divided by their maximum values and the lengths have been divided by a suitable reference length such as d for pipe diameter, b channel length or δ for boundary layer thickness. The Reynolds number is characterized by the mean flow $\mathrm{Re} = \frac{U_m d}{\nu}$ or Re $= \frac{U_m b}{\nu}$ or $\mathrm{Re} = \frac{U_m \delta}{\nu}$.

8.3.3 Orr-Sommerfeld Eigenvalue Problem

The Orr-Sommerfeld equation is a fourth order linear homogeneous ordinary differential equation. With this equation the linear stability problem has been reduced to an *eigenvalue problem* with the following boundary conditions at the wall and in the freestream:

$$\begin{aligned} y &= 0: \tilde{u} = \tilde{v} = 0 : \phi = 0, \phi' = 0 \\ y &= \infty: \tilde{u} = \tilde{v} = 0 : \phi = 0, \phi' = 0. \end{aligned} \quad (8.21)$$

Equation (8.20) contains the main flow velocity distribution $U(y)$ which is specified for the particular flow motion under investigation, the Reynolds number, and the parameters α, c_r, and c_i.

Before we proceed with the discussion of Orr-Sommerfeld equation, we consider the shear stress at the wall that generally can be written as:

$$\tau_W = \mu \left(\frac{\partial U}{\partial y}\right)_{y=0}. \quad (8.22)$$

If the flow is subjected to an adverse pressure gradient, the slope $(\partial U/\partial y)_{y=0}$ may approach zero and the wall shear stress disappears. This requires the velocity profile to have a pint of inflection as shown in Fig. 8.4. In this particular case the flow close to the wall behaves like an inviscid flow with the Reynolds number approaching infinity (Re $\to \infty$). For this spacial case the Orr-Sommerfeld stability equation reduces to the following Rayleigh equation:

$$(U - c)(\phi'' - \alpha^2\phi) - U''\phi = 0. \tag{8.23}$$

Equation (8.23) is a second order linear differential equation and need to satisfy only two boundary conditions:

$$\begin{aligned} y &= 0: \phi' = 0 \\ y &= \infty: \phi = 0. \end{aligned} \tag{8.24}$$

The Orr-Sommerfeld equation (8.20) is an eigenvalue problem with the boundary conditions (8.21). To solve this differential equation, first of all the velocity distribution $U(y)$ must be specified. As an example, the velocity distribution for plane Poisseule flow can be prescribed. In addition, Eq. (8.21) contains four more parameters, namely α, Re, c_r and c_i. We assume that Reynolds number and the wavelength $\lambda = 2\pi/\alpha$ are given. For each pair of given α and *Re* Eq. (8.20) with the boundary conditions (8.21) provide one eigenfunction $\phi(y)$ and one complex eigenvalue $c = c_r + ic_i$ with

$$c_r = \frac{\beta_r}{\alpha} == \frac{\beta_r \lambda}{2\pi} \tag{8.25}$$

as the phase velocity of the prescribed disturbance. For a given value of α disturbances are damped if $c_i \prec 0$ and stable laminar flow persists, whereas $c_i \succ 0$ indicates a disturbance amplification leading to instability of the laminar flow. The neutral stability is characterized by $c_i = 0$. For a prescribed laminar flow with a given $U(y)$ the results of a stability analysis is presented schematically in an α, R diagram Fig. 8.4, where every point of the diagram corresponds to a pair of c_r and c_i.

The curve of the neutral stability separates the region of stable laminar flow from that of unstable disturbances. The vertical line that tangents the stability curve constitutes the critical Reynolds number, below which all disturbances die out. Inside the stability curve the flow is unstable, whereas outside fully stable. The figure also show schematically the effect of velocity profile on the stability. A flow with the velocity profile described by (a) with a point of inflection is more sensitive to disturbances, whereas the one with the profile (b) has a smaller range of instability. These two profiles represents two different flow conditions. The profile (a) represents a boundary layer flow at positive pressure gradient, which is close to separation. In contrast, profile (b) may represent a boundary layer flow at negative pressure gradient. This explains why an accelerated laminar flow is more stable compared to a decelerated laminar flow we described in Chap. 7.

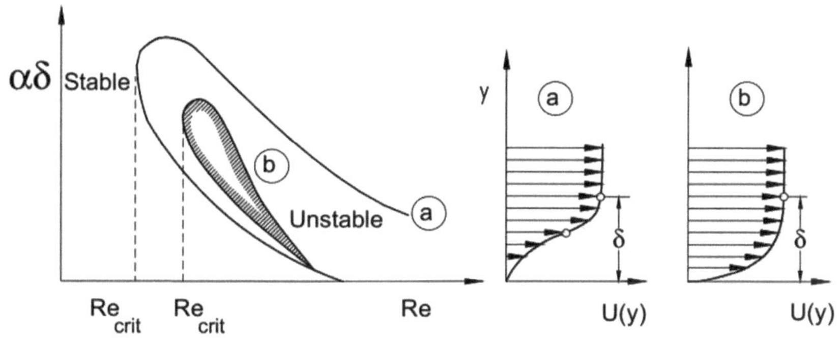

Fig. 8.4 Neutral stability curves for two-dimensional boundary layer with twodimensional disturbances, **a** frictionless Rayleigh stability for velocity profile with inflection point Re $\to \infty$, **b** viscous instability for velocity profile without inflection point

8.3.4 Solution of Orr-Sommerfeld Equation

As an example, we solve the Orr-Sommerfeld stability equation for the case of a plain Poiseuille flow between two parallel plates. The method used herein is based on the study by Orszag [59] who expanded the assumed solution and the boundary conditions in terms of linear combinations of Chebyshev orthogonal function of the first kind $T_n(y)$. For the particular case of a pure Poiseuille flow between parallel plates, we rearrange Eq. (8.20) and arrive at the following dimensionless result:

$$c(-\alpha\varphi'' + \alpha^3\varphi) + U(\alpha\varphi'' - \alpha^3\varphi) - \alpha\varphi U'' + \frac{1}{\mathrm{Re}}[-2i\alpha\varphi'' + i\varphi'''' + i\alpha^4\varphi] = 0. \tag{8.26}$$

The boundary conditions are:

$$\begin{aligned} \varphi(1) = 0; \varphi'(1) = 0 \\ \varphi(-1) = 0; \varphi'(-1) = 0 \end{aligned} \tag{8.27}$$

and the plane Poiseuille flow is given by the dimensionless profile:

$$U(y) = 1 - y^2. \tag{8.28}$$

In Eq. (8.28) the independent dimensionless variable y represents the ratio of the physical coordinate in lateral direction and the half-width with the value of unity between the plates. Likewise, $U(y)$ is the ratio of the undisturbed velocity distribution and the maximum velocity in the middle of the plates, and Re is based on the half-width between plates and is $Re = 1/\nu$. As reported in the open literature, there is no exact solution known for this set of equations. Therefore, we use numerical methods in order to solve the problem. One of the possible method, which is a very common practice is to expand the assumed solution in terms of a series of some type

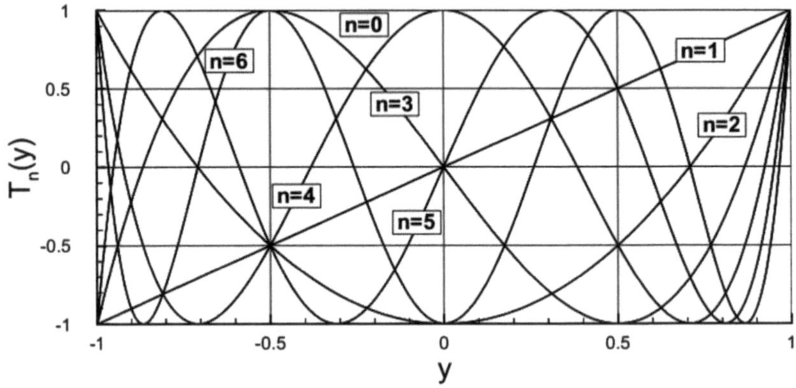

Fig. 8.5 The first six Chebyshev polynomials

of functions such as Taylor, Fourier, Chebyshev, Legendre, etc. For this particular problem, it was decided to expand the assumed solution in terms of Chebyshev orthogonal polynomials of the first kind $T_n(y)$. Chebyshev polynomial of the first kind is defined by:

$$T_n(\cos\Theta) = \cos n\Theta \tag{8.29}$$

for all non-negative integer n. Examples of these functions are:

$$\begin{aligned} T_0(y) &= 1 \\ T_1(y) &= y \\ T_2(y) &= 2y^2 - 1 \\ T_n(y) &= 2yT_{n-1}(y) - T_{n-2}(y). \end{aligned} \tag{8.30}$$

A plot of the first six polynomials is shown in Fig. 8.5. It can be seen that this set of polynomials is composed of both even and odd functions.
Another interesting characteristic is that all polynomials are orthogonal and non-singular in the interval $[-1, 1]$, i.e. the inner product of two polynomials is given by:

$$\langle T_i(y)T_j(y)\rangle = \int_{-1}^{+1} \frac{T_i(y)T_j(y)}{c_j\pi w(y)}dy = 1 \text{ if } i = j \text{ and } = 0 \text{ for } i \not\equiv j \tag{8.31}$$

with T_i as the Chebyshev polynomials, C_j a constant ($C_0 = 2$ and $C_n = 1$ for $n > 0$), and $w(y)$ is the weighting function defined as:

$$w(y) = \sqrt{1 - y^2}. \tag{8.32}$$

The orthogonality condition makes the Chebyshev polynomials particularly appropriate for solving the Orr-Sommerfeld problem. To solve the Orr-Sommerfeld differential equation we assume that the solution can be expressed in terms of Chebyshev polynomials $T_n(y)$, as shown:

$$\phi = \sum_{k=0}^{\infty} a_n T_n(y) \tag{8.33}$$

where the coefficient a_n can be determined from the inner product as (orthogonal property):

$$a_n = \frac{2}{\pi c_n} \int_{-1}^{1} \frac{\phi T_n(y)}{\sqrt{1-y^2}} dy. \tag{8.34}$$

ϕ from Eq. (8.33) is then introduced into the differential equation (8.26) and in the boundary conditions (8.27). It must be noted however, for this particular case, that the presence of only even derivatives in the differential equation and the symmetry of the boundary conditions reduce our solution to the combination of even polynomials only. The number of equations is then reduced to $k/2 + 1$ (one from each inner product), where k is the maximum degree used in the expansion. Since the boundary conditions must be satisfied, the last set of equations obtained from the inner product (related to the high frequency terms), are substituted by the boundary condition equations. At this point, the nontrivial solution to our set of unknowns is obtained finding the values of the complex number c that nulls the determinant of the matrix associated with the system of equations. In other words, we must solve an eigenvalue problem.

The problem of solving the Orr-Sommerfeld differential equation, which was in the past the subject of several dissertations can now be assigned as a routine homework problem. Using the symbolic capabilities and the library of built-in functions from several software (Maple, Mathematica, Matlab) it is possible to produce highly accurate expressions of the characteristic polynomial by increasing the order of Chebyshev polynomials. However, increasing the orde requires larger memory and computational time that are associated with the inner product (integration) between ϕ and the Chebyshev polynomials $T_i(y)$. The analysis showed that the results from the inner products were related.

Two aspects are worth noting: First, the polynomial matrix shown in Table 8.1, is always lower triangular. This means that almost half of the internal product between functions is already known without the need to perform the integration. Secondly but more important, each term of the table can be generated as a linear combination of the other terms. The constants for the combination are identical to the coefficients of the Chebyshev polynomials of the same degree as the column where the term of interest is located. Using these two simple properties, it is possible to generate the results of all internal products needed to form the set of equations.

Table 8.1 Example of a Chebyshev matrix used

	T_0	T_2	T_4	T_6	T_8	T_{10}	T_{12}	T_{14}
y_0	π	0	0	0	0	0	0	0
y_2	$\frac{1}{2}\pi$	$\frac{1}{4}\pi$	0	0	0	0	0	0
y_4	$\frac{3}{8}\pi$	$\frac{1}{4}\pi$	$\frac{1}{16}\pi$	0	0	0	0	0
y_6	$\frac{5}{16}\pi$	$\frac{15}{64}\pi$	$\frac{3}{32}\pi$	$\frac{1}{64}\pi$	0	0	0	0
y_8	$\frac{35}{128}\pi$	$\frac{7}{32}\pi$	$\frac{7}{64}\pi$	$\frac{1}{32}\pi$	$\frac{1}{256}\pi$	0	0	0
y_{10}	$\frac{63}{256}\pi$	$\frac{105}{512}\pi$	$\frac{15}{128}\pi$	$\frac{45}{1024}\pi$	$\frac{5}{512}\pi$	$\frac{1}{1024}\pi$	0	0
y_{12}	$\frac{231}{1024}\pi$	$\frac{99}{512}\pi$	$\frac{495}{4096}\pi$	$\frac{55}{1024}\pi$	$\frac{33}{2048}\pi$	$\frac{3}{1024}\pi$	$\frac{1}{4096}\pi$	0
y_{14}	$\frac{429}{2048}\pi$	$\frac{3003}{16384}\pi$	$\frac{1001}{8192}\pi$	$\frac{1001}{16384}\pi$	$\frac{91}{4096}\pi$	$\frac{91}{16384}\pi$	$\frac{7}{8192}\pi$	$\frac{1}{16384}\pi$

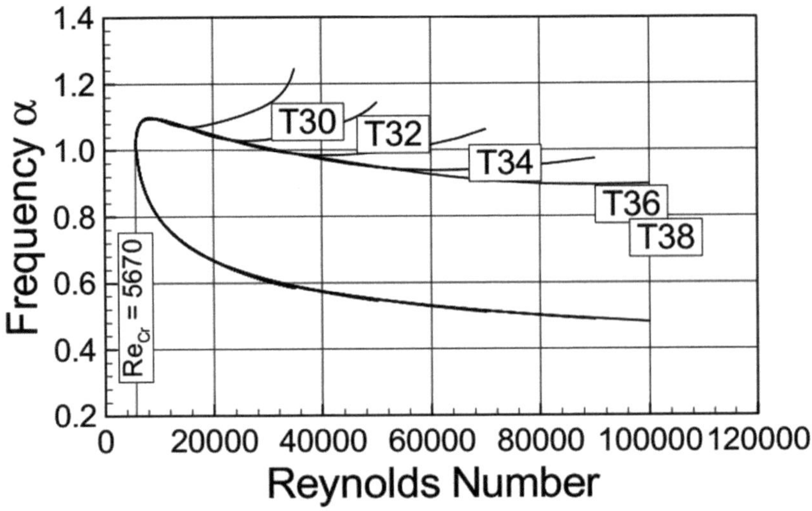

Fig. 8.6 Effect of degree of Chebychev polynomial on the numerical solution

8.3.5 *Numerical Results*

The accuracy of the results improves by using higher order Chebyshev polynomials in the expansion of the solution. However, a practical limit must be found since the computer resources (time, memory, etc.) required to solve the problem also increases. Figure 8.6 shows the effect of the use of five different Chebyshev Polynomials over the accuracy of the neutral stability curve. It can be seen that the location of the critical Reynolds number as well as the lower branch of the stability curve is not much affected if the order of the polynomials is greater than 30. On the other hand, the upper branch shows a great dependence on the degree selected. However within low ranges of Reynolds numbers, a low order polynomial may be used with a certain degree of confidence. Figure 8.7 contains the plots corresponding to the stability

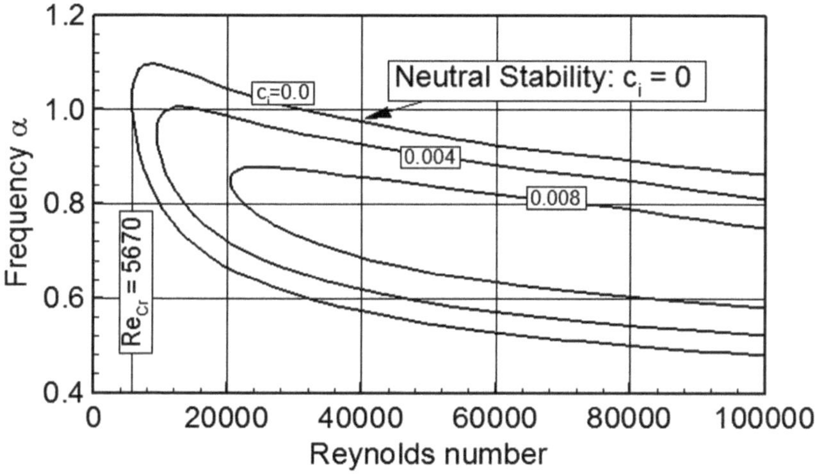

Fig. 8.7 Stability map for a plane Poiseulle flow

maps for the plane Poiseuille flow between parallel plates. The figure exhibits the frequency of the disturbance wave as a function of Reynolds number with c_i as the parameter. The tangent to the neutral stability curve with $c_i = 0$ predicts a Reynolds number of 5670 within 0.1% accuracy. Figure 8.7 contains three stability curves with $c_i = 0, 0.004, 0.008$. The neutral stability curve characterized by the zero damping $c_i = 0$ separates the stable outer region from the unstable inner region

The linearized stability theory presented above mathematically describes the basic physics of the change of flow state from laminar to turbulent. The theory is applicable to simple steady flows at low turbulence intensity levels. As the study by Morkovin [60] shows, the linearized Orr-Sommerfeld equation is not applicable to flows with high free-stream turbulence intensity (more than 10%), where the Tollmien-Schlichting waves we discussed above are completely *bypassed*.

8.4 Physics of an Intermittent Flow, Transition

As discussed in the preceding sections, the amplification/damping factor of the disturbance β_i determines the flow pattern. For $\beta_i \prec 0$, disturbances are damped and stable laminar flow persists. On the other hand, disturbances are amplified if $\beta_i \succ 0$. In this case instability may drastically change the flow pattern from laminar to turbulent. This change, however, does not occur suddenly. The instability triggers a transition process, which is characterized by its intermittently laminar-turbulent nature.

To better understand the physics of an intermittently laminar-turbulent flow, we consider a flat plate, Fig. 8.8, with a smooth surface placed within a wind tunnel with statistically steady flow velocity $\bar{\mathbf{V}}$ and a low turbulence fluctuation velocity

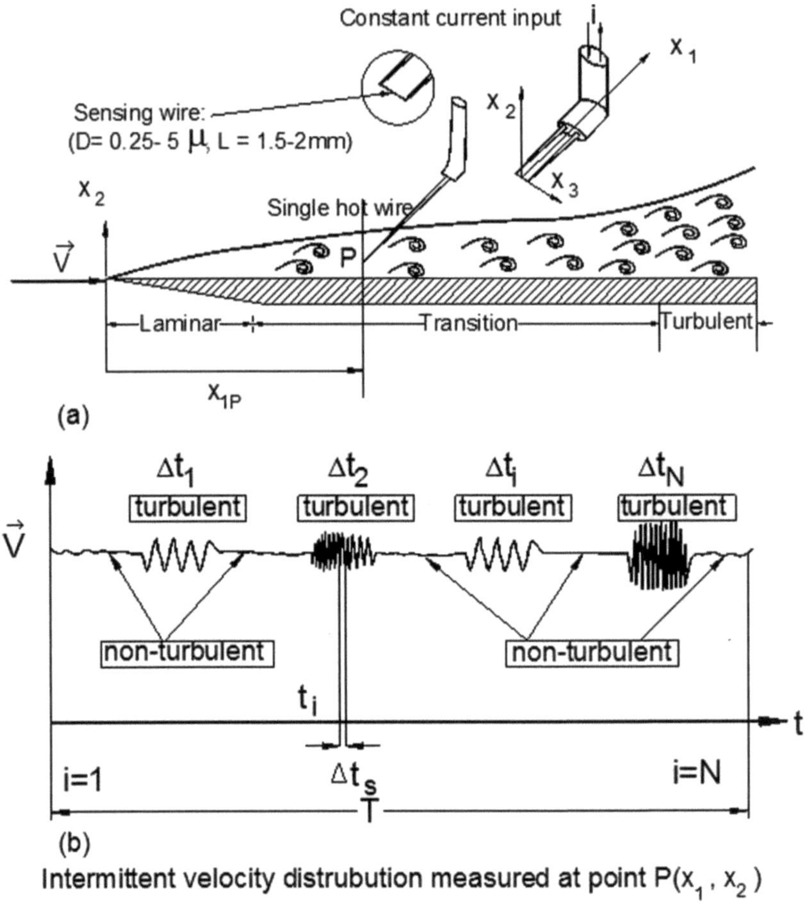

Fig. 8.8 Measurement of an intermittently laminar-turbulent flow, **a** positioning a hot wire sensor with the transitional portion of the boundary layer, **b** high frequency velocity signals acquired at point *P*

$\mathbf{V}'$.[1] It should be noted that, in contrast to the theoretical assumption we made for a fully laminar flow, the real flow using in wind tunnel tests always contains certain degree of turbulence fluctuations superimposed on the main flow velocity $\bar{\mathbf{V}}$. This is expressed in terms of turbulence intensity defined as $Tu = \sqrt{\overline{V'^2}}/\bar{V}$. Thus, from a practical point of view, it is more appropriate to use the term *non-turbulent* flow rather than laminar one.

[1] The superscript "′" pertains to stochastic fluctuations in contrast to "∼" used in Sect. 8.3 that stands for deterministic disturbance.

Downstream of the laminar region, we place a miniature *hot wire* sensor at an arbitrary point P within the boundary layer to measure the velocity (see Chap. 11 for detailed flow measurement).

The position of the sensor relative to the plate such that the axis of the sensing wire coincides with x_3-axis which is perpendicular to the $x_1 - x_2$-plane, Fig. 8.8a. The wire and the associated *anemometer* electronics provide a virtually instantaneous response to any high frequency incoming flow. Figure 8.8b schematically reflects the time dependent velocity of an otherwise statistically steady flow. As seen, the anemometer provides a sequence of signals that can be categorized as non-turbulent characterized by a time independent, non-turbulent pattern followed by a sequence of time dependent highly random signals that reflect turbulent flow. Since in a transitional flow regime, sequences of non-turbulent signals are followed by turbulent ones, we need to establish certain criteria that must be fulfilled before a sequence of signals can be called non-turbulent or turbulent. This is issue is treated in the following Section.

8.4.1 *Identification of Intermittent Behavior of Statistically Steady Flows*

To identify the laminar and turbulent states, Kovasznay et al. [65] introduced the intermittency function $I(\mathbf{x}, \mathbf{t})$. The value of I is unity for a turbulent flow regime and zero otherwise:

$$I(\mathbf{x}, \mathbf{t}) = \begin{cases} 1 & \text{for turbulent flow} \\ 0 & \text{for non-turbulent flow.} \end{cases} \tag{8.35}$$

Fig. 8.9 schematically exhibits an intermittently laminar-turbulent velocity with the corresponding intermittency function $I(\mathbf{x}, \mathbf{t})$ for a statistically steady flow at a given position vector $\mathbf{x}$ and an arbitrary time t.

Following Kovasznay et al. [65], the time-averaged value of $I(\mathbf{x}, \mathbf{t})$ is the intermittency factor γ, which gives the fraction of the time that a highly sensitive probe spends in turbulent flow in a sufficiently long period of time T:

$$\gamma(\mathbf{x}) = \frac{\sum_{\mathrm{i}=1}^{\mathrm{N}} \Delta \mathbf{t_i}}{\mathbf{T}} \tag{8.36}$$

which is equivalent to

$$\gamma(x) \equiv \bar{I} = \frac{\mathbf{1}}{T} \int_{t}^{(t+T)} I(x, t)\mathbf{dt}. \tag{8.37}$$

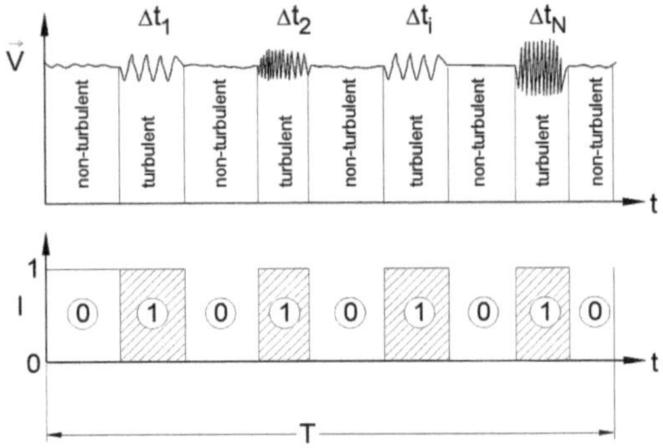

Fig. 8.9 Identification of non-turbulent ($I = 0$) and non-turbulent flow ($I = 1$)

Experimentally, the intermittency factor γ is determined from a set of N experimental data. This requires that the integral in Eq. (8.37) be replaced by Eq. (8.38):

$$\gamma(\mathbf{x}) = \frac{\mathbf{1}}{\mathbf{N}} \sum_{\mathrm{i}=1}^{\mathbf{N}} \mathbf{I}(\mathbf{x}, \mathbf{t}_\mathrm{i}). \tag{8.38}$$

The hatched areas in Fig. 8.9 labeled with ① indicate the portion of the velocity with random fluctuations, whereas, the blank areas point to signals lacking random fluctuations.

8.4.2 Turbulent/Non-turbulent Decisions

To make an instantaneous decision about the non-turbulent/turbulent nature of a flow it is possible to use a simple probe, such as a hot-wire, for measuring the velocity fluctuations and to identify the fine-scale structure in the turbulent fluid, as shown in Fig. 8.10.

Since the velocity fluctuation is not sufficient for making instantaneous decisions for or against the presence of turbulence, the velocity signals need to be sensitized to increase discriminatory capabilities. The commonly used method of sensitizing is to differentiate the signals. The sensitizing process generates some zeros inside the fully turbulent fluid. These zeros influence the decision process for the presence of turbulence or non-turbulence. The process of eliminating these zeros is to integrate the signal over a short period of time T, which produces a criterion function $S(t)$. After short term integration, a threshold level C is applied to the criterion function to distinguish between the true turbulence and the signal noise. Applying the threshold level results in an indicator function consisting of zeros and 1's satisfying:

$$I(\mathbf{x}, \mathbf{t}) = \begin{cases} 1 & \text{when } S(x,t) \geq C \\ 0 & \text{when } S(x,t) < C. \end{cases} \tag{8.39}$$

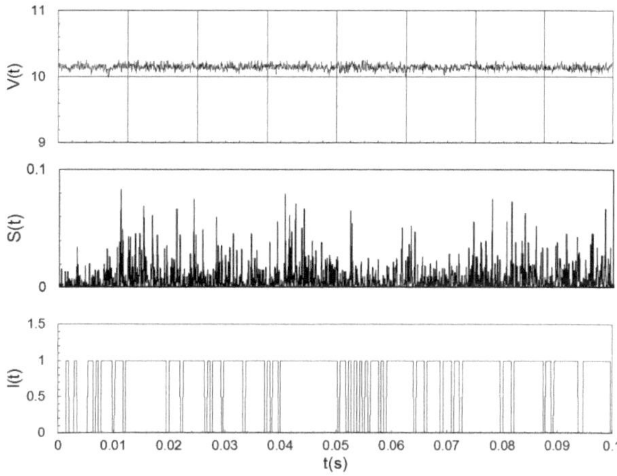

Fig. 8.10 Processing the instantaneous velocity signals forintermittency calculation for a statistically steady flow along a turbine blade. $V(t)$ = velocity signals, $S(t)$ = Detector function, $I(t)$ = indicator function, for non-turbulent $I = 0$, for turbulent flow $I = 1$, measurement*TPFL*

The resulting random square wave, $I(\mathbf{x}, \mathbf{t})$, along with the original signal is used to condition the appropriate averages using the equations above.

Performing the averaging process using Eqs. (8.37) or (8.38) for the statistically steady flow shown in Fig. 8.10, we find an intermittency factor $0 \le \gamma(\mathbf{x}) \le \mathbf{1}$. For the case $0 < \gamma(\mathbf{x}) < \mathbf{1}$ this means that flow is *transitional*. For a statistically steady flow, the time averaged intermittency at any point along the surface in streamwise direction can be obtained that reflects the intermittent behavior of the flow under investigation. As an example, Fig. 8.11 exhibits the intermittency distribution along the concave surface of a curved plate at zero longitudinal pressure gradient.
Using an entire set of velocity distributions along the concave surface of a curved plate at zero longitudinal pressure gradient, a detailed picture of the intermittency behavior of the boundary is presented in Fig. 8.11. It exhibits the intermittency contour within the boundary layer along the concave side of a curved plate under statistically steady flow condition at zero pressure gradient. Close to the surface, the intermittency starts from zero and gradually approaches its maximum value. The dark area with $\bar{\gamma}(\mathbf{x}) \approx \mathbf{1}$ encloses locations with the maximum turbulent fluctuations. Moving from the surface toward the edge of the boundary layer, the intermittency factor decreases approaching the non-turbulent freestream.

Figure 8.12 presents a more quantitative picture of the intermittency distribution with normal distance y as a parameter. Substantial changes of $\bar{\gamma}(\mathbf{x})$ occur within a range of $y = 0.0$ to 1.3 mm with the maximum intermittency $\bar{\gamma}(\mathbf{x}) \approx \mathbf{0.96}$ which means that the velocity has not reached a fully turbulent state. In fact in many engineering applications, for instance, turbomchinery aerodynamics, the flow is neither fully laminar $\bar{\gamma}(\mathbf{x}) \approx \mathbf{0}$ nor fully turbulent $\bar{\gamma}(\mathbf{x}) \approx \mathbf{1}$. It is *transitional* with $0 < \bar{\gamma}(\mathbf{x}) < \mathbf{1}$. The change of $\bar{\gamma}(\mathbf{x})$ in normal direction reflects the distribution of spots cross section in y-direction that decreases toward the edge of the boundary layer. The knowledge of $\bar{\gamma}(\mathbf{x})$-distribution is crucial in assessing the computational results of CFD-code, understanding the development of spot structure and the flow

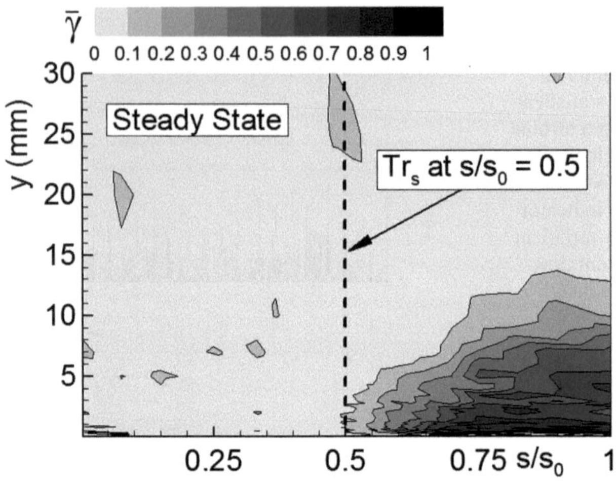

Fig. 8.11 Time-averaged intermittency contour for steady flow along the concave surface S of a curved plate at zero streamwise pressure gradient, S_0 is the arc length of the curved plate, Measurement*TPFL*

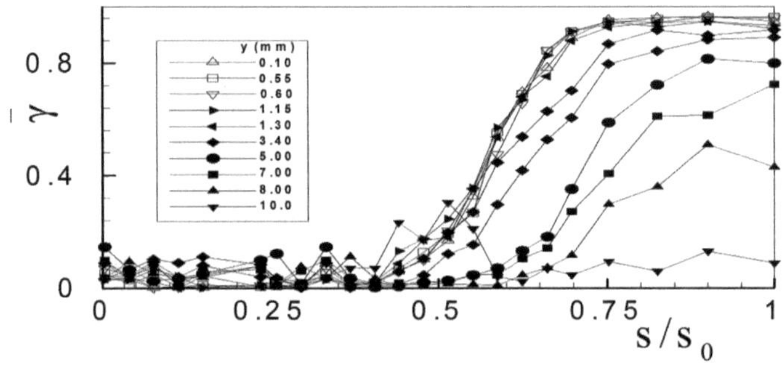

Fig. 8.12 Time-averaged intermittency distribution along the concave surface of a curved plate at zero streamwise pressure gradient with normal distance y as a parameter, measurement*TPFL*

situation within a transitional boundary layer. For calculating the transition boundary layer characteristics, the values close to the surface are used.

Figure 8.13 exhibits the γ-distribution along the concave side of the curved plate mentioned above at $y = 0.1$ mm above the surface as a function of Re-number in streamwise direction s, Re=Us/ ν. Up to $\mathrm{Re}_S \approx 176{,}000$, the boundary layer is fully non-turbulent with $\gamma = 0$. This point marks the start of the transition Re_S. Similarly, the end of the transition is marked with $\mathrm{Re}_E \approx 400{,}000$. The locations of transition start and end depend strongly on pressure gradient in streamwise direction and the inlet flow condition. The latter includes the free-stream turbulence intensity for steady inlet flow condition. For a periodic unsteady flow condition as is present in many engineering applications such as in turbomachinery fluid mechanics, periodic disturbances with specific characteristics play a key role in determining the start and end of the transition.

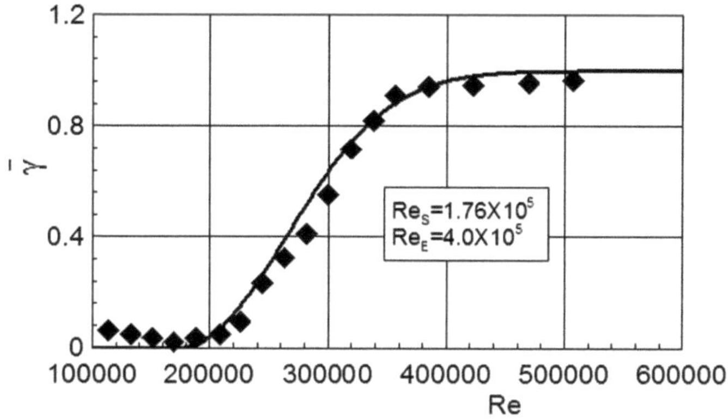

Fig. 8.13 Intermittency as a function of Re along the concave surface of a curved plate at $y = 0.1$ mm from, experiment ♦, solid line Eq. 8.42

8.4.3 Intermittency Modeling for Steady Flow at Zero Pressure Gradient

The transition process was first explained by Emmons [61] through the turbulent spot production hypothesis. Adopting a sequence of assumptions, Emmon arrived at the following intermittency relation:

$$\gamma(\mathbf{x}) = \mathbf{1} - e^{\frac{-sgx^3}{3U}} \tag{8.40}$$

with σ as the turbulent spot propagation parameter, g the spot production parameter, x the streamwise distance, and U the mean stream velocity. While the Emmon's spot production hypothesis is found to be correct, Eq. (8.40) does not provide a solution compatible with the experimental results. As an alternative, Schubauer and Klebanoff [62] used the Gaussian integral curve to fit the γ-distribution measured along a flat plate. Synthesizing the Emmon's hypothesis with the Gaussian integral, Dhawan and Narasimha [63] proposed the following empirical intermittency factor for natural transition:

$$\gamma(\mathbf{x}) = \mathbf{1} - e^{-A\xi^2} \tag{8.41}$$

with $\xi = (x - x_s)/\lambda$, $\lambda = (x)_{\gamma=0.75} - (x)_{\gamma=0.25}$ and x_S as the streamwise location of the transition start and A as constant. The solution of Eq. (8.41) requires the knowledge of λ which contains two unknowns and the location of transition start x_S. In [64] the constant A was set equal to 0.412. Thus, we are dealing with three unknowns, namely x_S, and the two streamwise positions at which the intermittency factor assumes values of 0.75 and 0.25. While the transition start x_S can be estimated, the two streamwise positions $(x)_{\gamma=0.75}$ and $(x)_{\gamma=0.25}$ are still unknown. Further more, the quantity A which was set equal to 0.412, may be itself a function of several

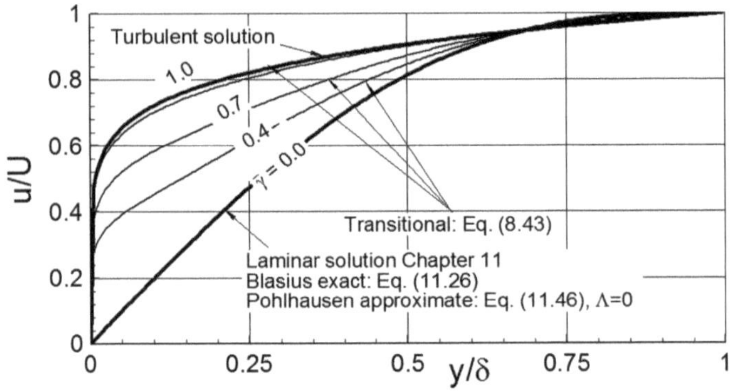

Fig. 8.14 Velocity profile in transition region using the intermittency function (8.42)

parameters such as the pressure gradient and the free-stream turbulence intensity. As we discuss in the following section, a time dependent universal unsteady transition model was presented in [64] for curved plate channel under periodic unsteady flow condition and generalized in [65] for turbomachinery aerodynamics application. The intermittency model for steady state turned out to be a special case of the unsteady model presented in [65, 66], it reads:

$$\bar{\gamma} = C_1 \left[1.0 - e^{-\left(C_2 \frac{Re - Re_{x,t}}{Re_{x,t} - Re_{x,e}} \right)^2} \right] \tag{8.42}$$

with $C_1 = 0.95$, $C_2 = 1.81$. With the known intermittency factor, the averaged velocity distribution in a transitional region is determined from:

$$\bar{\mathbf{V}} = (1 - \bar{\gamma})\mathbf{V}_L + \gamma \mathbf{V}_T. \tag{8.43}$$

with $\mathbf{V}_\mathrm{L}$ and $\mathbf{V}_\mathrm{T}$ as the solutions of laminar and turbulent flow, respectively. As an example, we take the Blasius solution for the laminar and the Prandtl-Schlichting solution for the turbulent portion of a transitional flow (for details see Chap. 11) along a flat long plate at zero streamwise pressure gradient and construct the transitional velocity distribution using Eq. (8.43). The results are plotted in Fig. 8.14, where the non-dimensional velocity $\mathbf{v}/\mathbf{V}$ is plotted versus the non-dimensional variable y/δ with δ as the boundary layer thickness. Two distinctively different curves mark the start and end of the transition denoted by $\bar{\gamma} = 0$ and $\bar{\gamma} = 1$. As seen, within the two $\bar{\gamma}$-values the velocity profile changes significantly resulting in boundary layer parameters and particularly and skin friction that are different from those pertaining to laminar or turbulent flow (see Chap. 11).

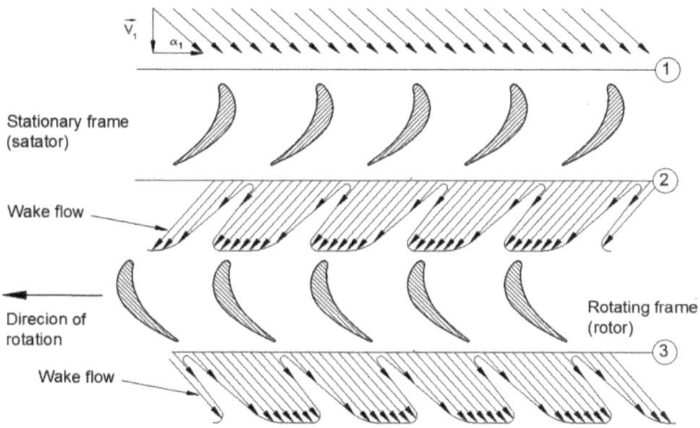

Fig. 8.15 Unsteady flow interaction within a turbine stage

Fig. 8.16 Comparison of boundary layer thickness developments, **a** steady state, **b** instantaneous image of the boundary layer thickness

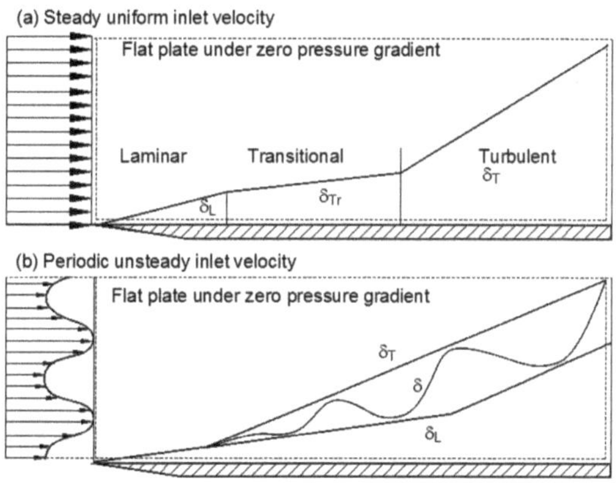

8.4.4 *Identification of Intermittent Behavior of Periodic Unsteady Flows*

The flow through a significant number of engineering devices is of periodic unsteady nature. Steam and gas turbine power plants, jet engines, turbines, compressors and pumps are a few examples. Within these devices unsteady interaction between individual components takes place. Figure 8.15 schematically represents the unsteady flow interaction between the stationary and rotating frame of a turbine stage.

A stationary probe traversing downstream of the stator at station (2) records a *spatially periodic* velocity distribution . Another probe placed on the rotor blade leading edge that rotates with the same frequency as the rotor shaft, registers the incoming velocity signals as a *temporally periodic*. The effect of this periodic unsteady inlet flow on the blade boundary layer is qualitatively and quantitatively different from those we discussed in the preceding section. The difference is shown in a simplified sketch presented in Fig. 8.16.

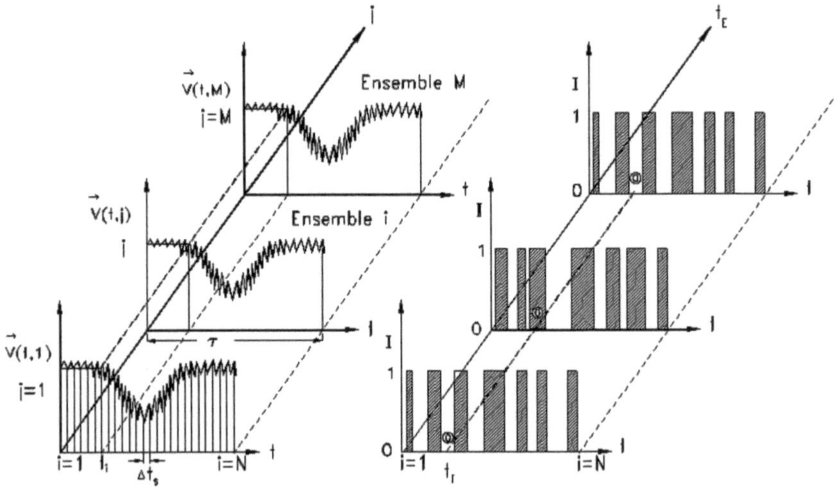

Fig. 8.17 Periodic unsteady flow velocity with the corresponding distribution of $I(x, t)$ at a particular position x

While the boundary layer thickness δ in case (a) is temporally independent, the one in case (b) experiences a temporal change. To predict the transition process under unsteady inlet flow condition using the intermittency approach, we first consider Fig. 8.17.

Figure 8.17 includes three sets of unsteady velocity data taken at three different times but during the same time interval Δt (corresponding to the sequence $i = 0$ to $i = N$). Each of these sets is termed an *ensemble*. Considering the velocity distribution at an arbitrary position vector **x** and at an ensemble j such as **V**(t,j), we use the same procedure we applied to the statistically steady flow discussed above to identify the nature of the periodic unsteady boundary layer flow. The corresponding intermittency function **I**(t,j) at a given position vector **x** is shown in Fig. 8.17. For a particular instant of time identified by the subscript i for all ensembles, the ensemble average of **I**(t,j) over N number of ensembles results in an ensemble averaged intermittency function $< \gamma(\mathbf{x}, \mathrm{t}) >$. This is defined as:

$$< \gamma(\mathbf{x}, \mathrm{t}) > = \frac{\mathbf{1}}{\mathrm{M}} \sum_{\mathrm{j}=\mathbf{1}}^{\mathrm{M}} \mathrm{I}(\mathrm{t_i}, \mathrm{j}). \tag{8.44}$$

In Eq. (8.44), M refers to the total number of ensembles and t_i the time at which the corresponding signal was acquired. In contrast to the intermittency factor $\bar{\gamma}(\mathbf{x})$, the ensemble averaged intermittency function $< \gamma(\mathbf{x}, \mathrm{t}) >$ is a time dependent quantity. Figure 8.18 shows the steps necessary to process the instantaneous velocity data to obtain the ensemble averaged intermittency. The periodic unsteady velocity $u(t)$ is

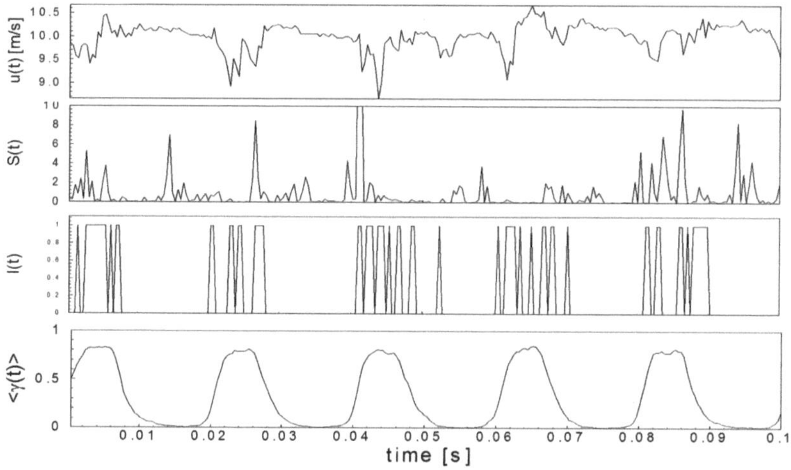

Fig. 8.18 Processing of instantaneous velocity signals to calculate ensemble averaged intermittency function for a periodic unsteady flow at $y = 0.1$ mm, $S/S_0 = 0.5235$ and a reduced wake frequency $\Omega = 3.443$

produced by moving a set of cylindrical rods with the diameter of 2 mm in front of a curved plate placed in the mid height of a curved channel (for details see [65, 66]). For each ensemble, the velocity derivatives are obtained leading to a time dependent intermittency function I(**x**, t). Taking the ensemble average of I(**x**, t) as defined by Eq. (8.44) results in an ensemble averaged $< \gamma(\mathbf{x}, t) >$, shown in Fig. 8.18. Repeating the same procedure for all velocity signals taken at $y = 0.1$ mm along the concave surface of the curved plate from leading to trailing edge, Fig. 8.19 shows a contour plot that reflects the intermittent behavior of the boundary layer under unsteady inlet flow condition. The contour variable γ is plotted for two unsteady wake passing periods. Figure 8.19 also includes the transition start for steady state case.

The areas with lower intermittency mark the non-turbulent flow within the wake external region, whereas dark areas indicate the regions with higher turbulent fluctuations. Upstream of the steady state transition start at $S/S_0 \approx 0.5$ there exists a stable laminar boundary layer region. This region is periodically disturbed by the wake strips that impinge on the surface. The wake strips are bound by two lines L1 and L2 that mark the leading and trailing edge with velocities $V_{LE} \approx 0.8 < V_0 >$ and $V_{TE} \approx 0.5 < V_0 >$, respectively. As seen, whenever the wake impinges on the surface, the boundary layer becomes turbulent ($\gamma \approx 1$). It returns to its previous stable laminar state, as soon as the wake strip has passed leaving behind a *calmed region.* The calmed region bound by L3 with a velocity approximately $V_{\text{Calm}} \approx 0.3 < V_0 >$. The calm region denoted by a triangle ABCA is characterized by $< \gamma(t) > \approx 0$. A comparison of Figs. 8.19 and 8.11 shows that the existence of the wakes has caused the non-turbulent region becomes larger, thus for this particular case, delaying the transition process. This is true as long as the wakes are not mixed with each other.

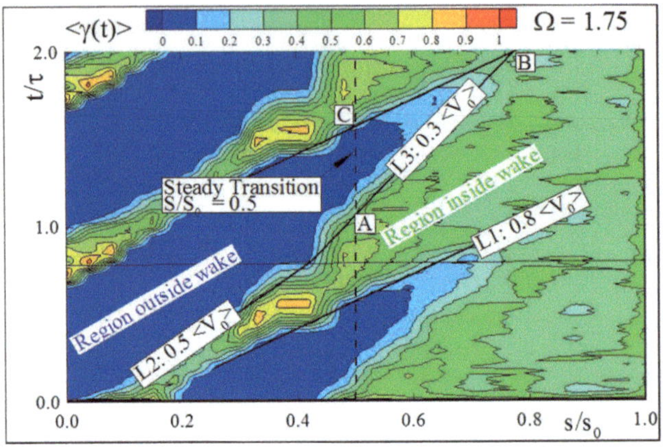

Fig. 8.19 Contour plot of intermittency function along the normalized streamwise distance s/s_0 for a reduced wake frequency $\Omega = 1.75$, dashed vertical line denotes the start of steady state case with ($\Omega = 0.0$), the ensemble averaged free stream velocity is denoted by $< V_0 >$

Increasing the reduced wake frequency of incoming wakes, causes mixing of the wakes associated with higher averaged turbulence intensity and a shift of transition toward the leading edge. The phenomenon of calming behind turbulent spots first mentioned by Schubauer and Klebanoff [53] was quantitatively determined by Herbst [66] and Pfeil et al. [67], Schobeiri and Radke [68] and Halstead et al. [69].

8.4.5 *Intermittency Modeling for Periodic Unsteady Flow*

The effect of periodic unsteady wake flow on boundary layer transition is discussed more in detail in Chap. 11, Sect. 8.2. The specific problematic of the transition, however, are discussed in this section. To establish an intermittency based transition model that accounts for the periodic unsteady inlet flow impinging on a flat plate, a curved plate, a compressor or turbine blade, we first introduce a dimensionless parameter that characterizes periodic nature of the incoming flow:

$$\zeta = \frac{U_w t}{b} \equiv= \frac{y}{b} \text{ with } b = \frac{1}{\sqrt{\pi}} \int_{-\infty}^{+\infty} \Gamma d\xi_2. \tag{8.45}$$

Equation (8.45) relates the passing time t of a periodic flow that impinges on the surface with the passing velocity in lateral direction U_w and the intermittency width b. The latter is directly related to the wake width introduced by Schobeiri et al. [70]. We define the relative intermittency function Γ in Eq. (8.45) as:

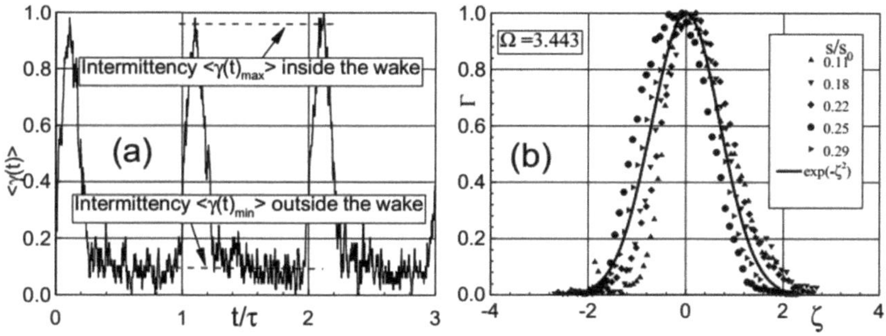

Fig. 8.20 Ensemble averaged intermittency, **a** marks the maximum values of the intermittency inside the wake and the minimum range outside the wake, **b** shows the relative intermittency and its description by a Gaussian function

$$\Gamma = \frac{<\gamma_i(t_i)> - <\gamma_i(t_i)>_{\min}}{<\gamma_i(t_i)>_{\max} - <\gamma_i(t_i)>_{\min}}. \tag{8.46}$$

In Eq. (8.46), $<\gamma_i(t_i)>$ is the time dependent ensemble-averaged intermittency function which determines the transitional nature of an unsteady boundary layer. The maximum intermittency $<\gamma_i(t_i)>_{\max}$, shown in Fig. 8.20a, exhibits the time dependent ensemble averaged intermittency value inside the wake vortical core. Finally, the minimum intermittency $<\gamma_i(t_i)>_{\min}$, represents the ensemble averaged intermittency values outside the wake vortical core.
Experimental results presented in Fig. 8.20b show that the relative intermittency function Γ closely follows a Gaussian distribution, which is given by:

$$\Gamma = e^{-\zeta^2}. \tag{8.47}$$

Here, ζ is the non-dimensionalized lateral length scale. The validity of Eq. (8.47) has been verified for different cases [65, 71, 72], suggesting it is a universal unsteady intermittency function. Using this function as a universally valid intermittency relationship for zero and non-zero pressure gradient cases [71], the intermittency function $<\gamma_i(t_i)>$ is completely determined if additional information about the minimum and maximum intermittency functions $<\gamma_i(t_i)>_{\min}$ and $\gamma_i(t_i)>_{\max}$ are available. The distribution of $\gamma_i(t_i)>_{\min}$ and $\gamma_i(t_i)>_{\max}$ in the streamwise direction are plotted in Fig. 8.21 (a). The steady case shown in Fig. 8.21 (b) serves as the basis of comparison for these maximum and minimum values. In the steady case, the intermittency starts to rise from zero at a streamwise Reynolds number $\mathrm{Re}_{x,s} = 2 \times 10^5$, and gradually approaches the unity corresponding to the fully turbulent state. This is typical of natural transition and follows the intermittency function (8.42). The distributions of maximum and minimum turbulence intermittencies $\gamma_i(t_i)>_{\min}$ and $<\gamma_i(t_i)>_{\max}$ in the streamwise direction are shown in Fig. 8.21a. For each particular streamwise location on the plate surface with a streamwise Reynolds number, for example $\mathrm{Re}_{x,s} = 1 \times 10^5$, two corresponding, distinctively different intermittency states are

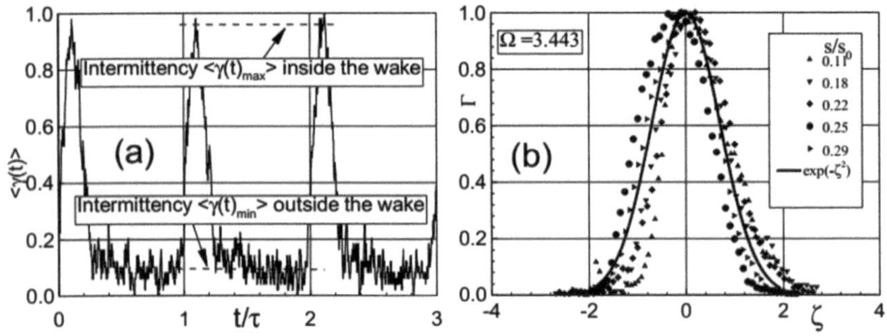

Fig. 8.21 **a** Maximum, minimum and averaged intermittency distribution along a curved plate under periodic unsteady flow condition compared to the intermittency at steady inlet floe condition (**b**)

periodically present. At this location, $< \gamma_i(t_i) >_{\max}$ corresponds to the condition where the wake with the high turbulence intensity core impinges on the plate surface at a particular instant of time. Once the wake has passed over the surface, the same streamwise location is exposed to a low turbulence intensity flow regime with an intermittency state of $< \gamma_i(t_i) >_{\min}$, where no wake is present. As seen, $< \gamma_i(t_i) >_{\min}$ has the tendency to follow the course of the steady (no-wake) intermittency distribution exhibited in Fig. 8.21b, with a gradual increase from an initial *non-turbulent* state with a value of zero approaching a final state of 0.8. This was expected as $< \gamma_i(t_i) >_{\min}$ is calculated outside the wake region where the turbulence intensity is relatively small. On the other hand, $< \gamma_i(t_i) >_{\max}$ reveals a fundamentally different behavior that needs to be discussed further. As Fig. 8.21a shows, the wake flow with an intermittency close to 1 (see also Fig. 8.19) impinges on the blade surface. By convecting downstream, its turbulent fluctuations undergo a strong damping by the wall shear stress forces (stable laminar flow). The process of damping continues until $< \gamma_i(t_i) >_{\max}$ reaches a minimum. At this point, the wall shear forces are not able to further suppress the turbulent fluctuations. As a consequence, the intermittency again increases to approach the unity, showing the combined effect of *wake induced* and *natural transition* due to an increased turbulence intensity level. The damping process of the high turbulence intensity wake flow discussed above explains the phenomena of *the becalming* effect of a wake induced transition observed by several researchers including [56–58]. Figure 8.21a also shows the average intermittency, which is a result of the integral effect of periodic wakes with respect to time. The maximum and minimum intermittency functions are described by:

$$< \gamma(t) >_{\max} = \left(1.0 - f_1 e^{-\left(\frac{Re_x - Re_{x,s}}{Re_{x,s} - Re_{x,e}}\right)^2}\right) \tag{8.48}$$

and

$$< \gamma(t) >_{\min} = f_2 \left(1.0 - e^{-\left(\frac{Re_x - Re_{x,s}}{Re_{x,s} - Re_{x,e}}\right)^2}\right) \tag{8.49}$$

Table 8.2 f_i-values for different Ω-values

	Reduced wake frequency Ω			
f_i	1.03	1.72	3.443	5.166
f_1	0.57	0.22	0.50	0.35
f_2	0.80	0.85	0.86	0.88
f_3	1.00	0.82	0.80	0.80
f_4	0.85	0.92	0.92	0.94

where the constants f_1 and f_2 may depend, among others, on the freestream turbulence intensity Tu, the wake parameter Ω and surface roughness. The time averaged intermittency is described by:

$$\bar{\gamma} = f_4\left(1.0 - f_3 e^{-\left(\frac{Re - Re_{x,t}}{Re_{x,t} - Re_{x,e}}\right)^2}\right). \tag{8.50}$$

The combined effect of $< \gamma_i(t_i) >_{max}$ and $< \gamma_i(t_i) >_{min}$ can be seen in the expression for $\bar{\gamma}$ through the constants f_3 and f_4. These constants are directly related to the constants f_1 and f_2. For the given Ω-values, the f_i-values are given in Table 8.2.

8.5 Implementation of Intermittency into Navier Stokes Equations

8.5.1 Reynolds-Averaged Equations for Fully Turbulent Flow

In most engineering applications, the flow quantities such as velocity, pressure, temperature, and density are generally associated with certain time dependent fluctuations. These fluctuations may be of deterministic or stochastic nature. Turbulent flow is characterized by random fluctuations in velocity, pressure, temperature, and density. Figure 8.22 schematically shows the time dependent turbulent velocity vector as a function of time for statistically steady, statistically unsteady, and periodic unsteady flows. It exhibits three representative cases encountered in engineering application. Case (a) represents a statistically steady flow through a duct (pipe, nozzle, diffuser etc.). Case (b) reveals the statistically unsteady velocity at the exit of a storage facility during a depressurizing process. Case (c) depicts a periodic unsteady turbulent flow (almost sinusoidal) with a time dependent mean that is encountered in combustion engines. Periodic unsteady flows are also found in all sorts of turbines and compressors.

Any turbulent quantity can be decomposed in a mean and a fluctuation part, where the mean may be time dependent itself as we saw in the ensemble averaging process. For a statistically steady flow, the velocity vector is decomposed in a mean and fluctuation term:

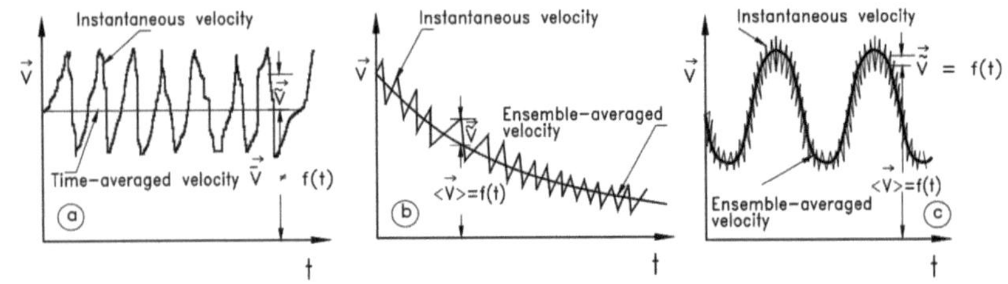

Fig. 8.22 Schematic representation of **a** statistically steady turbulentflow with a time independent mean, **b** statistically unsteady turbulentflow with a time dependent mean, and **c** periodic unsteady turbulentflow with a time dependent periodic mean

$$\mathbf{V}(\mathbf{x}, t) = \bar{\mathbf{V}}(\mathbf{x}) + \mathbf{V}'(\mathbf{x}, t). \tag{8.51}$$

The velocity components are obtained from Eq. (8.51) as:

$$V_i(x_j, t) = \bar{V}_i(x_j) + V_i'(x_j, t). \tag{8.52}$$

For a statistically unsteady flow, the flow velocity

$$\mathbf{V}(\mathbf{x}, t) =< \bar{\mathbf{V}}(\mathbf{x}, t) > +\mathbf{V}'(\mathbf{x}, t) \tag{8.53}$$

with $< \mathbf{V}(\mathbf{x}, \mathrm{t}) >$ as the ensemble averaged velocity according to Eq. (8.54).

$$< \mathbf{V}(\mathbf{x}, \mathrm{t}) >= \frac{\mathbf{1}}{\mathrm{M}} \sum_{\mathrm{j}=\mathbf{1}}^{\mathrm{M}} \mathbf{V}(\mathrm{t_i}, \mathrm{j}) \tag{8.54}$$

where the flow is realized M times and each time the velocity $\mathbf{V}(\mathbf{x}, \mathrm{t})$ is determined at the same position $\mathbf{x}$ and the same instant of time t. The velocity components are obtained from Eq. (8.53):

$$V_i(x_j, t) =< V_i(x_j, t) > +V_i'(x_j, t). \tag{8.55}$$

For a statistically steady flow, we define the time averaged quantity Q of a turbulent flow as:

$$\bar{Q} = \frac{1}{T} \int_0^T Q dt \tag{8.56}$$

with T as a time interval, over which the quantity is averaged. For a statistically steady and highly turbulent flow, the averaged quantity $\bar{Q}$ is time independent. However, if the turbulent flow is periodically unsteady with $\tilde{T}$ as the period, the averaging duration T must be an order of magnitude smaller than the period of the mean unsteady flow which means $\tilde{T} \gg T$. From Eq. (8.56), it immediately follows that:

$$\bar{\bar{Q}} = \bar{Q}, \qquad \bar{Q}' = 0. \tag{8.57}$$

The quantity Q may be a zeroth order tensor such as the temperature, pressure, density or a first order tensor such as the velocity. The spatial differentiation of the quantity is obtained from:

$$\overline{\frac{\partial Q}{\partial s}} = \frac{\partial \bar{Q}}{\partial s}. \tag{8.58}$$

For a velocity measured at an arbitrary position vector $\mathbf{x}$, the mean is expressed as:

$$\bar{\mathbf{V}}(\mathbf{x}) = \frac{1}{T}\int_0^T \mathbf{V}(\mathbf{x}, t)dt : \tag{8.59}$$

For further consideration and for the sake of simplicity, we may abandon the ensemble averaged parenthesis pair and instead use the over bar $< \mathbf{V}(\mathbf{x}, t) > \equiv \bar{\mathbf{V}}(\mathbf{x}, t) \equiv \bar{\mathbf{V}}$. However, we will resort to the parenthesis whenever there is a reason for confusion. We start from the Navier-Stokes equation for incompressible flow in an absolute frame of reference:

$$\frac{\partial \mathbf{V}}{\partial t} + \mathbf{V} \cdot (\nabla \mathbf{V}) = -\frac{1}{\rho}\nabla p + \nu \Delta \mathbf{V} \tag{8.60}$$

and assume that the flow quantities are associated with certain fluctuations. We replace the velocity vector $\mathbf{V}$ by $\mathbf{V} = \bar{\mathbf{V}} + \mathbf{V}'$, the pressure $p = \bar{p} + p'$ with $\bar{\mathbf{V}}$, $\bar{p}$ as the ensemble averaged velocity vector, ensemble averaged pressure and $\mathbf{V}'$, $\mathbf{p}'$ the fluctuation velocity vector and pressure which are inherently time dependent. A simple order of magnitude estimation shows that the density fluctuations may be neglected. Taking the average of Eq. (8.60) gives:

$$\overline{\frac{\partial(\bar{\mathbf{V}} + \mathbf{V}')}{\partial t} + (\bar{\mathbf{V}} + \mathbf{V}') \cdot \nabla(\bar{\mathbf{V}} + \mathbf{V}')} = \overline{-\frac{1}{\rho}\nabla(\bar{p} + p') + \nu\Delta(\bar{\mathbf{V}} + \mathbf{V}')}. \tag{8.61}$$

Further treatment of Eq. (8.61) in conjunction with the averaging rules Eqs. (8.56), (8.57) and (8.58) results in

$$\frac{\partial \bar{\mathbf{V}}}{\partial t} + \bar{\mathbf{V}} \cdot \nabla\bar{\mathbf{V}} + \nabla \cdot (\overline{\mathbf{V}'\mathbf{V}'}) = -\frac{1}{\rho}\nabla\bar{p} + \nu\Delta\bar{\mathbf{V}} + \mathbf{g}. \tag{8.62}$$

This equation is referred to as the *Reynolds averaged Navier-Stokes* (RANS) equation of motion for incompressible flow with constant viscosity. For the statistically steady state, $\frac{\partial \bar{\mathbf{V}}}{\partial t} = 0$, accordingly for statistically unsteady flow, we have the time derivative $\frac{\partial \bar{\mathbf{V}}}{\partial t} \equiv \frac{\partial < \bar{\mathbf{V}} >}{\partial t} \neq 0$, because $\bar{\mathbf{V}} =< \mathbf{V} >$. The term $(\overline{\mathbf{V}'\mathbf{V}'})$ is the *Reynolds stress tensor* and its divergence $\nabla \cdot (\overline{\mathbf{V}'\mathbf{V}'})$ is the "eddy force" acting on the fluid particle due to the turbulent fluctuations. It should be pointed out that the decomposition steps performed above were in order to find approximate solutions for the Navier-Stokes equation, whose direct numerical solution until very recently, appeared to be out of reach. Since the Reynolds stress tensor can not be expressed uniquely in terms of mean flow quantities, it must be modeled. This, however, is the subject of turbulence modeling that is treated in Chap. 9. Before starting with the implementation of intermittency function into the Reynolds equations, we rearrange Eq. (8.62) as follows:

$$\frac{\partial \bar{\mathbf{V}}}{\partial t} + \bar{\mathbf{V}} \cdot \nabla \bar{\mathbf{V}} = -\frac{1}{\rho} \nabla \bar{p} + \nabla \cdot (\nu \nabla \bar{\mathbf{V}} - \overline{\mathbf{V}'\mathbf{V}'}) + \mathbf{g}. \tag{8.63}$$

In Eq. (8.63) we assumed that the kinematic viscosity is constant throughout the flow field. In Cartesian coordinate systems the index notation of Eq. (8.63) is:

$$\frac{\partial \bar{V}_i}{\partial t} + \bar{V}_j \frac{\partial \bar{V}_i}{\partial x_j} = -\frac{1}{\rho} \frac{\partial \bar{p}}{\partial x_i} + \frac{\partial}{\partial x_j} \left(\nu \frac{\partial \bar{V}_i}{\partial x_j} - \overline{V'_i V'_j} \right) + g_i \tag{8.64}$$

and is decomposed in x_1-direction:

$$\begin{aligned} \frac{\partial \bar{V}_1}{\partial t} + \bar{V}_1 \frac{\partial \bar{V}_1}{\partial x_1} + \bar{V}_2 \frac{\partial \bar{V}_1}{\partial x_2} + \bar{V}_3 \frac{\partial \bar{V}_1}{\partial x_3} &= -\frac{1}{\rho} \frac{\partial p}{\partial x_1} + \nu \left(\frac{\partial^2 \bar{V}_1}{\partial x_1^2} + \frac{\partial^2 \bar{V}_1}{\partial x_2^2} + \frac{\partial^2 \bar{V}_1}{\partial x_3^2} \right) \\ &- \left(\frac{\partial}{\partial x_1} (\overline{V_1' V_1'}) + \frac{\partial}{\partial x_2} (\overline{V_1' V_2'}) + \frac{\partial}{\partial x_3} (\overline{V_1' V_3'}) \right) + g_1 \end{aligned} \tag{8.65}$$

in x_2-direction:

$$\begin{aligned} \frac{\partial \bar{V}_2}{\partial t} + \bar{V}_1 \frac{\partial \bar{V}_2}{\partial x_1} + \bar{V}_2 \frac{\partial \bar{V}_2}{\partial x_2} + \bar{V}_3 \frac{\partial \bar{V}_2}{\partial x_3} &= -\frac{1}{\rho} \frac{\partial p}{\partial x_2} + \nu \left(\frac{\partial^2 \bar{V}_2}{\partial x_1^2} + \frac{\partial^2 \bar{V}_2}{\partial x_2^2} + \frac{\partial^2 \bar{V}_2}{\partial x_3^2} \right) \\ &- \left(\frac{\partial}{\partial x_1} (\overline{V_2' V_1'}) + \frac{\partial}{\partial x_2} (\overline{V_2' V_2'}) + \frac{\partial}{\partial x_3} (\overline{V_2' V_3'}) \right) + g_2 \end{aligned} \tag{8.66}$$

and finally in x_3-direction, we have:

$$\begin{aligned} \frac{\partial \bar{V}_2}{\partial t} + \bar{V}_1 \frac{\partial \bar{V}_3}{\partial x_1} + \bar{V}_2 \frac{\partial \bar{V}_3}{\partial x_2} + \bar{V}_3 \frac{\partial \bar{V}_3}{\partial x_3} &= -\frac{1}{\rho} \frac{\partial p}{\partial x_3} + \nu \left(\frac{\partial^2 \bar{V}_3}{\partial x_1^2} + \frac{\partial^2 \bar{V}_3}{\partial x_2^2} + \frac{\partial^2 \bar{V}_3}{\partial x_3^2} \right) \\ &- \left(\frac{\partial}{\partial x_1} (\overline{V_3' V_1'}) + \frac{\partial}{\partial x_2} (\overline{V_3' V_2'}) + \frac{\partial}{\partial x_3} (\overline{V_3' V_3'}) \right) + g_3. \end{aligned} \tag{8.67}$$

The Reynolds averaged Navier-Stokes (RANS) Eq. (8.63), its index notation (8.64), and the component decomposition (8.65)–(8.67) are derived for a fully turbulent flow, which inherently includes the Reynolds stress tensor that has nine components whose divergence is shown in Eq. (8.64). This equation cannot be applied to a transitional flow, which is intermittently laminar and turbulent as is common in many engineering applications. To account for the intermittent nature of a transitional flow, RANS-equations require a conditioning as detailed below.

8.5.2 *Intermittency Implementation in RANS*

Following Eq. (8.51), we first re-arrange the velocity vector:

$$\mathbf{V}'(\mathbf{x}, t) = \mathbf{V}(\mathbf{x}, t) - \bar{\mathbf{V}}(\mathbf{x}, t). \tag{8.68}$$

For a non-turbulent flow, the left-hand side of Eq. (8.66) becomes zero:

$$\mathbf{V}(\mathbf{x}, t) - \bar{\mathbf{V}}(\mathbf{x}, t) = 0 \tag{8.69}$$

and for a fully turbulent flow, we have

$$\mathbf{V}(\mathbf{x}, t) - \bar{\mathbf{V}}(\mathbf{x}, t) = \mathbf{V}'(\mathbf{x}, t) \neq 0. \tag{8.70}$$

Thus, Eqs. (8.69) and (8.70) can be summarized as:

$$\mathbf{V}(\mathbf{x}, t) - \bar{\mathbf{V}}(\mathbf{x}, t) = I\mathbf{V}'(\mathbf{x}, t) \tag{8.71}$$

with $I = 1$ for fully turbulent flow and $I = 0$ otherwise. To arrive at a conditioned Navier-Stokes equation for implementation of intermittency function, it is easier to modify first the Navier-Stokes equations by adding the continuity equation for incompressible flow. This results in

$$\frac{\partial \mathbf{V}}{\partial t} + \nabla \cdot (\mathbf{V}\mathbf{V}) = -\frac{\mathbf{1}}{\rho}\nabla \mathbf{p} + \nu \Delta \mathbf{V}. \tag{8.72}$$

Inserting the velocity from Eq. (8.71) into the Eq. (8.72), we receive:

$$\frac{\partial(\bar{\mathbf{V}} + I\mathbf{V}')}{\partial t} + \nabla \cdot [(\bar{\mathbf{V}} + I\mathbf{V}')(\bar{\mathbf{V}} + I\mathbf{V}')] = -\frac{1}{\rho}\nabla(p + p') + \nu\Delta(\bar{\mathbf{V}} + I\mathbf{V}'). \tag{8.73}$$

Performing the multiplication, Eq. (8.73) is prepared for ensemble averaging:

$$\begin{aligned} &\overline{\frac{\partial \bar{\mathbf{V}}}{\partial t} + \frac{\partial (I\mathbf{V}')}{\partial t} + \nabla \cdot [(\bar{\mathbf{V}}\bar{\mathbf{V}} + 2I\bar{\mathbf{V}}\mathbf{V}' + I^2\mathbf{V}'\mathbf{V}')]} \\ &= \overline{-\frac{1}{\rho}\nabla(p + p') + \nu\Delta\bar{\mathbf{V}} + \nu\Delta(I\mathbf{V}')}. \end{aligned} \tag{8.74}$$

Carrying out the procedure of ensemble averaging, the following terms identically disappear:

$$\overline{I\mathbf{V}'} \equiv 0, \quad \overline{2I\bar{\mathbf{V}}V'} \equiv 0. \tag{8.75}$$

Utilizing the $I^2 \equiv I$, we arrive at the Reynolds stress tensor:

$$I\overline{\mathbf{V}'\mathbf{V}'} = \gamma\overline{\mathbf{V}'\mathbf{V}'} = \frac{1}{T}\int_{t}^{t+T} I\mathbf{V}'\mathbf{V}'dt. \tag{8.76}$$

With Eq. (8.76) and the ensemble averaged (8.74), we obtain the conditioned Reynolds equations for incompressible flow that describe non-turbulent, transitional, and fully turbulent flows as well:

$$\frac{\partial \bar{\mathbf{V}}}{\partial t} + \nabla \cdot (\bar{\mathbf{V}}\bar{\mathbf{V}} + \gamma\overline{\mathbf{V}'\mathbf{V}'}) = -\frac{1}{\rho}\nabla\bar{p} + \nu\Delta\bar{\mathbf{V}}. \tag{8.77}$$

Rearranging Eq. (8.77) leads to:

$$\frac{\partial \bar{\mathbf{V}}}{\partial t} + \bar{\mathbf{V}} \cdot \nabla\bar{\mathbf{V}} = -\frac{1}{\rho}\nabla\bar{p} + \nu\Delta\bar{\mathbf{V}} - \nabla \cdot (\gamma\overline{\mathbf{V}'\mathbf{V}'}). \tag{8.78}$$

The turbulent shear stress associated with the intermittency function, $\gamma\overline{\mathbf{V}'\mathbf{V}'}$, plays a crucial role in affecting the solution of the Navier-Stokes equations. This is particularly significant for calculating the wall friction and the heat transfer coefficient distribution. Two quantities have to be accurately modeled. One is the intermittency function γ, and the other is the Reynolds stress $\overline{\mathbf{V}'\mathbf{V}'}$ tensor with its nine components. Inaccurate modeling of these two quantities leads to a multiplicative error of their product $\gamma\overline{\mathbf{V}'\mathbf{V}'}$. This error particularly affects the accuracy of the total pressure losses, efficiencies, and heat transfer coefficient calculations. Equation (8.78) is coordinate invariant and can be transformed to any curvilinear coordinate system within an absolute frame of reference. Its component in x_1-direction is:

$$\begin{aligned}&\frac{\partial \bar{V}_1}{\partial t} + \bar{V}_1\frac{\partial \bar{V}_1}{\partial x_1} + \bar{V}_2\frac{\partial \bar{V}_1}{\partial x_2} + \bar{V}_3\frac{\partial \bar{V}_1}{\partial x_3} = -\frac{1}{\rho}\frac{\partial p}{\partial x_1} + \nu\left(\frac{\partial^2 \bar{V}_1}{\partial x_1^2} + \frac{\partial^2 \bar{V}_1}{\partial x_2^2} + \frac{\partial^2 \bar{V}_1}{\partial x_3^2}\right)\\ &\quad - \left(\frac{\partial(\gamma\overline{V_1'V_1'})}{\partial x_1} + \frac{\partial(\gamma\overline{V_2'V_1'})}{\partial x_2} + \frac{\partial(\gamma\overline{V_3'V_1'})}{\partial x_3}\right) + g_1\end{aligned} \tag{8.79}$$

in x_2-direction,

$$\begin{aligned}&\frac{\partial \bar{V}_2}{\partial t} + \bar{V}_1\frac{\partial \bar{V}_2}{\partial x_1} + \bar{V}_2\frac{\partial V_2}{\partial x_2} + \bar{V}_3\frac{\partial \bar{V}_2}{\partial x_3} = -\frac{1}{\rho}\frac{\partial p}{\partial x_2} + \nu\left(\frac{\partial^2 \bar{V}_2}{\partial x_1^2} + \frac{\partial^2 \bar{V}_2}{\partial x_2^2} + \frac{\partial^2 \bar{V}_2}{\partial x_3^2}\right)\\ &\quad - \left(\frac{\partial(\gamma\overline{V_1'V_2'})}{\partial x_1} + \frac{\partial(\gamma\overline{V_2'V_2'})}{\partial x_2} + \frac{\partial(\gamma\overline{V_3'V_2'})}{\partial x_3}\right) + g_2\end{aligned} \tag{8.80}$$

in x_3-direction

$$\frac{\partial \bar{V}_3}{\partial t} + \bar{V}_1 \frac{\partial \bar{V}_3}{\partial x_1} + \bar{V}_2 \frac{\partial \bar{V}_3}{\partial x_2} + \bar{V}_3 \frac{\partial \bar{V}_3}{\partial x_3} = -\frac{1}{\rho}\frac{\partial p}{\partial x_3} + \nu \left(\frac{\partial^2 \bar{V}_3}{\partial x_1^2} + \frac{\partial^2 \bar{V}_3}{\partial x_2^2} + \frac{\partial^2 \bar{V}_3}{\partial x_3^2} \right)$$
$$- \left(\frac{\partial(\gamma \overline{V_1' V_3'})}{\partial x_1} + \frac{\partial(\gamma \overline{V_2' V_3'})}{\partial x_2} + \frac{\partial(\gamma \overline{V_3' V_3'})}{\partial x_3} \right) + g_3. \tag{8.81}$$

The conditioning procedure discussed above and the subsequent decomposition lead to a turbulent shear stress expression that contains turbulent and non-turbulent terms. Detailed flow measurements by Chavary and Tutu [72], however, clearly show that the shear stress is carried completely by the turbulent region. In the case of the conditioning process we derived above, the non-turbulent term is embedded in the ensemble averaged terms, implying that the intermittent shear stress term $\gamma \overline{\mathbf{V}'\mathbf{V}'}$ in Eq. (8.76) is carried completely by the turbulent region. This exhibits a substantial improvement in terms of formulating the RANS-equations for unsteady intermittent flows.

8.6 Problems and Projects

Project 8.1 Derive the Orr-Sommerfeld differential equation (8.20). Using the boundary conditions (8.27), solve the Orr-Sommerfeld equation for a plane Poiseuille flow.

Project 8.2 Write a computer program for calculating the Blasius equation [see Chap. 11, laminar flow, Eq. (11.26)]. Add a subroutine for calculating the fully turbulent velocity profile using the Prandtl 1/7th law [see Chap. 11, Eq. (11.17)]. Add a subroutine for transition model and compute the flow velocity profiles from laminar state to turbulent state.

Problem 8.3 (1) Using Matlab or Maple random data generator tool, generate a set of random fluctuation over a period of 1 s with a frequency of 1kHz. The fluctuation amplitude should be around 3 m/s. The two consecutive data points should fluctuate around zero (positive, negative), so the mean of the entire fluctuating data is always zero. (2) Generate 50 events (or observation) of above data, arrange the data randomly. (3) Using a sinusoidal velocity distribution generate a periodic velocity distribution with a mean of 30 m/s, an amplitude of 10 m/s and a period of 1 s. Superimpose (1) on (3) get another set of data (1000 points) for each event. (4) Perform the ensemble averaging and find the fluctuation RMS and the ensemble average of the velocity.

Chapter 9
Turbulent Flow, Modeling

9.1 Fundamentals of Turbulent Flows

The preceding chapter dealt with stability of laminar flows, their perturbation and transition to the turbulent state. In discussing the transition process, we prepared the essentials for better understanding the basic physics of the more complex turbulent flow, which is still an unresolved and extremely challenging problem in fluid mechanics.

Using the intermittency function as a parameter to describe the flow state under consideration, two distinct flow regimes were distinguished in Chap. 8: (a) laminar flow regime characterized by the absence of irregular or random fluctuations with $\gamma = 0$ and, (b) turbulent flow state characterized by $\gamma = 1$ with irregularities expressed in terms of random variations in time and space. While the randomness is an inherent quality of a turbulent flow, it does not completely define the turbulent flow. In many engineering applications, however, turbulent flow can be described statistically by determining averaged values for flow quantities. Descriptions of averaged velocity, kinetic energy, and Reynolds stress tensor (see Chap. 8) distributions of wakes, free jets and jet boundaries are a few examples from *free turbulent flow* where the effect of molecular viscosity compared to the turbulence viscosity is neglected. For all engineering flow applications such as flows through pipes, nozzles, diffuseres, combustion engines, turbines and compressors, blade channels or flow around aircraft wings, averaged quantities are described utilizing different turbulence models that relate the Reynolds stress tensor to the mean flow quantities. In these cases, we deal with the wall turbulence which is generated by the presence of a solid wall. In contrast to the free turbulence, the wall turbulence is subjected to both molecular and turbulence viscosity associated with energy dissipation. Taking into consideration the complexity of turbulent flows encountered in engineering and a multitude of definitions found in literature, among others, by Taylor [73], von Kármán [74], Hinze [75], and Rotta [76], we term a flow regime turbulent that has the following characteristics:

M. T. Schobeiri, *Advanced Fluid Mechanics and Heat Transfer for Engineers and Scientists*, https://doi.org/10.1007/978-3-030-72925-7_9

1. Turbulent flows are generally irregular and their properties continuously undergo stochastic spatial and temporal changes. As a result, no reproducible turbulence data with stochastic distribution can be obtained.
2. Despite its stochastic nature, using statistical tools, time or ensemble averaged values can be constructed that are perfectly reproducible.
3. Turbulent flows are rotational motions (vorticity) with a wide variety of vortices with different sizes and vorticities.
4. Turbulent flows are generally unsteady and three-dimensional.

The above characteristics are implicitly indicative of the following features that are inherent in turbulence:

(a) There is no analytical exact solution for any type of turbulent flows, even for the simplest one. In the case of free turbulent flows, as we will see in the following, only approximate solutions that are based on an inductive approach are found.
(b) The inherent three-dimensional unsteady nature of turbulence associated with the velocity fluctuations is responsible for an intense mixing of fluid particles causing an enhanced mass, momentum, and energy transfer between the fluid particles. This process is called diffusion. The enhanced diffusivity is due to the existence of Reynolds stress which is, in general, several order of magnitudes larger than the viscous stresses. Exceptions are flows very close to the wall, where the viscosity has a predominant effect.
(c) The high level of spatial and temporal fluctuations of velocity, pressure, and temperature causes fluctuating vortices, also called eddies, of different sizes. The eddies convect, rotate, stretch, decompose in smaller eddies, overlap and coalesce, as Fig. 9.1 schematically shows. Figure 9.1 schematically summarizes the energy cascade process taking place within a turbulent boundary layer. As seen, a flat plate is exposed to a constant, steady, non-turbulent ($\nabla \times \boldsymbol{V} = 0$) flow that is separated from the rotational boundary layer ($\nabla \times \boldsymbol{V} \neq 0$) by a sharp interface. The averaged boundary layer thickness is shown as a dashed curve. Large eddies with a size l continuously extract energy from the main flow and transfers it to smaller eddies. The process of energy cascading leads to the smallest eddies whose energy dissipates as heat. The specific issues dealing with Kolmogorov's hypothesis, energy cascade, eddy structure, length, and time scale, are treated in more detail in the following sections.
(d) During the cascade process, the size of these eddies change from large to small. In a boundary layer flow, as shown in Fig. 9.1, the size of the largest eddy has the same order of magnitude as the local boundary layer thickness δ. It receives its energy from the mean flow. The larger eddies continuously transfer their kinetic energy to the smaller eddies. Once the eddy size is reduced to a minimum, its kinetic energy is dissipated by the viscous diffusion. A state of universal equilibrium is reached when the rate of energy received from larger eddies is nearly equal to the rate of energy of the smallest eddies that dissipate into heat. The process of transferring energy from the largest eddy to the smallest is called

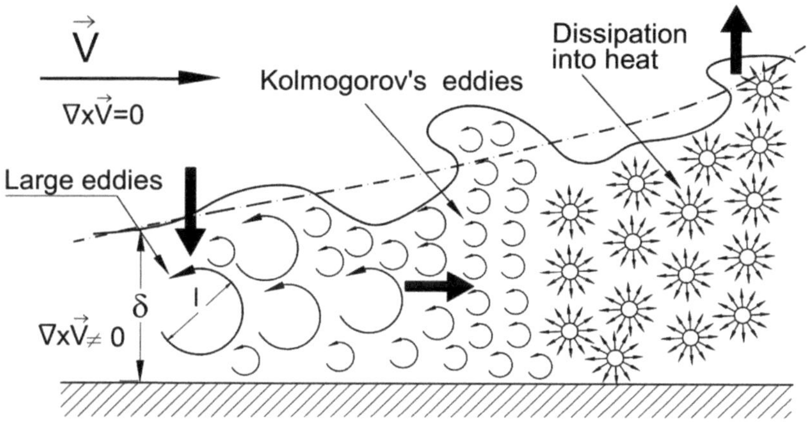

Fig. 9.1 Schematics of an instantaneous energy cascade in turbulent boundary layer. The arrows indicate energy extraction, transfer and dissipation

energy cascade process, introduced by Richardson [77]. While the statistics[1] of larger eddies change, Kolmogorov [78] introduced a hypothesis that enables quantifying the scale of the smallest eddy on the basis of isotropy of those eddies.

(e) Turbulent flow occurs at high Reynolds numbers. For engineering applications where a solid wall is present (boundary layer, wall turbulence), the order of magnitude for the Re-number to become fully turbulent depends on the pressure gradient along the surface, as well as the perturbation of the boundary layer by any incoming periodic unsteady disturbances such as wake impingements.

Note that in the course of the above introduction we utilized the rather vague term "eddy", which is used in the literature in context of turbulence research. In contrast to the precisely defined term "vortex" with a descriptive circulation Γ and its direct relation to the vorticity vector $\nabla \times \boldsymbol{V}$, the term eddy lacks a precise definition and is loosely used for any individual turbulent structure with some length-scale.

9.1.1 Type of Turbulence

A turbulent flow is called *homogeneous*, if its statistical quantities (or short: statistics) are independent of the spatial position vector $\boldsymbol{x}$. This requires that the mean velocity field described by Eq. (8.51), $\bar{\boldsymbol{V}}(\boldsymbol{X}, t)$, must also be independent of $\boldsymbol{x}$. If we assume a constant pressure flow field ($\nabla \bar{p} = 0$) and neglect the contribution of the body forces (gravitational, electromagnetic, electrostatic), the Reynolds equation (8.63) will reduce to $\partial \bar{\boldsymbol{V}} / \partial t = 0$ because all spatial derivatives disappear. As a consequence,

[1] The averages of a random quantity are called statistics. This includes mean and the rms (root-mean-square) of that quantity.

the resulting velocity field will be independent of time $\bar{V} \neq f(t)$, meaning that the velocity $\bar{V} = \text{const.}$ and the flow is *statistically stationary*.

A turbulent flow is called *isotropic* if there is no preference for any specific direction, i.e. $\overline{V_1'^2} = \overline{V_2'^2} = \overline{V_3'^2} = \overline{V_i'V_i'}/3$. In reality, these averages exhibit certain directional dependency, making the isotropy a hypothetical case. Despite this fact, the isotropic turbulence is significant for fundamental study of turbulent flow. From experimental point of view, flow regions can be found, where the fine structure of actual nonisotropic turbulent flows can be approximated as isotropic. This approximation associated with the major simplifications makes the complex turbulent flow more amenable to fundamental theoretical treatment than any other type of turbulent flow.

9.1.2 Correlations, Length and Time Scales

As we saw in Chap. 8, the Reynolds-averaging procedure has created an *apparent stress tensor* $\overline{\boldsymbol{V}'\boldsymbol{V}'}$ with nine components from which, for symmetric reasons, six are distinct. Thus, the creation of this tensor has added six more unknowns to Navier–Stokes equations. In order to find additional equations to close the equation set that consists of continuity, momentum, and energy balances, we need to construct additional equations. This is done by multiplying the ith component of the Navier–Stokes equation with the jth one. Thereby we expect to find turbulence models that establish relations between the new equations and the set of equations mentioned above. It should be pointed out that this purely mathematical manipulation does not represent any new conservation law with a physical background. However, it helps in providing additional tools that are necessary for turbulence modeling. In this context, correlations are indispensable tools for providing additional insight into turbulence. As we know from Navier–Stokes equations, the second order tensor $\overline{\boldsymbol{V}'\boldsymbol{V}'}$ is the mean product of the fluctuation components at a single point in space; it is called a *single point correlation*. It does not give any further information about the turbulence structure, such as the length and time scale of eddies. We obtain this information from a *two-point correlation*. It is a second order tensor of the mean product of fluctuation components at two different points in space and time, namely $(\boldsymbol{x}, t)$ and $(\boldsymbol{x} + \boldsymbol{r}, i + \tau)$. For a purely spatial correlation with $\tau = 0$, the same fluctuating quantity is measured at two different spatial positions x and $x + r$. Figure 9.2 shows the position of the fluctuation components for (a) single point correlation and several two-point correlations. For a general two-point correlation, we construct the second order tensor

$$e_i e_j R_{ij}(\boldsymbol{x}, t, \boldsymbol{r}, \tau) = e_i e_j \overline{V_i'(\boldsymbol{x}, t) V_j'(\boldsymbol{x} + \boldsymbol{r}, t + \tau)}, \text{ with}$$
$$R_{ij}(\boldsymbol{x}, t, \boldsymbol{r}, \tau) = \overline{V_i'(\boldsymbol{x}, t) V_j'(\boldsymbol{x} + \boldsymbol{r}, t + \tau)} \tag{9.1}$$

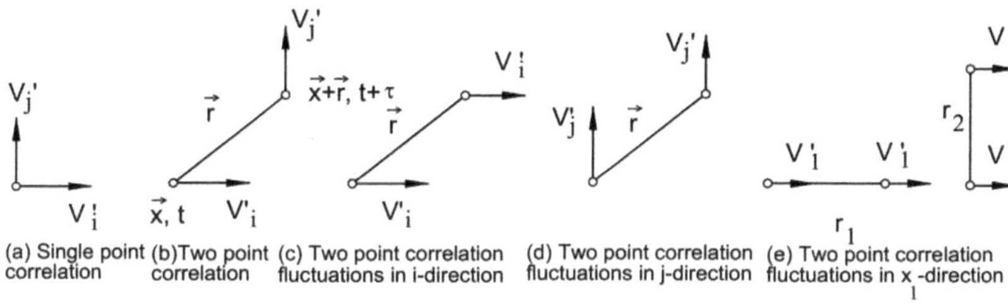

Fig. 9.2 Single- and two-point correlations

with $\boldsymbol{r} = e_i r_i$ and τ as the spatial and temporal distance between the two points. For $|r| \rightarrow \infty$ or $|\tau| \rightarrow \infty$, the fluctuation components V_i' and V_j' are independent from each other, leading to $R_{ij} = 0$. For a stationary or homogeneous process, the correlation tensor is symmetric and we may write:

$$R_{ij}(\boldsymbol{x}, t, \boldsymbol{r}, \tau) = R_{ji}(\boldsymbol{x} + \boldsymbol{r}, t + \tau, -\boldsymbol{r}, \tau). \tag{9.2}$$

Normalizing the correlation Eq. (9.1), we obtain the dimensionless correlation, also called *correlation coefficient* as:

$$\rho_{ij}(\boldsymbol{x}, t, \boldsymbol{r}, \tau) = \frac{\overline{V_i'(\boldsymbol{x}, t) V_j'(\boldsymbol{x} + \boldsymbol{r}, t + \tau)}}{[\overline{V_i'^2(\boldsymbol{x}, t)}\, \overline{V_j'^2(\boldsymbol{x} + \boldsymbol{r}, t + \tau)}]^{\frac{1}{2}}} \tag{9.3}$$

For a stationary or homogeneous field, the tensor $R_{ij}(\boldsymbol{x}, t, \boldsymbol{r}, \tau)$ is independent of t and x. We can construct an *auto-correlation* when the fluctuation components are measured at the same position but at different times and have the same direction ($i = j$). It is defined as

$$R_{ij}(t, \tau) = \overline{V_i'(t) V_j'(t + \tau)}. \tag{9.4}$$

Note that by setting $i = j$, we do not sum over the indices i and j. As an example, the auto-correlation for the fluctuation component V_1', is written as

$$R_{11}(t, \tau) = \overline{V_1'(t) V_1'(t + \tau)} \tag{9.5}$$

with the fluctuation V_1' at the same spatial position but at two different times t and $t + \tau$. On the other hand, the spatial correlation is obtained by setting in Eq. (9.1) $\tau = 0$.

$$R_{ij}(\boldsymbol{x}, \boldsymbol{r}) \equiv R_{ij}(\boldsymbol{r}) = \overline{V_i'(\boldsymbol{x}) V_j'(\boldsymbol{x} + \boldsymbol{r})}. \tag{9.6}$$

The corresponding correlation coefficient is

$$\rho_{ij}(\boldsymbol{r}) = \frac{\overline{V_i'(\boldsymbol{x}) V_j'(\boldsymbol{x} + \boldsymbol{r})}}{[\overline{V_i'^2(\boldsymbol{x})}\, \overline{V_j'^2(\boldsymbol{x} + \boldsymbol{r})}]^{\frac{1}{2}}} \tag{9.7}$$

For the component in x_1-direction, Eq. (9.6) is simplified as

$$R_{11}(r_1, 0, 0) = \overline{V_1'(x_1)V_1'(x_1 + r_1)}. \tag{9.8}$$

In Eq. (9.8), the reference position vector x_1 can be set $x_1 = 0$, resulting in

$$R_{11}(r_1, 0, 0) = \overline{V_1'(0)V_1'(r_1)}. \tag{9.9}$$

The right-hand side of Eq. (9.9) is called the *covariance* of the V_1'-component. The correlation coefficient then is obtained by setting in Eq. (9.7) $i = j$

$$\rho_{11}(r_1, 0, 0) = \frac{\overline{V_1'(0)V_1'(r_1)}}{[\overline{V_1'^2(0)}\,\overline{V_1'^2(r_1)}]^{\frac{1}{2}}}. \tag{9.10}$$

In most turbulence related literature, the term $\overline{V_1'^2(r_1)}$ is replaced by $\overline{V_1'^2(0)}$, thus, the modified coefficient is:

$$\rho_{11}(r_1, 0, 0) = \frac{\overline{V_1'(0)V_1'(r_1)}}{\overline{V_1'^2(0)}}. \tag{9.11}$$

In a similar manner, the coefficients in r_2 and r_3 may be constructed

$$\rho_{11}0, (r_2, 0) = \frac{\overline{V_1'(0)V_1'(r_2)}}{\overline{V_1'^2(0)}}. \tag{9.12}$$

The correlation functions are used to determine the length and time scales. The general definition of the integral length scale is

$$L_{ij,k}(\boldsymbol{x}, t) = \frac{\int_{-\infty}^{+\infty} R_{ij}(\boldsymbol{x}, t, r_k, 0)dr_k}{2\overline{V_i'(\boldsymbol{x}, t)V_j'(\boldsymbol{x}, t)}}, \tag{9.13}$$

with r_k as a specific spatial displacement in x_1, x_2 or x_3 direction as Fig. 9.3 shows for r_1, r_2 and r_3.
Likewise, the time scale is defined as

$$T_{ij}(\boldsymbol{x}, t) = \frac{\int_{-\infty}^{+\infty} R_{ij}(\boldsymbol{x}, t, 0, \tau)d\tau}{2\overline{V_i'(\boldsymbol{x}, t)V_j'(\boldsymbol{x}, t)}}. \tag{9.14}$$

For the special cases discussed above, the length scale is schematically plotted in Fig. 9.3 for longitudinal $\rho_{11}(r_1, 0, 0)$, as well as lateral $\rho_{11}(0, r_2, 0)$ correlation coefficients. In both cases, the length scale is simply the area underneath the coefficient curves. Using the hot wire anemometry for measuring the velocity fluctuation, it is necessary to use two parallel wires separated either by r_1 in a longitudinal or by r_2 in a lateral direction. In the longitudinal case, Fig. 9.3a, the second wire is in velocity

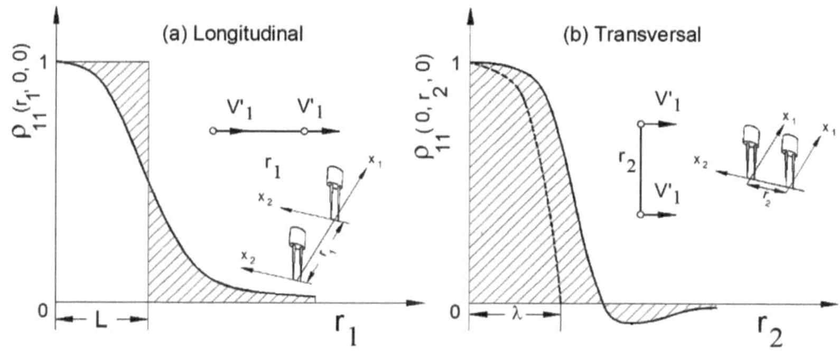

Fig. 9.3 Correlation coefficients with their *osculating* parabolas

and thermal wakes of the first wire located upstream of the second wire. This configuration leads to an erroneous longitudinal length scale. The lateral length scale can be measured more accurately using the two hot wire probes as arranged in Fig. 9.3b. Re-arranging Eq. (9.13), we find the longitudinal length scale

$$L_{\text{long}} = \int_{-\infty}^{+\infty} \rho_{11}(r_1, 0, 0) dr_1 = \frac{1}{2\overline{V_1'^2(0)}} \int_{-\infty}^{+\infty} \overline{V_1'(0) V_1'(r_1)} dr_1 \tag{9.15}$$

and the lateral length scale

$$L_{\text{lat}} = \int_{-\infty}^{+\infty} \rho_{11}(0, r_2, 0) dr_2 = \frac{1}{2\overline{V_1'^2(0)}} \int_{-\infty}^{+\infty} \overline{V_1'(0) V_1'(r_2)} dr_2. \tag{9.16}$$

Although measuring the lateral length scale delivers a more accurate result than the measured longitudinal one, from an experimental point of view, both are not practical. This following hypothesis offers a practical alternative.

Taylor Hypothesis: An alternative method to estimate the length scale is the utilization of a *frozen turbulence hypothesis* proposed by Taylor [73]. Considering a large scale eddy with sufficiently high energy content, Taylor proposed an hypothesis that the energy transport contribution of small size eddies that are carried by a large scale eddy, as shown in Fig. 9.4, compared with the one produced by a larger eddy, is negligibly small. In such a situation, the transport of a turbulence field past a fixed point is due to the larger energy containing eddies. It states that in certain circumstances, turbulence can be considered as "frozen" as it passes by a "sensor".
This statement is illustrated in Fig. 9.4. It shows a large eddy moving with an averaged constant velocity of $\bar{V}_1$ in the x_1-direction, carrying a number of smaller eddies with fluctuating velocity V_1'. The hypothesis is considered valid only if the condition $V_1'^2 << V_1^2$ holds. Despite this constraint, the Taylor hypothesis delivers a reasonable

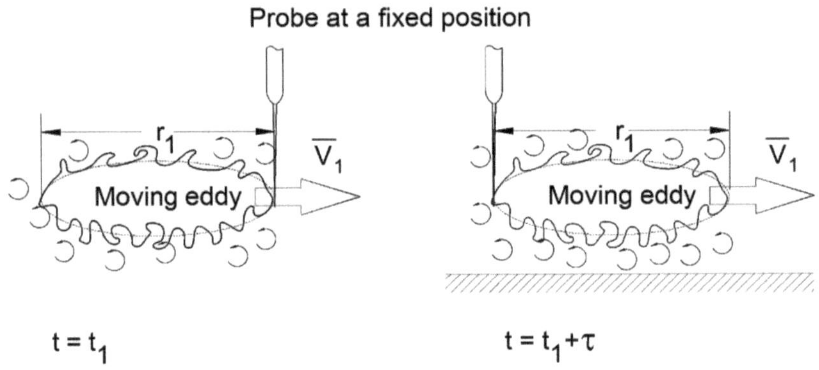

Fig. 9.4 Explaining Taylor "frozen" turbulence

approximation for the variations of fluctuating eddies that are carried along by larger scale eddies. Taylor established his hypothesis using a spatial (Eulerian) description rather than a material (Lagrangian) one (see Chap. 3). The hypothesis relates the spatial variation to the temporal variation measured at a single point. From an experimental point of view, this approach exhibits a substantial reduction in efforts to determine the turbulence length and time scales. Mathematically speaking, the Taylor hypothesis implies that the substantial change of the velocity $\boldsymbol{V} = \bar{\boldsymbol{V}} + \boldsymbol{V}'$ must vanish. Utilizing the Taylor's assumption of constant mean velocity, we have

$$\frac{D(\bar{\boldsymbol{V}} + \boldsymbol{V}')}{Dt} = \frac{\partial \boldsymbol{V}'}{\partial t} + \bar{\boldsymbol{V}} \cdot \nabla \boldsymbol{V}' = 0. \tag{9.17}$$

Equation (9.17) is the mathematical formulation of the Taylor hypothesis. The x_1 component of (9.17) reads

$$\frac{\partial V_1'}{\partial x_1} = -\frac{1}{\bar{V}_1}\frac{\partial V'}{\partial t}. \tag{9.18}$$

Approximating the differentials by differences leads to:

$$\frac{V_1'(x_1 + r_1) - V_1'(x_1)}{r_1} = -\frac{V_1'(t + \tau)V_1'(t)}{\bar{V}_1 \tau}. \tag{9.19}$$

Equation (9.19) implies that the spatial separation shown in Fig. 9.3a can be expressed in terms of a temporal separation. As seen, Eq. (9.17) is the left-hand side of the Navier–Stokes equation, with the right side $-1/\rho\nabla(\bar{p} + p') + \nu\Delta(\bar{\boldsymbol{V}} + \boldsymbol{V}') = 0$. This hypothesis is only valid if we assume that $\boldsymbol{V}'/\bar{V} << 1$ and $\bar{p} = \text{const}$. As a consequence of this assumption, the pressure fluctuation p', which has the order of V'^2, can be neglected. With $r_1 = \bar{V}_1 \tau$ in Eq. (9.19), we obtain

$$V_1(t) - V_1(t + \tau) = V_1(x_1 + r_1) - V_1(x_1). \tag{9.20}$$

Thus, the auto-correlation coefficient (9.11) becomes

$$\rho_{11}(r_1, 0, 0) \equiv \rho_{11}(\tau) = \frac{\overline{V_1'(t)V_1'(t+\tau)}}{\overline{V_1'^2}} \tag{9.21}$$

and the corresponding integral time scale follows from

$$T_1 = \frac{1}{2}\int_{-\infty}^{+\infty} \rho_{11}(\tau)d\tau \tag{9.22}$$

thus, the length scale results from

$$L = \bar{V}_1 T_1. \tag{9.23}$$

Shifting the time origin results in $\rho(\tau) = \rho(-\tau)$, meaning that Eq. (9.21) is an even function. Furthermore, Eq. (9.22) has the property that at $\rho(\tau) = 1$ and $\tau = 0$ for $\tau \Rightarrow \infty$.

An alternative method to determine the time scale of small dissipating eddies uses Eq. (9.21). For this purpose we first expand the corresponding correlation coefficient (9.21) about $\tau = 0$ with respect to time, and truncate beyond the quadratic term; as a result, we arrive at

$$\rho_{11}(\tau) = 1 + \frac{\tau^2}{2}\left(\frac{\partial^2 \rho_{11}}{\partial \tau^2}\right)_{\tau=0} + \cdots \tag{9.24}$$

This crude approximation allows constructing an *osculating parabola* with the same vertex value and the derivative at $\tau = 0$ as the exact $\rho_{11}(\tau)$-curve. The parabola is described by

$$\rho_{11}(\tau) \approx 1 - \left(\frac{\tau}{\tau_1}\right)^2. \tag{9.25}$$

The intersection of this parabola with the τ-axis delivers the Taylor time scale τ_1, from which the Taylor micro length scale can be inferred. Equating (9.24) and (9.25) gives the Taylor micro time scale

$$\tau_1 = \sqrt{\frac{-2}{\left(\frac{\partial^2 \rho_{11}}{\partial \tau^2}\right)_{\tau=0}}} \tag{9.26}$$

and the Taylor micro length scale

$$\lambda = \bar{V}_1 \tau. \tag{9.27}$$

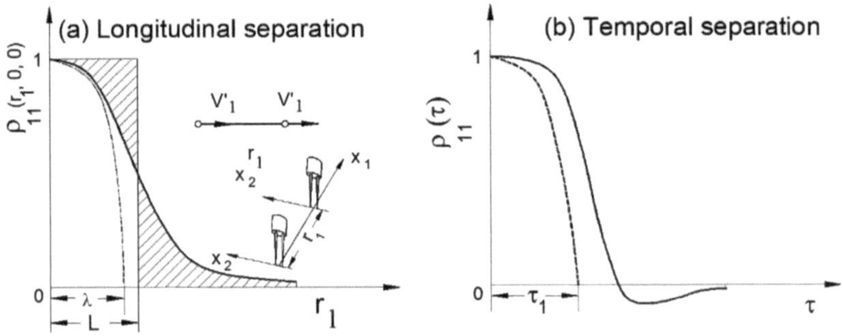

Fig. 9.5 Approximation of Taylor length and time scales by osculating parabolas

The two scales are shown in Fig. 9.5 with the correlation coefficients for (a) spatial and (b) temporal separations. It also includes the osculating parabola with the length scale and time scale intersects. The Taylor micro length scale can be found directly by using a similar procedure that leads to Eq. (9.26). In this case, we expand the correlation coefficient (9.11) about $r_1 = 0$ and truncate beyond the quadratic term. We may then approximate the result by the following parabola

$$\rho_{11}(r_1, 0, 0) \approx 1 - \left(\frac{r_1}{\lambda}\right)^2 \tag{9.28}$$

and arrive directly at the Taylor length scale

$$\lambda = \sqrt{\frac{-\overline{V_1'^2}}{\left(\frac{\partial^2 \rho_{11}}{\partial r_2^2}\right)}}. \tag{9.29}$$

It should be pointed out that the Taylor micro length scale is only an approximate length scale. It does not represent the length scale of large energy containing eddies or the smallest dissipating eddies. However, for a homogeneous isotropic turbulence, λ provides a useful tool to estimate the turbulence dissipation (Sect. 9.2.1.4, Eq. (9.71)). For this purpose, we first use the following length scale definition

$$\lambda = \sqrt{\frac{\overline{V_1'^2}}{\overline{\left(\frac{\partial V_1}{\partial x_1}\right)^2}}} \tag{9.30}$$

and then expand the dissipation equation defined in Sect. 9.2.1, Eq. (9.71) for isotropic turbulence and introduce (9.30). As a result, we find a relationship between the dissipation and the length scale.[2]

[2] Detailed derivations of (9.31) is found in Hinze [75], p. 179 and Rotta [76], p. 80.

$$\varepsilon = \nu \overline{\left(\frac{\partial V_i'}{\partial x_j} \frac{\partial V_i'}{\partial x_j} \right)} = \nu \overline{\left(\frac{\partial V_1'}{\partial x_1} \right)^2} (2\delta_{ii}\delta_{jj} - \delta_{ij}\delta_{ij})$$

$$= 15\nu \overline{\left(\frac{\partial V_1'}{\partial x_1} \right)^2} = 15\nu \left(\frac{\overline{V_1'}}{\lambda} \right)^2 . \tag{9.31}$$

Using dimensional analysis, Taylor established the following relationship for dissipation

$$\varepsilon \propto \frac{k^{3/2}}{l}. \tag{9.32}$$

Another important aspect is that length, time, and velocity scales describe the dissipative character of Kolmogorov's eddies as a result of energy cascading, as shown in Fig. 9.1. Using dimensional analysis, Kolmogorov arrived at his length scale (η), time scale (τ) and the velocity scale (υ) scales:

$$\eta = \left(\frac{\nu^3}{\varepsilon} \right)^{1/4}, \tau = \left(\frac{\nu}{\varepsilon} \right)^{1/2}, \upsilon = \left(\frac{\nu}{\varepsilon} \right)^{1/4} \tag{9.33}$$

which we will discuss in some details in the following section.

9.1.3 Spectral Representation of Turbulent Flows

As Fig. 9.1 shows, the scales of turbulence eddies distributed over a range of scales extend from the largest scales which interact with the mean flow, from which they extract their energy, to the smallest scales where their energy dissipates as heat. Utilizing a transformation from physical space into a wavenumber space, the energy of eddies can be expressed in terms of a spectral distribution represented by the function $E(k)dk$, which is the energy of eddies from k to $k + dk$ with k as the wavenumber. Since the wavenumber is expressed in terms of the wave length $\lambda = 2\pi/k$, the dimension of wavenumber is, L^{-1} in M, L, t dimension systems. If we assume that the eddy's length scale l is proportional to the wave length λ, then the wavenumber can be thought of as proportional to the inverse of an eddy's length l, i.e $k \propto 1/l$. Figure 9.6 exhibits the energy spectral distribution $E(k)$ as a function of the wavenumber k. This energy spectrum corresponds to the formation and the scales of eddies within a boundary layer shown in Fig. 9.1. Kolmogorov introduced three distinct length scale/wavenumber regions that are marked in Fig. 9.6.

The first region is occupied by large eddies that contain most of the energy. These eddies interact with the mean flow from which they extract their energy (downward arrow) and transfer it to smaller scale eddies. The large eddies affected by the flow boundary conditions are anisotropic. According to Kolmogorov, they loose their directional preference in the energy cascade process by which energy is transferred

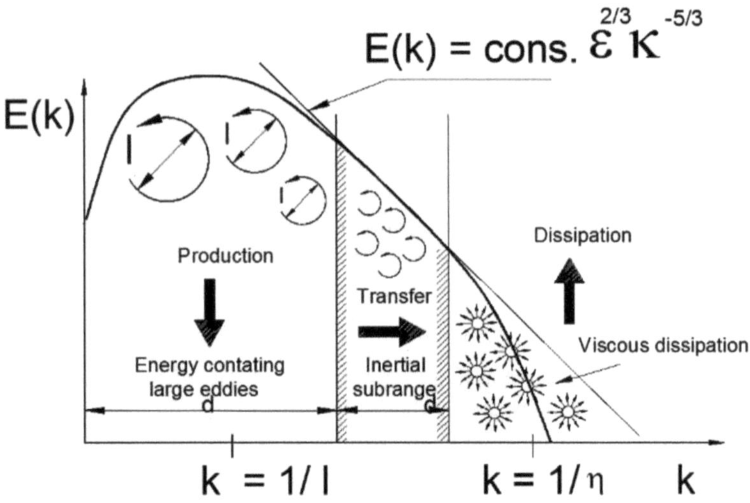

Fig. 9.6 Schematic of Kolmogorov energy spectrum as a function of wavenumber

to successively smaller and smaller eddies. The second region is the *inertial subrange* (Kolmogorov).

In this region, a transport of energy takes place from the large eddies to the eddies that are in dissipation range (horizontal arrow). Since in this subrange, the energy transfer is accomplished by inertial forces, it is called the inertial subrange. As shown in Fig. 9.6, the existence of this region requires that the Reynolds number must be high to establish a fully turbulent flow. The third region is the dissipation range where the eddies are small and isotropic and their kinetic energy dissipates as heat. The scales of the eddies are described by the Kolmogorov scales, Eq. (9.33).

Kolmogorov Hypotheses: Utilizing the above scale decomposition, Kolmogorov established his universal equilibrium theory based on two similarity hypotheses for turbulent flows. The first hypothesis states that for a high Reynolds number turbulent flow, the small-scale turbulent motions are isotropic and independent of the detailed structure of large scale eddies. Furthermore, there is a range of high wavenumbers where the turbulence is statistically in equilibrium and uniquely determined by the energy dissipation $\varepsilon[L^2T^{-3}]$ and the kinematic viscosity $\nu[L^2T^{-1}]$. With this hypothesis and in conjunction with dimensional reasoning, Kolmogorov arrived at length (η), time (τ) and the velocity (υ) scales which have already been presented in Eq. (9.33). Considering the Kolmogorov's length and velocity scales, the corresponding Kolmogorov's equilibrium Reynolds number is

$$\mathrm{Re} = \frac{\upsilon\eta}{\nu} = 1. \tag{9.34}$$

To define the range of the equilibrium, we first introduce a dissipation wavenumber to emphasize the strong effects of viscosity

$$K_d = \frac{1}{\eta} \tag{9.35}$$

and the wavenumber of energy containing eddies with the length scale l that may be interpreted as the average size of the energy containing eddies

$$K_e = \frac{1}{l}. \tag{9.36}$$

The equilibrium range contains wavenumbers for which $k >> k_e$ with $k_e << k_d$. Thus, the equilibrium wavenumber must satisfy the condition

$$K_e << k << k_d. \tag{9.37}$$

The range defined by the above condition is exactly the Kolmogorov inertial subrange within which the turbulence is independent of the energy containing eddies and of the range of strong dissipation. Utilizing the dimensional analysis, we find for the energy spectrum $E(k)[L^3T^{-2}]$ the following relationship

$$E(k) = \nu^{5/4}\varepsilon^{1/4}F(k\eta) \tag{9.38}$$

with the function $F(k\eta)$ to be determined.

The second hypothesis states that when the Reynolds number is large enough for the energy containing eddies, there exists a subrange of wavenumbers in which the condition (9.37) is satisfied, then the energy spectrum is independent of ν and is determined by dissipation parameter ε only. In this hypothesis, within the inertial subrange and by virtue of dimensional analysis, where the function $F(k\eta)$ becomes

$$F(k\eta) = \alpha(k\eta)^{-5/3} \tag{9.39}$$

with $\eta = \left(\frac{\nu^3}{\varepsilon}\right)^{1/4}$ from (9.33), Kolmogorov found the final equation for energy spectrum within the inertial subrange as

$$E(k) = \text{const.}\, \varepsilon^{2/3}k^{-5/3} \tag{9.40}$$

with const. $\equiv C_k$ as the universal Kolmogorov's constant shown in Fig. 9.6. Using tidal waves for measuring the spectrum, Grant et al. [79] found the values for $\alpha = 1.44 \pm 0.01$ and $C_k = 1.89 \pm 0.08$. Equation (9.40) is called the Kolmogorov spectrum, which is based on Kolmogorov's second hypothesis. Onsager [80] and Weizsäcker [81] arrived at the same equation independent of Kolmogorov and each other.

9.1.4 Spectral Tensor, Energy Spectral Function

As the energy spectrum schematically plotted in Fig. 9.6 reveals, in an energy cascade process, eddies with different length, time, and velocity scales interact with each other. Energy is continuously transferred from larger eddies to smaller and smaller ones reaching the dissipation as the final stage of the cascade process. To account for different scales in a more quantitative way, the Fourier analysis, as an appropriate tool, is utilized. To directly apply the Fourier analysis to the issues we discussed in the preceding section, we consider the two-point velocity correlation Eq. (9.1). To start with the simplest case, we assume that (a) the velocity field is spatially homogeneous, meaning that the two-point correlation is independent of the position vector **x** and, (b) there is no temporal separation between the two points measurement, then Eq. (9.1) reduces to

$$R_{ij}(t, \mathbf{r}) = \overline{\mathbf{V}'_\mathbf{i}(\mathbf{t})\mathbf{V}'_\mathbf{j}(\mathbf{r}, \mathbf{t})}. \tag{9.41}$$

We can now construct a second order velocity spectral tensor $\Phi(\mathbf{k}, \mathbf{t}) = \mathbf{e_i e_j}\Phi_{\mathbf{ij}}(\mathbf{k}, \mathbf{t})$ in terms of *wavenumber spectrum* as the Fourier transform of the two point correlation (9.41)

$$\Phi_{ij}(\mathbf{k}, \mathbf{t}) = \frac{1}{(2\pi)^3}\int\limits_{\mathbf{V(r)}} \mathbf{e}^{-\mathbf{ik}\cdot\mathbf{r}}\mathbf{R_{ij}}(\mathbf{t}, \mathbf{r})\mathbf{dr} \tag{9.42}$$

with the correlation

$$R_{ij}(t, \mathbf{r}) = \int\limits_{\mathbf{V(k)}} \mathbf{e}^{\mathbf{ik}\cdot\mathbf{r}}\Phi_{\mathbf{ij}}(\mathbf{k}, \mathbf{t})\mathbf{dk} \tag{9.43}$$

and

$$e^{i\mathbf{k}\cdot\mathbf{r}} = \cos(\mathbf{k}\cdot\mathbf{r}) + \mathbf{i}\sin(\mathbf{k}\cdot\mathbf{r}) \tag{9.44}$$

where $\mathbf{k} = \mathbf{e_i k_i}$ is the wavenumber vector which is related to the wave length by $l = 2\pi/|\mathbf{k}|$. Since we transformed the physical space into the wavenumber space, the integral boundaries in (9.42) and (9.43) constitute the volume in the wavenumber space. Furthermore, since the correlation tensor $R_{ij}(t, \mathbf{r})$ is real, the velocity spectrum tensor $\Phi_{ij}(\mathbf{k}, \mathbf{t})$ is, in general, of a complex nature. It also has the symmetry property

$$\Phi_{ij}(\mathbf{k}, \mathbf{t}) = \Phi_{\mathbf{ji}}(-\mathbf{k}, \mathbf{t}) \tag{9.45}$$

and satisfies the orthogonality condition

$$k_i\Phi_{ij}(\mathbf{k}, \mathbf{t}) = \mathbf{k_j}\Phi_{\mathbf{ij}}(\mathbf{k}, \mathbf{t}) = 0. \tag{9.46}$$

One-Dimensional Spectral Function: Of practical interest is the one-dimensional version of Eq. (9.42), where the physical component in r_1 and the wavenumber component in k_1 are considered; the one-dimensional case of Eq. (9.42) reads

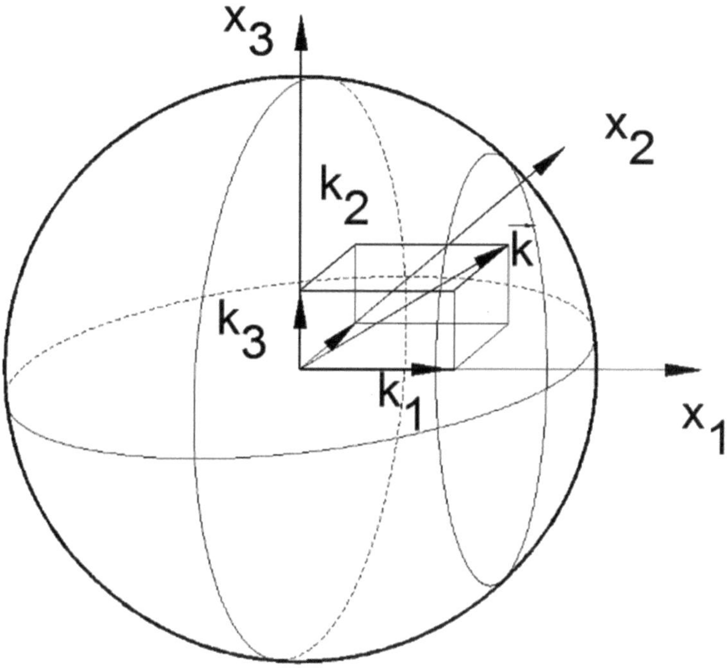

Fig. 9.7 Wavenumber vector and its components, from Rotta [4]

$$\Theta_{ij}(k_1, t) = \frac{1}{2\pi} \int_{-\infty}^{+\infty} R_{ij}(t, r_1) e^{-ik_1 r_1} dr_1. \tag{9.47}$$

The same result is obtained by integrating Eq. (9.42) over the other two wavenumber components

$$\Theta_{ij}(k_1, t) = \int \int_{-\infty}^{\infty} \Phi_{ij}(\mathbf{k}, \mathbf{t}) \mathbf{dk}_2 \mathbf{dk}_3. \tag{9.48}$$

Of particular practical interest is a spectral function which depends only on the magnitude of the wavenumber $k = |\mathbf{k}|$. It can be calculated as a surface integral over the surface of a sphere with the radius k in the wavenumber space as shown in Fig. 9.7.
The corresponding spectral function read

$$\psi_{ij}(k, t) = k^2 \oint \Phi_{ij}(\mathbf{k}, \mathbf{t}) \mathbf{d}\Omega \tag{9.49}$$

with $d\Omega$ as an infinitesimal solid angle. Equation (9.49) allows one to compute the energy spectral function as the half trace of (9.49), [76].

$$E(k,t) = \frac{1}{2}\psi_{ii}(k,t) = \frac{1}{2}k^2 \oint \Phi_{ii}(\mathbf{k},\mathbf{t})\mathbf{d}\Omega \tag{9.50}$$

with $E(k,t)$ as the specific kinetic energy of all wavenumbers with the magnitude $k \leq |\mathbf{dk}| \leq \mathbf{k} + \mathbf{dk}$. For isotropic turbulence, $\overline{V_1'^2} = \overline{V_2'^2} = \overline{V_3'^2} = \overline{V_i'V_i'}/3$, Eq. (9.50) is integrated and reduced to

$$E(k) = 4\pi k^2 \frac{1}{2}\Phi_{ii}(k) \tag{9.51}$$

with $\Phi_{ii}(k)$ as a scalar quantity. Without presenting the mathematical proof, the spectral tensor $\Phi(k) = e_i e_j \Phi_{ij}(k)$ can be reconstructed using Eq. (9.51)

$$\Phi(k) = \frac{E(k)}{4\pi k^4}(k^2 I - \mathbf{kk}) \tag{9.52}$$

with $I = e_i e_j \delta_{ij}$ as the unit tensor and $\mathbf{kk} = \mathbf{e_i e_j k_i k_j}$ as the second order wavenumber tensor. Thus, Eq. (9.52) can be rewritten as

$$\Phi_{ij}(k) = \frac{E(k)}{4\pi k^{42}}(k^2\delta_{ij} - k_i k_j). \tag{9.53}$$

The integration of the energy spectral function $E(k,t)$ over the entire wavenumber space leads to the total turbulent kinetic energy

$$\frac{1}{2}\overline{V_i'V_i'} = \int_0^\infty E(k,t)dk \tag{9.54}$$

9.2 Averaging Fundamental Equations of Turbulent Flow

In this section, we present the fundamental equations that describe turbulent flows. For the sake of completeness, we also represent some of those equations that were already presented in the preceding chapter.

Turbulent flow is characterized by random fluctuations in velocity, pressure, temperature, density, and etc. Any turbulent quantity can be decomposed in a mean and fluctuating part. Experimental observations revealed that average values with respect to time and space exist because distinct flow patterns are repeated regularly in time and space. In Chap. 8, an averaging procedure applied to a single quantity leading to Eqs. (8.57) and (8.58) was given. Before preceding further, we apply the averaging formalism to two arbitrary quantities Q1 and Q2 with $Q_1 = \bar{Q}_1 + Q_1'$ and $Q_2 = \bar{Q}_2 + Q_2'$. If we deal with a statistically steady flow, re-applying the averaging procedure results in:

$$
\begin{aligned}
&\overline{Q}_1 = \overline{\overline{Q_1} + Q'_1} = \overline{\overline{Q_1}} + \overline{Q'_1}, \text{ and } \overline{Q'_1}, = 0\\
&\overline{\overline{Q_1}\,\overline{Q_2}} = \overline{\overline{Q_1}}\,\overline{\overline{Q_2}} = \overline{Q_1}\,\overline{Q_2}\\
&\overline{\overline{Q_1}\,Q'_2} = \overline{\overline{Q_1}}\,\overline{Q'_2} = \overline{Q_1}\,\overline{Q'_2} = 0\\
&\overline{\overline{Q_2}\,Q'_1} = \overline{\overline{Q_2}}\,\overline{Q'_1} = \overline{Q_2}\,\overline{Q'_1} = 0\\
&\overline{Q_1 Q_2} = \overline{Q_1}\,\overline{Q_2} + \overline{Q'_1 Q'_2}.
\end{aligned}
\tag{9.55}
$$

Applying operators to a tensor valued function results in:

$$
\begin{aligned}
&\overline{\nabla(\bar{Q} + Q')} = \nabla\bar{Q} + \nabla\overline{Q'} = \nabla\bar{Q}\\
&\overline{\frac{D}{Dt}(\bar{Q} + Q')} = \frac{D\overline{Q}}{Dt} + \overline{V'\cdot\nabla Q'}.
\end{aligned}
$$

9.2.1 Averaging Conservation Equations

In this section, we apply the averaging procedure (9.55) to the conservation equations of continuity, motion, mechanical energy, and thermal energy presented in Chap. 4.

9.2.1.1 Averaging the Continuity Equation

Averaging the continuity equation reads:

$$
\frac{\partial\bar{\rho}}{\partial t} + \nabla\cdot(\bar{\rho}\,\bar{\mathbf{V}} + \overline{\rho'\mathbf{V}'}) = 0 \tag{9.56}
$$

with the index notation

$$
\frac{\partial\bar{\rho}}{\partial t} + \frac{\partial}{\partial x_i}(\bar{\rho}\,\bar{V}_i + \overline{\rho' V'_i}) = 0. \tag{9.57}
$$

9.2.1.2 Averaging the Navier–Stokes Equation

First, we decompose the velocity vector in the Navier–Stokes equation and find:

$$
\begin{aligned}
&\frac{\partial\bar{\mathbf{V}}}{\partial t} + \frac{\partial\mathbf{V}'}{\partial t} + \bar{\mathbf{V}}\cdot\nabla\bar{\mathbf{V}} + \bar{\mathbf{V}}\cdot\nabla\overline{V'} + \mathbf{V}'\cdot\nabla\bar{\mathbf{V}} + \mathbf{V}'\cdot\nabla\mathbf{V}'\\
&\quad = -\frac{1}{\rho}\nabla\bar{p} - \frac{1}{\rho}\nabla p' + \nu\Delta\bar{\mathbf{V}} + \nu\Delta\mathbf{V}' + \mathbf{g}
\end{aligned}
\tag{9.58}
$$

then we apply the averaging rules (9.55) to Eq. (9.58) and arrive at the Reynolds averaged Navier–Stokes Eq. (8.63):

$$\frac{\partial \bar{\mathbf{V}}}{\partial t} + \bar{\mathbf{V}} \cdot \nabla \bar{\mathbf{V}} + \nabla \cdot (\overline{\mathbf{V}'\mathbf{V}'}) = -\frac{1}{\rho} \nabla \bar{p} + \nu \Delta \bar{\mathbf{V}} + \mathbf{g}. \tag{9.59}$$

The index notation gives:

$$\frac{\partial \bar{V}_i}{\partial t} + \bar{V}_j \frac{\partial \bar{V}_i}{\partial x_j} = -\frac{1}{\rho} \frac{\partial \bar{p}}{\partial x_i} + \nu \frac{\partial^2 \bar{V}_i}{\partial x_j \partial x_j} - \frac{\partial (\overline{V_i' V_j'})}{\partial x_j} + g_i. \tag{9.60}$$

9.2.1.3 Averaging the Mechanical Energy Equation

The mechanical energy equation for a turbulent flow is obtained using Eq. (4.71), which includes the compressibility term ($\nabla \cdot \mathbf{V} \not\equiv 0$). Dividing the involved flow quantities into the mean and the fluctuating parts and applying the averaging procedure outlined in Sect. 9.2, results in a complex equation. To reduce the degree of complexity, we consider an incompressible flow with the mechanical energy equation given by Eq. (4.72) and also below:

$$\frac{D}{Dt}\left(\frac{V^2}{2}\right) = \nabla \cdot \left(-\frac{p}{\rho}\mathbf{V} + 2\nu \mathbf{V} \cdot \mathbf{D}\right) - 2\nu \mathbf{D} : \mathbf{D} + \mathbf{V} \cdot \mathbf{g}. \tag{9.61}$$

Using the identity $\mathbf{V} \cdot \nabla (V^2) = \nabla \cdot (\mathbf{V} V^2)$ for an incompressible flow and its index notation $V_i \frac{\partial}{\partial x_i}(V_j V_j) = \frac{\partial}{\partial x_i}(V_i V_j V_j)$, we find the index notation of Eq. (9.61):

$$\begin{aligned} \frac{\partial}{\partial t}\left(\frac{V_j V_j}{2}\right) &= -\frac{\partial}{\partial x_i} V_i \left(\frac{p}{\rho} + \frac{V_j V_j}{2}\right) + \nu \frac{\partial}{\partial x_i} V_j \left(\frac{\partial V_i}{\partial x_j} + \frac{\partial V_j}{\partial x_i}\right) \\ &\quad - \nu \left(\frac{\partial V_i}{\partial x_j} + \frac{\partial V_j}{\partial x_i}\right) \frac{\partial V_j}{\partial x_i} + V_i g_i. \end{aligned} \tag{9.62}$$

We introduce the following decompositions:

$$\begin{aligned} &\mathbf{V} = \overline{\mathbf{V} + \mathbf{V}'} \text{ with } \mathbf{V_i} = \bar{\mathbf{V}}_\mathbf{i} + \mathbf{V}'_\mathbf{i}, \\ &V^2 = V_i V_i = \bar{V}_i \bar{V}_i + 2\bar{V}_i V_i' + V_i' V_i' \\ &p = \bar{p} + p' \end{aligned} \tag{9.63}$$

and substitute the quantities in Eq. (9.61) by Eq. (9.63) and average the results, we find:

$$\begin{aligned}
&\frac{\partial}{\partial t}\left(\frac{\bar{V}_i\bar{V}_i}{2}\right)+\frac{\partial}{\partial t}\left(\frac{\overline{V_i'V_i'}}{2}\right)=-\frac{\partial}{\partial x_i}\bar{V}_i\left(\frac{\overline{p}}{\rho}+\frac{\overline{V_jV_j}}{2}\right)+\nu\frac{\partial}{\partial x_i}\bar{V}_j\left(\frac{\partial\bar{V}_i}{\partial x_j}\frac{\partial\bar{V}_j}{\partial x_i}\right)\\
&\quad-\nu\left(\frac{\partial\bar{V}_i}{\partial x_j}+\frac{\partial\bar{V}_j}{\partial x_i}\right)\frac{\partial\bar{V}_j}{\partial x_i}-\frac{\partial}{\partial x_i}\overline{V_i'\left(\frac{p'}{\rho}+\frac{V_jV_j}{2}\right)}\\
&\quad-\frac{\partial}{\partial x_i}\left(\bar{V}_j\overline{V_i'V_j'}\right)-\frac{\partial}{\partial x_i}\left(\bar{V}_i\frac{\overline{V_j'V_j'}}{2}\right)\\
&\quad+\nu\frac{\partial}{\partial x_i}\overline{V_j'\left(\frac{\partial V_i'}{\partial x_j}+\frac{\partial V_j'}{\partial x_i}\right)}-\nu\overline{\left(\frac{\partial V_i'}{\partial x_j}+\frac{\partial V_j'}{\partial x_i}\right)\frac{\partial V_j'}{\partial x_i}}+\bar{V}_i g_i.
\end{aligned}\tag{9.64}$$

Equation (9.64) is the mechanical energy equation for turbulent flow, where the energy of the Reynolds stress tensor appears on the right-hand side. In deriving Eq. (9.64), we considered the gravitation force as the only field force with the component in the x_3-direction. If other field forces such as electromagnetic or electrostatic forces are present, they are added to Eq. (9.64) in the same way as the above gravitational work.

9.2.1.4 Averaging the Thermal Energy Equation

A thermal energy equation can be expressed in terms of specific internal energy u or specific static enthalpy h. In both cases, the specific internal energy and specific static enthalpy can be expressed in terms of temperature $u = c_v T$ and $h = c_p T$. Both forms are fully equivalent and one can be converted into the other by $h = u + pv$, which is the defining equation for the specific static enthalpy. For averaging the thermal energy equation in terms of specific static enthalpy which we replace by the static temperature, we resort to Eq. (4.94)

$$c_p\frac{DT}{Dt}=\frac{k}{\varrho}\nabla^2T+\frac{1}{\varrho}\frac{Dp}{Dt}+\frac{1}{\rho}\mathbf{T}:\mathbf{D}\tag{9.65}$$

with the friction stress tensor $\mathbf{T}$ from Eq. (4.36):

$$\mathbf{T}=\lambda(\nabla\cdot\mathbf{V})\mathbf{I}+2\mu\mathbf{D}.\tag{9.66}$$

Decomposing, in Eq. (9.65), the temperature and pressure T and p as well as the friction and deformation tensors $\mathbf{T}$ and $\mathbf{D}$ while neglecting the density fluctuation, we find:

$$\begin{aligned}
c_p\frac{D(\bar{T}+T')}{Dt}&=\frac{k}{\varrho}\nabla^2(\bar{T}+T')+\frac{1}{\varrho}\frac{D(p+p')}{Dt}\\
&\quad+\frac{1}{\varrho}(\bar{\mathbf{T}}+\mathbf{T}'):(\bar{\mathbf{D}}+\mathbf{D}').
\end{aligned}\tag{9.67}$$

Averaging the entire Eq. (9.67) and considering the rule for averaging the spatial and substantial derivatives of tensor valued functions listed in Eq. (9.55), we arrive at:

$$c_p\left(\frac{D\bar{T}}{Dt}+\overline{V'\cdot\nabla T'}\right)=\frac{k}{\varrho}\nabla^2\bar{T}+\frac{1}{\varrho}\frac{D\bar{p}}{Dt}+\frac{1}{\varrho}(\bar{\mathbf{T}}:\bar{\mathbf{D}}+\overline{\mathbf{T}':\mathbf{D}'}). \tag{9.68}$$

Further expansion of Eq. (9.68) and considering Eq. (9.66) leads to:

$$c_p\left(\frac{\partial\bar{T}}{\partial t}+\bar{V}\cdot\nabla\bar{T}+\overline{V'\cdot\nabla T'}\right)=\frac{k}{\varrho}\nabla^2\bar{T}+\frac{1}{\varrho}\frac{D\bar{p}}{Dt}$$
$$+\frac{\lambda}{\varrho}(\nabla\cdot\bar{V})^2+2\nu\bar{\boldsymbol{D}}:\bar{\boldsymbol{D}}+\frac{\lambda}{\varrho}\overline{(\nabla\cdot\boldsymbol{V}')^2}+2\nu\overline{\boldsymbol{D}':\boldsymbol{D}'}. \tag{9.69}$$

Following Eqs. (4.72)–(4.74), the viscous dissipation of the mean flow is:

$$\bar{\Phi}=\varrho\left(\frac{\lambda}{\varrho}(\nabla\cdot\bar{\mathbf{V}})^2+2\nu\bar{\mathbf{D}}:\bar{\mathbf{D}}\right)\equiv\varrho\bar{\varepsilon}\text{ with }\bar{\varepsilon}=\frac{\lambda}{\varrho}(\nabla\cdot\bar{\mathbf{V}})^2+2\nu\bar{\mathbf{D}}:\bar{\mathbf{D}}. \tag{9.70}$$

Correspondingly, the turbulent dissipation reads

$$\Phi_{\text{tur}}=\lambda\overline{(\nabla\cdot\mathbf{V}')}^2+2\mu\overline{\mathbf{D}'}:\overline{\mathbf{D}'}\equiv\varrho\varepsilon$$
$$\text{with }\varepsilon=\frac{\lambda}{\varrho}\overline{(\nabla\cdot\mathbf{V}')}^2+2\nu\overline{\mathbf{D}'}:\mathbf{D}'. \tag{9.71}$$

Thus, the total dissipation reads

$$\Phi_{\text{total}}=\bar{\Phi}+\Phi_{\text{tur}}=\varrho(\bar{\varepsilon}+\varepsilon). \tag{9.72}$$

In Eqs. (9.70)–(9.72) $\bar{\varepsilon}$ and ε are the specific dissipations (dissipation per unit of mass) of the mean flow and the turbulent flow, respectively. The latter is also called the turbulent dissipation. Following Eq. (4.74), the index notation of Eq. (9.71) reads:

$$\Phi_{\text{tur}}=\lambda\overline{\left(\frac{\partial V_i'}{\partial x_i}\right)^2}+\frac{2}{4}\mu\overline{\left(\frac{\partial V_i'}{\partial x_j}+\frac{\partial V_j'}{\partial x_i}\right)\left(\frac{\partial V_i'}{\partial x_j}+\frac{\partial V_j'}{\partial x_i}\right)} \tag{9.73}$$

and its expansion is:

$$\Phi_{\text{tur}}=\lambda\overline{\left(\frac{\partial V_1'}{\partial x_1}+\frac{\partial V_2'}{\partial x_2}+\frac{\partial V_3'}{\partial x_3}\right)^2}+2\mu\left[\overline{\left(\frac{\partial V_1'}{\partial x_1}\right)^2}+\overline{\left(\frac{\partial V_2'}{\partial x_2}\right)^2}+\overline{\left(\frac{\partial V_3'}{\partial x_3}\right)^2}\right]$$
$$+\mu\left[\overline{\left(\frac{\partial V_1'}{\partial x_2}+\frac{\partial V_2'}{\partial x_1}\right)^2}+\overline{\left(\frac{\partial V_1'}{\partial x_3}+\frac{\partial V_3'}{\partial x_1}\right)^2}+\overline{\left(\frac{\partial V_2'}{\partial x_3}+\frac{\partial V_3'}{\partial x_2}\right)^2}\right]. \tag{9.74}$$

Generally, the total dissipation expresses the conversion of mechanical energy into heat and causes the system to heat up. Comparing the specific dissipation of the mean flow $\bar{\varepsilon}$ with the turbulent flow ε shows that the order of magnitude of ε by far surmounts the one of $\bar{\varepsilon}$. The reason is that, in spite of the fact that $|\mathbf{V}'| << \bar{\mathbf{V}}$, the changes of the fluctuating velocity $|\partial V_i'/\partial x_j|$ is much larger than changes of the mean flow velocity $|\partial V_i'/\partial x_j| >> \partial \bar{V}_i/\partial x_j$. This circumstance is an inherent characteristic of all turbulent flows and allows the mean flow dissipation to drop in all turbulence equations. Thus, with Eqs. (9.69) and (9.70), Eq. (9.68) becomes:

$$c_p\left(\frac{\partial \bar{T}}{\partial t} + \bar{V}\cdot\nabla\bar{T} + \overline{V'\cdot\nabla T'}\right) = \frac{k}{\varrho}\nabla^2 + \bar{T} + \frac{1}{\varrho}\frac{D\bar{p}}{Dt} + \bar{\varepsilon} + \varepsilon. \tag{9.75}$$

9.2.1.5 Averaging the Total Enthalpy Equation

The quantities in total enthalpy Eq. (4.95), also given below,

$$\rho\left(\frac{\partial H}{\partial t} + \mathbf{V}\cdot\nabla H\right) = \kappa\nabla^2 T + \frac{\partial p}{\partial t} + \nabla\cdot(\mathbf{T}\cdot\mathbf{V}) + \rho\mathbf{V}\cdot\mathbf{g} \tag{9.76}$$

with $\mathbf{T}$ as the temperature and $\mathbf{T} = \lambda(\nabla\cdot\mathbf{V})\mathbf{I} + 2\mu\mathbf{D}$ as the friction stress tensor. Before further treating the total enthalpy equation, we re-arrange the term $\nabla\cdot(\mathbf{T}\cdot\mathbf{V})$ as:

$$\nabla\cdot(\mathbf{T}\cdot\mathbf{V}) = (\nabla\cdot\mathbf{T})\cdot\mathbf{V} + \mathbf{T}:\nabla\mathbf{V} = (\nabla\cdot\mathbf{T})\cdot\mathbf{V} + \mathbf{T}:(\Omega + \mathbf{D}). \tag{9.77}$$

Since the friction tensor $\mathbf{T}$ is symmetric and the rotation tensor Ω is antisymmetric, their double dot product identically disappears leading to

$$\nabla\cdot(\mathbf{T}\cdot\mathbf{V}) = (\nabla\cdot\mathbf{T})\cdot\mathbf{V} + \mathbf{T}:\nabla\mathbf{V} = (\nabla\cdot\mathbf{T})\cdot\mathbf{V} + \mathbf{T}:\mathbf{D}. \tag{9.78}$$

The friction tensor $\mathbf{T}$ in Eqs. (9.76)–(9.78) includes $\nabla\cdot\mathbf{V}$, which is nonzero for compressible flows. In the context of turbulent flow treatment, its contribution is insignificant and brings only additional complexity to a topic which is complex anyway. Setting $\nabla\cdot\mathbf{V} = 0$ reduces Eq. (9.78) to

$$\nabla\cdot(\mathbf{T}\cdot\mathbf{V}) = 2\mu((\nabla\cdot\mathbf{D})\cdot\mathbf{V} + \mathbf{D}:\mathbf{D}). \tag{9.79}$$

Implementing Eq. (9.79) into Eq. (9.76), we find:

$$\rho\frac{DH}{Dt} = \kappa\nabla^2 T + \frac{\partial p}{\partial t} + 2\mu((\nabla\cdot\mathbf{D})\cdot\mathbf{V} + \mathbf{D}:\mathbf{D}) + \rho\mathbf{V}\cdot\mathbf{g}. \tag{9.80}$$

Thus, Eq. (9.80) is identical with

$$\rho\left(\frac{\partial H}{\partial t}+\mathbf{V}\cdot\nabla H\right)=\kappa\nabla^2 T+\frac{\partial p}{\partial t}+2\mu\nabla\cdot(\mathbf{D}\cdot\mathbf{V})+\rho\mathbf{V}\cdot\mathbf{g}. \tag{9.81}$$

Decomposing the quantities in Eq. (9.81) gives:

$$\begin{aligned}
&\varrho\frac{\partial}{\partial t}\left(\bar{h}+h'+\frac{1}{2}(\bar{V}+V')^2\right)+\varrho(\bar{\mathbf{V}}+\mathbf{V}')\cdot\nabla\left(\bar{h}+h'+\frac{1}{2}(\bar{V}+V')^2\right)\\
&=\kappa\nabla^2(\bar{T}+T')+\frac{\partial(\bar{p}+p')}{\partial t}+2\mu\nabla\cdot((\bar{\mathbf{D}}+\mathbf{D}')\cdot(\bar{\mathbf{V}}+\mathbf{V}'))\\
&+\rho(\bar{\mathbf{V}}+V')\cdot\mathbf{g}.
\end{aligned} \tag{9.82}$$

Comparing the order of magnitude of the fluctuation kinetic energy with the one of the mean flow shows that $V'^2/2 \ll V^2/2$. This is true even for flow situations with relatively high turbulence intensities of 10% and above. This order of magnitude comparison can directly be related to the square of turbulence intensity defined as Tu$=\sqrt{\overline{V'^2}}/V$. For a large turbulence intensity of Tu $=10\%$, we obtain a ratio of $V'^2/V^2=0.01$. This comparison allows neglecting the fluctuation kinetic energy. After averaging Eq. (9.82), we find:

$$\begin{aligned}
\varrho\left(\frac{\partial\bar{H}}{\partial t}+\bar{\mathbf{V}}\cdot\nabla\bar{H}\right)&=\kappa\nabla^2\bar{T}-\varrho\overline{\mathbf{V}'\cdot\nabla h'}-\varrho\overline{\mathbf{V}'\cdot\nabla(\bar{V}\cdot V')}\\
&+\frac{\partial\bar{p}}{\partial t}+2\mu(\nabla\cdot(\bar{\mathbf{D}}\cdot\bar{V})+\nabla\cdot(\overline{\mathbf{D}'\cdot V'}))+\varrho\bar{\mathbf{V}}\cdot\mathbf{g}.
\end{aligned} \tag{9.83}$$

For steady case and neglecting the gravitational work, we obtain:

$$\begin{aligned}
\varrho(\bar{\mathbf{V}}\cdot\nabla\bar{H})&=\kappa\nabla^2\bar{T}-\varrho\overline{\mathbf{V}'\cdot\nabla h'}-\varrho\overline{\mathbf{V}'\cdot\nabla(\bar{V}\cdot V')}\\
&+2\mu(\nabla\cdot(\bar{\mathbf{D}}\cdot\bar{V})+\nabla\cdot(\overline{\mathbf{D}'\cdot V'})).
\end{aligned} \tag{9.84}$$

In Eq. (9.84) we replace the averaged static temperature with static enthalpy, add and subtract the kinetic energy to introduce the total enthalpy, and introduce the Prandtl number $\mathrm{Pr}=\mu c_p/\kappa$. Furthermore, for the sake of practicability, we modify the third term in Eq. (9.84) by adding a zero $\nabla\cdot\mathbf{V}'=0$:

$$\overline{\mathbf{V}'\cdot\nabla(\bar{V}\cdot V')}+\overline{(\nabla\cdot V')\bar{\mathbf{V}}\cdot\mathbf{V}'}=\nabla\cdot[\overline{(\bar{\mathbf{V}}\cdot\mathbf{V}')\mathbf{V}'}] \tag{9.85}$$

as a result, we find:

$$\begin{aligned}
\varrho(\bar{\mathbf{V}}\cdot\nabla\bar{H})&=\frac{\mu}{Pr}\nabla^2\bar{H}-\frac{\mu}{Pr}\nabla^2\left(\frac{1}{2}\bar{V}^2\right)-\varrho\overline{\mathbf{V}'\cdot\nabla h'}-\varrho\nabla\cdot\overline{[(\bar{\mathbf{V}}\cdot\mathbf{V}')\mathbf{V}']}\\
&+2\mu[\nabla\cdot(\bar{\mathbf{D}}\cdot\bar{V})+\nabla\cdot(\overline{\mathbf{D}'\cdot V'})].
\end{aligned} \tag{9.86}$$

Equation (9.86) written in Cartesian index notation is:

$$\varrho\left(\bar{V}_i\frac{\partial\bar{H}}{\partial x_i}\right)=\frac{\mu}{Pr}\frac{\partial^2\bar{H}}{\partial x_i\partial x_i}-\frac{\mu}{P}\left[\frac{\partial}{\partial x_i}\left(\bar{V}_m\frac{\partial\bar{V}_m}{\partial x_i}\right)\right]-\varrho\overline{V_i'\frac{\partial h'}{\partial x_i}}-\varrho\frac{\partial(\bar{V}_m\overline{V_m'V_i'})}{\partial x_i}$$
$$+\mu\left(\frac{\partial\bar{V}_i}{\partial x_j}\frac{\partial\bar{V}_j}{\partial x_i}\right)+\mu\left[\frac{\partial}{\partial x_i}\left(\bar{V}_m\frac{\partial\bar{V}_m}{\partial x_i}\right)\right]$$
$$+\mu\left(\overline{\frac{\partial V_i'}{\partial x_j}\frac{\partial V_j'}{\partial x_i}}+\overline{\frac{\partial^2V_j'}{\partial x_i\partial x_i}V_j'}+\overline{\frac{\partial V_j'}{\partial x_i}\frac{\partial V_j'}{\partial x_i}}\right). \tag{9.87}$$

Combining the second and the sixth term on the right hand side of Eq. (9.87) results in a more compact version:

$$\varrho\left(\bar{V}_i\frac{\partial\bar{H}}{\partial x_i}\right)=\frac{\mu}{Pr}\frac{\partial^2\bar{H}}{\partial x_i\partial x_i}+\frac{\partial}{\partial x_i}\left[\mu\left(1-\frac{1}{Pr}\right)\left(\bar{V}_m\frac{\partial\bar{V}_m}{\partial x_i}\right)\right]$$
$$-\varrho\overline{V_i'\frac{\partial h'}{\partial x_i}}-\varrho\frac{\partial(\bar{V}_m\overline{V_m'V_i'})}{\partial x_i}+\mu\left(\frac{\partial\bar{V}_i}{\partial x_j}\frac{\partial\bar{V}_j}{\partial x_i}\right)$$
$$+\mu\left(\overline{\frac{\partial V_i'}{\partial x_j}\frac{\partial V_j'}{\partial x_i}}+\overline{\frac{\partial^2V_j'}{\partial x_i\partial x_i}V_j'}+\overline{\frac{\partial V_j'}{\partial x_i}\frac{\partial V_j'}{\partial x_i}}\right). \tag{9.88}$$

To apply Eq. (9.88) to boundary layer problems discussed in Chap. 11, it is more appropriate to deal with the correlation $\overline{\partial(V_i'h')/\partial x_i}$ rather than $\overline{(V_i'\partial h')\partial x_i}$. This requires a modification of (9.88) by introducing the following identity for incompressible flows

$$\varrho\overline{\mathbf{V}'\cdot\nabla h'}=\varrho\nabla\cdot(\overline{\mathbf{V}'h'})-\varrho\nabla\cdot\overline{\mathbf{V}'h'}$$
$$\varrho\overline{V_i'\frac{\partial h'}{\partial x_i}}=\varrho\overline{\frac{\partial(V_i'h')}{\partial x_i}}+0. \tag{9.89}$$

With (9.89), Eq. (9.88) becomes

$$\varrho\left(\bar{V}_i\frac{\partial\bar{H}}{\partial x_i}\right)=\frac{\mu}{Pr}\frac{\partial^2\bar{H}}{\partial x_i\partial x_i}+\frac{\partial}{\partial x_i}\left[\mu\left(1-\frac{1}{Pr}\right)\left(\bar{V}_m\frac{\partial\bar{V}_m}{\partial x_i}\right)\right]$$
$$-\varrho\overline{\frac{\partial(V_i'h')}{\partial x_i}}-\varrho\frac{\partial(\bar{V}_m\overline{V_m'V_i'})}{\partial x_i}+\mu\left(\frac{\partial\bar{V}_i}{\partial x_j}\frac{\partial\bar{V}_j}{\partial x_i}\right)$$
$$+\mu\left(\overline{\frac{\partial V_i'}{\partial x_j}\frac{\partial V_j'}{\partial x_i}}+\overline{\frac{\partial^2V_j'}{\partial x_i\partial x_i}V_j'}+\overline{\frac{\partial V_j'}{\partial x_i}\frac{\partial V_j'}{\partial x_i}}\right). \tag{9.90}$$

Equation (9.88) (or (9.90)) is the complete equation of the total enthalpy for steady incompressible three-dimensional flows. Summing over the range of indices, Eq. (9.88) can easily be expanded. The expanded version contains several terms that are insignificant for a two-dimensional flow and may be deleted altogether as shown in Chap. 11, when dealing with the boundary layer theory.

9.2.1.6 Quantities Resulting from Averaging to be Modeled

In addition to the viscous and turbulent dissipation terms, Eq. (9.86) includes a new correlation $-\overline{V_i' \frac{\partial h'}{\partial x_i}}$ and a transport term $\frac{\partial(\bar{V}_m \overline{V_m' V_i'})}{\partial x_i}$ as a result of averaging the enthalpy equation. As a result of the averaging procedure, the Reynolds stress tensor $\overline{\mathbf{V}'\mathbf{V}'}$ was created in Eq. (9.59) with nine components from which six are distinct:

$$-\rho\overline{V_i'V_j'} = \begin{pmatrix} \overline{V_1'V_1'} & \overline{V_1'V_2'} & \overline{V_1'V_3'} \\ \overline{V_2'V_1'} & \overline{V_2'V_2'} & \overline{V_2'V_3'} \\ \overline{V_3'V_1'} & \overline{V_3'V_2'} & \overline{V_3'V_3'} \end{pmatrix} = \mathbf{T}' = \begin{pmatrix} \tau_{11}' & \tau_{12}' & \tau_{13}' \\ \tau_{21}' & \tau_{22}' & \tau_{23}' \\ \tau_{31}' & \tau_{32}' & \tau_{33}' \end{pmatrix} \tag{9.91}$$

with $\tau_{12}' = \tau_{21}'$, $\tau_{13}' = \tau_{31}'$ and $\tau_{23} = \tau_{32}$. Considering the molecular friction tensor Eq. (4.73) for an incompressible fluid, the total friction tensor of a turbulent flow $\mathbf{T}_{\text{total}}$ consists of the molecular friction stress tensor $\bar{\mathbf{T}}$ and the turbulent stress tensor $\mathbf{T}'$:

$$\mathbf{T}_{\text{total}} = \bar{\mathbf{T}} + \mathbf{T}' = 2\mu\bar{\mathbf{D}} - \rho\overline{\mathbf{V}'\mathbf{V}'}. \tag{9.92}$$

Experimental results show that close to a solid wall, the order of magnitude of the Reynolds stress is comparable with the molecular stress. In free turbulent flow cases such as wake flow, jet flow and jet boundary, where the flow is not affected by a solid wall, the order of magnitude of $\mathbf{T}'$ can be much higher than $\bar{\mathbf{T}}$ such that the latter can be neglected.

The elements of the tensor $\overline{\mathbf{V}'\mathbf{V}'}$ in Eqs. (9.59) or (9.60) have added six more unknowns to the Navier–Stokes Eq. (8.62). With three velocity components, the pressure and six Reynolds stress terms, we have totally ten unknowns with only four differential equations resulting from Eq. (9.59) together with the continuity equation. Additional unknowns such as $\overline{\mathbf{V}' \cdot \nabla H'}$ and static temperature $\overline{\mathbf{T}' \cdot \mathbf{V}'}$, are added to the system of differential equations for solving the energy equation. In order to find a solution, one has to provide additional equations that relate the Reynolds stress tensor (9.60) with the quantities of the main flow. Likewise, empirical correlations need to be found that relate $\overline{\mathbf{V}' \cdot \nabla H'}$ and $\overline{\mathbf{T}' \cdot \mathbf{V}'}$ to the quantities of the main flow. Such correlations can be constructed by mathematically manipulating the equations of motion and by establishing empirical models.

These additional equations are called closure equations. To obtain these equations, in the following we perform certain time consuming, yet mathematically simple operations to drive new equations from the already existing ones. As we will see, these new equations contain additional unknowns that need to be determined. It should be pointed out that these new equations do not have any new physical background and are just simple mathematical manipulations. The purpose of these manipulations is to find some empirical correlations to close our new system of equations. To easily follow the sequence of operations that generates the new equation, we introduce a new operator $\mathbf{N(V)}$, which we call the Navier–Stokes operator, where the velocity is assumed to be a function of time and space. This assumption is valid for statistically stationary/non-stationary, with a constant time dependent mean and stochastic fluctuations. Resorting to Eq. (4.43), we define

$$N(\mathbf{V}) = \frac{\partial \mathbf{V}}{\partial t} + \mathbf{V} \cdot \nabla \mathbf{V} + \frac{1}{\rho} \nabla p - \nu \Delta \mathbf{V} - \mathbf{g} = 0. \tag{9.93}$$

With $\mathbf{N}$ as the operator and $\mathbf{V}$ the tensor valued argument, upon which the operator acts and builds the Navier–Stokes equation. The argument may be a vector such as $\mathbf{V} = \bar{\mathbf{V}} + \mathbf{V}'$ or a component of a vector such as V_i. If the argument is the component V_i, then $\mathbf{N}(V_i)$ describes the ith component of the Navier–Stokes equation. In case the vector is decomposed into a mean and a fluctuation part, then the argument of the operator is replaced by $\mathbf{V} = \bar{\mathbf{V}} + \mathbf{V}'$ leading to $\mathbf{N}(\bar{\mathbf{V}} + \mathbf{V}')$. If the entire Navier–Stokes equation is averaged, we replace the operator argument by $\overline{(\bar{\mathbf{V}} + \mathbf{V}')}$. Before discussing different turbulence models, we present equations of turbulent kinetic energy and its dissipation as the two major closure equations. Similarly, we may write Eq. (9.93) in index form

$$\mathbf{N}(V_i) = \frac{\partial V_i}{\partial t} + V_j \frac{\partial V_i}{\partial x_j} + \frac{1}{\rho} \frac{\partial p}{\partial x_i} - \nu \frac{\partial^2 V_i}{\partial x_j \partial x_j} - g_i = 0. \tag{9.94}$$

Equation (9.94) describes the ith component of the Navier–Stokes equation. We may also obtain $\mathbf{N}(\bar{\mathbf{V}} + \mathbf{V}')$ and $\overline{\mathbf{N}(\bar{\mathbf{V}} + \mathbf{V}')}$. In the course of the following derivations, we encounter cases where second order tensors such as, $\overline{V_j' \mathbf{N}(\bar{V_i} + V_i')}$, the jth derivative of the ith component such as, $\overline{\frac{\partial}{\partial x_j}(\mathbf{N}(\bar{V_i} + V_i'))}$, or a second order tensor product such as, $\overline{\frac{\partial' V_i'}{\partial x_j} \frac{\partial}{\partial x_j}(\bar{\mathbf{N}}(\bar{V_i} + V_i'))}$, are necessary to close the equation system.

9.2.2 *Equation of Turbulence Kinetic Energy*

To arrive at the equation of turbulence kinetic energy for an incompressible turbulent flow, we first subtract Eq. (9.59) from Eq. (9.58):

$$\frac{\partial \mathbf{V}'}{\partial t} + \bar{\mathbf{V}} \cdot \nabla \mathbf{V}' + \mathbf{V}' \cdot \nabla \bar{\mathbf{V}} + \mathbf{V}' \cdot \nabla \mathbf{V}'' = -\frac{1}{\rho} \nabla p' + \nu \Delta \mathbf{V}' - \nabla \cdot (\overline{\mathbf{V}'\mathbf{V}'}) \tag{9.95}$$

and scalarly multiply Eq. (9.95) with $\mathbf{V}'$:

$$\mathbf{V}' \cdot \frac{\partial \mathbf{V}'}{\partial t} + \mathbf{V}' \cdot (\bar{\mathbf{V}} \cdot \nabla \mathbf{V}') + \mathbf{V}' \cdot (\mathbf{V}' \cdot \nabla \bar{\mathbf{V}}) + \mathbf{V}' \cdot \mathbf{V}' \cdot \nabla \mathbf{V}'$$
$$= -\frac{1}{\rho \mathbf{V}'} \cdot \nabla p' + \nu \mathbf{V}' \cdot \Delta \mathbf{V}' - \mathbf{V}' \cdot \nabla \cdot (\overline{\mathbf{V}'\mathbf{V}'}) \tag{9.96}$$

and rearrange the Reynolds stress tensor $\nabla \cdot (\overline{\mathbf{V}'\mathbf{V}'})$ in Eq. (9.96) by subtracting the continuity equation:

$$\nabla \cdot (\overline{\mathbf{V}'\mathbf{V}'}) = \overline{\mathbf{V}' \cdot \nabla \mathbf{V}'} + \overline{\mathbf{V}' \nabla \cdot \mathbf{V}'} = \overline{\mathbf{V}' \cdot \nabla \mathbf{V}'}. \tag{9.97}$$

Inserting Eq. (9.97) into Eq. (9.96) results in:

$$\mathbf{V}' \cdot \frac{\partial \mathbf{V}'}{\partial t} + \mathbf{V}' \cdot (\bar{\mathbf{V}} \cdot \nabla \mathbf{V}') + \mathbf{V}' \cdot (\mathbf{V}' \cdot \nabla \bar{\mathbf{V}}) + \mathbf{V}' \cdot (\mathbf{V}' \cdot \nabla \mathbf{V}') + \mathbf{V}' \cdot \overline{\mathbf{V}' \cdot \nabla \mathbf{V}'}$$
$$= -\frac{1}{\rho \mathbf{V}'} \cdot \nabla p' + \nu \mathbf{V}' \cdot \Delta \mathbf{V}'. \tag{9.98}$$

Now we average Eq. (9.98) by considering the following identities for the second term on the left-hand-side:

$$\mathbf{V}' \cdot (\bar{\mathbf{V}} \cdot \nabla \mathbf{V}') = \bar{\mathbf{V}} \cdot \nabla \left(\frac{\mathbf{V}' \cdot \mathbf{V}'}{2} \right). \tag{9.99}$$

Using the index notation, it can be shown that the third term on the left-hand-side of Eq. (9.98) is:

$$\mathbf{V}' \cdot (\mathbf{V}' \cdot \nabla \bar{\mathbf{V}}) = (\mathbf{V}'\mathbf{V}') : \nabla \bar{\mathbf{V}}. \tag{9.100}$$

Since the gradient of the mean velocity $\nabla \bar{\mathbf{V}}$ is the sum of the deformation and rotation tensor $\nabla \bar{\mathbf{V}} = \bar{\mathbf{D}} + \bar{\mathbf{\Omega}}$, Eq. (9.100) can be modified as:

$$\mathbf{V}' \cdot (\mathbf{V}' \cdot \nabla \bar{\mathbf{V}}) = (\mathbf{V}'\mathbf{V}') : (\bar{\mathbf{D}} + \bar{\mathbf{\Omega}}). \tag{9.101}$$

Since the product $(\mathbf{V}'\mathbf{V}') : \bar{\mathbf{\Omega}} = 0$, Eq. (9.101) can be modified as:

$$\mathbf{V}' \cdot (\mathbf{V}' \cdot \nabla \bar{\mathbf{V}}) = (\mathbf{V}'\mathbf{V}') : \bar{\mathbf{D}}. \tag{9.102}$$

Now we define the *turbulent kinetic energy* as:

$$k = \frac{1}{2}\overline{\mathbf{V}' \cdot \mathbf{V}'} = \frac{1}{2}\overline{V_i' V_i'} = \frac{1}{2}(\overline{V_1'^2} + \overline{V_2'^2} + \overline{V_3'^2}) = \frac{1}{2}\overline{V'^2} \tag{9.103}$$

and insert into Eq. (9.98) and average:

$$\frac{\partial k}{\partial t} + \overline{\bar{\mathbf{V}} \cdot \nabla k} + \overline{(\mathbf{V}'V') : \bar{\mathbf{D}}} + \overline{\mathbf{V}' \cdot (\mathbf{V}' \cdot \nabla \mathbf{V}')} = -\frac{1}{\rho}\overline{\mathbf{V}' \cdot \nabla p'} + \nu \overline{\mathbf{V}' \cdot \Delta \mathbf{V}'}. \tag{9.104}$$

The forth term on the left-hand-side can be written as

$$\overline{\mathbf{V}' \cdot (\mathbf{V}' \cdot \nabla \mathbf{V}')} = \overline{\mathbf{V}' \cdot \nabla(\mathbf{V}' \cdot \mathbf{V}'/2)}. \tag{9.105}$$

Considering Eq. (9.105), the equation of turbulence kinetic energy (9.104) becomes:

$$\frac{\partial k}{\partial t} + \overline{\bar{\mathbf{V}} \cdot k} + \overline{(\mathbf{V}'\mathbf{V}') : \bar{\mathbf{D}}} + \overline{\mathbf{V}' \cdot \nabla k} = -\frac{1}{\rho}\overline{\mathbf{V}' \cdot \nabla p'} + \nu \overline{\mathbf{V}' \cdot \Delta \mathbf{V}'}. \tag{9.106}$$

A simple rearrangement of Eq. (9.106) yields:

$$\frac{\partial k}{\partial t} + \overline{\bar{\mathbf{V}} \cdot \nabla k} + \overline{(\mathbf{V}'\mathbf{V}') : \bar{\mathbf{D}}} = -\overline{\mathbf{V}' \cdot \left(\nabla k + \frac{\nabla p'}{\rho}\right)} + \nu \overline{\mathbf{V}' \cdot \Delta \mathbf{V}'}. \tag{9.107}$$

We add to the argument in the parenthesis on the right-hand side of Eq. (9.107) the following zeros:

$$\overline{k\nabla \cdot V'} = 0, \; \frac{\overline{p'\nabla \cdot V'}}{\rho} = 0 \tag{9.108}$$

and obtain:

$$\frac{\partial k}{\partial t} + \overline{\bar{\mathbf{V}} \cdot \nabla k} + \overline{(\mathbf{V}'\mathbf{V}') : \bar{\mathbf{D}}} = -\overline{\mathbf{V}' \cdot \nabla k} + \overline{k\nabla \cdot \mathbf{V}'} + \frac{\overline{\mathbf{V}' \cdot \nabla p'}}{\rho} + \frac{\overline{p'\nabla \cdot V'}}{\rho}$$
$$+ \nu \overline{\mathbf{V}' \cdot \Delta \mathbf{V}'}. \tag{9.109}$$

Rearranging the terms in the parentheses on the right-hand-side of Eq. (9.107) results in the final equation of turbulence kinetic energy for incompressible flow in a coordinate invariant form:

$$\frac{\partial k}{\partial t} + \overline{\bar{\mathbf{V}} \cdot \nabla k} + \overline{(\mathbf{V}'\mathbf{V}') : \bar{\mathbf{D}}} = -\nabla \cdot \left(\overline{V'k} + \frac{\overline{\mathbf{V}'p'}}{\rho}\right) + \nu \overline{\mathbf{V}' \cdot \Delta \mathbf{V}'}. \tag{9.110}$$

The Cartesian index notation is:

$$\frac{\partial k}{\partial t} + \bar{V}_i \frac{\partial k}{\partial x_i} = -\overline{V_i' V_j'}\frac{\partial \bar{V}_i}{\partial x_j} - \frac{\partial}{\partial x_i}\left(\overline{V_i'(k + \frac{p'}{\rho}}\right) + \nu \overline{V_i' \frac{\partial^2 V_i'}{\partial x_j \partial x_j}}. \tag{9.111}$$

Equation (9.110) with its index notation (9.111) is the balance equation of the turbulence kinetic energy per unit of mass. Before interpreting the individual terms in Eq. (9.111), we first modify the last term on the right-hand side. The modification is aimed at providing a detailed mathematical explanation that describes the dissipative nature of this term. We use the following identity

$$\nu\overline{\mathbf{V}'\cdot\Delta\mathbf{V}'} = 2\nu\left\{\overline{\nabla\cdot(\mathbf{V}'\cdot\mathbf{D}')} - \overline{\mathbf{D}':\nabla\mathbf{V}'}\right\} \quad (9.112)$$

with $\mathbf{D}'$ as the deformation tensor of the turbulence fluctuation and its components $D_{ij'} = \frac{1}{2}\left(\frac{\partial V_j'}{\partial x_i} + \frac{\partial V_i'}{\partial x_j}\right)$. The first term on the right-hand side of Eq. (9.112) written in index notation:

$$2\nu\{\overline{\nabla\cdot(\mathbf{V}'\cdot\mathbf{D}')}\} = \nu\frac{\partial}{\partial x_i}\left\{\overline{V_j'\left(\frac{\partial V_i'}{\partial x_j} + \frac{\partial V_j'}{\partial x_i}\right)}\right\} \text{ differentiating gives:}$$

$$2\nu\{\overline{\nabla\cdot(\mathbf{V}'\cdot\mathbf{D}')}\} = \nu\overline{\frac{\partial V_j'}{\partial x_i}\left(\frac{\partial V_i'}{\partial x_j} + \frac{\partial V_j'}{\partial x_i}\right)} + \nu\overline{V_j'\frac{\partial^2 V_j'}{\partial x_i \partial x_i}} \quad (9.113)$$

with $\nu\overline{V_j'\frac{\partial^2 V_i'}{\partial x_i \partial x_j}} = \nu\overline{V_j'\frac{\partial}{\partial x_j}\left(\frac{\partial V_i'}{\partial x_i}\right)} = 0$ in Eq. (9.113) as a consequence of the incompressibility requirement. The second term on the right-hand side of Eq. (9.112) written in index notation is:

$$-2\nu\overline{\mathbf{D}':\nabla\mathbf{V}'} = -\nu\overline{\left(\frac{\partial V_i'}{\partial x_j'} + \frac{\partial V_j'}{\partial x_i}\right)\frac{\partial V_j'}{\partial x_i}} \quad (9.114)$$

with the velocity gradient in that can be set as $\nabla\mathbf{V}' = \mathbf{D}' + \Omega'$ and since $\mathbf{D}' : \Omega' = 0$, Eq. (9.114) is rearranged as:

$$\varepsilon_c \equiv 2\nu\overline{\mathbf{D}':\mathbf{D}'} = \nu\overline{\frac{\partial V_i'}{\partial x_j}\frac{\partial V_j'}{\partial x_i}} + \nu\overline{\frac{\partial V_j'}{\partial x_i}\frac{\partial V_j'}{\partial x_i}}. \quad (9.115)$$

Equation (9.115) exhibits the complete turbulence dissipation as found, among others, in Hinze [75] and Rotta [76]. The above definition of dissipation differs from the definition of the dissipation we will use in conjunction with the modeling, which is defined as

$$\varepsilon \equiv \nu\overline{\frac{\partial V_j'}{\partial x_i}\frac{\partial V_j'}{\partial x_i}}. \quad (9.116)$$

Thus, Eq. (9.116) is expressed in terms of Eq. (9.115) through

$$\varepsilon = \varepsilon_c - \nu\overline{\frac{\partial V_i'}{\partial x_j}\frac{\partial V_j'}{\partial x_i}}. \quad (9.117)$$

Bradshaw and Pitt [82] have shown that for cases with strong velocity gradients such as shock waves, the maximum difference $\Delta\varepsilon = \varepsilon_c - \varepsilon$ is about 2% and is for other flow situations negligibly small. Returning to Eq. (9.112), the sum of (9.113) and (9.114) yields:

$$\nu\overline{\mathbf{V}'\cdot\Delta\mathbf{V}'} = \nu\frac{\partial}{\partial x_i}\left\{\overline{V_j'\left(\frac{\partial V_i'}{\partial x_j}+\frac{\partial V_j'}{\partial x_i}\right)}\right\} - \nu\overline{\left(\frac{\partial V_i'}{\partial x_j}\frac{\partial V_j'}{\partial x_i}\right)\frac{\partial V_j'}{\partial x_i}}. \tag{9.118}$$

Expressing the left-hand side of (9.118) in index notation, we get

$$\nu\overline{V_i'\frac{\partial^2 V_i'}{\partial x_j\partial x_j}} = \nu\frac{\partial}{\partial x_i}\left\{\overline{V_j'\left(\frac{\partial V_i'}{\partial x_j}+\frac{\partial V_j'}{\partial x_i}\right)}\right\} - \nu\overline{\left(\frac{\partial V_i'}{\partial x_j}+\frac{\partial V_j'}{\partial x_i}\right)\frac{\partial V_j'}{\partial x_i}}. \tag{9.119}$$

We replace in Eq. (9.110) the last term on the right-hand side by Eq. (9.112) and obtain:

$$\begin{aligned}\frac{\partial k}{\partial t} + \bar{\mathbf{V}}\cdot\nabla k = &-\overline{(\mathbf{V}'\mathbf{V}')}:\bar{\mathbf{D}} - \nabla\cdot\left(\overline{\mathbf{V}'k}+\frac{\overline{\mathbf{V}'p'}}{\rho}\right)\\ &+2\nu\{\overline{\nabla\cdot(\mathbf{V}'\cdot\mathbf{D}')} - \overline{\mathbf{D}':\nabla\mathbf{V}'}\}.\end{aligned} \tag{9.120}$$

The index notation of Eq. (9.120) is:

$$\begin{aligned}\frac{\partial k}{\partial t} + \bar{V}_i\frac{\partial k}{\partial x_i} = &-\overline{V_i'V_j'}\frac{\partial \bar{V}_j}{\partial x_i} - \frac{\partial}{\partial x_i}\left(\overline{V_i'\left(k+\frac{p'}{\rho}\right)}\right)\\ &+\nu\frac{\partial}{\partial x_i}\left\{\overline{V_j'\left(\frac{\partial V_i'}{\partial x_j}+\frac{\partial V_j'}{\partial x_i}\right)}\right\} - \nu\overline{\left(\frac{\partial V_i'}{\partial x_j}+\frac{\partial V_j'}{\partial x_i}\right)\frac{\partial V_j'}{\partial x_i}}.\end{aligned} \tag{9.121}$$

Equation (9.121) expresses the same physical content as Eq. (9.111), thus, it does not represent a new physical relationship that can be used to reduce the number of unknowns. Using some mathematical manipulations, we merely decomposed the last term of Eq. (9.111) to explicitly introduce the dissipation process into the turbulence kinetic energy balance.

Interpretation of Individual Terms in Eq. (9.121): The two terms on the left-hand side of Eqs. (9.120) and (9.121) describe the substantial change of the turbulence kinetic energy per unit of mass consisting of the local and convective changes of the kinetic energy. The first term on the right-hand side is the energy transferred from the mean flow through the turbulent shear stress. This term is also called the *production of turbulence energy*. This is explicitly expressed in terms of the double scalar product of the mean flow deformation tensor $\bar{\mathbf{D}}$ and the second order Reynolds stress tensor $\overline{\mathbf{V}'\mathbf{V}'}$. The second term is the spatial change of the work by the total pressure of the fluctuating motion. It exhibits the *convective diffusion by turbulence of the total turbulence energy*. The third term on the right-hand side is the spatial change of the specific work (work per unit mass) by the viscous shear stress of the turbulent motion. The last term expresses the *viscous dissipation by the turbulent motion*.

Introducing the Dissipation: There are a variety of alternative forms for turbulence kinetic energy. The following alternative form is used for the purpose of turbulence modeling. We further re-arrange the first term on the right-hand side of Eq. (9.119):

$$\nu\overline{\mathbf{V}'\cdot\Delta\mathbf{V}'} = \nu\frac{\partial}{\partial x_i}\left(\overline{V'_j\frac{\partial V'_i}{\partial x_j}} + \overline{V'_j\frac{\partial V'_j}{\partial x_i}}\right) - \nu\left(\overline{\frac{\partial V'_i}{\partial x_j}} + \overline{\frac{\partial V'_j}{\partial x_i}}\right)\overline{\frac{\partial V'_j}{\partial x_i}}. \tag{9.122}$$

The second term within the first parentheses of Eq. (9.122) is the spatial change of the kinetic energy

$$\nu\overline{\mathbf{V}'\cdot\Delta\mathbf{V}'} = \nu\frac{\partial}{\partial x_i}\left(\overline{V'_j\frac{\partial V'_i}{\partial x_j}} + \frac{\partial k}{\partial x_i}\right) - \nu\left(\overline{\frac{\partial V'_i}{\partial x_j}} + \overline{\frac{\partial V'_j}{\partial x_i}}\right)\overline{\frac{\partial V'_j}{\partial x_i}}. \tag{9.123}$$

We differentiate the expression in the first parentheses of Eq. (9.123) with respect to x_i and obtain:

$$\nu\overline{\mathbf{V}'\cdot\Delta\mathbf{V}'} = \nu\left\{\overline{\frac{\partial V'_j}{\partial x_i}\frac{\partial V'_i}{\partial x_j}} + \overline{V'_j\frac{\partial}{\partial x_i}\left(\frac{\partial V'_j}{\partial x_j}\right)} + \frac{\partial^2 k}{\partial x_i \partial x_i}\right\}$$
$$- \nu\left(\overline{\frac{\partial V'_i}{\partial x_j}\frac{\partial V'_j}{\partial x_i}} + \overline{\frac{\partial V'_j}{\partial x_i}\frac{\partial V'_j}{\partial x_i}}\right). \tag{9.124}$$

Because of the continuity requirement for an incompressible flow, the second term in the first parenthesis of Eq. (9.124) identically vanishes. Moreover, the first terms within the first parenthesis and the second parenthesis cancel each other out reducing Eq. (9.124) to:

$$\nu\overline{\mathbf{V}'\cdot\Delta\mathbf{V}'} = \nu\overline{V'_i\frac{\partial^2 V'_i}{\partial x_j \partial x_j}} = \nu\frac{\partial^2 k}{\partial x_i \partial x_i} - \nu\left(\overline{\frac{\partial V'_i}{\partial x_j}\frac{\partial V'_i}{\partial x_j}}\right). \tag{9.125}$$

With Eqs. (9.125) and (9.116), Eq. (9.111) reads:

$$\frac{\partial k}{\partial t} + \bar{V}_i\frac{\partial k}{\partial x_i} + \frac{1}{2}\overline{V'_iV'_j}\left(\frac{\partial \bar{V}_i}{\partial x_j} + \frac{\partial \bar{V}_j}{\partial x_i}\right) = -\frac{\partial}{\partial x_i}\left(\overline{V'_i\left(k + \frac{p'}{\rho}\right)}\right) + \nu\frac{\partial^2 k}{\partial x_j \partial x_j} - \varepsilon. \tag{9.126}$$

Equation (9.126) establishes a relationship between the substantial change of the turbulence kinetic energy and its dissipation. In Eq. (9.126), ε can be replaced by the complete dissipation ε_c, leading to:

$$\frac{\partial k}{\partial t} + \bar{V}_i \frac{\partial k}{\partial x_i} + \frac{1}{2}\overline{V_i' V_j'}\left(\frac{\partial \bar{V}_i}{\partial x_j} + \frac{\partial \bar{V}_j}{\partial x_i}\right) = -\frac{\partial}{\partial x_i}\left(\overline{V_i'\left(k + \frac{p'}{\rho}\right)}\right)$$
$$+ \nu \frac{\partial^2 k}{\partial x_j \partial x_j} + \nu \overline{\frac{\partial V_i'}{\partial x_j}\frac{\partial V_j'}{\partial x_i}} - \varepsilon_c. \tag{9.127}$$

9.2.3 *Equation of Dissipation of Kinetic Energy*

As we will see later in turbulence modeling, besides the equations of continuity, motion, and energy, the equation of dissipation is also used. To arrive at this equation, we write Eq. (9.95) in index notation

$$\frac{\partial V_i'}{\partial t} + V_k' \frac{\partial \bar{V}_i}{\partial x_k} + \bar{V}_k \frac{\partial V_i'}{\partial x_k} + V_k' \frac{\partial V_i'}{\partial x_k} = -\frac{1}{\rho}\frac{\partial p'}{\partial x_i} + \nu \frac{\partial^2 V_i'}{\partial x_k \partial x_k} + \frac{\partial(\overline{V_i' V_k'})}{\partial x_k}. \tag{9.128}$$

We differentiate Eq. (9.128) with respect to x_j and scalarly multiply the result with $2\nu \frac{\partial V_i'}{\partial x_j}$. After averaging, we arrive at the following exact dissipation equation by Launder et al. [83]

$$\frac{\partial \varepsilon}{\partial t} + \bar{V}_k \frac{\partial \varepsilon}{\partial x_k} = -2\nu \left\{ \frac{\partial^2 \bar{V}_j}{\partial x_1 \partial x_k}\left(\overline{\frac{V_k' \partial V_j'}{\partial x_l}}\right) + \frac{\partial \bar{V}_j}{\partial x_k}\left(\overline{\frac{\partial V_j'}{\partial x_l}\frac{\partial V_k'}{\partial x_l}} + \overline{\frac{\partial V_i'}{\partial x_j}\frac{\partial V_i'}{\partial x_k}}\right)\right.$$
$$+ \overline{\frac{\partial V_j'}{\partial x_k}\frac{\partial V_j'}{\partial x_l}\frac{\partial V_k'}{\partial x_l}} + \frac{1}{2}\frac{\partial}{\partial x_k}\left(\overline{V_k' \frac{\partial V_j'}{\partial x_l}\frac{\partial V_j'}{\partial x_l}}\right)$$
$$\left. + \frac{1}{\rho}\frac{\partial}{\partial x_k}\left(\overline{\frac{\partial V_k' \partial p'}{\partial x_l \partial x_l}}\right) + \nu \overline{\left(\frac{\partial^2 V_j}{\partial x_k \partial x_l}\right)^2}\right\} + \nu \frac{\partial^2 \varepsilon}{\partial x_k \partial x_k}. \tag{9.129}$$

Equation (9.129) exhibits an exact derivation of the dissipation equation and is more complicated than Eq. (9.126) for the turbulence kinetic energy. For modeling purposes, an empirical relation proposed by Launder and Spalding [84] is used as the standard model equation, which we present in the following section.

9.3 Turbulence Modeling

Equation (9.59) indicates that in order to obtain solutions for the Reynolds averaged Navier–Stokes equations (RANS), it is necessary to provide further information about the Reynolds stress Tensor $T' = \rho \overline{V'V'}$ which has generally nine components from which six are distinct. Many studies investigated the possibilities to establish a relationship between $T' = \rho \overline{V'V'}$ and the mean velocity field. This approach is called turbulence modeling. Tremendous amount of papers published in the last three decades show that none of the existing turbulence models can be universally applied to arbitrary types of turbulent flows. Very recent direct Navier–Stokes simulations (DNS) performed successfully for different flow situations exhibit a major breakthrough, making the turbulence modeling and its use superfluous. However, the computational effort to perform DNS makes it, at least for the time being, not attractive. This situation certainly will change in the near future. Until then, one has to work with several turbulence models, each of which is appropriate for certain types of flows. In the context of this chapter, we intend to make students familiar with the most representative models that are being used. In the following, we discuss models that are based on turbulent-viscosity models. Analogous to the Stokes material equation, this model is based on the assumption that the Reynolds stress tensor might be correlated with the mean velocity gradient. Boussinesq [85] was the first to set a relationship between the Reynold stress tensor and the mean velocity gradient such as $\tau'_{12} = A\overline{V_1}/dx_2$, with A as the mixing coefficient. For a three-dimensional flow, the Boussinesq relationship reads:

$$\mathbf{T}' \equiv -\rho\overline{\mathbf{V}'\mathbf{V}'} = -\mu_t \nabla\bar{\mathbf{V}} = -\mu_t(\bar{\mathbf{D}} + \bar{\Omega}) \tag{9.130}$$

with μ_t as the unknown turbulence viscosity, also called *eddy viscosity*, that needs to be determined. As seen, the Boussinesq relation is an analogon to the viscous stress relation. With Eq. (9.130), the problem of determining the unknown Reynolds stress is shifted to the problem of finding the unknown eddy viscosity. It can be argued that the Boussinesq formulation (9.130) is not compatible with the material objectivity principal that requires frame indifference. Since the only part of $\nabla\bar{\mathbf{V}}$ that fulfills the objectivity principle is the mean deformation tensor $\bar{\mathbf{D}}$, it is obvious to set the Reynolds stress tensor in relation with the deformation tensor of the mean flow

$$\rho\overline{\mathbf{V}'\mathbf{V}'} = \mu_t\bar{\mathbf{D}}. \tag{9.131}$$

The turbulence viscosity μ_t also called eddy viscosity can be set as

$$\mu_t = \rho l_t V_t \tag{9.132}$$

with l_t and V_t as the turbulence length and velocity scales.

9.3.1 Algebraic Model: Prandtl Mixing Length Hypothesis

Prandtl [86] was the first to present a working algebraic turbulence model that is applied to wakes, jets and boundary layer flows. The model is based on *mixing length hypothesis* deduced from experiments and is analogous, to some extent, to the mean free path in kinetic gas theory. For better understanding the basic physics of the mixing length hypothesis, we utilize the Prandtl approximation that uses the

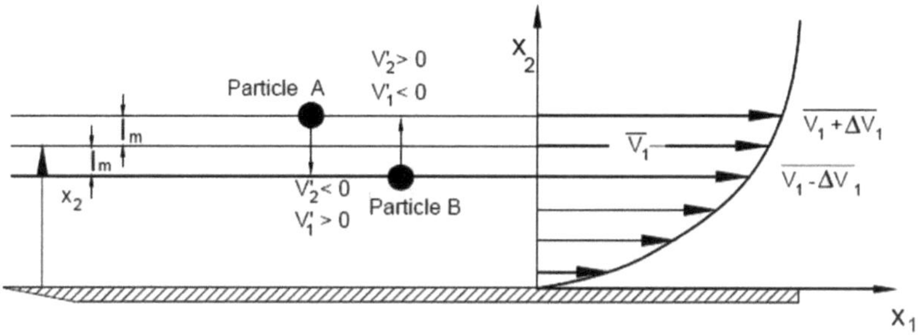

Fig. 9.8 Explaining the mechanism of Prandtl mixing length theory

simplest case of a parallel flow, where the flow velocity has only one component that changes in normal direction with $\overline{V_1} = \overline{V_1}(x_2)$; $\overline{V_2} = 0$; $\overline{V_3} = 0$, as shown in Fig. 9.8. Consider the boundary layer flow along a flat plate in X_1-direction. The fluid particle A with the mass dm located at the position $X_2 + l_m$ and has the longitudinal velocity component $\overline{V_1} + \overline{\Delta V_1}$ is fluctuating and moving downward with the lateral velocity $V_2' < 0$ and the fluctuation momentum $d\mathbf{I}'_{x_2} = dm\mathbf{V}'_2$. It arrives at the layer X_2 which has a lower velocity $\overline{V_1}$. According to the Prandtl hypothesis, this macroscopic momentum exchange most likely gives rise to a positive fluctuation $V_1' > 0$. This results in a negative non-zero correlation $\overline{V_1'V_2'}$. Inversely, the fluid particle B moving upward with the velocity $V_2' > 0$ from the layer $X_2 - l_m$, where a lower longitudinal velocity $\overline{V_1} - \overline{\Delta V_1}$ prevails, causes a negative fluctuation $V_1' < 0$. In both cases, the particles experience a velocity difference which can be approximated as:

$$\overline{\Delta V_1} = l_m \overline{\Delta V_1}/\Delta x_2 \approx l_m \overline{dV_1}/dx_2 \tag{9.133}$$

by using the Taylor expansion and neglecting all higher order terms. The distance between the two layers l_m is called mixing length. Since $|\overline{\Delta V_1}|$ has the same order of magnitude as $|V_1'|$, we may replace in Eq. (9.133) $|\overline{\Delta V_1}|$ by $|V_1'|$ and arrive at:

$$|V_1'| = l_m \left| \frac{d\overline{V_1}}{dx_2} \right|. \tag{9.134}$$

Note that the mixing length l_m is still an unknown quantity. Since, by virtue of the Prandtl hypothesis, the longitudinal fluctuation component V_1' was brought about by the impact of the lateral component V_2', it seems reasonable to assume that $|V_2'| \propto |V_1'|$ such that with Eq. (9.134), we may find for the lateral fluctuation component $|V_2'| = C_1 \cdot l_m |\frac{d\overline{V_1}}{dx_2}|$ with C_1 as a constant. Thus, the $\overline{V_1'V_2'}$ component of the Reynolds stress tensor becomes:

$$\overline{V_1'V_2'} = -C_1 l_m^2 |\overline{V_1'}|\,|\overline{V_2'}| = -C_1 l_m^2 \left(\frac{d\overline{V_1}}{dx_2}\right)^2. \tag{9.135}$$

Since the constant C_1 as well as the mixing length l_m are unknown, the constant C_1 may be included in the mixing length such that we may write

$$\overline{V_1'V_2'} = -l_m^2 \left(\frac{d\overline{V}_1}{dx_2}\right)^2. \tag{9.136}$$

Considering Eqs. (9.91) and (9.130), the shear stress component becomes:

$$\tau_{12}' = -\rho\overline{V_1'V_2'} = \rho l_m^2 \left(\frac{d\overline{V}_1}{dx_2}\right)^2. \tag{9.137}$$

Equation (9.137) does not take into account that the sign of the shear stress component τ_{12} changes with $d\overline{V}_1/dx_2$. To correct this, we may write

$$\tau_{12}' = -\rho\overline{V_1'V_2'} = \rho l_m^2 \left|\frac{d\overline{V}_1}{dx_2}\right| \left(\frac{d\overline{V}_1}{dx_2}\right) \equiv \mu_t \left(\frac{d\overline{V}_1}{dx_2}\right) \tag{9.138}$$

with μ_t as the eddy viscosity. This is the Prandtl mixing length hypothesis. From Eq. (9.138) we deduce that the eddy kinematic viscosity $\nu_t = \mu_t/\rho$ can be expressed as:

$$\nu_t = l_m^2 \left|\frac{d\overline{V}_1}{dx_2}\right|. \tag{9.139}$$

To find an algebraic expression for the mixing length l_m, several empirical correlations were suggested that are discussed by Schlichting [87] and summarized by Wilcox [88]. The mixing length l_m does not have a universally valid character and changes from case to case, therefore it is not appropriate for three-dimensional flow applications. However, it is successfully applied to boundary layer flow (for details see Chap. 11) and particularly to free turbulent flows. Utilizing the two-dimensional boundary layer approximation by Prandtl, and for the sake of simplicity, we use the boundary layer nomenclature with the mean-flow component, $\overline{V}_1 \equiv u$ as the significant velocity in $x_1 \equiv x$-direction, the distance from the wall $x_2 \equiv y$, the dimensionless velocity $u^+ = u/u_\tau$ and the dimensionless distance from the wall $y^+ = u_\tau y/\nu$. The *wall friction velocity* u_τ is related to the wall shear stress τ_w by the relation $\tau_w = \rho u_\tau^2$. Figure 9.9 exhibits the non-dimensionalized flow velocity distribution u^+ of a fully turbulent boundary layer as a function of the non-dimensionalized normal wall distance y^+. The plot with the log-scale for y^+ reveals three distinct layers: the *viscous sublayer* ranging from $y^+ = 0$ to 5, followed by a *buffer layer* that is tangent to *the logarithmic layer* at about $y^+ = 200$. The buffer layer extends from $y^+ = 5$ to 200. The viscous sublayer is approximated by *the linear wall function:*

$$u^+ = y^+ \tag{9.140}$$

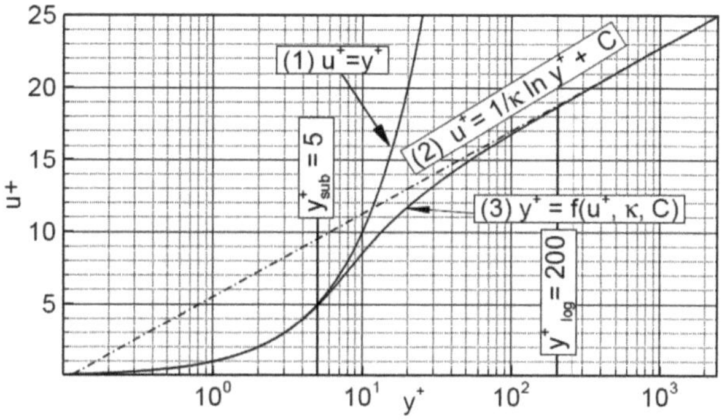

Fig. 9.9 Approximation of velocity distribution for a fully turbulent flow by its decomposition into a laminar sublayer curve (1), a logarithmic layer (2) and the buffer layer (3) extending from $y^+ = 5$ to 200

followed by the logarithmic layer which is approximated by

$$u^+ = \frac{1}{k} \ln \left(\frac{u_\tau y}{\nu} \right) + C. \tag{9.141}$$

For a *fully developed turbulent flow*, the constants in Eq. (9.141) are experimentally found to be $k = 0.41$ and $C = 5.0$, Fig. 9.9, Curve 2.

For *transitional boundary layer flows*, these constants change significantly, as detailed in a study by Müller [89]. A third layer, the *outer layer*, tangents the logarithmic layer described by a so-called *wake function*, is discussed in Chap. 10. Outside the viscous sublayer marked as the logarithmic layer, the mixing length is approximated by a simple linear function

$$l_m = ky. \tag{9.142}$$

Accounting for viscous damping, the mixing length for the viscous sublayer is modeled by introducing a damping function D into Eq. (9.142). As a result, we obtain

$$l_m = kDy \tag{9.143}$$

with the damping function D proposed by van Driest [90] as

$$D = 1 - e^{-y^+/A^+} \tag{9.144}$$

with the constant $A^+ = 26$ for a boundary layer at zero-pressure gradient. As we will discuss in Chap. 11, based on experimental evaluation of a large number of velocity profiles, Kays and Moffat [91] developed an empirical correlation for A^+ that accounts for different pressure gradients and boundary layer suction/blowing. For zero suction/blowing this correlation reduces to:

$$A^+ = \frac{26}{abP^+ + 1.0} \tag{9.145}$$

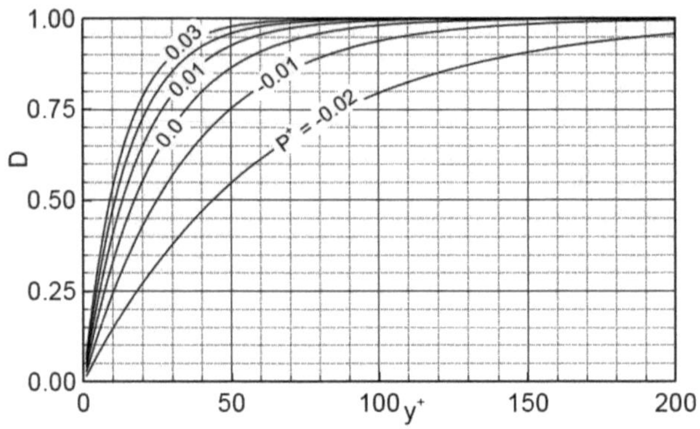

Fig. 9.10 Van Driest damping function with $p+$ as a parameter

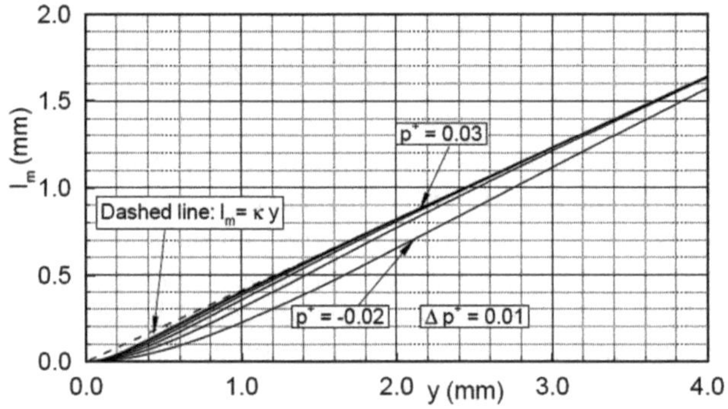

Fig. 9.11 Details of the mixing length from Fig. 9-11

with $a = 9.0$, $b = 4.25$ if $P^+ < 0$ and $b = 2.29$ if $P^+ > 0$. The dimensionless pressure gradient P^+ in Eq. (9.145) is defined in as

$$P^+ = \frac{\mu(d\bar{p}/dx)}{\rho^{1/2}\tau_w^{3/2}}. \tag{9.146}$$

Introducing Eq. (9.145) into Eq. (9.144), the Van Driest damping function is plotted in Fig. 9.10.

Figure 9.10 exhibits the damping function D as a function of y^+ with the pressure gradient P^+ as a parameter. The implementation of the damping function into the mixing length accounts for the non-linear distribution of the mixing length in the lateral direction as shown in Fig. 9.11. For comparison purposes, the linear distribution $l_m = ky$ is also plotted as a dashed line. For $P^+ \succeq 0$ and $y \succeq 6$ the curves pertaining to $l_m = kDy$ asymptotically approach the linear distribution. Inside the viscous sublayer significant differences are clearly visible, as shown in Fig. 9.12 which is a partial enlargement of Fig. 9.11. It exhibits the mixing length distribution very close to the wall. For comparison purposes, the linear distribution $l_m = ky$ is also plotted.

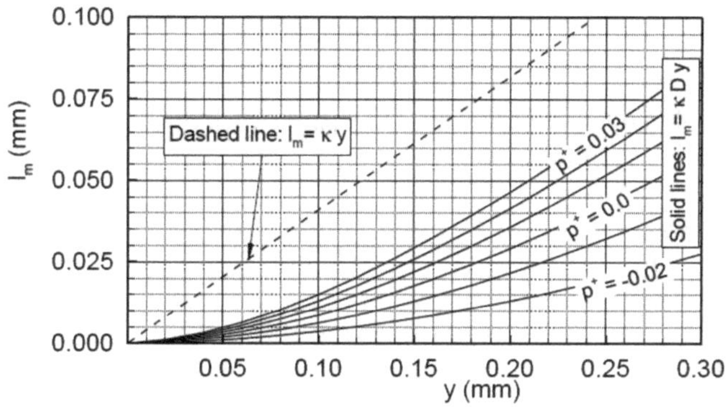

Fig. 9.12 Mixing length in lateral wall-direction with $p+$ as parameter

Considering these differences and the asymptotic approach mentioned above, it seems that Eq. (9.143) is suitable for describing the mixing length within a boundary layer for which the von Kármán constant k assumes the value $k = 0.41$. This value, as previously mentioned, is valid only for fully turbulent flows at zero-pressure gradient. It is not valid for transitional boundary layers, where κ changes significantly. Moreover, Eq. (9.139) implies that whenever $\overline{dV_1/dx_2} = 0$, the eddy kinematic viscosity approaches zero, which contradicts the experimental data. In addition, the mixing length concept does not apply to boundary layer flow cases where a flow separation occurs. Furthermore, it is not suitable for three-dimensional flow calculation as mentioned before. Concluding the discussion about the mixing length approach, it can be stated that the turbulence model based on the Prandtl mixing length theory, despite its shortcomings, is still used for boundary layer calculation and delivers satisfactory results, as seen in Chap. 11.

For the sake of completeness, in what follows, we present a system of equations that includes the algebraic model. The system can be used to solve steady incompressible free turbulent flow, as well as boundary layer problems, where no separation occurs. Utilizing the two-dimensional boundary layer flow assumption by Prandtl, for the sake of compatibility with his convention, we use the Prandtl nomenclature with $\overline{V_1} \equiv \bar{U}$, $\overline{V_2} \equiv \bar{V}$, $V_1' \equiv u$ and $V_2' \equiv v$ as the mean flow velocities and the fluctuation components in $x_1 \equiv x$ and $x_2 \equiv y$-direction, thus the continuity and momentum equations are reduced to

$$\frac{\partial \bar{U}}{\partial x} + \frac{\partial \bar{V}}{\partial y} = 0 \tag{9.147}$$

$$\bar{U}\frac{\partial \bar{U}}{\partial x} + \bar{V}\frac{\partial \bar{U}}{\partial y} = -\frac{1}{\rho}\frac{d\bar{p}}{dx} + \frac{\partial}{\partial y}\left(\nu\frac{\partial \bar{U}}{\partial y} - \overline{uv}\right). \tag{9.148}$$

According to the Prandtl boundary layer assumptions (see for details Chap. 11), the pressure gradient outside the boundary layer may be approximated by the Bernoulli equation , where the flow is assumed isentropic

$$-\frac{d\bar{p}}{dx} = \rho U_\infty \frac{dU_\infty}{dx} \tag{9.149}$$

with a known velocity distribution $U_\infty(x)$ outside the boundary layer. The turbulent shear stress in Eq. (9.148) becomes

$$-\rho\overline{uv} = \rho l_m^2 \left|\frac{\partial \bar{U}}{\partial y}\right| \frac{\partial \bar{U}}{\partial y} \tag{9.150}$$

with ℓ_m, from Eq. (9.143) in conjunction with Eqs. (9.147)–(9.149), a solution can be obtained for the main-velocity field in terms of $\bar{U}$ and $\bar{V}$.

9.3.2 Algebraic Model: Cebeci–Smith Model

Another algebraic model is the Cebeci–Smith [92] which has been used primarily in external high speed aerodynamics with attached thin boundary layer. It is a two-layer algebraic zero-equation model which gives the eddy viscosity by separate expressions in each layer, as a function of the local boundary layer velocity profile. The model is not suitable for cases with large separated regions and significant curvature/rotation effects. The turbulent kinematic viscosity for the inner layer is calculated from

$$\nu_{t_i} = l_m^2 \left[\left(\frac{\partial U}{\partial y}\right)^2 + \left(\frac{\partial U}{\partial x}\right)^2\right]^{\frac{1}{2}}. \tag{9.151}$$

For the outer layer kinematic viscosity is

$$\nu_{t_0} = \alpha U_e \delta_1 F_{Kl}(y;\delta) \tag{9.152}$$

with

$$F_{Kl}(y;\delta) = \left[1 + 5.5\left(\frac{y}{\delta}\right)^6\right]^{-1} \text{ and } \delta_1 = \int_0^\delta (1 - U/U_e)dy \tag{9.153}$$

$\alpha = 0.0168$, U_e the velocity at the edge of the boundary layer, δ_1 the boundary layer displacement thickness and F_{Kl} as the Klebanoff intermittency function [93]. The mixing length in Eq. (9.151) is determined by combining Eqs. (9.143) and (9.144)

$$l_m = \kappa y(1 - e^{-y^+/A^+}) \tag{9.154}$$

with $\kappa = 0.4$ and $A^+ = 26\left(1 + y\frac{dp/dx}{\rho u_\tau^2}\right)^{-1/2}$.

Table 9.1 Closure coefficients of Eqs. (9.157) through (9.159)

k	α	$A+$	C_{cp}	C_{Kleb}	C_w
0.4	0.0168	26	1.6	0.3	1

9.3.3 Baldwin–Lomax Algebraic Model

The third algebraic model is the Baldwin–Lomax model [94]. The basic structure of this model is essentially the same as the Cebeci–Smith model with the exception of a few minor changes. Similar to Cebeci–Smith, this model is a two-layer algebraic zero-equation model which gives the eddy kinematic viscosity ν_t as a function of the local boundary layer velocity profile. The model is suitable for high-speed flows with thin attached boundary-layers, typically present in aerospace and turbomachinery applications. While this model is quite robust and provides quick results, it is not capable of capturing details of the flow field. Since this model is not suitable for calculating flow situations with separation, its applicability is limited. We briefly summarize the structure of this model as follows. The kinematic viscosity for the inner layer is

$$\nu_{t_i} = l_m^2 |\Omega| \tag{9.155}$$

with

$$l_m = \kappa y(1 - e^{-y^+/A_0}) \tag{9.156}$$

and $\Omega = e_i e_j \omega_{ij}$ as the rotation tensor. The outer layer is described by

$$\upsilon_{t_0} = \alpha C_{cp} F_{wake} F_{kl}(y, y_{max}/C_{Kleb}) \tag{9.157}$$

with the wake function F_{wake}

$$F_{wake} = \min(y_{max} F_{max}; C_{wk} y_{max} U_{diff}/F_{max}) \tag{9.158}$$

and F_{max} and y_{max} as the maximum of the function

$$F(y) = y|\Omega|(1 - e^{-y^+/A_0}). \tag{9.159}$$

The velocity difference U_{diff} is defined as the difference of the velocity at y_{max} and y_{min}:

$$U_{diff} = \text{Max}(\sqrt{U_i U_i}) - \text{Min}(\sqrt{U_i U_i}) \tag{9.160}$$

with the closure coefficients listed in Table 9.1.

The above zero-equation models are applied to cases of free turbulent flow such as wake flow, jet flow, and jet boundaries.

9.3.4 One-Equation Model by Prandtl

A *one-equation* model is an enhanced version of the algebraic models we discussed in previous sections. This model utilizes one turbulent transport equation originally developed by Prandtl. Based on purely dimensional arguments, Prandtl proposed a relationship between the dissipation and the kinetic energy that reads

$$\varepsilon = C_D k^{3/2}/l_t \tag{9.161}$$

where the turbulence length scale ℓ_{t} is set proportional to the mixing length, ℓ_{m}, the boundary layer thickness δ or a wake or a jet width. The velocity scale in Eq. (9.132) is set proportional to the turbulent kinetic energy $V_t \propto k^{1/2}$ as suggested independently by Kolmogorov [95] and Prandtl [96]. Thus, the expression for the turbulent viscosity becomes:

$$\mu_t = C_\mu \ell_m k^{0.5} \tag{9.162}$$

with the constant C_μ to be determined from the experiment. The turbulent kinetic energy, k, as a transport equation is taken from Sect. 9.2.2 in the form of Eqs. (9.111) or (9.126) where the dissipation is implemented. For simple two-dimensional flows where no separation occurs, with the mean-flow component $\overline{V_1} \equiv \bar{U}$ as the significant velocity in $x_1 \equiv x$-direction, and the distance from the wall $x_2 \equiv y$, the following approximation by Launder and Spalding [97] may be used

$$\rho \frac{\mathrm{D}k}{\mathrm{D}t} = \mu_t \left(\frac{\partial \bar{U}}{\partial y}\right)^2 + \frac{\partial}{\partial y}\left(\frac{\mu_t}{\sigma_k}\frac{\partial k}{\partial y}\right) - C_D \frac{\rho k^{3/2}}{\ell_m}, \tag{9.163}$$

where $\sigma_k = 1$ and $C_D = 0.08$ are coefficients determined from experiments utilizing simple flow configurations. The one-equation model provides a better assumption for the velocity scale V_{t} than $\ell_m|\partial\bar{U}/\partial y|$. Similar to the algebraic model, the one-equation one is not applicable to the general three-dimensional flow cases since a general expression for the mixing length does not exist. Therefore the use of a one-equation model does not offer any improvement compared with the algebraic one. The one-equation models discussed above are based on kinetic energy equations. There are a variety of one-equation models that are based on Prandtl's concept and discussed in [88].

9.3.5 Two-Equation Models

Among the many two-equation models, three are the most established ones: (1) the standard $k - \varepsilon$ model, first introduced by Chou [98] and enhanced to its current form by Jones and Launder [99], (2) $k - \omega$ model first developed by Kolmogorov and enhanced to its current version by Wilcox [88] and (3) the shear stress transport

Table 9.2 Closure coefficients of Eq. (9.165)

C_μ	σ_k	σ_ε	$C_{\varepsilon 1}$	$C_{\varepsilon 2}$
0.09	1	1.3	1.44	1.92

(SST) model developed by Menter [100], who combined $\kappa - \varepsilon$ and $\kappa - \omega$ models by introducing a blending function with the objective to get the best out of these two models. All three models are built-in models of commercial codes that are used widely. In the following, we present these models and discuss their applicability.

9.3.5.1 Two-Equation $k - \varepsilon$ Model

The two equations utilized by this model are the transport equations of kinetic energy k and the transport equation for dissipation ε. These equations are used to determine the turbulent kinematic viscosity ν_t. For fully developed high Reynolds number turbulence, the exact transport equations for k (9.126) can be used. The transport equation for ε (9.129) includes triple correlations that are almost impossible to measure. Therefore, relative to ε, we have to replace it with a relationship that approximately resembles the terms in Eq. (9.129). To establish such a purely empirical relationship, dimensional analysis is heavily used. Launder and Spalding [98] used the following equations for kinetic energy

$$\frac{Dk}{Dt} = \frac{1}{\rho}\frac{\partial}{\partial x_j}\left(\frac{\mu_t}{\sigma_k}\frac{\partial k}{\partial x_j}\right) + \frac{\mu_t}{\rho}\left(\frac{\partial \bar{V}_i}{\partial x_j} + \frac{\partial \bar{V}_j}{\partial x_i}\right)\frac{\partial \bar{V}_i}{\partial x_j} - \varepsilon \tag{9.164}$$

and for dissipation

$$\frac{D\varepsilon}{Dt} = \frac{1}{\rho}\frac{\partial}{\partial x_j}\left(\frac{\mu_t}{\sigma_\varepsilon}\frac{\partial \varepsilon}{\partial x_j}\right) + C_{\varepsilon 1}\frac{\mu_t}{\rho}\frac{\varepsilon}{k}\left(\frac{\partial \bar{V}_i}{\partial x_j} + \frac{\partial \bar{V}_j}{\partial x_i}\right)\frac{\partial \bar{V}_i}{\partial x_j} - \frac{C_{\varepsilon 2}\varepsilon^2}{k}, \tag{9.165}$$

and the turbulent viscosity, μ_t, can be expressed as

$$\mu_t = \nu_t \rho = \frac{C_\mu \rho k^2}{\varepsilon}. \tag{9.166}$$

The constants σ_k, σ_ε, C_{ε_1}, C_{ε_2} and C_μ listed in Table 9.2 are calibration coefficients that are obtained from simple flow configurations such as grid turbulence. The models are applied to such flows and the coefficients are determined to make the model simulate the experimental behavior. The values of the above constants recommended by Launder and Spalding [83] are given in Table 9.2.

As seen, the simplified Eqs. (9.165) and (9.166) do not contain the molecular viscosity. They may be applied to free turbulence cases where the molecular viscosity is negligibly small compared to the turbulence viscosity. However, one cannot expect to obtain reasonable results by simulation of the wall turbulence using these equations. This deficiency is corrected by introducing the *standard* $k - \varepsilon$ *model*. This model uses the wall functions where the velocity at the wall is related to the wall shear stress τ_w by the logarithmic law of the wall. Jones and Launder [98] extended the original $k - \varepsilon$ model to the low Reynolds number form, which allows calculations right up to a solid wall. In the recent three decades, there have been many two-equation models, some of which Wilcox has listed in his book [88]. In general, the modified $k-$ and ε-equations, setting $v = \mu/\rho$ and $v_t = \mu_t/\rho$, are expressed as

$$\frac{Dk}{Dt} = \frac{\partial}{\partial x_j}\left\{\left(\nu + \frac{\nu_t}{\sigma_k}\right)\frac{\partial k}{\partial x_j}\right\} - \overline{V_i'V_j'}\frac{\partial \bar{V}_i}{\partial x_j} - \varepsilon \tag{9.167}$$

$$\frac{D\varepsilon}{Dt} = \frac{\partial}{\partial x_j}\left\{\left(\nu + \frac{\nu_t}{\sigma_\varepsilon}\right)\frac{\partial \varepsilon}{\partial x_j}\right\} - C_{\varepsilon 1}\frac{\varepsilon}{k}\overline{V_i'V_j'}\frac{\partial \bar{V}_i}{\partial x_j} - C_\varepsilon 2\frac{\varepsilon^2}{k}. \tag{9.168}$$

The closure coefficients are listed in Table 9.2. The Reynolds stress, $-\rho\overline{V_i'V_j'}$, can be expressed as

$$\nu_t = C_\mu f_\mu k^{1/2}\ell_t = \frac{C_\mu f_\mu k^2}{\varepsilon}, \tag{9.169}$$

where $\ell_t = k^{3/2}/\varepsilon$ is the eddy length scale.

Using the $k - \varepsilon$ model, successful simulations of a large variety of flow situations have been reported in a number of papers that deal with internal and external aerodynamics, where no or minor separation occurs. However, no satisfactory results are achieved whenever major separation is involved, indicating the lack of sensitivity to adverse pressure gradient. The model tends to significantly overpredict the shear-stress levels and thereby delays (or completely prevents) separation. This exhibits a major shortcoming, which Rodi [101] attributes to the overprediction of the turbulent length-scale in the near wall region. Menter [102] pointed to another shortcoming of the $k - \varepsilon$ model which is associated with the numerical stiffness of the equations when integrated through the viscous sublayer.

9.3.5.2 Two-Equation $k - \omega$-Model

This model replaces the ε-equation with the ω-transport equation, first introduced by Kolmogorov. It combines the physical reasoning with dimensional analysis. Following the Kolmogorov hypotheses, two quantities, namely ε and κ, seem to play a central role in his turbulence research. Therefore, it seemed appropriate to establish a transport equation in terms of a variable that is associated with the smallest eddy and includes ε and κ. This might be a ratio such as $\omega = \varepsilon/\kappa$ or $\omega = \kappa/\varepsilon$. Kolmogorov postulated the following transport equation

$$\frac{\partial \omega}{\partial t} + \bar{V}_i \frac{\partial \omega}{\partial x_i} = -\beta \omega^2 + \frac{\partial}{\partial x_i}\left(\sigma \nu_t \frac{\partial \omega}{\partial x_i}\right) \tag{9.170}$$

with β and σ as the two new closure coefficients. As seen, unlike the k-equation, the right-hand-side of Eq. (9.170) does not include the production term. This equation has undergone several changes where different researchers tried to add additional terms. The most current form developed by Wilcox [103], reads

$$\frac{\partial \omega}{\partial t} + \bar{V}_i \frac{\partial \omega}{\partial x_i} = \alpha \frac{\omega}{\kappa} \tau_{ij} \frac{\partial \bar{V}_j}{\partial x_i} - \beta \omega^2 + \frac{\partial}{\partial x_i}\left(\left(\nu + \sigma \frac{k}{\omega}\right) \frac{\partial \omega}{\partial x_i}\right) + \frac{\sigma_d}{\omega} \frac{\partial k}{\partial x_i} \frac{\partial \omega}{\partial x_i} \tag{9.171}$$

with $\tau_{ij} = -\overline{V_i' V_j'}$ as the specific Reynolds stress tensor. Wilcox also modified the k-equation as

$$\frac{\mathrm{D}k}{\mathrm{D}t} = \frac{\partial}{\partial x_i}\left\{\left(\nu + \sigma^* \frac{\kappa}{\omega}\right) \frac{\partial k}{\partial x_i}\right\} - \overline{V_i' V_j'} \frac{\partial \bar{V}_i}{\partial x_j} - \beta^* k \omega. \tag{9.172}$$

He also introduced the kinematic turbulent viscosity

$$\nu_t = \frac{k}{\tilde{\omega}}, \text{ with } \tilde{\omega} = \max\left(\omega, C_{\lim} \sqrt{\frac{2 D_{ij} D_{ij}}{\beta^*}}\right), \text{ and } C_{\lim} = \frac{7}{8}. \tag{9.173}$$

With $\overline{D_{ij}} = \frac{1}{2}\left(\frac{\partial \bar{V}_i}{\partial x_j} + \frac{\partial \bar{V}_j}{\partial x_i}\right)$ as the matrix of the mean deformation tensor. Wilcox defined the following closure coefficients and auxiliary relations

$$\begin{aligned} &\alpha = \frac{13}{25}, \beta = \beta_0 f_\beta, \\ &\beta^* = \frac{9}{100}, \sigma = \frac{1}{2}, \\ &\sigma^* = \frac{3}{5}, \sigma_{do} = \frac{1}{8} \\ &\sigma_d = 0, \text{ if } \frac{\partial k}{\partial x_i} \frac{\partial \omega}{\partial x_i} \leq 0 \\ &\sigma_d = \sigma_{do}, \text{ if } \frac{\partial k}{\partial x_i} \frac{\partial \omega}{\partial x_i} \succ 0. \end{aligned} \tag{9.174}$$

Furthermore,

$$\begin{aligned} &\beta_0 = 0.0708, f_\beta = \frac{1 + 85\chi_\omega}{1 + 100\chi_\omega}, \chi_\omega \equiv \left| \frac{\Omega_{ij} \Omega_{jk} \mathrm{Ski}}{(\beta^* \omega)^3} \right|, \\ &\varepsilon = \beta^* \omega \kappa, l_m = k^{1/2}. \end{aligned} \tag{9.175}$$

The $k-\omega$ model performs significantly better under adverse pressure gradient conditions than the $k-\varepsilon$ model. Another strong point of the model is the simplicity of its formulation in the viscous sublayer. The model does not employ damping functions, and has straightforward Dirichlet boundary conditions. This leads to significant advantages in numerical stability, Menter [100]. A major shortcoming of the $k-\omega$ model is its strong dependency on freestream values. Menter investigated this problem in detail, and showed that the magnitude of the eddy-viscosity can be changed by more than 100% just by using different values for $\tilde{\omega}$.

9.3.5.3 Two-Equation $k-\omega$-SST-Model

Considering the strength and the shortcomings of $\kappa-\varepsilon$ and $\kappa-\omega$ models briefly discussed in the previous two sections, Menter [104–106] introduced a *blending function* that combines the best of the two models. He modified the Wilcox $k-\omega$ model to account for the transport effects of the principal turbulent shear-stress. The resulting *SST*-model (Sear Stress Transport model) uses a $\kappa-\omega$ formulation in the inner parts of the boundary layer down to the wall through the viscous sublayer. Thus, the $SST-k-\omega$ model can be used as a low-Re turbulence model without any extra damping functions. The SST formulation also switches to a $k-\varepsilon$ mode in the free-stream and thereby avoids the common $k-\omega$ problem that the model is too sensitive to the turbulence free-stream boundary conditions and inlet free-stream turbulence properties. For the sake of completeness, we present the Menter's SST-model in terms of ω-equation with the blending function F_1

$$\frac{\partial(\rho\omega)}{\partial t}+\frac{\partial(\rho V_i\omega)}{\partial x_i}=\frac{\alpha}{\upsilon_t\tilde{P}_k}-\beta\rho k\omega^2+\frac{\partial}{\partial x_i}\left[(\mu+\sigma_\omega\mu_t)\frac{\partial\omega}{\partial x_i}\right]+2(1-F_1)\rho\sigma_{\omega 2}\frac{1}{\omega}\frac{\partial k}{\partial x_i}\frac{\partial\omega}{\partial x_i} \tag{9.176}$$

and the turbulence Kinetic Energy

$$\frac{\partial(\rho k)}{\partial t}+\frac{\partial(\rho V_i k)}{\partial x_i}=\tilde{P}_k-\beta^*\rho k\omega+\frac{\partial}{\partial x_i}\left[(\mu+\sigma_k\mu_t)\frac{\partial k}{\partial x_i}\right]. \tag{9.177}$$

The term $\tilde{p}_k$ in Eqs. (9.176) and (9.177) is a production limiter and is defined in Eq. (9.183). The blending function F_1 is determined from

$$F_1=\tanh(arg_1^4) \tag{9.178}$$

with the argument arg_1

$$arg_1=\min\left(\max\left(\sqrt{\frac{k}{\beta^*\omega y}\frac{500\nu}{y^2\omega}}\right),\frac{4\rho\sigma_{\omega 2}k}{CD_{k\omega}y^2}\right) \tag{9.179}$$

$$CD_{k\omega} = \max\left(2\rho\sigma_{\omega 2}\frac{1}{\omega}\frac{\partial k}{\partial x_i}\frac{\partial \omega}{\partial x_i}, 10^{-10}\right) \tag{9.180}$$

and $F1$ is equal to zero away from the surface (k-model), and switches over to one inside the boundary layer (k-model). The turbulent eddy viscosity is defined as follows:

$$v_t = \frac{a_1 k}{\max(a_1 \omega S F_2)} \tag{9.181}$$

with $S = \sqrt{D_{ij} D_{ij}}$ and $F2$ as a second blending function defined by:

$$F_2 = \tanh\left\{\left[\max\left(\frac{2\sqrt{k}}{\beta^* \omega y}, \frac{500 v}{y^2 \omega}\right)\right]^2\right\}. \tag{9.182}$$

A production limiter in Eq. (9.183) is used in the SST model to prevent the build-up of turbulence in stagnation regions. It is defined as

$$\tilde{P}_k = \min(P_k, 10\beta^* \rho k \omega) \tag{9.183}$$

with

$$P_k = \mu_t \frac{\partial V_i}{\partial x_j}\left(\frac{\partial V_i}{\partial x_j} + \frac{\partial V_j}{\partial x_i}\right). \tag{9.184}$$

All constants are computed by a blend from the corresponding constants of the $k - \varepsilon$ and the $k - \omega$ model via $\alpha = \alpha_1 F + \alpha_2(1 - F)$, etc. The constants for this model are

$$\begin{aligned}
&\beta^* = 0.09, \alpha_1 = 5/9, \beta_1 = 3/40,\\
&\sigma_{k1} = 0.85, \sigma_{\omega 1} = 0.5, \alpha_2 = 0.44,\\
&\beta_2 = 0.0828, \sigma_{k2} = 1, \sigma_{\omega 2} = 0.856.
\end{aligned}$$

According to [107], the above version of the $k - \varepsilon$ and $k - \omega$ equations, including constants listed above, is the most updated version.

9.4 Grid Turbulence

Calibration of closure coefficients and a proper model assessment require accurate definition of boundary conditions for experiments as well as computation. These include, among other things, information about inlet turbulence such as the turbulence intensity, length, and time scales. This information can be provided by using *turbulence grids*, Fig. 9.13.

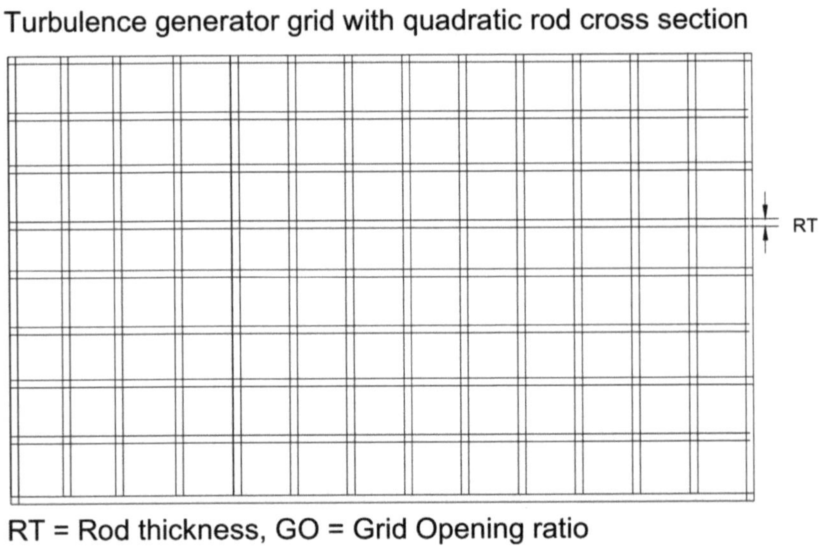

Fig. 9.13 A section of a turbulence grid with quadratic rod cross section, RT = rod thickness

Table 9.3 Turbulence grids: geometry, turbulence intensity and length scale

Turbulence grid	Grid opening GO (%)	Rod thickness RT (mm)	Turbulence intensity Tu (%)	Length scale Λ (mm)
No grid	100	0	1.9	41.3
TG1	77	6.35	3.0	32.5
TG2	55	9.52	8.0	30.1
TG3	18	12.7	13.0	23.4

The grids may consist of an array of bars with cylindrical or quadratic cross sections. The thickness of the grid bars and the grid openings determine the turbulence intensity, length, and time scale of the flow downstream of the grid. Immediately downstream of the grid, a system of discrete wakes with vortex streets are generated that interact with each other. Their turbulence energy undergoes a continuous decay process leading to an almost homogeneous turbulence. The grid is positioned at a certain distance upstream of the test section in such a way that it generates homogeneous turbulence. Table 9.3 shows the data of three different turbulence grids for producing inlet turbulence intensities Tu = 3.0, 8.0, and 13.0%. The grids consist of square shaped aluminum rods with the thickness RT and opening GO. The turbulence quantities were measured at the test section inlet with a distance of 130 mm from the grid.

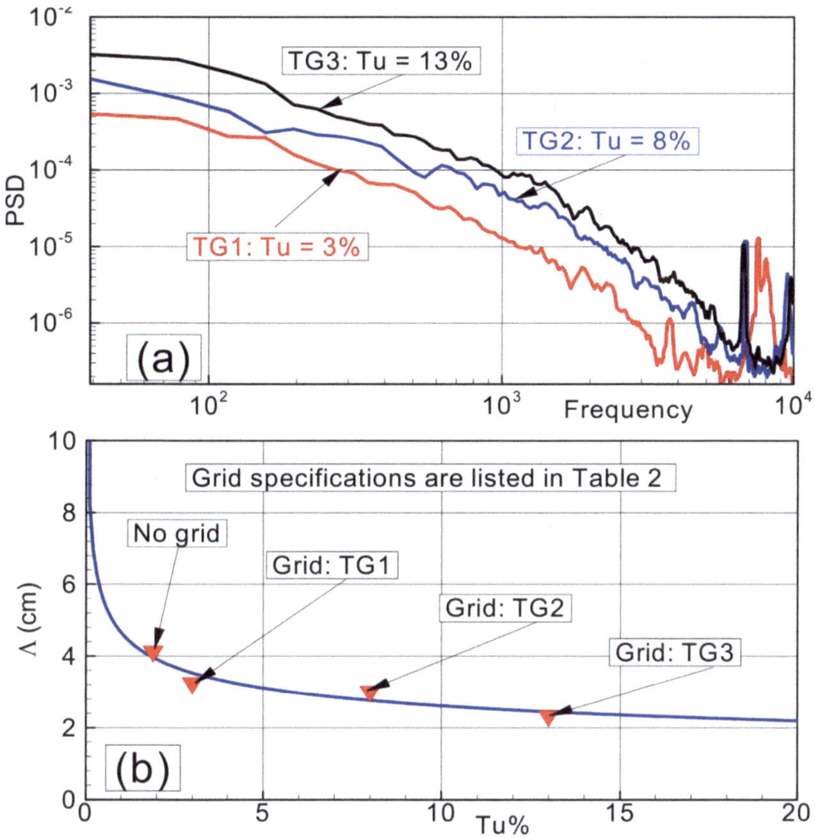

Fig. 9.14 **a** Power spectral distribution PSD as a function of frequency for three different grids described in Table 9.2. The results from **a** is used to generate the turbulence length scales as a function of turbulence intensity (**b**)

Figure 9.14a shows the power spectral density of the velocity signals from a hot wire sensor as a function of signal frequency. The length scale is calculated from $\Lambda = \bar{V} E_{(f=0)}/v_{rms}^2$ [mm], Fig. 9.14b.

9.5 Numerical Simulation Examples

In this section flow situations with engineering relevance are presented that demonstrate the application of the $k - \omega$ and standard $k - \varepsilon$ to steady and unsteady flow cases. While the steady flow simulation cases deal with flow in stationary frame of reference, the unsteady one simulate the flow in a rotating frame.

Fig. 9.15 Simulation with $k - \varepsilon$ model

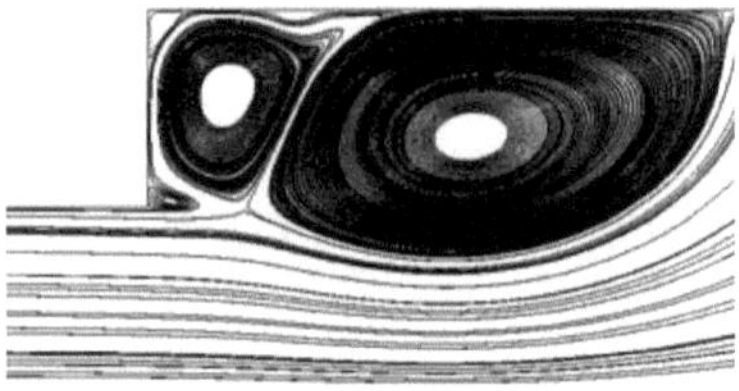

Fig. 9.16 Simulation with $k - \omega$ model

9.5.1 Examples of Steady Flow Simulations with Two-Equation Models

Internal Flow, Sudden Expansion: The following representative examples should illustrate the substantial differences between the two-equation models we presented above. The flow through a sudden expansion is appropriate for comparison purposes for two reasons: (1) It has a flow separation associated with a circulation zone and (2) it is very easy to obtain experimental data from this channel.

Figures 9.15 and 9.16 show flow simulations through a channel with a sudden expansion ratio of 2/1 using $k - \omega$ and standard $k - \varepsilon$ models. The purpose was to simulate the flow separation. The basic features of this flow can be briefly summarized as follows: The oncoming flow through the inlet section enters the sudden expansion, it separates and forms a free shear layer, which grows in flow direction. Closed to the sudden expansion wall a recirculation region is formed which contains a system of vortices. Durst et al. [108] presented a detailed velocity measurements for the flow through a plane nominally two-dimensional duct with a symmetric sudden expansion of area ratio 1:2. Their experiments show one long and one short separation zone.

The $k - \varepsilon$ simulation, Fig. 9.15 delivers a single large corner vortex, while the $k - \omega$ simulation shown in Fig. 9.16 exhibits a system of two vortices with the smaller vortex occupying the corner region juxtaposing the larger vortex. In context of a an accurate numerical prediction, it is essential to obtain flow details. These require accurately predicting the correct size and positions of the vortices, separation and re-attachment. Without a proper transition model, accurately predicting the above quantities is not possible.

Internal Flow, Turbine Cascade: Flow simulation with CFD has a wide application in aerodynamic design of turbines, compressors, gas turbine inlet nozzles and exit diffusers. As an example, Fig. 9.17 shows contour plots of velocity and pressure distributions in a high efficiency turbine blade using SST-turbulence model. On the

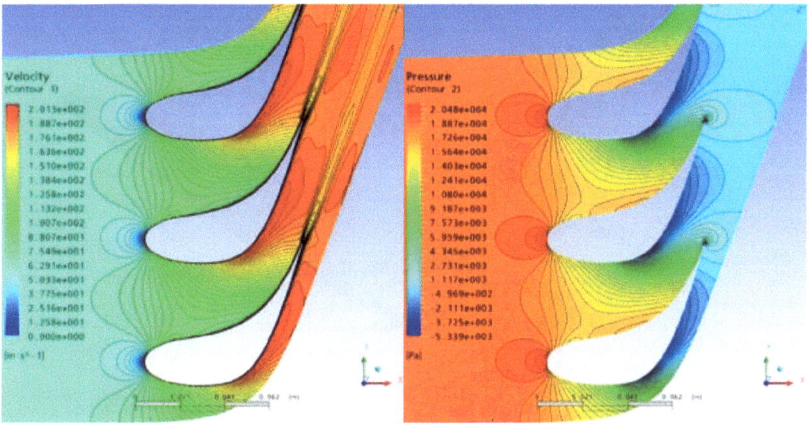

Fig. 9.17 Flow simulation through a turbine cascade, TPFL-design

convex surface (suction surface), the flow is initially accelerated at a slower rate from the leading edge and exits the channel at a higher velocity close to the trailing edge. The acceleration process is reflected in pressure contour. The cascade was designed to have high efficiency without separation. Thus, the flow simulation using either turbulence models delivered satisfactory results. For cascades with separation, neither $k - \varepsilon$ nor $k - \omega$ can provide satisfactory results. However, the combination of both in form of SST-model in conjunction with a working transition model delivers results that, to some extent, resemble experimental findings. This case is discussed in Chap. 11.

External Flow, Lift-Drag Polar Diagram: This example presents two test cases to predict the lift-drag polar diagram of an aircraft without and with engine integration, Fig. 9.18.

Figure 9.19 shows the predicted lift-drag *polar diagram* for the geometries presented in Fig. 9.18. The computation was performed using the SST-turbulence model and the results compared with the experiments. The lift and drag coefficients plotted in Fig. 9.19 are integral quantities that represents the lift and drag forces acting on the entire aircraft. The polar diagram is obtained by varying the angle of attack and measuring and computing the lift and drag forces. These forces are then non-dimensionalized with respect to a constant reference force, which is a product of a constant dynamic pressure and a characteristic area of the aircraft. Once a complete set of data for a given range of angle of attack is generated, then for each angle the lift coefficient is plotted against the drag coefficient as shown in Fig. 9.19.

Fig. 9.18 Geometry with engine integration for predicting the polar diagram, from [32]

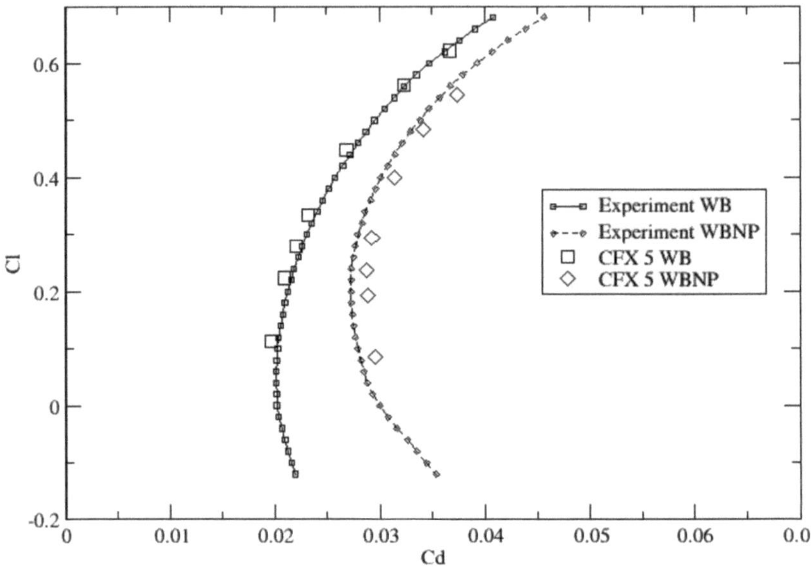

Fig. 9.19 Lift-drag polar diagram for an aircraft model without engine (WB) and with engines (WVBN), from [31]

9.5.2 Case Study: Flow Simulation in a Rotating Turbine

The above examples were dealing with relatively simple flow cases, where a flow simulation using $k - \varepsilon$, $k - \omega$ or the combination thereof render satisfactory qualitative results. However, simulating a complex flow situation, where different parameters are involved, leads to results, whose proper interpretations require the understanding of the model shortcomings. As an example, we discuss in detail the flow through a high pressure (HP) turbine (readers interested in turbomachinery aerodynamics and design are referred to a new textbook by Schobeiri [109]). The parameters involved in this case are, among others, change of frame of reference from stationary to rotating frame, unsteady inlet flow condition, impingement of periodic wakes from preceding rotating blade rows on the succeeding ones, changes of pressure gradient, presence of secondary vortices and laminar-turbulent transition. The case study presented in the following deals with the numerical simulation of the aerodynamic behavior of a two-stage high pressure (HP) turbine. HP-turbine blades have relatively small aspect ratios which develop major secondary flow regions at the hub and tip. For the shrouded HP-blades, close to the hub and casing, the fluid particles move from the pressure to the suction side and generate a system of vortices. In unshrouded blades, a system of tip clearance vortices are added to the tip endwall vortices. These vortices induce drag forces that are the cause of the secondary flow losses. Focusing on the secondary flow loss mechanisms, the fluid particles within the endwall boundary layers are exposed to a pitchwise pressure gradient in the blade channel. In addition, their interaction with the main flow causes angle deviation inside and outside the blade channel, resulting in additional losses. Thus, in HP-turbines with small aspect ratios, the secondary flow loss of almost 40–50% is the major loss contributor. The degree of accuracy in simulating the complex flow physics of the unsteady, transitional and highly vortical flow of a HP-turbine is an indicator of the predictive capability of any computational method including the steady Reynolds averaged Navier–Stokes (RANS) and unsteady Reynolds averaged Navier–Stokes (URANS). Numerous papers published over the past three decades show that, there are still substantial differences between the experimental and the numerical results pertaining to the individual flow quantities. These differences are integrally noticeable in terms of major discrepancies in aerodynamic losses, efficiency and performance of turbomachines. As a consequence, engine manufacturers are compelled to frequently calibrate their simulation package by performing a series of experiments before issuing efficiency and performance guaranties. In the following an attempt is made at identifying the quantities, whose simulation inaccuracies are preeminently responsible for the aforementioned differences. This task requires (a) a meticulous experimental investigation of all individual thermo-fluid quantities and their interactions resulting in an integral behavior of the turbomachine in terms of efficiency and performance, (b) a detailed numerical investigation using appropriate grid densities based on simulation sensitivity, and (c) steady and unsteady simulations to ensure their impact on the final numerical results. To perform the above experimental and numerical tasks, a two-stage high pressure axial turbine rotor has been designed

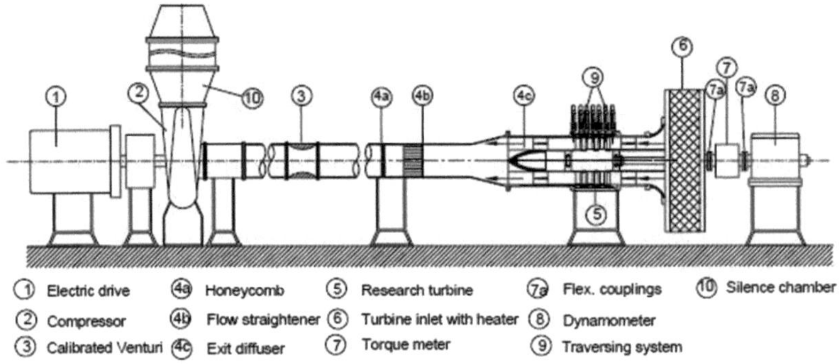

Fig. 9.20 The *TPFL* research turbine facility with its components, the circumferential traversing system (9) is driven by another traversing system sitting on a frame and is perpendicular to this plane

for generating benchmark data and to compare with the numerical results. Detailed interstage radial and circumferential traversing presents a complete flow picture of the second stage. Performance measurements were carried out for design and off-design rotational speed. For comparison with numerical simulation, the turbine was numerically modeled using a commercial code. An extensive mesh sensitivity study was performed to achieve a grid-independent accuracy for both steady and transient analysis. In discussing the results, a close look is taken at the physics behind the models. This helps to better understand the deficiency of the existing ε-equation and thus the associated ω-equation and explains, why these models are not capable of a priory predicting any arbitrary flow situation and require calibration for each single flow case.

Experimental Research Facility: The facility shown in Figs. 9.20 and 9.21 is described in [110–112]. The generation is a two-stage turbine shown in Fig. 9.21. Particular attention was paid to the stator blade design which has a symmetric compound lean, Fig. 9.22. The rotor blades, Fig. 9.23, however, are fully three-dimensional and are tapered from pub to tip. The turbine rotor pertaining to this investigation includes two stages as described below.

Turbine Blading: The turbine rotor specifically designed for the benchmark data Interstage flow measurements were conducted at stations 3, 4 and 5, Fig. 9.21. For this purpose, three five-hole probes were utilized using a non-nulling calibration method. Traversing the flow field in radial and circumferential direction was accomplished by a seven-axes traversing system with step motor controller with encoder and decoder on each axis.

Numerical Treatment: Starting with the mesh generation, numerical simulations are conducted using Reynolds averaged Navier–Stokes (RANS) simulation first. As seen in the following, the comparison of the numerical results with the experimental ones reveals discrepancies in quantities such as velocities, pressures, temperatures and particularly row loss coefficients and turbine efficiency. To find the cause of

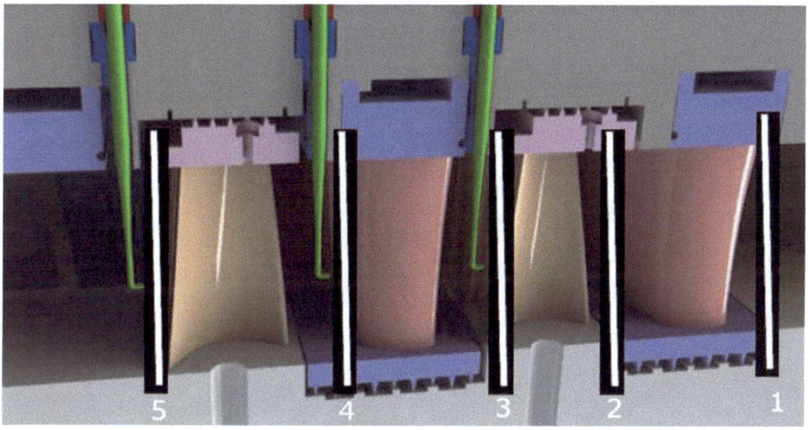

Fig. 9.21 Positions of three five-hole probes for radial and circumferential traverse of flow field. The probes are positioned in such a way that their wakes do not interact with each other

Fig. 9.22 Stator blade

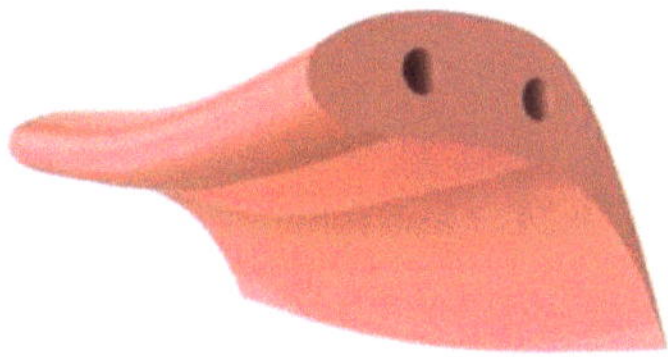

Fig. 9.23 Rotor blade

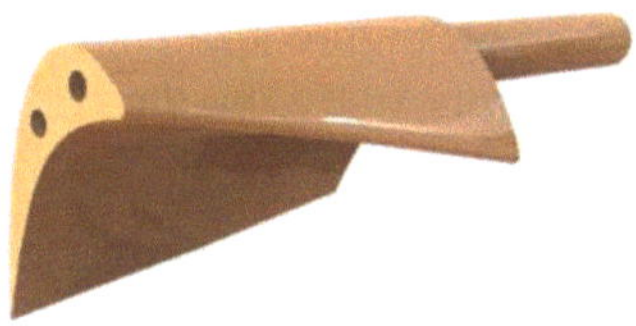

these discrepancies, before expediting the physics behind the turbulence and transition models that may cause the discrepancies, a rather time consuming unsteady simulation using unsteady RANS, the URANS was conduted. The results of both simulations are presented in the section Results and Discussions. The flow solver used in this study is a commercially available finite volume solver [113] with the SST-turbulence model publically known. Both simulation cases were run on the Texas A&M university supercomputing facility.

Mesh Specifications: Stator and seal domains were meshed using a multi-block meshing technique to produce a structured hexahedral mesh. During the meshing process the blade surfaces were surrounded with an O-grid to create a boundary layer mesh, which used 16 node points normal to the wall. Five mesh densities were created for mesh sensitivity study that requires a y^+ values close to 1. The results of several sensitivity studies showed that a mesh with the density of 4.8×10^6 can be

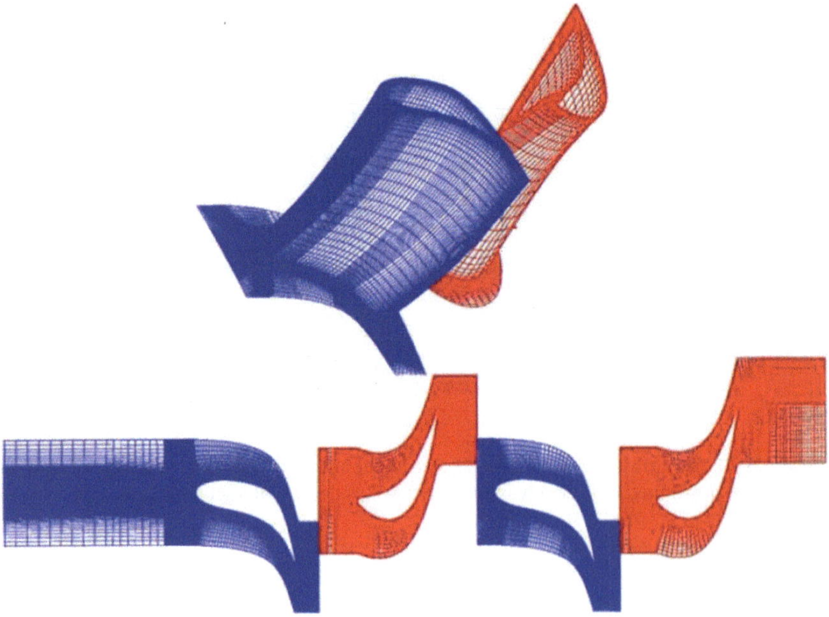

Fig. 9.24 Mesh for numerical analysis

deemed a good compromise between resource requirements and the required results accuracy. Figure 9.24 shows one of the numerical grids that were used for the mesh sensitivity study and the subsequent numerical analyses.

Boundary Conditions: The boundary conditions used for CFD-simulation were taken from turbine rig experiments conducted at various rotational speed, including 1800 and 3000 rpm. A list of all the boundary conditions used for the two simulation cases are shown in Table 9.4. The global machine inlet and exit stations are labeled 1 and 5 respectively. Referring to Fig. 9.21, Stations 3, 4 and 5 are the interstage measurement stations between the rotor exits and the stator inlets.

Simulations: In the following, a RANS and a URANS simulation procedure are presented. For the RANS simulation a specific row-to-row interface model is used which spatially averages velocity and pressure data from one row and passes this onto to the next row as boundary conditions. In the case of URANS simulation, the flow field is assumed to be unsteady and, therefore the rotor row is actually rotated when compared to the stator rows via the use of a sliding mesh interface. The sliding mesh allows for a full model of the interface between each row, without the need of any kind of averaging. This provides a full time-dependent solution and accounts for various interactions between stator and rotor rows. The main disadvantage of using a transient analysis are the greater computing resources required.

RANS: Steady State Simulation: For steady state cases, each turbine row was defined as a separate domain within the solver workspace, and specific interfaces

Table 9.4 Boundary conditions for three RPMs a pressure ratio of 1.38

Boundary condition	1800 RPM	2600 RPM	3000 RPM
Reference pressure (kPa)	101.328	101.382	101.382
Total pressure inlet (kPa)	100.744	100.826	100.357
Static pressure outlet (kPa)	70.0	69.7	69.4
Total temperature inlet (K)	317.2	318.6	318.3
Total temperature outlet (K)	293.4	291.4	290.7
Inlet turbulence intensity (%)	5	5	5

were defined for the seals' entrance and exit points for both the rotor and stator stages. Between the stator and rotor domains a "Stage" interface was used which performs a circumferential average of the fluxes through bands on the interface between the stator and rotor, this is the mixing-plane method introduced by [114]. Using this interface model each fluid zone (blade row) is solved and two adjacent frames are coupled by exchanging flow variables at the prescribed interface location. The spatial averaging at the interfaces removes any unsteady effects that would arise due to changes from one blade row to the next blade row, example of such effects include, wakes, shock waves and separated flow. This method allows for a steady-state calculations of multistage turbomachinery flows. In addition, circumferential periodic boundary conditions are placed on the upper and lower surfaces of both the stator and rotor domains. The use of the stage interface condition allows for unequal blade numbering between stator and rotor domains, in this case the stator blade number is 66 and the rotor blade number is 63 as listed in Table 9.4, which translates to a ratio of 1.047 at the stator to rotor interface which is lower than the 1.2.

Each simulation case was iterated until the global root mean square error residuals for the RANS-equations reached values below 1×10^{-6}. Convergence was achieved when the residuals' target value was reached, and stability was observed in the monitored variables. For extreme off-design cases with low expansion pressure ratios close to 1.15 and low shaft speeds $\omega/\omega_{\mathrm{D}}$ close to 60%, a 30% reduction in the false-time step was required, which in turn increased the computer wall time to achieve convergence. All cases used a second order spatial discretization for the velocity components V_{i}, the pressures, the energy and turbulence equations k and ω.

URANS Simulation: For the URANS-unsteady analysis, the stage interfaces were changed to a fully unsteady-stator-rotor interface which creates a sliding mesh interface between the stator and rotor patches. Using sliding mesh, URANS accounts for the unsteady flow interaction effects at the frame change locations. Simulation with URANS allows capturing the time dependent interactions between the stator and rotor rows, particularly the wake interaction with the subsequent row. This interaction, however, affects the boundary layer development, onset of the boundary transition process, separation and its re-attachment as evidenced by a series of studies published by Schobeiri and his co-workers [115–124]. A RANS-simulation is based on the steady-state mixing plane approach, where the time dependent interactions are averaged out. In contrast, the URANS unsteady simulation provides a means of simulating time-dependent stator rotor interactions. Considering the unsteady results from the above studies, conducting a URANS-simulation was expected to deliver more accurate results than RANS.

Turbulence Model: The turbulence model used in this study was the Shear Stress Transport model (SST) treated in Sect. 9.3.5. As mentioned, this model combines $k - \varepsilon$ and $k - \omega$ models by introducing a blending function with the objective to get the best out of these two models. In addition to the turbulence model a transition model was also employed.

9.5.3 Results, Discussion

Interstage Flow Traversing, Total, Static Pressure Distributions: Interstage measurements were carried at stations 3, 4 and 5, Fig. 9.25, with a rotational speed of 3000 rpm and the results were compared with RANS and URANS simulation. At the outset, it should be emphasized that the experiments were repeated at least two to three times to ensure the reproducibility of the entire data set. For numerical simulation, the boundary conditions were taken from the experiment as it was without any modification. Total and static pressure distributions are shown in Fig. 9.25 as a function of the dimensionless immersion $R = (r - r_h)/(r_t - r_h)$. In contrast to many papers that use dimensionless quantities, whenever experimental and numerical results are compared, all quantities, with the exception of a few, presented in this study are dimensional. This allows a direct quantitative assessment of the numerical results, when compared with the experimental ones. For the same reason, presentation of contour plots is limited to a few ones, just to show the qualitative differences.

Total Pressure: Measurement, Numerical Simulations: As Fig. 9.25 top shows, the measured total pressures up- and downstream of the stator (Station 3 and 4, ▼, ♦) differ from each other only slightly confirming the positive effect of the circumferential lean of the stator blades on reducing the secondary flow zones at the hub and tip. The comparison of the total pressure measured at the stator inlet (▼), station 3, with the steady RANS-results (—) shows a substantial difference of more than 3 kPa. Downstream of the stator at station 4, the difference between the measured and RANS-steady results (♦, —), is more than 3.5 kPa. Considering the pressure uncertainty of 0.19%, which amounts to 35.1 Pa, the above differences are more than 100 times the measurements uncertainty. The comparison of the measured and the RANS-results downstream of the rotor row (station 5) shows similar differences. The extraction of mechanical energy from the rotor causes a reduction in total pressures (top) and static pressure (bottom) at stations 3, 4 and 5 (▼ , ♦, ▲) with the RANS and URANS computational results, (solid, dashed curves, red, black, blue) pressure as shown in Fig. 9.25 (station 5, ▲). Compared to the RANS-simulation (—) a difference of more than 3.4 kPa is observed. The above comparison shows that we are dealing with significant quantitative differences between the experiment and RANS-simulation. Differences are also observed in course of both distributions particularly at the exit of the stator, rotor rows and close to the hub and tip. While the trend prediction outside the hub and tip regions is satisfactory, major differences are observed in those regions as illustrated in Fig. 9.25. However, from boundary layer and secondary flow development point of view, its is vital for any CFD-simulation to predict the same trend at above regions. Considering the above, we are dealing with two simulation deficiencies: (1) a substantial quantitative underprediction of total pressures and (2) an inaccurate prediction of the total pressure trend close to the hub and tip. Being aware of the RANS-simulation deficiencies, the differences encountered above were of much higher magnitude that we expected. To expedite the cause of these differences, we resorted to URANS simulation, whose results are also presented in Fig. 9.25 as dashed curves. After performing the URANS calculations, the data were averaged at any single point from hub to tip in radial, circumferential and axial direction. The purpose of this procedure was to account for the stator-rotor wake interaction and to preserve integrally the impact of the wakes on the boundary layer of the subsequent blade rows. As seen, in case of the stator inlet, Station 3, the URNAS simulation has reduced the total pressure differences from 3 to 2.3 kPa which exhibits a marginal improvement only. At the stator exit, station 4, which is identical with the rotor inlet station, the total pressure difference has reduced from 3.5 to 2.84 kPa which is again a marginal improvement. The total pressure difference between the stator inlet and exit calculated by URANS is almost 1 kPa which results in a significant total pressure loss coefficient. The rotor row URANS-simulation presents a slightly different picture. The pressure difference has been reduced substantially from 3.4 to 1.1 kPa. Although URANS has reduced the differences, particularly in the rotor exit, the remaining differences are so large that cannot be accepted as satisfactory. Furthermore, despite the improvement in rotor exit total pressure, because of differences at the stator exit (rotor inlet), the rotor total pressure loss coefficient is significantly higher than the measured one.

Static Pressure: Measurement, Numerical Simulations: The static pressure as the difference between the total and dynamic pressure reveals much larger differences between the experiment and RANS simulation as shown in Fig. 9.25 bottom. Here, as in the above case, the calculated static pressures at the inlet and exit of the stator row are substantially underpredicted. Similar to the total pressure discussed above, URANS simulation brings an improvement only at the exit of the rotor, while the pressure difference at the inlet has become much larger than the total pressure difference discussed above. The increase in static pressure differences is attributed to

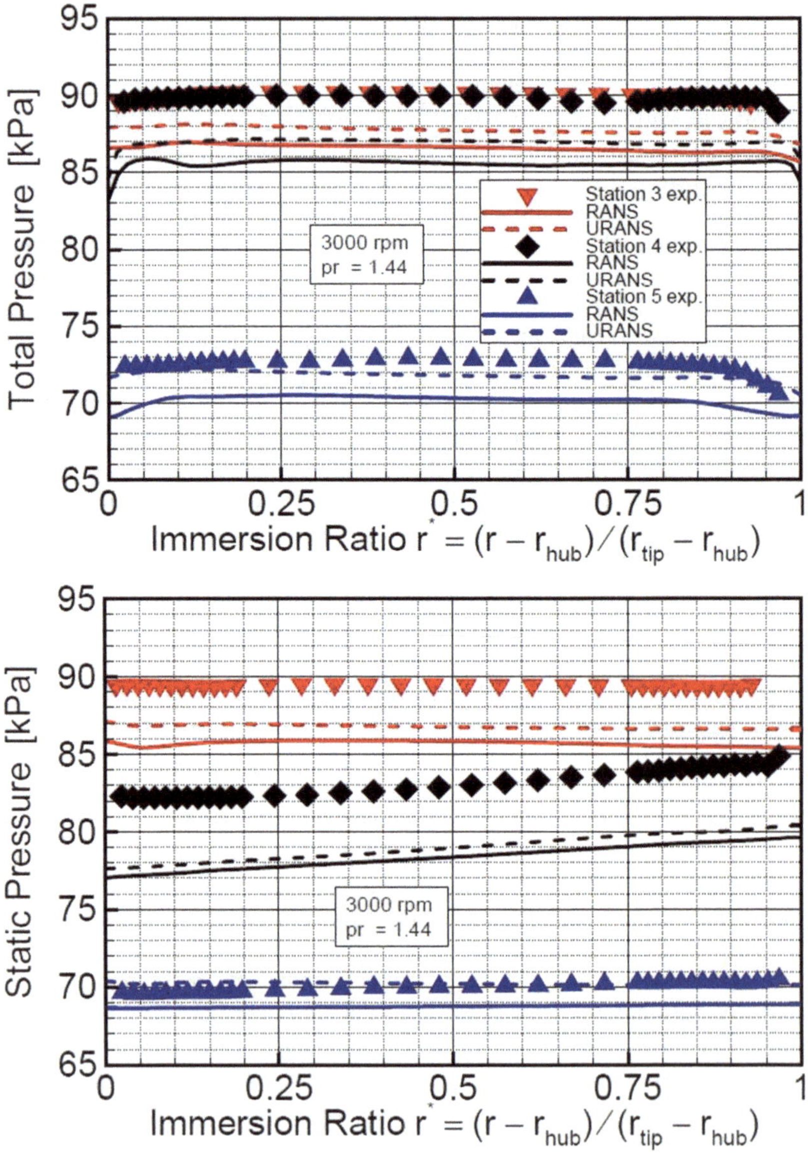

Fig. 9.25 Comparison of the measured total pressures (top) and static

an inaccurate calculation of the velocity distribution within the turbine blade flow path. In this context, an accurate prediction of the boundary layer development and transition is the prerequisite for accurately calculating the velocity components and thus the velocity vector that constitutes the dynamic pressure. This issue will be treated later in this section.

Absolute and Relative Velocities: Measurement, Numerical Simulations: The absolute velocity is the vector sum of the axial, radial and circumferential components, measured by the five hole probes locates at station 3, 4 and 5 and computed by the Reynolds averaged Navier–Stokes equations. Measured absolute velocities at station 3, 4, and 5 are shown in Fig. 9.26 (top) and compared with the RANS and URANS simulation results. At the stator inlet, Station 3, the experimental results differ substantially from RANS and URANS-results, particularly at the hub and tip regions, where the secondary flow is predominant. While within the immersion range of 0.25–0.85, the numerical simulations follow the experimental trend, they differ substantially at the hub and tip secondary flow regions. The velocity at the stator exit, Station 4, exhibits a similar picture. As Fig. 9.26 (top) shows, substantial differences up to 20 m/s exist at the rotor exit, Station 5. Here, as in case of the stator, the velocity distribution close to the hub and tip is neither predicted quantitatively nor qualitatively by both RANS and URANS. Interestingly, while the velocity up and downstream of the stator are overpredicted, the one downstream of the rotor is substantially underpredicted. The relative velocity is obtained by subtracting the constant rotational velocity vector from the absolute velocity, Fig. 9.26 (bottom). Thus, the results are directly related to each other and the differences are the same.

Absolute, Relative and Meridional Flow Angles: Measurement, Numerical Simulations: Before discussing the angle calculation results, it should be emphasized that the degree of accuracy of angle calculation is instrumental in determining the blade loss coefficient. The inaccuracy in angle calculation leads to inaccurate incidence angles and loss calculation. In this regard, the blade geometry, particularly its leading edge radius plays a crucial role. While the stator blades with a relatively large leading edge radius, Fig. 9.22, are less sensitive to flow incidence changes, the rotor blades, Fig. 9.23 have a much smaller leading edge radius causing higher sensitivity to incidence changes resulting in higher total pressure losses.

Figure 9.27 shows the absolute flow angle α (top), relative angle β (middle) and the meridional flow angle γ with the angle convention shown in the figure. The results of RANS and URANS calculations are compared with the measurements. At station 3, upstream of stator 2, both RANS and URANS results follow the experimental trend for $r* = 0.15$–0.75 with a $\Delta\alpha$ of almost 9°. Much larger differences are exhibited for station 5, downstream of the second rotor. At station 4, downstream of the second stator, the absolute flow angle α, shows small differences because of the diagram α-scale. However reducing the α-range from (0–175°) to (5–25°), reveals at the hub and tip differences of $\Delta\alpha$ of more than 5°. Similar pattern, however with smaller differences with a maximum $\Delta\beta$ of almost 5° are observed for relative flow angles, Fig. 9.27 (middle). Trend wise URANS delivers a quasi similar trend as the experiment, however with a displaced position of minima and maxima close to the

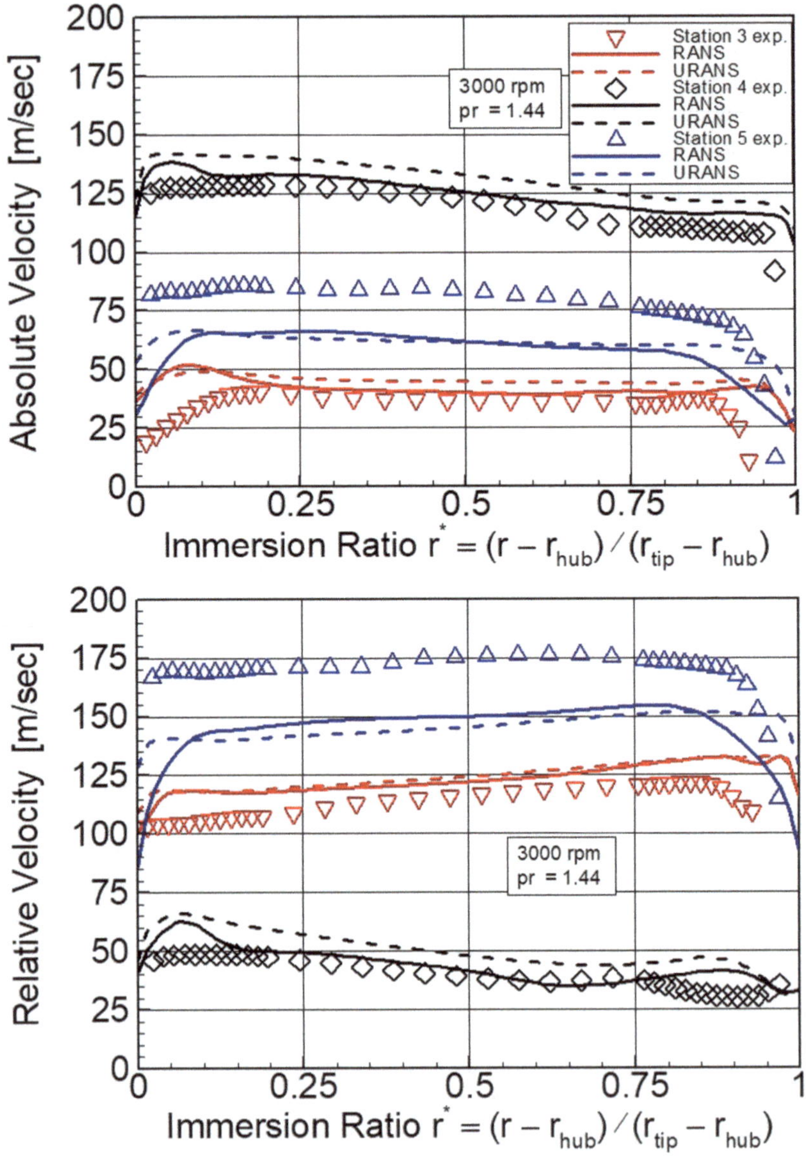

Fig. 9.26 Comparison of the measured absolute (top) and relative velocities (bottom) with the RANS and URANS computational results, (solid, dashed curves, red, black, blue)

hub and tip. This is also true for the meridional angle γ. These differences affect the calculation of the losses as discussed below.

Total Pressure Loss Coefficients: The interstage traversing provides the entire flow quantities, from which the absolute and relative total pressures and thus the total pressure loss coefficient for the stator and the rotor are determined. For this purpose, first the traversed quantities are averaged in the circumferential direction. The averaged quantities must be consistent with conservation laws of thermodynamics and fluid mechanics. The significance of an appropriate averaging technique and its impact on the results has been very well known in the turbomachinery community for more than four decades and does not need to be reviewed here (see extensive review by Traupel [125]). A simple *consistent averaging technique* that yields consistent results at a plane under inhomogeneous flow conditions was introduced by Dzung [126]. This averaging method is considered as a standard averaging technique and is used on routine basis by many researchers. For a given control volume, the averaging technique by Dzung yields integral quantities that are inherently consistent with the conservation laws. Using the averaged quantities, the total pressure loss coefficient for stator and rotor is defined as:

$$\zeta_{\text{Stator}} = \frac{P_3 - P_4}{P_3 - p_4}, \ \zeta_{\text{Rotor}} = \frac{P_{4r} - P_{5r}}{P_{4r} - p_5} \tag{9.185}$$

with P_3, P_4 as the absolute total pressure up- and downstream of the stator, Stations 3 and 4, $P_{4r} - P_{5r}$ as the relative total pressure up- and downstream of the rotor, and p_4, p_5 as the static pressure at station 4 and 5.

Figure 9.28 (top) and its enlarged portion (middle) show the stator and rotor loss coefficients. As seen, the experimental results for the stator row are below $\zeta = 0.1$. Close to the hub and tip, where usually increased secondary flows dictate the flow pattern, the loss coefficient is characterized by zones of reduced losses. This reduction is due to the re-configuration of the blades by introducing a symmetric circumferential lean at the hub and tip. As Fig. 9.28 (middle) shows, neither RANS nor URANS predict the stator loss coefficient accurately. Significant differences in loss coefficient with an order of magnitude $\Delta\zeta \approx 0.1$ for RANS and URANS show that both simulation methods cannot even approximately determine the loss coefficient. For the rotor, the results of RANS-simulation exhibits much higher differences. While RANS significantly overpredict the rotor loss coefficient, URANS overpredict the losses within 50% of the blade height from $r* = 0.25$ to $r* = 0.75$ and from $r* = 0.75$ onward it does underpredict the losses.

Efficiency Measurement, Prediction. The most important parameter for evaluating the performance of any power generating or power consuming engine is the efficiency. It integrally reflects all total pressure losses and thus the entropy changes caused by individual components involved in that engine. In case of the current two-stage turbine, two stators and two rotors are involved. The interstage measurement presented above provides the flow quantities of the second stage only. The performance measurement, however, delivers the efficiency of the entire turbine including

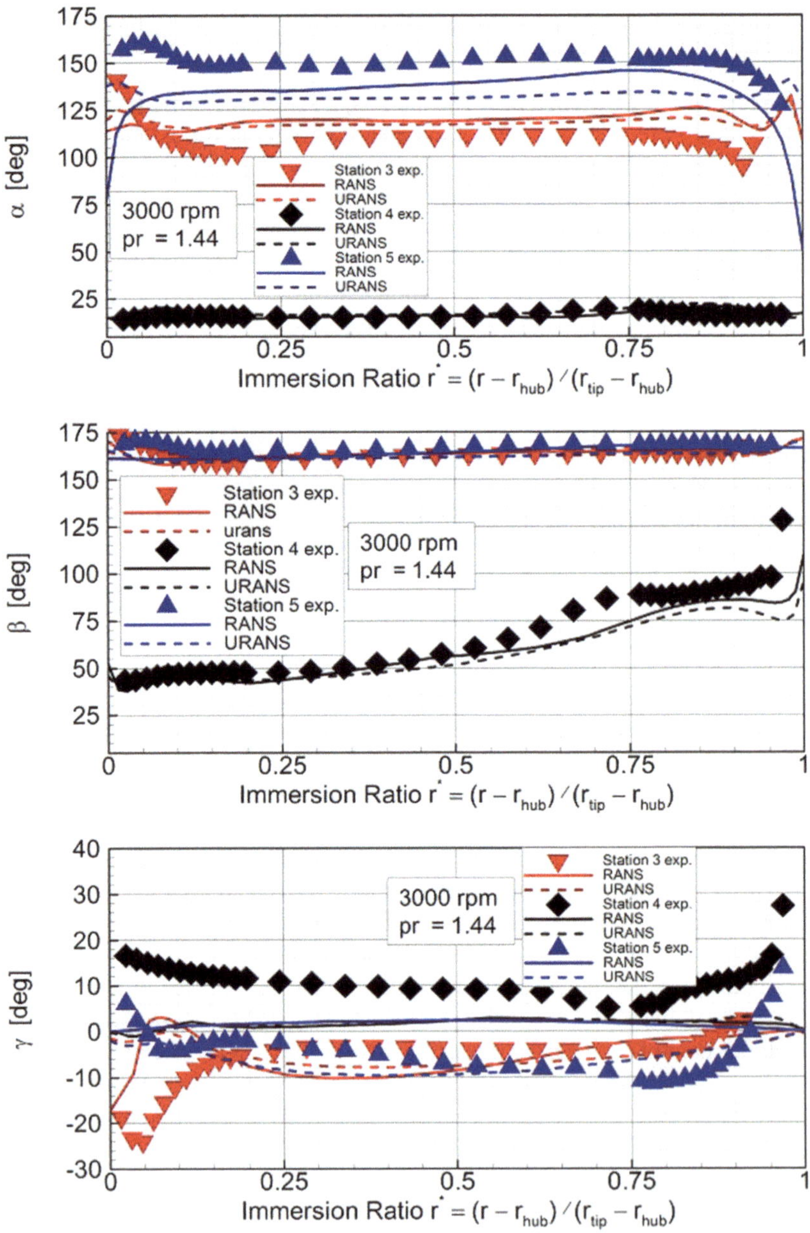

Fig. 9.27 Comparison of the measured absolute (top) and relative flow angles α and β and the meridional angle λ with the RANS and URANS computational results, (solid, dashed curves, red, black, blue)

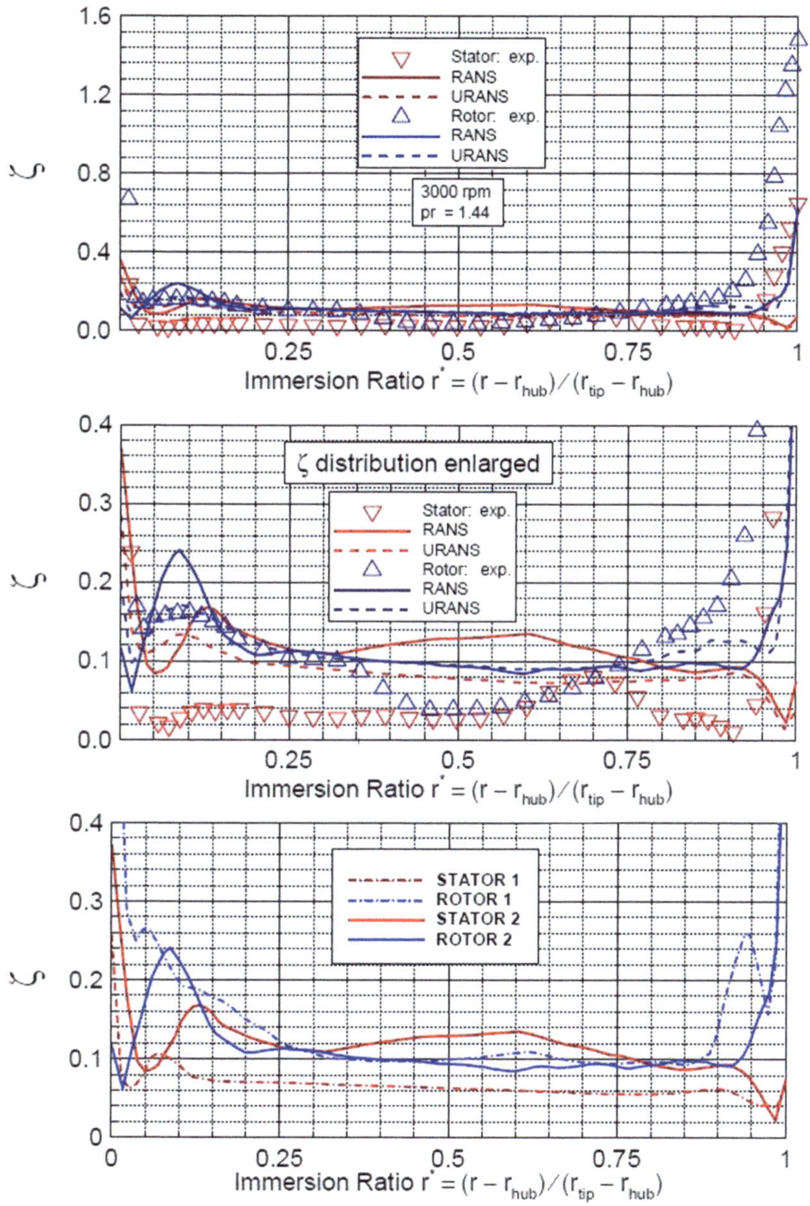

Fig. 9.28 Comparison of the measured stator and rotor loss coefficients (top) and their enlargement (middle) with RANS and URANS computational results. Bottom: RANS calculations of stator 1, 2, and rotor 1, 2 loss coefficients

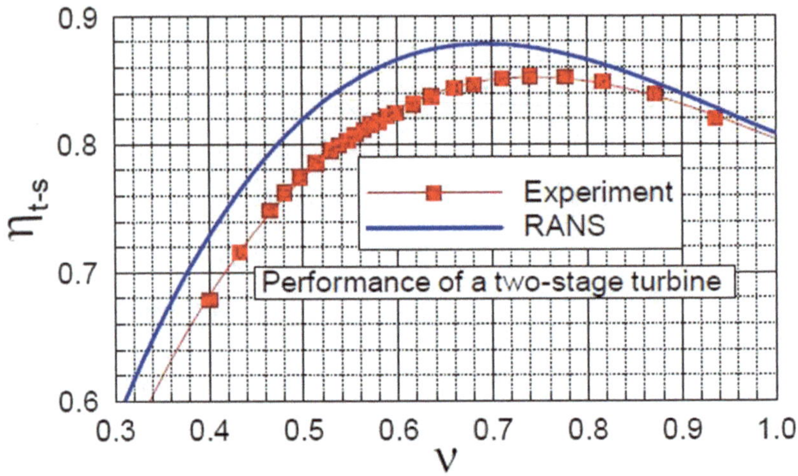

Fig. 9.29 Total to static efficiency of the two-stage turbine as a function of the dimensionless circumferential velocity

the contributions of the first stator and first rotor. For the numerical efficiency prediction of the entire turbine, it is essential to have the knowledge of the loss behavior of the first stator and first rotor row. This is shown in Fig. 9.28 (bottom), where the stator and rotor loss coefficients are calculated using RANS simulation. As seen, stator 1 has a substantially lower loss coefficient than stator 2. This is attributed to a very uniform velocity distribution at the turbine inlet, where stator 1 is located and there is no wake upstream of the first stator. The rotor 2, however has a slightly higher loss coefficient than rotor 1. This situation is reflected in performance curve, Fig. 9.29.

It represents the results of the performance experiments, where the total-to-static efficiency η_{t-s} is plotted versus the velocity ratio ν defined as

$$\eta_{t-s} = \frac{H_{\text{in}} - H_{\text{out}}}{H_{\text{in}-h_{\text{out}}}}, \; \nu = \frac{u}{\sqrt{2\Delta h_s}} \tag{9.186}$$

with H and h as the total and static enthalpy, respectively. Using RANS only, at the maximum points it exhibits an efficiency difference of 2.5%. Given the number of points calculated, this difference might be slightly smaller using URANS, which would have taken significantly longer computation time. Figure 9.29 integrally summarizes the deficiencies in calculating the individual flow quantities discussed above. As seen, the efficiency distribution over a large range of ν shows a strong overprediction that, at first glance, contradicts the loss calculation results in Fig. 9.28 (middle). This contradiction is explained by the fact that the strong underprediction of the first stator losses has shifted the entire efficiency prediction to higher level.

9.5.4 RANS-Shortcomings, Closing Remark

The multitude of the closure constants in the above discussed turbulence models have been calibrated using different experimental data. Since the geometry, Re-number, Mach number, pressure gradient, boundary layer transition and many more flow parameters differ from case to case, the constants may require new calibrations. The question that arises is this: can any of the models discussed above a priori predict an arbitrary flow situation? The answer is a clear no. Because all turbulence models are of purely empirical nature with closure constants that are not universal and require adjustments whenever one deals with a completely new case. As we saw, in implementing the exact equations for k and e that constitute the basis for $k - \varepsilon$ as well as $k - \omega$ model, major modifications had to be performed. Actually, in the case of ε-equation, the exact equation is surgically modified beyond recognition. Under this circumstance, none of the existing turbulence models can be regarded as universal. Considering this situation, however, satisfactory results can be obtained if the closure constants are calibrated for certain groups of flow situations. Following this procedure, numerous papers show quantitatively excellent results for groups of flow cases that have certain commonalities. Examples are flow cases at moderate pressure gradients and simple geometries. More complicated cases where the sign of the pressure gradient changes, flow separation and re-attachment occur and boundary layer transition plays a significant role still not adequately predicted.

The models presented above are just a few among many models published in the past three decades and summarized in [88]. In selecting these models, efforts have been made to present those that have been improved over the last three decades and have a longer lasting prospect of survival before the full implementation of DNS that makes the use of turbulence models unnecessary.

9.6 Problems and Projects

Problem 9.1: Given a second order tensor $\Phi = e_i e_j \phi_{ij} = e_i e_j V_i V_j$ with nine components, show that of these nine components only six are distinct. Also given is a third order tensor $\Psi = e_i e_j e_k \psi_{ijk} = e_i e_j e_k V_i V_j V_k$ that has 27 components, show that of these 27 components only 10 are distinct.

Problem 9.2: Using the dissipation Eq. (9.31) for a fully isotropic turbulence flow field verify that $\varepsilon = \nu \left(\overline{\frac{\partial V_i'}{\partial x_j} \frac{\partial V_i'}{\partial x_j}} \right) = 15\nu \overline{\left(\frac{\partial V_1'}{\partial x_1} \right)^2}$.

Problem 9.3: For a fully isotropic turbulence field the dissipation is given by Eq. (9.31) $\varepsilon = 15\nu \left(\frac{\overline{V_1'}}{\lambda} \right)^2$, the Kolmogorov time scale by Eq. (9.33) $\tau_K = \left(\frac{\nu}{\varepsilon} \right)^{1/2}$ and the Taylor micro time scale by Eq. (9.2) $\tau_T = \lambda / V_1$ show that these time scales are a related by the turbulence intensity.

Problem 9.4: Using the product $2\nu\overline{\frac{\partial V_i'}{\partial x_j}\frac{\partial}{\partial x_j}\left(\mathbf{N}(V_i)\right)}$ derive the exact solution for ε given in Eq. (9.129).

Problem 9.5: Correlations: Generate random velocities as a function of time with different frequencies and amplitudes. Using the correlation tensor Eq. (9.6), set the reference position vector, $x_i = x_1 = 0$. Find (1) $R_{11}(r_1, 0, 0)$, $R_{11}(0, r_2, 0)$, (b) the correlation coefficient $\rho_{11}(r_1, 0, 0)$, $\rho_{11}(0, r_2, 0)$, (c) the osculating parabolas, (d) length and time scales.

Problem 9.6: Expand the total enthalpy Eq. (9.87) and simplify the result for a two-dimensional boundary layer application.

Problem 9.7: For the thermal energy Eq. (9.75), Give (a) the index notation and (b) expand Eq. (9.75) in Cartesian coordinate system.

Problem 9.8: For the coordinate invariant averaged Navier–Stokes equation (9.59) give (a) the index notation for a general orthogonal coordinate system, (b) decompose it into three component and (c) use the corresponding relationships for metric coefficients and Christoffel symbols and express the three components in a cylindrical coordinate system.

Project 9.1: Using the index notation from Problem 9.8 and applying the results to the two dimensional orthogonal curvilinear coordinate given by Eq. (7.3) in Chap. 7:

$$w = -\frac{2}{a+ib}\ln z \text{ with } z = x + iy \text{ and } w = \xi_1 + i\xi_2$$

assume a uni-directional flow and substitute the Reynolds stress by the Prandtl mixing length. Formulate an appropriate velocity distribution at the inlet and numerically calculate the flow velocity distribution within (a) curved nozzle and (b) curved diffuser.

Project 9.2: For a free jet flow (for details see Chap. 10) using Prandtl mixing length model, determine the velocity and the turbulent shear stress distribution.

Chapter 10
Free Turbulent Flow

10.1 Types of Free Turbulent Flows

In Chap. 9 we primarily discussed the type of turbulent flow which is termed wall turbulence emphasizing the effect of wall shear stress on the turbulence, its production and dissipation. This chapter deals with the type of turbulence which is not confined by solid walls. We distinguish three different free turbulent flows: free jets, free wakes and jet boundaries shown in Fig. 10.1.

Free jets, Fig. 10.1a, are encountered in a variety of engineering applications. Hot gas jet exiting from the thrust nozzle of a jet engine, water jet exiting from a diffuser of a hydraulic turbine and the fluid discharged from an orifice are a few examples of how a fluid forms free jets. As Fig. 10.1a shows, the velocity profile of a free jet changes in longitudinal direction. The jet width b increases in lateral direction, while its velocity decreases. Furthermore, at the jet boundary, there is an exchange of mass, momentum and energy with the surrounding fluid at rest, which causes a partial mixing of the jet with the surrounding fluid. As we will discuss in more detail in the following, downstream of the nozzle, at some x/d-ratio, the non-dimensionalized velocity and turbulence quantities exhibit a similarity pattern.

Free wakes are generated behind any solid body that is exposed to a fluid flow. Figure 10.1b shows a two-dimensional free wake downstream of a cylindrical rod. Two quantities define the free wake development and decay in terms of wake *velocity defect* and *wake width*. The wake structure consists of a wake vortical core, within which there are intensive longitudinal and lateral fluctuations and the wake external region, where no major turbulence activities take place. Jet boundaries are formed between two streams that move parallel to each others with different velocities. They may be separated by a thin surface discontinuity as Fig. 10.1c shows or discharged into a an environment, where the flow is at rest, Fig. 10.1d. In all four cases illustrated in Fig. 10.1, the width changes in streamwise direction within a *mixing zone*. While free jets, free wakes and jet boundaries are frequently encountered in external aerodynamics, wake flow development within channels has a particular significance

M. T. Schobeiri, *Advanced Fluid Mechanics and Heat Transfer for Engineers and Scientists*, https://doi.org/10.1007/978-3-030-72925-7_10

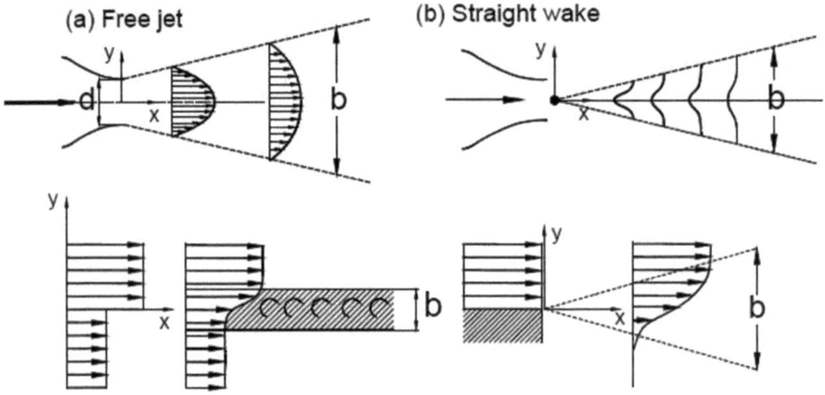

Fig. 10.1 Four different turbulent flows; the mixing range characterized by the width $b = b(x)$ separates the disturbed flow zone from the undisturbed flow regime

in internal aerodynamics such as the wake flow through a turbine or compressor blade channels. Unlike the free wakes that are subjected to zero streamwise pressure gradient, the channel wakes may experience positive, zero or negative pressure gradients in longitudinal as well as lateral directions. Most importantly, the channel wakes play a significant role in affecting the turbulence structure and the boundary layer development along a surface that is impinged by the incoming wakes.

10.2 Fundamental Equations of Free Turbulent Flows

The free turbulent flows briefly introduced above share the same flow characteristics, namely that their turbulent shear stress compared to the molecular shear stress is much larger. Further more, the surroundings in which they develop has a constant pressure (zero pressure gradient). As a result, at some downstream distance, where fully developed turbulent flow is established, molecular shear stress can be completely neglected and the pressure gradient term can be set equal to zero. Furthermore, we assume a two-dimensional flow and replace the velocity components $V_i = \bar{V}_i + V_{i'}$ by $V_1 = \bar{V}_1 + V_{1'} \equiv U = \bar{U} + u$, $V_2 = \bar{V}_2 + V_{2'} \equiv V = \bar{V} + v$ with $V_{1'} \equiv u$ and $V_{2'} \equiv v$ as the fluctuation components. Building the longitudinal, the lateral and the mixed velocity momenta, after averaging we obtain

$$\begin{aligned} \overline{U^2} &= \bar{U}^2 + \overline{u^2} \\ \overline{V^2} &= \bar{V}^2 + \overline{v^2} \\ \overline{UV} &= \bar{U}\,\bar{V} + \overline{uv} \end{aligned} \tag{10.1}$$

with $\bar{V}_{1'} \equiv \bar{u} = 0$, $\bar{V}_{2'} \equiv \bar{v} = 0$. In Eq. (10.1), the terms $\bar{U}\,\bar{V}$ and $\overline{UV}$ are referred to as the partial and total velocity momenta, respectively. With the above assumptions for free turbulent flows, the Reynolds equations (Eq. 8.65) can be substantially simplified leading to:

$$\frac{\partial}{\partial x}\left(\frac{\bar{p}}{\rho} + \bar{U}^2 + \overline{u^2}\right) = -\frac{\partial}{\partial y}(\bar{U}\,\bar{V} + \overline{uv})$$
$$\frac{\partial}{\partial y}\left(\frac{\bar{p}}{\rho} + \bar{V}^2 + \overline{v^2}\right) = -\frac{\partial}{\partial x}(\bar{U}\,\bar{V} + \overline{uv}). \tag{10.2}$$

The relation between the velocity component is given by the continuity equation:

$$\frac{\partial U}{\partial x} = -\frac{\partial V}{\partial y}. \tag{10.3}$$

For a fully developed free turbulent flow at some distance downstream of the turbulence origin, the pressure fluctuations compared to the constant static pressure outside the mixing zone are so small that they can be neglected. As the experimental results show, the longitudinal fluctuation velocity $|u|$ is much smaller than the mean velocity $\bar{U}$. The lateral fluctuation velocity $|v|$, however, has the same order of magnitude as the mean lateral velocity $\bar{V}$, while it is negligible compared with $\bar{U}$. This comparison leads to the conclusion that the contributions of the fluctuation velocity momenta are negligibly small compared to the contribution of the longitudinal mean velocity momentum $\bar{U}^2$.

10.3 Free Turbulent Flows at Zero-Pressure Gradient

As presented in Chap. 9, the Prandtl mixing length model was based on the his mixing length hypothesis. Likewise, Kolmogorov based his original $\kappa - \omega$-model on his hypotheses. Each turbulence model presented has its own shortcomings implying that none of them can be considered as universal. The type of approach that is based on hypotheses is called deductive approach. In treating the free turbulence, we use the inductive approach introduced by Reichardt [127]. This approach uses detailed experimental results, from which general conclusions are derived. The inductive approach which is distinctively different from the deductive one is very effective in predicting free turbulent flow cases we categorized above. This approach will be used in this chapter.

A free turbulent flow is established at some distance downstream of the turbulence origin and is characterized by the similarity of its velocity and momentum profiles. In order to solve Eqs. (10.1) and (10.2), we assume that from a definite distance x/d downstream of the wake origin, the velocity and the momentum defect profiles are similar. This distance will have to be experimentally verified, as is discussed in the following. The similarity assumption implies that for arbitrary points on the longitudinal coordinate x, there is a width $b = b/(x)$ as the corresponding length scale on the lateral coordinate y within which the mixing process takes place, for which we define a dimensionless variable:

$$\zeta = \frac{y}{b}. \tag{10.4}$$

Furthermore, we define a *velocity defect* $\bar{U}_1$ as the difference between the *undisturbed potential* velocity U_p which would exist outside the mixing zone and the actual velocity $\bar{U}$. Thus, the actual velocity can be expressed as:

$$\bar{U} = U_p - \bar{U}_1. \tag{10.5}$$

Correspondingly, we define a momentum defect $\bar{U}_I^2$, which exhibits the difference between the momentum of the potential velocity U_p^2 and the momentum of the actual velocity $\bar{U}^2$. Thus, the velocity momentum can be written as:

$$\bar{U}^2 = U_p^2 - \bar{U}_I^2, \text{ with } \bar{U}_I^2 = 2U_p\bar{U}_1 + \bar{U}_1^2. \tag{10.6}$$

In Eq. (10.5), $\bar{U}_1$ and $\bar{U}_I^2$ represent the time-averaged velocity and momentum defects within the mixing range. We also define the following dimensionless velocity, as well as the momentum defect functions:

$$\varphi_1 = \frac{\bar{U}_1}{\bar{U}_{1m}}, \varphi^2 = \frac{\bar{U}_I^2}{\bar{U}_{Im}^2}. \tag{10.7}$$

Here, $\bar{U}_{1m}$ and $\bar{U}_{Im}^2$ represent the maximum values of velocity and momentum defects within the mixing range. The above defined dimensionless variables and the value of U_p will be implemented into Eqs. (10.1) and (10.2), (10.13) to obtain general expressions to solve the above free turbulent flows.

The lateral velocity component is obtained by implementing Eq. (10.4) into Eq. (10.2) as follows

$$\bar{V} = -\int \frac{\partial \bar{U}_1}{\partial x} dy + c \tag{10.8}$$

with $\bar{U}_1 = \bar{U}_{1m}\varphi_1$ we obtain

$$\bar{V} = -\int \frac{\partial(\bar{U}_{1m}\varphi_1)}{\partial x} dy = -\int \left(\varphi_1 \frac{\partial \bar{U}_{1m}}{\partial x} dy + \bar{U}_{1m} \frac{\partial \varphi_1}{\partial x}\right) dy. \tag{10.9}$$

Since $\bar{U}_{1m}$ depends only on x and with $\zeta = y/b$ we may write

$$\frac{\partial \varphi_1}{\partial x} = \frac{\partial \varphi_1}{\partial \zeta}\frac{\partial \zeta}{\partial x} = -\frac{db}{dx}\frac{\zeta}{b}\frac{\partial \varphi_1}{\partial \zeta}. \tag{10.10}$$

Inserting Eq. (10.9) into Eq. (10.8), we obtain

$$\bar{V} = \bar{U}_{1m}\frac{db}{dx}\int \zeta \frac{\partial \varphi_1}{d\zeta} d\zeta - b\frac{d\bar{U}_{1m}}{dx}\int \varphi_1 d\zeta. \tag{10.11}$$

The mixing width b, the maximum velocity defect $\bar{U}_{1m}$ and therefore their product $\bar{U}_{1m}b$ are either a function of x or a constant. Assuming a constant static pressure downstream of the turbulence origin and replacing in x-component of Eq. (10.2) $\bar{U}^2 = U_p^2 - \bar{U}_I^2$ with U_p as the constant pressure outside the mixing range, the x-component of Eq. (10.2) is simplified as

$$\frac{\partial \bar{U}_I^2}{\partial x} = -\frac{\partial}{\partial y}(\bar{U}\,\bar{V} + \overline{uv}) = -\frac{\partial \overline{UV}}{\partial y}. \tag{10.12}$$

For a fully developed free turbulent flow in accord with Eq. (10.7) we replace the differential argument in Eq. (10.12) by $\bar{U}_I^2 = \bar{U}_{Im}^2\varphi^2$ and integrate Eq. (10.12) to arrive at the mixed velocity momentum:

$$\overline{UV} = \bar{U}_{Im}^2\frac{db}{dx}\zeta\varphi^2 - \frac{d(\bar{U}_{Im}^2 b)}{dx}\int \varphi^2 d\zeta. \tag{10.13}$$

With Eqs. (10.11) and (10.13) we are now able to find the solutions for the partial momentum $\bar{U}\,\bar{V}$, the mixing momentum $\overline{UV}$ and the turbulent shear stress $\overline{uv}$ provided that detailed experimental information about the mixing width b, the similarity functions $\varphi, \varphi_1, \bar{U}_{1m}$ and $\bar{U}_{Im}$ are available. Furthermore, appropriate reference velocity and reference velocity momentum must be found such that the resulting dimensionless partial momentum $\bar{U}\,\bar{V}$, total momentum $\overline{UV}$ and the shear stress momentum $\overline{uv}$ are functions of ζ only. Resorting to the idea of Prandtl [128] that the velocity fluctuations are proportional to the local mean velocity, implying that the ratios

$$\frac{\sqrt{\overline{u^2}}}{V_{1m}}, \frac{\sqrt{\overline{v^2}}}{V_{1m}}, \frac{\sqrt{\overline{uv}}}{V_{1m}} \tag{10.14}$$

are dependent upon ζ only, Reichardt argued that since $|\bar{U}\,\bar{V}|$ and $|\overline{UV}|$ are of the same order of magnitude as $|\overline{uv}|$, the ratios

$$\frac{\overline{UV}}{f_1(x)} = g_1(\zeta), \frac{\bar{U}\,\bar{V}}{f_2(x)} = g_2(\zeta) \tag{10.15}$$

Table 10.1 Characteristics of free turbulent flows

Characteristic quantities of free turbulent flows		
Plane free jet	Plane free wake	Plane free jet boundary
$b \propto x$	$b \propto x^{1/2}$	$b \propto x$
$\bar{U}_{1m} \propto x^{-1/2}$	$\bar{U}_{1m} \propto x^{-1/2}$	$\bar{U}_{1m} = \text{const.}$
$\varphi_1 = e^{-\frac{1}{2}\zeta^2}$	$\varphi_1 = e^{-\zeta^2}$	$\varphi_1 = \frac{1}{2}[1 - \tanh(\kappa_2 \zeta_1)]$
$\varphi^2 = e^{-\zeta^2}$	$\varphi^2 = e^{-\zeta^2}$	$\varphi^2 = \frac{1}{4}[1 - \tanh(\kappa_2 \zeta_1)]^2$
$\overline{UV} = -\frac{1}{2}\frac{db}{dx}\frac{\partial \bar{U}_I^2}{\partial \zeta}$	$\overline{UV} = -\frac{1}{2}\frac{db}{dx}\frac{\partial \bar{U}_I^2}{\partial \zeta}$	$\overline{UV} = -\frac{1}{2}\frac{db}{dx}\frac{\partial \bar{U}_I^2}{\partial \zeta}$
$\frac{\bar{U}\,\bar{V}}{\alpha \bar{U}_{Im}^2} = -\frac{1}{4}\left(\frac{3}{2}\frac{\partial \varphi^2}{\partial \zeta} - \frac{\partial \varphi_1}{\partial \zeta}\right)$	$\frac{\bar{U}\,\bar{V}}{\alpha \bar{U}_{Im}^2} = -\frac{1}{2}\left(\frac{3}{2}\frac{\partial \varphi^2}{\partial \zeta} - \frac{\partial \varphi_1}{\partial \zeta}\right)$	$\frac{\bar{U}\,\bar{V}}{\alpha \bar{U}_{Im}^2} = -\frac{1}{4}\left(\frac{3}{2}\frac{\partial \varphi^2}{\partial \zeta} - \frac{\partial \varphi_1}{\partial \zeta}\right)$
$\frac{\overline{uv}}{\alpha \bar{U}_{Im}^2} = -\frac{1}{4}\left(\frac{1}{2}\frac{\partial \varphi^2}{\partial \zeta} + \frac{\partial \varphi_1}{\partial \zeta}\right)$	$\frac{\overline{uv}}{\alpha \bar{U}_{Im}^2} = -\frac{1}{6}\left(\frac{1}{2}\frac{\partial \varphi^2}{\partial \zeta} + \frac{\partial \varphi_1}{\partial \zeta}\right)$	$\frac{\overline{uv}}{\alpha \bar{U}_{Im}^2} = -\frac{1}{4}\left(\frac{1}{2}\frac{\partial \varphi^2}{\partial \zeta} + \frac{\partial \varphi_1}{\partial \zeta}\right)$

must be functions of ζ for free turbulent flows with similar flow conditions. The functions $f_1(x)$ and $f_2(x)$ contain all free stream quantities that change in flow direction and have the same dimensions as $|\bar{U}\,\bar{V}|$ and $|\overline{UV}|$. For $f_1(x)$ or $f_2(x)$ we may set:

$$f_1(x) = \bar{V}_{1m}^2 \frac{db}{dx}, \quad \text{or}\, f_2(x) = \bar{V}_{Im}^2 \frac{db}{dx}. \tag{10.16}$$

The task of the inductive approach is to provide detailed experimental information necessary to find the necessary relationship for calculating the momenta $|\bar{U}\,\bar{V}|$ and $|\overline{UV}|$. Based on detailed free turbulence measurements, boundary conditions and similarity assumptions, Reichardt [127] derived a set of equations that accurately represent the free turbulent flows. To find a set of unifying equations that describe the free turbulent flows, Eifler [129] simplified the *exact* equations derived by Reichardt and arrived at a unifying set of equations presented in Table 10.1. As shown, all equations describing the dimensionless partial momentum, total momentum and the shear stress are related to the Gaussian function, its integrals or differentials.

Table 10.1 contains the characteristics of plane free jets, free wakes and the jet boundaries. The same equations are valid for axisymmetric free turbulent cases, however, the continuity equation as well as equation of motion must be written in polar coordinates.

10.3.1 Plane Free Jet Flows

The characteristics of the plane free jet flow are listed in Table 10.1. They contain the expressions for the velocity defect $\phi_1 = \bar{U}/\bar{U}_m$, the total momentum $\overline{UV}$, the partial momentum $\bar{U}\,\bar{V}$ and the turbulent shear stress $\overline{uv}$. One of the important characteristics of all free turbulent flows is the product of the local mixing width and the local maximum velocity defect $b\bar{U}_{1m}$. For free jet flows, the experiments by Reichardt

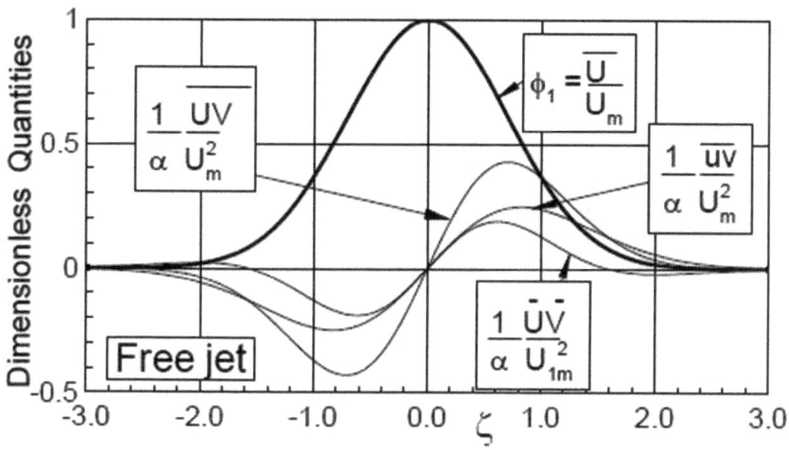

Fig. 10.2 Plane free jet quantities as a function of dimensionless coordinate, profiles of the total momentum, partial momentum and shear stress

show that this product is dependent upon the streamwise direction x, that means $b\bar{U}_{1m} = f(x)$.

Figure 10.2 shows the above quantities made dimensionelss with $\alpha\bar{U}^2_{Im}$ as the denominator with $\alpha = db/dx$. Extensive experiments by Reichardt shows that the velocity defect exactly $\phi_1 = \bar{U}/\bar{U}_m$ follows the Gaussian distribution. He also measured the ratio $\bar{U}^2_I/\bar{U}^2_{Im} \equiv \varphi^2$ and found that it also follows the Gaussian distribution. As seen from Fig. 10.2, the total velocity momentum $\overline{UV}$ is the sum of the partial momentum $\bar{U}\,\bar{V}$ and the shear stress momentum $\overline{uv}$. As expected, the shear stress is zero at the jet boundaries and changes its sign from negative to positive at the jet center.

10.3.2 Straight Wake at Zero Pressure Gradient

Before proceeding with the straight wake flow as an important free turbulent flow that is encountered in external and internal aerodynamics, we need to know how the wake width develops in streamwise direction. In order to eliminate the secondary effects of entrainment of ambient fluid particles into the wake, cylindrical rods of different diameters can be inserted into a two-dimensional channel, where the relative position x/d of the probe that measures the wake turbulence quantities is varied, Fig. 10.3. The zero-pressure gradient environment is established by slightly opening the side walls. This compensates for the cross section blockage caused by the boundary layer displacement thickness. Straight wake studies by Reichardt [127] and Eifler

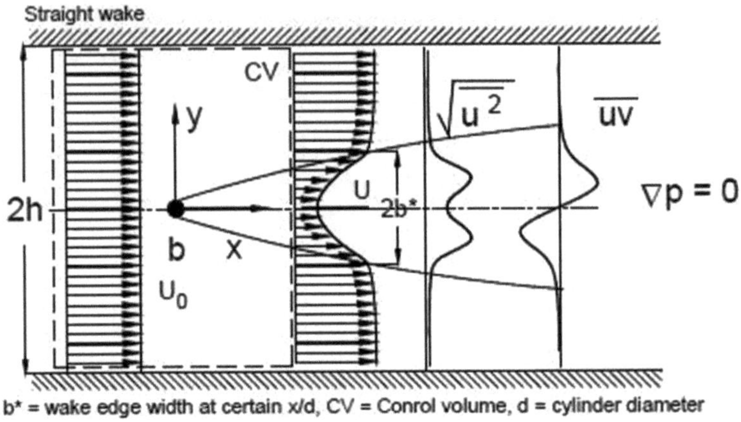

Fig. 10.3 Wake development downstream of a cylinder, the top and bottom walls slightly diverge to maintain $\nabla p = 0$

[129] have shown that the wake development in the longitudinal direction primarily depends upon the ratio x/d regardless of the separate variation of the parameters involved in the ratio. To define the wake width b developed within a straight channel with the height $2h$ and the width of unity, we integrate the area under $\bar{U}_1$-distribution

$$\bar{U}_{1m}b = \int_{-h}^{+h} \bar{U}_1 dy, b = \frac{1}{\bar{U}_{1m}} \int_{-h}^{+h} \bar{U}_1 dy \tag{10.17}$$

with h as the half height of the side walls shown in Fig. 10.3.
To relate the integral in Eq. (10.17) to the corresponding integral with infinity as the boundary, we find

$$\bar{U}_{1m}b_{y=\infty} = \int_{-\infty}^{+\infty} \bar{U}_1 dy, b_{y=\infty} = \frac{1}{\bar{U}_{1m}} \int_{-\infty}^{+\infty} \bar{U}_1 dy. \tag{10.18}$$

Introducing in Eq. (10.18) $\zeta = y/b$ results in

$$b_{y=\infty} = b \int_{-\infty}^{+\infty} \frac{\bar{U}_1}{\bar{U}_{1m}} \frac{dy}{b} = b \int_{-\infty}^{+\infty} \varphi d\zeta = b \int_{-\infty}^{+\infty} e^{-\zeta^2} d\zeta = b\sqrt{\pi}. \tag{10.19}$$

Thus, with Eq. (10.19) the two widths are interrelated as

$$b_{y=\infty} = b\sqrt{\pi} \equiv 2\gamma b \tag{10.20}$$

with $\gamma = \sqrt{\pi}/2$. We insert Eq. (10.20) into Eq. (10.18) and find

$$\bar{U}_{1m} b 2\gamma = \int_{-\infty}^{+\infty} \bar{U}_1 dy. \tag{10.21}$$

Rearranging Eq. (10.21) results in the wake with that can be determined as

$$b = \frac{1}{2\gamma} \int_{-\infty}^{+\infty} \frac{\bar{U}_1}{\bar{U}_{1m}} dy. \tag{10.22}$$

The wake width b is easily found by numerically integrating Eq. (10.22) and using the distribution of $\bar{U}_1$ given by experiments. To find simple algebraic relationships for the wake width as a function of x/d, the local drag coefficient C_D is calculated using the experimental data. For the determination of C_D, a control volume CV is placed inside the channel, Fig. 10.3, that includes the undisturbed inlet velocity U_0 and the velocity profile at the position x/d. Applying the continuity and linear momentum equation as presented in Chap. 5, we find the relationship for the drag coefficient C_D as:

$$C_D = \frac{2W}{\rho U_0^2 d} \approx \frac{4}{U_0 d} \int_0^{b*} \bar{U}_1 dy. \tag{10.23}$$

Eifler [129] introduced a ratio C_D/β^2 with $\beta = \bar{U}_{1m}/U_0$ and U_0 as the constant undisturbed inlet velocity. This ratio allows collapsing the experimental data on two straight lines that represents the *near wake* $0 < x/d < 60$ region, the transition region $60 < x/d < 130$ and the *far wake* $x/d > 130$ region as shown in Fig. 10.4. While the far wake region is characterized by a constant product $\bar{U}_{1m} b$ =const., the near wake region influenced by the von Kárman vortex street shows a dependency of $\bar{U}_{1m} b$ upon x/d. Figure 10.5 shows the implementation of the results in Fig. 10.4 for calculating the wake width distribution. Figure 10.6 shows, the velocity defect $\varphi_1 = \bar{U}_1/\bar{U}_{1m}$ as a function of dimensionless ζ for different x/d ratios. As shown, the experimental results follow the Gaussian distribution, $\varphi_1 = e^{-\zeta^2}$.

Figure 10.6 exhibits the measured and the predicted turbulent shear stresses. At the wake center, the velocity in the longitudinal direction has a maximum, while in the lateral direction as well as the wake boundaries it diminishes. This is reflected in Fig. 10.7, where the shear stress values at the wake center and the boundaries approach zero.

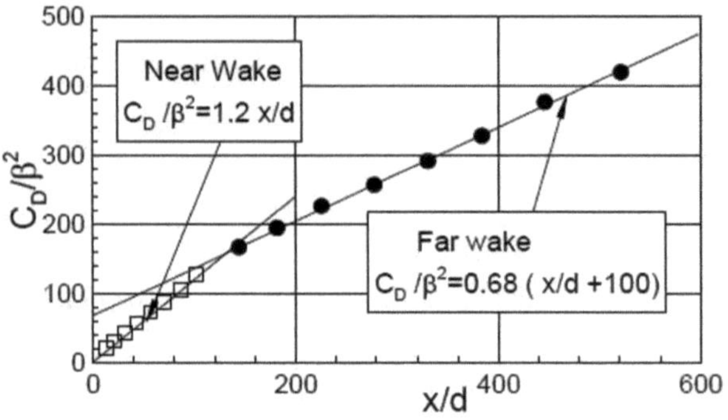

Fig. 10.4 Relative drag coefficient as a function of dimensionless distance x/d

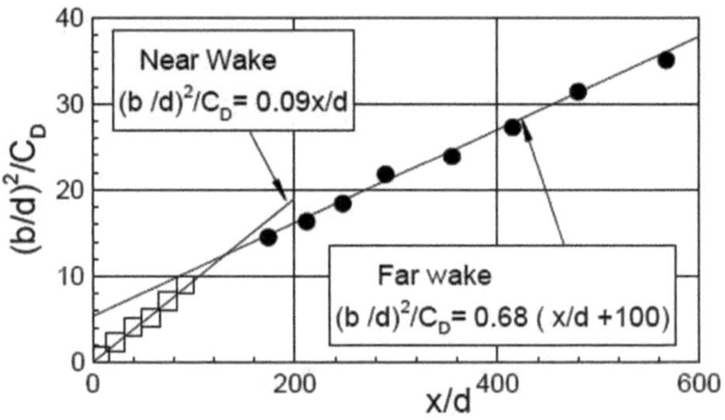

Fig. 10.5 Relative wake width as a function of dimensionless distance x/d with the equations for near and far wakes

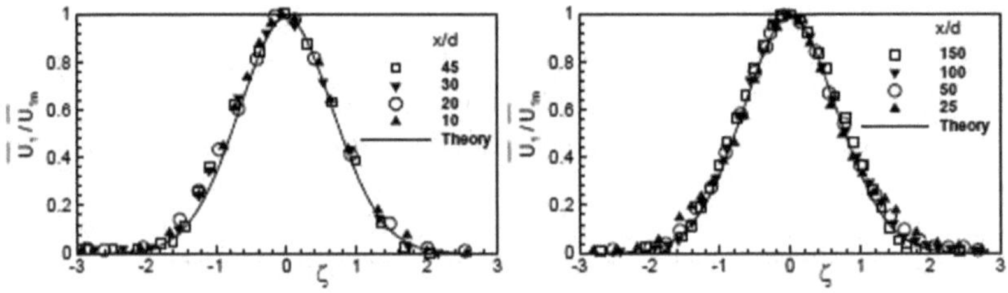

Fig. 10.6 Dimensionless wake velocity defect φ_1 as a function of ζ experiments from Eifler [129]

Figure 10.8 summarizes the wake characteristics in terms of velocity defect $\phi_1 = \bar{U}_1/\bar{U}_{1m}$, the total momentum $\overline{UV}$, the partial momentum $\bar{U}\,\bar{V}$ and the turbulent shear stress $\overline{uv}$. These quantities made dimensionless with $\alpha\bar{U}_{Im}^2$ as the denominator with $\alpha = db/dx$. As seen, the total velocity momentum $\overline{UV}$ is the sum of the partial momentum $\bar{U}\,\bar{V}$ and the shear stress momentum $\overline{uv}$. The shear stress is zero at the wake boundaries and changes its sign from negative to positive at the wake center. Free wake flow exhibits similar free turbulence characteristics as free jets. For far

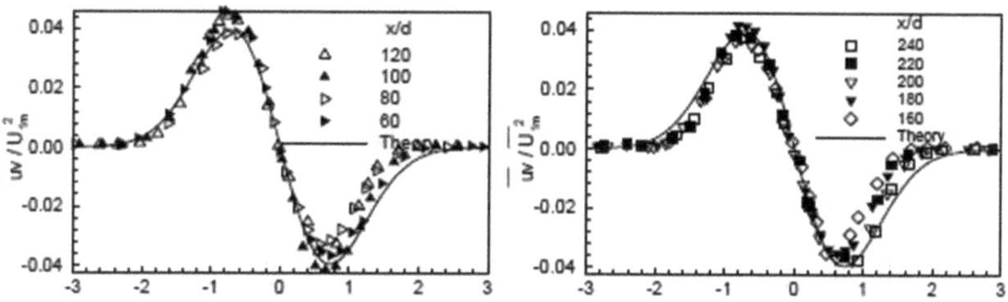

Fig. 10.7 Dimensionless turbulent shear stress as a function of ζ experiments from Eifler [129]

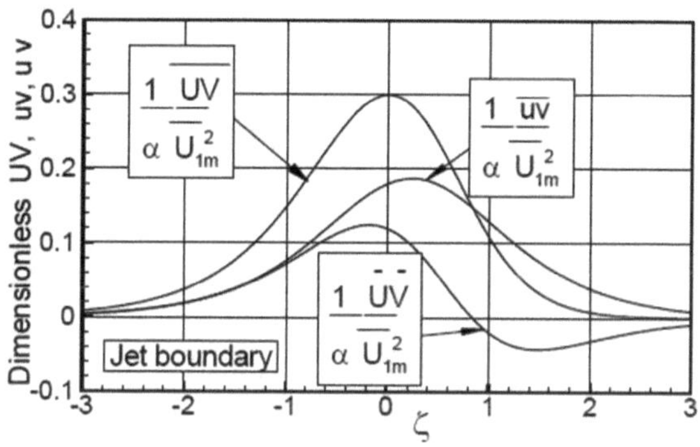

Fig. 10.8 Plane free wake quantities as a function of dimensionless coordinate ζ, profiles of the total momentum, partial momentum ans shear stress

wake region, the experiments in [127, 129] show that the product $b\bar{U}_{1m}$ is independent from the streamwise direction x. Comparing Figs. 10.2 and 10.8 reveals a striking similarity. While in free jet flow, the dimensionless velocity $\bar{U}/\bar{U}_m$ is described by the Gaussian function, it is the velocity defect in free wake flow that is described by the Gaussian function. Furthermore, the other turbulence quantities in both cases behave similarly.

10.3.3 Free Jet Boundary

The treatment of the free jet boundary is very similar to the two cases we discussed in previous sections using the inductive approach. For the dimensionless quantity $\varphi_1 = \bar{U}_1/\bar{U}_{1m}$ Reichardt [127] found the following relation:

$$\varphi_1 = \frac{1}{\sqrt{2}}\left(1 - \frac{2}{\sqrt{2}}\int e^{-\zeta_1^2} d\zeta_1\right)^{1/2} \tag{10.24}$$

with $\zeta_1 = \kappa_1(\zeta - \zeta_0)$, $\kappa = 0.815$ and $\zeta_0 = 0.477$. Equation (10.24) can be approximated as

$$\varphi_1 = \frac{1}{2}[1 - \tanh(\kappa_2\zeta_1)]. \tag{10.25}$$

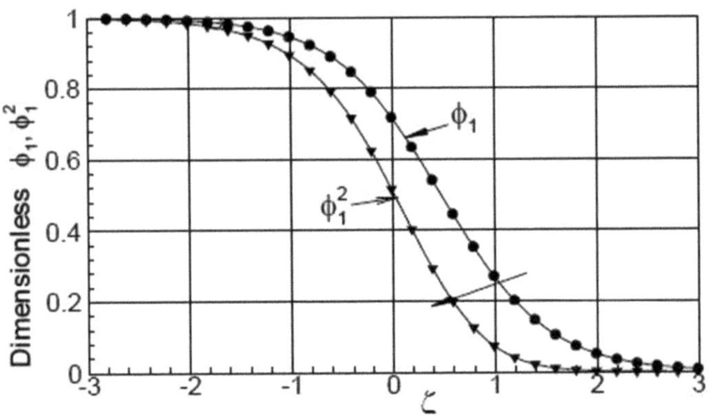

Fig. 10.9 Plane free jet boundary, distribution of the longitudinal velocity ratio $\varphi_1 = \bar{U}_1/\bar{U}_{1m}$ and the momentum ratio φ_1^2 as a function of dimensionless coordinate ζ

Equations (10.24) and (10.25) are plotted in Fig. 10.9 as solid lines that coincide with each other.

The symbols in Fig. 10.9 representing the experimental data by Reichardt [127] seem to compare very well with the both functions. This is also true for the momentum equation:

$$\varphi_1^2 = \frac{1}{2}\left(1 - \frac{2}{\sqrt{2}}\int e^{-\zeta_1^2} d\zeta_1\right). \tag{10.26}$$

For the free jet boundary the velocity $\bar{U}_{1m} = \bar{U}_{um} = \bar{U}_0$ =const. is constant. As a result, the second integral in Eq. (10.11) disappears leading to

$$\bar{V} = \bar{U}_{1m}\frac{db}{dx}\int \zeta \frac{\partial \varphi_1}{d\zeta} d\zeta. \tag{10.27}$$

The partial velocity momentum is obtained by multiplying Eq. (10.27) with $\bar{U}_1 = \varphi_1 \bar{U}_{1m}$ that leads to:

$$\bar{U}\,\bar{V} = \bar{U}_{1m}^2 \frac{db}{dx}\varphi_1 \int \zeta \frac{\partial \varphi_1}{d\zeta} d\zeta. \tag{10.28}$$

The turbulent shear stress results as the difference between the total velocity momentum, Eq. (10.13) and the partial momentum Eq. (10.27) leading to:

$$\overline{uv} = \frac{\bar{U}_{1m}^2}{2}\left(\frac{1}{2}\frac{\partial \varphi^2}{\partial \zeta} + \frac{\partial \varphi_1}{\partial \zeta}\right). \tag{10.29}$$

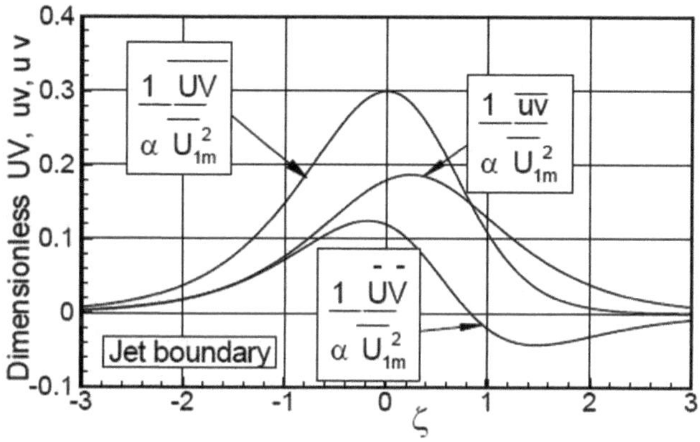

Fig. 10.10 Free jet boundary quantities as functions of dimensionless coordinate ζ, profiles of the dimensionless total momentum, partial momentum and shear stress

The characteristic quantities of this flow are also listed in Table 10.1 from which the dimensionless total momentum, partial momentum and the shear stress are plotted in Fig. 10.10.

10.4 Wake Flow at Non-zero Lateral Pressure Gradient

The wake flow at non-zero pressure gradient constitutes a special case of free turbulent flows with a broad range of general engineering applications. In the field of internal aerodynamics, the wake development under the influence of curvature and pressure gradient is a common feature found in several components of an aircraft gas turbine engines, power generation gas turbines and steam turbines in steady and periodic unsteady forms. The wake flow caused by the turbomachinery blades is associated with inherent unsteadiness. The periodic unsteady wake flow is induced by mutual interaction between the stator and rotor blades of a turbomachine and influences the boundary layer transition behavior and heat transfer characteristics of the turbine or compressor blades positioned downstream of the wake. Because of the significant impact of the wakes in internal aerodynamics, particularly in the area of turbomachinery flow, efforts have been made to describe the fundamental physics of wake development and decay at non-zero pressure gradient environment very similar to the one encountered in turbomachinery internal aerodynamics. The following treatment deals with the fundamental physics of wake development and decay in a curved channel at non-zero lateral pressure gradient. More detailed theoretical and experimental investigations by Schobeiri and his co-researchers at *TPFL* [131–133] study the phenomena of steady and periodic unsteady wake development and decay within curved channel at zero, positive and negative longitudinal pressure gradient as well as non-zero lateral pressure gradients.

10.4.1 Wake Flow in Engineering, Applications, General Remarks

Turbomachines are devices without which no modern society can perform its daily activities. Within turbomachines conversion of total energy of a working medium into mechanical energy and vice versa takes place. Turbomachines are generally divided into two main categories. The first category is used primarily to produce power. It includes, among others, steam turbines, gas turbines, and hydraulic turbines. The main function of the second category is to increase the total pressure of the working fluid by consuming power. The conversion of total energy into shaft work or vice versa is based on exchange of momentum between the blading and the working fluid. This category includes compressors, pumps, and fans. Figure 10.11 shows the rotor of a heavy duty gas turbine engine.

The multi-stage compressor (left: 21 stages) raises the total pressure of air from inlet pressure to a required exit pressure of about 16 bar. Fuel is added in the combustion chamber, where the total temperature raises. Hot combustion gas enters a multis-stage turbine (right: 5 stages) that drives the compressor and the generator. Turbine and compressor stages consist of a stator and rotor rows. The function of the stator row is to provide the necessary velocity and incidence angle for the following turbine rotor. A detailed treatment of turbomchinery theory, design and nonlinear dynamic performance is found in the recent textbook by Schobeiri [130].

Fig. 10.11 Rotor unit of a heavy duty gas turbine with multi-stage compressor and turbine, compressor pressure ratio 15:1 (BBC-GT13E2)

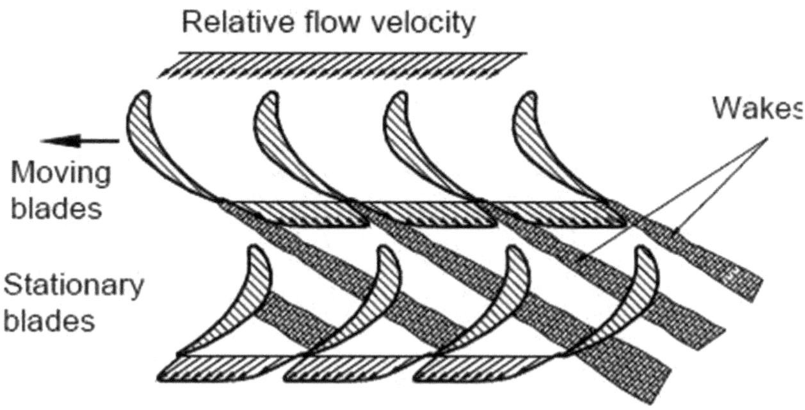

Fig. 10.12 Schematic of moving wakes from rotating rotor blades that impinge on stationary stator blades

Figure 10.12 shows the wake development originating from a turbine stator blade row that impinges on the subsequent rotating rotor blades. Similarly, the wakes generated by the rotor blades impinge on the succeeding stator blades. The interaction of the wake with the succeeding blades is always between a stationary and a rotating frame, regardless of their sequential position.

The turbulence structure of a wake is defined in terms of the velocity defect, the turbulence characteristics and the drag coefficient C_D. The drag coefficient can be used as a similarity parameter for comparing the wakes that originate from turbine or compressor blades with those that originate form a set of cylindrical rods. Thus, if wakes generated by a given set of blades or by a set of cylindric rods have the same drag coefficient C_D, they have approximately the same turbulence structure.
Figure 10.13 shows a cascade of cylindrical bars moving with the translational velocity vector **U**. The bars are subjected to a relative inlet flow with the velocity vector **W** which constitutes the difference between the absolute velocity vector **V** and the translational vector is vector $W = \mathbf{V} - \mathbf{U}$.

Measuring the wake structure essential for understanding the basic physics of wake development under the turbomachinery condition is extremely difficult in a rotating turbomachine. However, producing wakes downstream of a cylindrical rods through a curved channel at positive, zero or negative pressure gradient yield information very similar to the wakes through a turbine or compressor blade channels. Prerequisite for the similarity is that the diameter of the cylindrical rod has the same drag coefficient C_D as the blade. Figure 10.14 schematically shows the wake development through curved channels under negative, zero, and positive longitudinal pressure gradients. In all three cases, there exist a lateral pressure gradient that stems from the channel curvature. Figure 10.15 shows the experimental setup for investigating the wake development within a curved channel that consists of constant curvature inner and outer walls as well as two side walls. It has a wake generator that

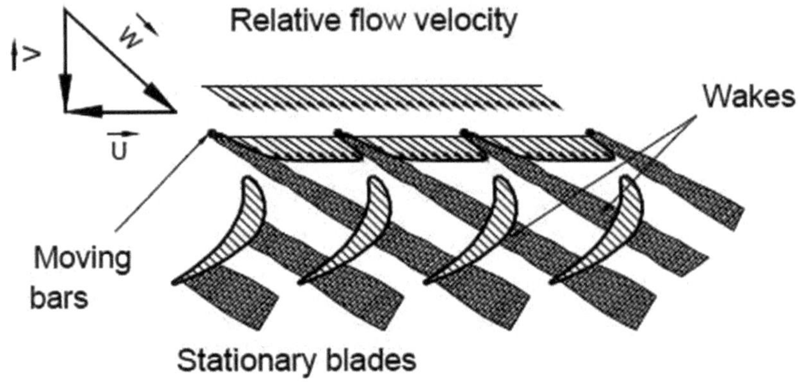

Fig. 10.13 Simulation of wakes by a set of moving cylindrical rods

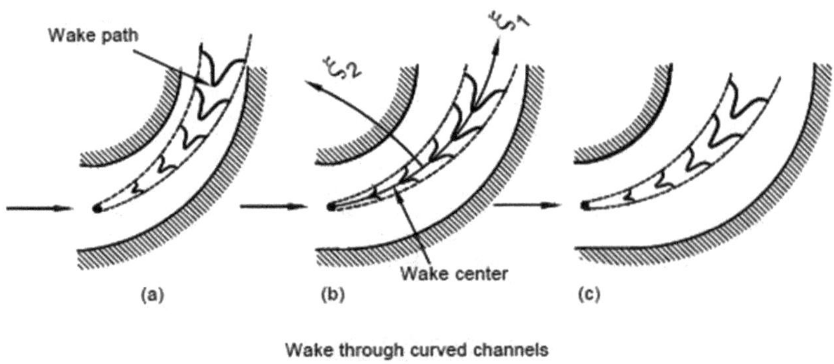

Fig. 10.14 Wake development through curved channels, **a** $\partial p/\partial \xi_1 < 0$, **b** $\partial p/\partial \xi_1 = 0$ and **c** $\partial p/\partial \xi_1 > 0$

can generate steady as well as periodic unsteady wakes. More details on steady and unsteady wake development and decay are found in [131, 132].

In the following the theoretical framework for predicting the wake development under the effect of curvature and zero longitudinal pressure gradient is presented. More details are found in [131]. The theoretical framework is also extended to negative and positive pressure gradient which is detailed in [133].

10.4.2 Theoretical Concept, an Inductive Approach

This section deals with deriving expressions for wake characteristics that describe the steady wake development under the influence of pressure gradient and curvature. The wake characteristics include dimensionless mean velocity defect, mean longitu-

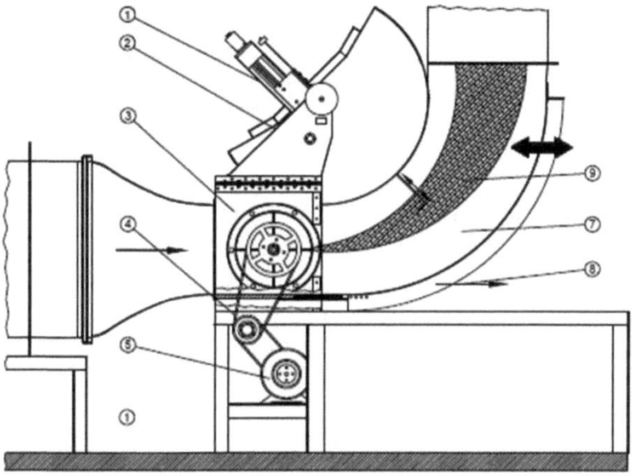

Fig. 10.15 Research facility at *TPFL* [132] for measuring the wake development and decay under different longitudinal and lateral pressure gradients. Different longitudinal pressure gradients are established by moving the outer wall in and out

dinal and lateral velocities, and total and partial momenta that lead to an expression for Reynolds shear stress. The first step in the development of the theory is to transform the coordinate invariant equations of motion and continuity into curvilinear coordinate systems. For the present theoretical considerations, an incompressible turbulent flow through a two-dimensional curved channels is assumed. Further, it has been assumed that the velocity vector has a temporal and spatial dependency and can be decomposed into a time-independent mean and a time-dependent turbulent fluctuation vector. Based on the experimental observations, the flow regime under investigation can be divided into three distinct zones: (1) a highly vortical wake core characterized by the mean velocity components that are asymmetric about the wake centerline. (2) The wake external zone where the velocity distribution approximately corresponds to that of a potential flow. In this connection, it should be noted that the wake region is highly rotational where a potential flow assumption does not apply. (3) The third zone is the boundary layer at the convex and concave channel walls, where the viscosity effect causes a boundary layer displacement and thus a slight flow acceleration. To compensate for blockage, the exit cross section is slightly increased to ensure a constant longitudinal pressure distribution as detailed in [131].

Conservation laws are first presented in a coordinate invariant form and then transformed into an orthogonal curvilinear coordinate system (ξ_i). In this coordinate system, ξ_1 is the direction along a streamline near the wake center and ξ_2 is the direction normal to it. Starting with the conservation law of mass, the equation of continuity in coordinate invariant form is:

$$\nabla \cdot \mathbf{V} = \mathbf{0}. \tag{10.30}$$

Following the argument presented in (10.1) that in a free turbulent flow the turbulent shear stress compared to the molecular shear stress is much larger, we assume that the viscosity effects can be neglected. Under this assumption the equation of motion in a coordinate invariant form is:

$$\mathbf{V} \cdot \nabla \mathbf{V} = -\frac{\mathbf{1}}{\text{æ}} \nabla \mathbf{p}. \tag{10.31}$$

Hereafter, Eq. (10.31) is referred to as the *version 1* of the equation of motion. Combining Eqs. (10.30) and (10.31) results in a modified, more appropriate version (referred to as *version 2*) of the equation of motion as:

$$\nabla \cdot (\mathbf{V}\,\mathbf{V}) = -\frac{\mathbf{1}}{\text{æ}} \nabla \mathbf{p}. \tag{10.32}$$

Equation (10.32) is particularly useful for comparing the order of magnitude of individual terms and their contributions. For further treatment of conservation laws, the velocity vector is decomposed into a time-averaged mean and a time-dependent fluctuation as:

$$\mathbf{V} = \bar{\mathbf{V}} + \tilde{\mathbf{V}}. \tag{10.33}$$

In Eq. (10.33) we introduced tilde ($\tilde{}$) instead of ($\prime$) since the latter would interfere with the contra variant superscripts that we use in the following orthogonal curvilinear components. Introducing Eq. (10.33) into Eq. (10.32) and time averaging the entire expression, we arrive at:

$$\nabla \cdot (\overline{\mathbf{V}\,\mathbf{V}}) = \overline{\nabla \cdot [(\bar{\mathbf{V}} + \tilde{\mathbf{V}})(\bar{\mathbf{V}} + \tilde{\mathbf{V}})]} = \nabla \cdot (\bar{\mathbf{V}}\,\bar{\mathbf{V}} + \overline{\tilde{\mathbf{V}}\,\tilde{\mathbf{V}}}). \tag{10.34}$$

To keep the above introduced nomenclature, the time averaged second order tensor $\overline{\mathbf{V}\,\mathbf{V}}$ in Eq. (10.34) is called the *total velocity momentum*, the expression $\bar{\mathbf{V}}\,\bar{\mathbf{V}}$ is termed the *partial velocity momentum* and the expression $\overline{\tilde{\mathbf{V}}\,\tilde{\mathbf{V}}}$ is the Reynolds stress tensor. As seen from Eq. (10.34), the Reynolds stress tensor is the difference between the total and the partial velocity momenta. In a three-dimensional flow, the above tensors have generally nine components, from which, due to the symmetry, only six are distinct. For the two-dimensional flow assumption of this study, the number of distinct components reduces to three. For analytical treatment, it is appropriate to transform Eqs. (10.31) and (10.32) into the wake orthogonal curvilinear coordinate system ξ_1 and ξ_2 shown in Fig. 10.11. Transforming Eqs. (10.30)–(10.32) into the wake curvilinear coordinate system and using Eqs. (10.33) and (10.34), the corresponding index notation for continuity equation reads:

$$\bar{V}^i_i + \bar{V}^i \Gamma^j_{ij} = 0 \tag{10.35}$$

the version 1 of equation of motion is

$$\bar{V}^j \bar{V}^i_{,j} + \bar{V}^j \bar{V}^k \Gamma^i_{kj} = -\frac{1}{\rho} g^{ij} \bar{p}_j - (\overline{\tilde{V}^m \tilde{V}^i})_m - \overline{\tilde{V}^m \tilde{V}^i} \Gamma^j_{mj} - \overline{\tilde{V}^m \tilde{V}^n} \Gamma^i_{mn} \tag{10.36}$$

and the version 2 index notation reads

$$\begin{aligned}(\bar{V}^m \bar{V}^j_{,m} + \bar{V}^m \bar{V}^j \Gamma^i_{mi} + \bar{V}^m \bar{V}^n \Gamma^j_{mn} = &-\frac{1}{\rho} g^{ji} \bar{p}_i - (\overline{\tilde{V}^m \tilde{V}^j})_m \\ &- \overline{\tilde{V}^m \tilde{V}^j} \Gamma^i_{mi} + \overline{\tilde{V}^m \tilde{V}^n} \Gamma^j_{mn}.\end{aligned} \tag{10.37}$$

In Eqs. (10.36) and (10.37) the comma before the subscripts indicates the partial differentiation with respect to the subscript that follows the comma. The metric coefficients and Christoffel symbols for the current curvilinear coordinate system are:

$$\begin{aligned} g_{ij} &= \begin{pmatrix} \left(\frac{R+\xi_2}{R}\right)^2 & 0 \\ 0 & 1 \end{pmatrix}; \; g^{ij} = \begin{pmatrix} \left(\frac{R}{R+\xi_2}\right)^2 & 0 \\ 0 & 1 \end{pmatrix} \\ \Gamma^1_{ij} &= \begin{pmatrix} 0 & \frac{1}{R+\xi_2} \\ \frac{1}{R+\xi_2} & 0 \end{pmatrix}; \; \Gamma^2_{ij} = \begin{pmatrix} -\frac{R+\xi_2}{R^2} & 0 \\ 0 & 1 \end{pmatrix}. \end{aligned} \tag{10.38}$$

In Eq. (10.38), R represents the radius of curvature of the wake centerline at $\xi_2 = 0$, taken to be positive if convex in the positive ξ_2 direction. For further treatment, the co- and contra-variant components in Eqs. (10.35)–(10.37) must be replaced by the physical components. Introducing $\bar{U}$ and $\bar{V}$ for the time-averaged physical velocity components and $\bar{u}^2$, $\bar{v}^2$, $\overline{uv}$ for the three distinct time-averaged physical components of the Reynolds stress tensor into Eqs. (10.35)–(10.37), the time-averaged version of continuity equation (10.35), in the wake curvilinear coordinates is:

$$\bar{U}_1 + \left[\left(1 + \frac{\xi_2}{R}\right)\bar{V}\right]_2 = 0. \tag{10.39}$$

The subscripts ",1" and ",2" refer to the derivatives in ξ_1 and ξ_2 directions, respectively. The version 1 equation of motion, Eq. (10.36), decomposed into longitudinal direction ξ_1 is:

$$\begin{aligned}\frac{R}{R+\xi_2}\bar{U}\,\bar{U}_{,1} + \bar{V}\,\bar{U}_{,2} + \frac{\bar{U}\,\bar{V}}{R+\xi_2} = -\frac{1}{\rho}\frac{R}{R+\xi_2}\bar{p}_1 - \frac{R}{R+\xi_2}(\bar{u}^2)_1 - (\overline{uv})_{,2} \\ - \frac{2}{R+\xi_2}\overline{uv},\end{aligned} \tag{10.40}$$

and lateral direction ξ_2 is:

$$\frac{R}{R+\xi_2}\bar{U}\,\bar{V}_{,1}+\bar{V}\,\bar{V}_{,2}-\frac{\bar{U}^2}{R+\xi_2}=-\frac{1}{\rho}\bar{p}_2-\frac{R}{R+\xi_2}(\overline{uv})_{,1}-(\bar{v}^2)_{,2}-\frac{\bar{v}^2}{R+\xi_2}$$
$$+\frac{\bar{u}^2}{R+\xi_2}. \tag{10.41}$$

Similarly, version 2 of the equation of motion (10.37) decomposed into ξ_1, ξ_2 components is:

$$\frac{R}{R+\xi_2}\left(\frac{\bar{p}}{\rho}+\bar{U}^2+\bar{u}^2\right)_{,1}+(\bar{U}\,\bar{V}+\overline{uv})_{,2}+\frac{2}{R+\xi_2}(\bar{U}\,\bar{V}+\overline{uv})=0 \tag{10.42}$$

$$\frac{R}{R+\xi_2}(\bar{U}\,\bar{V}+\overline{uv})_{,1}+\left(\frac{\bar{p}}{\rho}+\bar{V}^2+\bar{v}^2\right)_{,2}-\frac{1}{R+\xi_2}(\bar{U}^2-\bar{V}^2+\bar{u}^2-\bar{v}^2)=0. \tag{10.43}$$

Equations (10.42) and (10.43) are of practical interest for estimating the order of magnitude of each individual term compared with the others. As the experimental results show, the longitudinal fluctuation velocity $|u|$ is much smaller than the mean velocity $\bar{U}$. The lateral fluctuation velocity $|v|$, however, has the same order of magnitude as the mean lateral velocity $\bar{V}$, while it is negligible compared with $\bar{U}$. This comparison leads to the conclusion that the contributions of the fluctuation velocity momenta are negligibly small compared to the contribution of the longitudinal mean velocity momentum $\bar{U}^2$. Equations (10.42) and (10.43) describe the wake development through a curved channel under the influence of pressure gradients. The next step is to introduce non-dimensional parameters aimed at verifying the dynamic similarity assumptions by properly defining the local length and velocity scales.

10.4.3 Nondimensional Parameters

In order to solve Eqs. (10.39)–(10.43), we assume that from a definite distance ξ_1/d downstream of the wake origin, the velocity and the momentum defect profiles are similar. This distance will have to be experimentally verified, as is discussed in the following. Similar to Eq. (10.4) we introduce a dimensionless parameter

$$\zeta=\frac{\xi_2}{b} \tag{10.44}$$

with the lateral coordinate ξ_2, and the wake width $b=b(\xi_1)$. Furthermore, similar to Eq. (10.5) we define a *wake velocity defect* $\bar{U}_1$ as the difference between the *hypothetical potential* velocity U_p which would exist without the cylinder and the actual velocity $\bar{U}$ as shown in Fig. 10.16. Similar to the straight wake, the actual wake velocity can be expressed as:

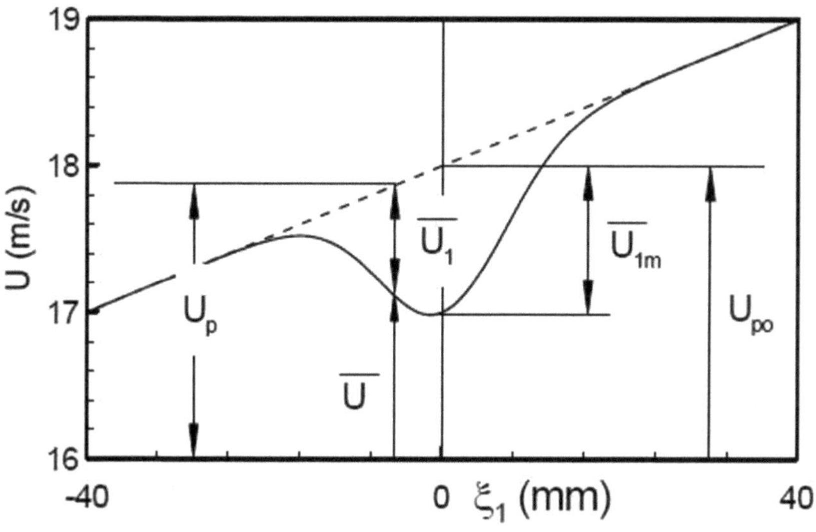

Fig. 10.16 Asymmetric wake velocity distribution, definitions of $\bar{U}$, $\bar{U}_1$, $\bar{U}_{1m}$, U_p and U_{p0} with U_p as the hypothetical potential flow velocity

$$\bar{U} = U_p - \bar{U}_1. \tag{10.45}$$

Correspondingly, the wake momentum defect $\bar{U}_I^2$ is the difference between the momentum of the potential velocity U_p^2 and the momentum of the actual velocity $\bar{U}^2$. Thus, the wake velocity momentum can be written as:

$$\overline{U^2} = U_p^2 - \overline{U_I^2}. \tag{10.46}$$

In Eqs. (10.45) and (10.46), U_p represents the hypothetical velocity distribution that is an extension of the undisturbed wake-external velocity into the wake. The hypothetical potential velocity U_p in Fig. 10.16 can easily be determined by neglecting in Eq. (10.40) all turbulence quantities. Since this section deals with the channel wake at zero-longitudinal pressure gradient, we set in Eq. (10.40) $\partial\bar{p}/\partial\xi_1 \equiv \bar{p}_{,1} = 0$. As a result we have:

$$\frac{R}{R+\xi_2} U_p \frac{\partial U_p}{\partial \xi_1} + V_p \frac{\partial U_p}{\partial \xi_2} + \frac{U_p V_p}{R+\xi_2} = 0. \tag{10.47}$$

Since $\partial U_p/\partial\xi_1 \ll \partial U_p/\partial\xi_2$ the first term in Eq. (10.47) can be neglected leading to:

$$\frac{\partial U_p}{U_p} + \frac{\partial \xi_2}{R+\xi_2} = 0. \tag{10.48}$$

Integrating Eq. (10.48) and determining the integration constant by setting $(U_p)_{\xi_2=0} = U_{p_0}$, we find

$$U_p = U_{p^0}\left(1+\frac{\xi_2}{R}\right)^{-1}. \tag{10.49}$$

Expanding the expression in the parenthesis as a Taylor series and neglecting the higher order terms, the final expression for U_p as a linear function of ξ_2 is:

$$U_p = U_{p^0}\left(1-\frac{\xi_2}{R}\right). \tag{10.50}$$

Here, U_{p^0} is the hypothetical potential velocity at wake center, $\xi_2 = 0$, Fig. 10.13. Thus, the potential velocity U_p outside the wake is a function of ξ_2 only. Similar to the straight wakes, the similarity assumption requires the following dimensionless wake velocity, as well as the momentum defect functions:

$$\varphi_1 = \frac{\overline{U_1}}{\overline{U_{1m}}},\ \varphi^2 = \frac{\overline{U_I^2}}{\overline{U_{Im}^2}} \tag{10.51}$$

with $\overline{U_1}$, $\overline{U_{1m}}$ and $\overline{U_I^2}$, $\overline{U_{Im}^2}$ as the time-averaged velocity and momentum defects and their maximum values within the wake region.

10.4.4 Near Wake, Far Wake Regions

To estimate the influence region of the wake generating cylinder, also referred to as "near and far wake regions," we use the nondimensional momentum defect ratio or the drag coefficient $C_D = 2\delta_2/d$ and the wake shape factor $H_{12} = \delta_1/\delta_2$, where the velocity defect function φ_1 is introduced:

$$\delta_2 = b\frac{\bar{U}_{1m}}{U_{p^0}}\left(1-\frac{1}{\sqrt{2}}\frac{\bar{U}_{1m}}{U_{p^0}}\right) \tag{10.52}$$

$$C_D = \frac{2\delta_2}{d} = 2\frac{b}{d}\frac{\bar{U}_{1m}}{U_{p^0}}\left(1-\frac{1}{\sqrt{2}}\frac{\bar{U}_{1m}}{U_{p^0}}\right) \tag{10.53}$$

$$H_{12} = \frac{\delta_1}{\delta_2} = \frac{1}{1-\frac{1}{\sqrt{2}}\frac{\bar{U}_{1m}}{U_{p^0}}}. \tag{10.54}$$

Figure 10.17 shows the C_D-distribution for the zero longitudinal pressure gradient cases for a wide range of ξ_1/d locations. The C_w-distribution does not exhibit any major changes, however, considering the wake shape factor H_{12}, shown in Fig. 10.17,

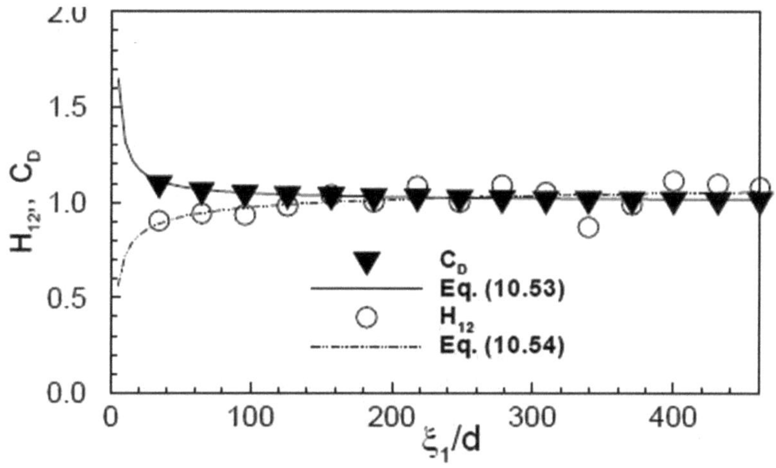

Fig. 10.17 Drag coefficient C_D and shape parameter H_{12} as functions of dimensionless longitudinal distance ξ_1/d with d as that bar diameter

as an alternative indicator, a transition zone $\lambda_{tr} = \xi_1/d < 40$ may be defined, for which the nondimensional wake velocity defect φ_1 indicates a certain dependency upon ξ_1/d. This dependency diminishes for $\lambda_{tr} = \xi_1/d \succeq 40$.

10.4.5 Utilizing the Wake Characteristics

Introducing the wake velocity defect, Eq. (10.45) in connection with Eq. (10.49), into the continuity equation (10.39) and integrating the resulting equation, we obtain an expression for mean lateral velocity $\bar{V}$ as:

$$\left(1 + \frac{\xi_2}{R})\right) \bar{V} = -\int \frac{d}{d\xi_1}\left[U_{p^0}\left(1 - \frac{\xi_2}{R}\right) - \bar{U}_{1m}\varphi_1\right] bd\zeta + c. \tag{10.55}$$

After some further rearrangements of terms in Eq. (10.55) we obtain:

$$\left(1 + \frac{\xi_2}{R}\right) \bar{V} = \frac{d(\bar{U}_{1m}b)}{d\xi_1}\int \varphi_1 d\zeta - \frac{db}{d\xi_1}\bar{U}_{1m}\varphi_1\zeta - \frac{dU_{p^0}}{d\xi_1}b\int\left(1 - \frac{\zeta b}{R}\right)$$
$$d\zeta + c. \tag{10.56}$$

Equation (10.56) and shows that the mean lateral velocity is determined by the turbulent mixing and decay process in longitudinal direction characterized by the longitudinal changes of the velocity-width product, $\bar{U}_{1m}b$, and by the longitudinal changes of potential velocity at wake center U_{p^0}. Note that the longitudinal changes of

U_{p^0} are closely related to the pressure gradient. Since the lateral velocity component $\bar{V}$ is zero at the wake center, the integration constant c in Eq. (10.56) must identically vanish. Thus, the general expression for $\bar{V}$ after evaluating the integrals is:

$$\left(1+\frac{\xi_2}{R}\right)\bar{V} = \frac{d(\bar{U}_{1m}b)}{d\xi_1}\frac{\sqrt{\pi}}{2}erf(\zeta) - \frac{db}{d\xi_1}\bar{U}_{1m}\varphi_1\zeta - \frac{dU_{p^0}}{d\xi_1}b\zeta \left(1-\frac{b}{2R}\zeta\right). \tag{10.57}$$

In Eq. (10.57), $erf(\zeta)$ stands for error function, which is the integral of Gaussian distribution. With Eq. (10.57), the distribution of the mean lateral velocity component can be found provided the wake velocity defect function φ_1, the distribution of the wake width $b = b(\xi_1)$, as well as longitudinal distributions of $\bar{U}_{1m}$ and U_{p^0} are known. The information regarding the distributions of b, $\bar{U}_{1m}$, and U_{p^0} are obtained from the experiment. Similar to the straight wake, a length and a velocity scale are chosen such that the nondimensional wake velocity defect φ_1 is a function of ξ_2/b, i.e.,

$$U\frac{\bar{U}_1}{\bar{U}_{1m}} = \varphi_1 = f\left(\frac{\xi_2}{b}\right) = f(\zeta). \tag{10.58}$$

Similar solution for $\bar{U}_1/\bar{U}_{1m}$ is found by using Eq. (10.40) in conjunction with the order of magnitude analysis of Eq. (10.42). This procedure delivers an ordinary second order differential equation that can be solved numerically. The numerical solution of the resulting ordinary differential equation follows the Gaussian distribution

$$\varphi_1 = e^{-\zeta^2}. \tag{10.59}$$

Experimental results presented in Fig. 10.18 (symbols) over a wide range of ξ_1/d show that for far wake all experimental results collapse to a single curve (solid line) that is described by Eq. (10.59). As seen, the mean nondimensional velocity defect profiles are symmetric and identical to profiles obtained for straight wakes. Comparing the straight wake results presented in Fig. 10.6 and those of curved wakes, Fig. 10.18, leads to the conclusion that after transforming the governing equations into an appropriate wake coordinate system, the assumption of similarity in wake velocity defect profiles is valid. This statement is also valid for positive and negative pressure gradient cases as shown in [133].

Using Eq. (10.50) and considering Eq. (10.59) in conjunction with the experimentally verified assumption of $(\bar{U}_{1m}b) \simeq$ const., the longitudinal velocity component is obtained from:

$$\bar{U} = U_{p^0}\left(1-\frac{b}{R}\zeta\right) - \bar{U}_{1m}\varphi_1. \tag{10.60}$$

Figure 10.19 shows the distribution of the experimental (symbols) and the theoretical (Eq. (10.60)) mean longitudinal velocity component plotted for different longitudinal locations as a function of the lateral distance from the wake center. As shown in Fig. 10.19, the velocity distributions are strongly asymmetric with higher velocities at the positive side of ξ_2 that corresponds to the location closer to the convex wall with $\xi_2 = 0$ as the geometric location of the wake center. Setting in Eq. (10.57) $d(\bar{U}_{1m}b)/d\xi = 0$ and $dU_{p^0}/d\xi_1 \simeq 0$, the lateral velocity component is approximated by:

$$\left(1 + \frac{\xi_2}{R}\right)\bar{V} = -\bar{U}_{1m}\frac{db}{d\xi_1}\zeta\varphi_1. \tag{10.61}$$

It should be noted that, the lateral velocity component is very small compared to the longitudinal one and the accuracy of its measurement falls into the accuracy range of the cross wire probe, with which the velocity components are measured [131].

The partial momentum is the product of Eqs. (10.60) and (10.61) which in conjunction with Eq. (10.50) gives

$$\left(1 + \frac{\xi_2}{R}\right)\frac{\bar{U}\,\bar{V}}{\bar{U}_{1m}^2} = -\frac{U_{p^0}}{\bar{U}_{1m}}\frac{db}{d\xi_1}\left(1 - \frac{b}{R}\zeta\right)\zeta\varphi_1 + \frac{db}{d\xi_1}\zeta\varphi_1^2. \tag{10.62}$$

Since the mean longitudinal turbulent fluctuation in comparison with the mean flow can be neglected, $\bar{u}^2 \ll \bar{U}^2$ and also the variation of the potential velocity at the wake center in ξ_1-direction is very small for the case of zero longitudinal pressure gradient,

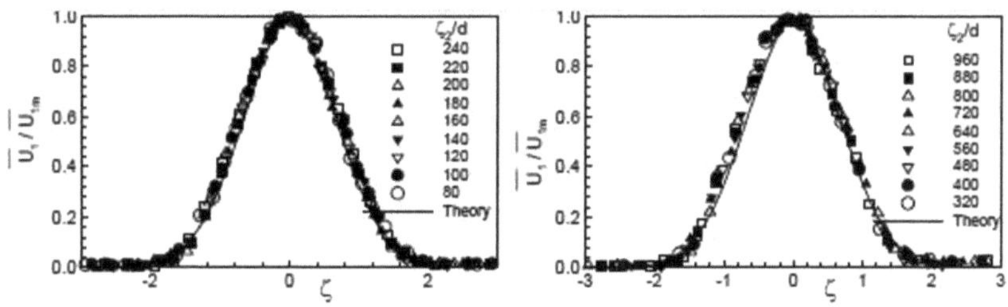

Fig. 10.18 Nondimensional mean velocity defect as a function of dimensionless lateral distance from Schobeiri et al. [132]

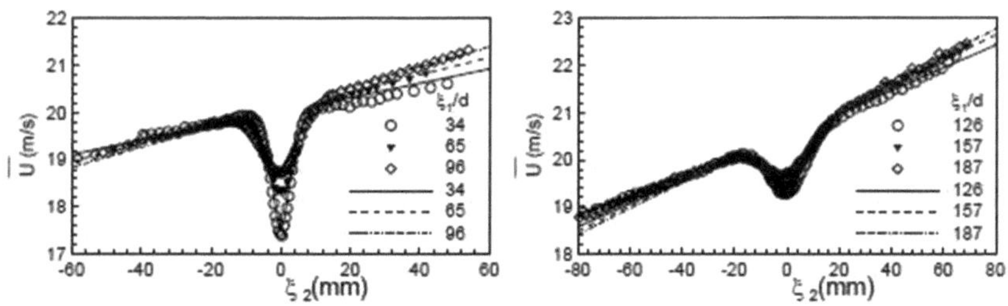

Fig. 10.19 Mean longitudinal velocity distribution in lateral direction, solid lines, Eq. (10.60), experiments form Schobeiri et al. [132]

Eq. (10.42) in connection with Eqs. (10.45) and (10.46) can be simplified as:

$$\frac{R}{R+\xi_2}\frac{\partial}{\partial \xi_1}(-\bar{U}_I^2)+(\overline{UV})_{,2}+\frac{2}{R+\xi_2}(\overline{UV})=0. \tag{10.63}$$

A further comparison of order of magnitude shows that

$$\frac{\overline{UV}}{R}\ll\frac{\partial}{\partial \xi_2}\left[\left(1+\frac{\xi_2}{R}\right)\overline{UV}\right] \tag{10.64}$$

with Eq. (10.64), a further rearrangement and the subsequent integration of the results, the total momentum yields

$$\left(1+\frac{\xi_2}{R}\right)\overline{UV}=\int\frac{\partial}{\partial \xi_1}(\bar{U}_I^2)bd\zeta+c=\int\frac{\partial}{\partial \xi_1}(\bar{U}_{Im}^2\varphi^2)bd\zeta+c. \tag{10.65}$$

In the second integral of Eq. (10.65) $\bar{U}_I^2$ is replaced by the product $\overline{U_I^2}=\overline{U_{Im}^2}\varphi^2$ and $\overline{U_{Im}^2}$ by $\overline{U_{Im}^2}=2U_{p^0}\overline{U_{1m}}-\overline{U_{1m}^2}$. The approximate equality of dimensionless wake velocity and momentum defects, i.e., $\varphi^2\cong\varphi_1$, has been experimentally verified for different locations downstream of the wake generating body. Using this approximation, a further rearrangement of the individual terms in Eq. (10.65) results in:

$$\left(1+\frac{\xi_2}{R}\right)\frac{\overline{UV}}{\bar{U}_{1m}^2}=\frac{U_{p^0}}{\bar{U}_{1m}}\frac{db}{d\xi_1}\left[-2\zeta\varphi_1+\frac{b\varphi_1}{R}(1+2\zeta^2)\right]+\frac{db}{d\xi_1}\zeta\varphi_1+\frac{c}{\bar{U}_{1m}^2}. \tag{10.66}$$

From a physical point of view, its is of interest to determine the order of magnitude of the individual terms involved in Eq. (10.66). The computation of individual terms showed that:

$$\left|\frac{b\varphi_1}{R}(1+2\zeta^2)\right|\ll|2\zeta\varphi_1|,\ \text{and}\ \left|\frac{db}{d\xi_1}\zeta\varphi_1\right|\ll|2\zeta\varphi_1|. \tag{10.67}$$

Equation (10.67) shows that the second term in the bracket as well as the second term on the right-hand side of Eq. (10.67) can be neglected. Despite this fact, these terms were not neglected, when computing Eq. (10.66) to avoid oversimplification. The constant c in Eq. (10.66) is evaluated from experimental results at $\zeta=0$. For near wake its value is zero, however, for $\xi_1/d>100$ it changes slightly in ξ_1-direction but still remains close to zero.

The expression for the turbulent shear stress can be obtained from the difference of total and partial momenta, i.e. Eqs. (10.62) and (10.66)

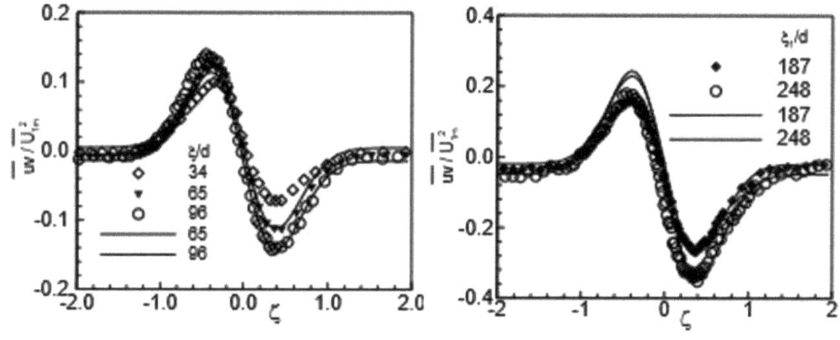

Fig. 10.20 Dimensionless Reynolds shear stress $\overline{uv}/\overline{U_{1m}^2}$ as a function of dimensionless lateral distance ζ with longitudinal distance ξ_1/d as a parameter, curves are calculation results, symbols are experiments from Schobeiri et al. [132]

$$\left(1+\frac{\xi_2}{R}\right)\frac{\overline{uv}}{\bar{U}_{1m}^2}=\frac{U_{p^0}}{\bar{U}_{1m}}\frac{db}{d\xi_1}\left[-\zeta\varphi_1+\frac{b\varphi_1}{R}(1+\zeta^2)\right]+\frac{db}{d\xi_1}\zeta\varphi_1(1-\varphi_1)+\frac{c}{\bar{U}_{1m}^2}. \quad (10.68)$$

The results of calculating Eq. (10.68) are plotted in Fig. 10.20. Figure 10.20 shows the dimensionless Reynolds shear stress distribution at five longitudinal locations. As shown, the shear stress is non-zero at the wake center because of the curvature effect that causes a pressure gradient in the lateral direction, resulting in a highly asymmetric distribution of shear stress profiles. It is interesting to note that at the wake center, where $\partial\bar{U}/\partial\xi_2 = 0$, the shear stress is not necessarily zero. This is in accord with the findings by Raj and Lakshminarayana [134]. They also observed a non-zero value of Reynolds shear stress at the wake center. They also concluded that the mixing length hypothesis is not valid for predicting the mean and turbulent quantities in such a region. The Reynolds shear stress in the hypothetical potential flow outside the wake is not exactly equal to zero due to the turbulence existing in that region. Also the $\overline{uv}$ outside the wake has a higher absolute value near the concave side of the wall ($\xi_2 < 0$). Measurements at selected longitudinal locations without the wake showed a lateral gradient of $\overline{uv}$ with a negative value near the concave wall and a positive value near the convex wall. The radial position where $\overline{uv} = 0$ was located between the convex wall and mean radius of the channel. A similar distribution of Reynolds shear stress has been observed in turbulent flows in curved channels, as reported by Wattendorf [135] and Eskinazi and Yeh [136].

As shown in Fig. 10.20, the shear stress distribution is strongly asymmetric, which can be attributed to the asymmetry of the mean longitudinal velocity component. Generally, in a curved shear flow, the positive velocity gradient in a positive radial direction suppresses turbulence (stabilizing effect) while a negative velocity gradient in a positive radial direction promotes turbulence (destabilizing effect). From the mean longitudinal velocity distributions it is apparent that the velocity gradient is negative in the positive radial direction on the inner half of the wake (the concave side of the trajectory of the wake centerline). Thus, higher values of Reynolds shear stress are expected on the inner half of the wake. The opposite trend is true for outer half of the wake, which results in lower values of Reynolds shear stress. It appears that the Reynolds shear stress in the outer half of the wake is more closer to self-preservation than the inner half of the wake.

The experimentally determined shear stress distributions (symbols) shown in Fig. 10.20, are compared with the developed inductive theory (lines). As shown, the shear stress was calculated as the difference between the total and partial momenta $\overline{uv} = \overline{UV} - \bar{U}\,\bar{V}$ by integrating the conservation equations. The integration constants in the corresponding expressions were evaluated from experimental measurements corresponding to the values at the wake center.

Wake Flow, Concluding Remarks: The inductive approached developed for predicting the wake decay and development in straight and curved channels at zero-longitudinal pressure gradient presented in this chapter was further extended to cases with negative and positive pressure gradients [133]. It also was extended to periodic unsteady wakes [132]. This approach is an alternative for predicting the free wakes as well as channel wakes characteristics based on experimental findings. Simple relationships for wake velocity distribution were found to derive the longitudinal and lateral velocity distributions as well as the turbulent shear stress. It is of course possible to use the turbulence models discussed in Chap. 9 to predict the wake characteristics. This task is presented as Problems at the end of this chapter.

10.5 Computational Projects

Project 1: Air exits through a subsonic two-dimensional nozzle into the atmospheric environment. The nozzle has the width of 200.00 mm and a height of 50.00 mm. Using the inductive approach discussed in Sect. 10.3, write (1) a computer code to calculate free jet turbulence quantities. (2) Utilize an existing CFD-platform for calculating the same quantities using (a) the mixing length model, (b) $k - \varepsilon$ model, (c) $k - \omega$ model and (d) SST-model. Critically analyze the results.

Project 2: Given is a straight duct with the width of 1000.00 mm, the height of 500.00 mm and a length of 1000 mm. Establish a zero pressure gradient in longitudinal direction by slightly opening the top and bottom channel walls. At $x = 100.00$ mm insert a cylindrical rod of 2 mm diameter. Assume an inlet velocity of 10 m/s and a static pressure at the inlet which is equal to the difference of atmospheric total pressure and the inlet dynamic pressure. Write a computer program for calculating the wake velocity defect, longitudinal and lateral velocity components and the Reynolds shear stress. Use (1) the inductive approach discussed in this chapter. (2) Utilize an existing CFD-platform for calculating the same quantities using (a) the mixing length model, (b) $k - \varepsilon$ model, (c) $k - \omega$ model and (d) SST-model. Critically analyze the results.

Project 3: Given is a curved channel similar to the one shown in Fig. 10.15. The walls are arranged concentric. The convex top wall has a radius of $R_I = 500.00$ mm, whereas the bottom concave wall radius is $R_0 = 920.00$ mm. Assuming an inlet velocity of 10.0 m/s, write a computer code to verify the results of the inductive approach presented Sect. 10.3.

Chapter 11
Boundary Layer Aerodynamics

In Chap. 9 we have shown that using the computational fluid dynamics (CFD), flow details in and around complex geometries can be predicted accuracy. The flow field calculation includes details very close to the wall, where the viscosity plays a significant role. In the absence of random fluctuations the (laminar) flow can be calculated with high accuracy. For predicting turbulent flows, however, turbulence models were required to be implemented into the Navier-Stokes equations to account for turbulence fluctuations. One of the more important tasks in engineering fluid mechanics is to predict the drag forces acting on the surfaces of components, among others, pipes, diffusers, nozzles, turbines, compressors, or wings of aircrafts. As seen in Chap. 5, the drag forces are produced by the fluid viscosity which causes the shear stress acting on the surface. The question that arises is how far from the surface the viscosity dominates the flow field. Prandtl [137] was the first to answer this question. Combining his physical intuition with experiments, he developed the concept of the boundary layer theory. In what follows the concept of the boundary layer theory for two dimensional flow is presented. Utilizing the two-dimensional boundary layer approximation by Prandtl, and for the sake of simplicity, we use the boundary layer nomenclature with the mean-flow component, $V_1 \equiv u$, $V_2 \equiv v$, as the significant velocities in $x_1 \equiv x$, and $x_2 \equiv y$-direction.

Based on his experimental observations, Prandtl found that effect of the viscosity is confined to a thin viscous layer that he called, the *boundary layer*. Prandtl estimated that at any longitudinal position x the boundary layer thickness $\delta(x)$ compared to the position x is small, meaning that $\delta \ll x$. For the flat plate under zero pressure gradient shown in Fig. 11.1 with the length L, we have $\delta_L/L \ll 1$. If we assume that $x \propto L$ and $y \propto \delta$, then we may estimate the changes in longitudinal direction compared to the normal one. Furthermore, based on Prandtl's experimental findings, following order of magnitude comparison holds:

M. T. Schobeiri, *Advanced Fluid Mechanics and Heat Transfer for Engineers and Scientists*, https://doi.org/10.1007/978-3-030-72925-7_11

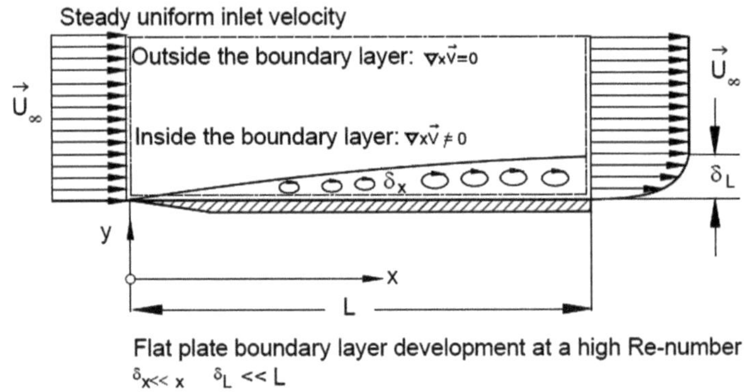

Fig. 11.1 Development of boundary layer along a flat plate at zero pressure gradient and high Re-number

$$\frac{\partial}{\partial x} \ll \frac{\partial}{\partial y} \sim \frac{1}{\delta} \text{ and } \frac{\partial^2}{\partial x^2} \ll \frac{\partial^2}{\partial y^2} \sim \frac{1}{\delta^2}. \tag{11.1}$$

The above order of magnitude estimation enables a substantial simplification of the Navier-Stokes equations that can be solved relatively easily. Furthermore, the concept of the boundary layer theory allows the separation of a flow field into the boundary layer region where the viscous forces play a dominant role and a region outside the boundary layer, where the convective forces dominate the flow field.

Thus, it is admissible to treat the region outside the boundary layer as a quasi-inviscid flow field that is described by the Euler equation of motion. Thus, the approximate solution of the flow field is composed of the viscous solutions, described by the boundary layer theory and the non-viscous Euler solution. To combine these two solutions, the boundary layer solution has to satisfy the no-slip condition at the wall and at the edge of the boundary layer, where the two solutions tangent each others they must share the same values and the same slopes. The combination of the viscous-inviscid solutions can exhibit a very fast alternative for providing information about the distribution of the drag forces. The major shortcoming of the boundary layer theory is that it is not capable of handling the flow separation. It also does not account for the boundary layer transition.

11.1 Boundary Layer Approximations

The theoretical structure of boundary layer theory is based on the Navier-Stokes equations for incompressible steady and two dimensional flows. Assuming a flow with a large Reynolds number and the order of magnitude comparison in Eq. (11.1) for the velocity component in x-direction and a dimensionless boundary layer thickness $\delta/L \equiv \delta$ with $L = 1$ the following order of magnitude estimations hold:

$$\frac{\partial u}{\partial x} \sim 1, \frac{\partial u}{\partial y} \sim \frac{1}{\delta} \text{ and } \frac{\partial^2 u}{\partial x^2} \sim 1, \frac{\partial^2 u}{\partial y^2} \sim \frac{1}{\delta^2}. \quad (11.2)$$

To estimate the order of magnitude of all terms involved in a two-dimensional incompressible Navier-Stokes equations, we first introduce the following dimensionless quantities:

$$x^* = \frac{x}{L}, V^* = \frac{V}{V_\infty}, P^* = \frac{p}{\rho V_\infty^2}, t^* = \frac{tV_\infty}{L}, Re = \frac{\varrho V_\infty L}{\mu} = \frac{V_\infty L}{\nu}. \quad (11.3)$$

Introducing the dimensionless quantities (11.3) into the Navier-Stokes equations, we encounter the Re-number defined in Eq. (11.3), for which we need to find the order of magnitude. This is done by establishing a ratio between the convective forces that dominate the flow outside the boundary layer and the viscous forces inside the boundary layer. We assume that within the boundary layer the viscous forces have the same order of magnitude as the convective forces. This assumption leads to:

$$\frac{u}{\nu} \frac{\partial u/\partial x}{\partial^2 u/\partial y^2} \sim 1. \quad (11.4)$$

Equation (11.4) can be used to estimate the order of magnitude of the boundary layer thickness δ and to relate it to the Re-number. Using U_∞ as the undisturbed velocity and the L as the reference length scale in x-direction, we find the order of magnitude for the numerator of Eq. (11.4):

$$u\frac{\partial u}{\partial x} \sim \frac{U_\infty^2}{L}. \quad (11.5)$$

The length scale in the y direction is the boundary layer thickness δ, so that the following relationship holds

$$\nu\frac{\partial^2 u}{\partial y^2} \sim \nu\frac{U_\infty}{\delta^2}. \quad (11.6)$$

Using Equations (11.5) and (11.6) we estimate the order of magnitude of:

$$\frac{U_\infty^2/L}{\nu U_\infty/\delta^2} \sim 1. \quad (11.7)$$

Inserting $Re = U_\infty L/\nu$ into Eq. (11.7) we obtain

$$\frac{\delta}{L} \sim Re^{-\frac{1}{2}}, \quad \text{or} \quad \left(\frac{\delta}{L}\right)^2 = \frac{1}{Re}. \quad (11.8)$$

Implementing Eqs. (11.4)–(11.6) into the Navier-Stokes equations, we obtain its dimensionless version:

$$\frac{\partial \mathbf{V}^*}{\partial t*} + \mathbf{V}^* \cdot \nabla \mathbf{V}^* = \nabla p^* + \frac{1}{\mathbf{Re}} \Delta \mathbf{V}^*. \tag{11.9}$$

Decomposing Eq. (11.9), the x-component with its order of magnitude estimation yields:

$$\begin{array}{ccccccc} \dfrac{\partial u^*}{\partial t*} & + u^* \dfrac{\partial u^*}{\partial x^*} & + v^* \dfrac{\partial u^*}{\partial y^*} & = -\dfrac{\partial p}{\partial x^*} & + \dfrac{1}{Re} \Bigg(& \dfrac{\partial^2 u^*}{\partial x^{*^2}} & + \dfrac{\partial^2 u^*}{\partial y^{*^2}} \Bigg) \\ 1 & 1 \quad 1 & \delta^* \dfrac{1}{\delta^*} & & \delta^{*^2} & 1 & \dfrac{1}{\delta^{*^2}} \end{array} \tag{11.10}$$

similarly for the y-component we have

$$\begin{array}{ccccccc} \dfrac{\partial v^*}{\partial t*} & + u^* \dfrac{\partial v^*}{\partial x^*} & + v^* \dfrac{\partial v^*}{\partial y^*} & = -\dfrac{\partial p^*}{\partial y^*} & + \dfrac{1}{Re} \Bigg(& \dfrac{\partial^2 v^*}{\partial x^2} & + \dfrac{\partial^2 v^*}{\partial y^2} \Bigg) \\ \delta^* & 1 \quad \delta^* & \delta^* \; 1 & & \delta^{*^2} & \delta^* & \dfrac{1}{\delta^*} \end{array} \tag{11.11}$$

with the dimensionless boundary layer thickness $\delta^* \equiv \delta/L$. Since $\delta^* \ll 1$ all terms with the order of magnitude of δ^* can be neglected compared to those with the magnitude of 1.

Going back to the dimensional Navier-Stokes equations and assuming a steady flow, the consequence of the order of magnitude estimation in Eq. (11.10) is that the only term that can be omitted is the shear stress term $\nu \partial^2 u/\partial x^2$. Thus, the x component of the Navier-Stokes equations reduces to

$$u\frac{\partial u}{\partial x} + v\frac{\partial u}{\partial y} = -\frac{1}{\varrho}\frac{\partial p}{\partial x} + \nu\frac{\partial^2 u}{\partial y^2} \tag{11.12}$$

with u and v as the velocity components in x and y-directions. On the other hand, the only term in the y-component that survives is

$$0 + 0 = -\frac{1}{\varrho}\frac{\partial p}{\partial y} + 0 + 0. \tag{11.13}$$

With the continuity equation

$$\frac{\partial u}{\partial x} + \frac{\partial v}{\partial y} = 0 \tag{11.14}$$

the system of three differential equations is complete that allows to calculate the boundary layer. The y-component, Eq. (11.13), indicates that at any arbitrary x-position the pressure inside the boundary layer including the boundary layer edge at

$y = \delta$ remains constant, meaning that $p = p(x)$. This implies that pressure inside the boundary layer $p(x)$ has the same value as outside it. This value is known from the inviscid solution, where it can be obtained by differentiating the Bernoulli equation:

$$-\frac{1}{\varrho}\frac{\partial p}{\partial x} = U\frac{\partial U}{\partial x} \tag{11.15}$$

with the velocity U outside the boundary layer. The solution of differential equations (11.12)–(11.15) require boundary conditions that must be formulated from case to case as we will discuss in the following sections. For certain cases of laminar boundary layer problems, the system of partial differential equations can be transformed into a system of ordinary differential equations that can be solved using either finite difference or Runge-Kutta integration method. In the following sections we distinguish between laminar and turbulent boundary layers that we treat separately.

11.2 Exact Solutions of Laminar Boundary Layer Equations

Exact solutions for a class of laminar boundary problems are presented in this section. In the context of boundary layer theory, a solution is considered exact when it is a complete solution of the boundary layer equation, irrespective of whether it is obtained analytically or numerically. However, it should be noted that even for the simplest turbulent boundary layer problem, no analytical solution has been found. For a certain class of boundary layer cases, the streamwise pressure distribution outside the boundary layer can be explained as simple power functions. Figure 11.2 shows two representative cases of accelerated flow through a turbine cascade and a nozzle and decelerated flow through a compressor cascade and a diffuser. In these cases, the velocity distributions outside the boundary layer can be approximated by simple power laws allowing the partial differential equations (11.12)–(11.15) reduce to ordinary differential equations that can be solved using initial and boundary conditions. Considering cases, where the velocity distributions can be described by a simple power law:

$$U(x) = Cx^m \tag{11.16}$$

where the exponent m represents different flow types. As an example $m = 0$ represents the boundary layer flow along a flat plate at zero pressure gradient. For stagnation point flow the exponent is $m = 1$. Flow past a wedge with different angles that are directly related to the exponent m is another interesting example.

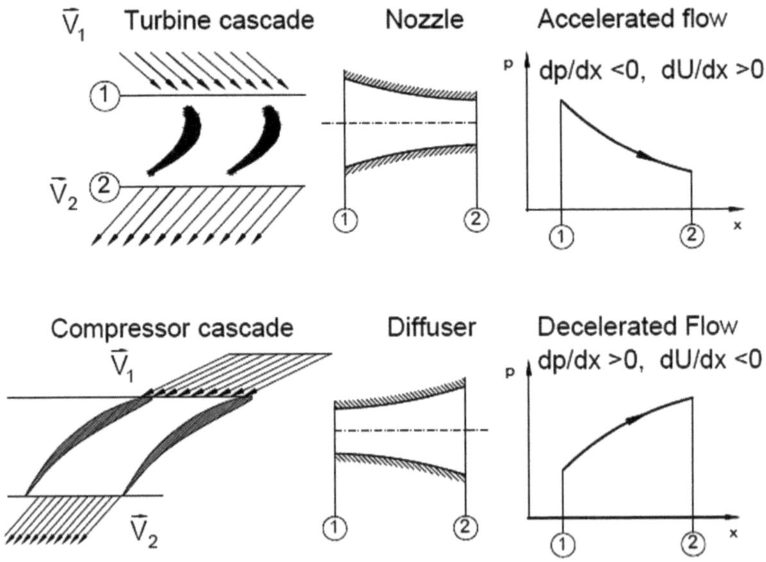

Fig. 11.2 Velocity distributions outside the boundary layer for accelerated flows through a turbine cascade and a nozzle, decelerated flow through a compressor cascade and a diffuser

11.2.1 Laminar Boundary Layer, Flat Plate

The laminar flow along a flat plate at zero pressure gradient with a constant velocity outside the boundary layer $U = U_\infty$ =const. exhibits the first application of Prandtl boundary layer theory. Staring from the Prandtl boundary layer Eq. (11.12) we set $\partial p/\partial x = 0$ that leads to:

$$u\frac{\partial u}{\partial x} + v\frac{\partial u}{\partial y} = \nu\frac{\partial^2 u}{\partial y^2}. \tag{11.17}$$

To find an exact solution for the velocity distribution, Blasius [138] introduced the following dimensionless coordinates within the laminar boundary layer along the flat plate:

$$\xi = \frac{x}{L}, \text{ and } \eta = \frac{y}{\delta} \tag{11.18}$$

with L as the plate length and $\delta = \delta(x)$ as the boundary layer thickness. From Eq. (11.8) we may set $\delta\sqrt{\nu x/U_\infty}$ and introduce a stream function

$$\psi = f(\eta)\sqrt{\nu x U_\infty} \tag{11.19}$$

with $\eta = y/\delta = y/\sqrt{\nu x/U_\infty}$ and the velocity component as defined in Eq. (6.17)

$$u = \frac{\partial \psi}{\partial y} = \frac{\partial \psi}{\partial \eta}\frac{\partial \eta}{\partial y}, v = -\frac{\partial \psi}{\partial x}. \tag{11.20}$$

Thus we obtain for u and v the following relationships:

$$u = \left(\frac{\partial f}{\partial \eta}\sqrt{\nu x U_\infty}\right)\sqrt{\frac{U_\infty}{\nu x}} = f'(\eta)U_\infty \tag{11.21}$$

and

$$v = -\frac{\partial \psi}{\partial x} = -\frac{\partial}{\partial x}(f(\eta)\sqrt{\nu x U_\infty}) = -\sqrt{\nu x U_\infty}\frac{\partial f(\eta)}{\partial x} - \frac{f(\eta)}{2}\sqrt{\nu U_\infty/x}. \tag{11.22}$$

Further differentiation leads to

$$v = -\sqrt{\nu x U_\infty}\frac{\partial f(\eta)}{\partial \eta}\frac{\partial \eta}{\partial x} - \frac{f(\eta)}{2}\sqrt{\nu U_\infty/x} = \frac{1}{2}\sqrt{\frac{\nu U_\infty}{x}}(\eta f'(\eta) - f(\eta)) \tag{11.23}$$

with$f'(\eta) = \partial f/\partial \eta$. Using the same differentiation procedure as in Equations (11.20) and (11.23) we arrive at the velocity derivatives in x- and y-directions:

$$\frac{\partial u}{\partial x} = -\frac{\eta}{2x}U_\infty f''(\eta) \tag{11.24}$$

and in y-direction

$$\frac{\partial u}{\partial y} = U_\infty\sqrt{\frac{U_\infty}{\nu x}}f''(\eta). \tag{11.25}$$

Inserting these terms into Eq. (11.17) we arrive at a nonlinear third order ordinary differential equation

$$2f'''(\eta) + f(\eta)f''(\eta) = 0. \tag{11.26}$$

Equation (11.26) developed by Blasius has to satisfy the no-slip condition at the wall, namely:

$$y = \eta = 0 :=> u = v0, \text{ with } u = U_\infty f'(\eta = 0) = 0 => f'(\eta = 0) = 0 \tag{11.27}$$

and at some far distance from the boundary layer

$$y \gg \delta : u = U_\infty = U_\infty f', => f'(\eta \gg \delta) = 1. \tag{11.28}$$

Actually, the Blasius equation with the above boundary conditions exhibits a boundary value problem. However, using an iterative method, similar to the one discussed

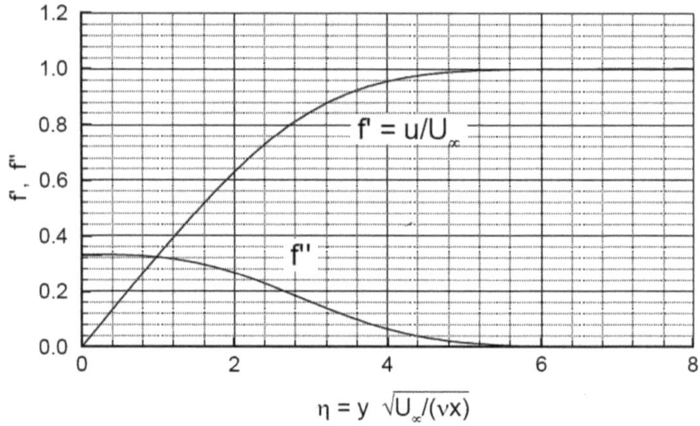

Fig. 11.3 Velocity $f' = u/U_\infty$ and its slope f'' as a function of dimensionless parameter η

in Chap. 7, it can be treated as an initial value problem. Assuming a certain initial value for $f''(\eta = 0)$, Eq. (11.26) can be solved using Runge-Kutta or Predictor- Corrector method to calculate, among others, $f'(\eta \gg \delta)$. Figure 11.3 exhibits a plot of the Blasius profiles for $f' = u/U_\infty$ and the velocity slope f''. As shown, the velocity asymptotically approaches the unity, while the velocity gradient f'' approaches zero. The velocity slope at the wall indicates the order of magnitude of the wall shear stress τ_w which can be calculated from

$$\tau_w = \mu \frac{\partial u}{\partial y}\bigg|_{y=0} = \mu U_\infty \sqrt{\frac{U_\infty}{\nu x}} f''(0). \tag{11.29}$$

For a flat plate with a length L, the friction force per unit of depth is calculated by integrating Eq. (11.29)

$$F = b \int_{x=0}^{L} \tau_w dx. \tag{11.30}$$

Inserting Eq. (11.29) into Eq. (11.30), we find

$$F = \mu b U_\infty f''(0) \sqrt{\frac{U_\infty}{\nu}} \int_{x=0}^{L} \frac{dx}{\sqrt{x}} = 2bf''(0)\varrho U_\infty^2 L \left(\frac{U_\infty L}{\nu}\right)^{-1/2} \tag{11.31}$$

with b as the depth of the plate. The drag coefficient is calculated from

$$c_f = \frac{F}{\varrho/2U_\infty^2 bL} = \frac{4f''(0)}{\sqrt{Re}} = \frac{1.328}{\sqrt{Re}} \tag{11.32}$$

with $f''(0) = 0.332$ as shown in Fig. 11.3.

11.2.2 Wedge Flows

As discussed in the preceding section, to arrive at the Blasius solution for a flat plate at zero pressure gradient, we assumed a constant velocity $U = U/_{\infty}$ =const. outside the boundary layer. This assumption in conjunction with Eq. (11.15) has lead to $\partial U/\partial x = 0$ and the subsequent elimination of the pressure gradient, $\partial p/\partial x = 0$ from the Navier-Stokes equations. Introducing a dimensionless coordinate η and a stream function $\Psi = \Psi(\eta)$, the partial differential equations of motion were reduced to an ordinary differential equation, which we solved as an initial value problem by iteratively obtaining $f''(0) = 0.332$ as shown in Fig. 11.3. The solution that we obtained satisfied the similarity requirement namely that the velocities at any two locations are "similar", meaning that all velocity profiles became identical when u/U_{∞} is plotted against η. In this section, we treat a flow past a wedge with an angle of 2β as shown in Fig. 11.4. Varying the wedge angle causes the pressure distribution outside the boundary layer to change. A positive wedge angle is associated with a positive value of m in Eq. (11.16) implying that the flow outside of the boundary layer is accelerated.

We assume a non-zero-pressure gradient outside the boundary layer and require that the solutions must satisfy the similarity condition:

$$\frac{u(x_1, \eta_1)}{U(x_1)} = \frac{u(x_2, \eta_2)}{U(x_2)} \tag{11.33}$$

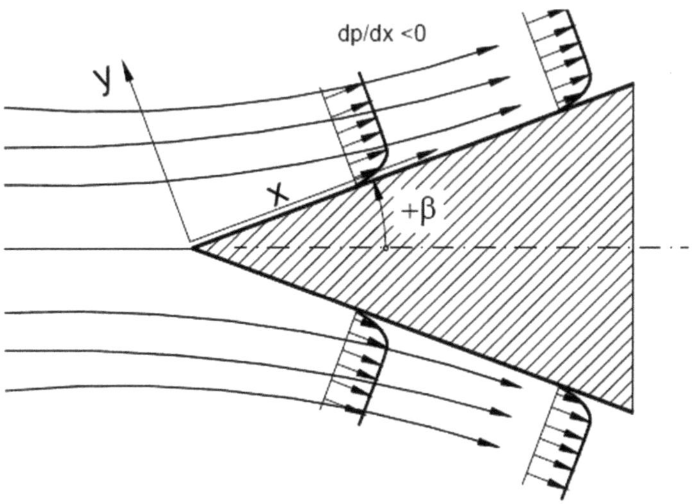

Fig. 11.4 Boundary layer development along a wedge

with $\eta = y/\delta(x)$. Now we consider a set of differential equations consisting of continuity and Navier-Stokes equations with a non-zero pressure gradient term in the x-component:

$$\frac{\partial u}{\partial x} + \frac{\partial v}{\partial y} = 0 \tag{11.34}$$

$$u\frac{\partial u}{\partial x} + v\frac{\partial u}{\partial y} = U\frac{dU}{dx} + \nu\frac{\partial^2 u}{\partial y^2}. \tag{11.35}$$

Introducing a stream function $\psi = \psi(x, y)$ with

$$u = \frac{\partial \psi}{\partial y} = \frac{\partial \psi}{\partial \eta}\frac{\partial \eta}{\partial y}, v = -\frac{\partial \psi}{\partial x} \tag{11.36}$$

the equation of motion in x-direction becomes

$$\frac{\partial \psi}{\partial y}\frac{\partial^2 \psi}{\partial x \partial y} - \frac{\partial \psi}{\partial x}\frac{\partial^2 \psi}{\partial y^2} = U\frac{dU}{dx} + \nu\frac{\partial^3 \psi}{\partial y^3} \tag{11.37}$$

which has to satisfy the boundary conditions $u = \partial\psi/\partial y = 0$ and $v = \partial\psi/\partial x = 0$ for $y = 0$ and $\partial\psi/\partial y = U$ at $y = \infty$. Furthermore, we require that the velocity distribution outside the boundary layer follows a simple power function

$$U(x) = Cx^m, U\frac{dU}{dx} = C^2 m x^{2m-1}. \tag{11.38}$$

Introducing the same relationship for the stream function as in Eq. (11.19), namely

$$\psi = f(\eta)\sqrt{\nu x U_\infty} \tag{11.39}$$

with the same dimensionless similarity coordinate as in the preceding section $\eta = y/\delta = y/\sqrt{\nu x/U_\infty}$, Eq. (11.37) is transformed into an ordinary differential equation

$$f''' + \frac{m+1}{2}ff'' + m(1 - f'^2) = 0. \tag{11.40}$$

This is the so-called *Falkner-Skan equation* [139] which describes a laminar flow past a wedge as shown in Fig. 11.4. The solution of this equation was provided by Hartree [140] and is presented in Fig. 11.5. As discussed by Schlichting [142] and Spurk [143], the velocity exponent m is related to the wedge angle by the following equation:

$$\beta = \pi\frac{m}{m+1}. \tag{11.41}$$

The special case of $m = 0$ delivers the Blasius equation discussed in the preceding section. Equation (11.40) is a nonlinear ordinary differential equation that can be

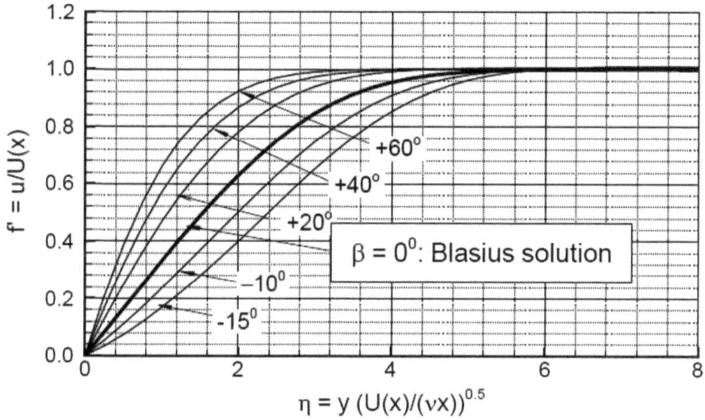

Fig. 11.5 Boundary layer velocity distributions in a wedge flow, accelerated flow: $\beta > 0$, decelerated flow: $\beta < 0$

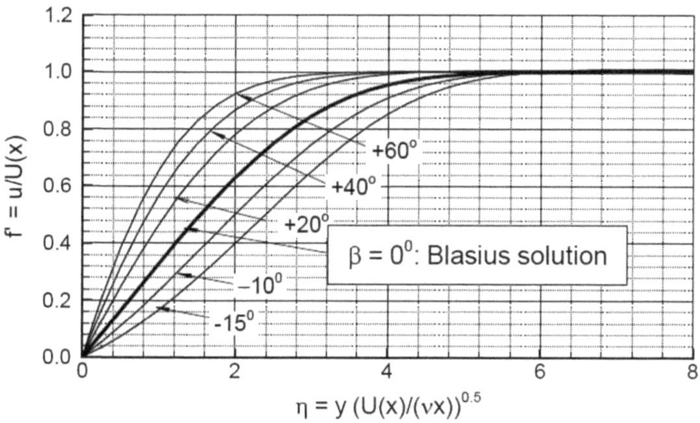

Fig. 11.6 Flow along a convex corner ($\beta < 0$)

reduced to an initial value problem by iteratively determining the derivative $f''(\eta = 0)$ such that the boundary condition $f' = u/U_\infty = 1$ for $\eta = \infty$ is satisfied. The solution of Falkner-Skan equation is plotted in Fig. 11.5, where the dimensionless velocity $f' = u/U(x)$ is plotted against the dimensionless coordinate η with β as a parameter. The special case of $\beta = 0$ represents the laminar flow along a flat plate with $m = 0$. Increasing the wedge angle β cause the flow outside the boundary layer to accelerate. The cases with negative β that correspond to negative $m < 0$ are taught of flows past a convex corner.

Once the potential flow passes over a convex corner, the streamlines diverge causing a flow deceleration that is associated with a pressure increase in flow direction as sketched in Fig. 11.6.

In this case the two forces, namely the viscous force and the pressure force co-act against the movement of the fluid particle. Thus, very close to the wall, the slope of the velocity profile becomes zero showing a typical inflection pattern. This is indicative of the beginning of a flow separation as shown in Fig. 11.6. Given the inherent susceptibility of the laminar boundary layer to even very small positive pressure gradients, the results for $\beta < -10^0$ do not seem to be plausible.

11.2.3 Polhausen Approximate Solution

Considering the exact solutions of Blasius and Falkner-Skan flows, one may conclude that for these type of flows approximate solutions can be obtained by using simple polynomial equations. In case of a laminar flow with pressure gradient as a parameter, Pohlhausen [143] assumed a forth order polynomial for the dimensionless velocity distribution as a function of $\eta = y/\delta(x)$ with the pressure gradient

$$\Lambda = -\frac{dp}{dx}\frac{\delta}{\mu U/\delta} = \frac{\delta^2}{\nu}\frac{dU}{dx} \tag{11.42}$$

as the parameter. To obtain an approximate solution, Polhausen set:

$$\frac{u}{U} = \Sigma_{n=1}^{4} a_i \eta^i = a_1\eta + a_2\eta^2 + a_3\eta^3 + a_4\eta^4 \tag{11.43}$$

with a_i as free constants that have to satisfy the continuity and the Navier-Stokes boundary conditions for exact solutions:

$$\begin{aligned} &\text{at } y = 0\text{: } \nu\frac{\partial^2 u}{\partial y^2} = \frac{1}{\rho}\frac{dp}{dx} = -U\frac{dU}{dx} \\ &\text{at } y = \delta\text{: } u = U,\ \frac{\partial u}{\partial y} = 0,\ \frac{\partial^2 u}{\partial y^2} = 0. \end{aligned} \tag{11.44}$$

The above boundary conditions are sufficient to find the coefficients a_i. There is no need to define explicitly the no-slip condition at the wall, since Eq. (11.43) inherently satisfies this requirement. Using the boundary conditions (11.44), following expressions are obtained for the coefficients in Eq. (11.43)

$$a_1 = 2 + \frac{\Lambda}{6};\ a_2 = -\frac{\Lambda}{2};\ a_3 = -\frac{\Lambda}{2};\ a_4 = 1 - \frac{\Lambda}{6} \tag{11.45}$$

and hence the velocity profile can be expressed in terms of

$$\frac{u}{U} = (2\eta - 2\eta^3 + \eta^4) + \frac{\Lambda}{6}(\eta - 3\eta^2 + 3\eta^3 - \eta^4). \tag{11.46}$$

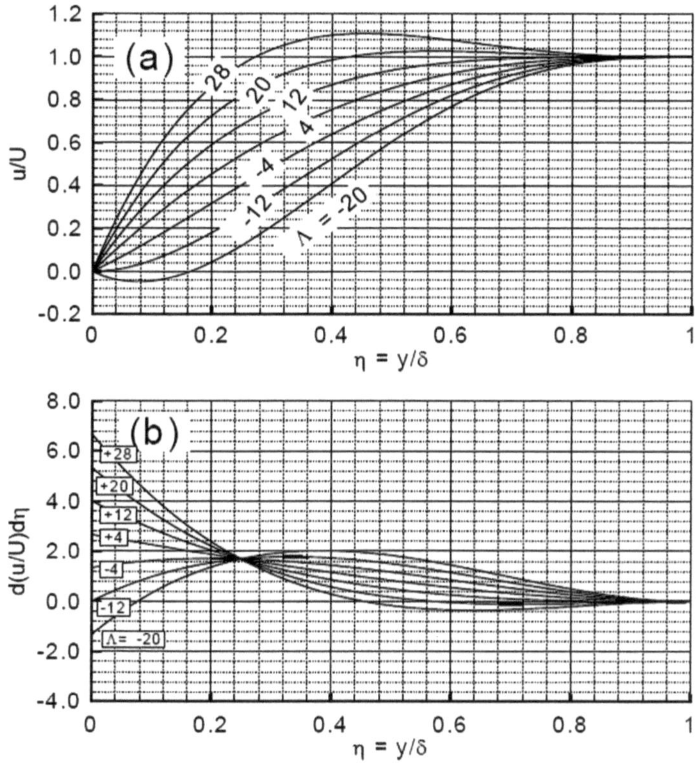

Fig. 11.7 Pohlhausen profiles: **a** velocity profiles, **b** slope of the velocity profiles with Λ as parameter

The velocity profiles and their slopes are plotted in Fig. 11.7a, b. Figure 11.7a exhibits the velocity profiles as a function of η with the dimensionless pressure gradient Λ as a parameter. Accelerated flows are denoted by $\Lambda > 0$, while decelerated flows are characterized by $\Lambda < 0$. The slopes of the velocities are plotted in Fig. 11.7b. Of particular interest are the slopes at the wall since they determine the curvature of the velocity profiles near the wall. The curves with $\Lambda_{\eta=0} = -12$ with zero-slope at the wall, Fig. 11.7a ,b indicate the point of inflection at the wall. The curves with $\Lambda_{\eta=0} < -12$ pertaining to separated flow situations are also plotted. However, they are not compatible with the concept of the boundary layer theory, which excludes flow separation.

11.3 Boundary Layer Theory Integral Method

As we saw in the preceding sections, in order to solve laminar boundary layer problems, Navier-Stokes equations were drastically simplified. Detailed velocity distribution within the boundary layer were presented that allowed the calculation of the friction forces caused by the wall shear stress acting on the surface under investigation. Accurate calculation of the wall shear stress is of primary importance for calculation of the total pressure loss and thus the efficiency of any engineering device, within which a fluid dynamic, heat transfer or energy conversion takes place. The integral method presented in this section offers an alternative to determine the wall shear stress. It is based on continuity, momentum and energy equations in integral form as treated in Chap. 5. Applying the integral balances to a boundary layer problem, we find the boundary layer thicknesses that are part of the boundary layer integral equation derived in the following.

11.3.1 Boundary Layer Thicknesses

Fig. 11.8 shows the different nature of the boundary layer developed along a flat plate at zero streamwise pressure gradient as we discussed in more details in Chap. 8. The application of integral balances to the control volume shown in Fig. 11.8 delivers the boundary layer *displacement thickness* δ_1, the boundary layer *momentum deficiency thickness* δ_2 and the energy *deficiency thickness* δ_3.

Figure 11.8 shows the boundary layer development along a flat plate. To calculate the above thicknesses, different control volumes may be placed on the flat plate. Regardless of the choice, they must include the inlet velocity and the velocity distribution at any arbitrary position in x-direction. The two control volumes delivers the same results. While the control surface BC in Fig. 11.8a is parallel to the plate, the BC-surface in Fig. 11.8b is a streamline that goes through the edge of the boundary layer at point C and intersects with the surface AB at the inlet. Applying the integral balances of continuity, momentum and energy to Fig. 11.8a requires the mass flow calculations along BC, which is immediately found by subtracting the mass flow through CD from the one at AB. Using Fig. 11.8 (b), the mass flow balance needs to apply to the inlet surface AB and the exit surface CD. The stream surface BC act as a solid surface, where no mass flow can cross. Furthermore Fig. 11.8b shows, the *displacement* δ_1 of the streamline from the height h at the inlet to the height $h+\delta_1$. Application of the continuity balance Eq. (5.4) to Fig. 11.8b results in:

$$\int_S \rho \mathbf{V}\cdot\mathbf{ndS} = \int_{S_{AB}} \rho \mathbf{V}\cdot\mathbf{ndS} + \int_{S_{CD}} \rho \mathbf{V}\cdot\mathbf{ndS} = \mathbf{0}. \tag{11.47}$$

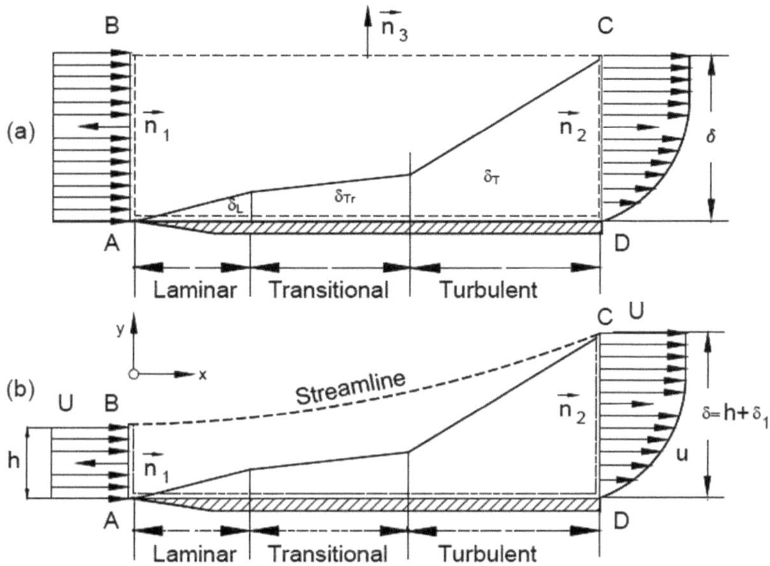

Fig. 11.8 Boundary layer development along a flat plate at zero pressure gradient

Assuming incompressible flow and considering the continuity balance per unit of depth, Eq. (11.47) reduces to

$$-\int_0^h U dy + \int_0^\delta (u - U + U)dy = 0; -Uh + U\delta + \int_0^\delta (u - U)dy = 0. \quad (11.48)$$

Replacing in Eq. (11.48) δ by $\delta = h + \delta_1$, we find immediately the boundary layer displacement thickness:

$$\delta_1 = \int_0^\delta (1 - \frac{u}{U})dy. \quad (11.49)$$

Applying the linear momentum Eq. (5.25) to the same control volume while neglecting the shear stress integrals, we find the drag force per unit of depth:

$$D = \int_0^h U\dot{dm} - \int_0^\delta u\dot{dm} = U\int_0^\delta \varrho u dy - \int_0^\delta \varrho u^2 dy$$

$$D \equiv \varrho \int_0^\delta (Uu - u^2)dy = \varrho \int_0^\delta u(U - u)dy. \quad (11.50)$$

Non-dimensionalizing the drag force with a reference force per unit of depth $\frac{1}{2}\varrho U^2 L$ gives:

$$C_D = \frac{2}{L}\int_0^\delta a\frac{u}{U}\left(1-\frac{u}{U}\right)dy = \frac{2\delta_2}{L}. \tag{11.51}$$

With L as the plate length an δ_2 as the momentum deficiency thickness or short *momentum thickness*. In an analogous way we find the energy deficiency thickness δ_3:

$$\delta_3 = \int_0^\delta \frac{u}{U}\left(1-\frac{u^2}{U^2}\right)dy \tag{11.52}$$

In case of compressible flow, density must be included in the thicknesses we derived above.

$$\delta_1 = \int_0^\delta \left(1-\frac{\varrho}{\varrho_1}\frac{u}{U}\right)dy$$

$$\delta_2 = \int_0^\delta \frac{\varrho}{\varrho_1}\frac{u}{U}\left(1-\frac{u}{U}\right)dy$$

$$\delta_3 = \int_0^\delta \frac{\varrho}{\varrho_1}\frac{u}{U}\left(1-\frac{u^2}{U^2}\right)dy \tag{11.53}$$

with ϱ_1 as the reference density. We also define the form parameter H_{12} and H_{23}:

$$H_{12} = \frac{\delta_1}{\delta_2};\, H_{32} = \frac{\delta_3}{\delta_2};\, H_{23} = \frac{\delta_2}{\delta_3}. \tag{11.54}$$

The above thicknesses and parameters are the characteristics of boundary layer and will be implemented into the integral equations of the boundary layer in the following section.

11.3.2 Boundary Layer Integral Equation

The complexity of solving the boundary layer equations and the lack of high performance computational device of any sort motivated Prandtl and his co-researchers to find a solution for the boundary layer problem that could be handled with the tools available in the early twenties, namely slide rules (straight, circular and cylindrical). Using the boundary layer differential equations (11.12), (11.14) and considering

Eq. (11.15), we integrate Eq. (11.12) in lateral direction from $y = 0$ at the wall to $y = h$ at the edge of the boundary layer:

$$\int_{y=0}^{h}\left(u\frac{\partial u}{\partial x}+v\frac{\partial u}{\partial y}-U\frac{dU}{dx}\right)dy=\int_{y=0}^{h}\frac{\mu}{\varrho}\frac{\partial^2 u}{\partial y^2}. \tag{11.55}$$

The integration of the right hand side is:

$$\int_{y=0}^{h}\frac{\mu}{\varrho}\frac{\partial^2 u}{\partial y^2}=\int_{y=0}^{h}\frac{1}{\varrho}\frac{\partial \tau}{\partial y}\approx\frac{1}{\varrho}(\tau_{y=h}-\tau_{y=0})=-\frac{\tau_W}{\varrho}. \tag{11.56}$$

In Eq. (11.56), in accord with the boundary layer concept, the shear stress at the edge of the boundary layer was set $\tau_{y=h}\equiv 0$ and $\tau_{y=0}=\tau_W$ represents the wall shear stress. Thus, Eq. (11.55) is re-arranged as:

$$\int_{y=0}^{h}\left(u\frac{\partial u}{\partial x}+v\frac{\partial u}{\partial y}-U\frac{dU}{dx}\right)dy=-\frac{\tau_W}{\varrho}. \tag{11.57}$$

Equation (11.57) is valid for both laminar and turbulent flows provided that the velocity components are time averaged. For further treatment, we replace the v-component in Eq. (11.57) by the integration of the continuity equation $v=-\int^{y}0\frac{\partial u}{\partial x}dy$:

$$\int_{y=0}^{h}\left(\frac{\partial u}{\partial x}-\frac{\partial u}{\partial y}\int_{0}^{y}\frac{\partial u}{\partial x}dy-U\frac{dU}{dx}\right)dy=-\frac{\tau_W}{\varrho}. \tag{11.58}$$

The partial integration of the second term within the parenthesis gives:

$$\int_{y=0}^{h}\left(\frac{\partial u}{\partial y}\int_{0}^{y}\frac{\partial u}{\partial x}\right)dy=U\int_{0}^{h}\frac{\partial u}{\partial x}dy-\int_{0}^{h}u\frac{\partial u}{\partial x}dy. \tag{11.59}$$

Inserting Eq. (11.59) into Eq. (11.58) results in

$$\int_{y=0}^{h}\left(2u\frac{\partial u}{\partial x}-U\frac{\partial u}{\partial x}-U\frac{dU}{dx}\right)dy=-\frac{\tau_W}{\varrho} \tag{11.60}$$

which can be modified as:

$$\int_{0}^{h}\frac{\partial}{\partial x}[u(U-u)]dy+\frac{dU}{dx}\int_{0}^{h}(U-u)dy=-\frac{\tau_W}{\varrho}. \tag{11.61}$$

Now we introduce the displacement thickness δ_1 and the momentum thickness δ_2 from Eq. (11.53) into Eq. (11.61) and obtain the final integral equation of boundary layer:

$$\frac{d}{dx}(U^2\delta_2) + \delta_1 U \frac{dU}{dx} = \frac{\tau_w}{\varrho}. \tag{11.62}$$

Equation (11.62) is the momentum integral equation of boundary layer theory developed by von Kármán [144]. Expanding the first term and dividing the results by U^2 reads:

$$\frac{d\delta_2}{dx} + \frac{1}{U}\frac{dU}{dx}(2\delta_2 + \delta_1) = \frac{\tau_w}{\varrho U^2} = \frac{C_f}{2}. \tag{11.63}$$

Further re-arrangement can be performed by expressing $\delta_1 = H_{12}\delta_2$

$$\frac{d\delta_2}{dx} + (2 + H_{12})\frac{\delta_2}{U}\frac{dU}{dx} = \frac{C_f}{2}. \tag{11.64}$$

With Eq. (11.64), we reduced the combined partial differential equations (11.12), (11.14) to an ordinary one. In order to solve this differential equation for δ_2, besides the initial condition, the pressure distribution outside the boundary layer in terms of velocity distribution $U = U(x)$, the form parameter H_{12}, and the skin friction coefficient C_f must be known. The accurate determination of the skin friction, however, has been the subject of many research works of purely empirical nature. For turbulent boundary layers, Ludwieg and Tillman [145] presented an empirical correlation with an acceptable accuracy. It reads:

$$C_f = 0.246\frac{10^{0.678H_{12}}}{Re_{\delta_2}^{0.268}} \tag{11.65}$$

with $Re_{\delta_2} = U\delta_2/\nu$ and $H_{12} = \delta_1/\delta_2$. Coles [146] modified Eq. (11.65) and arrived at

$$C_f = 0.3\frac{e^{-1.33H_{12}}}{\log(Re_{\delta_2})^{1.74+0.31H_{12}}}. \tag{11.66}$$

Figure 11.9 shows the Ludwieg-Tillman and Coles C_f-distribution as a function of Re_{δ_2} with H_{12} a parameter.

As seen the two correlations are almost identical with the exception of the range $Re_{\delta_2} < 430$, where the relative difference $(C_{f_{LT}} - C_{f_{\text{Coles}}})/C_{f_{LT}}$ is about $\pm 30\%$. The question arises, which of these two formulas is more appropriate for implementation into the von Kármán integral Eq. (11.64). White [147] argues that the accuracy of the Ludwieg-Tillman formula is only about $\pm 10\%$, while the Coles formula is accurate to about $\pm 3\%$. Investigations by Lehman [148] show much higher relative differences with respect to both Ludwieg-Tillman and Coles correlations. However at lower Re_{δ_2} they are closer to Ludwieg-Tillmans results. This suggests, that none

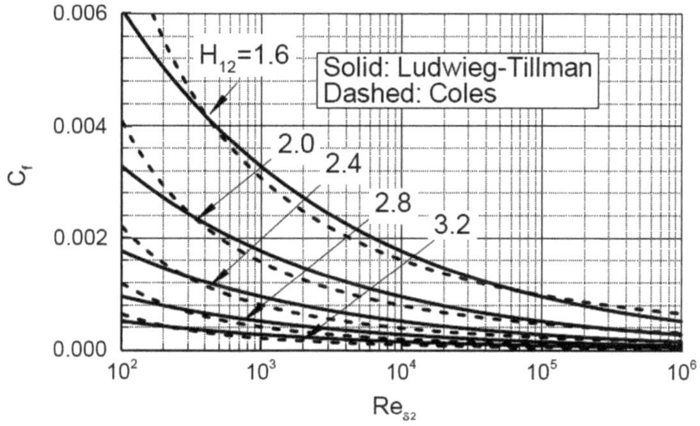

Fig. 11.9 Skin friction coefficient C_f as function of momentum thickness Reynolds number with the H_{12} as a parameter

of the above correlations covers the entire range of Re_{δ_2} appropriately, therefore the selection should be made based on the range of the above parameters.

Prescribing $U = U(x)$, Eq. (11.64) in conjunction with Eq. (11.65) can be solved iteratively. Once the difference between the left and the right hand side of Eq. (11.64) is small enough, the iteration process can be stopped. It should be pointed out that (a) Eq. (11.64) follows the concept of boundary layer theory, which is only valid for cases without flow separation and (b) the outcome of the integration which is the boundary layer momentum thickness is an approximate solution associated certain degree of accuracy ($\pm 10\%$ and above). More elaborate calculation procedures introduced, among others, by Rotta [149], Truckenbrodt [150] and Pfeil and his co-researchers [151, 152]. These authors use the velocity distribution including the wake-function (see following section) as the point of departure.

11.4 Turbulent Boundary Layers

As already discussed in Chap. 8, once the transition process has been completed, and the intermittency factor has reached its asymptotic value of $\gamma = 1$, the boundary layer becomes fully turbulent. Its motion is described by the Reynolds averaged Navier-Stokes equations (8.76), where the Reynolds stress tensor $\gamma \overline{\mathbf{V}'\mathbf{V}'}$ is replaced by $\overline{\mathbf{V}'\mathbf{V}'}$. The index notation also given as Eq. (9.60) is:

$$\frac{\partial \bar{V}_i}{\partial t} + \bar{V}_j \frac{\partial \bar{V}_i}{\partial x_j} = -\frac{1}{\rho}\frac{\partial \bar{p}}{\partial x_i} + \nu \frac{\partial^2 \bar{V}_i}{\partial x_j \partial x_j} - \frac{\partial (\overline{V_i' V_j'})}{\partial x_j} + g_i. \tag{11.67}$$

Assuming a two dimensional statistically steady boundary layer flow and neglecting the gravitational force, the component of Eq. (11.67) in x_1-direction is

$$\bar{V}_1 \frac{\partial \bar{V}_1}{\partial x_1} + \bar{V}_2 \frac{\partial \bar{V}_1}{\partial x_2} = -\frac{1}{\varrho}\frac{\partial \bar{p}}{\partial x_1} + \nu\left(\frac{\partial^2 \bar{V}_1}{\partial x_1^2} + \frac{\partial^2 \bar{V}_1}{\partial x_2^2}\right) - \left(\frac{\partial(\overline{V_1'V_1'})}{\partial x_1} + \frac{\partial(\overline{V_1'V_2'})}{\partial x_2}\right) \tag{11.68}$$

and in x_2-direction reads

$$\bar{V}_1 \frac{\partial \bar{V}_2}{\partial x_1} + \bar{V}_2 \frac{\partial \bar{V}_2}{\partial x_2} = -\frac{1}{\varrho}\frac{\partial \bar{p}}{\partial x_2} + \nu\left(\frac{\partial^2 \bar{V}_2}{\partial x_1^2} + \frac{\partial^2 \bar{V}_2}{\partial x_2^2}\right) - \left(\frac{\partial(\overline{V_2'V_1'})}{\partial x_1} + \frac{\partial(\overline{V_2'V_2'})}{\partial x_2}\right). \tag{11.69}$$

Following the boundary layer concept and the order of magnitude estimates in Equations (11.1), (11.10), and (11.11) for the mean velocity components and their changes, Eq. (11.68) in $x_1 \equiv x$-direction is:

$$\bar{u}\frac{\partial \bar{u}}{\partial x} + \bar{v}\frac{\partial \bar{u}}{\partial y} = -\frac{1}{\varrho}\frac{\partial \bar{p}}{\partial x} + \nu\frac{\partial^2 \bar{u}}{\partial y^2} - \left(\frac{\partial\overline{(u'^2)}}{\partial x} + \frac{\partial\overline{(u'v')}}{\partial y}\right) \tag{11.70}$$

and in $x_2 \equiv y$-direction reads:

$$0 = -\frac{1}{\varrho}\frac{\partial \bar{p}}{\partial y} + \nu\frac{\partial^2 \bar{v}}{\partial y^2} - \left(\frac{\partial\overline{(u'v')}}{\partial x} + \frac{\partial\overline{(v'^2)}}{\partial y}\right). \tag{11.71}$$

As seen from Eqs. (11.10) and (11.11), estimating the order of magnitudes has led to a drastic reduction of the mean velocity components and their derivatives in Eqs. (11.70) and (11.71). Similar order of magnitude estimation must be applied to the turbulence quantities. For this purpose we first extend the list of dimensionless parameters in Eq. (11.3) be introducing a dimensionless fluctuation velocity vector $\mathbf{V}'^*$ and the turbulence intensity Tu.

$$\begin{aligned} x^* &= \frac{x}{L}, V^* = \frac{V}{V_\infty}, P^* = \frac{p}{\rho V_\infty^2}, t^* = \frac{tV_\infty}{L} \\ Re &= \frac{\varrho V_\infty L}{\mu} = \frac{V_\infty L}{\nu}, Tu = \frac{\sqrt{\overline{V'^2}}}{U_\infty}. \end{aligned} \tag{11.72}$$

These dimensionless parameters inserted into Eq. (9.59) results in

$$\frac{\partial \boldsymbol{V}^*}{\partial t^*} + \boldsymbol{V}^* \cdot \nabla \boldsymbol{V}^* = \nabla p^* + \frac{1}{\mathrm{Re}}\Delta \boldsymbol{V}^* - \nabla \cdot \overline{(\boldsymbol{V}' * \boldsymbol{V}'^*)}. \tag{11.73}$$

Decomposing Eq. (11.73) into the x-component reads

$$\begin{array}{ccccccccc}\dfrac{\partial u^*}{\partial t*} & + u^*\dfrac{\partial u^*}{\partial x^*} & + v^*\dfrac{\partial u^*}{\partial y^*} & = -\dfrac{\partial p^*}{\partial x^*} & + \dfrac{1}{Re}\Bigg(& \dfrac{\partial^2 u^*}{\partial x^{*2}} & + \dfrac{\partial^2 u^*}{\partial y^{*2}}\Bigg) & - \Bigg(\dfrac{\partial \overline{(u'*^2)}}{\partial x^*} & + \dfrac{\partial \overline{(u'*v'*)}}{\partial y^*}\Bigg) \\ 1 & 1\ 1 & \delta^*\dfrac{1}{\delta^*} & & \delta^{*2} & 1 & \dfrac{1}{\delta^{*2}} & T_u^2 & \dfrac{Tu^2}{\delta^*} \end{array} \quad (11.74)$$

and in y-component we have:

$$\begin{array}{ccccccccc}\dfrac{\partial v^*}{\partial t*} & + u^*\dfrac{\partial v^*}{\partial x^*} & + v^*\dfrac{\partial v^*}{\partial y^*} & = -\dfrac{\partial p}{\partial y^*} & + \dfrac{1}{Re}\Bigg(& \dfrac{\partial^2 v^*}{\partial x^{*2}} & + \dfrac{\partial^2 v^*}{\partial y^{*2}}\Bigg) & - \Bigg(\dfrac{\partial \overline{(u'*v'*)}}{\partial x^*} & + \dfrac{\partial \overline{(v'*^2)}}{\partial y^*}\Bigg) \\ \delta^* & 1\ \delta^* & \delta^*\ 1 & & \delta^{*2} & \delta^* & \dfrac{1}{\delta^{*2}} & T_u^2 & \dfrac{Tu^2}{\delta^*} \end{array} \quad (11.75)$$

In Eqs. (11.74) and (11.75) all terms with the order of magnitude δ^* can be neglected as shown in Eqs. (11.11) and (11.12). The terms with the order of magnitude Tu^2 may be neglected also. This is admissible because in engineering applications, the turbulence intensity ranges from 2% (0.02) to 15% (0.15) or above. This means, the order of magnitude of Tu^2 may range from 0.0004 to 0.0225. On the other hand, the terms with the order of magnitude of Tu^2/δ^* may or may not be neglected. Since $O(Tu) \approx O(\delta^*)$, the order of magnitude of Tu^2/δ^* can be approximated as $O(Tu^2/\delta^*) \approx O(Tu)$. As a consequence, Eqs. (11.70) and (11.71) are further reduced to:

$$\bar{u}\frac{\partial \bar{u}}{\partial x} + \bar{v}\frac{\partial \bar{u}}{\partial y} = -\frac{1}{\varrho}\frac{\partial \bar{p}}{\partial x} + \nu\frac{\partial^2 \bar{u}}{\partial y^2} - \frac{\partial \overline{(u'v')}}{\partial y} \quad (11.76)$$

and in $x_2 \equiv y$-direction reads:

$$0 = -\frac{1}{\varrho}\frac{\partial \bar{p}}{\partial y} - \frac{\partial \overline{(v'^2)}}{\partial y}. \quad (11.77)$$

In solving turbulent boundary layer problems in differential form it is a common practice to set $\partial\overline{(v'^2)}/\partial y = 0$. This implies that the pressure changes only in x-direction. With this additional simplification, we have only one component of momentum equation to deal with, which is

$$\bar{u}\frac{\partial \bar{u}}{\partial x} + \bar{v}\frac{\partial \bar{u}}{\partial y} = -\frac{1}{\varrho}\frac{\partial \bar{p}}{\partial x} + \frac{\partial}{\partial y}\left(\nu\frac{\partial \bar{u}}{\partial y} - \overline{u'v'}\right) \quad (11.78)$$

with $\nu\partial\bar{u}/\partial y = \mu/\varrho\partial\bar{u}/\partial y = \tau_l/\varrho$ as the laminar shear stress and $-\overline{u'v'} = \tau_t/\varrho$ as the shear stress component of the Reynolds stress tensor. Thus we can define a total shear stress component as:

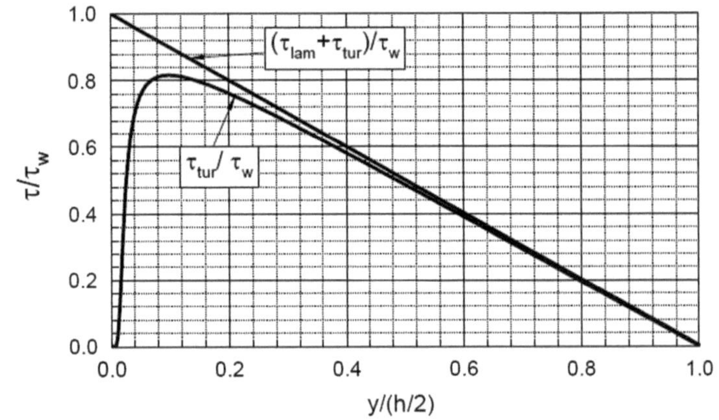

Fig. 11.10 Molecular and turbulence shear stress distribution in a channel flow with the height h

$$\left(\nu\frac{\partial\bar{u}}{\partial y}-\overline{u'v'}\right)=\frac{1}{\varrho}(\tau_l+\tau_t) \tag{11.79}$$

which is the sum of the molecular and the turbulence shear stress. While the molecular shear stress occupies a small region very close to the wall, the turbulence shear stress dominates the rest of the channel.

This is quite clearly shown in Fig. 11.10. Expressing the Reynolds stress tensor in terms of Eq. (9.129), $-\rho\overline{\mathbf{V}'\mathbf{V}'}=\mu_t\bar{\mathbf{D}}$, its shear components reads:

$$-\overline{u'v'}\equiv\nu_t\frac{\partial\bar{u}}{\partial y} \tag{11.80}$$

with ν_t as the kinematic turbulence viscosity. Implementing Eqs. (11.79) and (11.80) into Eq. (11.78), we have the defining equation of motion for turbulent boundary layer as:

$$\bar{u}\frac{\partial\bar{u}}{\partial x}+\bar{v}\frac{\partial\bar{u}}{\partial y}=-\frac{1}{\varrho}\frac{\partial\bar{p}}{\partial x}+\frac{\partial}{\partial y}\left[(\nu+\nu_t)\frac{\partial\bar{u}}{\partial y}\right]. \tag{11.81}$$

The continuity equation provides a relation between the longitudinal and lateral velocity components:

$$\frac{\partial\bar{u}}{\partial x}+\frac{\partial\bar{v}}{\partial y}=0. \tag{11.82}$$

Equations (11.81) and (11.82) are the basis for implementation of different turbulence models as we presented in Chap. 9. One of the significant parameter that is involved in the models we discussed is the wall shear stress, which is extracted from wall function that we discuss in the following section.

11.4.1 Universal Wall Functions

In the context of turbulence modeling discussed in Chap. 9, we tried to find a relationship between the mixing length and the distance from the wall. We introduced a linear function to describe the *linear sublayer:*

$$u^+ = y^+ \tag{11.83}$$

followed by the *logarithmic layer* which is described as a logarithmic function:

$$u^+ = \frac{1}{k}\ln\left(\frac{u_\tau y}{\nu}\right) + C == \frac{1}{k}\ln y^+ + C \tag{11.84}$$

with the dimensionless velocity $u^+ = u/u_\tau$ and the dimensionless distance from the wall $y^+ = u_\tau y/\nu$. The *wall friction velocity* u_τ is related to the wall shear stress τ_w by the relation $\tau_w = \rho u_\tau^2$. Equations (11.83) and (11.84) introduced by Prandtl [153] and [154] are universal laws of the wall and are found on dimensional basis. Figure 11.11 gives an overview of a typical turbulent velocity profile, which is described by several functions. The linear relationship for the sublayer accounts for the laminar friction within the range of $y^+ < 5$. In the logarithmic layer the influence of the turbulent friction outweighs the laminar one. The *overlap region* is characterized by the interaction between the molecular and turbulent viscosity.

The *inner layer* includes the laminar sublayer, the logarithmic layer and the overlap. Moving towards the edge of the boundary layer, the viscosity effect diminishes. At the same time the effects of parameters such as the pressure gradient, the curvature, the freestream turbulence intensity and the *inner layer unsteadiness* determine the *outer layer*.

Figure 11.12 shows the equations that define the laminar sublayer with its influence range of y^+ from 0 to 5 and the range of start of logarithmic layer is estimated at $y^+ = 30 - 200$. The accurate location of this point depends on how the overlap region

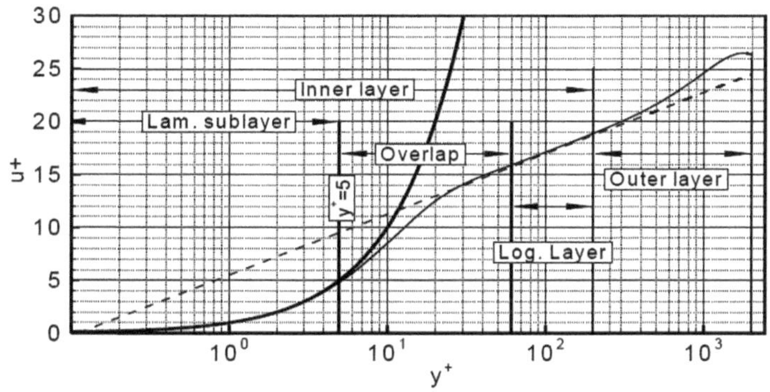

Fig. 11.11 Development of boundary layer along a flat plate

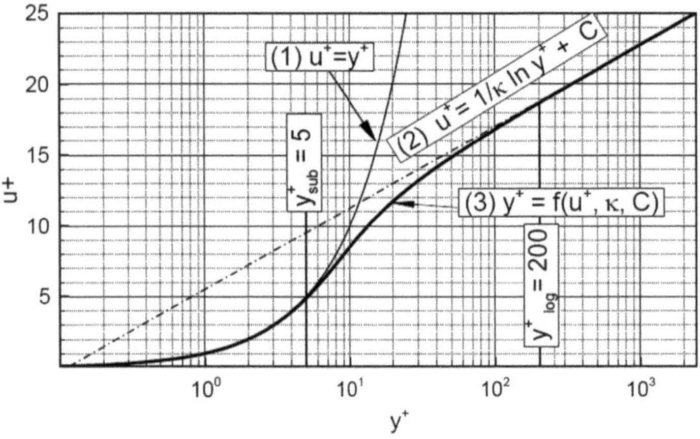

Fig. 11.12 Development of boundary layer along a flat plate

is defined. For the case of zero pressure gradient showed in Fig. 11.12, the location is $y^+ \approx 30$. For a *fully developed turbulent flow* in smooth pipes Nikuradse [155] experimentally determined the constants in Eq. (11.84) to be $k = 0.4$ and $C = 5.50$. As seen in Fig. 11.12, Equations (11.83) and (11.84) describe only two portions of a turbulent velocity profile. A major portion of the velocity profile, the *overlap layer*, that connects the viscous sub layer with the logarithmic layer is missing. Likewise the *outer layer* which merges the boundary layer with the mainstream is not described. For zero-pressure gradient turbulent flow, Spalding [156] introduced an implicit function that provides a single relation that covers the sublayer, the overlap layer and the logarithmic layer. It reads:

$$y^+ = u^+ + e^{\kappa C}\left(e^{\kappa}u^+ - 1 - \kappa u^+ - \frac{1}{2}(\kappa u^+)^2 - \frac{1}{6}(\kappa u^+)^3\right) \tag{11.85}$$

Equation (11.85) labeled as $y^+ = f(u^+, \kappa, C)$ is also plotted in Fig. 11.12. Though this equation adequately describes the velocity profile of a turbulent flow at zero pressure gradient, due to its implicit nature its practical applications is limited. Pfeil and Sticksel [151] developed an explicit equation that describes the velocity profiles of a turbulent flow at zero pressure gradient. The equation is based on Reichard's proposal [157] that the shear stress in the immediate vicinity of the wall must be at least proportional to y^3. The equation is:

$$u^+ = \frac{1}{\kappa}\ln(1 + a_1 y^+) + C1[1 - e^{-a_2 y^+}(1 + a_3 y^+)]. \tag{11.86}$$

The coefficients depend on the pressure gradient $p^+ = \frac{\nu}{\varrho u_\tau^3}\frac{dp}{dx}$. For a flat plate at $p^+ = 0$, the coefficients listed in [151] are:

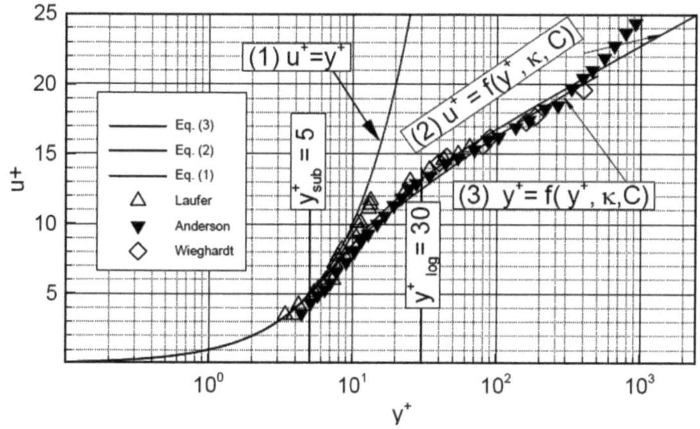

Fig. 11.13 Sublayer (1), implicit Spalding (2), explicit Pfeil and Sticksel (3) and experiments

$$a_1 = 0.215, a_2 = 0.174, \text{ and } a_3 = 0.127, C_1 = 9.22. \qquad (11.87)$$

Equations (11.85) and (11.86) are plotted in Fig. 11.13, which also includes the laminar sublayer.

Figure 11.13 compares Equations (11.85) and (11.86) with the experimental data by Laufer [158], Anderson et al. [159] and Wieghardt [160]. Both equations show a satisfactory agreement with the measurements. However, with the exception of the wake region, the experimental match Eq. (11.86) almost entirely. Also, systematic measurements by Sticksel [161] performed at different pressure gradients show that Eq. (11.86) satisfactorily covers the overlap region which is in general shorter than the one described by Eq. (11.85). These measurements are useful complements to those documented by Coles and Hirst [162].

11.4.2 Velocity Defect Function

Equations (11.83) and (11.84) cover major portions of the velocity that are under influence of the laminar and turbulent wall sear stresses. Moving towards the edge of the boundary layer, the influence of these stresses diminishes while the influence of the main stream becomes more important in shaping the velocity pattern. This suggests that in the outer layer, the wall turbulence may be replaced by the free turbulence (Chap. 10). One of the major parameters that shape the outer portion of the velocity profile is the pressure gradient. This parameter can be take into consideration by adding another term to the existing wall functions. It may have the following form:

$$\frac{\bar{u}}{u_\tau} = f(yu_\tau/\nu) + \frac{\Pi(x)}{k} W(y/\delta). \qquad (11.88)$$

Utilizing Eqs. (11.84), (11.88) gives:

$$u^+ = \frac{1}{k}\ln y^+ + C + \frac{\Pi(x)}{k}W(y/\delta) \tag{11.89}$$

with $\Pi(x)$ as the *wake parameter* that represents the pressure gradient. The additional term $W(y/\delta)$ is called wake function $W(y/\delta)$ introduced by Coles [163] in numerical format. It has the typical pattern of a jet boundary or half of a wake (see Chap. 10). Experimental data taken at non-zero pressure gradient pertaining to the outer layer show that they are well approximated by the function

$$W(y/\delta) = 2\sin^2\left(\frac{\pi}{2}\frac{y}{\delta}\right) \tag{11.90}$$

which satisfies the normalization requirement

$$\int_0^1 W(y/\delta)d(y/\delta) = 1. \tag{11.91}$$

Equation (11.90) is a purely curve fit and unlike Eqs. (11.83) and (11.84) it is not based on dimensional reasoning. To determine the parameter $\Pi(x)$, we set in Eq. (11.89) $y = \delta$ and with Eq. (11.90) we find:

$$U^+ = \frac{u_{y=\delta}}{u_\tau} = \frac{U}{u_\tau} = \frac{1}{k}\ln(\delta u_\tau/\nu) + C + 2\frac{\Pi}{k}. \tag{11.92}$$

Equation (11.92) is a direct relationship between the wall shear stress and wake parameter $\Pi(x)$. Subtracting Eq. (11.88) from Eq. (11.92), we arrive at a relationship for the wake *velocity defect*:

$$\frac{U - \bar{u}}{u_\tau} = -\frac{1}{k}\ln(y/\delta) + \frac{\Pi}{k}[2 - W(y/\delta)]. \tag{11.93}$$

Equation (11.93) includes the dimensionless wall distance, the pressure gradient and the wall velocity. It can be written as:

$$\frac{U - \bar{u}}{u_\tau} = f\left(\frac{y}{\delta}, \frac{\delta}{\tau_w}\frac{dp}{dx}\right). \tag{11.94}$$

The last equation is called the *velocity defect law.* Clauser [164] replaced the boundary layer thickness by the displacement thickness δ_1 and introduced the *Clauser equilibrium parameter*:

$$\beta = \frac{\delta_1}{\tau_w}\frac{dp}{dx}. \tag{11.95}$$

Clauser argued that for a boundary layer with variable pressure gradient but constant β all properties can be scaled with a single parameter. This type of boundary layer is called equilibrium boundary layer. Introducing the local friction coefficient and inserting $u_\tau = \sqrt{\tau_w/\varrho}$ into Eq. (11.95), Spurk [165] introduced the following relationship between the pressure parameter and the local friction coefficient:

$$\sqrt{\frac{2}{c_{f\text{local}}}} = \frac{U}{u_\infty} = \frac{1}{k}\ln\left(\frac{\delta U}{\nu}\sqrt{\frac{c_{f\text{local}}}{2}}\right) + C + 2\frac{\Pi}{k}. \tag{11.96}$$

Ignoring the effect of the viscous sublayer in integrating and using the definition of the displacement thickness δ_1 from Eq. (11.88) Spurk obtained the following relations for the displacement thickness

$$\delta_{\frac{1}{\delta}} = (1+\Pi)u_{\frac{\infty}{Uk}} = \sqrt{\frac{c_{f\text{local}}}{2}}\frac{1+\Pi}{k} \tag{11.97}$$

and for the momentum thickness

$$\delta_{\frac{2}{\delta}} = \sqrt{\frac{c_{f\text{local}}}{2}}\frac{1+\Pi}{k} - \frac{2+3.18\Pi+1.5\Pi^2}{k^2}\frac{c_{f\text{local}}}{2}. \tag{11.98}$$

In Eq. 11.98 the unknowns are $c_{f\text{local}}$, δ, δ_1, δ_2 and Π. With the integral momentum equation (11.64) and the following empirical relation by Mellor and Gibson [166]

$$\Pi \approx 0.8(\beta + 0.5)^{3/4}, \tag{11.99}$$

we have five equations available for the five unknowns. For a flat plate at zero pressure gradient, Eq. (11.99) the wake parameter value is $\Pi = 0.4757$. However, in the literature different values are suggested (Coles [147]: $\Pi = 0.55$, White [147] $\Pi = 0.5$). The equilibrium parameter can be re-arranged in terms of velocity gradient outside the boundary layer as

$$\beta = \frac{\delta_1}{\tau_w}\frac{dp}{dx} = -\frac{\delta_1}{\delta_2}\frac{2}{c_f'}\frac{\delta_2}{U}\frac{dU}{dx}. \tag{11.100}$$

With the given five equations for the five unknowns and the prescribed initial values, the turbulent boundary layer can be calculated by numerical methods. For a flat plate at zero-pressure gradient with $\beta = 0$ and $\Pi \approx 0.55$ we obtain from the momentum Eq. (11.64)

$$\frac{d\delta_2}{dx} = \frac{c_f}{2}. \tag{11.101}$$

Expressing δ_2 and x in terms of the corresponding Re-number $Re_{\delta_2} = U_\infty\delta_2/\nu$ and $Re_x = U_\infty x/\nu$, Eq. (11.101) is re-written as

$$\frac{dRe_{\delta_2}}{dRe_x} = \frac{c_f}{2}. \tag{11.102}$$

White [147] fitted the numerical values that he gained from the computation of *law of the wake* the following relationship between the friction coefficient and Re_{δ_2}:

$$c_f = 0.012 Re_{\delta_2}^{-1/6}. \tag{11.103}$$

Inserting Eq. (11.103) into Eq. (11.102) and integrating the result with the assumption that at $x = 0$: $\delta_2 = 0$ White [147] arrived at

$$Re_{\delta_2} = 0.0142 Re_x^{6/7}. \tag{11.104}$$

To find the friction coefficient c_f in Eq. (11.103) as a function of Re_x, the Reynolds number Re_{δ_2} is replaced by Eq. (11.104) resulting in

$$c_f = 0.025 Re_x^{-1/7}. \tag{11.105}$$

White also computed a flat plate case, where the wake was neglected, and arrived at a similar relation but with different multiplication factor:

$$c_f = 0.027 Re_x^{-1/7}. \tag{11.106}$$

In this context it is informative to compare Eq. (11.105) with the other correlations that are used in boundary layer calculation. Earlier correlation by Prandtl and Schlichting [167] reads:

$$c_f = 0.445(\log Re_x)^{-2.58}. \tag{11.107}$$

Falkner [168] suggested the following relation:

$$c_f = 0.0303 Re_x^{-1/7}. \tag{11.108}$$

Fig. 11.14 compares Eqs. (11.105)–(11.109).

As seen, correlations by Prandtl-Schlichting (11.107) and Falkner (11.108) are very close. Differences of about 17% are seen, when comparing White's correlation (11.105) with the other correlations. White argued that neglecting the wake effect increases the multiplication factor from 0.025 to 0.027 causing an 8% higher c_f. Plotting Eq. (11.106), however shows that for the no-wake case, there is a substantial difference between the results of Eq. (11.106), the Falkner's and Prandtl-Schlichting's correlations. If we assume that the difference of wake-no wake of $\Delta_{\text{wake-no wake}} = 0.025 - 0.027 = -0.002$ correctly reflects the wake effect on the multiplication factor, then we arrive at a modified Falkner equation:

$$c_f = 0.0283 Re_x^{-1/7}. \tag{11.109}$$

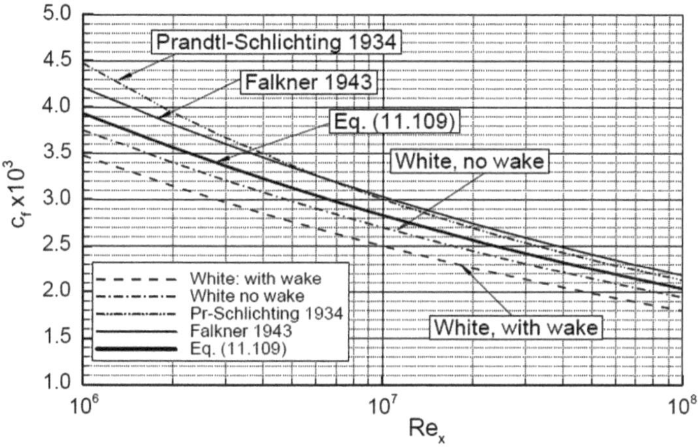

Fig. 11.14 Comparison of several friction factors

Equation (11.109) is also plotted in Fig. 11.14. These equations are valid for a fully turbulent flow with a Re_x-range of $10^5 < Re^9_{x<10}$.

For a few special cases of fully developed turbulent flows such as the flow through a pipe with a smooth surface or a fully turbulent boundary layer flow along a flat plate, the velocity distribution can be approximated by *power laws*. Based on Blasius [167] work, for a pipe with a radius R, Prandtl [154] introduced a power law to approximate the velocity ratio $u/U_{\max}$ within a pipe. Blasius [167] established the following empirical *coefficient of resistance* λ for smooth pipes:

$$\lambda = 0.3164 \left(\frac{\bar{u}D}{\nu}\right)^{-1/4} = \frac{0.3164}{Re^{0.25}} \tag{11.110}$$

with $\bar{u}$ as the averaged velocity that satisfies the continuity requirement. The Blasius Equation is valid for $Re^5_{x<10}$. The coefficient λ is related to the pressure loss coefficient of the pipe defined as:

$$\zeta = \frac{\Delta p}{\frac{\rho}{2}\bar{u}^2} \equiv \lambda \frac{L}{D} \tag{11.111}$$

with L as the length of the pipe, $\lambda = \lambda(Re, k/D)$ and k/D as the relative surface roughness. Setting the reaction force caused by the wall shear stress equal to the force by pressure drop, we obtain:

$$\frac{\Delta p}{L} = \frac{4}{D}\tau_w. \tag{11.112}$$

Inserting in Eq. (11.112) Δp from Eq. (11.111), we find

$$\tau_w = \frac{1}{8}\lambda\rho\bar{u}^2. \tag{11.113}$$

Substituting in Eq. (11.113) λ by Eq. (11.110) yields

$$\tau_w = 0.03955\rho\bar{u}^{7/4}\nu^{1/4}D^{-1/4} = 0.03325\rho\bar{u}^{7/4}\nu^{1/4}R^{-1/4} = \rho u_\tau^2. \tag{11.114}$$

From Eq. (11.114) immediately follows that

$$\left(\frac{\bar{u}}{u_\tau}\right)^{-\frac{7}{4}} = \frac{1}{0.03325}\left(\frac{u_\tau R}{\nu}\right)^{\frac{1}{4}} \text{ or } \left(\frac{\bar{u}}{u_\tau}\right) = 6.99\left(\frac{u_\tau R}{\nu}\right)^{\frac{1}{7}}. \tag{11.115}$$

This is the 1/7-th power law for the velocity distribution within a smooth pipe. Nikuradse [156] showed that the velocity distribution of a fully turbulent pipe flow can be described by

$$\frac{u}{U_{max}} = \left(\frac{y}{R}\right)^{1/7}. \tag{11.116}$$

With U_{max} as the as the maximum velocity at the pipe center and y the distance from the pipe wall. Comparing the velocity measurement of a fully turbulent flow along a smooth plate, Prandtl [154] argued that for the flat plate the following approximation can be made

$$\frac{\bar{u}}{U} = \left(\frac{y}{\delta}\right)^{1/7} \tag{11.117}$$

with U as the velocity at the edge of the boundary layer. The 1/7th power law can also be used to approximate a portion of the velocity profile that is described by the logarithmic law. It reads:

$$u^+ = 8.775y^{+1/7}. \tag{11.118}$$

As Fig. 11.15 shows, Eq. (11.117) tangents the logarithmic layer and delivers almost identical values from $y^+ = 40$ to $y^+ = 400$. It is interesting to note that Eq. (11.117) captures a portion of the outer region that is described by the wake function.

11.5 Boundary Layer, Differential Treatment

This section treats the boundary layer problem from a differential point of view. In contrast to the integral method, the differential method provides the distribution of flow quantities in a two dimensional coordinate system. It includes the equations of continuity and motion. For calculating the heat transfer aspects of the boundary layer, the energy equation that includes the heat flux must be added to the equations

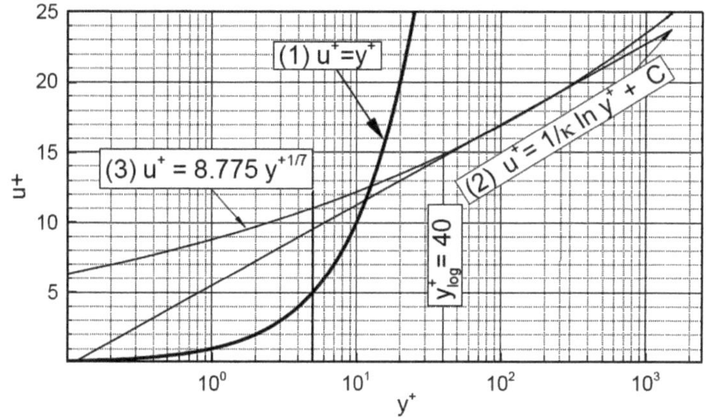

Fig. 11.15 Overlap section of power law with the logarithmic layer

mentioned above. This is treated in Chap. 12. Using the nomenclature introduced for two-dimensional boundary layer, the time averaged continuity equation is given by

$$\frac{\partial}{\partial x}(\varrho\bar{u}) + \frac{\partial}{\partial y}(\varrho\bar{v}) = 0 \tag{11.119}$$

and time averaged momentum equation in the x-direction reads:

$$\varrho\bar{u}\frac{\partial\bar{u}}{\partial x} + \varrho\bar{v}\frac{\partial\bar{u}}{\partial y} = -\frac{\overline{dp}}{dx} + \frac{\partial}{\partial y}\left[\left(\mu\frac{\partial\bar{u}}{\partial y} - \rho\overline{u'v'}\right)\right]. \tag{11.120}$$

11.5.1 Solution of Boundary Layer Equations

To solve the system consisting of continuity and momentum equations (11.119), (11.120), the turbulence models discussed in Chap. 9 may be utilized. The solution of this system provides the 2-D velocity distribution. For two-dimensional boundary layer problems, where a flow separation is not present, the Prandtl mixing length model has shown to deliver reasonable results as shown in Fig. 11.16.

If, in addition to the velocity distribution, the temperature distribution needs to be determined, the equation of energy must be added to the system. In this case, in addition to the turbulent shear stress, the turbulent heat flux must also be modeled. Thus, we obtain the heat transfer and friction coefficients as shown by Schobeiri et al. [173–175]. The combination of equations of energy, momentum and continuity will be discussed in detail in Chap. 12. Since at high Reynolds numbers, the boundary layer development undergoes a transition, the mixing length model must include the intermittency function as we discussed in more detail in Chap. 8. Figure 11.16

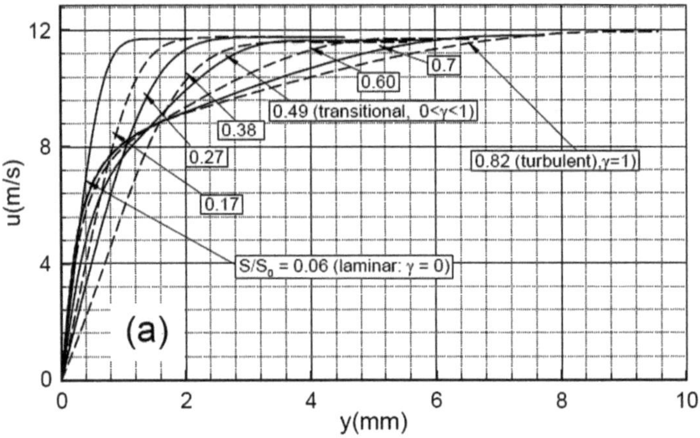

Fig. 11.16 Boundary layer development along the curved plate, Fig. 11.18, at zero-longitudinal pressure gradient and s/s_0 the parameter for relative arc length with $s_0 = 690$ mm

shows the velocity distribution along the curved plate, Fig. 11.18, at different s/s_0 locations. These locations refer to positions, where the flow is laminar $\gamma = 0.0$, transitional, $0 < \gamma < 1$, and fully turbulent with $\gamma = 1.0$. The results of the boundary layer velocity calculations for this plate are shown in Fig. 11.16. It should be pointed out that using any turbulence model without considering the boundary layer transition process delivers results that are applicable to the turbulent portion of a boundary layer only. In numerous engineering applications, however, the boundary layer is mostly transitional. Figure 11.16 shows the velocity distribution starting from $s/s_0 = 0.06$, which is very close to the leading edge. Laminar boundary layer extends from the leading edge to a local relative position of $s/s_0 = 0.49$. As seen, the major portions of the laminar profiles can be approximated by the wall function $u^+ = y^+$, which is equivalent to $u = \frac{\partial u}{\partial y} y$. This linear behavior is clearly reflected in Fig. 11.16. The boundary layer transition starts at about $s/s_0 = 0.49$ and is completed at about $s/s_0 = 0.82$. The transitional profiles reveal a very small portion of the logarithmic layer. The fully turbulent profile at $s/s_0 = 0.82$ has all the features we discussed in Sect. 11.5.1. As we saw in the preceding sections, using the Prandtl-mixing length model in conjunction with the transition model presented in Chap. 8 delivers satisfactory results for two-dimensional boundary layers. It can also be used for moderate adverse pressure gradient as the study by Schobeiri and Chakka [186] shows.

The Prandtl mixing length model may be replaced by any of the models described in Chap. 9. As an example, a channel flow case computed using $\kappa - \omega$-model by Wilcox [187] depicted in Fig. 11.17 shows a good agreement between the computation and the experiment in sublayer and logarithmic layer. As seen the linear sublayer and the logarithmic layer is reasonably well predicted.

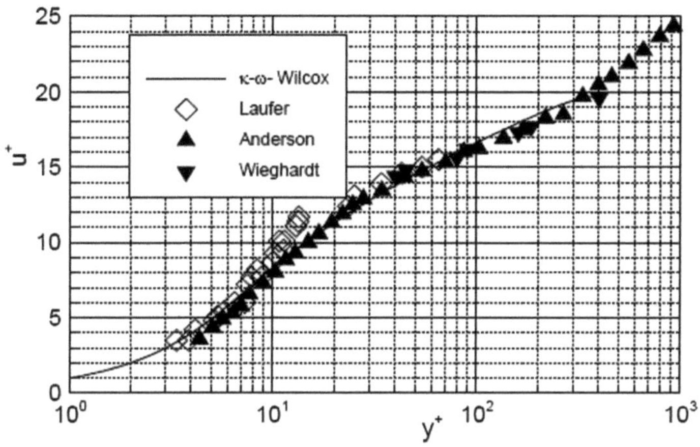

Fig. 11.17 Channel flow calculation using Wilcox $\kappa - \omega$ turbulence model, computation solid line, experiments: Symbols

11.6 Measurement of Boundary Layer Flow

The foregoing sections are dedicated to the boundary layer theory, its integral and differential treatments as well as the methods of solutions. Generally, in science and engineering any hypothesis, modeling or calculation method must undergo a critical experimental verification to substantiate its validity. At this juncture, it is necessary to provide the essentials for understanding the basic physics of the boundary layer development, transition, separation and re-attachment from an experimental point of view. In order to achieve the objectives of providing the detailed and accurate data of relevant boundary layer quantities, the researcher needs to establish: (1) facilities with the corresponding test sections that allow the generation of experimental data, (2) instrumentation of the test sections to deliver experimental data with the desired accuracy, (3) data acquisition system and (4) data analysis.

11.7 Design Facilities for Boundary Layer Research

In designing a research facility, generally, the researcher must have a clear idea about the scope of the parameters involved in experimental project and their variation through the facility. In case of the boundary layer research, the relevant parameters that define the behavior of boundary layer flow under steady and unsteady flow conditions are:

(1) Geometry including surface curvature, surface roughness,
(2) Flow unsteadiness,
(3) Turbulence intensity,

(4) Reynolds and Mach number,
(5) Pressure gradient.

Some of the above parameters are interrelated. For instance, a test object that has convex and concave surface, it inherently produces positive and negative pressure gradients.

This section provides the engineering researchers with an idea as to how to design facilities for fundamental and applied unsteady boundary layer research. The facilities produce steady and unsteady velocity and heat transfer data that can be used for comparison with calculation results. Focusing on obtaining the unsteady flow and heat transfer data with steady flow as a special case, we introduce facilities for fundamental steady and unsteady boundary layer flow research. The systematic experimental approach starts with investigating the boundary layer development along a flat plate inserted into the test section of a wind tunnel.

To investigate the impact of curvature on boundary layer development, a test section with curved walls is introduced, where parameters such as the pressure gradient and the degree of unsteadiness expressed in terms of *reduced frequency* Ω defined below, can be varied. These curved surfaces are particularly interesting, because the surface curvature changes the boundary layer pattern.

11.7.1 Facilities for Boundary Layer Research in Stationary Frame

The first facility designed for fundamental unsteady boundary layer research is shown in Fig. 11.18. Although this facility is designed for taking unsteady flow data, it allows operating in steady state mode as shown bellow.
It consists of a flat plate placed in a test section of a wind tunnel. The plate can be pivoted around the x_2 axis, thus creating zero, positive and negative pressure gradient $0 \geq \frac{\partial p}{\partial x_2} > 0$ with x_2 as the axis perpendicular to the drawing plane.

The periodic unsteady wake flow is generated by a *squirrel cage* type wake generator that consists of two parallel rotating circular disks. In order not to generate secondary flow, the discs rotate behind a cover. For creating periodic wake flow that impinge on the flat plate, cylindrical rods are attached to the circumference of the disks. The number of the rods translated into *Strouhal number* represents the degree of unsteadiness of the incoming wake flow. For obtaining experimental data at steady inlet flow condition, the rods are removed. The degree of unsteadiness or *reduced frequency* is defined by the rotational speed of the disks, spacing of the rods, the undisturbed inlet velocity and the length of the flat plate, summarized as: $\Omega = \frac{c}{s}\frac{U}{V_{ax}}$ with c as the chord length of the plate, the spacing $s = \frac{\pi D}{N}$ and N as the number of the rods. For no rod $N = 0$, the spacing becomes $s = \infty$ and $\Omega = 0$ resulting in steady flow. Detailed information are found in [176].

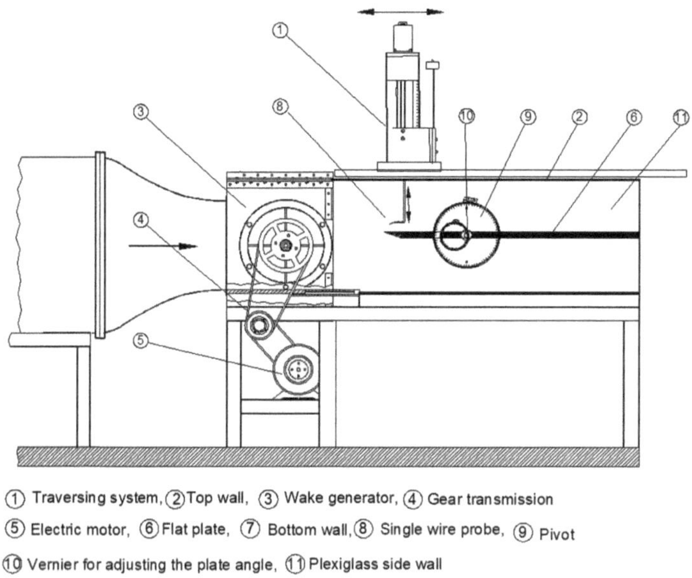

Fig. 11.18 Flat plate at zero longitudinal pressure gradient, adjustment of pressure gradient by pivoting the plate, velocity measurement by single hot wire probe

Unlike the flat plate boundary layer at zero-longitudinal pressure gradient, the boundary layer along the curved surfaces of a curved plate, for example, in Fig. 11.19, is subjected to a lateral pressure gradient, which is brought about by the plate curvature. The boundary layer development along this curved plate resembles the one along a turbine or compressor blade. The corresponding test section is shown in Fig. 11.19. Based on the position of the plate within the test section, negative, zero or positive pressure gradients can be established by varying the plate leading edge angle relative to the incoming flow. A single wire custom designed probe attached to a traversing system measures the boundary layer region from leading edge to trailing edge. As in Figs. 11.18 and 11.19, the *squirrel cage* type wake generator is used to generate periodic unsteady flow condition present at the inlet of the test section.

11.7.2 Facilities for Boundary Layer Research in Rotating Frame

One of the complex engineering systems, within which the unsteady aerodynamics and heat transfer play a central role is a turbomachinery system with the aircraft gas turbine engine shown in Fig. 11.20 as its representative.

As seen, the engine has several compressor and turbine *stages*. Each stage consists of a stationary part called *stator* and a rotating part called *rotor*. Within the

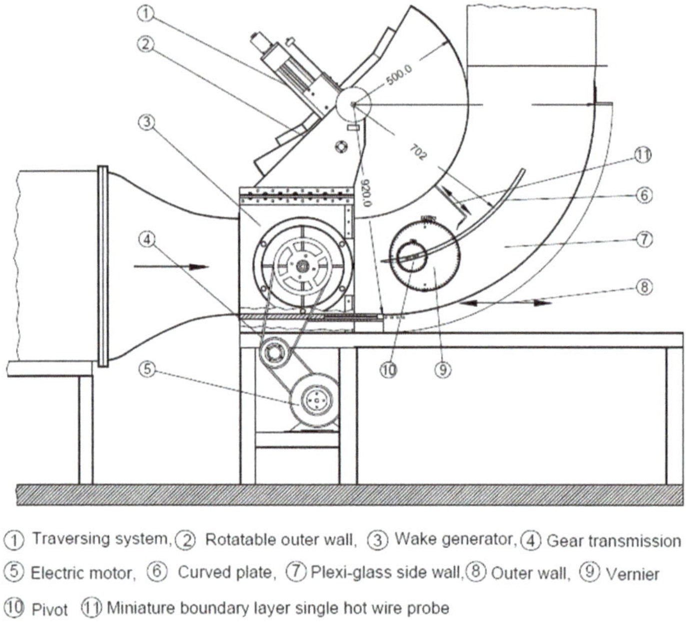

Fig. 11.19 Curved plate at zero longitudinal pressure gradient, adjustment of pressure gradient by pivoting the curved plate at the leading edge, velocity measurement by single hot wire probe

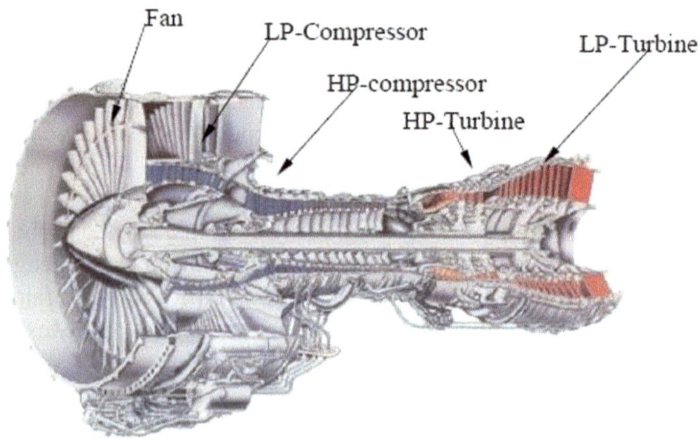

Fig. 11.20 An advanced aircraft gas turbine engine with fan stage, low pressure (LP) and high pressure (HP) compressor stages, combustion chamber, HP- and LP-turbine stages, several diffusers and nozzles

components of these systems, parameters such as pressure gradient, unsteady wake interaction, Re-number and turbulence intensity determine the development of the boundary layer, its separation and re-attachment. High operating temperature and the limited accessibility to the engine inner structure along with the component rotational speed make it impossible to measure any aerodynamic and heat transfer quantities. This circumstance compels aerodynamicists to design research facilities for extracting detail information about the particular parameters they wish to investigate. A research facility that emulates the unsteady flow within a complex system described above must have the following capabilities to:

(1) systematically vary the Re-and Mach number of the incoming flow
(2) generate clean 2-dimensional unsteady wakes
(3) systematically vary the frequency of the wakes
(4) systematically vary the thickness and the width of unsteady wakes
(5) systematically vary the turbulence intensity

The facility with the above capabilities is shown in Figs. 11.21 and 11.22. It consists of (1) a wake a generator unit with rods attached to two timing belts. (2) a linear cascade, where test objects such as compressor or turbine blades can be placed into it, and (3) the wind tunnel with the exit nozzle attached to the wake generator. The spacing of the rods can be arranged such that the facility can simulate different wake

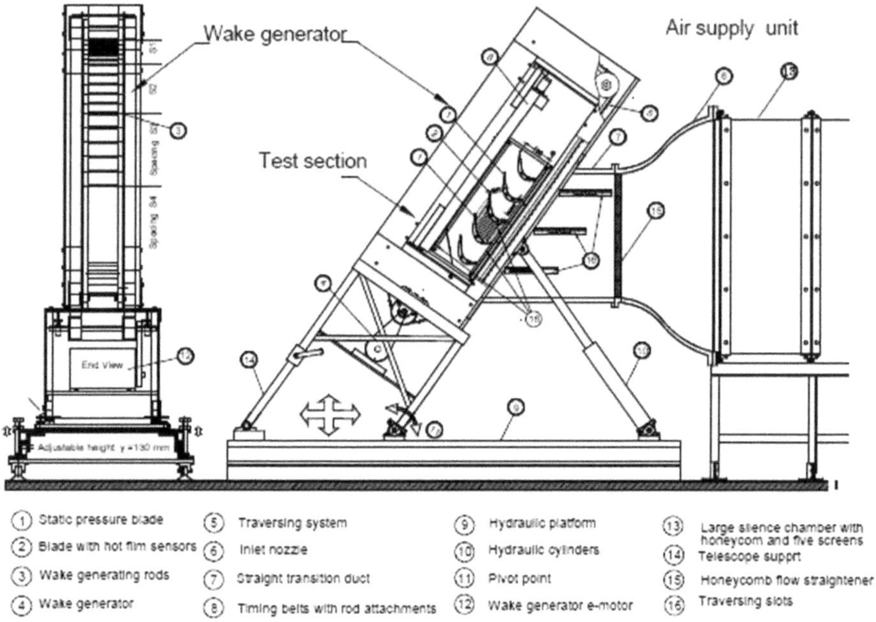

Fig. 11.21 Cascade facility for simulation of unsteady wake flow that impinges on the following blades, (a) rotor-stator interaction, wakes generated from the rotor row imping on the stator blades, (b) wakes generated by moving rods impinging on the stator blades

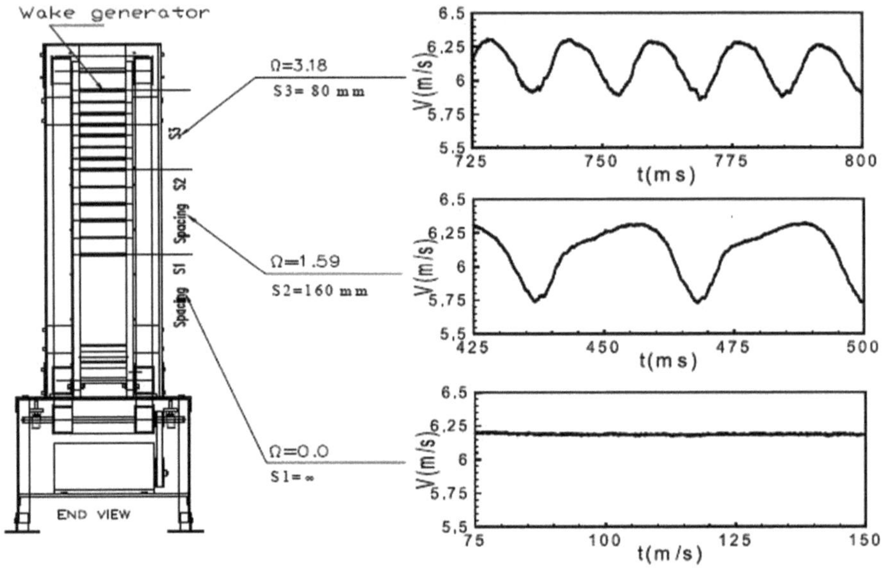

Fig. 11.22 Wake Generator with two timing belts with rods attached, velocity distributions generated with three different dimensionless frequencies $\Omega = 0$ (steady), 1.59, and 3.18. Location of the data measured: 30 mm upstream of the leading edge

reduced frequencies. To this end, the reduced frequency we defined in the previous section for Figs. 11.18 and 11.19 namely, $\Omega = \frac{c}{s}\frac{U}{V_{ax}}$, with c as the chord length of a flat or curved plate, s as the circumferential spacing of the rods, V_{ax} as the axial velocity component of the incoming flow, and U as the circumferential velocity of the rotating wake generator disk. This reduced frequency definition must be extended to include the spacing of the rods and the blades as well as the translational velocity of the moving rods. Thus the reduced frequency for this facility is defined as:

$$\Omega = \frac{c}{S_B}\frac{U}{V_{ax}}\frac{S_B}{S_R} = \frac{\sigma}{\varphi}\frac{S_B}{S_R} \tag{11.121}$$

with c as the blade cord, S_R the rod spacing, S_B the blade spacing, $\sigma = \frac{c}{S_B}$ and $\phi = \frac{V_{ax}}{U}$. On the left hand side of the Fig. 11.21 two sketches with captions (a) and (b) are shown. Sketch (a) qualitatively exhibits the interaction of the wakes originating from the first blade row that impinge on the downstream blade row. Sketch (b) also qualitatively exhibits the interaction of the wakes originating from a set of rods that impinge on the downstream blade row. The rod wakes can be so structured that they exactly resemble the wakes that originate from a blade row, [177].

Three sets of velocity distributions are shown in Fig. 11.21 with three different reduced frequencies. The largest spacing represents the steady state case with $\Omega = 0$. The side view of the wake generator facility in Fig. 11.21 shows the arrangement of the blade cascade. It also shows the rotated wake generator. This is to align the leading edge of the blades with the incoming flow from the wind tunnel. The rear view of the wake generator shows different distributions of rod clusters to simulate different reduced wake frequencies. The purpose of the rod cluster distribution becomes clear, when one looks at the wake velocity they generate as shown in Fig. 11.24. As seen, from Fig. 11.24 and referring to Eq. (11.121), because of $\frac{c}{S_R} \approx 0$ the resulting reduced frequency becomes $\Omega \approx 0$, which represents the velocity of a steady state case.

11.8 Instrumentation and Data Acquisition

The experimental techniques used for flow velocity measurement are, among other things, Laser Doppler anemometry (LDA), Particle Image Velocimetry (PIV) and hot wire anemometry (HWA). For boundary layer flow measurement, where the extraction of flow details close to the wall is of primary interest, optical methods like LDA and PIV have severe problems. Larger required measuring volume, noncontinuous signals and bias due to nonuniform distribution of seeding particles in the measuring volume close to the wall make LDA and PIV less appropriate for boundary layer measurement compared to HWA. Another major advantage is that the HWA is cheaper, it is relatively simple to use by novice users and easier to maintain than its LDA-or PIV competitors. Furthermore it delivers quite accurate results without excessive experimental efforts. Working with HWA, understanding its underlying physics and calibrating single- and cross wire probes is presented in Chap. 14. In the context of this Chapter, we briefly present the necessary features only and refer to Chap. 14 for in depth study.

11.8.1 How to Measure Boundary Layer Velocity with HWA

Details about different HW-probes, such as single wire, cross wire and three wire probes as well as operating modes are discussed in great extent in Chap. 14. Single-wire, cross-wire and three-wire probes are used to measure 1-,2-, and 3-dimensional flows. For capturing the statistics within the boundary layer, the cross-wire or three-wire probes cannot be used, because of the wire configuration and their tip geometry which are much larger than the boundary layer thickness itself. As a result, the single wire probe is the appropriate sensor for boundary layer measurement.[1] Figure 11.23 schematically shows a flat plate inserted in a test section, where the boundary layer

[1] A single wire probe can only measure the longitudinal velocity component within the boundary layer. Based on Prandtl experimental observations and dimensional analysis explained in Sect. 1.4, the velocity in lateral direction is small compared to the longitudinal one. This, however, is not the case for fluctuation components, they are of the same order of magnitude. Consider this fact, Prandtl proposed a relationship for calculating the lateral fluctuation component proportional to the longitudinal one.

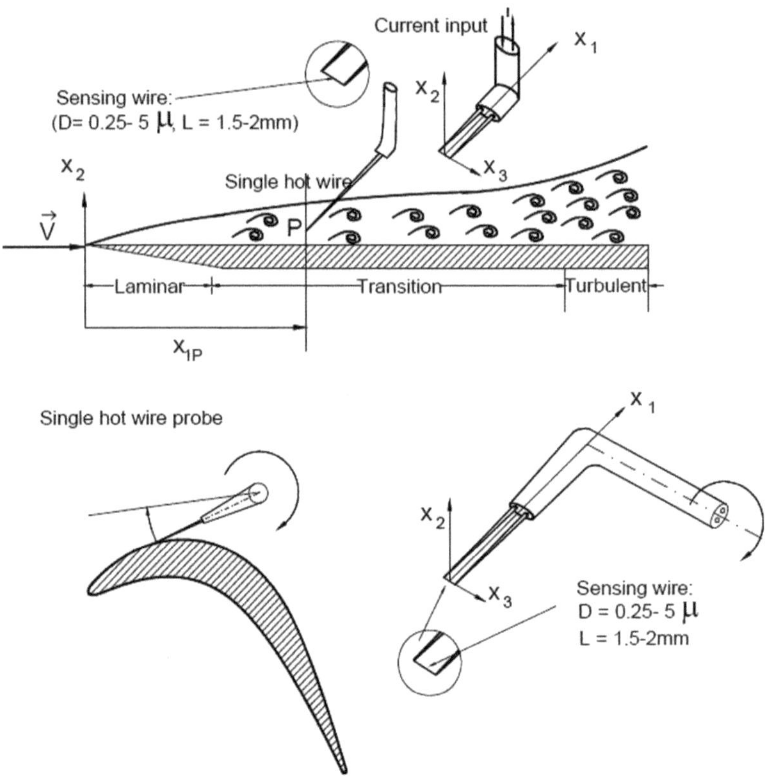

Fig. 11.23 Measurement of boundary layer flow using single hot wire

flow is measured using a single hot wire sensor. Underneath of the flat plate there is a test object with a convex and a concave surface and a round leading edge. It resembles a turbine blade.

The sensing wire materials are tungsten, platinum, platinum-rhodium or platinum iridium alloys. In the *Constant Temperature Anemometer CTA*-operating mode, the anemometer system provides the sensing wire with a current such that the wire resistance remains almost constant. The heated sensing element connected to a Wheatstone bridge is subjected to the incoming flow that removes the heat from the sensor wire in a convective heat transfer process. The change in flow condition causes a change in the bridge voltages and thus in the wire resistance. These voltages form the input to an amplifier generating an output current, which is inversely proportional to the resistance change of the sensing wire. This current is fed back to the bridge restoring the sensor resistance. Special boundary layer applications may require custom designed probes. In this case, attention must be paid to the position of the wire relative to the tips of the two prongs, on which the wire is soldered. The wire must be attached to the tips of the prongs and not in between. Also one has to make sure that the wire is perfectly parallel to the surface. The boundary layer probes, mostly custom designed, may be inserted with the probe holder shaft in x_2-direction as shown in Fig. 11.23 (top, flat plate). Boundary layer traverse in x_2-direction occurs while the probe is positioned at a fixed x_1-location. Traversing the boundary layer development along a convex or a concave surface, however, requires an elbow probe that turns around x_3-axis as depicted in Fig. 11.23 (bottom). In both cases, the probe tip is always parallel to x_3-axis.

It is highly recommended to calibrate the probe before taking data. The calibration can be done either in-situ or in a special calibration channel, where the flow velocity can be accurately varied in small increments. Detailed explanation of calibration procedure is found in Chap. 14. Exposing the probe to a pre-defined velocity range, for example $V = 1.0 - 20.01$ m/s with an increment of $\Delta V = 1.0$ m/s, the anemometer responds with a certain voltage for each velocity. Using the calibration equation, a set of voltage-velocity table is produced that can be spline fitted. The calibration equation is then implemented into a data acquisition and analysis system for further analysis. In addition to Chap. 14, calibration of single and x-wire probes are found in many publications listed by the hot wire system manufacturers (TSI, Dantec, etc.). A new and much simpler method than those found in TSI and Dantec is presented by John and Schobeiri [177] and Schobeiri et al. [178]

Another important issue that needs to be addressed is the wall influence on the results, when taking data close to the wall. The heat conductivity of the wall material has a significant influence on the results and must be taken into consideration. There are a number of empirical correlations that can be used, among others, by Durst et al. [181]

11.8.2 HWA Averaging, Sampling Data

Steady and unsteady data are averaged using the methods explained in Chap. 8. For statistical analysis of periodic unsteady flow, data must be ensemble averaged (see Chap. 8). The larger the number of ensembles, the better is the smoothing of the ensemble averaged data. However, from case to case, there is a limit for the ensemble number, beyond which, there is no noticeable improvement. In many cases including the ones discussed in the following, an ensemble number of 100 seems to be sufficient.

11.8.3 Data Sampling Rate

Another important parameter that affects the quality of the results and particularly the resolution of the details of flow statistics is the sampling rate. Consider a time interval T within which N samples are taken, Fig. 11.24. The sampling frequency f_S is defined as the ratio of the number of the samples divided by the total sample time T:

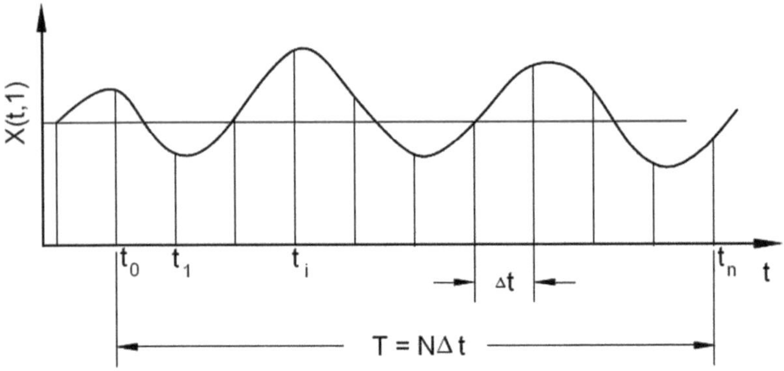

Fig. 11.24 Continuous record of a time dependent flow quantity $X(t, 1)$ for ensemble number 1

$$fs = \frac{N}{T}. \tag{11.122}$$

As shown in Fig. 11.24, dividing the sample time T by the number of samples N results in the time interval Δt between the samples. Consequently, we have $T = N\Delta t$. Thus, in Eq. (11.124) T can be replaced by $N\Delta t$ that leads to

$$fs = \frac{1}{\Delta t}. \tag{11.123}$$

Now, given a total sample time of T-seconds, the question that arises is: what is the appropriate sample frequency. The answer is: the selected sample frequency must exceed twice the analog signal highest frequency to avoid the *aliasing effect*. This frequency is called *Nyquist-frequency, or folding frequency.* Conversely, if one has chosen the sample frequency that satisfies the *Nyquist-criterion* and the number of the samples, the total sample time follows immediately. The analog signals in form of output voltage measured by an anemometer passes through a low pass filter that is used for removing high frequency electrical noise. The filtered voltage signal enters a *signal conditioner* that conditions the output signal to match the *analog/digital converter* input ranging from 0 to 10V. The digitized signals then enter the data acquisition computer, where a regularly updated calibration curve converts the voltage signals into velocity information. As mentioned previously, a detailed description of hot wire-anemometry is given in Chap. 14. A compilation of a broad range of knowledge and experience of many researchers using hot wire anemometry is found in an excellent book by Bruun [182].

11.9 Experimental Verification, Steady Flow

The calculation results presented in Fig. 11.16 in conjunction with the transition model required an experimental verification which is presented in Fig. 11.25. The figure shows the results of boundary layer measurement performed at three different longitudinal locations. It shows that the flow is laminar at $s/s_0 = 0149$, transitional at $s/s_0 = 0.514$ and turbulent at $s/s_0 = 0.823$ (solid lines). Satisfactory agreement with the experiment (symbols) shows that the mixing length model along with the transition model Eq. (8.42) is capable of capturing the details within the boundary layer. The laminar-turbulent transition process that takes place along the curved wall is suitably displayed in an intermittency contour plot as shown in Fig. 11.26. This figure exhibits details of the intermittency distribution inside and outside the boundary layer. Figure 11.26 displays two distinguished flow regions: (a) a transitional zone with $0 < \gamma < 1$ that starts at $s/s_0 \approx 0.5$ and extends up to the plate trailing edge and (b) a low turbulence flow region with $\gamma \approx 0$ that occupies the rest of the flow domain (deep blue). The intermittency picture revealed in Fig. 11.26 describes the transitional behavior of the boundary layer, which was shown in Figs. 11.25.

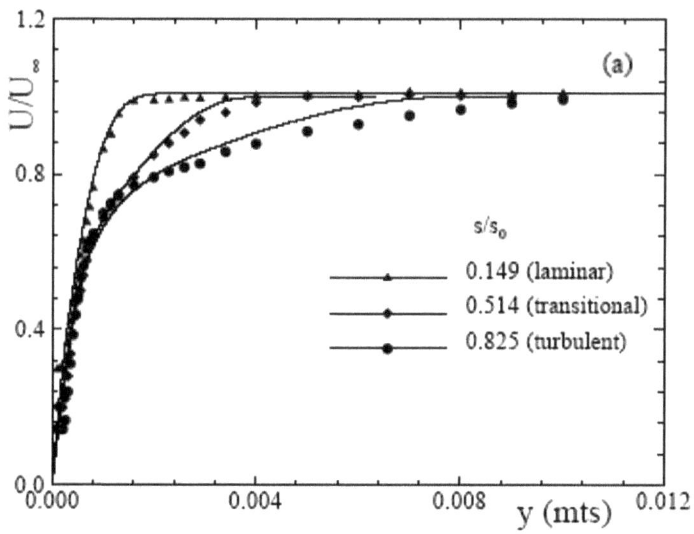

Fig. 11.25 Boundary layer velocity distribution at three different longitudinal position along the curved plate. Computation (solid lines), Experiments (Symbols), Chakka and Schobeiri [64]

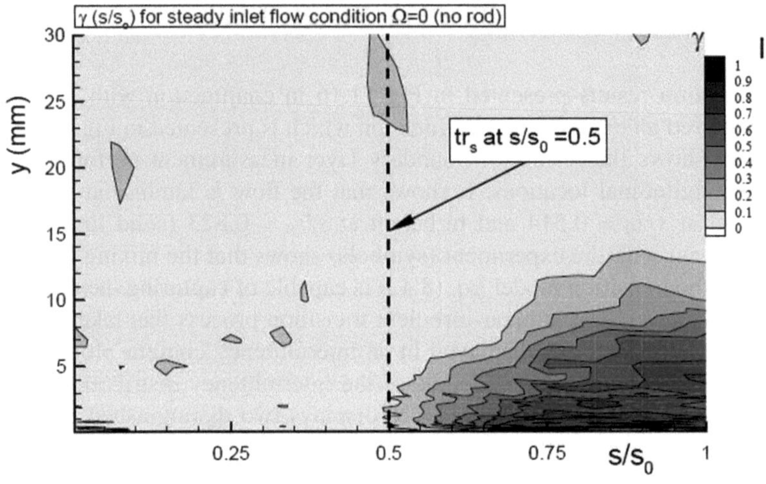

Fig. 11.26 Contour plot of intermittency function for steady inlet flow condition

11.10 Case Studies

In this section we present several cases from engineering that are aimed as providing the researchers as well practicing engineers with an in-depth knowledge of advanced fluid mechanics and its application to engineering.

11.10.1 Case Study 1: Curved Test Section

As the first case, we investigate the development of the boundary layer affected by a periodic unsteady inlet flow condition. For this case, we choose the curved channel depicted in Fig. 11.19. The unsteady flow produced by the wake generator is characterized by the reduced frequency Ω as defined in Sect. 11.7.1. For the experimental examples presented in this section, at each boundary layer position, samples are taken at a rate of 20 kHz for each of 100 revolutions of the wake generator. The data are ensemble averaged with respect to the rotational period of the wake generator. Figure 11.27 (top) shows a representative set of ensemble averaged velocity and Fig. 11.27 (bottom) depicts the fluctuation distributions inside the boundary layer along the curved plate of Fig. 11.19 (for more details see [185]).

Both figures are parts of the same set of unsteady flow measurement. Figure 11.27-top shows the velocity distribution as a function of dimensionless time t/τ for three time periods τ taken at different y-positions normal to the surface of the plate. Outside the boundary layer at about $y = 10$ mm, a wide portion of the velocity has a pronounced undisturbed region which we call external region. This region is characterized by a low level of random rms fluctuation of $<u> \approx 0.15 m/s$, Fig. 11.27

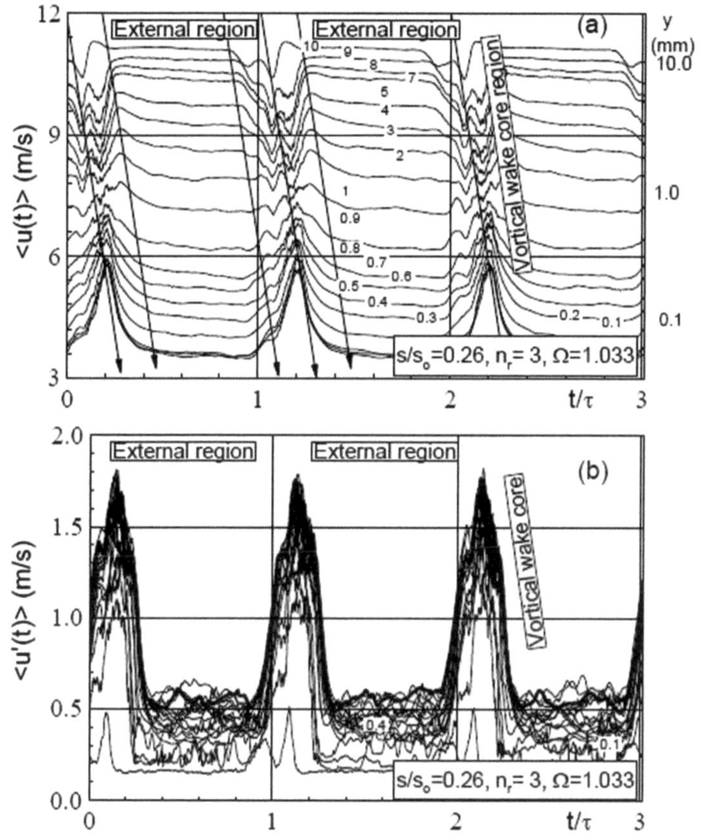

Fig. 11.27 Ensemble-averaged velocity and fluctuation-rms distribution as a function of non-dimensional time at different y-locations at $s/s0 = 0.26$ for $\Omega = 1.033$ corresponding to 3-rods attachment. Top: ensemble averaged velocity distribution $<U>$, bottom ensemble averaged fluctuation velocity $<u>$ both as a function of time

(bottom). Considering the velocity inside the external region, $<U> \approx 11.1m/s$, we find a turbulence intensity of about 1.3%. As the hot wire probe moves toward the plate surface, the velocity continuously reduces because of the increasing viscosity effect but maintains its pattern. A change of pattern is observed, when the probe is at a position $y < 1$mm.

While the velocity in external region reduces, the fluctuation velocity within the vortical core increases leading to a much higher turbulence intensity. Figure 11.27-bottom shows the ensemble averaged velocity fluctuation with a highly vortical wake core region which is occupied by small vortices. For a y-range between 0.1–10 mm, the velocity fluctuations within the vortical core have maximums that appear periodically. Very close to the wall at y =0.1 mm the maximum fluctuation rms is about 1.8m/s that corresponds to a turbulence intensity of about 52%. The increase in turbulence intensity caused by the impingement of the unsteady wake flow on the boundary layer is one of the mechanisms that suppresses the flow separation under unsteady wake flow condition as seen in the next section.

11.10.2 Unsteady Turbulence Activities, Calm Regions

Unsteady turbulence activities along the curved plate at $y = 0.1$ mm at different instant of time is shown in a time space diagram, Fig. 11.28. A set of ensemble-averaged data is utilized to generate the ensemble-averaged turbulence intensity contour plot for a lateral position of $y = 0.1$ mm presented in Fig. 11.28. As shown, for $\Omega = 1.033$, (3 rods), the boundary layer is periodically disturbed by the high turbulence intensity wake strips. These strips are contained between the wake leading edge and the wake trailing edge that move with two different velocities namely $0.88 < V_0 >$ and $0.5 < U_0 >$ as marked in the figure. Outside the wake strips undisturbed low turbulence regions are observed with significantly lower intensity levels indicating the absence of any visible wake interaction. As seen, whenever the wake strip with high turbulence intensity passes over the plate, the boundary layer becomes turbulent. However, the flow state changes from turbulent to laminar one as soon as the wake strip passes by. Thus, the flow state changes intermittently from laminar to turbulent and vice versa. The intermittent character of the boundary layer flow subjected to a periodic unsteady wake flow condition is shown in a time-space intermittency contour plot as shown in Fig. 11.29. The frequency as well as the y-locations along the plate correspond to those shown in Fig. 11.28.

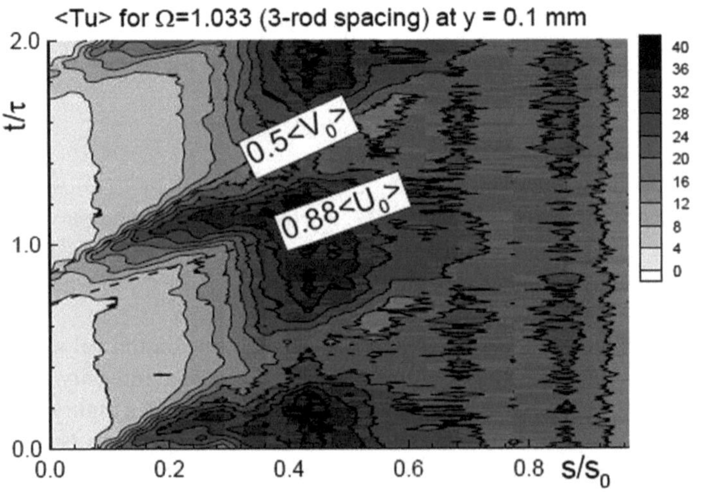

Fig. 11.28 Ensemble-averaged turbulence intensity $<TU>$ in temporal-spatial domain at $y = 0.1$mm, wake passing frequency $\Omega = 1.033$ (3 rods)

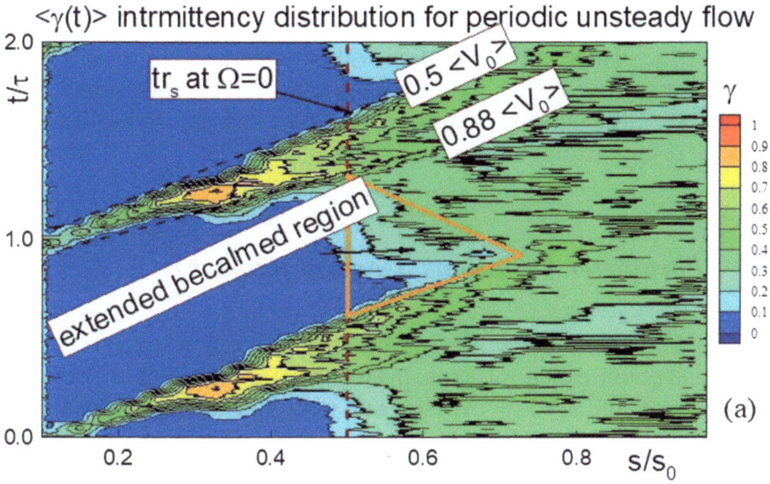

Fig. 11.29 Ensemble-averaged turbulence intermittency $<\gamma>$ in temporal-spatial domain at y =0.1mm, wake passing frequency Ω =1.033 (3 rods)

Figure 11.29 exhibits three distinguished flow zones: (a) a periodic laminar flow [2] zone with the intermittency close to zero, (b) a periodic turbulent zone occupied by the wake vortical core denoted by $<\gamma> \approx 1$ and (c) an extended *becalmed region* marked with a triangle. To highlight the effect of the unsteady wake flow impingement on the transition behavior, the vertical dashed line marks the position of the transition start of the boundary layer under steady inlet flow condition shown in Fig. 11.26. As Fig. 11.29 shows, the wake passing has caused a delay in transition start resulting in a becalmed region mentioned above.

11.11 Case Study 2: Aircraft Engine

This case study represents a challenging experimental case in advanced fluid mechanics. In this respect, it is noted that these type of case studies are not treated in any traditional fluid textbooks.

One of the areas of engineering applications, where several parameters interact with each other is the turbomachinery aerodynamics. As a representative example we refer to the aircraft engine displayed in Fig. 11.20 with the components listed in its caption. Within these components, the pressure gradient, the unsteady wake interaction, Re-number and turbulence intensity determine the development of the boundary layer, its separation and re-attachment on the blade surfaces. In an engine like the one shown in Fig. 11.20, as mentioned previously, it is very difficult, almost

[2] Using the term “laminar” is per definition not quite descriptive, perhaps the term “non-turbulent” would be more appropriate.

impossible to investigate the effect of these parameters individually. The facility that emulates the complex flow situation within an enigine shown in Fig. 11.20 is aleady presented in Figs. 11.21 and 11.22. Since the facility is described above and detailed in [188, 189], only the parameter variation capability of this facility and the experimental results acquired from the facility is discussed.

11.11.1 Parameter Variations, General Remarks

Referring to the facility in Sect. 11.7, it has the capability to measure the relevant parameters and perform their variation. In the following we present a brief discussion of each parameter followed by experimental results pertaining to the subject. The overarching research issue in this context is the variation of the inlet flow condition that includes:
(1) Flow unsteadiness, variation
(2) Turbulence intensity, length scale,
(3) Reynolds and Mach number,
(4) Pressure gradient.

11.11.2 Flow Unsteadiness: Kinematics of Periodic Wakes

In Sect. 8.5.1, different unsteady flow types were introduced. For this section, we consider the periodic unsteady flow case with the wake frequency as a parameter. The facility for this experimental research is already presented in Sect. 11.7.2 in Figs. 11.21 and 11.22. To simulate different frequencies, as shown in Fig. 11.22, rods are attached to the belts at different spacings. For variation of reduced frequency, Ω defined in Eq. (11.39) is used. The facility allows three variations of Ω, be carried out. The reduced frequency Ω given in Eq. (11.39) incorporates the rod spacing S_R and the blade spacing S_B, in addition to the inlet velocity and wake generator speed. For the rod spacings of $S_R = 80$ mm, $S_R = 160$ mm and $S_R = \infty$ (no rod, steady case), the corresponding dimensionless frequencies are $\Omega = 3.18$, 1.59 and 0.0. Figure 11.30 exhibits the cascade with a number of typical turbine blade installed in the facility 11.20. To understand the basics, we present just a case with $\Omega = 1.59$, where the contour plots of turbulence intensity and intermittency shown in Figs. 11.31 and 11.32, for detailed information we refer to [190, 192]. The boundary layer measurements were conducted on the concave surface with a single hot wire. The boundary layer was traversed in close proximity to the surface from 0.1mm to the edge of the respective boundary layer.

The periodic unsteady turbulence intensity contour is shown in Fig. 11.31. This figure shows two distinguished zones, one is the zone occupied by the high intensity wake core and the other characterized by low turbulence intensity zone. A more precise quantity that characterizes these zones is the intermittency distribution exhibited

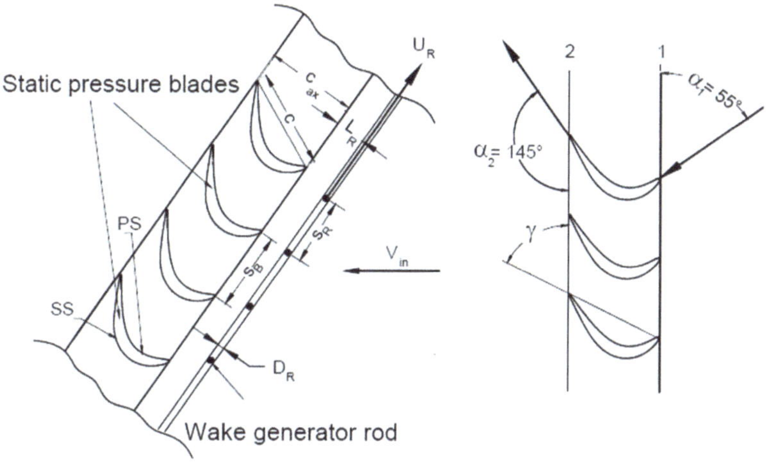

Fig. 11.30 Turbine cascade test section with the blade geometry and flow angles, SS= Suction Surface (convex), PS = Pressure Surface (concave), two blade instrumented with static pressure taps. For simulation of steady state case, the rods are removed

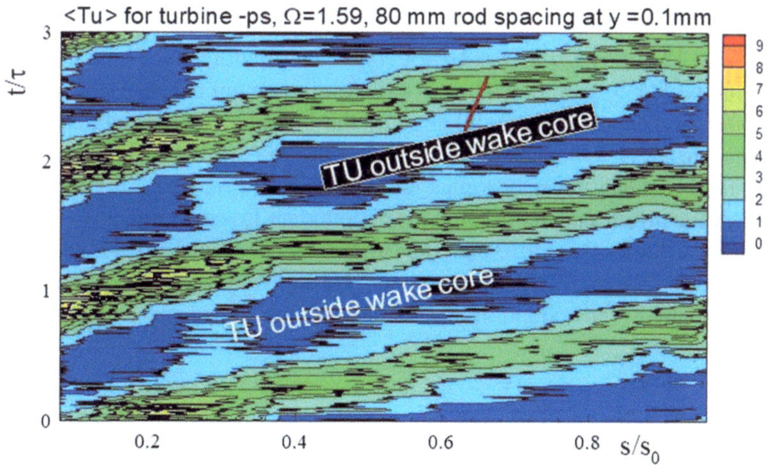

Fig. 11.31 Unsteady ensemble averaged turbulence intensity distribution at the lateral position of y=0.1 mm from the wall with S as the current location of the probe and S_o the arch length of the concave surface

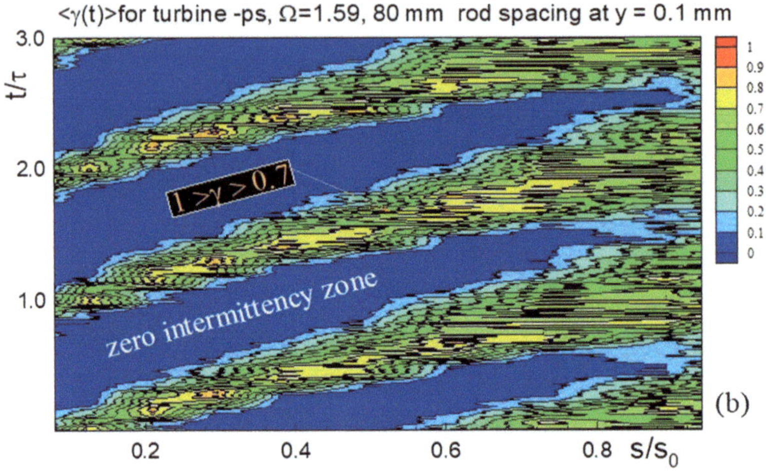

Fig. 11.32 Unsteady ensemble averaged intermittency $<\gamma>$ distribution at the lateral position of y = 0.1 mm from the wall with S as the current location of the probe and s S_o the arch length of the concave surface

in Fig. 11.32. The two *zero-intermittency zones* are separated by a high intermittency wake stripe with $1 > \gamma > 0.7$. The periodic impingement of the wake vortical core on the boundary layer, causes a change of the boundary layer characteristic from laminar to turbulent. Once a wake stripe has traveld over the location under investigation, the boundary layer becomes laminar again, [190, 192]. Figures 11.32 and 11.29 show the difference in the behavior of boundary layers under periodic unsteady inlet flow condition. This difference has significant implications for engine design and its performance prediction as quantified by the Navier-Stokes equations.

Revisiting the RANS-equation (8.78) that contains the viscous and turbulent shear stress:

$$\frac{\partial \bar{\mathbf{V}}}{\partial t} + \bar{\mathbf{V}} \cdot \nabla \bar{\mathbf{V}} = -\frac{1}{\rho} \nabla \bar{p} + \nu \Delta \bar{\mathbf{V}} - \nabla \cdot (\gamma \overline{\mathbf{V}'\mathbf{V}'}). \tag{11.124}$$

the term $\nabla \cdot (\gamma \overline{\mathbf{V}'\mathbf{V}'})$ shows the impact of intermittency on Reynolds shear stress. For $\gamma = 1$, it means fully turbulent flow while for $\gamma = 0$, it indicates that the flow is fully laminar. In fact in an engineering system the intermittency is always $1 > \gamma > 0$, which means the flow is transitional. As discussed in Sect. 8.5.2, the turbulent shear stress associated with the intermittency function, $\gamma \overline{\mathbf{V}'\mathbf{V}'}$, plays a crucial role in affecting the solution of the Navier-Stokes equations. This is particularly significant for calculating the wall friction and the heat transfer coefficient distribution. Two quantities have to be accurately modeled. One is the intermittency function γ, and the other is the Reynolds stress $\overline{\mathbf{V}'\mathbf{V}'}$ tensor with its nine components. Inaccurate modeling of these two quantities leads to a multiplicative error of their product $\gamma \overline{\mathbf{V}'\mathbf{V}'}$. This error particularly affects the accuracy of the total pressure losses, efficiencies, and heat transfer coefficient calculations. Equation (8.78) is coordinate invariant and can be transformed to any curvilinear coordinate system within an absolute frame of reference. Comparing the Fig. 11.29 with the Fig. 11.32 it becomes obvious that in case of the turbine blade, there is apparently no becalmed region as in curved plate. This fact indicates that the conventional assumption about a fully turbulent flow in engines like the one shown in Fig. 11.20 is from a physical point of view untenable.

11.11.3 Variation of Pressure Gradient

By choosing the special turbine blade shown in Fig. 11.30 as a representative example, an aerodynamic system is introduced that inherently generates a continuous distribution of negative, zero and positive pressure gradients. In steady operation mode the rods are removed thus the velocity at the cascade inlet is fully uniform. At the inlet, the velocity vector is tangent to the camberline of the blade such that an incidence free inlet flow condition is established. The Reynolds number $Re = s_0 V_{exit}/\nu$ is built with the suction surface length S_0 and the exit velocity.

Figure 11.33 displays the pressure coefficient $C_p = (p - p_1)/(p_t - p)_{inl}$ along the suction and pressure surfaces of the turbine blade with p_t as the total pressure and p_1 the static pressure of the first tap. As Fig. 11.33 shows, the suction surface (convex), exhibits a strong negative pressure gradient. The flow accelerates at a relatively steep rate and reaches its maximum surface velocity that corresponds to the minimum pressure with $c_p = -4.0$ at $s/s_0 = 0.48$. Passing through the minimum pressure, the fluid particles within the boundary layer encounter a positive pressure gradient that causes a deceleration until $s/s_0 = 0.55$ has been reached. This point signifies the beginning of the laminar boundary layer separation and the onset of a separation zone. As seen in the subsequent boundary layer discussion, the part of the separation zone characterized by a constant c_p-plateau extends up to $s/s_0 = 0.702$.

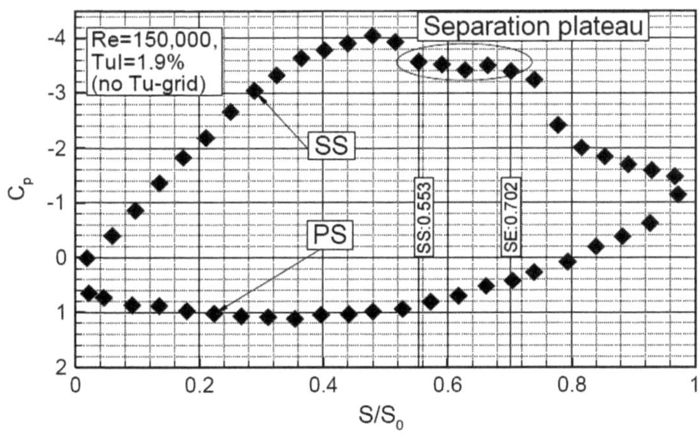

Fig. 11.33 Pressure distribution along the turbine blade, SS = Suction Surface, PS=Pressure Surface

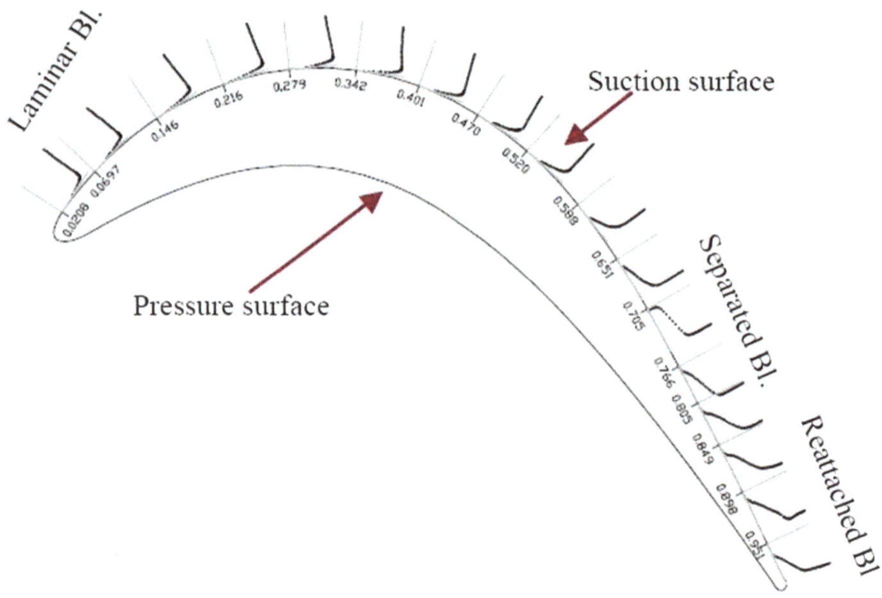

Fig. 11.34 Boundary layer development along the LP-turbine blade

11.11.4 Velocity Distribution

The pressure distribution around the blade shown in Fig. 11.33 directly affects the the development of the boundary layer velocity, its separation and re-attachment as seen in Fig. 11.34. It shows the measured velocity distribution normal to the surface its deceleration, separation and re-attachment. As seen from the figure, the separation starts around $s/s_0 = 0.65$ and manifests itself as a separation bubble at around $s/s_0 = 0.7$.

11.11.5 Velocity and Its Fluctuations

Generally, in engineering applications, the flow velocity is associated with certain fluctuations, whose degree can be expressed in terms of turbulence intensity defined as $Tu = \frac{\sqrt{v^2}}{V}$ with v as the fluctuation velocity and V as the mean velocity. This is true for statistically steady as well as unsteady flow situations. The fluctuations have usually high frequencies that require highly sensitive probes to capture them. For low temperature applications, hot wire anemometry is used, whose data acquisition frequency can be adjusted. For statistically steady flow, the time dependent velocity is measured, from which the time averaged and the fluctuation components can be extracted. The issue of hot-wire anemometry, the type of probes, the calibration and data acquisition is extensively treated in Chap. 14.

11.11.6 Time Averaged Velocity, Unsteadiness

The distribution of the time averaged velocity at different streamwise locations is presented in Figs. 11.35 and 11.36 for the turbulence intensity Tu-level of 1.9% (without grid)using a single hot-wire probes. The diagrams in include one steady state data for reference purposes, $\Omega = 0.0$ (SR = ∞) and two sets of unsteady data for $\Omega = 1.59$ (SR = 160 mm) and $\Omega = 3.18$ (SR = 80 mm). Note that at this relatively low turbulence intensity, the effect of unsteady wake is still noticeable particularly at the locations past the beginning of the separation process at $s/s_0 = 0.5$. The stable laminar boundary layer that starts from $s/s_0 = 0.0$ and extends to $S/S_0 = 0.5$ is not affected by the unsteady wake impingement. This is the result of the time averaging that we discussed in Sect. 11.10.1 earlier.

11.11.7 Time Averaged Fluctuation, Unsteadiness

As the Fig. 11.35 in conjunction with the Fig. 11.36 indicates, in upstream region of the separation bubble at $s/so = 0.49$, the flow is fully attached. The velocity distributions inside and outside the boundary layer experience slight decreases with increasing the reduced frequency. At the same positions, however, the time averaged fluctuations shown in Fig. 11.36 exhibits substantial changes within the boundary layer as well as outside it. The introduction of the periodic unsteady wakes with highly turbulent vortical cores and subsequent mixing has systematically increased the free stream turbulence intensity level from 1.9% in steady case, to almost 3% for $\Omega = 3.18$ (SR = 80 mm). This intensity level is obtained by dividing the fluctuation velocity at the edge of the boundary layer, Fig. 11.36, by the velocity at the same normal position.

Comparing the unsteady cases $\Omega = 1.59$ and 3.18, with the steady reference case $\Omega = 0.0$, indicates that, with increasing Ω, the lateral position of the maximum fluctuation shifts away from the wall. This is due to the periodic disturbance of the stable laminar boundary layer upstream of the separation bubble.

Convecting downstream, the initially stable laminar boundary layer flow experiences a change in pressure gradient from strongly negative to moderately positive causing the boundary layer to separate. This was discussed in Figs. 11.33 and 11.35. The inflectional pattern of the velocity distribution at $s/so = 0.57$ signifies the inception of a separation bubble that extends up to $s/so = 0.85$, resulting in a large sized closed separation bubble. As opposed to open separation zones that are encountered in flow decelerated systems such as compressor blades and diffuser boundary layers, the closed separation bubbles are characterized by a low velocity flow circulation within the bubble as shown in Fig. 11.35. Measurement of boundary layer also with single wire probes along the suction surface of the same blade is reported, among others, in [191, 193] reveals exactly the same pattern as shown in Fig. 11.35. In contrast, the single wire measurement in an open separation zone exhibits a pronounced

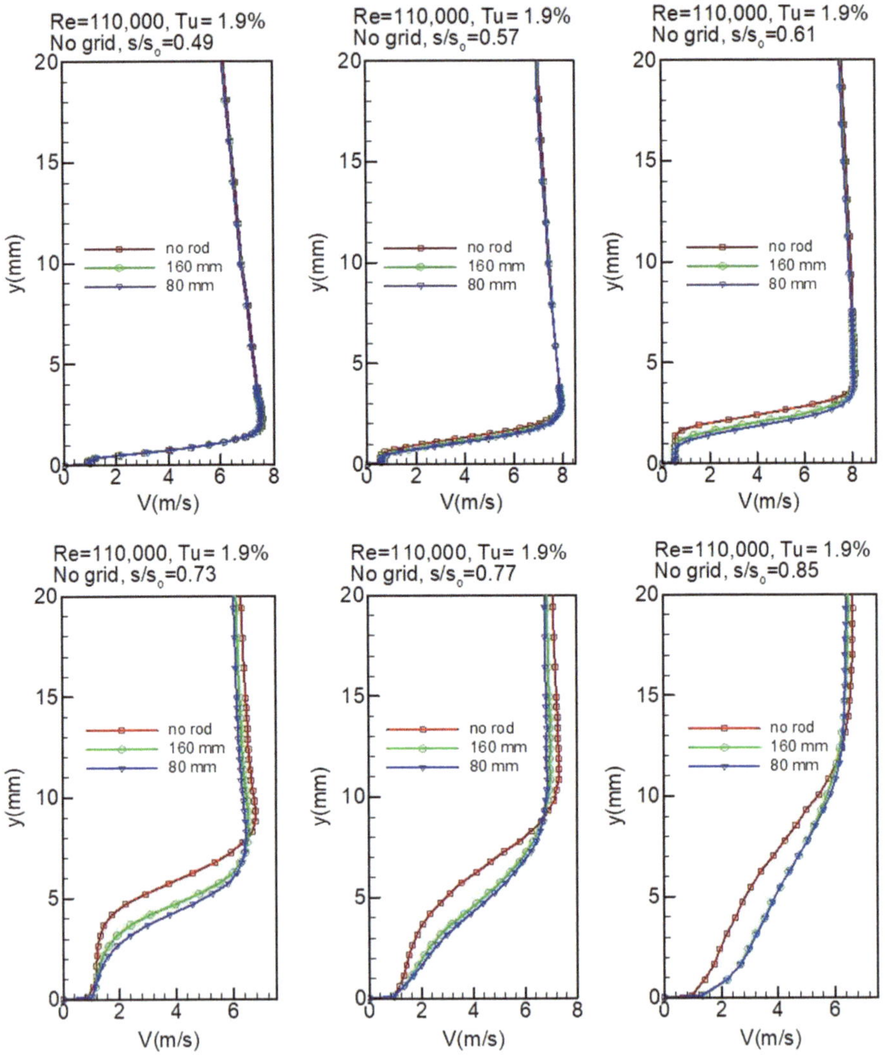

Fig. 11.35 Distribution of time-Averaged velocity at different location along the suction surface for steady case $\Omega = 0(S_R = \infty)$ and unsteady cases $\Omega = 1.59(S_R = 160$ mm) and $\Omega = 3.18(S_R = 80$ mm) at Re=110,000 and free-stream turbulence. intensity of 1.9% (without grid)

kink at the lateral position, where the reversed flow profile has its zero value. Despite the fact that a single wire probe does not recognize the flow direction, the appearance of a kink in a separated flow is interpreted as the point of reversal with a negative velocity.

The effect of unsteady wake frequency on boundary layer separation is distinctly illustrated in both Figs. 11.35 and 11.36 at s/so =0.61, 0.73, and 0.77. While the

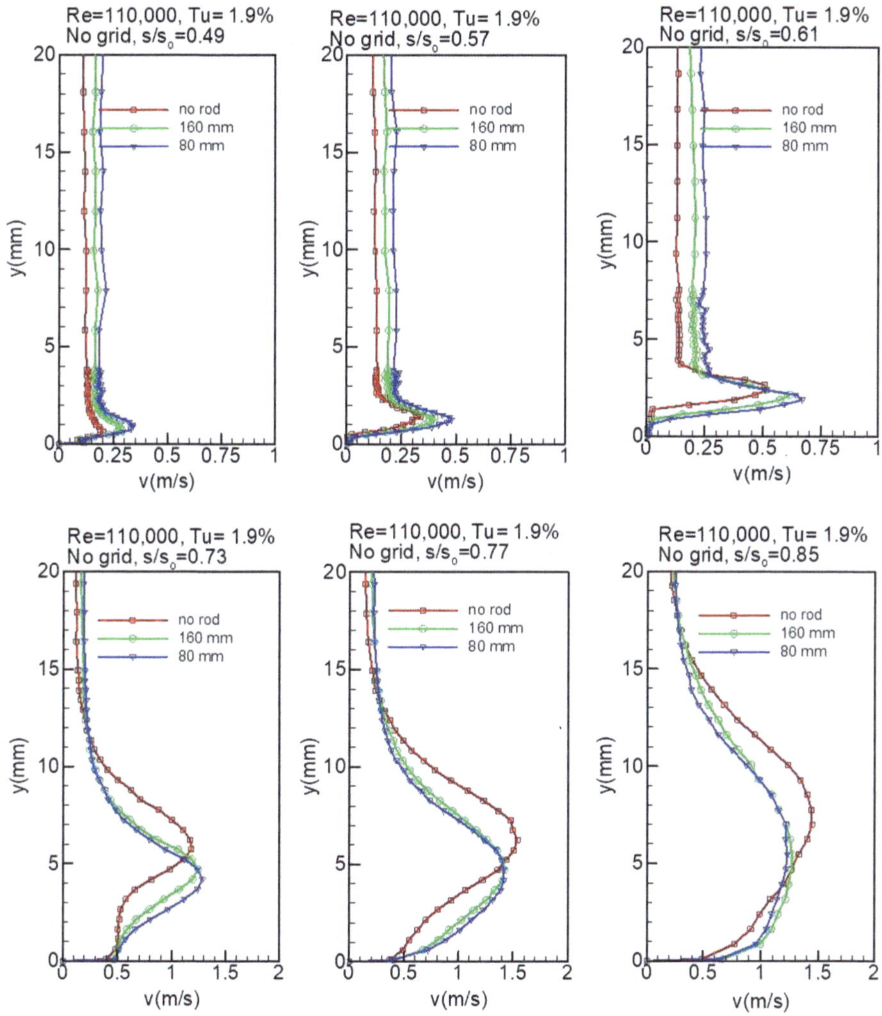

Fig. 11.36 Distribution of time-Averaged fluctuation velocity at different location along the suction surface for steady case $\Omega = 0(S_R = \infty)$ and unsteady cases $\Omega = 1.59(S_R = 160$ mm) and $\Omega = 3.18(S_R = 80$ mm) at Re=110,000 and free-stream turbulence. intensity of 1.9% (without grid)

steady flow case (no rod, $\Omega = 0.0$) is fully separated, the impingement of wakes with $\Omega = 1.59$ on the bubble has the tendency to reverse the separation causing a reduction of the separation height. This is due to the exchange of mass, momentum and energy between the highly turbulent vortical cores of the wakes and the low energetic fluid within the bubble as shown in Fig. 11.36. Increasing the frequency to $\Omega = 3.18$ has moved the velocity distribution further away from the separation, as seen in Fig. 11.35 at $s/so = 0.77$. Passing through the separation regime, the reattached flow still shows the unsteady wake effects on the velocity and fluctuation profiles.

The fluctuation profile, Fig. 11.36 at $s/so = 0.85$, depicts a decrease of turbulence fluctuation activities caused by unsteady wakes (Ω =1.59 and 3.18) compared to the steady case ($\Omega = 0$, no rod). This decrease is due to the calming phenomenon extensively discussed by several researchers [194–197]).

11.11.8 Combined Effects of Wakes and Turbulence

The time-averaged velocity, Fig. 11.35, as well as the fluctuation distribution Fig. 11.36 experience noticeable changes with increasing the dimensionless frequency from Ω =0.0 to 3.18. To demonstrate impact of turbulence level on both velocity as well as the fluctuations, we increase the turbulence level from 1.9 to 8%. To this end we use the turbulence grids detailed in Chap. 9. This reveals the dominant impact of the turbulence level on velocity and fluctuation distribution, which is confirmed by the Figs. 11.37, 11.38 and 11.39. It was generated by TG1 (from Chap. 9). It dictates the pattern of boundary layer development from leading edge to trailing edge. While the high frequency stochastic fluctuations of the incoming turbulence seem to overshadow the periodic unsteady wakes and the lateral extent of the separation bubble, they are not capable of completely suppressing the separation. A similar situation is encountered at higher turbulence intensity levels of 8% produced by grid TG2 (from Chap. 9), Fig. 11.38a, b. The time averaged velocity as well as fluctuation rms do not exhibit effects of unsteady wake impingement on the suction surface throughout. In contrast to the above 3% case, the 8% turbulence intensity case, exhibited in Fig. 11.38, seems to substantially reduce the separation bubble, where an inflection velocity profile.

The impact of the higher turbulence intensity on boundary layer development under unsteady inlet flow condition becomes visible at a turbulence intensity level of 3% shown in Fig. 11.37 at $s/s_0 = 0.61$ is still visible. An almost complete suppression is accomplished by utilizing the turbulence intensity of 13% that is produced by turbulence grid TG3 (from Chap. 9). In both turbulence cases of 8% and 13%, the periodic unsteady wakes along with their high turbulence intensity vortical cores seem to be completely submerged in the stochastic high frequency free-stream turbulence.

11.11.9 Impact of Wake Frequency on Flow Separation

Before explaining the physics of the flow separation, we clarify with an example, how the frequency of a periodic unsteady flow changes the dynamic pattern of a boundary layer flow separation. Considering the Fig. 11.40 with an $\Omega = 1.59 (S_R = 160$ mm) and a free stream turbulence level of 1.9%, it becomes clear that the presence of a low frequency of $\Omega = 1.59$ has not noticeably affected the separation dynamics. Even if we double the frequency to $\Omega = 3.18$, as shown in Fig. 11.41, will not significantly

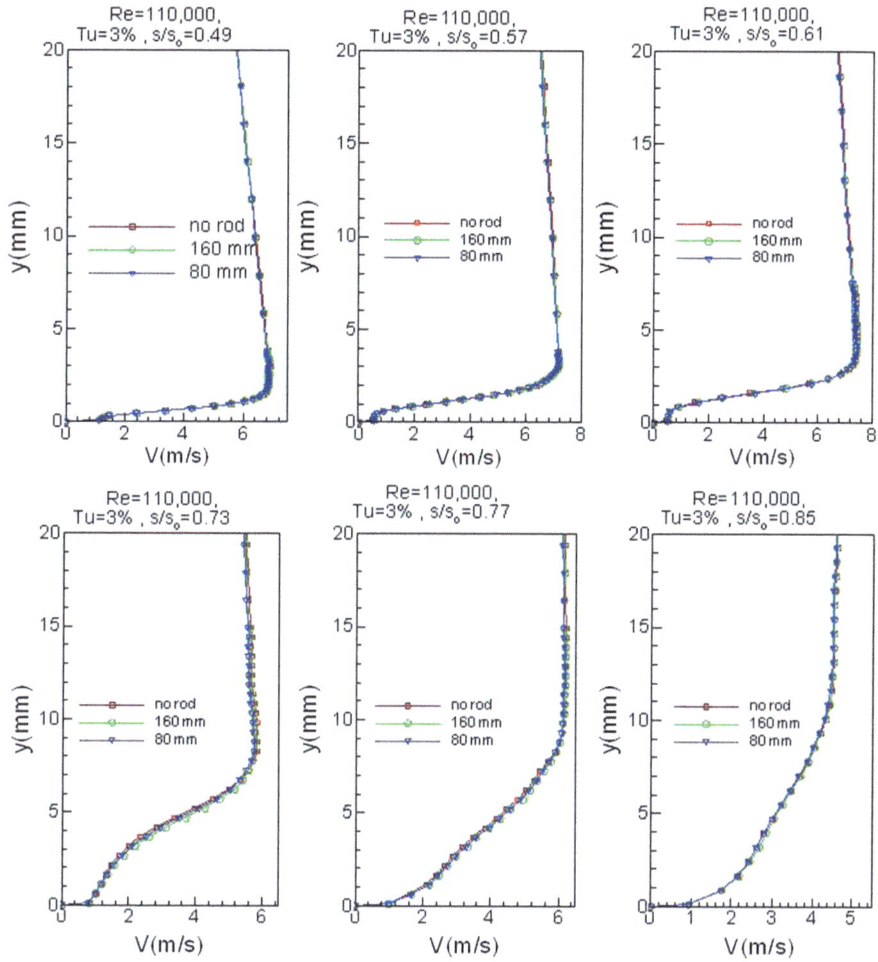

Fig. 11.37 Time-averaged velocity along the suction surface for different s/s_0 with $\Omega = 3.18(S_R =$ 40mm) at Re=110,000 and Tu=3.0 %

change the separation pattern. At this juncture the question, which arises is this: what criteria must be fulfilled in order for a periodic unsteady wake to substantially affect the separation pattern. To answer this question we have to revisit the ensemble averaged turbulence intensity $<Tu>$ as well as the ensemble averaged turbulence intermittency $<\gamma>$ shown in Figs. 11.31 and 11.32. The first criterium deduced from these figures is: as long as, the wakes are confined to narrow strips, they are not able to affect the separation pattern. As the Fig. 11.32 shows, the wakes join each other close to at $S/S_0 = 0.85$. In order to affect the separation pattern, they have to join each other at $S/S_0 < 0.5$. This constitute the second criterium.

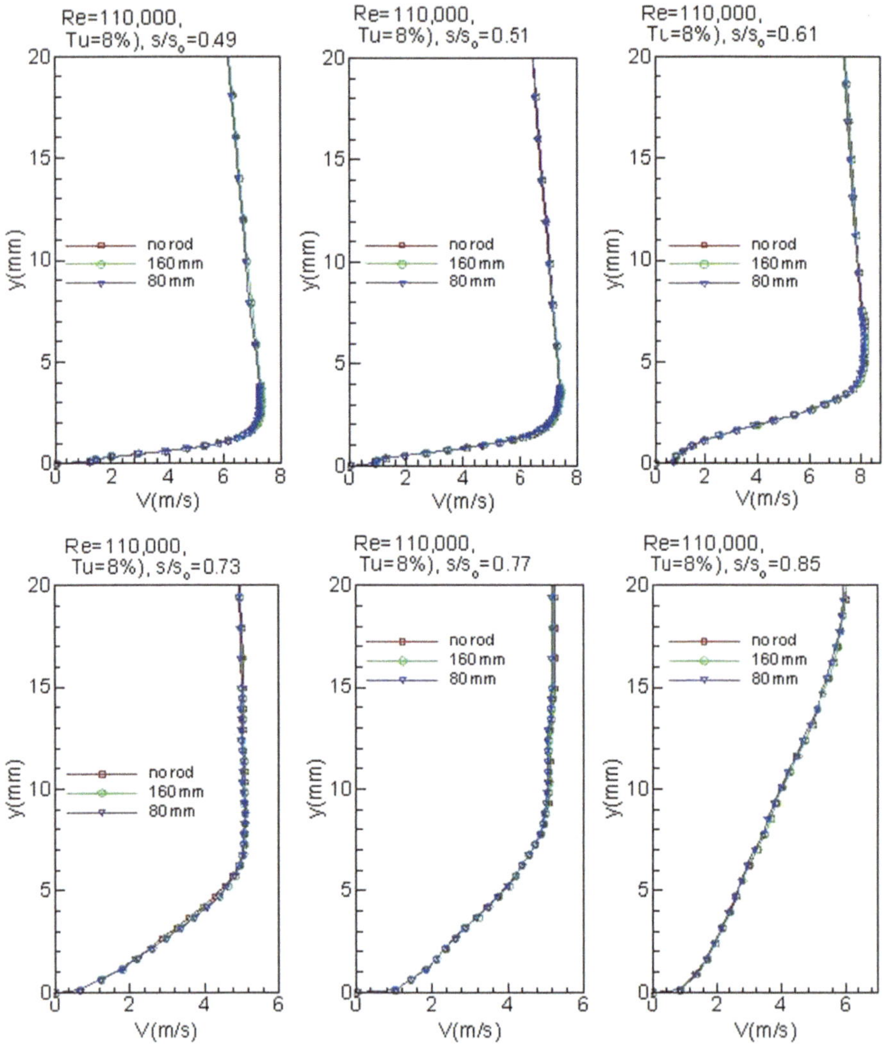

Fig. 11.38 Time-averaged velocity along the suction surface for different s/s_0 with $\Omega = 3.18(S_R = 40\text{mm})$ at Re=110,000 and Tu=8.0%

11.11.10 Dynamics of Separation Bubbles

For better understanding the physics, the ensemble averaged velocity contours are presented for Tu = 1.9%, 3%, 8% and 13.0%, respectively. The combined effects of the periodic unsteady wakes and high turbulence intensity on the onset and extent of the separation bubble are shown in Fig. 11.40 for the Reynolds number of 110,000.

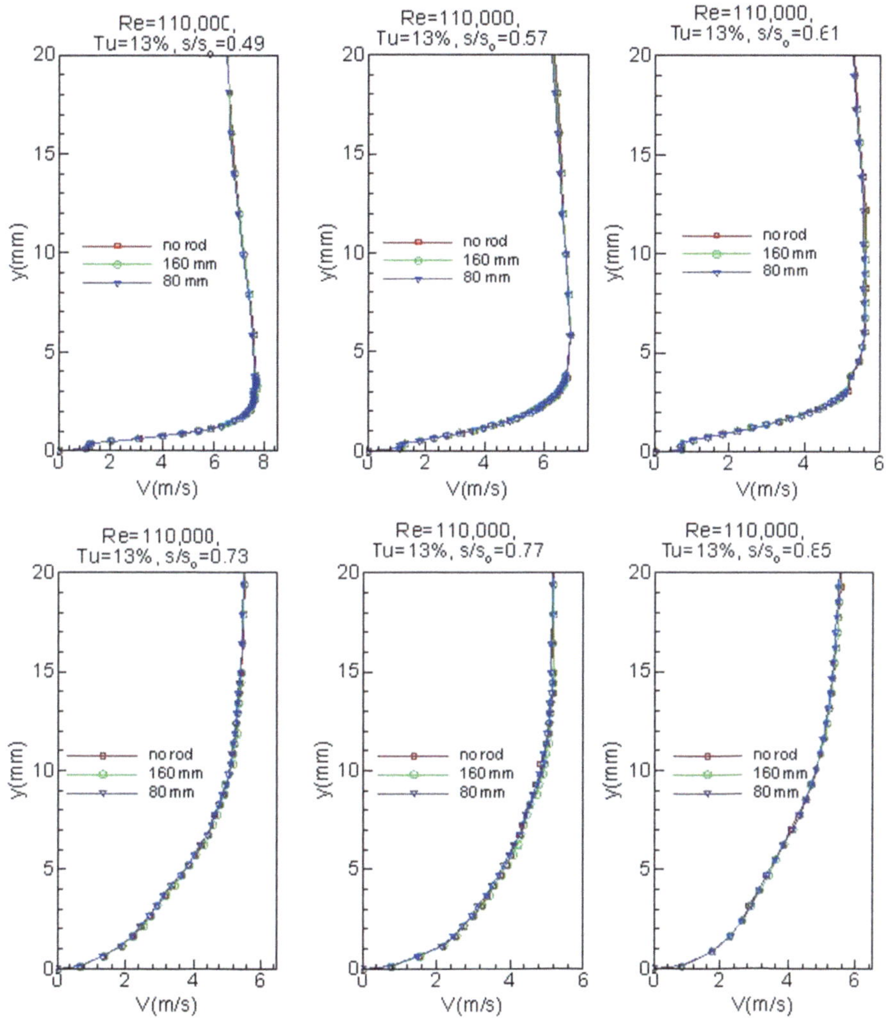

Fig. 11.39 Time-averaged velocity along the suction surface for different s/s_0 with $\Omega = 3.18(S_R =$ 40mm) at Re=110,000 and Tu=13.0%

These figures display the full extent of the separation bubble and its dynamic behavior under a periodic unsteady wake flow impingement at different t/τ. For each particular point s/s_0 on the surface, the unsteady velocity field inside and outside of the boundary layer is traversed in normal direction and ensemble-averaged at 100 revolution with respect to the rotational period of the wake generator. To obtain a contour plot for a particular $t\tau$, the entire unsteady ensemble-averaged data traversed from leading to trailing edge are stored in a large size file (of several giga bites) and sorted for the particular t/τ under investigation.

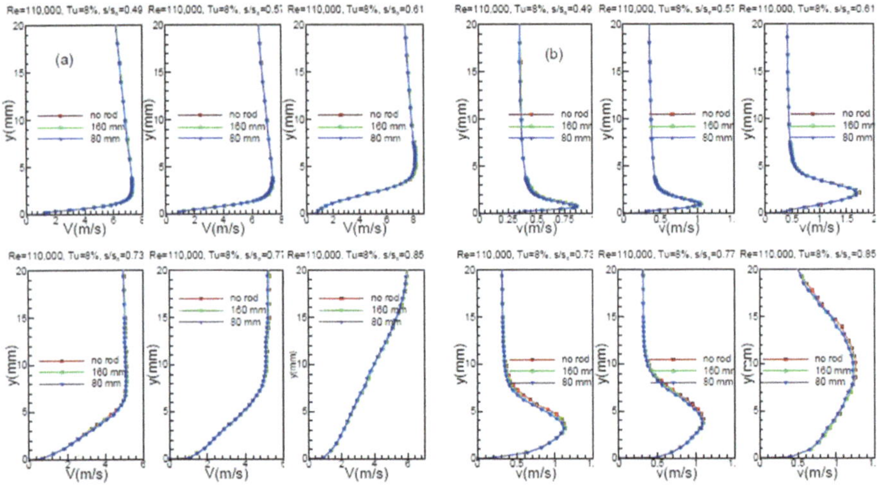

Fig. 11.40 Ensemble-averaged velocity contours along the suction surface for different s/s_0 with time t/τ as parameter for $\Omega = 1.59(S_R = 160\text{mm})$ at Re=110,000 and Tu=1.9% (no turbulence grid)

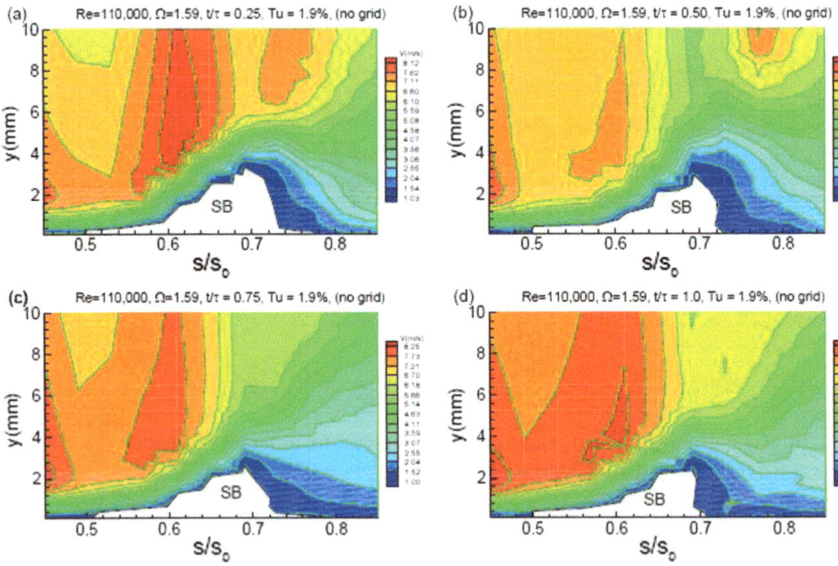

Fig. 11.41 Ensemble-averaged velocity contours along the suction surface for different s/s_0 with time t/τ as parameter for $\Omega = 3.18(S_R = 80\text{mm})$ at Re = 110,000 and 1.9 %

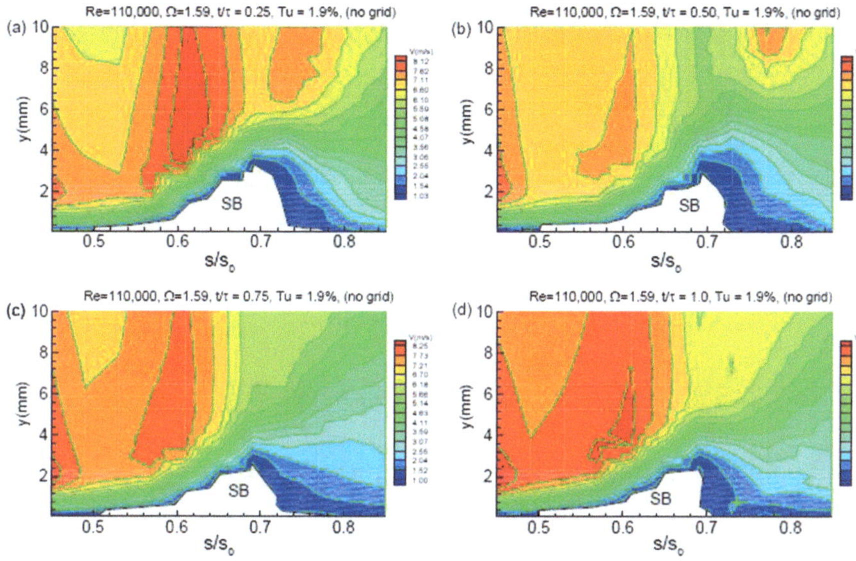

Fig. 11.42 Ensemble-averaged velocity contours along the suction surface for different s/s_0 with time t/τ as parameter for $\Omega = 1.59(S_R = 160$ mm) at Re = 110,000 and Tu = 1.9% (no turbulence grid)

11.11.11 Variation of Tu

Fig. 11.43 with $\Omega = 1.59$ and turbulence level of Tu = 1.9% exhibits the reference configuration for $\Omega = 1.59(S_R = 160\text{mm})$, where the bubble undergoes periodic contraction and expansion as extensively discussed in [198, 199]. During a rod passing period, the wake flow and the separation bubble undergo a sequence of flow states which are not noticeably different when the unsteady data are time-averaged. Starting with Re = 110,000 and $\Omega = 1.59$, Fig. 11.43 (a) exhibits the separation bubble in its full size at $t/\tau = 0.25$. At this instant of time, the incoming wakes have not reached the separation bubble (Fig. 11.42).

At a low turbulence intensity of Tu-1.95 the kinematics of the bubble is completely governed by the wake external flow which is distinguished by red patches traveling above the bubble. At $t/\tau = 0.5$, the wake with its highly turbulent vortical core passes over the blade and generates high turbulence kinetic energy. At this point, the wake turbulence penetrates into the bubble causing a strong mass, momentum and energy exchange between the wake flow and the fluid contained within the bubble. This exchange causes a dynamic suppression and a subsequent contraction of the bubble. As the wake travels over the bubble, the size of the bubble continues to contract at $t/\tau = 0.75$ and reaches its minimum size at, $t/\tau = 1.0$. At $t\tau = 1$, the full effect of the wake on the boundary layer can be seen before another wake appears and the bubble moves back to the original position. Increasing the turbulence level to 3%, 8%, and 13% by successively attaching the turbulence grids TG1, TG2, and TG3

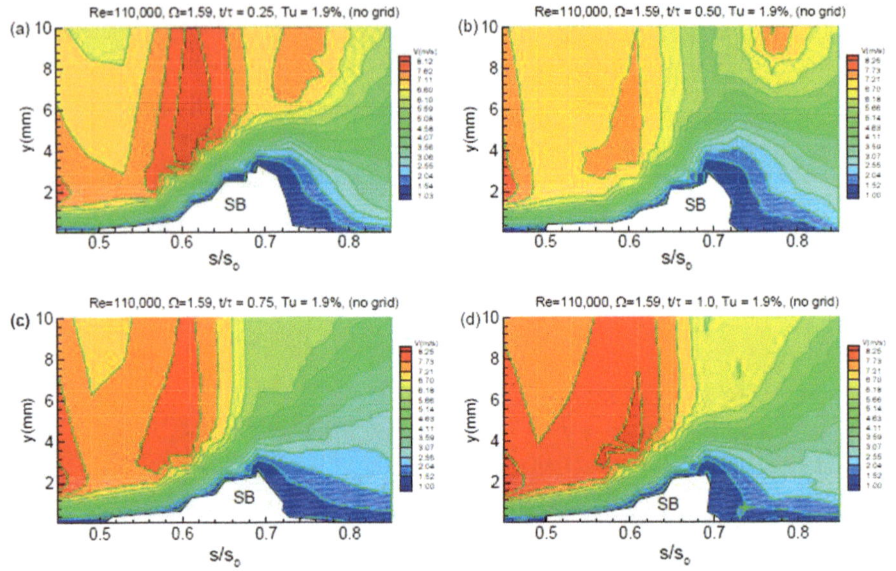

Fig. 11.43 Ensemble-averaged velocity contours along the suction surface for different s/s_0 with time t/τ as parameter for $\Omega = 3.18 (S_R = 80$ mm) at Re = 110,000 and Tu = 3.0%

(detail specifications are listed in Table 9.3 from Sect. 9.3 and keeping the same dimensionless frequency of $\Omega = 3.18$, has significantly reduced the lateral extent of the bubble. Starting with the case with Tu = 3%, Fig. 11.43, it shows the overall reduction of the bubble size as the result of higher turbulence intensity compared with Fig. 11.44.

As shown in Fig. 11.43, the size of the separation bubble though marginally reduced but not much diminished. This is because the turbulence intensity of 3% is not strong enough to substantially reduce the size of the bubble. The dynamic behavior of the bubble changes substantially by increasing the turbulence intensity to 8% and 13%, as shown in Figs. 11.44 and 11.45. The instance of the wake traveling over the separation bubble, which is clearly visible in Fig. 11.43, has diminished almost entirely. Increasing the turbulence intensity to 8%, Fig. 11.44a–d, and 13% respectively, Fig. 11.45 caused the bubble height to further reduce (the corresponding figure for 13% is very similar to the one with 8%). Although the higher turbulence level has, to a great extent, suppressed the separation bubble as Fig. 11.45 clearly shows, it was not able to completely eliminate it. There is still a small core of separation bubble remaining. Its existence is attributed to the stability of the separation bubble at the present Re-number level of 110,000.

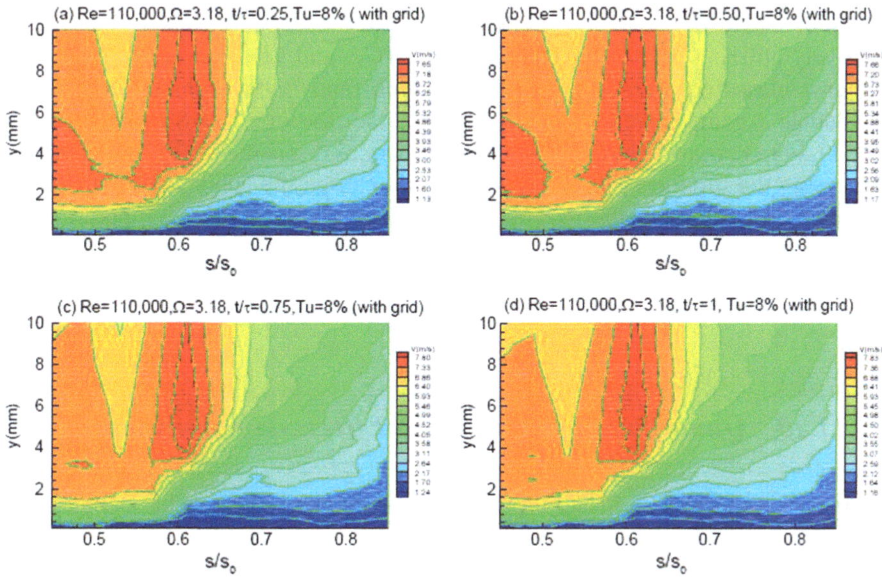

Fig. 11.44 Ensemble-averaged velocity contours along the suction surface for different s/s_0 with time t/τ as parameter for $\Omega = 3.18(S_R = 80$ mm) at Re = 110,000 and Tu = 8.0%

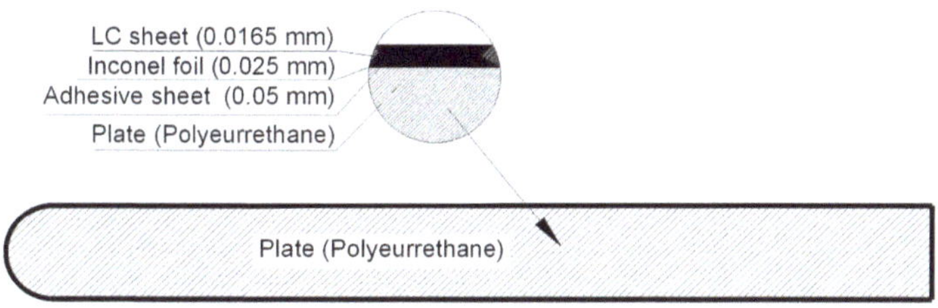

Fig. 11.45 Ensemble-averaged velocity contours along the suction surface for different s/s_0 with time t/τ as parameter for $\Omega = 3.18(S_R = 80\text{mm})$ at Re = 110,000 and Tu = 13.0% (with grid TG2)

11.11.12 Variation of Tu at Constant Wake Frequency

Figures represent the dynamic behavior of the separation bubble at Tu = 1.9%, but at a higher dimensionless frequency of $\Omega = 3.18$. Similar to Fig. 11.43, the case with the Tu = 1.9 exhibits the reference configuration for $\Omega = 3.18(S_R = 80\text{mm})$ where the bubble undergoes periodic contraction and expansion. The temporal sequence of events is identical with the case discussed in Fig. 11.43, making a detailed discussion unnecessary. In contrast to the events described in Fig. 11.43, the increased wake frequency in the reference configuration, Fig. 11.40, is associated with higher mixing and, thus, higher turbulence intensity that causes a more pronounced contraction and expansion of the bubble.

As in case with $\Omega = 1.59$, applying turbulence levels of 3, 8, and 13% by successively utilizing the turbulence grids TG1, TG2, and TG3 and keeping the same dimensionless frequency of $\Omega = 3.18$, has significantly reduced the lateral extent of the bubble. Again, as a representative example, the case with Tu = 8% is presented in Fig. 11.44 (a to d), which reveals similar behavior as discussed in Fig. 11.43. Further increasing the turbulence intensity to 13% has caused the bubble height to further reduce. Although the higher turbulence level has, to a great extent, suppressed the separation bubble, it was not able to completely eliminate it. There is still a small core of separation bubble remaining. As in Fig. 11.45, its existence is attributed to the stability of the separation bubble at the present Re-number level of 110,000.

11.11.13 Quantifying the Combined Effects on Aerodynamics

Figures 11.43, 11.44, 11.45 and 11.46 show the combined effects of turbulence intensity and unsteady wakes on the onset and extent of the separation bubble. Detailed information relative to propagation of the wake and the turbulence into the separation bubble is provided by Fig. 11.46a–d, where the time dependent ensemble averaged velocities and fluctuations are plotted for Re = 110,000 at a constant location $s/s_0 = 3.36$ mm inside the bubble for different intensities ranging from 1.9 to 13%. As Fig. 11.46a depicts, the wake has penetrated into the separation bubble, where its high turbulence vortical core and its external region is clearly visible.

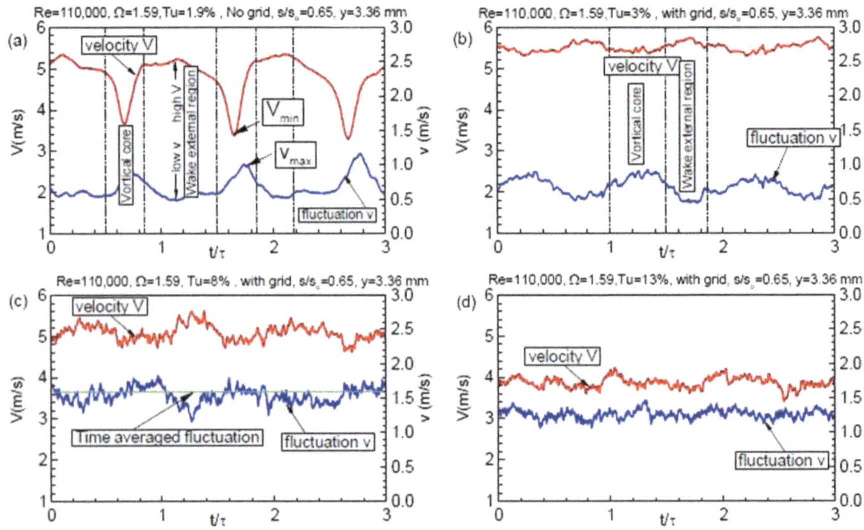

Fig. 11.46 (a, b, c, and d) Time dependent ensemble averaged velocities and fluctuations for Re=110,000 at a constant location s/s_0 =0.65 mm inside the bubble for different inlet turbulence intensities ranging from 1.9% to 13%

Lowest turbulence fluctuations occur outside the vortical core, whereas the highest is found within the wake velocity defect. Increasing Tu to 3%, Fig. 11.46b, reduces the velocity amplitude of the periodic inlet flow and its turbulence fluctuations. Despite a significant decay in amplitude, the periodic nature of the impinging wake flow is unmistakably visible. Further increase of Tu to 8%, Fig. 11.46c, shows that the footprint of a periodic unsteady inlet flow is still visible, however the deterministic periodicity of the wake flow is being subject to the stochastic nature caused by the high turbulence intensity. Further increase of turbulence to Tu =13% causes a degradation of the deterministic wake ensemble averaged pattern to a fully stochastic one. Comparing Figs. 11.46a, c leads to the following conclusion: The periodic unsteady wake flow definitely determines the separation dynamics as long as the level of the time averaged turbulence fluctuations is below the maximum level of the wake fluctuation ν_{max}, shown in Fig. 11.46a. In this case, this apparently takes place at a turbulence level between 3 and 8%. Increasing the inlet turbulence level above ν_{max} causes the wake periodicity to partially or totally submerge in the free-stream turbulence, thus, downgrading into stochastic fluctuation, as shown in Figs. 11.46c, d. In this case, the dynamic behavior of the separation bubble is governed by the flow turbulence that is responsible for the suppression of the separation bubble. One of the striking features this study reveals is, that the separation bubble has not disappeared completely despite the high turbulence intensity and the significant reduction of its size which is reduced to a tiny bubble. At this point, the role of the stability of the laminar boundary layer becomes apparent which is determined by the Reynolds number.

11.12 Numerical Simulation

As seen in previous sections, the boundary layer development along the suction surface of a low pressure turbine blade includes laminar, transitional and turbulent flow regimes with laminar separation and turbulent re-attachment. Thus, this particular case has all essential features relevant for assessing the predicting capability of any numerical method. To quantify the differences between the turbulence models discussed in Chap. 9, we utilize the low pressure turbine cascade in Fig. 11.21 with steady inlet flow condition. The models used are: $k-\varepsilon$, $k-\omega$, SST without transition and SST with transition model $y-\gamma-\theta$ from [200]. The numerical solutions are compared with the experimental results taken from [188] and presented in Fig. 11.47.

The laminar region from the leading edge up to $s/s_0 = 0.49$ are satisfactorily predicted with SST. It should be noted that in this region no transition takes place. As expected, close to the wall, the simulation with $k-\varepsilon$ results in major differences between the experiment and simulation. These differences are slightly less using $k-\omega$. Moving to $s/s_0 = 0.57$, the first signs of the velocity inflection at the wall becomes visible which signifies the onset of a separation zone, which grows downstream, $s/s_0 = 0.61$ and reaches its maximum size at about $s/s_0 = 0.73$. Further downstream

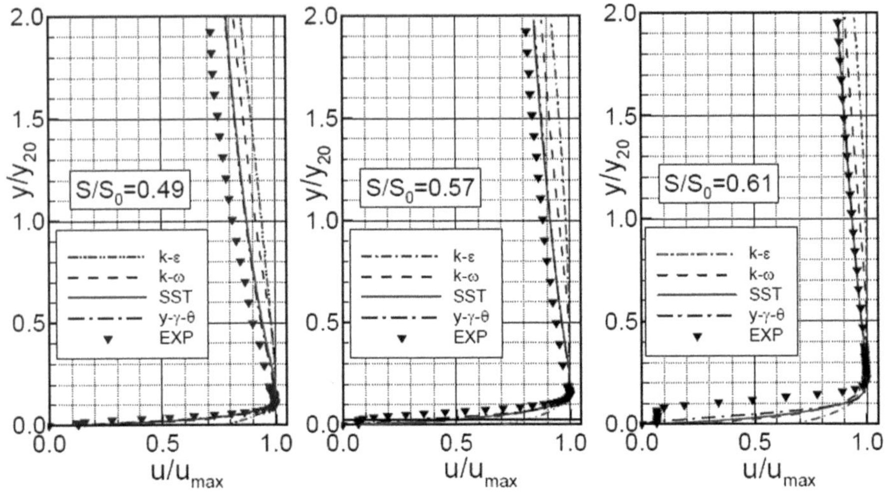

Fig. 11.47 Comparison of $k - g$, $k - \omega$, SST without transition and with $y - \gamma - \theta$ transition model and the experimental results

at $s/s_0 = 0.77$, the velocity profile starts reattaching and at $s/s_0 = 0.85$ the process of re-attachment is completed. As seen, non of the profiles resembles the experimental results. The profiles generated with SST in conjunction with transition model $y - \gamma - \theta$ seem to predict the separation, the onset and extension of the separation zone is not predicted. Likewise the re-attachment profile presented at $s/s_0 = 0.85$ does not correspond to the measured profile.

Chapter 12
Boundary Layer Heat Transfer

12.1 Introduction

In Chap. 11, we have treated the theoretical and experimental aspects of the boundary layer flow and introduced two facilities for conducting experimental investigations. Mean velocity and the fluctuation distributions within the boundary layer were presented for statistically steady as well as periodic unsteady inlet flow conditions. One important quantity, namely, the surface temperature was not mentioned in Chap. 11. In order to calculate the surface temperature distribution, the differential equation of energy must be added to the continuity and momentum equations of boundary layer. The energy equation is expressed in terms of averaged total enthalpy and the fluctuation enthalpy. Using thermodynamic relations,the enthalpy can easily be converted into temperature, from which the heat transfer coefficient can be calculated. To substantiate the theoretical basis, heat transfer experiments has been conducted and reported in this chapter. For this purpose, facilities in Chap. 11, Figs. 11.18 and 11.19 are modified to adopt the temperature measurement requirement. In this Chapter, for the sake of completeness, we first present the entire set of equations that are necessary for calculating the aerodynamics and heat transfer rather than referring the reader to the equations that are spread over several chapters. In conjunction with the above mentioned equation set, it should be noted that the boundary layer transition described by the intermittency function is an integral part of the boundary layer and heat transfer calculation. Since this topic, was extensively discussed in Chap. 8, we refer the reader to this chapter for an in-depth consultation. Instantaneous velocity signals are used to determine the intermittency throughout the boundary layer. The implementation of the transition model in the boundary layer code is explained only briefly.

M. T. Schobeiri, *Advanced Fluid Mechanics and Heat Transfer for Engineers and Scientists*, https://doi.org/10.1007/978-3-030-72925-7_12

12.2 Equations for Heat Transfer Calculation

Following Prandtl's boundary layer theory, all equations presented here are time-averaged. We start with the continuity equation for a two-dimensional boundary layer:

$$\frac{\partial}{\partial x}(\rho\bar{U}) + \frac{\partial}{\partial y}(\rho\bar{V}) = 0. \tag{12.1}$$

The time averaged momentum equation in the x-direction is:

$$\bar{U}\frac{\partial\bar{U}}{\partial x} + \bar{V}\frac{\partial\bar{U}}{\partial y} = -\frac{1}{\rho}\frac{d\bar{p}}{dx} + \frac{\partial}{\partial y}\left(\nu\frac{\partial\bar{U}}{\partial y} - \overline{uv}\right). \tag{12.2}$$

The energy equation for turbulent flow is:

$$\varrho\bar{U}\frac{\partial\bar{H}}{\partial x} + \varrho\left(\bar{V}\frac{\partial\bar{H}}{\partial y}\right) = \frac{\mu_{eff}}{Pr_{eff}}\frac{\partial^2\bar{H}}{\partial y^2} + \frac{\partial}{\partial y}\left[\mu\left(1-\frac{1}{Pr}\right)\bar{U}\frac{\partial\bar{U}}{\partial y}\right] - \varrho\frac{\overline{\partial(V'h')}}{\partial y}$$
$$-\varrho\frac{\partial(\bar{U}\,\overline{U'V'})}{\partial y} + \varrho\varepsilon. \tag{12.3}$$

now we introduce the turbulence kinematic viscosity, also called eddy kinematic viscosity ε_m, the eddy diffusivity of heat ε_h and the turbulent Prandtl number Pr_t

$$\nu_t \equiv \varepsilon_m = \frac{-\overline{u'\,v'}}{\partial u/\partial y}, \quad \varepsilon_h = \frac{-\overline{T'\,v'}}{\partial T/\partial y}, \quad Pr_t = \frac{\varepsilon_m}{\varepsilon_h} \tag{12.4}$$

and rewrite Eq. (12.3) using Eq. (12.4), we arive at a more compact form:

$$\varrho\bar{U}\frac{\partial\bar{H}}{\partial x} + \varrho\left(\bar{V}\frac{\partial\bar{H}}{\partial y}\right) = \frac{\partial}{\partial y}\left(\frac{\mu}{Pr} + \varrho\frac{\varepsilon_m}{Pr_t}\right)\frac{\partial\bar{H}}{\partial y} + \left(\frac{1}{Pr_t} - 1\right)$$
$$\times\left[\bar{u}\frac{\partial\bar{u}}{\partial y} + \varrho\varepsilon + \frac{\partial}{\partial y}\right]\mu\left(1-\frac{1}{Pr}\right) - \varrho\varepsilon_m. \tag{12.5}$$

The turbulent shear stress modeled by the eddy diffusivity for momentum ε_M

$$-\overline{u'v'} = \varepsilon_M\frac{\partial\bar{U}}{\partial y} = \frac{\mu_t}{\rho}\frac{\partial\bar{U}}{\partial y} \tag{12.6}$$

which is expressed in terms of the turbulent viscosity μ_t. This combined with the laminar viscosity to give

$$\mu_{\text{eff}} = (\mu + \mu_t) = \rho(\nu + \varepsilon_M). \tag{12.7}$$

Similarly, after introducing the concept of eddy diffusivity for heat, ε_H, the energy equation becomes

$$\varrho \bar{U}\frac{\partial \bar{H}}{\partial x} + \varrho\left(\bar{V}\frac{\partial \bar{H}}{\partial y}\right) = \frac{\mu_{eff}}{Pr_{eff}}\frac{\partial^2 \bar{H}}{\partial y^2} + \frac{\partial}{\partial y}\left[\mu\left(1-\frac{1}{Pr}\right)\bar{U}\frac{\partial \bar{U}}{\partial y}\right] - \varrho\frac{\overline{\partial(V'h')}}{\partial y}$$
$$- \varrho\frac{\partial(\bar{U}\,\overline{U'V'})}{\partial y} + \varrho\varepsilon \tag{12.8}$$

with $\bar{H}$, $\bar{U}$, $\bar{V}$, h' as the averaged total enthalpy, the averaged velocity component in x and y-direction, the static enthalpy fluctuation. The $\mathrm{Pr}_{\mathrm{eff}}$ is the effective Prandtl number [226] defined as

$$\mathrm{Pr}_{\mathrm{eff}} = \frac{\mu_{\mathrm{eff}}}{(k/c_p)_{\mathrm{eff}}} = \frac{1+\varepsilon_{\frac{M}{\nu}}}{\frac{1}{\mathrm{Pr}} + \epsilon_{\frac{M}{\nu}}\frac{1}{\mathrm{Pr}_t}}. \tag{12.9}$$

The effective thermal conductivity k as the sum of the material conductivity and the eddy (turbulent) conductivity:

$$\begin{aligned} k_{eff} &= k + k_t = k + \varrho\, c\, \varepsilon_h \\ \mu_{eff} &= \mu + \mu_t = \mu + \varrho\, \varepsilon_m \end{aligned} \tag{12.10}$$

and k as the thermal conductivity, c_p the specific heat at constant pressure, and Pr_t is the turbulent Prandtl number given by

$$\mathrm{Pr}_t = \frac{\varepsilon_M}{\varepsilon_h}. \tag{12.11}$$

Crawford and Kays [226] presented an empirical correlation for Pr_t for gases in terms of turbulent Peclet number $Pe_t = (\varepsilon_m / \nu)\, Pr$. Expressing this correlation in terms of ε_m^+ and $Pe_t^+ = \varepsilon_m^+\, Pr$, it reads:

$$Pr_t = \frac{1}{\frac{1}{2}\alpha^2 + \alpha\, c\, Pe_t^+ - (c\, Pe_t^+)^2\, (1. - e^{-\alpha/cPe_t^+})} \tag{12.12}$$

with $\alpha^2 = 1./0.86$, $c = 0.2$, and Pr_t as the turbulent Prandtl number.

The eddy viscosity term is modeled through the mixing length theory and the above intermittency model is implemented by

$$\varepsilon_M = \gamma l^2 \left|\frac{\partial U}{\partial y}\right| \tag{12.13}$$

where l is the mixing length and γ is the intermittency. In case of a steady flow, the intermittency is time averaged$\gamma \equiv \bar{\gamma}$ correspondingly for an unsteady flow it is ensemble averaged $\gamma \equiv <\gamma>$. The solution process uses Patankar-Spalding's

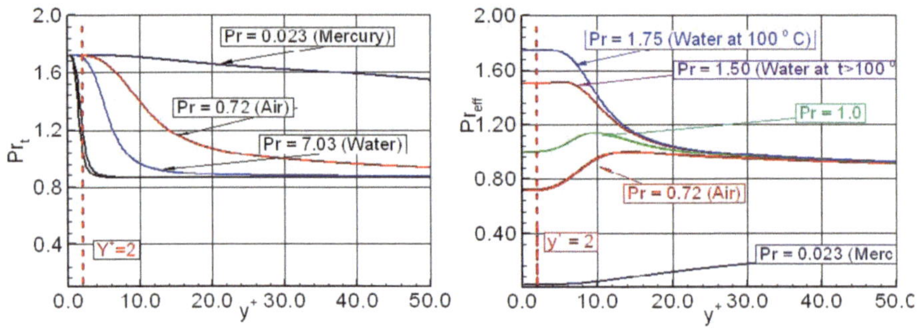

Fig. 12.1 Left: Turbulent Prandl number for different working media, right: effective Prandl number for different working media

omega (non-dimensional stream function) transformation, [223]. In stream function coordinates, the momentum equation without the body forces becomes

$$U\frac{\partial U}{\partial x} + U\frac{\partial}{\partial \psi}\left[U\nu_{\text{eff}}\frac{\partial U}{\partial \psi}\right] = -\frac{1}{\rho}\frac{dp}{dx}. \qquad (12.14)$$

The boundary layer equations are integrated after non-dimensionalizing the stream function and solved numerically.

12.3 Instrumentation for Temperature Measurement

To measure the surface temperature of a test object, four different instrumentations may be used (1) the thermocouples (TC), (2) infra red thermography (IR), (3) liquid crystal (LC) and (4) pressure/temperature sensitive paints (PSP/TSP). Currently, the measurement of temperature taught in heat transfer lab is conducted using the conventional thermo-couples (TC). Although it provides the students with the basic idea how the temperature is measured, it does not provide a picture about the temperature distribution on a surface. There are two problems using TC: first, a thermocouple is capable only of measuring the temperature at discrete points. Second, to obtain the temperature distribution on any surface, a number of TC-s must be applied in a way that they do not disturb the incoming flow. This requires an accurate treatment of the surface after the TC has been installed. This is to ensure that the tip of the TC is flush with the surface. To overcome these problems, we have prepared several practical and effective solutions. These are (1) Infrared Thermal Imaging (IR), (2) Liquid Crystal (LC) technique and (3) Pressure, Temperature Sensitive Paints (PSP, TSP). In what follows, we discuss these techniques in detail. With these three techniques, the temperature distribution on an entire surface can be measured. Chapter 15 is dedicated to these different temperature measurement techniques are discussed. For the heat transfer case studies presented in this chapter, we use the liquid crystal technique. It is easy to use and is very accurate.

12.4 Heat Transfer Calculation Procedures

In the following the governing equations that describe the boundary layer flow are coupled with the energy equation to calculate the transfer of mass, momentum and energy to the boundary layer. The unsteady transition model developed and discussed in Chap. 8 is implemented into the equation system. The time averaged continuity equation is given by

12.5 Experimental Heat Transfer

As we saw in Chap. 11, there is a multitude of parameters affecting the behavior of the boundary layer flow, among them the unsteadiness, pressure gradient and curvature were important to investigate. These parameters are also affecting the heat transfer process, which we treat in this chapter. For this purpose we resort to the plate and the cascade test facilities shown in Figs. 11.19 and 11.24. To measure the surface temperature of the test object heat must be added to the surface. This requires instrumentation of the surface of the test object and integration of a heating mechanism to the surface. This is accomplished by attaching a thin heater sheet of Inconel to the surface. For visualization of temperature distribution, we apply liquid crystal to the surface of the test object. The heating of the test surface is accomplished by using a power supply which is capable of varying the voltage and the current. Commercial liquid crystal sheets have different chemical compositions with a spectrum of color bands when they are exposed to different temperatures. Among the color spectrum, one is distinctly sharp. The type of liquid crystal used in these experiments has a sharp *yellow band* with the temperature $T_{yel} = 44.6\,^{\circ}\mathrm{C}$. For the case studies presented in this chapter, the plate and the cascade will be instrumented for temperature measurements. The convective portion of the energy equation is obtained using with the following equations:

$$h = \frac{Q_{conv}}{(T_{yl} - T_{\infty})} \tag{12.15}$$

Q_{conv} is the convective portion of the energy equation per unit area and is defined by:

$$Q_{conv} = Q_{foil} - Q_{rad} - Q_{cond} \tag{12.16}$$

where Q_{foil} is the heat flux of the Inconel foil and Q_{rad} is the radiation heat flux emitting from the surface of the liquid crystals. These two quantities are given by:

$$Q_{foil} = \frac{VI}{A_{foil}} \tag{12.17}$$

with V and I as the voltage and the current controlled by the power input device and A_{foil} is the area of the airfoil exposed to the incoming flow.

$$Q_{rad} = \varepsilon({T^4}_{yl} - {T^4}_{\infty}) \tag{12.18}$$

The emissivity ε of the liquid crystals has a value of 0.85. The amount of radiation reflected back from the top aluminum wall was found to be less than two percent of the input power and as a result was neglected. The third loss, the conduction losses, were found to be negligible based upon a two-dimensional finite difference nodal analysis method on a slice of the heat transfer plate. In context of heat transfer, we introduce two dimensionless parameters namely Nusselt number and Stanton number:

$$Nu_x = \frac{h_x}{k}, \text{ and } St = \frac{h}{\varrho U c_p} \tag{12.19}$$

The plate shown in Fig. 12.2 has an arc length of 690.0 mm, a radius of curvature of 702.5 mm, a leading edge radius of 1.0 and a thickness of 15.0 mm. Starting from the leading edge,the power is adjusted such that the position of the yellow band is identical with the target location on the test object, in this case the plate leading edge. The power is then reduced to move the yellow band away from the leading edge in a definite increment ΔS in longitudinal direction. For the channel the increment was set $\Delta S = 10$ mm. It must be noted that, in order to achieve steady state operation,

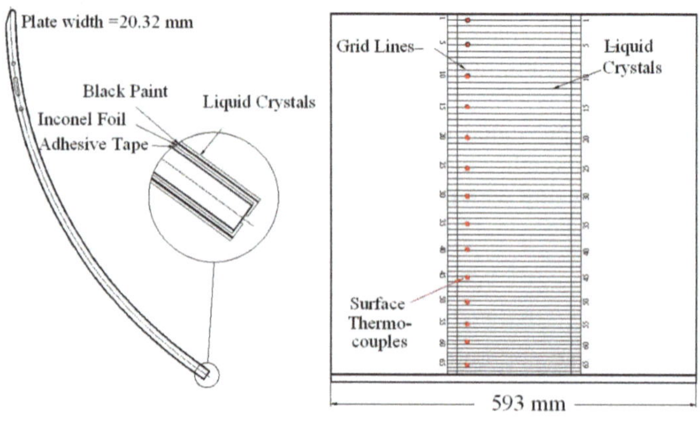

Fig. 12.2 Plate used in Chap. 11 modified for heat transfer measurement by attaching a liquid crystal sheet to the surface

certain time has to elapse before moving to the next point. This process is continued until the full longitudinal length of the plate for both surfaces is completely mapped. This process is the same for all wake passing frequencies. The collected data are reduced based upon a heat flux analysis which entails determining all the energy losses on a flux basis and subtracting them from the heat flux of the Inconel foil. To ensure that the yellow band temperature indicator corresponds to real temperature at the location of the measurement, thermocouples flush with surface are installed on the test object and also serve as the calibrators. Note the thermocouples arranged on the left hand side of the plate for simultaneous calibration of the liquid crystal temperature. The collected data are reduced based upon a constant heat flux analysis, which entails determining all the energy losses on a flux basis and subtracting them from the heat flux of the Inconel foil. For more details we refer to [224]. Details of heat transfer measurement including film cooling effectiveness along with a detailed description of different measurement techniques are presented in Chap. 15. Also heat transfer measurements using the liquid crystal technique are presented in a review documented in [226].

12.6 Local Heat Transfer Coefficient Distribution on Concave Surface

Figures 12.3 and 12.4 show the Nusselt number distribution on the concave and convex surface along the longitudinal length of the plate for $\Omega = 0.000$, 1.033, 1.725, 3.443, and 5.166. For the concave side of the plate, the effect of the wakes on the Nusselt number can clearly be seen. As shown in Fig. 12.3, the start and end of transition shifts towards the leading edge as the wake passing increases. The beginning of transition occurs for $\Omega = 0.0$, 1.033, 1.725, 3.443, and 5.166 around $s/s_0 = 0.35, 0.27, 0.26, 0.21$, and 0.17, respectively. For the aforementioned values of $\boldsymbol{\Omega}$, the end of transition occurs around s/s0 = 0.63, 0.60, 0.58, 0.38, and 0.34, respectively. average heat transfer coefficient occurs for higher values of $\boldsymbol{\Omega}$ mainly due to a higher freestream turbulence level generated by stronger wake mixing and turbulent activities inside the boundary layer. In full accord with the aerodynamic picture of the boundary layer transition discussed previously, the heat transfer is predominantly dictated by freestream turbulence.

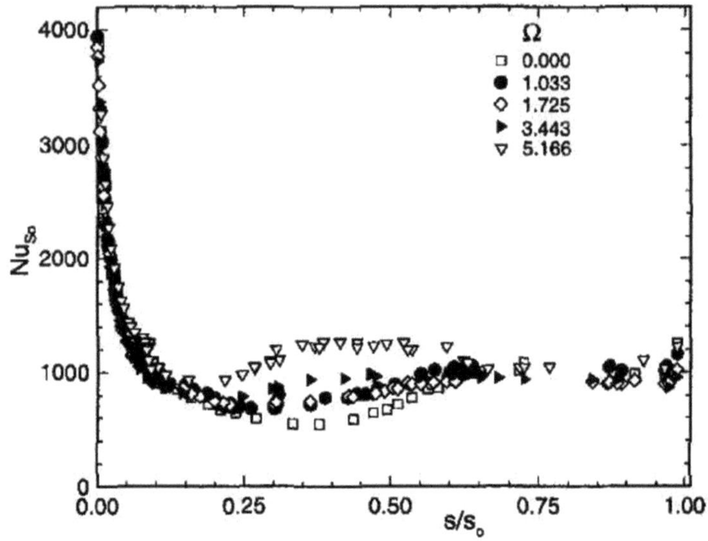

Fig. 12.3 Nusselt number distribution along the concave side of a curved plate in non-dimentional steamwise direction

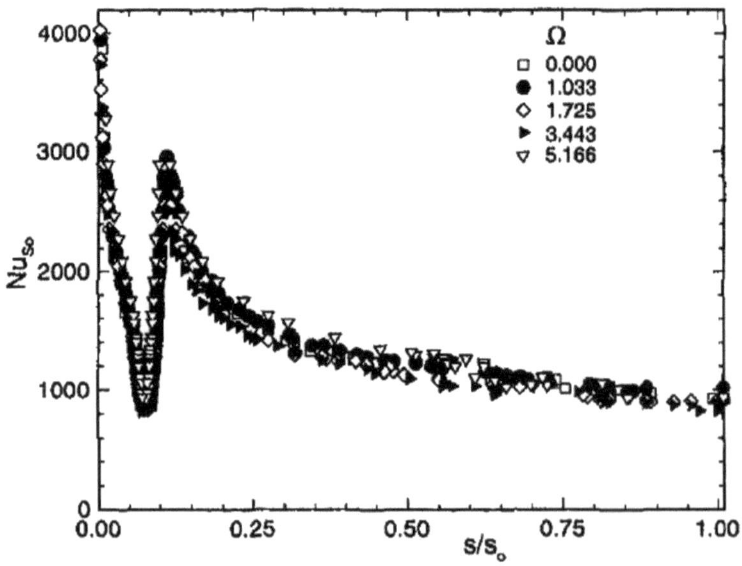

Fig. 12.4 Nusselt number distribution along the convex side of a curved plate in non-dimentional steamwise direction

12.6.1 Heat Transfer Coefficient Distribution

As indicated previously, we used liquid crystal technique developed by Hippensteele [225] for heat transfer measurement. This is being routinely applied by numerous researchers because it has the advantage of not affecting the turbulence structure at the surface, as thermocouples or surface mounted hot wire/film probes do. However, its slow response does not allow extracting valuable unsteady information. As a result, only time-averaged response can be acquired in unsteady cases. Stanton number distributions on the suction and pressure surfaces of the turbine blade for four different wake passing frequencies of $\Omega = 0$, 0.755, 1.51, and 3.02 that correspond to the spacings of 40 mm, 80 mm, 160 mm and ∞ are shown in Fig. 12.5. The uncertainties with the heat transfer coefficients are 6.8%. These results confirm the recent investigations on the effect of unsteady flows on heat transfer distribution of turbine blade by Han et al. [226]. The enhancement of the heat transfer coefficient with increase in the wake passing frequency is clearly apparent from these results. For the steady case on the suction surface, the transition starts near the trailing edge and the beginning of the transition point moves toward the leading edge as wake passing frequency is increased. There is also a consistent increase in the heat transfer coefficient with increase in wake passing frequency, but there is no apparent transition phenomena occurring on the pressure surface of the turbine blade, which is

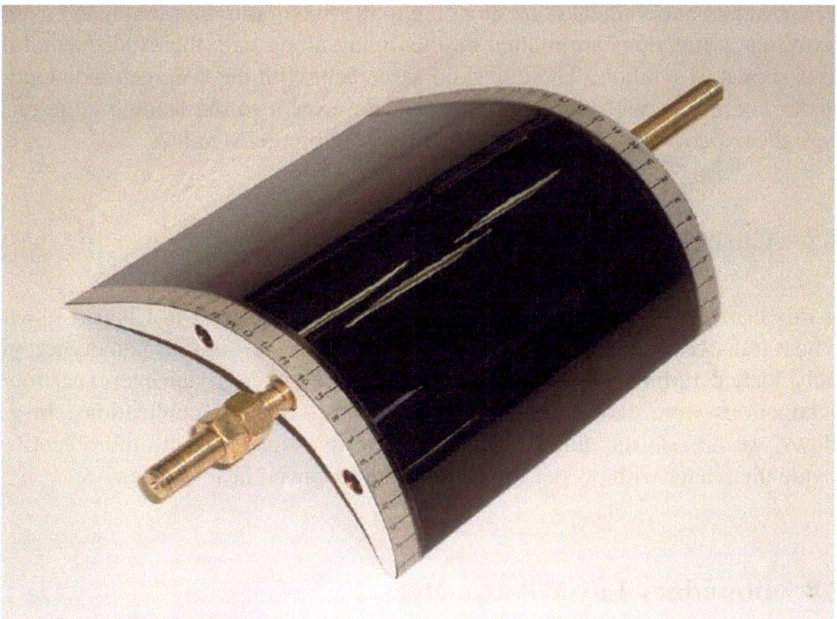

Fig. 12.5 Plate instrumented with a surface heater, liquid crystal sheet and thermo-couples for yellow band temperature calibration

also evident from the aerodynamic measurements shown in Fig. 12.4a, b. The expressions from the intermittency analysis of the aerodynamic data are implemented in the boundary layer code, TEXSTAN, and the results are compared with the experimental data. Figure 12.5a, b shows the heat transfer distribution on the suction surface for Ω values of 0.755 and 1.51. Three lines that predict the heat transfer coefficient corresponding to the maximum, minimum, and average intermittency functions are plotted in these figures along with the experimental data shown by symbols. Again, the three lines that correspond to the maximum, minimum, and average intermittency distributions are shown in these figures. The upper dashed curve represents the streamwise Stanton number distribution when the plate is subjected to an inlet flow intermittency state of $<\gamma(t)>_{max}$. On the other hand, when the plate is subjected to $<\gamma(t)>_{min}$, the lower point-dashed curve depicts its Stanton number distribution. However, because of the periodic character of the inlet flow associated with unsteady wakes, the plate would experience a periodic change of heat transfer represented by upper and lower Stanton number curves (dashed and point-dashed line) as an envelope. The liquid crystal responds to this periodic event with time averaged signals. This time-averaged result is reflected by the solid line, which is given by corresponding time-averaged intermittency. As seen, a reasonably good agreement is found for the entire laminar and transitional portions on the suction side of the turbine blade. Good agreement is seen between the with the average predicted heat transfer coefficient and the experimental data. Figure 12.5 shows the heat transfer distribution on the pressure surface for Ω value of 0.755. Three lines that predict the heat transfer coefficient corresponding to the maximum, minimum, and average intermittency functions are plotted on this figure along with the experimental data that is shown in symbols. There is good agreement with the average predicted heat transfer coefficient with the experimental data, except in the leading edge region where the experimental values are higher than the predicted values.

12.7 Case Study, Heat Transfer

For this case study, we utilize the cascade facility described in Chap. 11, where the heat transfer investigations were performed on the test object, namely a typical highly loaded turbine blade installed in aircraft engines. Extensive experimental investigations were carried out and reported in a series of publications. In what follows, we present the quint essential of that investigations with the objective to provide the reader with deeper insight into the advanced heat transfer.

12.8 Boundary Layer Parameters

The parameters defining the aerodynamic behavior of a boundary layer were discussed in Chap. 11. In this section we provide a detailed insight into the heat transfer behavior of a test object that is exposed to different flow parameters such as:

1. Pressure gradient,
2. Re-and Mach number,
3. Turbulence Intensity, and
4. Unsteadiness.

The pressure gradient is one of the major parameters that affects the boundary layer development from laminar to transitional to turbulent and separation. This parameter is an inherent part of the geometry of the test object. Examples of such test objects are: Nozzles and diffusors creating negative, positive pressure gradients. Turbine blades create on their section surface negative, while on the pressure surface positive pressure gradients. Therefor, as the test object, we choose one that inherently produces negative, zero and positive pressure gradients. To this end, we resort to the turbine blades that inherently create negative and positive pressure gradient. These blades, instrumented with liquid crystals are installed into the test section, where detailed heat transfer experiments were conducted. Details of heat transfer measurements using the liquid crystal technique are presented in an extensive literature review in [201]

The turbine blade prepared for heat transfer investigations is shown in Fig. 12.6. It is specially manufactured from a nonconductive material with a thin Inconel heater foil attached to its surface via a two-sided temperature resistant adhesive tape as shown in Fig. 12.6. On top of this tape, a sheet of liquid crystal is attached for temperature measurements. The power required for heating the blade is supplied by a direct current (DC) power supply system, which controls the positioning of the yellow band. Data are collected an reduced based upon a constant heat flux analysis. They determine all energy losses and subtract them from the heat flux of the Inconel foil.

The heat transfer experiment presented in this section investigates the individual and combined effects of steady and periodic unsteady wake flows and freestream turbulence intensity Tu on heat transfer behavior of the boundary layer including the laminar, the transtional, the turbulent and the flow separation.
Fig. 12.7 shows a composite picture that included the development of the flow and heat transfer quantities from the leading edge to trailing edge of the test object. It includes (a) the pressure distribution, (b) the velocity contour, (c) the fluctuation contour, and (d) a representative time averaged heat transfer coefficient (HTC)distribution for for a periodic unsteady wake flow at $\Omega = 3.18$(d). Furthermore, the diagrams in Fig. 12.7a–d deliver a coherent picture of the separation bubble. It depicts four distinct intervals that mark different events along the suction surface. An initially strong negative pressure gradient starting from the leading edge preserves the stable laminar boundary layer until the pressure minimum at $S/S_0 = 0.494$ has been reached. The laminar boundary layer characterized by the lack of lateral turbulence fluctuations, is not capable of transferring mass, momentum, and energy to the boundary layer, resulting in a steep drop of HTC from leading edge to $S/S_0 = 0.494$, where the pressure gradient changes the sign. The HTC drops further at a larger slope and assumes a minimum at the start of the separation bubble $S/S_0 = 0.583$. Passing through the pressure minimum, the initially stable laminar (or nonturbulent) bound-

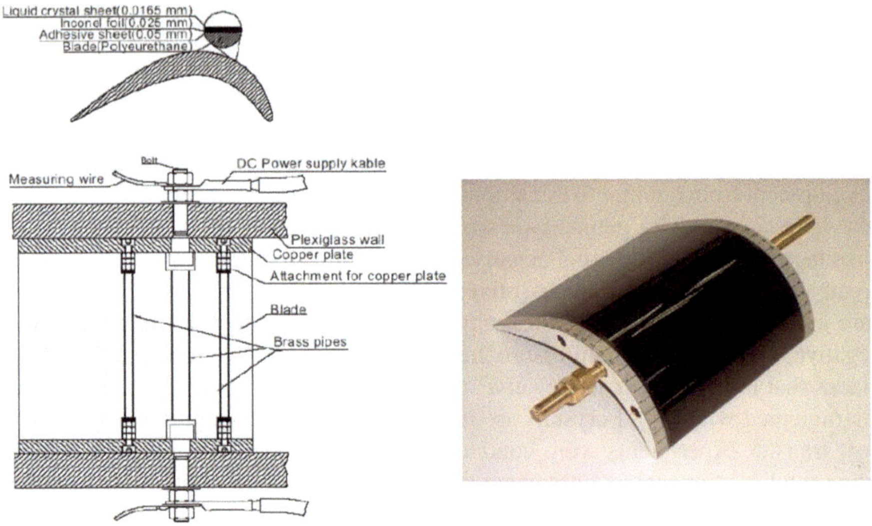

Fig. 12.6 Details of heat transfer blade with LC-Instrumentation

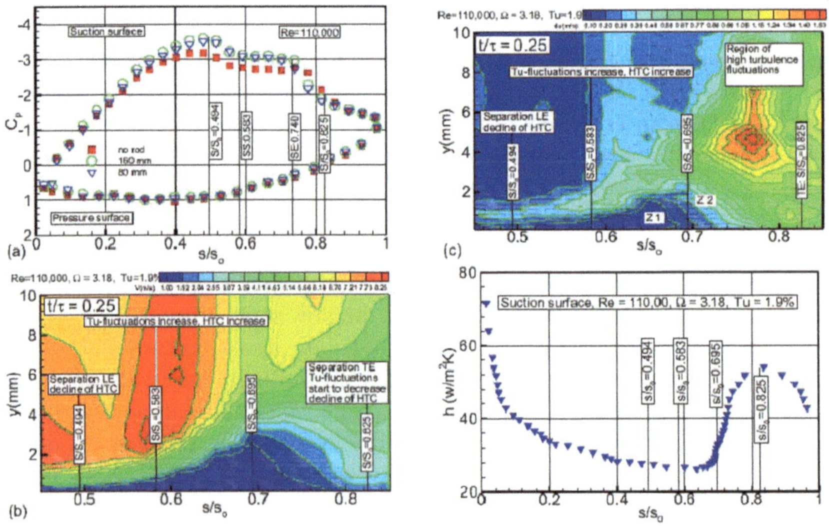

Fig. 12.7 Composite picture of **a** pressure distribution, **b** inception of separation bubble, **c** impact of the wake on separation bubble and **d** time averaged heat transfer coefficient

ary layer encounters a change in pressure gradient from negative to positive, causing it to become unstable and to separate at $S/S_0 = 0.583$. This point marks the leading edge of the separation bubble Fig. 12.7b, c. From this point on, the turbulence activities outside the bubble continuously increase, causing the heat transfer coefficient to rise. Further increase of HTC beyond $S/S_0 = 0.583$ occurs at a steep rate until the separation bubble trailing edge at $S/S_0 = 0.825$ has been reached. The steep increase of HTC within the separation bubble is due to an increased longitudinal and lateral turbulence fluctuation caused by the flow circulation within the bubble. As shown in Fig. 12.7c, the extent of low turbulence envelope is much smaller than the bubble size itself, Fig. 12.7b. This implies that the turbulence activity within the bubble is nonuniform and can be subdivided into two distinct zones, Z1 with lower fluctuation activities that occupies the bubble from the leading edge up to the location, where the bubble lateral extent reaches its maximum height at $S/S_0 = 0.695$ and Z2 the higher fluctuation zone beyond the maximum height, Fig. 12.7c.

12.8.1 Parameter Variation at Steady Inlet Flow Condition

Heat transfer experiments were carried out at Reynolds numbers of 110,000, 150,000, and 250,000 based on the suction surface length and the cascade exit velocity while keeping the turbulence intensity TU constant constant as depicted in Fig. 12.8. This figure shows the heat transfer distribution on suction side (SS) and pressure side (PS) with $S/S_0 = 0$ as the stagnation point. These diagrams exhibit the impact of the geometry on pressure distribution and thus the pressure gradient. They also reveal the significant influence of the Re-number and TU on heat transfer coefficient. Note that in each of these diagrams the turbulence intensity TU was kept constant. A comparison of the these two diagrams with each other shows the significant increase of heat transfer coefficient caused by higher turbulence intensity TU.

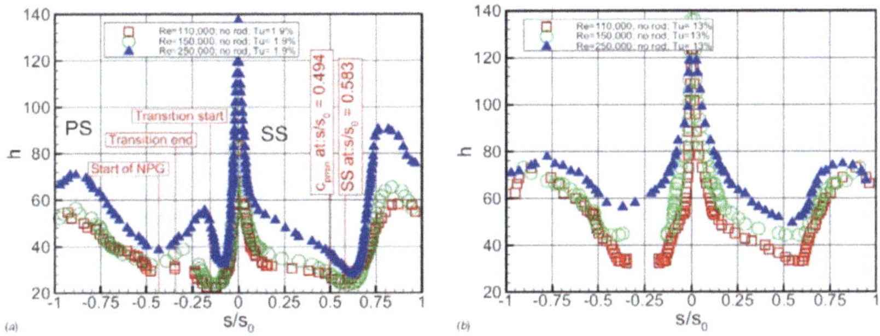

Fig. 12.8 Impact of the Reynolds number on heat transfer coefficient: diagram **a** has the turbulence intensity of 1.9%, while diagram **b** has a turbulence intensity of 13% for steady inlet flow condition

In presenting the heat transfer results, we prefer to use the plain heat transfer coefficient rather than the Nusselt number, which uses thermal conductivity and a constant characteristic length, such as the blade chord, to form the Nusselt number. Three sets of plots are generated to extract the effect of each individual parameter on heat transfer coefficients. Each figure includes the pressure as well as the suction surface heat transfer coefficient h as a function of dimensionless surface length s/s_0. While according to the equation of motion the boundary layer behavior is completely decoupled from thermal boundary layer behavior, the latter is through the equation of energy directly coupled with the boundary layer aerodynamics. Thus, a detailed description of heat transfer behavior is directly coupled with the aerodynamic results.

12.8.2 Parameter Variation at Unsteady Inlet Flow Condition

In this section we are using a periodic unsteady wake flow generated by a system of rods attached to the two timing belts as discussed in Chap. 11. The parameters are: (1) Re-number, (2) reduced frequency Ω and (3) the turbulence intensity TU Periodic Unsteady Inlet Flow Condition, Variation of Re Number at Constant Tu.

In Fig. 12.9, the periodic unsteady inlet flow conditions was generated for reduced frequencies $\Omega = 1.59$ that is created by a rod spacing $SR = 160$ mm.

Keeping the reduced frequency and the turbulence intensity constant, heat transfer measurements were carried out for Re = 110,000, 150,000, and 250,000. HTC distribution along the suction and the pressure surfaces is shown in Fig. 12.9 for $\Omega = 1.59$ and a turbulence intensity of $TU = 1.9\%$. The HTC distributions in Fig. 12.9 exhibit a systematic change for all three Re cases but a substantial increase for Re = 250,000. This is due to the fact that of the three Reynolds cases, the one with Re = 250,000 was able to influence the laminar portion of the boundary layer and, thus, the heat HTC behavior. Comparing the periodic unsteady wake flow case in Fig. 12.9 with the steady flow case shown in Fig. 12.8, only a marginal change in HTC can be observed. This is due to the fact that wakes generated by the translating rods with an $\Omega = 1.59$ are far apart from each other. Consequently, the turbulence activities of their vortical cores are not mutually interacting and, therefore, they are unable to substantially affect the total turbulence picture of the flow leading to almost the same HTC picture.

Keeping the reduced frequency $\Omega = 1.59$, $SR = 160$mm and increasing the turbulence intensity to $Tu = 3\%$ in Fig. 12.9 we observe only a minor increase in HTC for the suction surface compared to in Fig. 12.8 the pressure surface, however, a major increase in HTC is clearly visible, where the transition length almost completely disappeared. In this case, it seems that the wake unsteadiness is about to submerge in the stochastic high frequency freestream turbulence generated by Grid TG1. Further increasing the turbulence intensity by subsequently attaching Grids TG2 and TG3 whose specifications are listed in Tables 1 and 2 has not brought a substantial change compared to the case shown in Fig. 18b. Results are presented in Figs. 18c and 18d, where a systematic shift of HTC toward slightly higher values is shown for Re = 110,000 and 150,000 and substantially higher values for Re = 250,000.

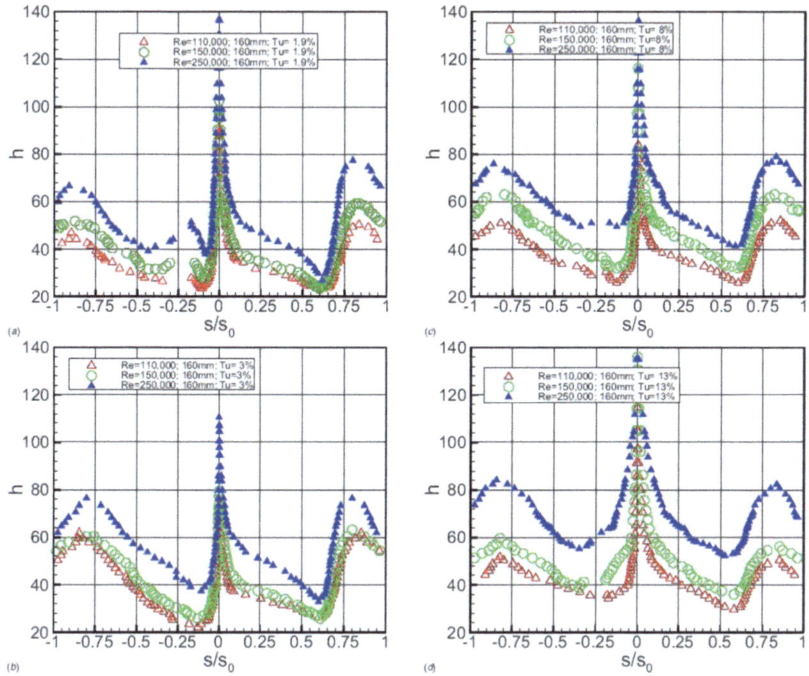

Fig. 12.9 Impact of the Reynolds number on heat transfer coefficient: diagram **a** has the turbulence intensity of 1.9%, while diagram **b** has a turbulence intensity of 13% for steady inlet flow condition

12.9 Temperature Measurement in Rotating Frame

In Sect. 12.8 we used liquid crystal method to measure surface temperature on a turbine blade in a stationary frame. This method is simple to use and delivers accurate results for measuring the surface temperature of test objects installed in a stationary test section, where the test object is neither moving nor rotating. However for the following reasons it is not appropriate for application in a rotating frame:

1. It requires installation of heaters to position the yellow line. However, it is not possible to position the yellow bands while the test object is rotating,
2. Attaching a heater- and a liquid crystal sheet changes the geometry of test object,
3. The attachment of the above sheets to a rotating test object, because of centrifugal force causing both sheets to detach from the surface,
4. It cannot be used for measuring the film cooling heat transfer because of surface holes.

An alternative technique is using the pressure/temperature sensitive paints (PSP/TSP) presented in this section. The measurement of surface temperature using Pressure/Temperature Sensitive Paint (PSP/TSP)is distinctively different from the methods discussed above. The temperature measurement capability of this method is far superior to the methods mentioned previously, however the data acquisition system is more complex than the two previously discussed methods. Considering this circumstance, the author decided to introduce the pressure/temperature sensitive paints technique for measuring the blade surface temperature in a high speed rotating turbine at Texas A&M Turbomachinery Performance and Flow Research Laboratory (TPFL). It is worth noting that the theoretical basis of this topic is neither taught in graduate experimental heat transfer classes nor mentioned in existing heat transfer textbooks.

12.9.1 PSP: Working Principle, Calibration

In order to provide the reader with the basic knowledge of this measurement technique, we first present the a brief description of this technique and for more details refer to Chap. 15.

The Pressure Sensitive Paint (PSP) consists of photo luminescent molecules held together by a binding compound as shown in Fig. 12.10. The luminous particles in the PSP emit light when excited, with the emitted light intensity being inversely proportional to the partial pressure of oxygen in the surroundings. The emitted light intensity can be recorded using a CCD-camera (Charged-Coupled Device) Fig. 12.11, and the corresponding oxygen partial pressures can be obtained by calibrating the emitted intensity against the partial pressure of oxygen. For details of PSP/TSP-measurement, data acquisition, calibration and data analysis we refer to Chap. 15 and to papers published by Schobeiri and his co-workers in the bibliography at the end of this Chapter. Moreover, the PSP-technique and its calibration exhibits a major section of Chap. 15.

The image intensity obtained from PSP by the camera during data acquisition is normalized with a reference image intensity taken under no-flow conditions. Background noise in the optical setup is removed by subtracting the image intensities with the image intensity obtained under no-flow conditions without excitation. The resulting intensity ratio can be converted to pressure ratio using the previously determined calibration curve and can be expressed as:

$$\frac{I_{ref} - I_{blk}}{I - I_{blk}} = f\left(\frac{(P_{O_2})_{air}}{(P_{O_2})_{ref}}\right) = f(P_{ratio}) \tag{12.20}$$

where I denotes the intensity obtained for each pixel and $f(P_{ratio})$ is the relation between intensity ratio and pressure ratio obtained after calibrating the PSP. Further details in using PSP for pressure measurements are given among others in McLachlan and Bell.

Similar to any other temperature measurement method, the results obtained by using PSP must be calibrated. The calibration for PSP is performed using a vacuum

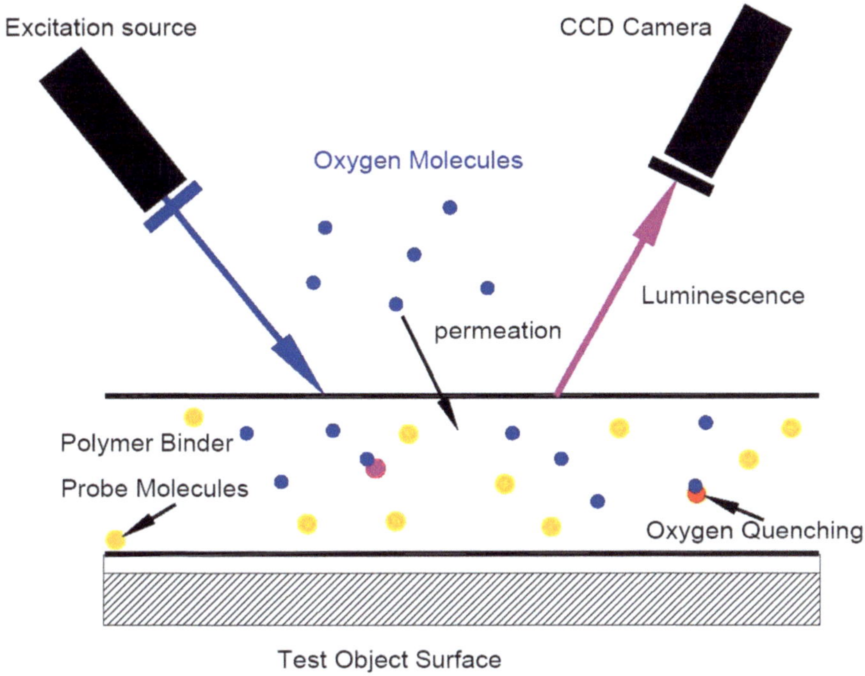

Fig. 12.10 Working principle of using PSP-measurement technique

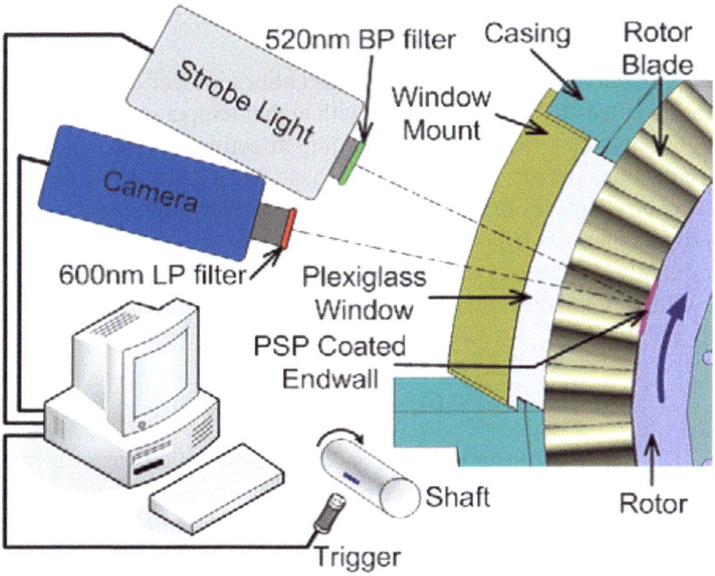

Fig. 12.11 Optical setup for PSP data acquisition

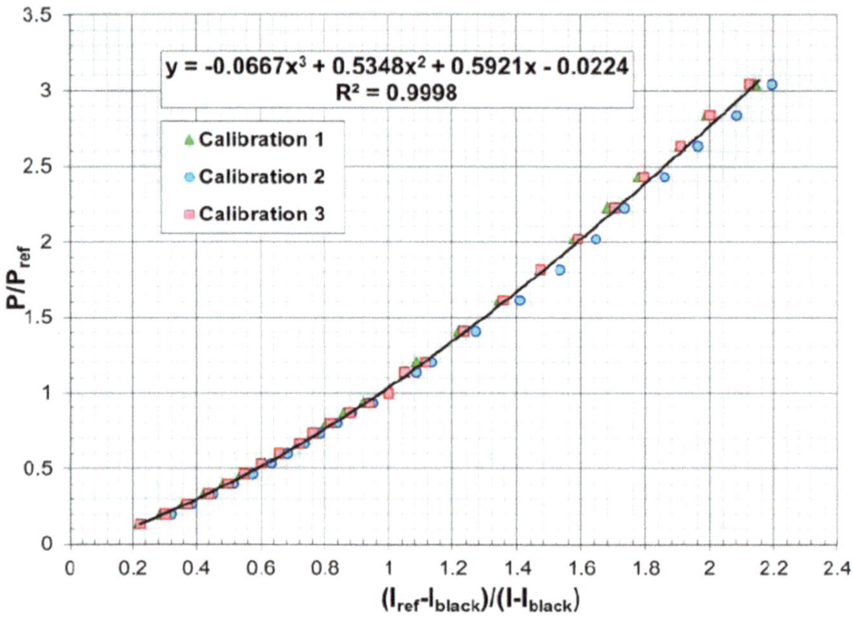

Fig. 12.12 Sample PSP-calibration curve with I as the pixel intensity of an image, I_{ref} as the reference image and I_{black} as the image without illumination

chamber, where its pressure is from 0 to 2 atm with corresponding emitted intensity that is recorded for each pressure setting.

The calibration curve, Fig. 12.12, is a typical calibration curve for PSP. Generally, the calibration is sensitive to temperature with higher temperatures resulting in lower emitted light intensities. To obtain film cooling effectiveness, air and nitrogen were used alternately as coolant. Nitrogen which has approximately the same molecular weight as the air displaces the oxygen molecules on the surface causing a change in the emitted light intensity from PSP. By noting the difference in emitted light intensity and subsequently the partial pressures between the air and nitrogen injection cases, the film cooling effectiveness can be determined using the following equation:

$$\eta = \frac{C_{air} - C_{mix}}{C_{N_2} - C_{air}} = \frac{(P_{O_2})_{air} - (P_{O_2})_{mix}}{(P_{O_2})_{air}} \tag{12.21}$$

where C_{air}, C_{mix} and C_{N2} are the oxygen concentrations of mainstream air, air/nitrogen mixture and nitrogen on the test surface respectively and are directly proportional to the partial pressure of oxygen. The definition of adiabatic film cooling effectiveness is:

$$\eta = \frac{T_f - T_m}{T_C - T_m} \tag{12.22}$$

indent with T_c, T_f and T_m as coolant temperature, local film temperature and mainstream temperature. In the following we present three case studies dealing with measuring film cooling effectiveness of turbine blades operating at high rpm up to 3000. The results show the capability of measuring the film cooling effectiveness.

12.10 Case Studies, Heat Transfer in Rotating Frame

Because of its engineering relevance, we present three representative cases that deal with film cooling heat transfer in rotating frame. These three cases include measurement of film cooling heat transfer on:

1. Turbine blade tip rotating at three different rpm,
2. Turbine leading edge film cooling at three different blowing ratios
3. Turbine end wall film cooling at one rpm and six blowing ratios.

The Case Studies reported in this and other chapters of this book are the results of several years of experimental research. In this context it should be noted that establishing an experimental facility, instrumentation, data acquisition and analysis is a very time consuming and expensive undertaking. In case of aerodynamic investigations, the RANS based Computational Fluid Dynamics (CFD) codes can be used that deliver reasonable results. Particularly, it can be utilized for systematic parameter studies. In case of temperature calculation, however, because of coupling the Navier-Sokes equations with energy equation and the lack of properly modeling the impact of the velocity fluctuation components on temperature calculation, errors ranging from 50% to over 100% are observed. It should be noted that, prior to undertaking the experiments that are presented here as Case Studies, extensive CFD-investigation were carried out. Comparing the results from CFD with the experiment revealed the same type of errors mentioned above.

12.10.1 Rotating Heat Transfer, Case I

Fig. 12.13 shows the film cooling effectiveness measured on tip of turbine blades with different geometries. Two tip geometries are presented. The first tip geometry of Fig. 12.13(a) shows the film cooling effectiveness contour for a plane tip with tip film cooling holes.
The second tip geometry belongs to a blade with squealer tip with film cooling holes shown in Fig. 12.13b. Film cooling heat transfer investigation for both blade tips were conducted at 3000.0, 2550., and 2000.0 rpm. The experiments were performed at *TPFL*.

The purpose of this experiment was to verify the validity of CFD-results. Particularly to determine whether the number of the holes, the locations of the cooling holes and the amount of injected cooling air would establish a full coverage of film

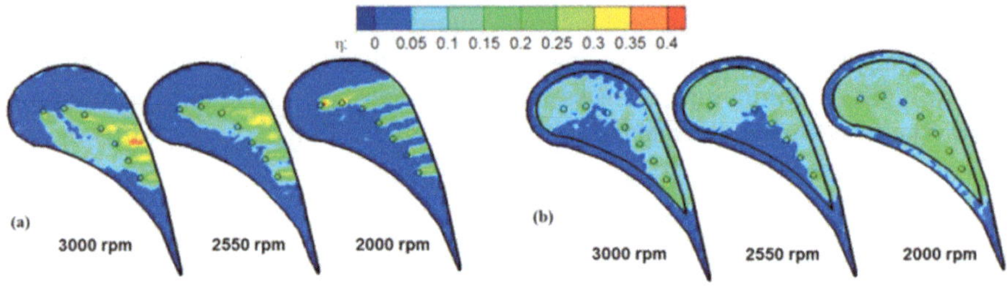

Fig. 12.13 Turbine blade geometries with:**a** plane tip and **b** squealer tip

for the surface to be cooled. As seen from Fig. 12.13, only the tip of the blade with squealer rotating at 2000 rpm delivers a full coverage. As a result, the location of the film cooling holes, their diameters, their distributions and the amount of injected coolant must be redesigned. Comparing the results from CFD with the experiment revealed the same type of errors mentioned above.

12.10.2 Rotating Heat Transfer, Case II

This case shows the film cooling effectiveness distribution on the surface of a turbine blade that is running at 2400 rpm. Three rows of film cooling holes, each having five holes are incorporated in and around the stagnation line. A borescope attached to a traversing system takes data from hub to tip and after analysis using Eqs. (12.21)–(12.22), the effectiveness picture is created as shown in Fig. 12.15. Once the image is obtained, the data has to be further analyzed to obtain quantitative results as the following diagram, Fig. 12.14 shows.
The significance of film cooling experiments for an engineering application is to achieve a uniform surface cooling through the cooling air injection. Looking at the Fig. 12.13a it becomes obvious that the film did not cover the entire surface of the turbine blade tip. This information compels the designer to modify the position of the cooling holes. The situation looks quite different, if the blade tip has a squealer geometry. In this case the film is trapped within the squealer, particularly at lower rpm.

The purpose of this experiment was to verify the sufficient film coverage around the leading edge of the validity of the CFD-prediction. The spectrum of the cooling effectiveness is shown in Fig. 12.14 with the minimum $\eta = 0.0$ and the maximum $\eta = 0.6$.
As seen in Fig. 12.15, traversing of the leading edge area from hub to tip of the blade is accomplished by a borescope attached to an Argon Laser system that serves as the excitation source. The emitted light from PSP transmitted via a camera to data acquisition system for further processing, which after data analysis delivers the contour plots Fig. 12.14. The three columns of film cooling pictures belong to the

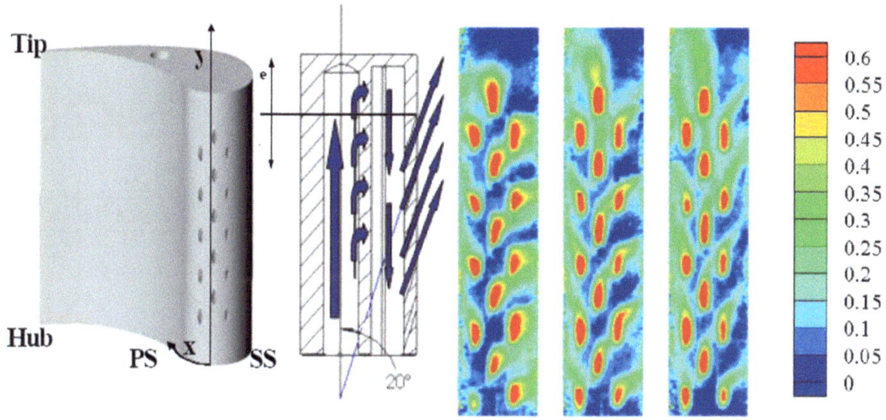

Fig. 12.14 Turbine blade with leading edge film cooling holes (left), ejection of coolant via internal channels through the cooling holes (middle). The contour plot of film cooling effectiveness for three blowing ratios $M = 0.5,\ 1.0\ 2.0$

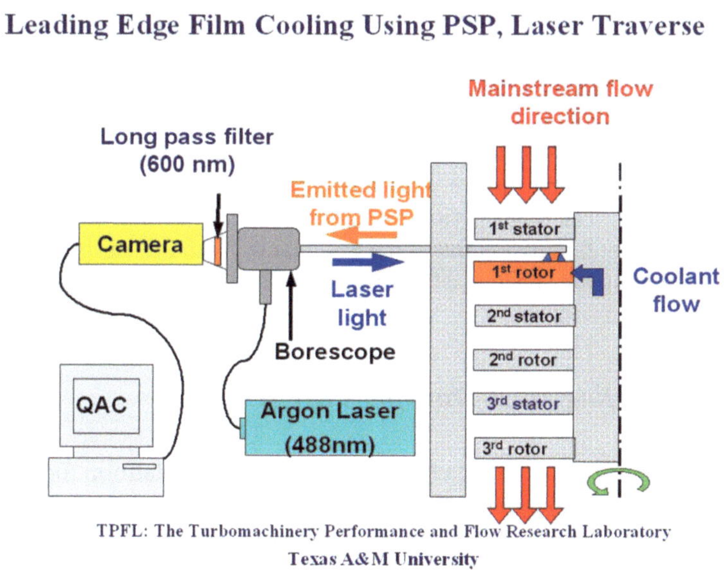

Fig. 12.15 Data acquisition setup and the control system for moving the borescope

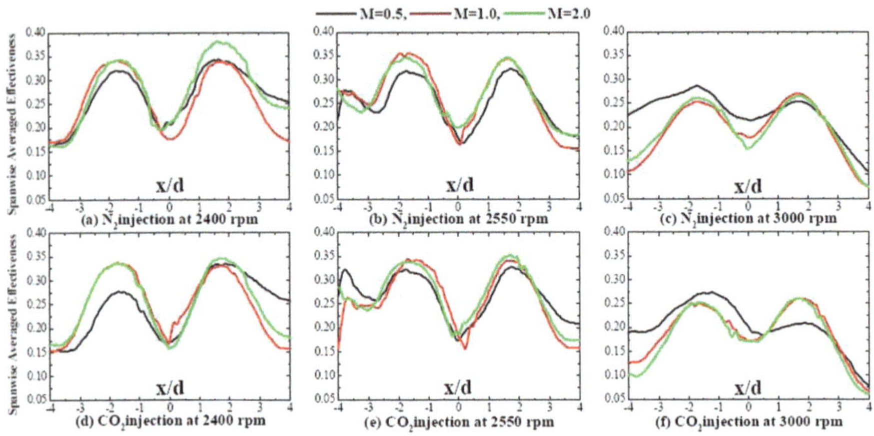

Fig. 12.16 Spanwise averaged quantitative results from the PSP measurement

blowing ratios $M = 0.5, 1.0, 2.0$ (from the left to right). As seen, the leading edge of the turbine blade is not completely covered. Here again, this information compels the designer to redesign the blade and place more holes at the leading edge.

Finally we have extracted the quantitative results from the PSP measurement and plotted in Fig. 12.16. As seen, the spanwise averaged effectiveness is plotted against the ratio x/D with x as the stream-wise distance from leading edge stagnation and D the hole diameter. The quasi periodic pattern with $\eta_{\max} \cong 0.35$ and $\eta_{\min} \cong 0.15$ shows a significant fluctuation that is not acceptable. With increasing the number of holes and reducing the spacing, a better coverage can be achieved.

12.10.3 Rotating Heat Transfer, Case III

Figure 12.17 shows film-cooling effectiveness on the rotor platform for downstream film cooling holes for the reference speed, 3000 rpm, and all mass flow ratios $M = \frac{\rho_c V_c}{\rho_\infty V_\infty}$. As expected film-cooling effectiveness is maximum near the coolant hole exit. As we proceed downstream of the holes, effectiveness magnitude diminishes as the coolant mixes with the mainstream flow. Peak effectiveness values occur as we approach $M_{holes} = 1$ and are approximately equal to $M_{holes} = 0.7$ exactly where the coolant ejects out of the holes. Effectiveness values and film distribution begin to decrease below and above $M_{holes} = 0.75$ and $M_{holes} = 1.25$, respectively. M_{holes} in this range provides good film-cooling protection on the platform covering most of the downstream passage surface. The contribution of each hole toward effective film coverage also varies depending on its location on the platform surface. Since coolant density is assumed to be the same as that of the mainstream, M_{holes} is dependent only on the exit velocity of the coolant gas. For $M_{holes} = 1.0$, the velocity of the

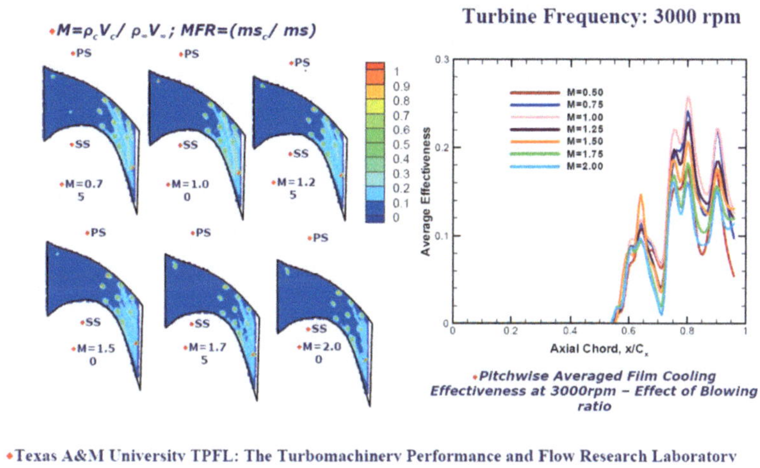

Fig. 12.17 Details of the cooling effectiveness on the blade platform

coolant ejecting out of the individual holes is approximately the same as that of the mainstream relative velocity at the first stage rotor exit. As the coolant flow velocity approaches the mainstream relative velocity, it appears that the ejected coolant has just the right momentum to adhere to the platform surface, displacing the mainstream boundary layer and minimizing the effects of the secondary flows.

This allows the coolant to provide better film coverage and higher effectiveness magnitudes as minimal coolant is dissipated into the mainstream flow before providing any protection. At $M_{holes} < 0.75$ the coolant quantity for film cooling is small and is incapable of providing any effective protection on the platform surface. The lower momentum prevents the coolant from penetrating the boundary layer on the platform surface, hindering the development of an effective thermal barrier. The low momentum coolant tends to get carried away by the higher momentum mainstream flow decreasing the effectiveness. On the contrary, the ejected coolant for $M_{holes} > 0.75$ possesses larger momentum and has a tendency to lift-off as it leaves the coolant holes.

Chapter 13
Compressible Flow

13.1 Steady Compressible Flow

As we discussed in Sect. 4.1.1, for an unsteady compressible flow, the density may generally vary as a function of space and time $\rho = \rho(\mathbf{x}, \mathbf{t})$. The necessary and sufficient condition for a flow to be characterized as compressible is that the substantial change of the density must not vanish. This statement is expressed by the relation:

$$\frac{D\rho}{Dt} = \frac{\partial \rho}{\partial t} + \mathbf{V} \cdot \nabla \rho \neq 0. \tag{13.1}$$

Steady compressible flow constitutes a special case where the density may vary throughout the flow field without changing with time at any spatial position. Thus, Eq. (13.1) reduces to:

$$\nabla \rho \neq 0. \tag{13.2}$$

In order to estimate the spatial changes of the density given by Eq. (13.2), we first establish the relationship between the change of the density with respect to pressure. This relationship is closely related to the speed of sound which enables us to define the flow Mach number as $M = V/c$, with c as the speed of sound. Using the basic conservation principles, we then derive a relationship between the density changes, the other thermodynamic properties, and the flow Mach number. To better understand the underlying physics, we assume an isentropic one dimensional flow, where we set $x_1 \equiv x$.

13.1.1 Speed of Sound, Mach Number

To calculate the speed of sound in a fluid which is contained in an open end channel with constant cross section, we generate an infinitesimal disturbance proceeding

M. T. Schobeiri, *Advanced Fluid Mechanics and Heat Transfer for Engineers and Scientists*, https://doi.org/10.1007/978-3-030-72925-7_13

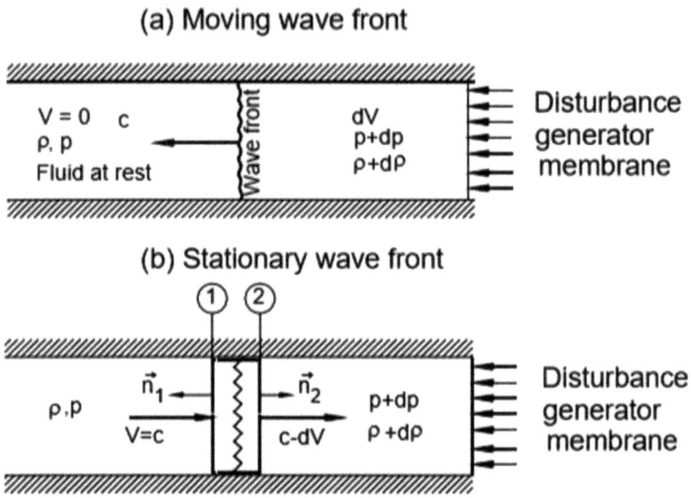

Fig. 13.1 Propagation of a pressure disturbance, **a** moving wave front, stationary frame of reference, **b** moving frame, stationary wave front

along the channel by moving a disturbance generator, Fig. 13.1. This weak disturbance causes a pressure wave which is then propagated with the speed of sound c. Upstream of the wave front, the fluid experiences an infinitesimal velocity dV at the pressure $p + dp$ and the density $\varrho + d\varrho$. Downstream of the wave front, the fluid is at rest with density ρ and pressure p. To obtain the speed of sound using the steady conservation laws, we simply change the frame of reference by placing an observer directly on the wave front, thus, moving with velocity c. Assuming an isentropic flow, we apply the conservation equations of mass, momentum, and energy to the control volume sketched in Fig. 13.1.
The continuity balance for steady flow, Eq. (5.16), applied to stations 1 and 2 results in:

$$\int_{S_1} \rho \mathbf{V}\cdot\mathbf{n}d\mathbf{S} + \int_{\mathbf{S}_2} \rho \mathbf{V}\cdot\mathbf{n}d\mathbf{S} = \mathbf{0}. \tag{13.3}$$

Substituting the velocities in Eq. (13.3) by those from Fig. 13.1, we have

$$\rho A c = (\rho + d\rho) A (c - dV). \tag{13.4}$$

Neglecting the second order terms, Eq. (13.4) reduces to:

$$dV = c\frac{d\rho}{\rho}. \tag{13.5}$$

Now we apply the linear momentum balance for steady flow, Eq. (5.47), to the control volume in Fig. 13.1. Because of the isentropic flow assumption, the shear stress terms identically vanish resulting in zero reaction force, thus, the momentum balance reduces to:

$$\mathbf{R} = \int_{S_{C1}} \mathbf{V}d\dot{\mathbf{m}} - \int_{S_{C2}} \mathbf{V}d\dot{\mathbf{m}} + \int_{S_{C1}} (-\mathbf{n}p)d\mathbf{S} + \int_{S_{C2}} (-\mathbf{n}p)d\mathbf{S} = \mathbf{0}. \tag{13.6}$$

Substituting the velocities in Eq. (13.6) by those from Fig. 13.1, we have:

$$\dot{m}c - \dot{m}(c - dV) + pA - (p + dp)A = 0. \tag{13.7}$$

With the mass flow $\dot{m} = \rho c A$, we arrive at:

$$dV = \frac{dp}{c\rho}. \tag{13.8}$$

Equating (13.5) and (13.8) results in:

$$c^2 = \frac{dp}{d\rho}. \tag{13.9}$$

We derived Eq. (13.9) under the assumption of isentropic flow. To underscore this assumption, we replace the ordinary derivative by partial derivative at constant entropy:

$$c^2 = \left(\frac{\partial p}{\partial \rho}\right)_s. \tag{13.10}$$

For an isentropic process we have:

$$pv^\kappa = \text{const. with } v = \frac{1}{\rho}, \text{ thus, } p = \text{const. } \rho^\kappa. \tag{13.11}$$

With κ as the isentropic exponent that can be set constant for a perfect gas. Taking the derivative

$$\left(\frac{\partial p}{\partial \rho}\right)_s = \frac{\kappa p}{\rho} = \kappa p v. \tag{13.12}$$

Using the equation of state for perfect gas $pv = RT$, we arrive at

$$c^2 = \left(\frac{\partial p}{\partial \rho}\right)_s = \kappa R T. \tag{13.13}$$

Thus, the speed of sound is directly related to the thermodynamic properties of the fluid:

$$c = \sqrt{\kappa R T}. \tag{13.14}$$

Equation (13.14) states that the speed of sound is a function of the substance properties. As mentioned above, the density change within a flow field is directly related to the Mach number. This statement will be explained more in detail in the following sections.

13.1.2 Fluid Density, Mach Number, Critical State

As we indicated earlier, the density and the flow Mach numbers are related to each other. To derive this relationship, we apply the energy equation for an adiabatic system to a large container, Fig. 13.2. The container is connected to a convergent nozzle with the exit diameter d that is negligibly small compared to the container diameter D.
The total enthalpy balance is written as

$$H \equiv h_t = h_0 + \frac{1}{2V_0^2} = h_1 + \frac{1}{2V_1^2} = \text{const.} \tag{13.15}$$

Since in this chapter we are dealing with one dimensional flow, the velocity subscripts refer to the stations and not to the velocity components as we had before. Thus, the subscript refers to the *stagnation point* where the velocity is assumed to be zero. Assuming a perfect gas, for enthalpy we introduce the temperature and divide the result by the static temperature. Thus, the dimensionless version of Eq. (13.15) in terms of total temperature reads:

$$\frac{T_t}{T} = 1 + \frac{1}{2}(\kappa - 1)M^2. \tag{13.16}$$

The specific heats at constant pressure and volume are related by the specific gas constant:

$$c_p - c_v = R, \ \frac{c_p}{c_v} = \kappa, \ \text{and } c = \sqrt{\kappa R T}. \tag{13.17}$$

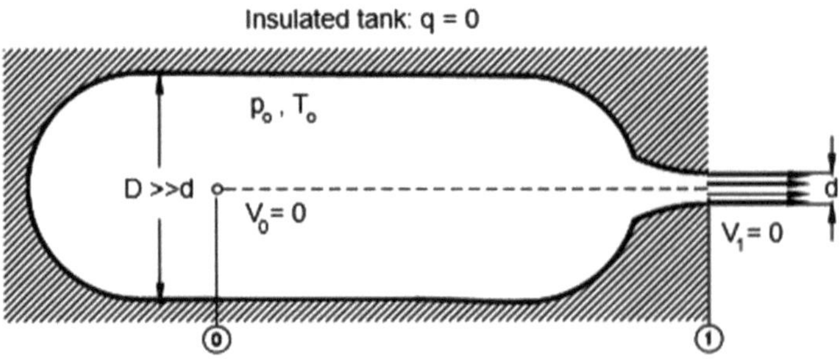

Fig. 13.2 Adiabatic system at a given total pressure and temperature

Using the above relations, the total temperature ratio is expressed in terms of the Mach number:

$$\frac{T_t}{T} = 1 + \frac{1}{2}(\kappa - 1)M^2. \tag{13.18}$$

To obtain a similar relationship for the density ratio, we assume an isentropic process described by:

$$pv^{\kappa} = p_t v_t^{\kappa} = \text{con.} \tag{13.19}$$

that we combine with the equation of state for ideal gases

$$v = \frac{p}{RT}. \tag{13.20}$$

Thus, eliminating the specific volume results in:

$$\frac{p_t}{p} = \left(\frac{T_t}{T}\right)^{\frac{\kappa}{\kappa-1}}. \tag{13.21}$$

Introducing Eq. (13.21) into Eq. (13.18) results in:

$$\frac{p_t}{p} = \left(1 + \frac{\kappa - 1}{2}M^2\right)^{\frac{\kappa}{\kappa-1}}. \tag{13.22}$$

Likewise, we obtain the density ratio as:

$$\frac{\rho_t}{\rho} = \left(1 + \frac{\kappa - 1}{2}M^2\right)^{\frac{1}{\kappa-1}}. \tag{13.23}$$

Equation (13.23) expresses the ratio of a stagnation point density relative to the density at any arbitrary point in the container including the exit area. Assuming air as a perfect gas with $\kappa = 7/5$ at a temperature of $T = 300°K$, the ratios $\Delta\rho/\rho_t = (\rho_t - \rho)/\rho_t$, $\Delta p/p_t = (p_t - p)/p_t$, and $\Delta T/T_t = (T_t - T)/T_t$ from Eqs. (13.18), (13.22) and (13.23) are plotted in Fig. 13.3. We find that for very small Mach numbers ($M \prec 0.1$), the density change $\Delta\rho/\rho_t$ is small and the flow is considered incompressible. Increasing the Mach number, however, results in a significant change of the density ratio. In practical applications, flows with $M \prec 0.3$ are still considered incompressible. Increasing the Mach number above $M \succ 0.3$ results in higher density changes that cannot be neglected, as Fig. 13.3 shows.

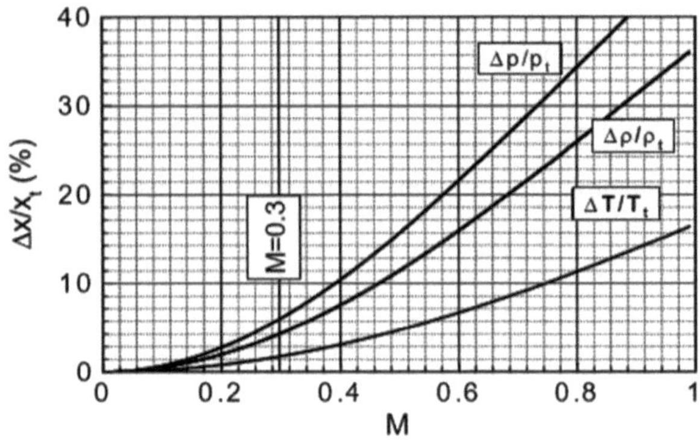

Fig. 13.3 Density, pressure and temperature changes as a function of flow Mach number

Thus, the flow is considered as compressible with noticeable change of density. If the velocity approaches the speed of sound, i.e. $V = c$ and $M = 1$, it is called the critical velocity and the flow state is called critical. In this case, the properties in Eqs. (13.18), (13.22) and (13.23) are calculated by setting $M = 1$. To distinguish this particular flow state, quantities are labeled with the superscript *. The critical temperature ratio is:

$$\frac{T_t^*}{T^*} = \frac{(\kappa+1)}{2} \text{ for } \kappa = 7/5 : \frac{T_t^*}{T^*} = 1.2. \tag{13.24}$$

The critical pressure ratio reads:

$$\frac{p_t^*}{p^*} = \left(\frac{\kappa+1}{2}\right)^{\frac{\kappa}{\kappa-1}} \quad \text{for } \kappa = 7/5 : \frac{p_t^*}{p^*} = 1.893 \tag{13.25}$$

and finally the critical density ratio is obtained from:

$$\frac{\rho_t^*}{\rho_*} = \left(\frac{\kappa+1}{2}\right)^{\frac{1}{\kappa-1}} \quad \text{for } \kappa = 7/5 : \frac{\rho_t^*}{\rho^*} = 1.577. \tag{13.26}$$

From Eq. (13.25), it is obvious that in order to achieve the sonic flow ($M = 1$), the critical pressure ratio must be established first. In a system like the one in Fig. 13.2 with a *convergent exit nozzle* and air as the working medium with $\kappa = 1.4$, the critical pressure ratio is $p_t^*/p^* = 1.893$. At this pressure ratio, the mass flow per unit area has a maximum, and the flow velocity at the exit nozzle equals to the speed of sound. Any increase in the pressure ratio above the critical one results in a *choking state* of the exit nozzle. In this case, the convergent nozzle produces its own exit pressure such that the critical pressure ratio is maintained. To calculate the mass flow of a calorically perfect gas through a convergent nozzle in terms of pressure ratios, we first replace the enthalpy in energy equation (13.15) by:

$$h = c_p T = \frac{k}{k-1} RT = \frac{k}{k-1} pv = \frac{k}{k-1} \frac{p}{\rho}. \tag{13.27}$$

Thus, the energy equation for a calorically perfect gas is:

$$\frac{V_0^2}{2} + \frac{\kappa}{\kappa - 1} \frac{p_0}{\varrho_0} = \frac{V_1^2}{2} + \frac{\kappa}{\kappa - 1} \frac{p_1}{\varrho_1}. \tag{13.28}$$

To eliminate the density at the exit, we now apply the isentropic relation to the right-hand side of Eq. (13.28) and arrive at:

$$\frac{V_0^2}{2} + \frac{\kappa}{\kappa - 1} \frac{p_0}{\varrho_0} = \frac{V_1^2}{2} + \frac{\kappa}{\kappa - 1} \frac{p_0}{\varrho_0} \left(\frac{p_1}{p_0} \right)^{\frac{\kappa-1}{\kappa}}. \tag{13.29}$$

We assume that inside the container, because of $D \gg d$, the velocity V_0 is negligibly small compared to the velocity at the nozzle exit V_1. In this case, the static pressure p_0 would represent the total pressure at the same position $p_0 \equiv p_t$. We now set $p_1 \equiv p_e$ and call it the nozzle exit or back pressure. If the actual pressure ratio is less than the critical one, $p_t/p_e < p_t^*/p_e^*$, and the mass flow exits into the atmosphere, then the nozzle exit pressure is identical with the ambient pressure and the nozzle is not choked. On the other hand, if $p_t/p_e > p_t^*/p_e^*$, the convergent nozzle is choked indicating that it has established a back pressure which corresponds to the critical pressure. With the above assumption, the mass flow through a convergent channel is calculated by

$$\dot{m} = V \rho A = \sqrt{\frac{2\kappa}{\kappa - 1} p_t \rho_t \left[\left(\frac{p_e}{p_t} \right)^{\frac{2}{\kappa}} - \left(\frac{p_e}{p_t} \right)^{\frac{\kappa+1}{\kappa}} \right]} = A\psi \sqrt{\frac{2\kappa}{\kappa - 1} p_t \rho_t} \tag{13.30}$$

where the *mass flow function* Ψ is defined as

$$\psi = \sqrt{\left(\frac{p_e}{p_t} \right)^{\frac{2}{\kappa}} - \left(\frac{p_e}{p_t} \right)^{\frac{\kappa+1}{\kappa}}} \tag{13.31}$$

thus, the mass flow through the nozzle is calculated by

$$\dot{m} = A\psi \sqrt{\frac{2\kappa}{\kappa - 1} p_t \rho_t}. \tag{13.32}$$

Fig. 13.4 shows Ψ as a function of the pressure ratio for different κ.
The maximum value of Ψ is obtained from:

$$\psi_{\max} = \sqrt{\frac{\kappa - 1}{\kappa + 1}} \left(\frac{2}{\kappa + 1} \right)^{\frac{1}{\kappa - 1}}. \tag{13.33}$$

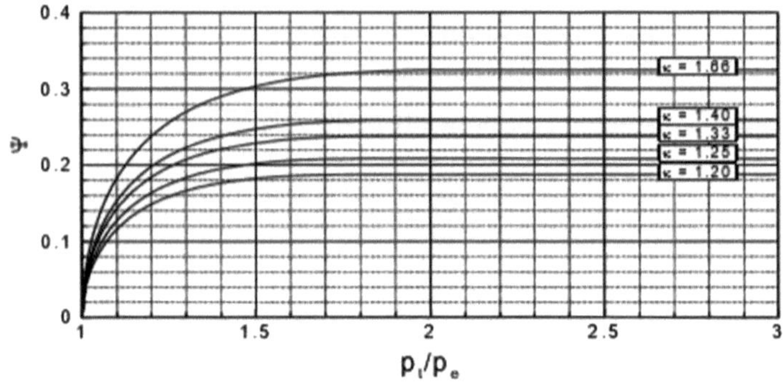

Fig. 13.4 Mass flow function Ψ for different κ-values

Fig. 13.4 shows that increasing the pressure ratio results in an increase of the mass flow function until Ψ_{max} has been reached. Further increase in pressure ratio results in a choking state where the flow function remains constant.

13.1.3 Effect of Cross-Section Change on Mach Number

As seen in the previous section, once the speed of sound has been reached at the exit of a convergent channel, the nozzle exit velocity can not exceed the speed of sound $V_1 = c$ which corresponds to $M = 1$. In order to establish an exit Mach number of $M > 1$, the nozzle geometry has to change. This is achieved by using the conservation of mass:

$$\varrho V A = \text{const.} \tag{13.34}$$

Differentiating Eq. (13.34) with respect to x-direction and dividing the result by Eq. (13.34), we obtain the expression

$$\frac{1}{V}\frac{dV}{dx} + \frac{1}{A}\frac{dA}{dx} + \frac{1}{\rho}\frac{d\varrho}{dx} = 0. \tag{13.35}$$

Introducing the speed of sound, Eq. (13.10) into Eq. (13.35), we find:

$$\frac{1}{V}\frac{dV}{dx} + \frac{1}{A}\frac{dA}{dx} + \frac{1}{c^2\rho}\frac{dp}{dx} = 0. \tag{13.36}$$

Applying the Euler equation of motion for one-dimensional flow, we obtain:

$$V\frac{dV}{dx} = -\frac{1}{\varrho}\frac{dp}{dx}. \tag{13.37}$$

Introducing Eq. (13.37) into Eq. (13.36), we have

$$\frac{1}{V}\frac{dV}{dx} + \frac{1}{A}\frac{dA}{dx} = \frac{V}{c^2}\frac{dV}{dx} \tag{13.38}$$

with the definition of Mach number, Eq. (13.38) reduces to:

$$\frac{1}{A}\frac{dA}{dx} = -\frac{1}{V}\frac{dV}{dx}\left(1 - M^2\right). \tag{13.39}$$

Introducing Eq. (13.37) into Eq. (13.39) results in:

$$\frac{1}{A}\frac{dA}{dx} = \frac{1}{\varrho V^2}\frac{dp}{dx}(1 - M^2). \tag{13.40}$$

With Eqs. (13.39) and (13.40) we have established two relationships between the cross section change, the velocity change, the pressure change and the Mach number. For a subsonic inlet flow condition $M < 1$, a decrease in cross-sectional area leads to an increase in velocity and consequently a decrease in pressure Fig. 13.5.
On the other hand, increasing the cross-section area ($dA > 0$) leads to decreasing the velocity ($dV < 0$) that is associated with an increase in pressure, Fig. 13.5(b). For $M = 1$ we obtain $dA/dx = 0$. For a supersonic inlet flow condition $M > 1$, Eqs. (13.39) and (13.40) show that if the cross-sectional area increases ($dA/dx > 0$), the velocity must also increase ($dV/dx > 0$), or if the cross-section decreases, so does the velocity. As a result, we obtain the geometries for supersonic nozzles and diffusers as shown in Fig. 13.6.
As shown in Figs. 13.5 and 13.6, the cross-section undergoes negative and positive changes to establish subsonic and supersonic flow regimes. The transition from a positive to a negative change requires that $dA/dx = 0$. This, however, means that the product, $dV/dx(1 - M^2)$, on the right-hand side of Eq. (13.39), must vanish. Since dV/dx has for both nozzle and diffuser flow cases a non-zero value, only the expression $(1 - M^2)$ has to vanish, which results in $M = 1$. As a consequence,

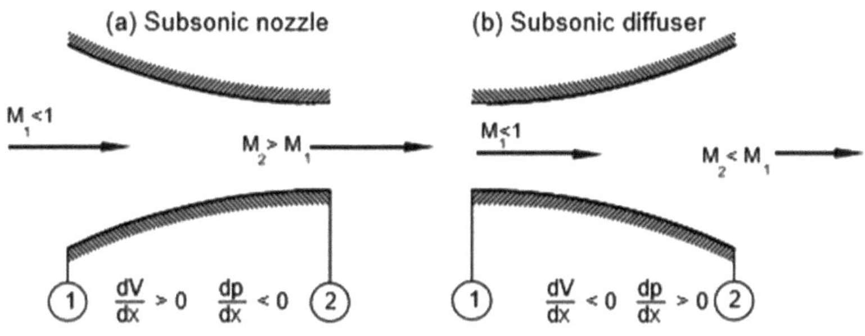

Fig. 13.5 **a** Subsonic nozzle with $dA < 0, dV > 0, dp < 0$, **b** Subsonic diffuser with $dA > 0, dV < 0, dp > 0$

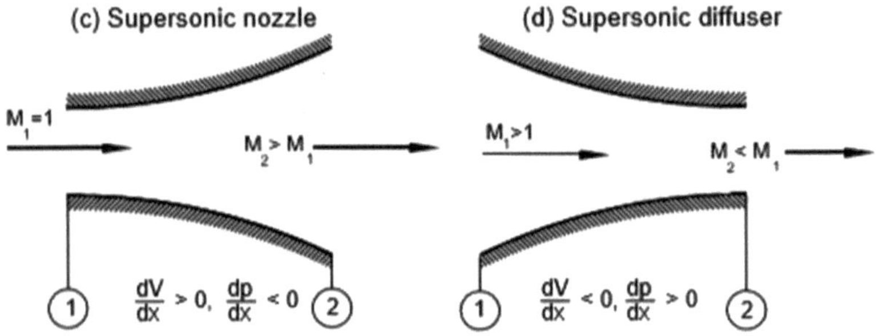

Fig. 13.6 **c** Supersonic nozzle with $dA > 0, dV > 0, dp < 0$, **d** Supersonic diffuser with $dA > 0, dV < 0, dp > 0$

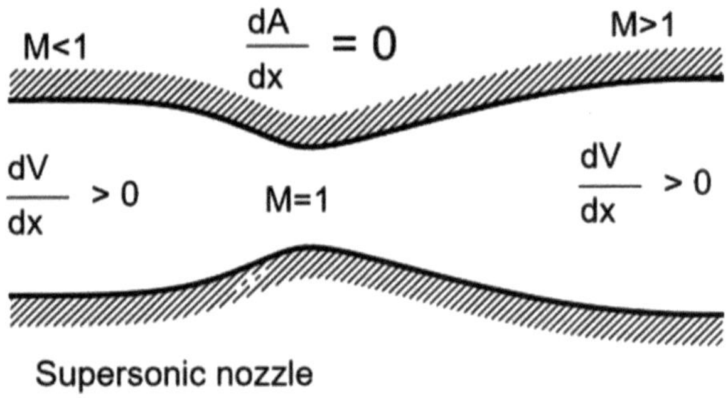

Fig. 13.7 A supersonic nozzle operating at super critical pressure ratio

Mach number $M = 1$ can be reached only at the position where the cross-section has a minimum. The above conditions provide a guideline to construct a *Laval nozzle* which is a convergent-divergent channel for accelerating the flow from subsonic to supersonic (*Laval nozzle*) Mach range. The condition for a supersonic flow to be established is that the pressure ratio along the nozzle from the inlet to the exit must correspond to the nozzle design pressure ratio which is far above the critical pressure ratio. In this case, the flow is accelerated in the convergent part, reaches the mach number $M = 1$ in the *throat* and is further accelerated in the divergent portion of the nozzle.

Fig. 13.7 shows the schematic of a Laval nozzle which is used in the first stage of power generation steam turbines, thrust nozzle of rocket engines, and in the after-burner of supersonic aircraft engines. If the channel pressure ratio is less than the critical pressure ratio, the flow in the convergent part is accelerated to a certain Mach number M<1 and then decelerates in the divergent part.

A supersonic diffuser is characterized by a convergent divergent channel, however, its inlet Mach number is supersonic ($M > 1$). Figure 13.8 shows a schematic of a supersonic diffuser. The incoming supersonic flow is decelerated from $M > 1$ to $M = 1$ at the throat where the sonic velocity has been reached. Further deceleration occurs at the divergent part of the supersonic diffuser.

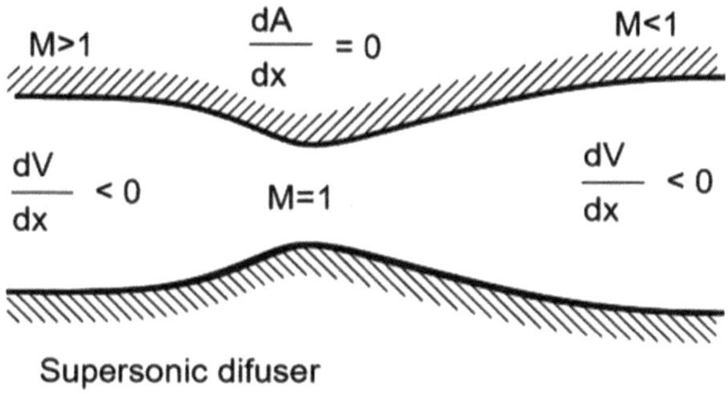

Fig. 13.8 Schematic of a supersonic diffuser

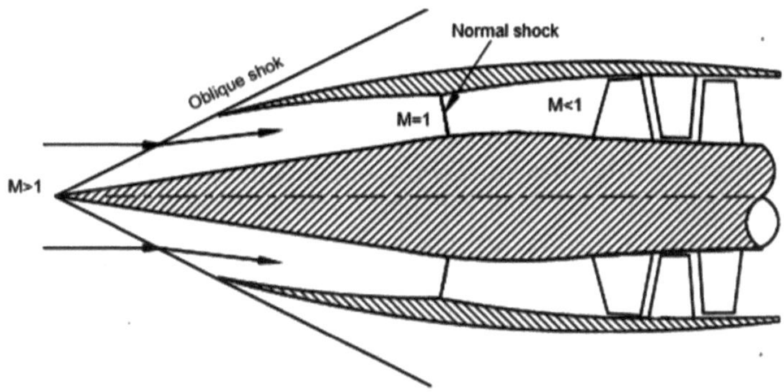

Fig. 13.9 Schematic of an inlet diffuser of a supersonic aircraft

This principle of supersonic flow deceleration is applied to the inlet of a supersonic aircraft, schematically shown in Fig. 13.9. The incoming supersonic inlet flow hits an oblique shock system that originates from the tip of the inlet cone and touches the cone casing. After passing though the oblique shock front, the flow is deflected and its velocity reduced. Passing through the convergent part of the supersonic diffuser, the velocity continuously decreases, and reaches the throat where a normal shock reduces the supersonic flow to a sonic one.

Further deceleration occurs in the divergent part of the diffuser where the flow is exiting into a multi-stage compressor. Equation (13.39) and its subsequent integration, along with the flow quantities listed in Table 13.1, indicate the direct relation between the area ratio and the Mach number. These relations can be utilized as useful tools for estimating the distribution of the cross-sectional area of a Laval nozzle, a supersonic stator blade channel, or a supersonic diffuser. If, for example, the Mach number distribution in the streamwise direction is given, $M(S) = f(S)$, the distribution of the cross-sectional area $A(S) = f(M)$ is directly calculated from Table 13.1. If, on the other hand, the cross-section distribution in the streamwise direction is prescribed, then the Mach number distribution, and thus, all other flow quantities can be calculated using an inverse function. Since we assumed the isentropic flow with calorically perfect gases as the working media, important features, such as flow

Table 13.1 Summary of the gas dynamic functions

Ratios for two arbitrary sections	Ratios relative to the throat
$\frac{A}{A_1} = \frac{1}{M}\left(\frac{1+\frac{\kappa-1}{2}M^2}{1+\frac{\kappa-1}{2}M_1^2}\right)^{\frac{1}{2}\frac{\kappa+1}{\kappa-1}}$	$\frac{A}{A^*} = \frac{1}{M}\left(\frac{1+\frac{\kappa-1}{2}M^2}{\frac{\kappa+1}{2}}\right)^{\frac{1}{2}\frac{\kappa+1}{\kappa-1}}$
$\frac{p}{p_1} = \left(\frac{1+\frac{\kappa-1}{2}M_1^2}{1+\frac{\kappa-1}{2}M^2}\right)^{\frac{\kappa}{\kappa-1}}$	$\frac{p}{p^*} = \left(\frac{\frac{\kappa+1}{2}}{1+\frac{\kappa-1}{2}M^2}\right)^{\frac{\kappa}{\kappa-1}}$
$\frac{T}{T_1} = \frac{1+\frac{\kappa-1}{2}M_1^2}{1+\frac{\kappa-1}{2}M^2}$	$\frac{T}{T^*} = \frac{\frac{\kappa+1}{2}}{1+\frac{\kappa-1}{2}M^2}$
$\frac{h}{h_1} = \frac{1+\frac{\kappa-1}{2}M_1^2}{1+\frac{\kappa-1}{2}M^2}$	$\frac{h}{h^*} = \frac{\frac{1+\kappa}{2}}{1+\frac{\kappa-1}{2}M^2}$
$\frac{V}{V_1} = \frac{M}{M_1}\left(\frac{1+\frac{k-1}{2}M_1^2}{1+\frac{\kappa-1}{2}M^2}\right)^{\frac{1}{2}}$	$\frac{V}{V^*} = M\left(\frac{\frac{1+k}{2}}{1+\frac{\kappa-1}{2}M^2}\right)^{\frac{1}{2}}$
$\frac{\rho}{\rho_1} = \frac{v_1}{v} = \left(\frac{1+\frac{\kappa-1}{2}M_1^2}{1+\frac{\kappa-1}{2}M^2}\right)^{\frac{1}{\kappa-1}}$	$\frac{\rho}{\rho^*} = \frac{v^*}{v} = \left(\frac{\frac{1+\kappa}{2}}{1+\frac{\kappa-1}{2}M^2}\right)^{\frac{1}{\kappa-1}}$

separation, as a consequence of the boundary layer development under adverse pressure gradient, will not be present. Therefore, in both cases, the resulting channel geometry or flow quantities are just rough estimations, no more, no less. Appropriate design of such channels, particularly transonic turbine or compressor blades, require a detailed calculation where the fluid viscosity is fully considered.

In order to represent the thermodynamic variables as functions of a Mach number, we use the continuity and energy equations in conjunction with the isentropic relation and the equation of state for the thermally perfect gases with $p = \varrho RT$. The isentropic flow parameters as a function of a Mach number are summarized in Table 13.1 which contains two columns. The first column gives the individual parameter ratios at arbitrary sections, whereas, the second one gives the ratios relative to the critical state. The gas dynamics relations presented in Table 13.1 are depicted in Fig. 13.10. Laval nozzles were first used in steam turbines, but many other applications for these nozzles have been found, for example, in rocket engines, supersonic steam turbines, and supersonic air craft engines. In the following, we briefly discuss the operational behavior of a generic Laval nozzle which is strongly determined by the pressure ratio. Detailed discussion of this topic can be found in excellent books by Spurk [229], Prandtl et al. [230], and Shapiro [231]. Starting with the design operating point where the exit pressure is set equal to the ambient pressure, curve ① in Fig. 13.11 corresponds to the design pressure ratio. In this case, the Mach number continuously increases from the subsonic at the inlet to the supersonic at the exit, Fig. 13.11 a. An increase in the ambient pressure results in an *overexpanded* jet depicted in Fig. 13.12, indicating that the flow in the nozzle expands above a pressure that does not correspond to its design pressure point ② with $p_e < p_a$. At this particular pressure condition, the flow pattern inside the nozzle does not change as curve ① indicates. However, outside the nozzle, the flow undergoes a system of *oblique shocks* (in the following section, shocks are treated in a detailed manner). The shocks originate at the nozzle exit rim, raising the nozzle exit pressure to the ambient pressure. Based on the magnitude of the ambient pressure, the transition from the

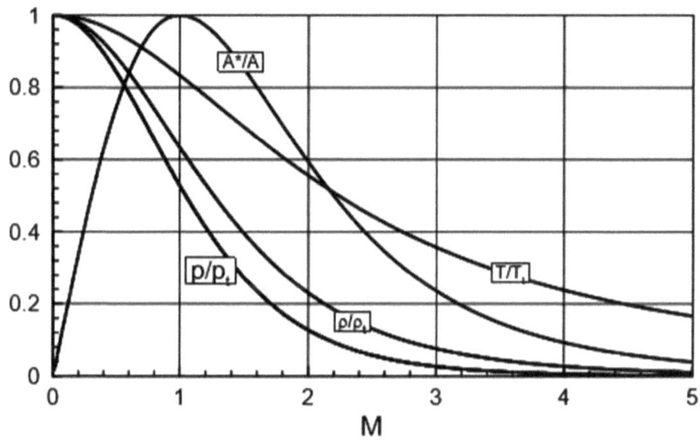

Fig. 13.10 Area ratio and the thermodynamic property ratios as a function of Mach number for $\kappa = 1.4$

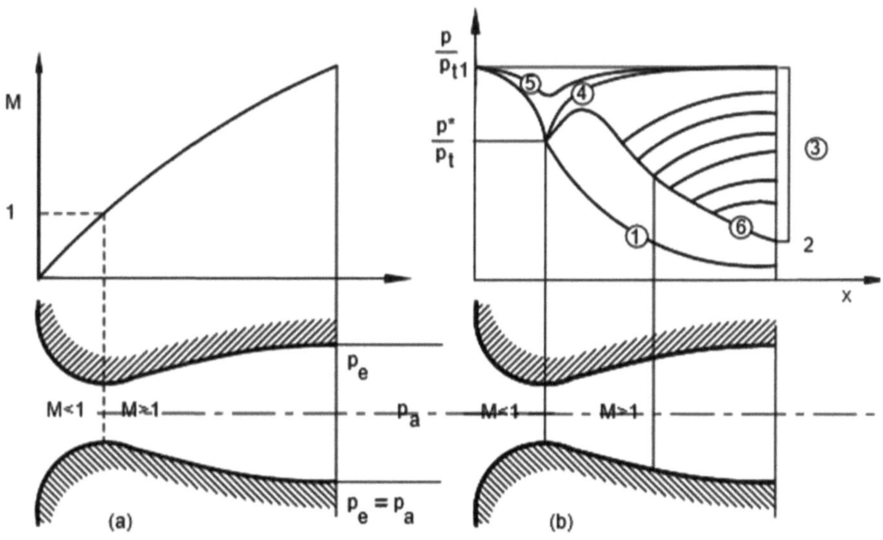

Fig. 13.11 Operational behavior of a generic Laval nozzle, **a** Expansion to the design exit pressure, **b** Overexpansion

nozzle design pressure p_e to the ambient pressure p_a is accomplished either by a system of oblique shock waves and their reflection on the jet boundary, as depicted in Fig. 13.12, or by a combination of oblique and normal shocks, as shown in Fig. 13.13. Further increase of the ambient pressure causes the shocks to move into the nozzle, forming a normal shock. The formation of the normal shock is associated with a discontinuous increase in pressure, shown in Fig. 13.11, curves ③. As seen in Fig. 13.11, upstream of the shock, the pressure follows the nozzle design pressure, curve ①, with

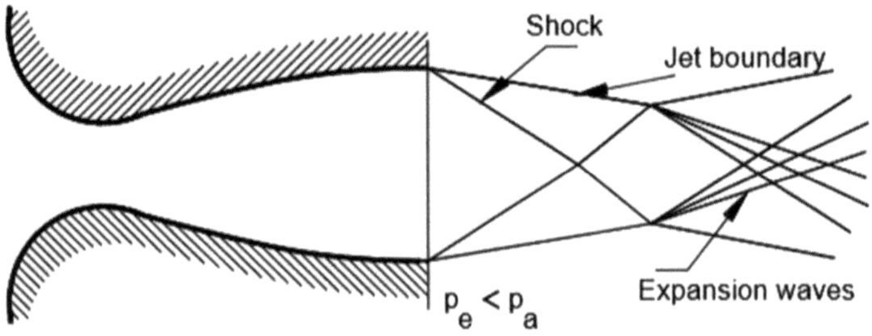

Fig. 13.12 Overexpanded jet—ambient pressure p_a, nozzle exit pressure p_e

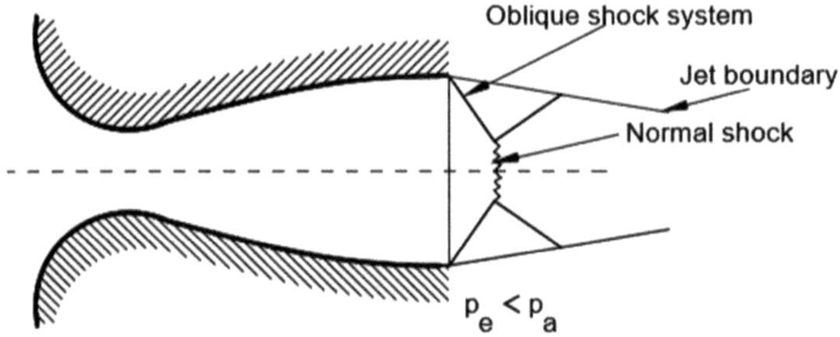

Fig. 13.13 Nozzle operating at overexpanded mode, formation of a system of normal shocks downstream of the nozzle exit

a supersonic velocity, while downstream of the shock, the flow is subsonic. If the ambient pressure is increased in such a way that the shock reaches the nozzle throat, curve ④, then the nozzle is not capable of producing a supersonic flow in its entire length. Any further increase in ambient pressure that corresponds to a nozzle pressure ratio below the critical one, causes the flow to accelerate within the convergent section to reach a maximum subsonic velocity in the throat and to decelerate within the divergent part of the nozzle.

A different flow pattern emerges when the ambient pressure p_a drops below the exit design pressure p_e. This occurs when a rocket engine ascends through the atmosphere. The lower ambient pressure causes the generation of a system of expansion and compression waves outside the nozzle, causing a pressure balance between the jet and the environment, as shown in Fig. 13.14. A similar picture is observed at the exit of a convergent nozzle that operates at a supercritical pressure ratio. The transition from higher exit pressure to a lower ambient pressure is achieved by a system of expansion waves and their reflection on the jet boundary as compression waves, Fig. 13.15.

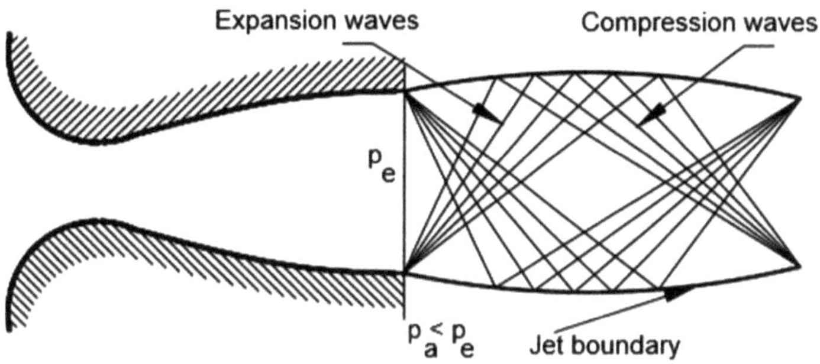

Fig. 13.14 Underexpanded with $p_1 < p_e$

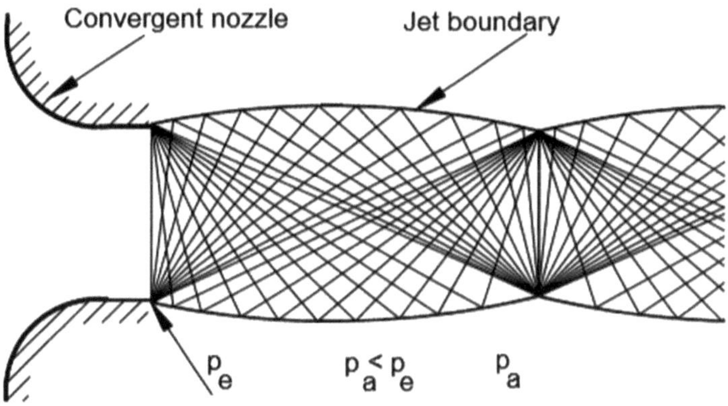

Fig. 13.15 Convergent nozzle with an after-expanding jet

13.1.3.1 Flow Through Channels with Constant Area

This type of flow is encountered in several engineering applications such as pipes, labyrinth seals of turbines and compressors, and to a certain degree of simplicity, in aircraft combustion. In the case of pipes and labyrinth seals, we are dealing with an adiabatic flow process where the total enthalpy remains constant. However, entropy increases are present due to the internal friction, shocks, or throttling. Combustion chamber flow can be approximated by a constant cross-section pipe with heat addition. The characteristic features of these devices are that the entropy changes are caused by heat addition such that the friction contribution to the entropy increase can be neglected. This assumption leads to a major simplification that we may add heat to a constant cross-section pipe and assume that the linear momentum remains constant. The constant total enthalpy is described by the *Fanno process*, whereas the constant linear momentum case is determined by the *Rayleigh* process.

Starting with the Rayleigh process, we will specifically consider the flow in a duct with a constant cross-section without surface or internal friction, but with heat transfer through the wall. In the absence of the shaft power, Eq. (5.73) is reduced to:

$$\frac{V_2^2}{2} + h_2 = \frac{V_1^2}{2} + h_1 + q. \tag{13.41}$$

In applying the momentum balance, we assume here that the contribution of the friction forces to the total entropy increase compared to the entropy increase by external heat addition, is negligibly small, thus, the resultant force in Eq. (5.26) may be set $\mathbf{F_r} = \mathbf{0}$. This results in:

$$\rho_2 V_2^2 + p_2 = \rho_1 V_1^2 + p_1 = \rho V^2 + p = \text{const.} \tag{13.42}$$

To find the flow quantities for the Rayleigh process, we present the calculation of a pressure ratio. The other quantities such as velocity ratio, temperature ratio, density ratios, and etc., are obtained using a similar procedure. We start with the calculation of the pressure ratio by utilizing the following steps.

Step 1: We combine the differential form of the momentum equation,

$$dp + 2\rho V dV + d\rho V^2 = 0 \tag{13.43}$$

with the differential form of a continuity equation for a constant cross-section,

$$d\rho V + \rho dV = 0 \tag{13.44}$$

and obtain the modified momentum equation. This equation can immediately be found from the 1-D Euler equation.

$$v dp + V dV = 0. \tag{13.45}$$

We rearrange the modified momentum Eq. (13.45) and introduce the Mach number

$$\frac{dp}{p} = -\frac{V dV}{vp} = -\frac{V dV}{RT} = -\kappa M^2 \frac{dV}{V}. \tag{13.46}$$

Step 2: We combine the differential form of the continuity equation (13.44) with the differential form of the equation of state for ideal gases. With this step, we eliminate the density from Eq. (13.44), and we have

$$\frac{dV}{V} + \frac{dp}{p} = \frac{dT}{T}. \tag{13.47}$$

To eliminate the velocity ratio in Eqs. (13.46) and (13.47), we use the definition of the Mach number. Its differentiation yields

$$\frac{dV}{V} = \frac{dM}{M} + \frac{1}{2}\frac{dT}{T}. \tag{13.48}$$

Step 3: Inserting the velocity ratio (13.48) into the momentum equation (13.46) and the equation of continuity (13.44), we obtain for the pressure ratio,

$$\frac{2}{\kappa M^2}\frac{dp}{p} - 2\frac{dM}{M} = \frac{dT}{T}. \tag{13.49}$$

We also replace in Eq. (13.47) the velocity ratio with Eq. (13.48) and find a second equation for temperature ratio,

$$\frac{2dp}{p} + \frac{2dM}{M} = \frac{dT}{T}. \tag{13.50}$$

Step 4: Equating (13.49) and (13.50), we find

$$\frac{dp}{p} = \frac{2\kappa M dM}{1 + \kappa M^2}. \tag{13.51}$$

Equation (13.51) can be integrated between any two positions, including the one where $M = 1$:

$$\frac{p_2}{p_1} = \frac{1 + \kappa M_1^2}{1 + \kappa M_2^2} \tag{13.52}$$

and for the critical state

$$\frac{p}{p^*} = \frac{1 + \kappa}{1 + \kappa M^2}. \tag{13.53}$$

In a similar manner, the temperature ratio calculated as,

$$\frac{dT}{T} = 2\frac{dM}{M}\left(\frac{1 - \kappa M_1^2}{1 + \kappa M_2^2}\right). \tag{13.54}$$

Considering the initial assumption of calorically perfect gas, the integration gives,

$$\frac{T_2}{T_1} = \frac{h_2}{h_1} = \frac{M_2^2}{M_1^2}\left(\frac{1 + \kappa M_1^2}{1 + \kappa M_2^2}\right)^2 \tag{13.55}$$

and relative to critical state, we have

$$\frac{T}{T^*} = \frac{h}{h^*} = M^2\left(\frac{1 + \kappa}{1 + \kappa M_2^2}\right)^2. \tag{13.56}$$

Table 13.2 Summary of gas dynamic equations for constructing the Rayleigh and Fanno curves

Constant Momentum, Rayleigh Process	Constant Total Energy, Fanno Process
$\frac{A}{A^*} = 1$	$\frac{A}{A^*} = 1$
① $\frac{p}{p^*} = \left(\frac{\kappa+1}{1+\kappa M^2}\right)$	$\frac{p}{p^*} = \frac{1}{M}\left(\frac{\kappa+1}{2\left(1+\frac{\kappa-1}{2}M^2\right)}\right)^{1/2}$
② $\frac{T}{T^*} = M^2\left(\frac{\kappa+1}{(1+\kappa)M^2}\right)^2$	$\frac{T}{T^*} = \frac{c^2}{c*2} = \frac{\frac{\kappa+1}{2}}{1+\frac{\kappa-1}{2}M^2}$
③ $\frac{h}{h^*} = M^2\left(\frac{\kappa+1}{(1+\kappa)M^2}\right)^2$	$\frac{h}{h^*} = \frac{\frac{1+\kappa}{2}}{1+\frac{\kappa-1}{2}M^2}$
④ $\frac{V}{V^*} = \frac{(k+1)M^2}{1+\kappa M^2}$	$\frac{V}{V^*} = M\left(\frac{\frac{1+k}{2}}{1+\frac{\kappa-1}{2}M^2}\right)^{\frac{1}{2}}$
⑤ $\frac{\rho^*}{\rho} = \frac{(k+1)M^2}{1+\kappa M^2}$	$\frac{\rho^*}{\rho} = M\left(\frac{\frac{1+k}{2}}{1+\frac{\kappa-1}{2}M^2}\right)^{\frac{1}{2}}$
⑥ $\Delta s = c_p \ln\left[\left(\frac{T}{T^*}\right)\left(\frac{p^*}{p}\right)^{\frac{\kappa-1}{\kappa}}\right]$	$\Delta s = c_p \ln\left[\left(\frac{T}{T^*}\right)\left(\frac{p^*}{p}\right)^{\frac{\kappa-1}{\kappa}}\right]$
⑦ $\frac{s-s^*}{c_p} = \ln M^2\left(\frac{k+1}{1+\kappa M^2}\right)^{\frac{\kappa+1}{\kappa}}$	$\frac{s-s^*}{c_p} = \ln M^2\left(\frac{\frac{k+1}{2}}{M^2\left(1+\frac{\kappa-1}{2}M^2\right)}\right)^{\frac{\kappa+1}{2\kappa}}$

Similarly, we find the pressure, temperature, velocity, and density ratios for the Fanno process as functions of a Mach number. These quantities are listed in Table 13.2. With these ratios as functions of Mach number, the entropy change is determined by using any of the two following equations, (13.57) or (13.58).

$$\Delta s = c_p \ln\left[\left(\frac{T_2}{T_1}\right)\left(\frac{p_1}{p_2}\right)^{\frac{\kappa-1}{\kappa}}\right] = c_v \ln\left[\left(\frac{T_2}{T_1}\right)\left(\frac{v_2}{v_1}\right)^{\kappa-1}\right]. \tag{13.57}$$

In terms of critical state:

$$\Delta s = s - s^* = c_p \ln\left[\left(\frac{T}{T^*}\right)\left(\frac{p^*}{p}\right)^{\frac{\kappa-1}{\kappa}}\right] = c_v \ln\left[\left(\frac{T}{T^*}\right)\left(\frac{v}{v^*}\right)^{\kappa-1}\right]. \tag{13.58}$$

Replacing the temperature and pressure ratios by the corresponding functions listed in Table 13.2, we find,

$$\frac{s - s^*}{c_p} = \ln M^2\left(\frac{k+1}{1+\kappa M^2}\right)^{\frac{\kappa+1}{\kappa}}. \tag{13.59}$$

The above properties can be determined by varying the Mach number. As mentioned previously, a Rayleigh curve is the locus of all constant momentum processes. It can be easily constructed by varying the Mach number and finding the corresponding, enthalpy, pressure, or entropy ratios.

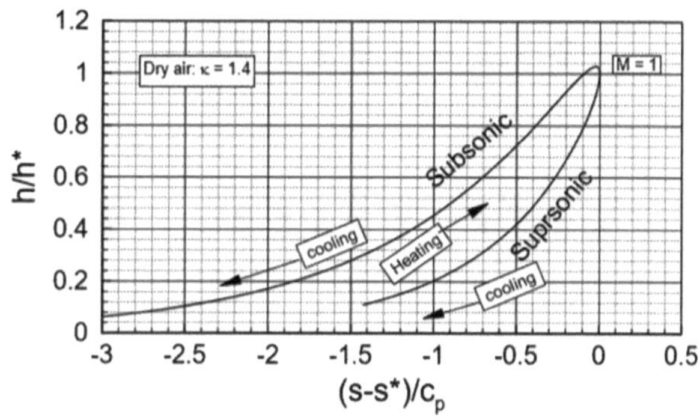

Fig. 13.16 Dimensionless h-s diagram for Rayleigh process for dry air assumed as a perfect gas with $\kappa = 1.4$, upper (subsonic acceleration) branch indicates the heat addition ($ds > 0$) to reach sonic speed ($M = 1$) followed by the lower branch (supersonic acceleration) caused by the heat rejection ($ds < 0$)

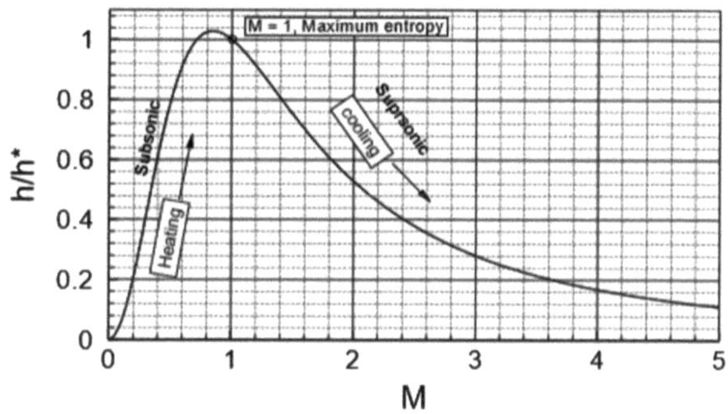

Fig. 13.17 Change of Mach number in a constant area channel with heat addition, rejection

Fig. 13.16 shows the Rayleigh curve in terms of enthalpy ratio as a function of dimensionless entropy difference for dry air as a calorically perfect gas. As seen, it has a subsonic upper branch with $M < 1$, a supersonic lower branch, with $M > 1$ joined by the sonic point with $M = 1$. Moving along the subsonic upper branch, the addition of heat causes the specific volume and, consequently, the velocity and the Mach number to increase until the speed of sound ($M = 1$) has been reached. Further increase of the Mach number requires cooling the mass flow. If the inlet mach number is supersonic (lower branch), a continuous addition of heat will cause a flow deceleration up to $M = 1$. Further deceleration required a continuous heat rejection along the subsonic upper branch.

Fig. 13.17 shows the enthalpy (or temperature) distribution as a function of a Mach number. The flow acceleration and deceleration as a result of heat addition/rejection is illustrated in a fictive channel shown in Fig. 13.18.

To emulate the Rayleigh process, we think of a channel that consists of two parts having the same type of heat conductive material and the same cross section. The parts are joined together by a thin, perfect heat insulating joint such that no heat can flow through it from either side. We assume that the streamwise location of the joint coincides with the streamwise location of the point with $M = 1$. Starting with a subsonic inlet Mach number, Fig. 13.18a, an amount of heat is added such that the increase in specific volume causes the velocity to increase, Fig. 13.16

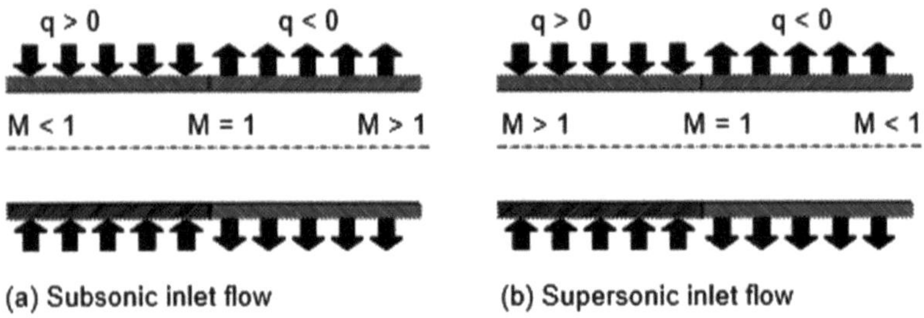

Fig. 13.18 A fictive channel for realization of Rayleigh process, **a** subsonic acceleration with heat addition and rejection, **b** supersonic deceleration with heat addition and rejection

(upper branch), and Fig. 13.17. To go beyond the speed of sound, heat needs to be rejected, Fig. 13.16 (lower branch), and 13.17. In the absence of the heat rejection, an increase of velocity is not possible. The channel will choke. Figure 13.18b shows the Rayleigh process that starts with a supersonic inlet To decelerate the flow, heat is added such that the speed of sound is reached. Further deceleration requires a reduction of specific volume which is established by rejecting the heat. As seen, the preceding Rayleigh process was characterized by reversibly adding and rejecting heat at a constant momentum that resulted in flow acceleration or deceleration. We consider now an adiabatic process through a channel with a constant cross sectional where internal and wall frictions are present. It is called the Fanno process and is characterized by constant total enthalpy. As a result, the static enthalpy experiences a continuous decrease, while the velocity increases. To construct the Fanno curve, first the flow quantities are expressed in terms of Mach number in a manner similar to the Rayleigh process presented above. The corresponding relations are summarized in Table 13.2.

Applying the energy, continuity, and impulse to an adiabatic constant cross-section duct flow, we find the pressure ratio from:

$$\frac{p}{p^*} = \frac{1}{M\left(\frac{\kappa+1}{2\left(1+\frac{\kappa-1}{2}M^2\right)}\right)}^{\frac{1}{2}}. \tag{13.60}$$

The other thermodynamic properties are calculated from:

$$\frac{T}{T^*} = \frac{\frac{\kappa+1}{2}}{1+\frac{\kappa-1}{2}M^2},\ \frac{h}{h^*} = \frac{\frac{1+\kappa}{2}}{1+\frac{\kappa-1}{2}M^2},\ \frac{\rho^*}{\rho} = M\left(\frac{\frac{1+k}{2}}{1+\frac{\kappa-1}{2}M^2}\right)^{\frac{1}{2}}. \tag{13.61}$$

Finally, the velocity ratio is given by

$$\frac{V}{V^*} = M\left(\frac{\frac{1+k}{2}}{1+\frac{\kappa-1}{2}M^2}\right)^{\frac{1}{2}}. \tag{13.62}$$

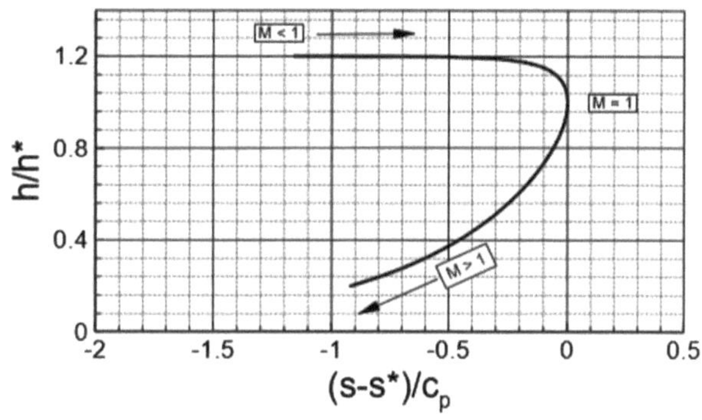

Fig. 13.19 Dimensionless h-s diagram for Fanno process for dry air with $\kappa = 1.4$ and internal friction

Taking the pressure and temperature ratios from Eqs. (13.60) and (13.61), we obtain the entropy difference from,

$$s - s^* = \Delta s = c_p \ln \left[\left(\frac{T}{T^*} \right) \left(\frac{p^*}{p} \right)^{\frac{\kappa - 1}{\kappa}} \right] \tag{13.63}$$

or in terms of Mach number, we obtain

$$\frac{\Delta s}{cp} = \frac{s - s^*}{c_p} = \ln M^2 \left(\frac{\frac{k+1}{2}}{M^2 \left(1 + \frac{\kappa - 1}{2} M^2\right)} \right)^{\frac{\kappa + 1}{2\kappa}} \tag{13.64}$$

the Fanno curve in an $h - s$ diagram plotted in Fig. 13.19.
This curve is valid for a duct flow without heating, independent of the wall and internal friction. The upper part of the curve is the subsonic, while the lower is the supersonic. Considering a flow through a long pipe, because of the entropy increase, the static enthalpy, the static pressure, and the density decreases. As a consequence, the velocity increases until the speed of sound with Mach number $M = 1$ has been reached. A further increase of the velocity pass the speed of sound resulting in $M > 1$, requires a decrease in entropy which violates the second law. Once the speed of sound has been reached, a normal shock will occur that reduces the velocity to subsonic. Therefore, the velocity cannot exceed the speed of sound.

A typical application of the Fanno process is shown in Fig. 13.20. The high pressure side of a steam turbine shaft is sealed against the atmospheric pressure. To reduce the mass flow that escapes from the process through the radial gap between the shaft and the casing, labyrinth seals are installed on the shaft and in the casing, Fig. 13.20a.
High pressure steam enters the gap and expands through the clearance C where its potential energy is converted into kinetic energy. By entering the cavity, the kinetic energy is dissipated causing a noticeable pressure drop. The process of expansion and dissipation repeats in the following cavities resulting in a relatively small mass flow that leaves the turbine. The end points of all expansions through the clearances are located on a Fanno line, which corresponds to a constant total enthalpy.

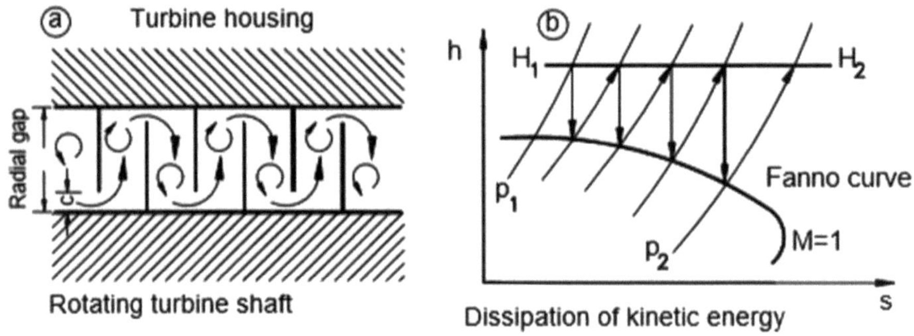

Fig. 13.20 Flow through a labyrinth seal of a turbomachine (**a**), Fanno process (**b**)

In Table 13.2, the equations are summarized and steps are marked that are necessary for constructing the Rayleigh and Fanno lines using the following steps. In step ① the Mach number is varied and the corresponding thermodynamic properties are calculated from steps ② to ⑤. With the temperature and pressure ratios calculated in steps ① and ②, the entropy can be calculated. These steps were performed to plot Fig. 13.19. Once the thermodynamic properties are calculated, different versions of Fanno and Rayleigh curves can be constructed easily.

13.1.3.2 The Normal Shock Wave Relations

The normal shock occurs when a supersonic flow encounters a strong perturbation. If a supersonic flow impinges on a blunt body, it will generate a normal shock in front of the body. Behind the shock, the flow velocity becomes subsonic causing the pressure, density and temperature to rise. The transition from supersonic to subsonic velocity occurs within a thin surface with a thickness that has an order of magnitude of the mean free path of the fluid. Thus, in gas dynamics it is approximated as a surface discontinuity with an infinitesimally small thickness.

Given the quantities in front of the shock, the quantities behind the shock can be determined using the conservation laws presented in Chap. 5. We assume that changes in flow quantities up- and downstream of the actual shock compared to the changes within the shock itself is negligibly small. Furthermore, we assume a steady adiabatic flow and, considering the infinitesimal thickness of the shock, we neglect the volume integrals. In addition, we assume that the inlet and exit control surfaces are approximately equal ($S_1 \approx S_2$) and the wall surface S_W is very small.
Using the control volume in Fig. 13.21 where position 1 and 2 refer to locations up and downstream of the shock, we apply the continuity equation,

$$\varrho_1 V_1 = \varrho_2 V_2 \tag{13.65}$$

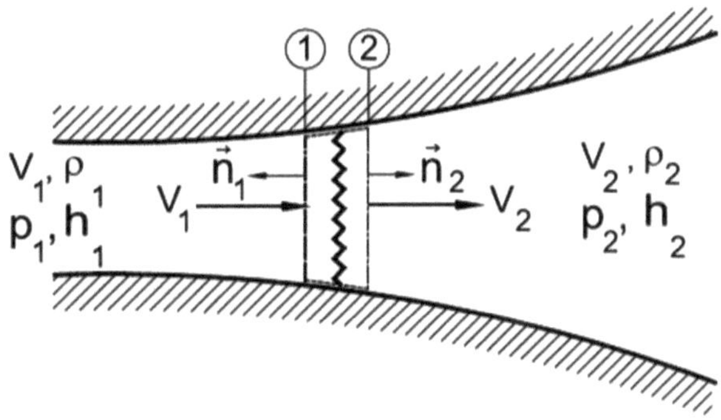

Fig. 13.21 Normal shock in a divergent part of a Laval nozzle

the balance of momentum,

$$\varrho_1 V_1^2 + p_1 = \varrho_2 V_2^2 + p_2 \tag{13.66}$$

and the balance of energy,

$$\frac{V_1^2}{2} + h_1 = \frac{V_2^2}{2} + h_2. \tag{13.67}$$

To close the systems of equations, we introduce the equation of state either in general p-v-T form

$$p = p(\varrho h) \tag{13.68}$$

or in particular for a perfect gas

$$p = \varrho R T = \varrho h \frac{\kappa - 1}{\kappa}. \tag{13.69}$$

With Eqs. (13.65)–(13.69) and known quantities in front of the shock, the unknown quantities behind the shock are determined. Inserting the continuity equation (13.65) into the balances of momentum (13.66) and of energy (13.67), we obtain:

$$p_2 - p_1 = \varrho_1 V_1^2 \left(1 - \frac{\varrho_1}{\varrho_2}\right) \tag{13.70}$$

and

$$h_2 - h_1 = \frac{V_1^2}{2} \left[1 - \left(\frac{\varrho_1}{\varrho_2}\right)^2\right]. \tag{13.71}$$

Eliminating the velocity V_1 from Eqs. (13.70) and (13.71), we obtain a relation between the thermodynamic quantities, the so-called *Hugoniot relation:*

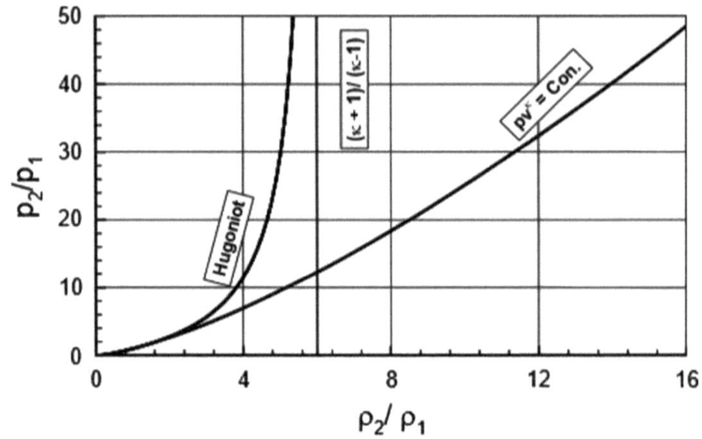

Fig. 13.22 Shock and isentropic compression

$$h_2 - h_1 = \frac{1}{2}(p_2 - p_1)\left(\frac{1}{\varrho_1} + \frac{1}{\varrho_2}\right). \tag{13.72}$$

Replacing the enthalpy in Eq. (13.72) by the pressure from Eq. (13.69), we find for a perfect gas the following relation

$$\frac{p_2}{p_1} = \frac{(\kappa+1)\varrho_2/\varrho_1 - (\kappa-1)}{(\kappa+1) - (\kappa-1)\varrho_2/\varrho_1}, \tag{13.73}$$

between the pressure and the density ratios. The maximum density ratio is obtained by setting in Eq. (13.73) $p_2/p_1 \to \infty$:

$$\left(\frac{\varrho_2}{\varrho_1}\right)_{max} = \frac{\kappa+1}{\kappa-1}. \tag{13.74}$$

Fig. 13.22 shows the pressure ratio for the normal shock as well as for the isentropic process.
As seen, the Hugoniot curve approaches an asymptotic value of $(\kappa+1)/(\kappa-1) = 6.0$ for diatomic with $\kappa = c_p/c_v = 7/5$. In contrast, the isentropic change of state results for $p_2/p_1 \to \infty$ in an infinitely large density ratio ϱ_2/ϱ_1.

Considering the upstream Mach number as the determining parameter for calculating the state downstream of the shock, the following relations are presented that directly relate the flow states up- and downstream of the shock as a function of the upstream Mach number. From Eq. (13.70), the velocity can be obtained as:

$$V_1^2 = \frac{p_1}{\varrho_1}\left(\frac{p_2}{p_1} - 1\right)\left(1 - \frac{\varrho_1}{\varrho_2}\right)^{-1}. \tag{13.75}$$

Introducing the speed of sound $c^2 = \kappa p/\varrho$ for calorically perfect gases, Eq. (13.75) can be modified as:

$$\left(\frac{V_1}{c_1}\right)^2 = M_1^2 = \frac{1}{\kappa}\left(\frac{p_2}{p_1} - 1\right)\left(1 - \frac{\varrho_1}{\varrho_2}\right)^{-1} \tag{13.76}$$

from which we can eliminate ϱ_1/ϱ_2. Using the Hugoniot relation (13.73), we obtain an equation for the pressure ratio

$$\left(\frac{p_2}{p_1} - 1\right)\left(\frac{p_2}{p_1} - 1 - 2\frac{\kappa}{\kappa+1}\left(M_1^2 - 1\right)\right) = 0. \tag{13.77}$$

Equation (13.77) is a product of two expressions that results in two solutions: $p_2/p_1 = 1$ and the following solution

$$\frac{p_2}{p_1} = 1 + 2\frac{\kappa}{\kappa+1}(M_1^2 - 1). \tag{13.78}$$

Equation (13.78) is an explicit relation between the pressure ratio across the shock and the upstream Mach number M_1. The density ratio is found by replacing p_2/p_1 in Eq. (13.78), the Hugoniot relation (13.73)

$$\frac{\varrho_2}{\varrho_1} = \frac{(\kappa+1)M_1^2}{2 + (\kappa-1)M_1^2}. \tag{13.79}$$

The temperature jump is obtained by using Eqs. (13.78) and (13.79):

$$\frac{T_2}{T_1} = \frac{p_2}{p_1}\frac{\varrho_1}{\varrho_2} = \frac{(2\kappa M_1^2 - (\kappa-1))(2 + (\kappa-1)M_1^2)}{(\kappa+1)^2 M_1^2}. \tag{13.80}$$

To find the Mach number behind the shock, we use the continuity equation and the speed of sound to get

$$M_2^2 = \left(\frac{V_2}{a_2}\right)^2 = V_1^2\left(\frac{\varrho_1}{\varrho_2}\right)^2 \frac{\varrho_2}{\kappa p_2} = M_1^2 \frac{p_1 \varrho_1}{p_2 \varrho_2}. \tag{13.81}$$

Introducing Eqs. (13.78) and (13.79) into Eq. (13.81), we finally find:

$$M_2^2 = \frac{\kappa + 1 + (\kappa - 1)\left(M_1^2 - 1\right)}{\kappa + 1 + 2\kappa(M_1^2 - 1)}. \tag{13.82}$$

We infer from this equation that in a normal shock wave, because $M_1 > 1$, the Mach number behind the shock is always lower than 1. In the case of a very strong shock M_2 takes on the limiting value

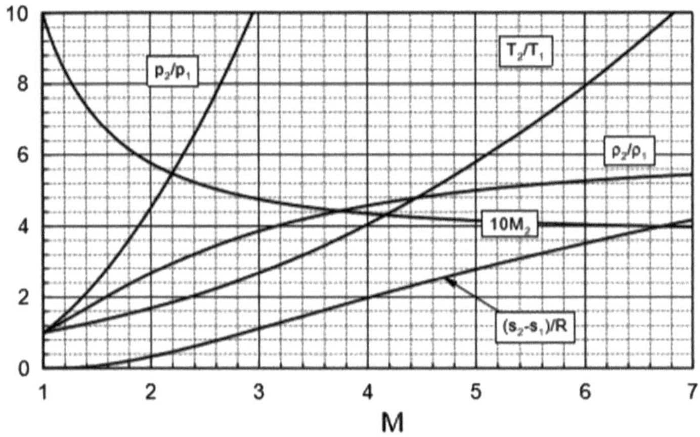

Fig. 13.23 Mach number and the thermodynamic properties behind the shock as a function of Mach number ahead of the shock for a diatomic gas with $\kappa = 7/5$

$$M_2\bigg|_{(M_1 \to \infty)} = \sqrt{\frac{1}{2}\frac{\kappa - 1}{\kappa}}. \tag{13.83}$$

The shock relations are shown in Fig. 13.23.
As seen, the curves for density ratio ϱ_2/ϱ_1 and downstream Mach number M_2 approach their asymptotes at 6.0 and 0.378. The change of the state from supersonic to subsonic due to the normal shock is associated with an entropy increase through the shock. This is explicitly expressed by the second law for perfect gases:

$$s_2 - s_1 = c_v \ln\left[\frac{p_2}{p_1}\left(\frac{\varrho_2}{\varrho_1}\right)^{-\kappa}\right]. \tag{13.84}$$

Replacing the pressure and density ration in Eq. (13.84) by Eqs. (13.78) and (13.79), we obtain the entropy equation as a function of Mach number M_1.

$$\frac{s_2 - s_1}{c_v} = \ln\left[\left(1 + \frac{2\kappa}{\kappa + 1}(M_1^2 - 1)\right)\left(\frac{2 + (\kappa - 1)M_1^2}{(\kappa + 1)M_1^2}\right)^{\kappa}\right]. \tag{13.85}$$

Eliminating the density ratio using the Hugoniot relation we find:

$$\frac{s_2 - s_1}{c_v} = \ln\left[\frac{p_2}{p_1}\left(\frac{(\kappa - 1)p_2/p_1 + \kappa + 1}{(\kappa + 1)p_2/p_1, +\kappa - 1}\right)^{\kappa}\right]. \tag{13.86}$$

For a strong shock $p_2/p_1 \to \infty$, the entropy difference tends to infinity. However, for a weak shock with a pressure ratio in the order of $p_2/p_1 = 1 + \varepsilon$ and small ε, the right-hand side of Eq. (13.85) is expanded resulting in:

$$\frac{s_2 - s_1}{c_v} = \frac{\kappa^2 - 1}{12\kappa^2}\left(\frac{p_2}{p_1} - 1\right)^3 - \frac{\kappa^2 - 1}{8\kappa^2}\left(\frac{p_2}{p_1} - 1\right)^4 + \cdots . \tag{13.87}$$

Defining the shock strength $\varepsilon = p_2/p_1 - 1$, Eq. (13.87) shows that for weak shocks (small ε) the entropy increase is of the third order with respect to ε. Therefore, the entropy increase may be neglected and the isentropic relation may be used for calculating the states on both sides of the shock. Figure 13.23 also shows the dimensionless entropy difference $(s_2 - s_1)/R$. Up to $M_1 = 1.4$, no noticeable change of entropy can be seen, which confirms the above statement.

13.1.4 Supersonic Flow

In Sect. 12.3, the transition from subsonic to supersonic flow regime was described by means of a convergent-divergent channel, the Laval nozzle, as an example of internal aerodynamics. The shock waves with the special case of normal shocks we treated, were generated by pressure disturbances across the shock. A similar situation occurs in external aerodynamics. Consider a stationary sound source that emits pressure disturbances in a fluid at rest. It generates sound waves with concentric spherical fronts. Now we suppose that the sound source, for instance a subsonic aircraft, moves with a subsonic speed ($u < c$) as shown in Fig. 13.24a. After a period of time t the source moved from point P_1 to point P_4 which is equivalent to the distance $\overline{P_1 P_4} = ut$. After the same period of time, the pressure disturbance propagated spherically with the speed of sound c. The spherical wave front has the radius $r_1 = ct$. Figure 13.24a also shows the locations of the source at distances $r_2 = 2/3ct$, $r_3 = 1/3ct$ and $r_4 = 0$. As seen, because of $u < c$, the disturbance source always remains within the respective spherical wave front.

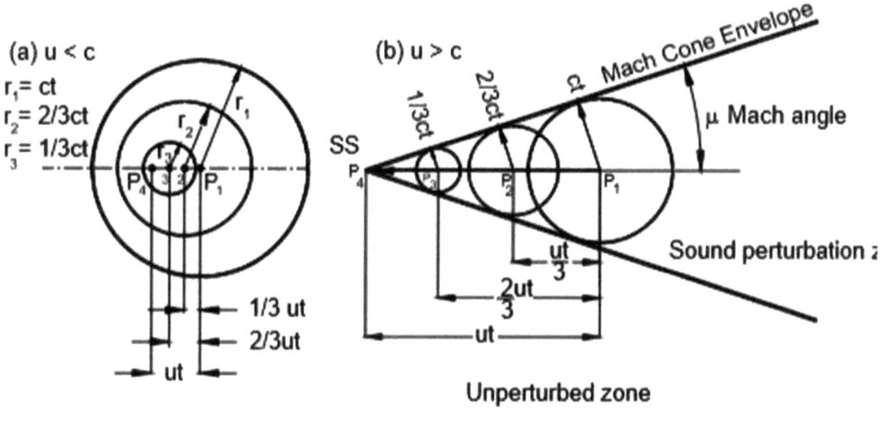

Fig. 13.24 Disturbance propagation in subsonic (**a**) and supersonic (**b**) flow

A different wave pattern arises, however, if the source of disturbance, for example a supersonic aircraft, moves with a supersonic speed ($u > c$) as Fig. 13.24b reveals. Similar to the previous case, after the period of time t, the spherical front of the pressure disturbance has reached a radius of $r_1 = ct$, while the source of the disturbance arrived at P_4 leaving the distance of $\overline{P_1P_4} = ut$ behind. The sound source moving with supersonic speed forms a conical envelop, the *Mach cone*, whose angle is calculated from

$$\sin\mu = \frac{ct}{ut} = \frac{1}{M}. \tag{13.88}$$

Figure 13.24 shows that the sound waves reaches the observer within the Mach cone described by the Mach angle μ. An observer positioned outside the Mach cone registers first the arrival of the supersonic aircraft and then its sound waves once the aircraft has passed overhead.

13.1.4.1 The Oblique Shock Wave Relations

In the previous section we treated the normal shock wave, a special type of shock, whose front is perpendicular to the flow direction. The more prevalent shocks encountered in engineering such as in transonic turbine or compressor blade channels, as well as supersonic aircrafts, are the oblique shocks. The basic mechanism of the oblique shock is shown in Fig. 13.25.
Supersonic flow with uniform velocity V_1 approaches a wedge with a sharp angle 2δ. A surface discontinuity characterized by an oblique shock wave is formed that builds an angle Θ with the flow direction. This particular shock is called the attached shock. Following a streamline by passing through the shock front, the streamline is deflected by an angle which corresponds to the half wedge angle δ_a. A different shock pattern is observed when the same supersonic flow approaches another wedge with $(2\delta)_b > (2\delta)_a$, as shown in Fig. 13.25b. Again, following an arbitrary streamline upstream of the leading edge, a *strong shock* is formed which is detached. Figure 13.25a, b

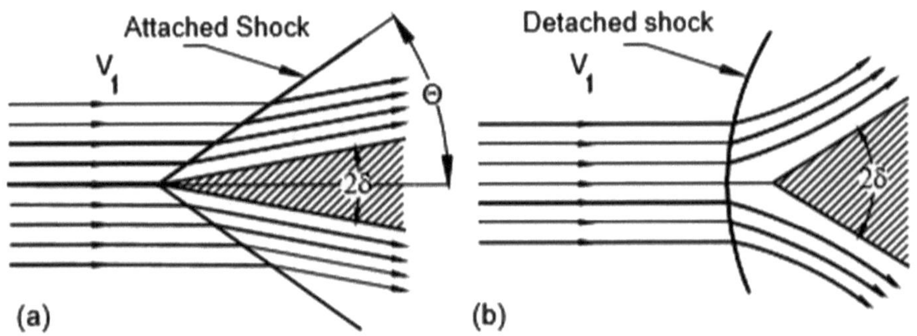

Fig. 13.25 A qualitative picture of two different shock patterns based on the same Mach number but different wedge angles

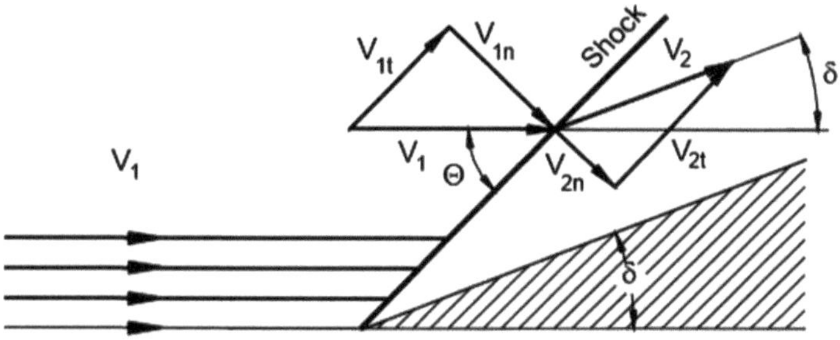

Fig. 13.26 Incoming velocity vector V_1 decomposed into normal and tangential components. $\Theta =$ shock angle, $\delta =$ half wedge angle

suggest that, depending on the magnitude of the incoming Mach number and the wedge angle or, generally body bluntness, attached or detached shocks may occur. To establish the corresponding relationships between the Mach number, the wedge angle, and the angle of the oblique shock, we use the same procedure that we applied to the normal shock waves. To do this, we decompose the velocity vector $\mathbf{V_1}$ in front of the shock into a component normal to the shock front V_{1n}, and a component tangential to the shock front V_{1t}, as shown in Fig. 13.26.
The tangential component is

$$V_{1t} = V_1 \cos \Theta \tag{13.89}$$

and the normal component follows from

$$V_{1n} = V_1 \sin \Theta . \tag{13.90}$$

Introducing the normal Mach number built with the normal component, we arrive at:

$$M_{1n} = \frac{V_{1n}}{c_1} = M_1 \sin \Theta . \tag{13.91}$$

The normal shock Eqs. (13.78)–(13.80) can then be carried over to the oblique shock wave by replacing M_1, with M_{1n}, from Eq. (13.91):

$$\frac{p_2}{p_1} = 1 + 2\frac{\kappa}{\kappa+1}(M_1^2 \sin^2 \Theta - 1) \tag{13.92}$$

$$\frac{\varrho_2}{\varrho_1} = \frac{(\kappa+1)M_1^2 \sin^2 \Theta}{2 + (\kappa-1)M_1^2 \sin^2 \Theta} \tag{13.93}$$

$$\frac{T_2}{T_1} = \frac{[2_\kappa M_1^2 \sin^2 \Theta - (\kappa-1)][2 + (\kappa-1)M_1^2 \sin^2 \Theta]}{(\kappa+1)^2 M_1^2 \sin^2 \Theta} . \tag{13.94}$$

Obtaining the normal component of the velocity behind the shock $V_{2n} = V_2 \sin(\Theta - \delta)$, the corresponding normal Mach number is

$$M_{2n} = \frac{V_{2n}}{c_2} = M_2 \sin(\Theta - \delta). \tag{13.95}$$

Relative to the shock front, the normal component M_{1n}, which might be supersonic, experiences a drastic deceleration resulting in a subsonic normal component M_{2n} behind the shock. The Mach number M_2, however, can be supersonic. If we again replace M_1 and M_2 with M_{1n} and M_{2n} in Eq. (13.82) from normal shock relations, we find:

$$M_2^2 \sin^2(\Theta - \delta) = \frac{\kappa + 1 + (\kappa - 1)(M_1^2 \sin^2 \Theta - 1)}{\kappa + 1 + 2(M_1^2 \sin^2 \Theta - 1)}. \tag{13.96}$$

Introducing the continuity equation (13.81), we find a relationship between the shock angle Θ and the wedge angle δ (Fig. 13.26):

$$\tan \delta = \frac{2 \cot \Theta (M_1^2 \sin^2 \Theta - 1)}{2 + M_1^2(\kappa + 1 - 2 \sin^2 \Theta)}. \tag{13.97}$$

Since the incoming Mach number M_1 and the wedge angle δ are supposed to be known, we have with Eqs. (13.96) and (13.97) two equations and two unknowns namely Θ and M_2.

Figure 13.27a shows the shock angle Θ as a function of the wedge angle δ with M_1 as parameter. For each Mach number, there is a maximum δ_{max}, beyond which the shock is detached. This maximum wedge angle is associated with a maximum shock angle Θ_{max}. The curve with the full circles is the locus of all δ_{max} and separates the upper Θ-branch from the lower one. As shown in Fig. 13.27a, b, for each given δ, two solutions, a strong shock and a weak shock, can be found based on the magnitude of the incoming M_1. A shock is called a *strong shock* if the shock angle Θ for a given Mach number M_1 is larger than the angle Θ_{max} (dashed curve in Fig. 13.27a, b) associated with the maximum deflection δ_{max}. A strong shock has distinguishing characteristics that the Mach number behind the shock, M_2, is always subsonic.

In contrast, the velocity downstream of a weak shock can lie in either the subsonic or the supersonic range. If the deflection angle δ is smaller than δ_{max}, there are then two possible solutions for the shock angle Θ. Which solution actually arises depends on the boundary conditions far behind the shock. Figure 13.27b displays the Mach number after the shock. Here again, strong shock leads to a subsonic mach number after the shock, whereas a week shock may maintain the supersonic character of the flow.

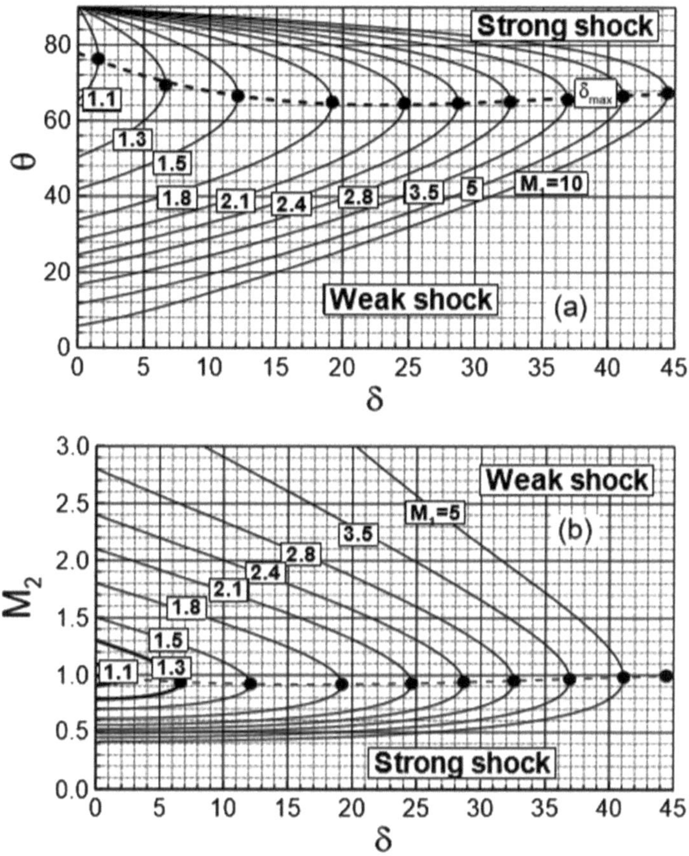

Fig. 13.27 Shock angle (**a**) and Mach number (**b**) as functions of wedge angle with incoming Mach number as parameter

13.1.4.2 Detached Shock Wave

Referring to Fig. 13.28, the wedge angle $\delta > \delta_{max}$ causes a strong detached shock. The stagnation streamline passes through a normal shock, where its initial supersonic Mach number is reduced to a subsonic one. Moving from the intersection of the stagnation streamline, the shock is deflected and its strength is reduced. Far downstream, the shock deteriorates into a Mach wave.

Detached shocks are frequently encountered in transonic and supersonic compressors operating at off-design conditions. To keep the shock losses at a minimum, the compressor blades are generally designed with a sharp leading edge, such that the shocks are always attached at the design operating point. Figure 13.29a, b show two profile families with the attached shocks. For transonic compressor stages with an inlet Mach number of $M = 0.95 - 1.05$, double-circular arc (DCA) profiles are used.

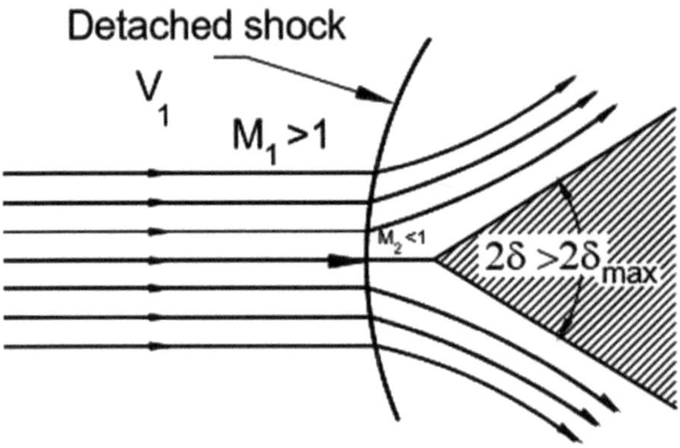

Fig. 13.28 Detached shock formation for $\delta > \delta_{max}$

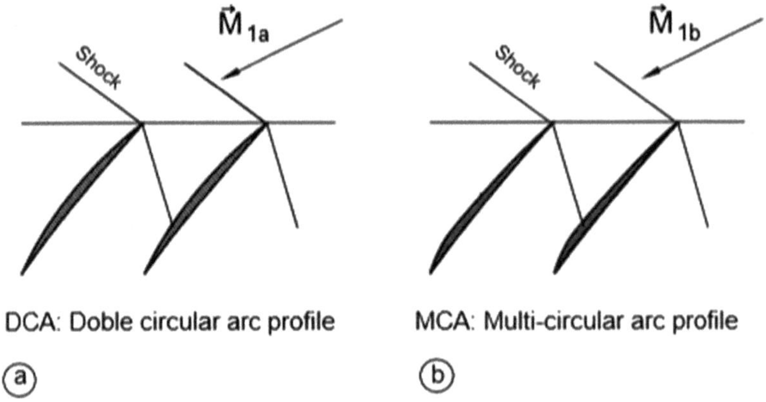

Fig. 13.29 Attached shock formation in front of a transonic compressor with DCA-profiles (**a**), and supersonic compressor with MCA-profiles (**b**)

Supersonic compressor stages require multi-circular arc (MCA) profiles. The profile shown in Fig. 13.29a belongs to the DCA family where the convex (suction side) and the concave (pressure side) surfaces are circular arcs of different diameters. In contrast, the suction side of the profiles illustrated in Fig. 13.29b consists of two or more arcs. The off-design operation affects the position of the shocks and may causing it to detach from the blade leading edge, [232].

Fig. 13.30 illustrates the impact of the variation of the back pressure on shock position. Beginning with a design point speed line, Fig. 13.30a, the operating point (a) is given by the inlet Mach number M_1 with a uniquely allocated inlet flow angle β_1. Increasing the back pressure from the design point back pressure to a higher level (b) causes the passage shock to move toward the cascade entrance.

By further increasing the back pressure from (b) to (c), a normal shock is established, which is still attached. The corresponding shock angle γ can be set equal to $\gamma_{lim} =$

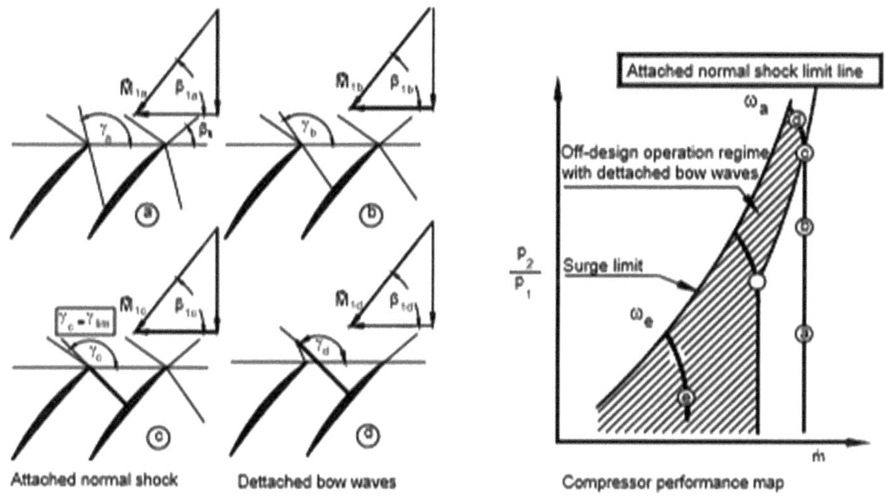

Fig. 13.30 Left, change of the shock angle for a given cascade geometry at different operation conditions, right: the effect of shock angle γ on compressor performance map, From [4]

γ_{att}. Decreasing the mass flow beyond this point causes the shock to detach from the leading edge, as shown in Fig. 13.30c. Reducing the rotational speed changes the incidence and may further move the shock from the leading edge as shown in Fig. 13.30d. These operating points are plotted schematically in a compressor performance map, shown in Fig. 13.30 (right), with a surge limit and an attached normal shock line.

During startup, shutdown and dynamic load change of a gas turbine engine, the compressor undergoes a change of rotational speed (rpm). One of these off-design speed lines is given in Fig. 13.30e. The changes of the rpm causes a change in the velocity diagram resulting in the detachment of the shock. Calculating the kinematics of the detached shock was a major research subject of NACA. The subject was treated among others by [233]. With today's computational capabilities attached and detached shockwaves are calculated with a reasonable accuracy.

13.1.4.3 Prandtl-Meyer Expansion

Unlike the supersonic flow along a concave surface, Fig. 13.26, which was associated with an oblique shock leading to a Mach number $M_2 < M_1$, a supersonic flow along a convex surface, Fig. 13.31, experiences an expansion process, Fig. 13.31a.
The parallel streamlines with the uniform Mach number M_1 pass through a system of expansion or Mach waves, thereby moving apart from each other and accelerate to a new Mach number $M_2 > M_1$. The expansion is associated with a deflection of the incoming supersonic flow with the Mach angle μ_1 to μ_2. To calculate the new Mach number, we first consider a supersonic flow around a corner of an infinitesimal

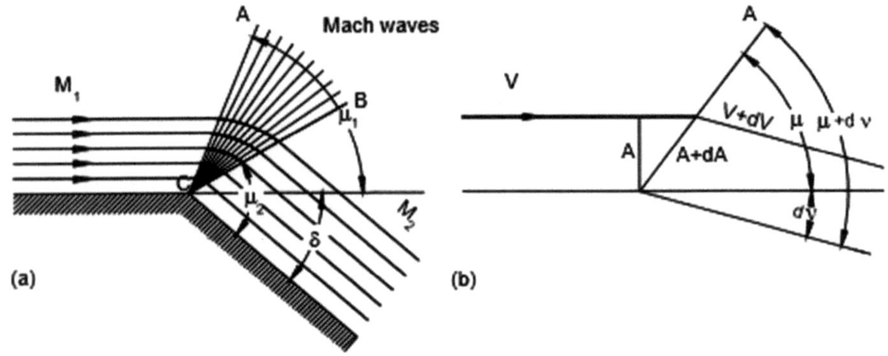

Fig. 13.31 Prandtl-Meyer Expansion around a convex corner

deflection, dA, as shown in Fig. 13.31b, and apply the continuity equation (13.38):

$$\frac{dA}{A} = -\frac{dV}{V}(1 - M^2). \tag{13.98}$$

The velocity ratio is expressed in terms of Mach number by utilizing the energy equation

$$\frac{dV}{V} = \frac{dM}{M(1 + \frac{\kappa-1}{2}M^2)}. \tag{13.99}$$

Inserting Eq. (13.99) into Eq. (13.98), we obtain:

$$\frac{dA}{A} = \frac{(M^2 - 1)dM}{M(1 + \frac{\kappa-1}{2}M^2)}. \tag{13.100}$$

The geometric relation from Fig. 13.31 reads:

$$\frac{A + dA}{A} = \frac{\sin(\mu + d\nu)}{\sin\mu} = \frac{\sin\mu\cos d\nu + \cos\mu\sin d\nu}{\sin\mu} = 1 + d\nu\cot\mu. \tag{13.101}$$

In Eq. (13.101), we assumed dv as infinitesimally small allowing to set $\cos dv = 1$ and $\sin dv = d\nu$. With this approximation, Eq. (13.101) becomes:

$$\frac{dA}{A} = d\nu\cot\mu = d\nu\sqrt{M^2 - 1} \tag{13.102}$$

with μ as the Mach angle that can be expressed as $\sin\mu = 1/M$. Equating (13.102) and (13.100) leads to:

$$d\nu = \frac{\sqrt{(M^2 - 1)}dM}{M(1 + \frac{\kappa-1}{2}M^2)} \tag{13.103}$$

and its subsequent integration gives:

$$\nu = \sqrt{\frac{\kappa+1}{\kappa-1}} \arctan\left(\sqrt{\frac{\kappa+1}{\kappa-1}}\sqrt{M^2-1}\right) - \arctan\sqrt{M^2-1}. \tag{13.104}$$

This deflection angle ν as well as the Mach angle μ equation are plotted in Fig. 13.32. As shown, each arbitrary supersonic Mach number is uniquely associated with a deflection angle ν. As an example, we assume that in Fig. 13.32 the flow has the Mach number $M_1 = 1.5$ and turns around a corner with an angle $\delta = 40°$. For this Mach number, the corresponding deflection angle $\nu_1 = 12.2$ is found. After turning around the corner, the deflection is $\nu_2 = \nu_1 + \delta = 52.2$, which results in a Mach number of $M_2 = 3.13$.

The Prandtl-Meyer expansion theory is widely used for design and loss calculation of transonic and supersonic compressor blades. Although this topic is treated in the corresponding chapter, in the context of this section, it is useful to point to a few interesting features from a turbomachinery design point of view. Figure 13.33 shows a supersonic compressor cascade with an inlet Mach number $M_\infty > 1$. The incoming supersonic flow impinges on the sharp leading edge and forms a weak oblique shock followed by an expansion fan. Passing through the shock front, the Mach number, although smaller, remains supersonic. Expansion waves are formed along the suction surface (convex side) of the blade from the leading edge L to the point e, where the subsequent Mach wave at point e intersects the adjacent blade leading edge. Since the angle Θ is known, the Mach number M_e at position e can easily be calculated from Prandtl-Meyer relation.

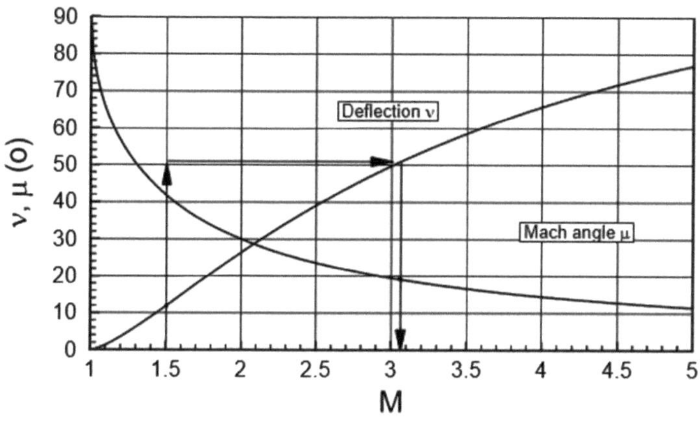

Fig. 13.32 Deflection angle ν and Mach angle μ as functions of Mach number

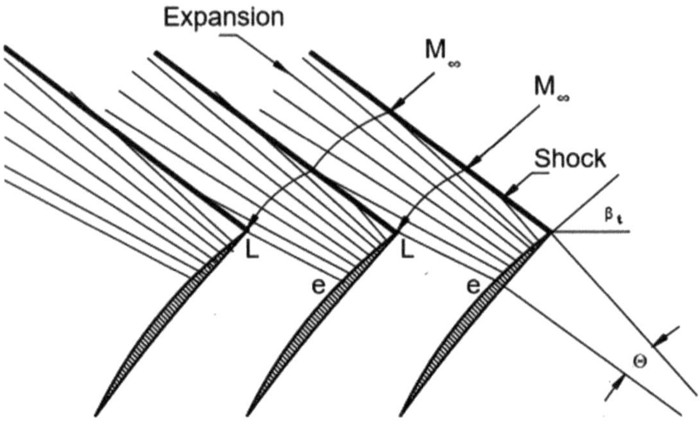

Fig. 13.33 A supersonic compressor cascade with supersonic inlet flow

13.2 Unsteady Compressible Flow

The following sections deal with the basic physics of unsteady compressible flow that is essential to predict unsteady flow and transient behavior of different engineering components. The flow in all engineering applications where mass, momentum, and energy transfer occurs within a stationary frame followed by a rotating one and vice versa, is periodic unsteady. Flows through turbines, compressors, internal combustion engines, and pumps are examples where periodic records of unsteady flow quantities characterize the flow situations. In contrast, a non-periodic unsteady flow situation is characterized by sets of non-periodic data records. The process of depressurizing a container under high pressure, non-periodic events within a shock tube, and pressurizing an air-storage cavern, are examples of non-periodic unsteady flow situations.

In the following sections, a system of nonlinear differential equations is presented that describes the basic physics of unsteady flow. A brief explanation of the numerical method for solution is followed by a detailed dynamic simulation of a shock tube.

13.2.1 One-Dimensional Approximation

The thermo-fluid dynamic processes that take place within engineering systems and components are mostly of the unsteady nature. The steady state, a special case, always originates from an unsteady condition during which the temporal changes in the process parameters have largely come to a standstill. For the purpose of the unsteady dynamic simulation of an engineering component, conservation laws presented and discussed in Chap. 4 are rearranged such that temporal changes of thermo-fluid dynamic quantities are expressed in terms of spatial changes. A summary of relevant

Table 13.3 Summary of thermo-fluid dynamic equations

Conservation equations in terms of local derivatives $\partial/\partial t$
Equation of continuity
$\frac{\partial \rho}{\partial t} = -\nabla \cdot (\rho \mathbf{V})$
Equation of motion, stress tensor decomposed
$\frac{\partial(\rho \mathbf{V})}{\partial t} + \nabla \cdot (\rho \mathbf{V}\mathbf{V}) = -\nabla p + \nabla \cdot \mathbf{T}$
Equation of mechanical energy including ρ
$\frac{\partial(\rho K)}{\partial t} = -\nabla \cdot (\rho K \mathbf{V}) - V \cdot \nabla p + \nabla \cdot (\mathbf{T} \cdot \mathbf{V}) + \rho \mathbf{V} \cdot \mathbf{g}$
Equation of thermal energy in terms of u for ideal gas
$\frac{\partial(\rho u)}{\partial t} = -\nabla \cdot (\rho u \mathbf{V}) - \nabla \dot{\mathbf{q}} - p \nabla \cdot \mathbf{V} + \mathbf{T} : \nabla \mathbf{V}$
Equation of thermal energy in terms of h for ideal gas
$\frac{\partial(\rho h)}{\partial t} = -\nabla \cdot (\rho h \mathbf{V}) - \nabla \dot{\mathbf{q}} + \frac{Dp}{Dt} + \mathbf{T} : \nabla \mathbf{V}$
Equation of thermal energy in terms of c_v and T
$\frac{\partial(\rho c_v T)}{\partial t} = -\nabla \cdot (\rho u \mathbf{V}) - \nabla \dot{\mathbf{q}} - p \nabla \cdot \mathbf{V} + \mathbf{T} : \nabla \mathbf{V}$
Equation of thermal energy in terms of h for ideal gases
$\frac{\partial(\rho c_p T)}{\partial t} = -\nabla \cdot (\rho h \mathbf{V}) - \nabla \dot{\mathbf{q}} + \frac{Dp}{Dt} + \mathbf{T} : \nabla \mathbf{V}$
Energy equation in terms of total pressure
$\frac{\partial P}{\partial t} = -k \nabla \cdot (\mathbf{V} P) - (\kappa - 1)[\nabla \cdot \dot{\mathbf{q}} + \nabla \cdot (\mathbf{V} \cdot T)] - (\kappa - 2)\left[\frac{\partial(\rho K)}{\partial t} + \nabla \cdot (\rho K \mathbf{V})\right]$
Energy equation in terms of total enthalpy
$\frac{\partial H}{\partial t} = -k \mathbf{V} \cdot \nabla H - (\kappa - 1)\left(\frac{1}{\rho} \nabla \cdot (\rho \mathbf{V})(H + K) + \frac{\mathbf{V} \cdot \partial(\rho \mathbf{V})}{\rho \partial t}\right) + \left(-\frac{\kappa \nabla \cdot \mathbf{q}}{\rho} + \frac{\kappa}{\rho} \nabla \cdot (\mathbf{V} \cdot \mathbf{T})\right)$

equations is presented in Table 13.3. They constitute the theoretical basis describing the dynamic process that takes place within an engineering component. In the context of the one-dimensional flow approximation, a one-dimensional time dependent calculation procedure provides a sufficiently accurate picture of a non-linear dynamic behavior of an engineering component. In the following, the conservation equations are presented in index notation. For the one-dimensional time dependent treatment, the basic equations are prepared first by setting the index $i = 1$. Thus, in the continuity equation of the Cartesian coordinate system the continuity reads:

$$\frac{\partial \rho}{\partial t} = -\frac{\partial}{\partial x_i}(\rho V_i). \tag{13.105}$$

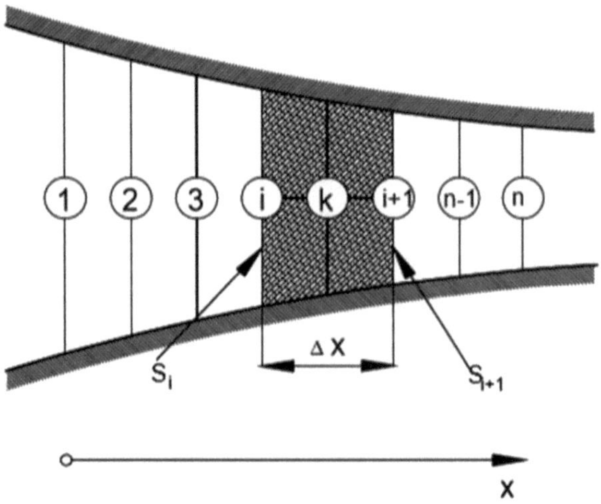

Fig. 13.34 Discretization of an arbitrary flow path with variable cross section $S = S(x)$

Equation (13.105) after setting $\rho V_1 = \dot{m}/S$ becomes:

$$\frac{\partial \rho}{\partial t} = -\frac{\partial}{\partial x_1}\left(\frac{\dot{m}}{S}\right) \tag{13.106}$$

with $x_1 \equiv x$ as the length in streamwise direction and $S = S(x)$ the cross-sectional area of the component under investigation. Equation (13.106) expresses the fact that the temporal change of the density is determined from the spatial change of the specific mass flow within a component. The partial differential Eq. (13.106) can be approximated as an ordinary differential equation by means of conversion into a difference equation. The ordinary differential equation can then be solved numerically with the prescribed initial and boundary conditions. For this purpose, the flow field is equidistantly divided into a number of discrete zones with prescribed length, ΔX, inlet and exit cross sections S_i and S_{i+1} as Fig. 13.34 shows.
Using the nomenclature in Fig. 13.34, Eq. (13.106) is approximated as:

$$\frac{\partial \rho_k}{\partial t} = -\frac{1}{\Delta x}\left(\frac{\dot{m}_{i+1}}{S_{i+1}} - \frac{\dot{m}_i}{S_i}\right) \tag{13.107}$$

with $\dot{m}_i$ and $\dot{m}_{i+1}$ as the mass flows at stations i and $i+1$ with the corresponding cross-sections. For a constant cross-section, Eq. (13.107) reduces to:

$$\frac{\partial \rho_k}{\partial t} = -\frac{1}{\Delta x S}(\dot{m}_{i+1} - \dot{m}_i) = -\frac{1}{\Delta V}(\dot{m}_{i+1} - \dot{m}_i) \tag{13.108}$$

with $\Delta V = \Delta x S$ as the volume of the element k enclosed between the surfaces i and $i+1$. The index k refers to the position at $\Delta x/2$, Fig. 13.34. The time dependent equation of motion in index notation of the momentum equation is:

$$\frac{\partial(\rho V_i)}{\partial t} = -\frac{\partial}{\partial x_j}(\rho V_i V_j) - \frac{\partial p}{\partial x_i} + \frac{\partial T_{ij}}{\partial x_j}. \tag{13.109}$$

In Eq. (13.109), $\nabla \cdot \mathbf{T} = e_i \partial T_{ij}/\partial x_j$ represents the shear force acting on the surface of the component. For a one-dimensional flow, the only non-zero term is $\partial\tau_{21}/\partial x_2$. It can be related to the wall shear stress τ_w which is a function of the friction coefficient c_f.

$$\tau_w = c_f \frac{\rho}{2} V^2. \tag{13.110}$$

In the near of the wall, the change of the shear stress can be approximated as the difference between the wall shear stress τ_W and the shear stress at the edge of the boundary layer, which can be set as $\tau_e \approx 0$

$$\left(\frac{\partial \tau_{12}}{\partial x_2}\right)_{x_2=0} = \frac{\tau_e - \tau_w}{\Delta x_2} = -\frac{\tau_w}{\Delta x_2}. \tag{13.111}$$

The distance in Δx_2 can be replaced by a characteristic length such as the hydraulic diameter D_h. Expressing the wall shear stress in Eq. (13.111) by the skin friction coefficient

$$\left(\frac{\partial \tau_{12}}{\partial x_2}\right)_{x_2=0} = -c_f \frac{\rho}{D_h} \frac{V^2}{2} = -c_f \frac{\dot{m}^2}{2 D_h \rho S^2} \tag{13.112}$$

and inserting Eq. (13.112) into the one-dimensional version of Eq. (13.109), we obtain

$$\frac{\partial \dot{m}}{\partial t} = -\frac{\partial}{\partial x_1}(\dot{m} V_1 + pS) + (\dot{m} V_1 + pS)\frac{1}{S}\frac{\partial S}{\partial x_1} - c_f \frac{\dot{m}^2}{2 D_h \rho S}. \tag{13.113}$$

Equation (13.113) relates the temporal change of the mass flow to the spatial change of the velocity, pressure and shear stress momentum. As we will see in the following sections, mass flow transients can be accurately determined using Eq. (13.113). Using the nomenclature from Fig. 13.34, we approximate Eq. (13.113) as:

$$\begin{aligned} \frac{\partial \dot{m}_k}{\partial t} &= -\frac{1}{\Delta x}(\dot{m}_{i+1} V_{i+1} - \dot{m}_i V_i + p_{i+1} S_{i+1} - p_i S_i) \\ &+ \left(\frac{\dot{m}_k V_k + P_k S_k}{S_k}\right)\left(\frac{S_{i+1} - S_i}{\Delta x}\right) - c_f \frac{m_k^2}{2 D_k^k \rho_k S_k}. \end{aligned} \tag{13.114}$$

For a constant cross-section, Eq. (13.114) is modified as:

$$\frac{\partial \dot{m}_k}{\partial t} = -\frac{1}{\Delta x}[\dot{m}_{i+1} V_{i+1} - \dot{m}_i V_i + (p_{i+1} - p_i)S] - c_f \frac{m_k^2}{2 D_{h_k} \rho_k S_k}. \tag{13.115}$$

The energy equation in terms of total enthalpy, is written in index notation

$$\frac{\partial H}{\partial t} = -kV_i\frac{\partial H}{\partial x_i} - \frac{\kappa - 1}{\rho}\left[(H + K)\frac{\partial(\rho V_i)}{\partial x_i} + \frac{V_i \cdot \partial(\rho V_i)}{\partial t}\right] - \frac{\kappa}{\rho}\left[\frac{\partial \dot{q}_i}{\partial x_i} - \frac{\partial(V_j T_{ij})}{\partial x_i}\right] \tag{13.116}$$

with K as the specific kinetic energy. Expressing the total enthalpy, Eq. (13.116), in terms of total temperature results in:

$$\frac{\partial(c_p T_0)}{\partial t} = -kV_i\frac{\partial(c_p T_0)}{\partial x_i} - \frac{\kappa - 1}{\rho}\left[(c_p T_0 + K)\frac{\partial(\rho V_i)}{\partial x_i} + \frac{V_i \cdot \partial(\rho V_i)}{\partial t}\right] - \frac{\kappa}{\rho}\left[\frac{\partial \dot{q}_i}{\partial x_i} - \frac{\partial(V_j T_{ij})}{\partial x_i}\right]. \tag{13.117}$$

For calculating the total pressure, the equation of total energy is written in terms of total pressure which is presented for the Cartesian coordinate system as:

$$\frac{\partial P}{\partial t} = -\kappa\frac{\partial}{\partial x_i}(PV_i) - (\kappa - 1)\left(\frac{\partial \dot{q}_i}{\partial x_i} - \frac{\partial}{\partial x_i}(V_j T_{ij})\right) (\kappa - 2)\left(\frac{\partial(\rho K V_i)}{\partial x_i} + \frac{\partial(\rho K)}{\partial t}\right). \tag{13.118}$$

Before treating the energy equation, the shear stress work Eq. (13.118) needs to be evaluated:

$$\nabla \cdot (\mathbf{T} \cdot \mathbf{V}) = \delta_{ij}\delta_{km}\frac{\partial(\tau_{jk} V_m)}{\partial x_i} = \frac{\partial(\tau_{ij} V_j)}{\partial x_i}. \tag{13.119}$$

For a two-dimensional flow, Eq. (13.119) gives

$$\nabla \cdot (\mathbf{T} \cdot \mathbf{V}) = \frac{\partial(\tau_{ij} V_j)}{\partial x_i} = \frac{\partial(\tau_{11} V_1 + \tau_{12} V_2)}{\partial x_1} + \frac{\partial(\tau_{21} V_1 + \tau_{22} V_2)}{\partial x_2}. \tag{13.120}$$

Assuming a one-dimensional flow with $V_2 = 0$, the contribution of the shear stress work Eq. (13.120) is reduced to

$$\nabla \cdot (\mathbf{T} \cdot \mathbf{V}) = \frac{\partial(\tau_{11} V_1)}{\partial x_1} \approx \frac{(\tau_{11\,\text{inlet}} V_{\text{inlet}} - \tau_{11\,\text{exit}} V_{1\,\text{exit}})}{\Delta x_1}. \tag{13.121}$$

The differences in τ_{11} at the inlet and exit of the component under simulation stem from velocity deformation at the inlet and exit. Its contribution, however, compared to the enthalpy terms in the energy equation, is negligibly small. Thus, the one-dimensional approximation of total energy equation in terms of total enthalpy reads:

$$\frac{\partial H}{\partial t} = -\frac{\kappa \dot{m}}{\rho S}\frac{\partial H}{\partial x_1} - \frac{\kappa - 1}{\rho}\left[(H + K)\frac{\partial}{\partial x_1}\left(\frac{\dot{m}}{S}\right) + \frac{1}{2\rho S^2}\frac{\partial \dot{m}^2}{\partial t}\right] - \frac{\kappa}{\rho}\frac{\partial \dot{q}_i}{\partial x_i}. \tag{13.122}$$

For a steady state case, without changes of specific mass $\dot{m}/S$, Eq. (13.122) leads to:

$$\frac{\partial H}{\partial x_1} = -\frac{S}{\dot{m}}\frac{\partial \dot{q}_i}{\partial x_i}. \tag{13.123}$$

For a given constant cross-section and constant mass flow, Eq. (13.123) gives

$$\frac{\partial H}{\partial x_1} = -\frac{\partial}{\partial x_1}\left(\frac{S\dot{q}_i}{\dot{m}}\right). \tag{13.124}$$

Integrating Eq. (13.124) in streamwise direction results in:

$$H_{\text{out}} - H_{\text{in}} = -\left(\frac{S}{\dot{m}}\right)\Delta\dot{q}. \tag{13.125}$$

For Eq. (13.125) to be compatible with the energy equation discussed in Chap. 5, Eq. (5.75) is presented:

$$H_{\text{Out}} - H_{\text{In}} = q + w_{\text{Shaft}}. \tag{13.126}$$

Equating (13.126) and (13.125) in the absence of a specific shaft power, the following relation between the heat flux vector and the heat added or rejected from the element must hold:

$$q = -\left(\frac{S}{\dot{m}}\right)\Delta\dot{q}. \tag{13.127}$$

From Eq. (13.127) it immediately follows that

$$\Delta\dot{q} = -\frac{q\dot{m}}{S} = -\frac{\dot{Q}}{S} \tag{13.128}$$

where $\dot{Q}$ (kJ/s) is the thermal energy flow added to or rejected from the component. In the presence of shaft power, the specific heat $q = \dot{Q}/\dot{m}$ (kJ/kg) in Eq. (13.128) may be replaced by the sum of the specific heat and specific shaft power:

$$\Delta\dot{q} = -\frac{\dot{m}q + \dot{m}l_m}{S} = -\left(\frac{\dot{Q} + L}{S}\right) \tag{13.129}$$

with $L = \dot{m}l_m$ (kJ/s) as the shaft power. Equation (13.129) in differential form in terms of $\dot{Q}$ and $\mathbf{L}$ is

$$\frac{\partial \dot{q}}{\partial x} = -\frac{\partial}{\partial x}\left(\frac{\dot{m}q + \dot{m}l_m}{S}\right) = -\frac{\partial}{\partial x}\left(\frac{\dot{Q} + L}{S}\right). \tag{13.130}$$

With Eq. (13.122), we find:

$$\frac{\partial H}{\partial t} = -\frac{\kappa \dot{m}}{\rho S}\frac{\partial H}{\partial x_1} - \frac{\kappa - 1}{\rho}\left[(H+K)\frac{\partial}{\partial x_1}\left(\frac{\dot{m}}{S}\right) + \frac{1}{2\rho S^2}\frac{\partial \dot{m}^2}{\partial t}\right] - \frac{\kappa}{\rho}\frac{\partial}{\partial x}\left(\frac{\dot{Q}+L}{S}\right). \tag{13.131}$$

Using the nomenclature in Fig. 13.34, Eq. (13.131) is written as:

$$\begin{aligned}\frac{\partial H}{\partial t} &= \kappa_k \frac{\dot{m}_k}{\rho_k S_k}\left(\frac{H_{i+1}-H_i}{\Delta x}\right)\\ &- \left(\frac{\kappa-1}{\rho}\right)_k \left[\left(\frac{H_k + K_k}{\Delta x}\right)\left(\frac{\dot{m}_{i+1}}{S_{i+1}} - \frac{\dot{m}_i}{S_i}\right) + \frac{\dot{m}_k}{\rho_k S_k^2}\frac{\partial \dot{m}_{i+1}}{\partial t}\right]\\ &- \frac{\kappa_k}{\rho_k}\left(\frac{\Delta \dot{Q} + \Delta L}{\Delta V}\right).\end{aligned} \tag{13.132}$$

In terms of total temperature, Eq. (13.132) is rearranged as:

$$\begin{aligned}\frac{\partial c_p T_0}{\partial t} &= -\kappa_k \frac{\dot{m}_k}{\rho_k S_k}\left(\frac{c_p T_{0_{i+1}} - c_p T_{0_i}}{\Delta x}\right)\\ &- \left(\frac{\kappa-1}{\rho}\right)_k \left[\left(\frac{c_p T_{0_k} + K_k}{\Delta x}\right)\left(\frac{\dot{m}_{i+1}}{S_{i+1}} - \frac{\dot{m}_i}{S_i}\right) + \frac{\dot{m}_k}{\rho_k S_k^2}\frac{\partial \dot{m}_{i+1}}{\partial t}\right]\\ &\frac{\kappa_k}{\rho_k}\left(\frac{\Delta \dot{Q} + \Delta L}{\Delta V}\right).\end{aligned} \tag{13.133}$$

In terms of total pressure, the energy equation reads:

$$\begin{aligned}\frac{\partial P}{\partial t} &= -\kappa \frac{\partial}{\partial x_1}(P V_1) - (\kappa - 1)\left(\frac{\partial \dot{q}_1}{\partial x_1} - \frac{\partial}{\partial x_1}(V_j T_{1j})\right)\\ &\qquad - (\kappa - 2)\left(\frac{\partial(\rho K V_1)}{\partial x_1} + \frac{\partial(\rho K)}{\partial t}\right).\end{aligned} \tag{13.134}$$

which is approximated as:

$$\begin{aligned}\frac{\partial P_k}{\partial t} &= -\frac{\kappa_k}{\Delta x}\left(\frac{\dot{m}_{i+1} p_{i+1}}{\rho_{i+1} S_{i+1}} - \frac{\dot{m}_i p_i}{\rho_i S_i}\right) - (\kappa_k - 1)\left(\frac{\dot{m}_k q_k}{\Delta V} + c_{f_k}\frac{\dot{m}_{i+1}\dot{m}_k^2}{2 D_{h_{i+1}} S_{i+1} \rho_k^2}\right)\\ &- (\kappa_k - 2)\frac{\dot{m}_k}{\rho_k S_k^2}\left(\frac{1}{2}\frac{\dot{m}_k}{\rho_k}\frac{1}{\Delta x}\left(\frac{\dot{m}_{i+1}}{S_{i+1}} - \frac{\dot{m}_i}{S_i}\right) + \frac{\partial \dot{m}_{i+1}}{\partial t}\right)\\ &- \frac{(\kappa_k - 2)}{2\Delta x}\left(\frac{\dot{m}_{i+1}^3}{\rho_{i+1}^2 S_{i+1}^3} - \frac{\dot{m}_i^3}{\rho_i^2 S_i^3}\right)\end{aligned} \tag{13.135}$$

with:

$$\rho_k = \frac{1}{R}\left(\frac{p_{i+1} + p_i}{T_{i+1} + T_i}\right),\; c_{p_k} = \frac{H_{i+1} - H_i}{T_{i+1}^* - T_i^*},\; \kappa_k = \frac{c_{p_k}}{c_{p_k} - R}.$$

13.3 Numerical Treatment

The above partial differential equations can be reduced to a system of ordinary differential equations by a one-dimensional approximation. The simulation of a complete aero-thermodynamics system is accomplished by combining individual components that have been modeled mathematically. The result is a system of ordinary differential equations that can be dealt with numerically. For weak transients, Runge-Kutta or Predictor-Corrector procedures may be used for the solution. When strong transient processes are simulated, the time constants of the differential equation system can differ significantly so that difficulties must be expected with stability and convergence with the integration methods. An implicit method avoids this problem. The system of ordinary differential equations generated in a mathematical simulation can be represented by:

$$\frac{d\mathbf{X}}{dt} = G(\mathbf{X}, t) \tag{13.136}$$

with $\mathbf{X}$ as the state vector sought. If the state vector $\mathbf{X}$ is known at the time t, it can be approximated as follows for the time $t + dt$ by the trapezoidal rule:

$$\boldsymbol{X}_{t+\Delta t} = \boldsymbol{X}_t + \frac{1}{2}\Delta t(G_{t+\Delta t} + G_t). \tag{13.137}$$

Because the vector $\boldsymbol{X}$ and the function G are known at the time t, i.e., $\boldsymbol{X}_t$ and G_t are known, Eq. (13.137) can be expressed as:

$$\boldsymbol{X}_{t+\Delta t} - \boldsymbol{X}_t - \frac{1}{2}\Delta t(G_{t+\Delta t} + G_t) = F(\boldsymbol{X}_{t+\Delta t}). \tag{13.138}$$

As a rule, the function F is non-linear. It can be used to determine $\boldsymbol{X}_{t+dt}$ by iteration when $\boldsymbol{X}_t$ is known. The iteration process is concluded for the time $t + dt$ if the convergence criterion

$$\frac{\boldsymbol{X}_i^{(k+1)} - \boldsymbol{X}_i^{(k)}}{\boldsymbol{X}_i^{(k+1)}} < \varepsilon \tag{13.139}$$

is fulfilled. If the maximum number of iterations, $k = k_{\max}$, is reached without fulfilling the convergence criterion, the time interval Δt is halved, and the process of iteration is repeated until the criterion of convergence is met. This integration process, based on the implicit one-step method described by Liniger and Willoughby [234] is reliable for the solution of stiff differential equations. The computer time required depends, first, on the number of components in the system and, second, on the nature of the transient processes. If the transients are very strong, the computer time can be 10 times greater than the real time because of the halving of the time interval. For weak transients, this ratio is less than 1.

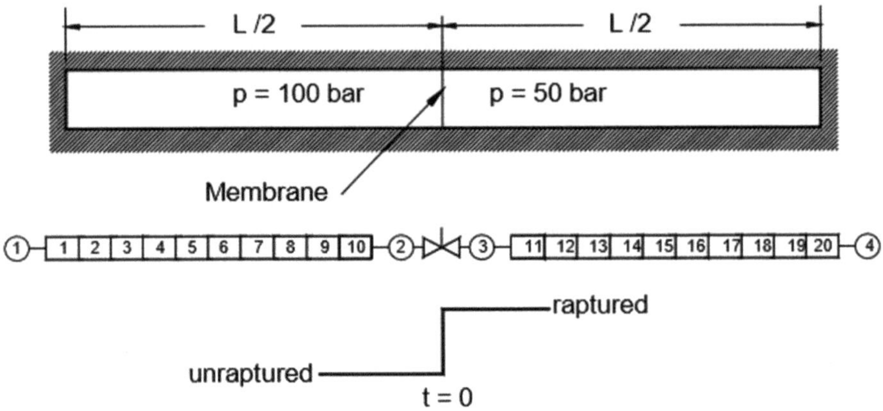

Fig. 13.35 Simulation schematic of a shock tube with a membrane separating the two pressure regions

13.3.1 Unsteady Compressible Flow: Example: Shock Tube

Dynamic Behavior of a shock tube exhibits a representative example of a compressible unsteady flow situation. The shock tube, Fig. 13.25, under investigation has a length of $L = 1$ m and a constant diameter $D = 0.5$ m. The tube is divided into two equal length compartments separated by thin a membrane. The left compartment has a pressure of $p_l = 100bar$, while the right one has a pressure of $p_r = 50bar$. Both compartments are under the same temperature of $T_l = T_r = 400K$.

The working medium is dry air, whose thermodynamic properties, specific heat capacities, absolute viscosity, and other substance quantities change during the process and are calculated using a gas table integrated into the computer code. The pressure ratio of 2 to 1 is greater than the critical pressure ratio and allows a shock propagation with the speed of sound. As shown in Fig. 13.35, each half of the tube is subdivided into 10 equal pieces. The corresponding coupling *plena* 1, 2, and thus, the left half, of the tube are under pressure of 100 bar, while the right half with the plena 3 and 4 are under the pressure of 50 bar. The membrane is modeled by a throttle system with a ramp that indicates the cross-sectional area shown underneath the throttle. The sudden rupture of the membrane is modeled by a sudden jump of the ramp.

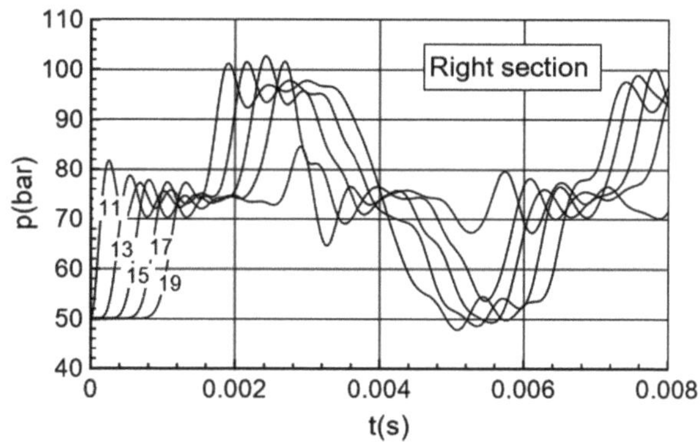

Fig. 13.36 Pressure transients within the shock tube. Right section includes all tube sections initially under high pressure of 50 bar, while the left section includes those initially at 100 bar

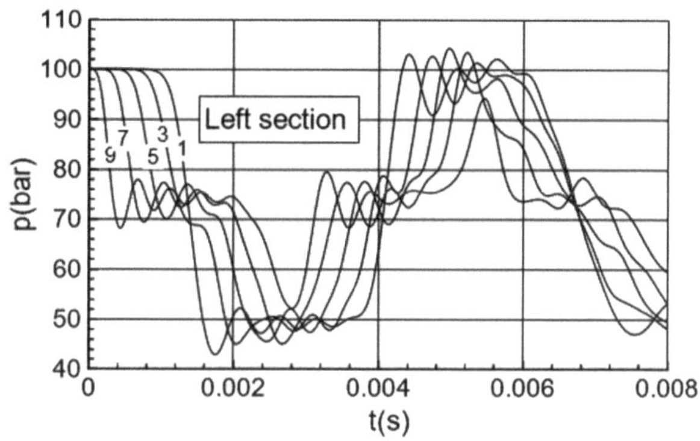

Fig. 13.37 Pressure transients within the shock tube. Left section includes all tube sections initially under high pressure of 100 bar, while the right section include those initially at 50 bar

13.3.2 Shock Tube Dynamic Behavior

13.3.2.1 Pressure Transients

The process of expansion and compression is initiated by suddenly rupturing the membrane. At time $t = 0$, the membrane is ruptured which causes strong pressure, temperatures, and thus, mass flow transients. Since the dynamic process is primarily determined by pressure, temperature, and mass flow transients, only a few representative results are discussed, as shown in Figs. 13.36, 13.37, 13.38, 13.39, 13.40 and 13.41.

Fig. 13.36 shows the pressure transients within the left sections 1–9. As curve 9 shows, the section of the tube that is close to the membrane reacts with a steep expansion wave. On the other hand, the pressure within the pipe section ahead of the shock, Fig. 13.37, curve 11, increases as the shock passes through the section. Oscillatory behavior is noted as the shock strength diminishes. The pipe sections that are farther away from the membrane, represented by curves 7, 5, 3, and 1 on the left and curves 13, 15, 17, and 19 on the right section, follow the pressure transient with certain time lags. Once the wave fronts have reached the end wall of the tube, they are reflected

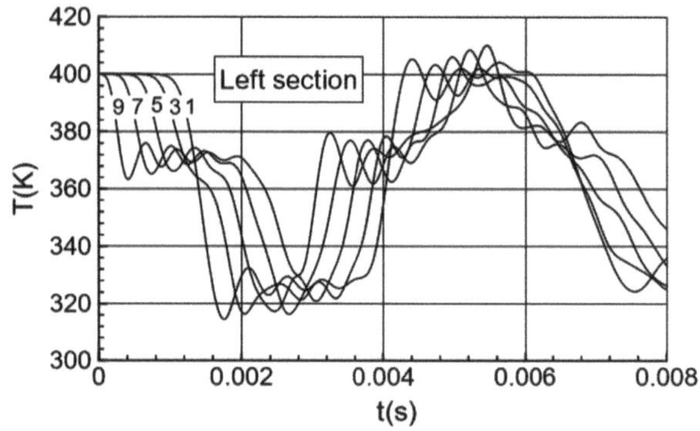

Fig. 13.38 Temperature transients within the left sections of the tube. Left and right sections includes all tube sections initially under temperature of 400 K

as compression waves. The aperiodic compression-expansion process is associated with a propagation speed which corresponds to the speed of sound. The expansion and compression waves cause the air, which was initially at rest, to perform an aperiodic oscillatory motion. Since the viscosity and the surface roughness effects are accounted for by introducing a friction coefficient, the transient process is of dissipative nature.

13.3.2.2 Temperature Transients

Fig. 13.38 shows the temperature transients within the left sections 1–9. As curve 9 shows, the section of the tube that is close to the membrane reacts with a steep temperature decrease. The pipe sections that are farther away from the membrane, represented by curves 7, 5, 3, and 1 on the left and curves 13, 15, 17, and 19 on the right section, follow the temperature transient with certain time lags. Once the shock waves have reached the end wall of the tube, they are reflected as compression waves where the temperature experiences a continuous increase.

Slightly different temperature transient behavior of the right sections are revealed in Fig. 13.39. Compared to the temperature transients of the left sections, the right sections temperature transients seem to be inconsistent. However, a closer look at the pressure transients explains the physics underlying the temperature transients. For this purpose we consider the pressure transient curve 11, in Fig. 13.39. The location of this pressure transient is in the vicinity of the membrane's right side with the pressure of 50 bar. Sudden rapture of the membrane simulated by a sudden ramp (Fig. 13.35) has caused a steep pressure rise from 50 bar to slightly above 80 bar. This pressure rise is followed by a damped oscillating wave that hits the opposite wall and reflects back with an initially increased pressure followed by a damped oscillation. This behavior is in temperature distribution where the pressure rise causes a temperature increase and vice versa. The temperature transients at downstream locations 12–20 follow the same trend.

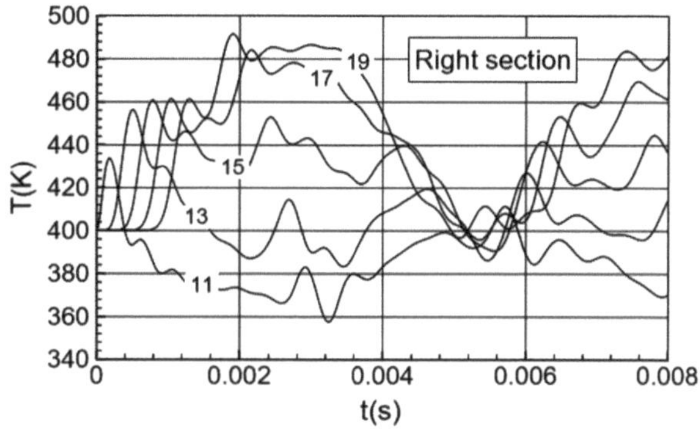

Fig. 13.39 Temperature transients within the right sections of the tube. Left and right sections includes all tube sections initially under temperature of 400 K

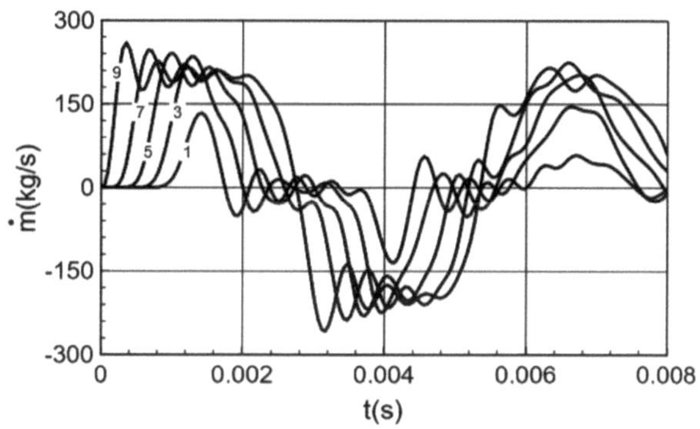

Fig. 13.40 Mass flow transients within left section of shock tube. The part includes all tube sections initially under high pressure of 100 bar, while the right part includes those initially at 50 bar

13.3.2.3 Mass Flow Transients

Figure 13.40 shows the mass flow transients within the left section of the tube. The steep negative pressure gradient causes the mass contained within the tube to perform aperiodic oscillatory motions. During the expansion process, curve 1, mass flows in the positive x-direction. It continues to stay positive as long as the pressure in individual sections are above their minimum. This means that the shock front has not reached the right wall yet. Once the shock front hits the right wall, it is reflected initiating a compression process that causes the mass to flow in the negative x-direction.

Figs. 13.36, 13.37, 13.38, 13.39, 13.40 and 13.41 clearly show the dissipative nature of the compression and expansion process that results in diminishing the wave amplitudes and damping the frequency. The degree of damping depends on the magnitude of the friction coefficient c_f that includes the Re-number and surface roughness effects. For a sufficiently long computational time, the oscillations of pressure, temperature, and mass flow will decay. For $c_f = 0$, the a-periodic oscillating motion persists with no decay.

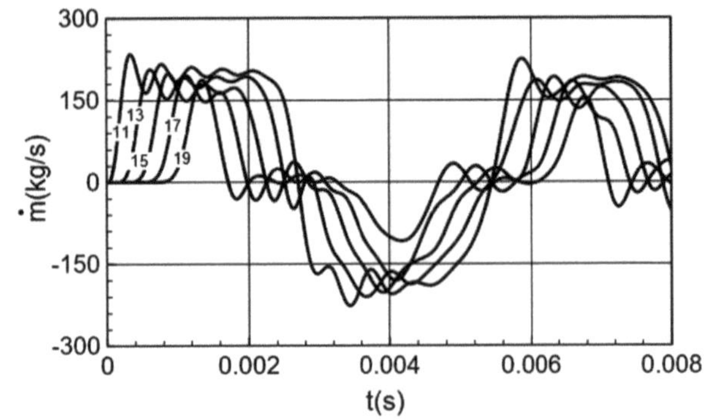

Fig. 13.41 Mass flow transients within the shock tube. The right section include all tube sections initially under pressure of 50 bar, while the left section include those initially at 100 bar

13.4 Problems and Projects

Problem 13.1 The supersonic flow at the inlet of a plane channel, Fig. 13.42, generates two crossing oblique shocks of equal strengths as shown below. The shocks are not reflected at the corners of the convergent part of the inlet (deflection angle $\delta = 10°$). The undisturbed Mach number is $M_1 = 3$, the undisturbed pressure $p_1 = 1$ bar. The working medium is air considered as an ideal gas with $\kappa = 1.4$, $R = 287$ J/(kg K)).

Problem 13.2 A wedge with a thin plate in front of it, Fig. 13.43, has an angle of $16°$ and is subjected to a plane supersonic air flow. The inlet incoming flow is parallel to the plate such that the plate's leading edge causes only a small perturbation.
(a) The angle between the thin plate upper surface and the Mach wave is $45°$. Determine the flow Mach number M_1.
(b) Find the shock angle Θ, the Mach number M_2 downstream of the first oblique shock, the pressure ratio p_2/p_1, and the temperature ratio T_2/T_1.
(c) Sketch the streamlines.

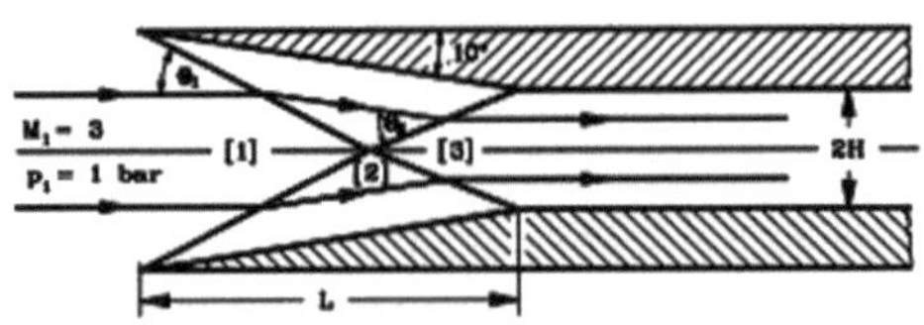

Fig. 13.42 Inlet of a plane channel

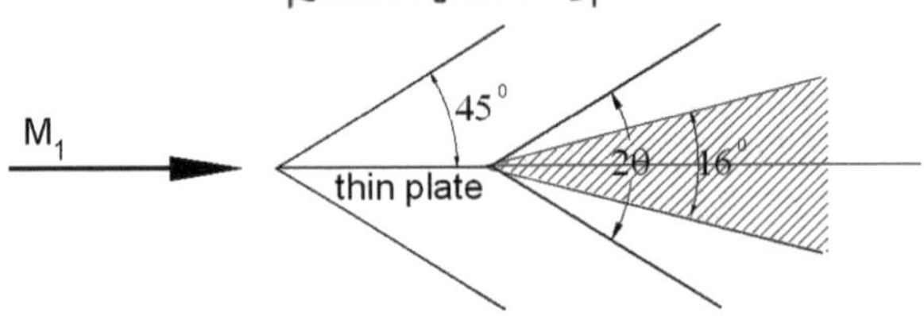

Fig. 13.43 Wedge with a thin plate in front of it

(a) Determine the shock angle Θ_1 of the weak shocks before crossing and the Mach number M_2 in the region between the shocks.
(b) Find the shock angle Θ_2 of the weak shocks after crossing and the Mach number M_3 downstream of the shocks.
(c) Find the pressure at station [3] behind the shocks.
(d) Calculate the entropy increase.
(e) Find the ratio L/H such that the sketched flow pattern can be established.

Problem 13.3 The lower wall of a plane channel turns at [A] and [B] reducing the channel height from h_1 to h_3 (see also Problem 10.4–13 for incompressible channel flow). The working medium is an ideal gas ($\kappa = 1,4$) with the Mach number $M_1 = 5.0$ (Fig. 13.44).
(a) For a given h_1 determine the distance l between the points A and B such that the downstream flow at point [3] is parallel and uniform. Find the channel height h_3.
(b) Find the value of the downstream Mach number M_3.

Problem 13.4 Air as an ideal gas ($\kappa = 1.4$, $R = 287$ J/kg K) flows through a plane channel, whose upper contour is shaped like a streamline of a Prandtl-Meyer flow. The flow is deflected from a given state [1] ($M_1 = 1.6$, $p1 = 0.4$ bar, $T_1 = 250$ K) by a centered wave with a deflection angle of $\delta = 306o$. The channel height upstream of the deflection is $h_1 = 0.3$ m. Fig. 13.45
(a) Determine the flow velocity u_1 and the mass flux $\dot{m}$ (per unit of depth) through the channel.
(b) Give the coordinates of point B at which the curvature of the upper channel contour starts.
(c) Determine M_2, p_2, T_2, ϱ_2 and u_2.

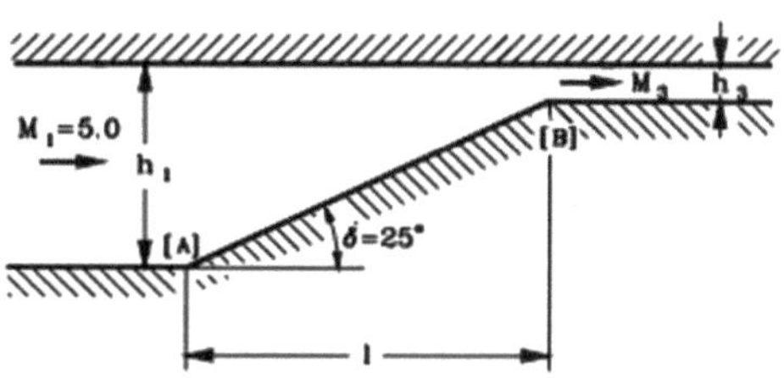

Fig. 13.44 Supersinic flow in convergent channel

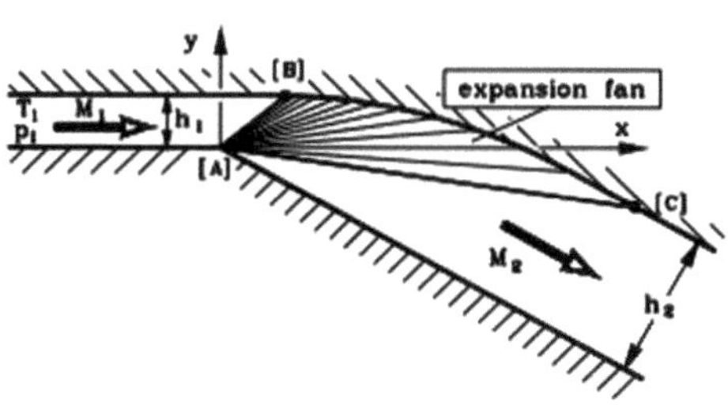

Fig. 13.45 Centered expansion wave around a convex corner

(d) Give the equation of the upper channel contour. Which end height h_2 has the channel? Examine the results using the continuity equation.

Project 13.1 A shock tube with the configuration shown in Fig. 13.46 has two separate compartments with the pressure and temperature of the left compartment greater than those of the right compartment. The pressure ratio is above the critical one. Using dry air as the working medium, for which the ideal gas equation holds, write a source code with pipe length, pipe diameter, pipe friction coefficient, pressure and temperature ratios as input parameter. Investigate (a) the effect of friction factor on the shock oscillation, (b) the effect of temperature ratio on the mixing process after the shock. As in Fig. 13.45, assume a sudden ramp for the membrane rapture. Furthermore, each compartment can be subdivided into 10 subsections that are joined together via plena 1–11 and 12–21. The volume of each plenum consists of half of the volume of each pipe attached to the plenum. Hint: The resulting set of differential equations is of *stiff* nature. Thus, the Runge-Kutta or Predictor-Corrector solvers may create numerical stability problems (Fig. 13.46).

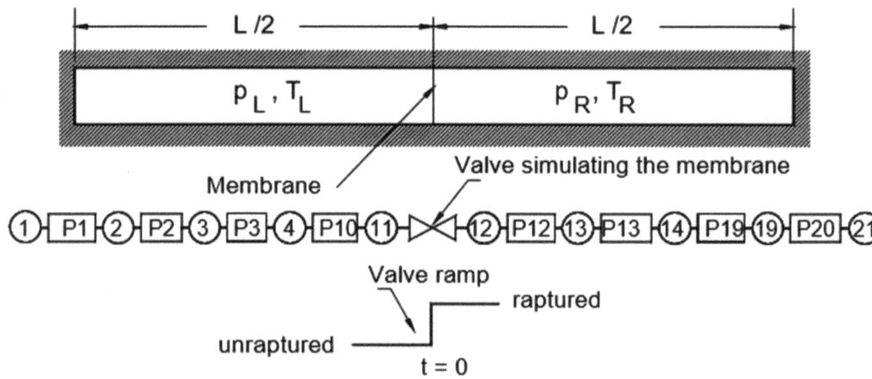

Fig. 13.46 Subdividing each compartment into 10 pipes with the corresponding plena volume

Chapter 14
Flow Measurement Techniques, Calibration

In the following sections, we introduce the basic essentials for flow measurement. As mentioned in previous chapters, the flow in engineering applications generally may be three-dimensional, transitional, turbulent and, in some cases, laminar. If the objective of a flow measurement is to obtain the time dependent quantities, techniques must be applied that are capable of capturing high frequency flow details. In this case, hot wire anemometry (HWA) or Laser Doppler Anemometry (LDA) can be used. If, on the other hand, obtaining the time averaged flow quantities is the objective of a flow measurement, pneumatic probes can be used. It is one thing to use either measurement techniques with the corresponding hardware for acquiring data and another thing to analyze the data. The latter requires detailed calibrations.

For the reasons discussed below, HWA is preferably used for measuring high frequency data. Regarding the pneumatic flow measurements, using five-hole probes is the standard method of acquiring the velocity components and total and static pressure distributions within a flow field. In the following sections, we present simple and effective calibration methods for HWA and five-hole probes.

14.1 Measurement of Time Dependent Flow Field Using HWA

HWA exhibits an effective and relatively simple measurement technique to acquire high frequency flow velocity data. The sensing element of HWA is the hot wire probe that may consist of one, two or three wires Fig. 14.1. Compared with the LDA, its handling is much easier, particularly for a novice, and its traversing system is less complicated. It is also relatively cheaper compared to LDA. Its frequency range is almost twice the range of LDA. Furthermore, the size of the hot-wire sensor and the measuring volume of LDA determines the accuracy of the data. While the diameter of a typical hot-wire sensor ranges between 5 and 15 μm with a length of 1–2 mm,

M. T. Schobeiri, *Advanced Fluid Mechanics and Heat Transfer for Engineers and Scientists*, https://doi.org/10.1007/978-3-030-72925-7_14

depending upon their applications, the size of a typical measuring volume of LDA is much larger (50 μm by 0.25 mm). For boundary layer measurement, it is imperative to be able to traverse the probe as close to the wall as possible (<0.1 mm). Another important aspect of using HWA is its signal-to-noise ratio which is substantially lower than the LDA.

While HWA has the above advantages, the following problems are associated with its application: It cannot be used in high temperature environments, particularly when there is a combustion involved. The sensing wire is sensitive to deposition of impurities that impact its frequency response, requiring frequent calibration. It is very delicate to handle and prone to breakage. Furthermore, the sensing wires are insensitive to a flow reversal. This restricts its applicability to flows at high turbulence intensity levels. These and other related issues are extensively discussed in an excellent book by Bruun [235]. A detailed treatment of hot wire-anemometry with a broad range of compiled knowledge and experience of many researchers using hot wire anemometry is found in [235].

14.1.1 Probe Type, Wire, Film Arrangements

Based on the individual flow field and its corresponding velocity components, sensors with a single wire, two wires (x-wire) and three wires can be used. As an example, Fig. 14.1 shows a single wire and an x-wire probe. Many different probe configurations are found in manufacturers' catalogs (TSI, DANTEC).

A single wire probe, Fig. 14.1a, consists of a shaft and two prongs, between which the sensing wire is soldered. According to [235], the wire material is generally tungsten with a temperature coefficient resistivity of 0.0036 ($°C^{-1}$), a strong tensile strength of 250,000 Ncm^{-2} but a poor resistance to oxidation above 350 °C. To increase the oxidation resistance, platinum may be used but it has a much weaker tensile strength (35,000 Ncm^{-2}) than tungsten. Figure 14.1b shows the 3-D sketch of the single wire including the velocity vector. The component sensed by the wire is the one which is perpendicular to the y-axis. This, in Fig. 14.1a, is the component $V\cos\alpha$. Figure 14.1c exhibits the position of this component relative to the wire. Figure 14.1d exhibits a miniature x-wire probe designed by TPFL and manufactured by TSI. The x-wire arrangement is shown schematically in Fig. 14.1e with the incoming velocity vector in Fig. 14.1f.

Hot film probes have the same geometry as hot wire ones but are more robust. The sensor is usually coated with quartz or aluminum oxide resulting in a sensor diameter of 25–70 μm compared to 5–15 μm for hot wire. The probe can be used in gases whenever finding the averaged and fluctuation velocity values are the objectives of the measurement. To utilize hot film probes for velocity measurement in water, the sensor must be heavily coated with quartz to provide electrical insulation.

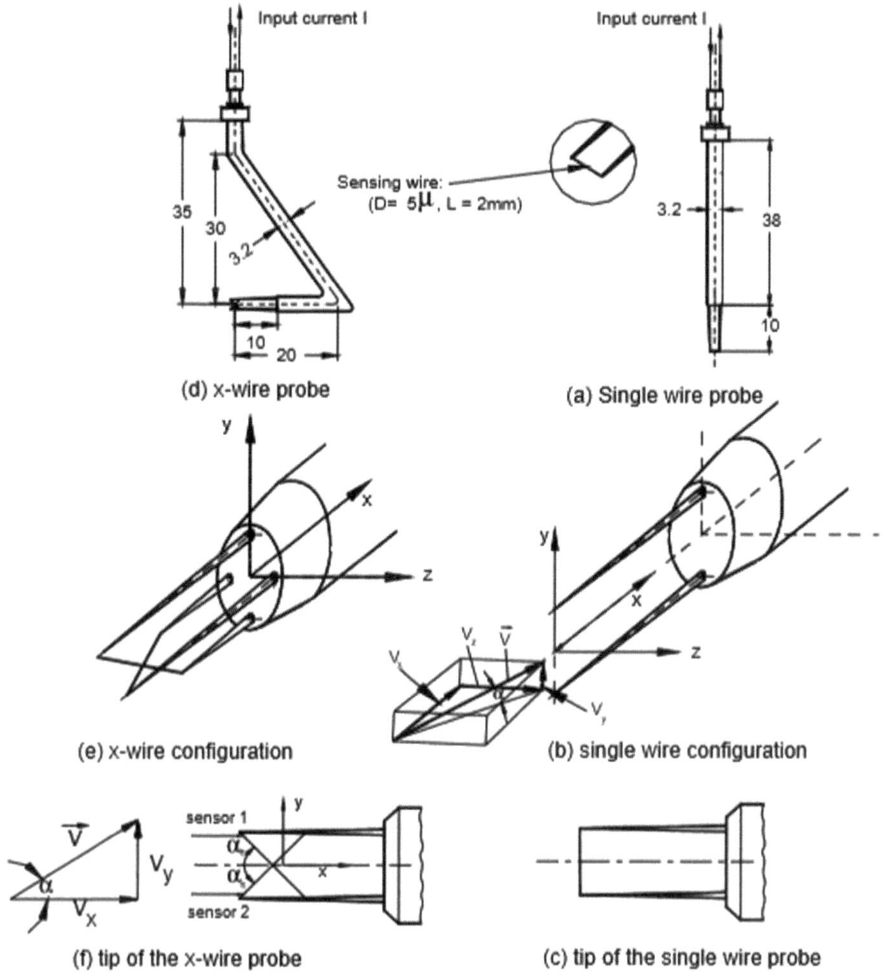

Fig. 14.1 Hot wire probes: single wire (**a**), (**b**) an (**c**); x-wire (**d**), (**e**) and (**f**)

14.1.2 Energy Balance of HW Probes

The working principle of a hot wire/film is based on the conversion of electrical energy into heat and the removal of the latter by means of heat transfer. Electric current I, Fig. 14.1, passes through the sensing wire and generates electric power equivalent to the mechanical energy flow (kJ/s) that follows the energy equation (5.51):

$$\dot{Q} + \dot{W} = \frac{DE}{Dt}. \tag{14.1}$$

Equation (14.1) is the power balance with $\dot{Q}$ and $\dot{W}$ as the thermal and mechanical energy flow (power), respectively. Since there is no macroscopic kinetic energy involved and the effect of gravity is neglected, the total energy equation (5.50) is reduced to

$$E = me = m\left(u + \frac{1}{2}V^2 + gz\right) = mu = mc_w\bar{T}_w \tag{14.2}$$

with m as the mass of wire, c_w the specific heat of the wire material, and $\bar{T}_w$ the averaged wire temperature. While $\dot{W}$ is the electric power added to the system, $\dot{Q}$ is heat rejected from the system, meaning that the sign of $\dot{Q}$ is inherently negative. With Eq. (14.2), the energy equation (14.2) becomes

$$-\dot{Q} + \dot{W} = mc_w\frac{D\bar{T}_w}{Dt} = mc_w\left(\mathbf{V}\cdot\nabla\bar{\mathbf{T}}_\mathbf{w} + \frac{\partial\bar{\mathbf{T}}_\mathbf{w}}{\partial\mathbf{t}}\right) = mc_w\left(\frac{\partial\bar{T}_w}{\partial t}\right) \tag{14.3}$$

with electric power input $\dot{W} = I^2R_w$ and $m = \varrho_w AL$ as the mass of the wire. For a steady state operation, the right-hand side of Eq. (14.3) becomes zero. For an infinitesimal wire length of dx, Eq. (14.3) is written as

$$d\dot{W} = d\dot{Q} + \varrho_w c_w\left(\frac{\partial T_w}{\partial t}\right)Adx. \tag{14.4}$$

Operating the HWA in a calibration facility, a wind tunnel or a test section, the wire will be subjected to convective heat removal, heat conduction, radiation and in case of an unsteady operation, certain amount of the thermal energy will be stored in the wire. Thus, the thermal power (heat rate) removed from the wire $\dot{Q}$ can be divided into

$$\dot{Q} = \dot{Q}_c + \dot{Q}_{cn} + \dot{Q}_r + \dot{Q}_s. \tag{14.5}$$

The convective part $\dot{Q}_c$ generally includes the heat rate due to a forced convection and a natural convection. The latter, however, is negligibly small whenever the test section Re-number is not too small. The convection equation in general is

$$\dot{Q}_c = h(T_w - T_F)A = \pi dh(T_w - T_F)L \tag{14.6}$$

with $h[\mathrm{J/(m^2\ s\ K)}]$ as the heat transfer coefficient, d the wire diameter, A the heat transfer area, L the wire length, T_w and T_F the temperatures of wire and fluid, respectively. For an infinitesimally small length of wire, Eq. (14.6) is written as

$$d\dot{Q}_c = \pi dh(T_w - T_F)dx. \tag{14.7}$$

The conduction portion $\dot{Q}_{cn}$ is derived from the energy equations in Chap. 4, Sect. 4.4.2.

$$d\dot{Q} = -\frac{\nabla\dot{\mathbf{q}}}{\varrho} = \frac{1}{\varrho}k\nabla^2 T \text{ with } \dot{\mathbf{q}} = -k\nabla T \tag{14.8}$$

$$d\dot{Q}_{cn} = -dm_w \frac{\nabla \dot{\mathbf{q}}}{\varrho} = A_w dx k_w \nabla^2 T = A_w dx k_w \frac{\partial^2 T_w}{\partial x^2}. \tag{14.9}$$

Equation (14.8) is heat rate/unit of mass. Considering the mass of the wire, we have with A_w as the wire cross-sectional area and $k_w[J/(msK)]$ the wire thermal conductivity. The radiative heat transfer is written as:

$$d\dot{Q}r = \pi d\sigma\varepsilon(T_w^4 - -T_F^4)dx. \tag{14.10}$$

In Eq. (14.10), $\sigma[\mathrm{J/(ms^2\ K^4)}]$ is the Stefan-Bolzman constant and ε is the emissivity of the wire defined as the total radiant energy emitted per unit area and unit of time divided by the corresponding rate of energy emission from a black body. Finally, the electrical power input is

$$\dot{W} = I^2 R_w \tag{14.11}$$

with $R_w = \varrho_W L/A_w[\Omega]$ as the wire resistance and $\varrho[\Omega\mathrm{m}]$ as its resistivity. Since ϱ is a function of temperature, it can be approximated by function

$$\varrho_w = \varrho_F[1 + \alpha(T_w - T_F)] \tag{14.12}$$

with ϱ_F as the resistivity at T_F and α the temperature coefficient of resistivity. With Eq. (14.12), the wire resistance becomes

$$R_w = R_F[1 + \alpha(T_w - T_F)]. \tag{14.13}$$

In Eq. (14.13), T_F is the wire resistance at a fixed fluid temperature. For a differential element, Eq. (13.11) is reduced to

$$d\dot{W} = I^2 dR_w = I^2 \varrho dx/A_w = I^2 \varrho_F[1 + \alpha(T_w - T_F)]dx/A_w. \tag{14.14}$$

Summarizing Eqs. (14.6), (14.9), (14.10) and (14.14) we have

$$I^2\varrho_F[1 + \alpha(T_w - T_F)]\frac{dx}{A_w} = \pi dh(T_w - T_F)dx - A_w k_w \frac{\partial^2 T_w}{\partial x^2}dx +$$
$$\pi d\sigma\varepsilon(T_w^4 - -T_F^4)dx + \varrho_w c_w \left(\frac{\partial T_w}{\partial t}\right) A dx. \tag{14.15}$$

The wire temperature in Eq. (14.15) is, in general, a function of time and length, $T_w = T_w(t, x)$. For steady state, it reduces to $T_w = T_w(x)$ leading to the following simplified relation for T_w

$$\frac{1}{A_w} I^2 \varrho_F[1 + \alpha(T_w - T_F)] = \pi dh(T_w - T_F) - A_w k_w \frac{d^2 T_w}{dx^2} + \pi d\sigma\varepsilon(T_w^4 - -T_F^4). \tag{14.16}$$

The last term in Eq. (14.16) compared to the remaining three terms is very small, thus, it can be neglected. As a result, we have

$$\frac{I^2\varrho_F[1+\alpha(T_w - T_F)]}{A_w} = \pi dh(T_w - T_F) - A_w k_w \frac{d^2 T_w}{dx^2}. \tag{14.17}$$

Rearranging Eq. (14.17) leads to

$$\frac{d^2 T_w}{dx^2} + \left(\frac{I^2\varrho_0\alpha}{A_w^2 k_w} - \frac{\pi dh}{A_w k_w}\right)(T_w - T_F) + \frac{I^2\varrho_0}{A_w^2 k_w} = 0. \tag{14.18}$$

Setting $T_w - T_F = T$, Eq. (14.18) reduces to

$$\frac{d^2 T}{dx^2} + K_1 T + K_2 = 0 \tag{14.19}$$

with the constants

$$K_1 = \left(\frac{I^2\varrho_0\alpha}{A_w^2 k_w} - \frac{\pi dh}{A_w k_w}\right) \text{ and } K_2 = \frac{I^2\varrho_0}{A_w^2 k_w}. \tag{14.20}$$

Equation (14.19) is an ordinary differential equation with constant coefficients. Its general solution is

$$T = C_1 e^{\sqrt{K_1}x} + C_2 e^{-\sqrt{K_1}x + \frac{K_2}{K_1}}. \tag{14.21}$$

The constants in Eq. (14.21) are determined from the boundary conditions at $BC1 : x = 0 : dT/dx = 0 \Rightarrow C_1 = C_2 = C$ and $BC2 : x = l/2 : T = 0$ While the boundary condition $BC1$ is exact, $BC2$ is an approximation, since at both prongs the temperature $T = T_w - T_F$ is not exactly zero. With $BC2$, we find the value for C and, thus, the solution for Eq. (14.19)

$$T = T_w - T_F = \frac{K_2}{K_1}\left[1 - \frac{\cosh(\sqrt{K_1}x)}{\cosh(\sqrt{K_1}l/2)}\right]. \tag{14.22}$$

As discussed by Bruun [235], Betchov [236] introduced a *cold length* l_c

$$\frac{1}{|K_1|^{1/2}} \equiv l_c = \frac{d}{2}\left(\frac{k_w}{k}\frac{R_w}{R_F}\frac{1}{Nu}\right)^{1/2} \tag{14.23}$$

where R_w, R_F are the sensor resistances at wire and fluid temperatures, respectively. Inserting Eq. (14.23) into Eq. (14.22), we find

$$T = T_w - T_F = \frac{K_2}{|K_1|}\left[1 - \frac{\cosh(x/lc)}{\cosh(l/2lc)}\right] \tag{14.24}$$

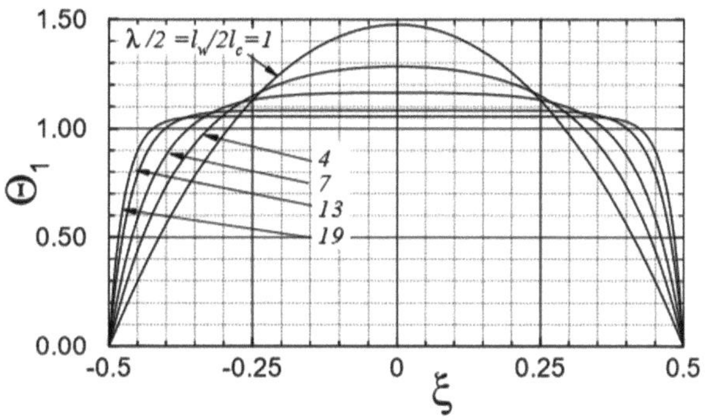

Fig. 14.2 Dimensionless wire temperature distribution Θ_1 relative to its mean as a function of dimensionless wire length with $\lambda/2$ as parameter

with l_c from Eq. (14.23). Integration of Eq. (14.24) along the length l delivers the mean wire temperature

$$\bar{T} = \frac{K_2}{|K_1|}\left[1 - \frac{\tanh(l/2lc)}{(l/2lc)}\right] \tag{14.25}$$

and the maximum temperature is

$$T_m = \frac{K_2}{|K_1|}\left[1 - \frac{1}{\cosh(l/2lc)}\right]. \tag{14.26}$$

Introducing the following dimensionless parameters

$$\xi = \frac{x}{l}, \lambda = \frac{l}{lc}, \frac{x}{l_c} = \xi\lambda, \theta_1 = \frac{T_w - T_F}{\bar{T}_w - T_F}, \theta_2 = \frac{T_w - T_F}{T_{\max} - T_F} \tag{14.27}$$

the results are plotted in Figs. 14.2 and 14.3. Bruun [235] found that for a 5 μm tungsten wire operated at an overheat ratio of $R_w/R_F = 1.8$, l_c will be about $30d$. The range of the overheat ratio is generally provided by the manufacturers (TSI, DANTEC, etc.). Bruun estimated that for a standard probe of an active length of 1.5 mm, the ratio for $\lambda/2$ will be about 4. Figure 14.2 shows the dimensionless wire temperature distribution relative to its mean. As seen, the highest temperature is concentrated in the middle of the wire.

Figure 14.3 shows the dimensionless wire temperature distributions relative to their corresponding maxima. This complementary figure shows that with a larger l/l_c, a more uniform temperature distribution can be achieved that covers a major part of the wire.

Fig. 14.3 Dimensionless wire temperature distribution Θ_1 relative to its maximum as a function of dimensionless wire length with $\lambda/2$ as parameter

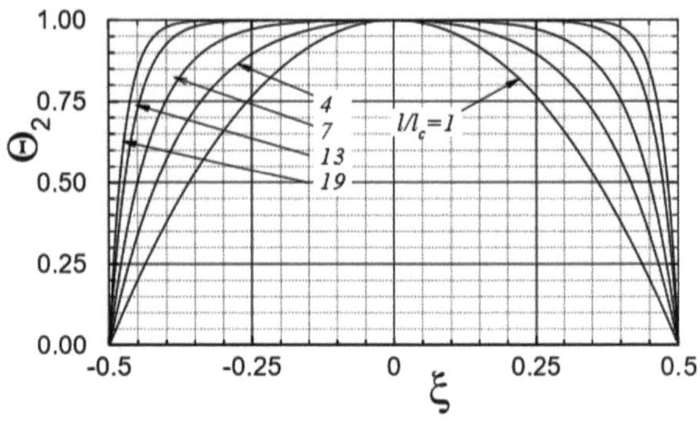

14.1.3 *Heat Transfer of HW Probes*

The heat transfer coefficient h in Eq. (14.20) can be expressed in terms of Nusselt number Nu. A functional relationship for Nu that includes a set of dimensionless parameters such as Reynolds number (Re), Prandtl number (Pr), Grasshof number (Gr) and Mach Number (M) may be written as

$$Nu = f(Re, Pr, Gr, M) \tag{14.28}$$

with

$$Nu = \frac{hd}{k}, Re = \frac{\varrho U d}{\mu}, Pr = \frac{c_p \mu}{k},$$
$$Gr = \frac{g\rho^2 d^3 (T_w - T_F)}{\mu^2}, M = \frac{U}{c} \tag{14.29}$$

with c_p as the specific heat at constant pressure, k the thermal conductivity of the fluid and c the speed of sound. In case of natural convection, the flow velocity is very low and the gravitational force has a major effect on heat transfer with the Grasshof number as the determining parameter. In case of a forced convection, the contribution of the Grasshof number compared to those of Re and Pr is negligible. Similarly, for an incompressible flow, the effect of Mach number can be neglected. Given the above conditions, the functional relation (14.28) is reduced to

$$Nu = f(Re, Pr). \tag{14.30}$$

King [237] was the first to propose the first correlation for HW-convective heat transfer Nusselt number as

$$Nu = A + BRe^{1/2} \tag{14.31}$$

with A and B as calibration coefficients experimentally determined for each fluid. These implicitly include the Prandtl number. Kramer [238] proposed the following correlation

$$Nu = 0.42Pr^{1/5} + 0.57Pr^{1/3}Re^{1/2} \tag{14.32}$$

that is valid for a Re-range of $0.01 < Re < 10{,}000$ and a Pr-range of $0.71 < Pr < 1000$. To replace the heat transfer coefficient h with the Nusselt number, we consider the steady case of Eq. (14.3)

$$\dot{Q} = \dot{W} = I^2 R_w = h(T_w - T_F)A \tag{14.33}$$

and replace the temperature difference $T_w - T_F$ with a resistance difference using Eq. (14.13) that is re-arranged as

$$T_w - T_F = \frac{R_w - R_F}{\alpha R_F}. \tag{14.34}$$

With Eq. (14.34) and the definition of Nusselt number, Eq. (14.33) assumes the form

$$\frac{I^2 R_w}{R_w - R_F} = \frac{hA}{\alpha R_F} = Nu\left(\frac{kA}{d\alpha_F}\frac{1}{R_F}\right). \tag{14.35}$$

For a given fluid, the expression in the above parentheses is a constant. It can be absorbed in the coefficients of Nusselt number as follows

$$A = 0.42\left(\frac{kA}{d\alpha_F}\frac{1}{R_F}\right)Pr^{1/5},\ B = 0.57\left(\frac{kA}{d\alpha_F}\frac{1}{R_F}\right)Pr^{1/3}. \tag{14.36}$$

Thus, Eq. (14.36) is re-written as

$$\frac{I^2 R_w}{R_w - R_F} = A + BRe^{1/2} \tag{14.37}$$

with the coefficients from Eq. (14.36). In practice, it is custom to use a modified version of Eq. (14.37), which is

$$\frac{I^2 R_w}{R_w - R_F} = A + BU^n. \tag{14.38}$$

14.1.4 Calibration Facility

The accurate calibration of sensors is a prerequisite for measurement of any physical quantity. Concerning the fluid mechanics, these quantities are static and total

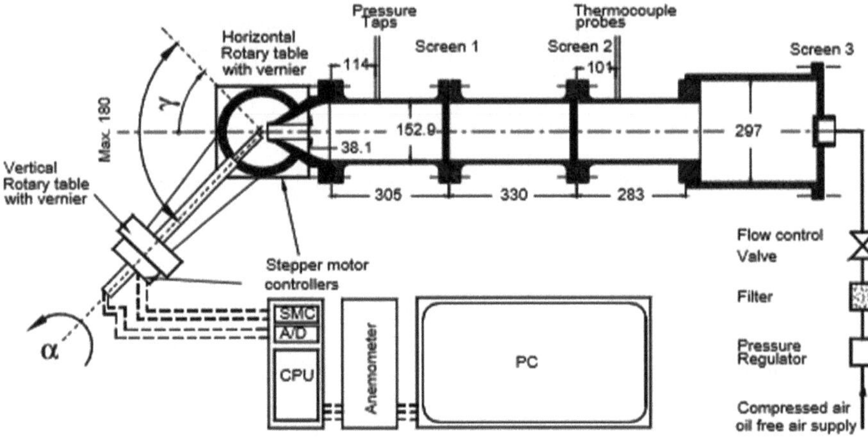

Fig. 14.4 Facility for calibrating hot-wire and five hole probes with stepper motor controller SMC, Analog-Digital Converter A/D and the CPU

pressures and temperatures as well as the velocity vectors with their three components with the corresponding angles. If the time averaged values of these quantities is the objective of the measurements, pneumatic probes are used. If, however, the information about turbulence intensity, velocity, pressure and temperature fluctuations is required, sensors must be used capable of acquiring high frequency signals. Regardless of the nature of the data, understanding the experimental environment, eliminating parasitic effects and properly designing a calibration facility are the first steps towards acquiring accurate data. The facility described below is used for calibrating single-hot wire and x-wire probes for high frequency data acquisition. It is also used to calibrate pneumatic probes. Analog signals from probes are voltage signals to be converted into real physical quantities by means of calibration. The calibration facility is shown in Fig. 14.4. This facility has been used to calibrate the hot wire and five−hole probes.

Compressed air is drawn from a reservoir which passes through a pressure regulator, filter and flow control valve before entering the calibration facility. It consists of a settling chamber followed by a pipe with three axisymmetric sections, each having a diameter of 152.9 mm. A nozzle with an exit diameter of 38.1 mm is attached to the end of the pipe. The inlet and outlet of the nozzle are parallel to the axis of the facility. Several thermocouples are placed in the first section of the calibration facility to measure the air temperature. Also, several pressure taps are placed in the last section of the pipe to measure the pressure difference between static pressure inside the pipe and the atmospheric pressure with a differential pressure transducer. Two lynx stepper motors with controllers are used to automatically vary the pitch and the yaw angles. The stepper motors are mounted on a traversing system with an arm at 90°. The system is designed to place the probe tip in the center of the exiting jet.

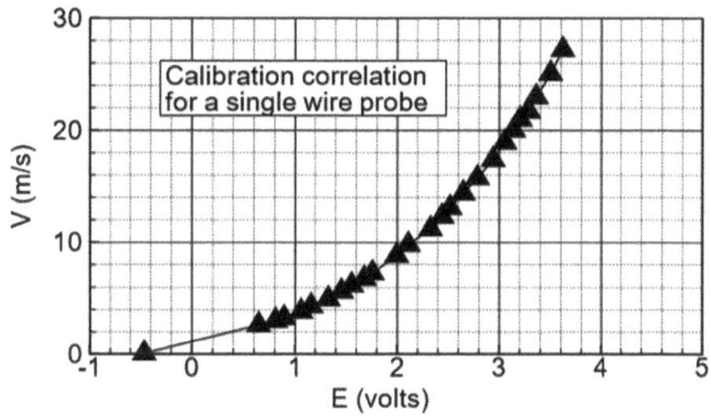

Fig. 14.5 Calibration curve for a single wire probe

14.1.5 *Calibration of Single Hot Wire Probes*

This single wire calibration provides the velocity as a function of the voltage across the wire. During the velocity calibration, a single hot wire probe is kept normal to the flow and the flow velocity is varied. The exit velocity is calculated by the differences between the static pressure at the inlet and the static pressure at the outlet of the nozzle. Using the Bernoulli−equation in conjunction with the continuity equation for incompressible flow, and considering the velocity magnitude inside the calibration tunnel compared to the exit velocity as negligible, we calculate exit velocity from:

$$V = \sqrt{\frac{2\Delta p}{\varrho} \frac{d_1^4}{d_1^4 - d_2^4}}. \tag{14.39}$$

Varying the exit velocity from 0 to the desired maximum, for each velocity, a set of data from constant temperature anemometer (CTA), temperature, and pressure voltage signals are taken. Since the flow velocity vector is normal to the wire axis, the output voltage is a function of the velocity, $E = f(V)$ only. Thus, a calibration equation can be established between the hot wire voltage E and the velocity V by a fourth order polynomial:

$$V = a_0 + a_1 E + a_2 E^2 + a_3 E^3 + a_4 E^4. \tag{14.40}$$

The coefficients are calculated using the least-square method. As an example, the results of a calibration is shown in Fig. 14.5.

14.1.6 *Calibration of Single Hot Wire Probes*

X-probes are widely used for measuring two components of the flow velocity. Various methods are employed for interpreting the signals from two sensor probes and almost all methods relate the output voltage to the effective cooling velocity V_e. The effective velocity is related to the actual flow velocity V in different forms. The most accurate representation of the directional response of a hot-wire or hot-film

$$V_e^2 = V_N^2 + k^2\, V_T^2 \tag{14.41}$$

was introduced by Hinze [239]. This equation takes into account the cooling due to both normal and tangential velocity components. Webster [240] experimentally found the values of k in Eq. (14.41) for various l/d ratios for a single hot-wire sensor. Jorgensen [241] investigated the dependence of k on a yaw angle and found that the value of k varies with the yaw angle. Bradshaw [242] introduced a calibration method based on an effective angle defined as:

$$V_e = V \cos \alpha_e. \tag{14.42}$$

Bruun et al. [244] used a conventional calibration method with a constant k-factor in Eq. (14.41) and compared it with Bradshaw's method, which is based on Eq. (14.42). As a result, Bruun showed that the conventional method based on Eq. (14.41) gives the smallest error over the complete angle range. Lekakis [245] obtained the k values at a given velocity from a yaw calibration and found that a constant value for k, determined by a least square-fit of all k values, provides a good representation of the probe angular response. In a similar way, he determined the value of k for different velocities. Using the above k values, Lekakis developed an analytical method for the calculation of the velocity and its components.

The use of constant k at a given velocity, which is common to the previously described methods, introduces significant errors particularly at higher yaw angles. Schröder [246] employed a simple calibration method by introducing an ideal flow angle, for which k was equal to zero, and applied the method to X-hot-wire probes. A yaw calibration covering the entire angle range was performed at a single velocity. Compared to constant k-factor based methods, the method by Schröder was more accurate to a wide range of yaw angles ($-40° < \alpha < +40°$). A more detailed description of this method is given later.

Another alternative to reduce calibration errors is to utilize a full-range velocity-angle calibration technique in conjunction with look-up tables. Such a calibration technique was first introduced by Willmarth and Bogar [247], further developed by Johnson and Eckelmann [248] and Leuptow et al. [249], and modified by Browne et al. [250]. The generation of such tables as a result of a full-range velocity-angle calibration, however, requires a significant amount of calibration time and effort. Furthermore, the implementation of the above tables into a corresponding data reduction and analysis program necessitates excessive computational overhead particularly in connection with unsteady flow measurements. The need to process high volume unsteady flow data motivated John and Schobeiri [251] to develop a simple method that significantly reduces the calibration time and accounts for high accuracy. The need for developing such a method was particularly indicated for flow situations where the angle as well as the velocity undergo significant changes. Such situations are frequently encountered in turbomachinery wake flows immediately downstream of the stator or rotor trailing edge plane. Similar flow situations are also observed in wakes immediately downstream of a stationary or rotating cylinder.

This section presents an improved calibration technique based on the ideal flow angle previously mentioned. A correction procedure is introduced which allows accurate measurement at lower velocities. The calibration technique is applicable to x-hot-wire and hot-film probes, however, the measurements presented in this section were performed with a hot-film probe.

14.1.6.1 Calibration at a Single Free Stream Velocity, Equation

Using the facility described in Sect. 13.1.3, calibration measurements were performed with an X-hot-film probe which was connected to the two channels of a TSI IFA 100 constant temperature anemometer (CTA) system. The diameter of the platinum hot-film sensor was 25 μm with the two sensors separated by a distance of approximately 1 mm. The sensors were operated at an overheat ratio of 1.5. The analog signals from the pressure transducer, thermocouple and hot-film anemometer were transferred to the input channels of an analog/digital converter (A/D) board plugged into the expansion slot of a personal computer, which is used for the data acquisition and analysis.

The probe geometry, the flow velocity V and direction α, as well as the components V_x and V_y, are shown in Fig. 14.1. Each of the sensors of the X-hot-film probe under investigation had an angle $\alpha_s = 45°$ to the X-axis. The components of the velocity V along and perpendicular to the probe axis are V_x and V_y respectively. The effective cooling velocity V_{ej} is approximated as a fourth-order polynomial function of the anemometer output voltage E_j

$$V_{ej} = a_{0j} + a_1 E_j^2 + a_{3j} E_j^3 + a_{4j} E_j^4 \tag{14.43}$$

where the coefficients a_{ij} are obtained by a least-squares fit. The angle response equations for sensors 1 and 2 are derived from Eq. (14.41) as

$$V_{e1}^2 = V^2[\sin^2(\alpha_s + \alpha) + k_1^2 \cos^2(\alpha_s + \alpha)] \tag{14.44}$$

$$V_{e2}^2 = V^2[\sin^2(\alpha_s - \alpha) + k_2^2 \cos^2(\alpha_s - \alpha)]. \tag{14.45}$$

V_{e1} and V_{e2} are the effective cooling velocities and k_1 and k_2 are the yaw coefficients for sensors 1 and 2. As mentioned in the introduction, Schröder [246] defined an ideal angle α_{id} for which k_j is equal to zero. Applying this definition to Eqs. (14.44) and (14.45) leads to

$$\alpha_{id} = \tan^{-1}\left(\frac{V_{e1} - V_{e2}}{V_{e1} + V_{e2}} \tan \alpha_s\right). \tag{14.46}$$

For probes with $\alpha_s = 45°$, Eq. (14.46) can be written as:

$$\alpha_{id} = \tan^{-1}\left(\frac{V_{e1}}{V_{e2}}\right) - 45. \tag{14.47}$$

Schröder also introduced a nondimensional parameter H that relates the effective cooling velocities to the actual velocity. It is defined by:

$$H = \frac{V^2}{V_{e1}^2 + V_{e2}^2}. \tag{14.48}$$

The yaw angle calibration determines the values of α_{id} and H for various α. H and α are represented by a fifth-order polynomial function of α_{id} by a least-squares fit, i.e.,

$$H = b_0 + b_1\alpha_{id} + b_2\alpha_{id}^2 + b_3\alpha_{id}^3 + b_4\alpha_{id}^4 + b_5\alpha_{id}^5 \tag{14.49}$$

$$\alpha = c_0 + c_1\alpha_{id} + c_2\alpha_{id}^2 + c_3\alpha_{id}^3 + c_4\alpha_{id}^4 + c_5\alpha_{id}^5. \tag{14.50}$$

In order to increase the curve fit accuracy, we introduced a new function H^* defined by:

$$H^* = \frac{V\cos\alpha}{\sqrt{V_{e1}^2 + V_{e2}^2}}. \tag{14.51}$$

Similar to H, the new function H^* is also represented by a fifth order polynomial function of α_{id}, i.e.,

$$H^* = d_0 + d_1\alpha_{id} + d_2\alpha_{id}^2 + d_3\alpha_{id}^3 + d_4\alpha_{id}^4 + d_5\alpha_{id}^5. \tag{14.52}$$

The variation of α with α_{id} is shown in Fig. 14.6. For comparison, the line $\alpha = \alpha_{id}$ is also plotted. It can be seen that the difference between α and α_{id} is very high at higher absolute values of α.

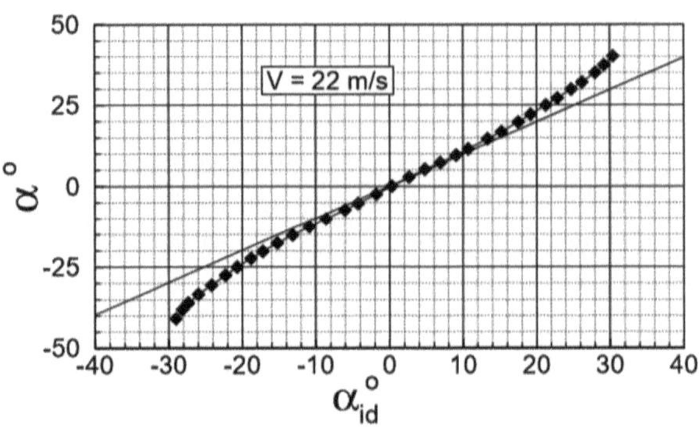

Fig. 14.6 Variation of α with α_{id} from yaw calibration at 22 m/s. For uncertainty estimates see Table 14.1

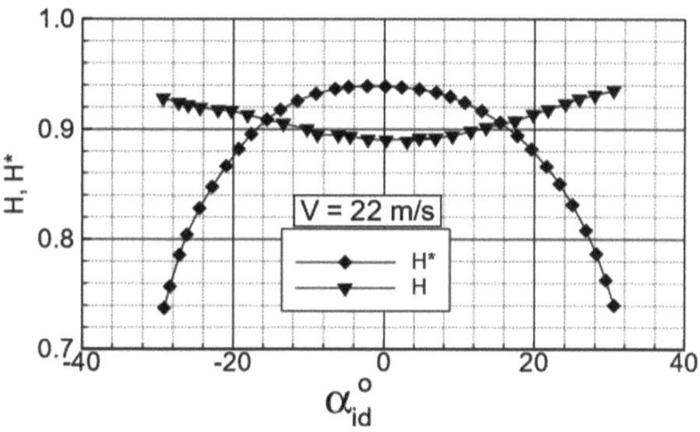

Fig. 14.7 Variation of H and H^* with α_{id} obtained from yaw calibration at 22m/s. For uncertainty estimates see Table 14.1

The values of H and H^* are plotted against α_{id} in Fig. 14.7. The solid lines in Figs. 14.6 and 14.7 show the fifth-order polynomial curve fit. As seen from Fig. 14.7, the function H^* gave a lower scatter of data points compared to the function H.

14.1.6.2 Calibration Procedure

We obtain the coefficients of Eq. (14.43) for each of the sensors by keeping the sensors normal to the flow and varying the velocity. The yaw angle calibration is carried out for α from $-40°$ to $40°$ at a constant velocity V. The functions H^* and α_{id} are calculated from Eqs. (14.51) and (14.47). The coefficients c_i in Eq. (14.50) and d_i in Eq. (14.52) are found by least-squares fit of α_{id} against and H^*, respectively. The values of α_{id} and H^* are found to have good reproducibility. As compared to α_{id} and H^*, the coefficients of velocity calibration in Eq. (14.43) can drift and frequent calibration is essential. It is always preferable to perform the calibration in the test section where actual measurements are carried out. Many times the specific geometry of the test section and the probes makes it impossible to yaw the probe such that the sensors are normal to the flow. However, the velocity calibration can be conducted simultaneously for both the sensors at any angle, provided c_i and d_i are already determined by yaw angle calibration and the velocity V is known and can be varied. For a known angle α, α_{id} is obtained from Eq. (14.50) using the Newton-Raphson method. The corresponding value of H^* is determined using Eq. (14.52). The effective cooling velocities are obtained from the following relations derived from Eqs. (14.47) and (14.51):

$$V_{e1} = \frac{V \cos\alpha \tan(\alpha_{id} + 45)}{H^* \sqrt{1 + \tan^2(\alpha_{id} + 45)}}$$

$$V_{e2} = \frac{V \cos\alpha}{H^* \sqrt{1 + \tan^2(\alpha_{id} + 45)}}. \tag{14.53}$$

Velocity calibrations with $\alpha = -45°$, 0 and $+45°$ are shown in Fig. 14.8.

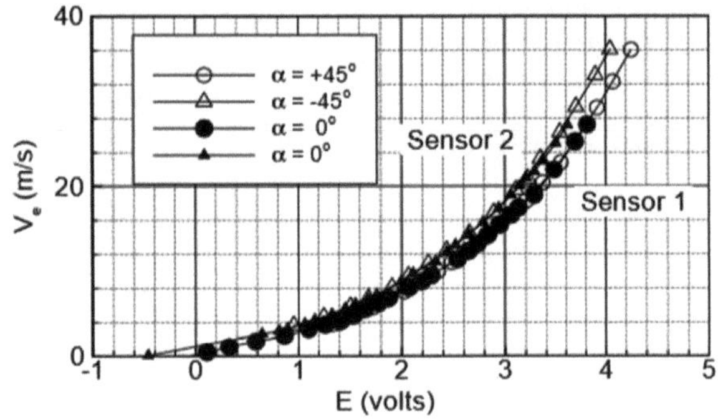

Fig. 14.8 Effective velocities of sensors 1 and 2. For uncertainty estimates see Table 14.1

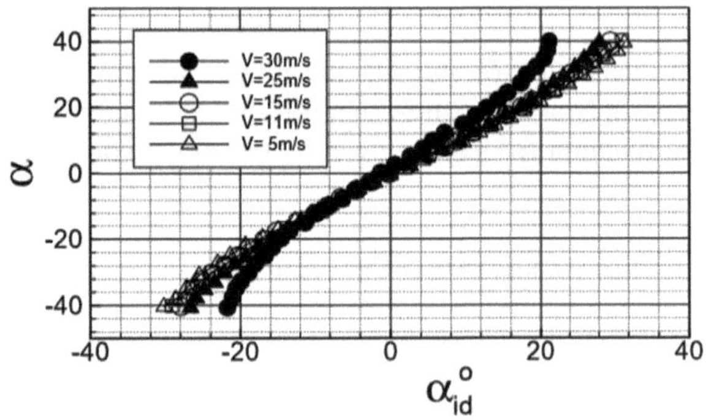

Fig. 14.9 Distribution of α_{id} versus α for different velocities. For uncertainty estimates see Table 14.1

The flow velocity V is norm to sensors 1 and 2 when α is $+45°$ and $-45°$, respectively. The two calibration curves for each of the sensors are almost identical, which proves the validity of the above mentioned calibration procedure.

14.1.6.3 Calibration at Different Velocities

To investigate the effect of velocity on the yaw response, a yaw angle calibration was carried out at different velocities and the results are plotted in Fig. 14.9.
Note that the value of α_{id} varies with velocity V. For a given α, the variation of α_{id} increases with decreasing velocity. The above variation is greater at a higher absolute value of α. This can be explained in terms of the k-factors of each sensor. A preliminary investigation showed that for a given α, the k remains almost constant for velocities above 17 m/s and for $-20° < \alpha < +20°$. Lekakis et al. [245] found that k is constant for flow velocities higher than approximately 15–20 m/s. The relation between α_{id} and k_1 and k_2 can be obtained from Eqs. (14.44), (14.45) and (14.47) i.e.,

$$\tan(\alpha_{id} + 45) = \sqrt{\frac{\sin^2(45+\alpha) + k_1^2\cos^2(45+\alpha)}{\sin^2(45+\alpha) + k_2^2\cos^2(45-\alpha)}}. \tag{14.54}$$

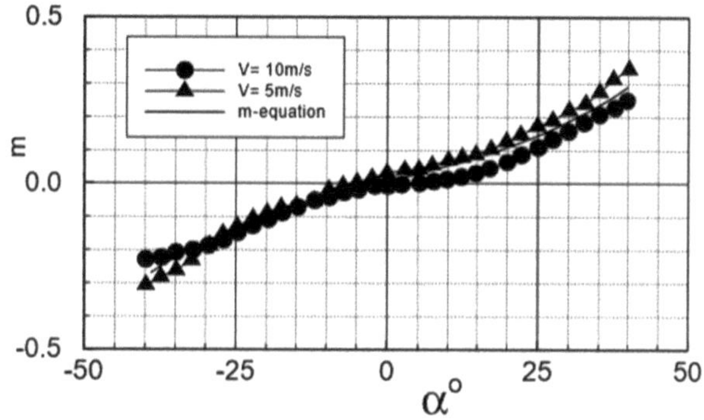

Fig. 14.10 Variation of correction parameter m with " For uncertainty estimates see Table 13.1

Since k_1 and k_2 are functions of V and α, the left-hand side of Eq. (14.54) is a function of V for a particular α. Based on the assumption of a power law relationship between Eq. (14.54) and V, Eq. (14.55) can be derived to compensate for the deviation of α_{id}.

When using the result of yaw angle calibration at a velocity above 20 m/s to any other velocity below 15 m/s, α_{id} was corrected using the following relation.

$$\frac{\tan(\alpha_{id}^* + 45)}{\tan(\alpha_{id} + 45)} = \left(\frac{V^*}{V}\right)^m. \tag{14.55}$$

The superscript * denotes the quantities at the reference velocity at which the yaw angle calibration was carried out. The power m is a function of α and can be least-square-fitted by a third-order polynomial as shown in Fig. 14.10, i.e.,

$$m = e_0 + e_1\alpha + e_2\alpha^2 + e_3\alpha^3. \tag{14.56}$$

Figure 14.10 was obtained using a reference velocity yaw angle calibration at 22 m/s and two other calibrations at 5 m/s and 10 m/s.

14.1.6.4 Reduction Method

This section briefly describes the steps to calculate the velocity components from the voltage outputs of the two sensors obtained during actual measurements. The first step is to compensate the anemometer output voltage E_{mj} of each sensor for the change in fluid temperature. An equation for temperature compensation is obtained from the modified form of King's law, $E_b^2 = (A + B.V^n)(T_s - T)$. As shown in [245], the temperature dependency of the constants A and B can be neglected for a hot-film sensor of 25 μm diameter at velocities above 1 m/s. This leads to a temperature compensation formula given by

$$E_j = E_{mj}\sqrt{\frac{T_s - T}{T_s - T_c}} \tag{14.57}$$

where T_s, T_c, and T are the sensor operating temperature, calibration temperature and the temperature of the fluid during actual measurement, respectively. As a second step, the instantaneous effective velocities V_{e1} and V_{e2} are calculated from instantaneous temperature-compensated voltages using Eq. (14.43). The instantaneous value of α_{id} is obtained from V_{e1} and V_{e2} using Eq. (14.46) while α and H^* are calculated using Eqs. (14.50) and (14.52). The magnitude of instantaneous velocity V is determined using Eq. (14.51) and its components V_x and V_y are calculated from its magnitude and direction.

If the velocity V is below 15 m/s, α_{id} is corrected by Eq. (14.55) by using the yaw correction parameter m obtained from Eq. (14.56). From the new α_{id}, instantaneous values of H^* and V are computed again using Eqs. (14.50), (14.52) and (14.51), respectively. The instantaneous velocity components V_x and V_y are calculated as before.

14.1.6.5 Calibration Uncertainty

The calibration uncertainties of the various quantities are estimated according to the method suggested by Yavuzkurt [252], which is based on the uncertainty analysis by Moffat [253]. The uncertainties for four different velocities are shown in Table 14.1. The main contribution to the uncertainty comes from the differential pressure transducer which has an uncertainty of ± 0.6 Pa. However, significantly lower uncertainties can be achieved by using transducers or manometers with lower uncertainties.

14.1.6.6 Correction Results, Comparison

In this section, the results of the single velocity calibration without the yaw correction are compared to those with the yaw correction. After the calibration was complete, 50 new sets of data independent from that of calibration were taken covering a velocity range of 10–30 m/s and angle $-30° < \alpha < +30°$. The actual values of V, V_x, and V_y are obtained from the measured pressure drop across the nozzle, temperature of the air and on value obtained from the vernier of the rotary table. The absolute percentage deviation between actual and calculated V is given by

Table 14.1 Calibration uncertainties

V	$\Delta V/V$	$\Delta V_e/V_e$	$\Delta \alpha_{id}/\alpha_{id}$	$\Delta H^*/H^*$
m/s	%	%	%	%
5	2	2.3	3.3	2.95
10	0.5	1.3	1.8	1.28
20	0.1	1.22	1.71	1.09
30	0.06	1.2	1.7	1.07

Table 14.2 Maximum deviation between actual and calculated values of velocities and angle for the case without yaw angle correction

Angle α	Velocity V	$\epsilon_{V\max}$	$\epsilon_{Vx,\max}$	$\epsilon_{Vy,\max}$	$\Delta\alpha_{\max}$
0	m/s	%	%	%	o
–20 to +20	20–30	1.3	1.2	3	0.5
±20–±30	20–30	2	2.7	3.8	0.9
–20 to +20	10	1.3	1.5	8	1.9
±20–±30	10	1.1	3.9	10	3.5

Table 14.3 Maximum deviation between actual and calculated values of velocities and angle for the case with yaw angle correction

Angle α	Velocity V	$\epsilon_{V\max}$	$\epsilon_{Vx,\max}$	$\epsilon_{Vy,\max}$	$\Delta\alpha_{\max}$
0	m/s	%	%	%	o
–20 to +20	10	1.25	1.27	3	0.74
±20–±30	10	1	1.7	3.6	1.1

$$\epsilon_V = \frac{|V_a - V_c|}{V_a} \times 100 \tag{14.58}$$

where V_a and V_c are the actual and calculated values of V. Similarly, percentage deviations of V_x and V_y, denoted by ϵ_{Vx} and ϵ_{Vy} are computed. The absolute difference between actual and calculated is denoted by $\Delta\alpha$.

Table 14.2 shows the maximum deviations of V, V_x, V_y, and α when the yaw angle calibration at 22 m/s is used. These deviations were expected from the results plotted in Figure 13.9. At the velocity of 10 m/s, $\Delta\alpha_{\max}$ is higher than values obtained for the velocity range of 20–30 m/s. These large angle deviations result in significant errors for velocity particularly for the y-direction component. Table 14.3 shows the result for 10 m/s velocity when the yaw correction is applied. The error decreases significantly and the deviations are of the same order as those obtained for the 20–30 m/s range.

14.2 Measurement of Time Averaged Flow Quantities Using Five-Hole Probes

Five-hole probes are pneumatic instruments that measure static and total pressure distributions in a three-dimensional flow-field. Using a calibration system, velocity components and their respective angles are extracted from the above mentioned pressures. The pneumatic nature of the signal transfer from the probe to the pressure transducers associated with a strong damping of the signal frequency does not allow measuring high frequency pressure fluctuations. Based on the low sampling frequency of pressure transducers, signals are time averaged. Thus, these probes are appropriate for measuring the time averaged total and static pressure distributions as well as the velocity components. For acquiring high frequency signals, sensors can be placed at the tip of the five-hole probes to measure time resolved high frequency pressure signals as reported in [254].

Five-hole probes are typically custom built of miniature stainless steel tubes that are fused together and machined to produce a symmetrically conical head, which is bent and positioned at a 90° angle from the probe shaft that houses all the tubes and extends outside the measurement domain. The probe tip can have the standard L-shape design or can be axially offset from the probe body as with the cobra-shape design. Figure 14.11 shows both probe configurations and the incoming flow velocity vector and its decomposition onto axial, circumferential and radial direction. Figure 14.11 also shows the design of a standard L-shaped probe.

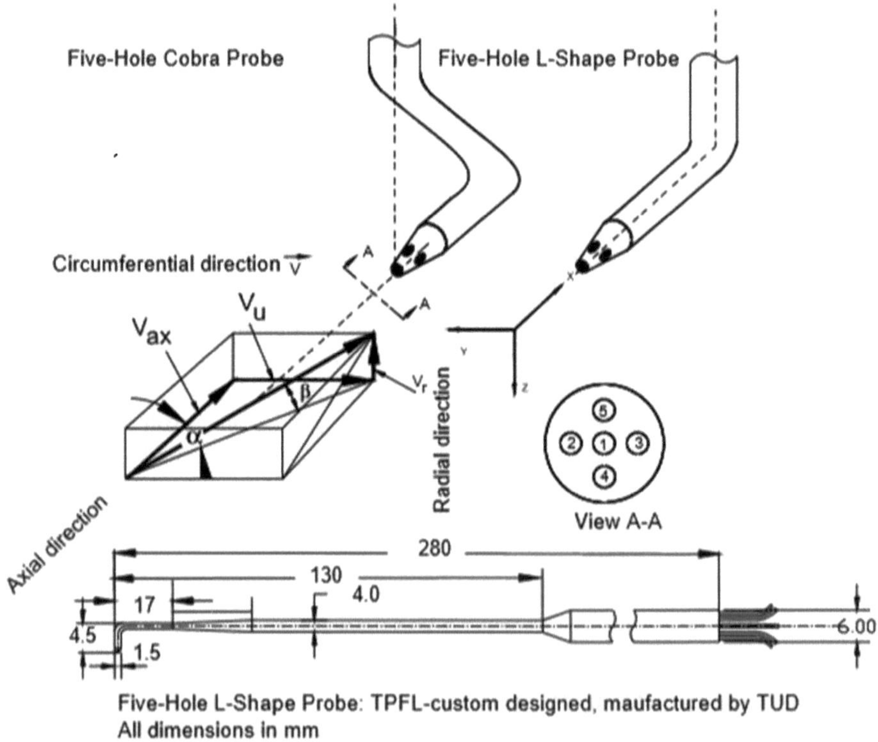

Fig. 14.11 Pneumatic five-hole probes: Cobra, L-probe

14.2.1 Flow Field Measurement Using Five Hole Probes

Five-hole probes are used wherever the flow velocity changes during a measurement. The flow angle may undergo changes in excess of $\pm 30°$. There are two methods of using five-hole probes for measuring static and total pressures as well as velocity components. They are termed *nulling* and *non-nulling*. In nulling mode the probe tip axis is adjusted to the flow angle such that the pressures of holes 2–3 and 5–6 are equalized. This requires changing angular position of the probe. Measuring the flow field in this mode is extremely time consuming and in most cases impractical. In a non-nulling mode, the probe orientation remains fixed as long as the flow variation occurs within the calibration range. In case that the actual flow angle exceeds the calibration range, a partial adjustment of β-angle (also called yaw) is possible as long as the variation of α-angle (also called pitch) occurs withing the calibration range. The non-nulling mode requires the probe to be calibrated to account for flow pressure and direction variations before being placed in the flow-field at a fixed position. This allows for performing extensive multi-point flow measurements where the probe is typically radially or axially traversed across the flow domain without the need to continuously adjust the probe body or tip angles.

Numerous non-nulling calibration methodologies have been developed over the last three decades, among others by [255–258]. A recent paper by Town and Camci [259] discusses the calibration issues of existing methods. An improved method of calibration was introduced by by TPFL [260]. The common feature of almost all existing calibration methods is the structure of calibration coefficients which include an algebraic average of the static pressures measured in the outer four holes. Implementing this average in calibration coefficient is admissible as long as the flow angle does not cause flow separation. For adverse flow angles, one or more outer holes measure pressure in a separated flow zone resulting in an erroneous average. To circumvent this deficiency, Rubner and Bohn [261] introduced a new calibration procedure which is deemed to be one of the most accurate and reliable procedures. The following presentation is based on Bohn's technique.

14.2.2 Calibration Procedure of Five-Hole Probes

To calibrate the probe, it is inserted into the automatic calibration facility shown in Fig. 14.4. The traversing system is designed to keep the position of the probe tip exactly in the center of the jet at any given α and β angle. Note that the α and β-planes in Fig. 14.12 are body fixed. The indexing mechanism varies the α and β angles through prescribed calibration range of $\pm 40°$. Exceeding this range may cause flow separation on probe holes, leading to erroneous results. The indexing mechanism controlled by two stepper motor controllers can be programmed to operate as very small angle increments. However, an increment of one degree delivers sufficiently accurate results. An example of a traversing schedule is shown in Fig. 14.4b.

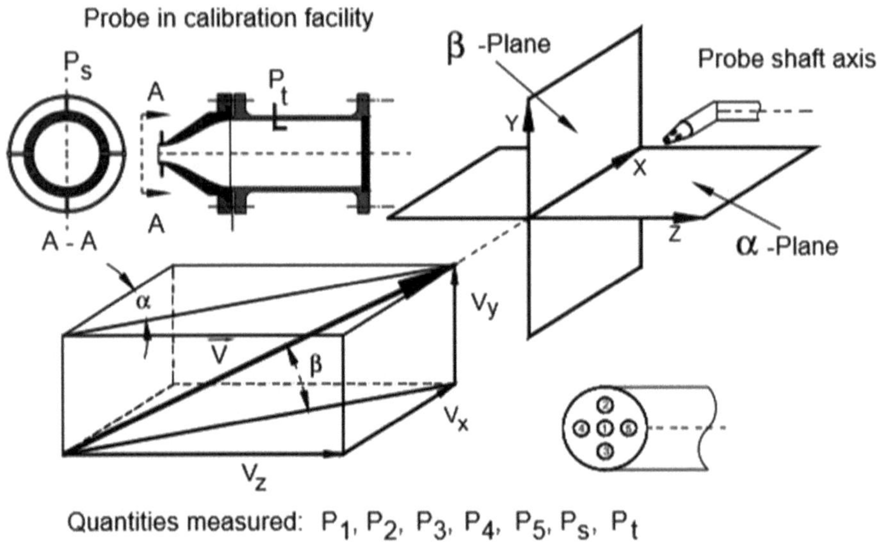

Fig. 14.12 Position of the probe tip and shaft axis relative to the calibration facility, angle definition

During the calibration process, the flow velocity exiting from the calibration facility is kept constant while pressure data from all five holes are recorded. Utilizing the five pressures, five flow coefficients are calculated. These coefficients relate the total and static pressures of the flow as well as the α and β angles to the five individual pressures of the probe. Using curve fit tools, three dimensional curves are computed that correlate the coefficients with the flow angles. These coefficients are

$$Q_1 = \frac{P_4 - P_5}{P_1 - P_s}, \; Q_2 = \frac{P_3 - P_2}{P_1 - P_s}, \; Q_3 = \frac{P_1 - P_s}{P_t - P_s}, \; Q_4 = \frac{P_1 - P_4}{P_t - P_s}, \; Q_5 = \frac{P_1 - P_5}{P_t - P_s} \tag{14.59}$$

with P_1 through and P_5 the pressures measured by the five holes and P_s and P_t as the static and total pressure measured by four static pressure taps close to the nozzle exit and a Pitot tube inside the calibration tunnel. These pressures, however, do not reflect the true static and total pressure measured by the probe itself. Their true values are crucial for the degree of accuracy of the calibration. An iterative method proposed by Rubner and Bohn [261] accurately determines the static and total pressure.

Starting with Q_1 and Q_2, the static pressure P_s is obtained using an iterative process by setting $P_s = (P_s)_{i=1} = 0.8P_1$. This step determines Q_1 and Q_2 and, thus, $\alpha = f_1(Q_1, Q_2)$ and $\beta = f_1(Q_1, Q_2)$. With the calculated α and β, the remaining coefficients Q_3, Q_4 and Q_5 are calculated. The iteration function given in [261] is

$$(P_s)_{i+1} = P_1 - Q_3 \left(\frac{P_1 - P_4}{2Q_4} + \frac{P_1 - P_5}{2Q_5} \right) \tag{14.60}$$

which in conjunction with the following convergence criterium

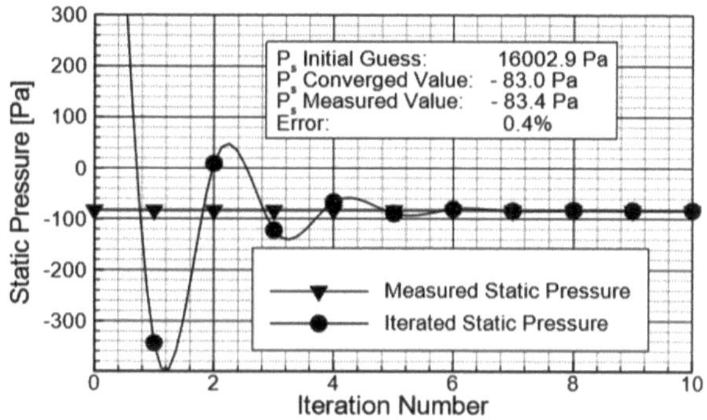

Fig. 14.13 Convergence diagram for static pressure

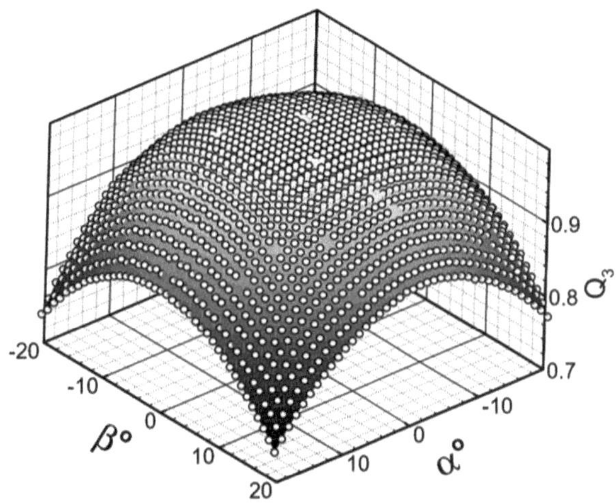

Fig. 14.14 Q3-surface versus α and β

$$-5Pa < |P_{s_{i+1}} - P_{s_i}| \leq +5Pa \tag{14.61}$$

gives an uncertainty for the static pressure of 5 Pa for the angles $\alpha, \beta = \pm 0.1°$ and for the velocity of 0.2 m/s. As an example, the convergence diagrams for static and total pressure are presented in Fig. 14.13. As seen, for both pressures, the convergence is achieved after six iterations for the above prescribed convergence limit with an error of 0.4% for static pressure and 0.02% for total pressure. Once the convergence has been reached, the coefficients Q_1 through Q_5 can be plotted. An example is given in Fig. 14.14, where Q_3 is plotted versus α and β. The surfaces can be described using optimized 3-D polynomial surface fit with standard least square minimization, Lorentzian minimization or any other appropriate fit. Once the probe has been calibrated, it can be inserted into the test section. With the five pressures measured from the probe using an appropriate data analysis, the coefficients Q_i are calculated and, thus, all flow parameters obtained.

Chapter 15
Heat Transfer

15.1 Heat Transfer Measurement, Calibration

To measure the surface temperature of a test object, four different instrumentations may be used (1) the thermocouples (TC), (2) infra red thermography (IR), (3) liquid crystal (LC) and (4) pressure/ temperature sensitive paints (PSP/TSP). Currently, the measurement of temperature taught in heat transfer lab is conducted using the conventional thermo-couples (TC). Although it provides the students with the basic idea how the temperature is measured, it does not provide a picture about the temperature distribution on a surface. There are two problems using TC: first, a thermocouple is capable only of measuring the temperature at discrete points. Second, to obtain the temperature distribution on any surface, a number of TC-s must be applied in a way that they do not disturb the incoming flow. This requires an accurate treatment of the surface after the TC has been installed. This is to ensure that the tip of the TC is flush with the surface. To overcome these problems, we have prepared several practical and effective solutions. These are (1) Infrared Thermal Imaging (IR), (2) Liquid Crystal (LC) technique and (3) Pressure, Temperature Sensitive Paints (PSP, TSP). In what follows, we discuss these techniques in detail. With these three techniques, the temperature distribution on an entire surface can be measured. These three techniques are discussed below.

15.1.1 Infrared Thermal Imaging

This technique can be used for temperature measurement of an object with or without being exposed to a flow. Using a hand-held IR-device, it delivers temperature distribution on the surface of an object. It is simple to operate and one can immediately obtain the temperature distribution on the surface of a test object. The use of Infra

M. T. Schobeiri, *Advanced Fluid Mechanics and Heat Transfer for Engineers and Scientists*, https://doi.org/10.1007/978-3-030-72925-7_15

Red Camera (IRC) enables the temperature measurement of an entire surface. The data acquired by an IRC can be used to obtain the heat transfer coefficient on the surface of a test object.

15.1.2 Film Cooling Effectiveness Measurement with Infrared Camera

IR also can be used for measuring the film cooling effectiveness, which requires a test section. The image of the temperature distribution on the surface of a test object can be obtained using an (IRC) mounted on top of the test section as shown in Figs. 15.1 and 15.2.

15.1.3 Film Cooling Effectiveness

The quantities involved in calculating the film cooling effectiveness are: (1) the temperature of the incoming flow $T\infty$, (2) the adiabatic wall temperature T_{av} and the temperature of the coolant ejecting from the holes T_C. With these quantities the effectiveness of surface cooling, the so called film cooling effectiveness is determined. Thus, the film cooling effectiveness is defined as:

$$\eta = \frac{(T_\infty - T_{av})}{(T_\infty - T_C)} \tag{15.1}$$

Film cooling performance is significantly affected by various parameters such as wall geometry, wall curvature, boundary layer influence, coolant velocity, Reynolds

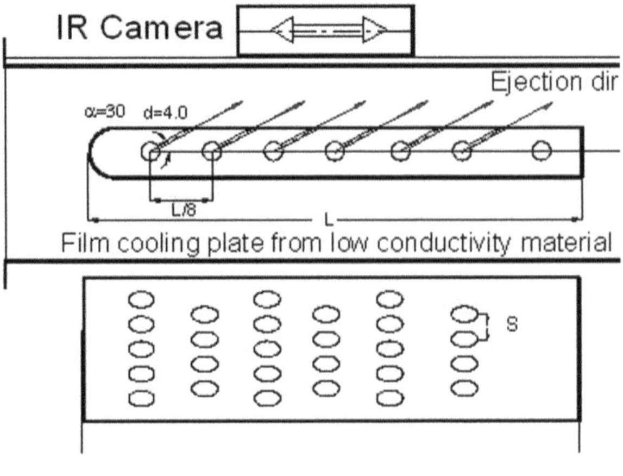

Fig. 15.1 Film cooling test object with IR-camera

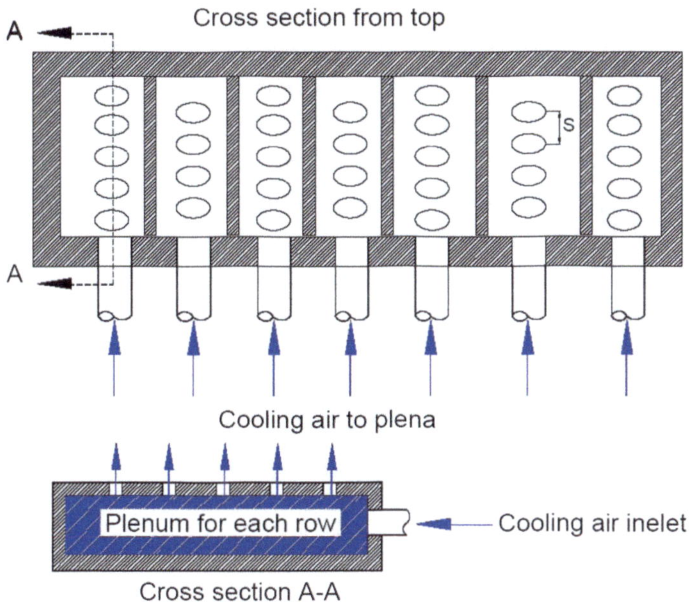

Fig. 15.2 Details of the film cooling plate with cooling plana

number, blowing ratio, momentum flux ratio, pressure gradient, Mach number, etc. As Eq. (15.1) shows, the effectiveness is a non-dimensional parameter used to characterize the film cooling performance (Fig. 15.3).

Conducting a cooling effectiveness test provides the information as to what extent a material, for example a turbine blade, that is exposed to a high temperature environment should be thermally protected.

15.1.4 Working Principle of IR-Thermography

As described by the IRC-manufacturer Fluke [262] All objects emit infrared energy, known as a heat signature. An infrared camera detects and measures the infrared energy of objects. The camera converts that infrared data into an electronic image that shows the apparent surface temperature of the object being measured. An infrared camera contains an optical system that focuses infrared energy onto a special detector chip (sensor array) that contains thousands of detector pixels arranged in a grid. Each pixel in the sensor array reacts to the infrared energy focused on it and produces an electronic signal. The camera processor takes the signal from each pixel and applies a mathematical calculation to it to create a color map of the apparent temperature of the object. Each temperature value is assigned a different color. The resulting matrix of colors is sent to memory and to the camera's display as a temperature picture

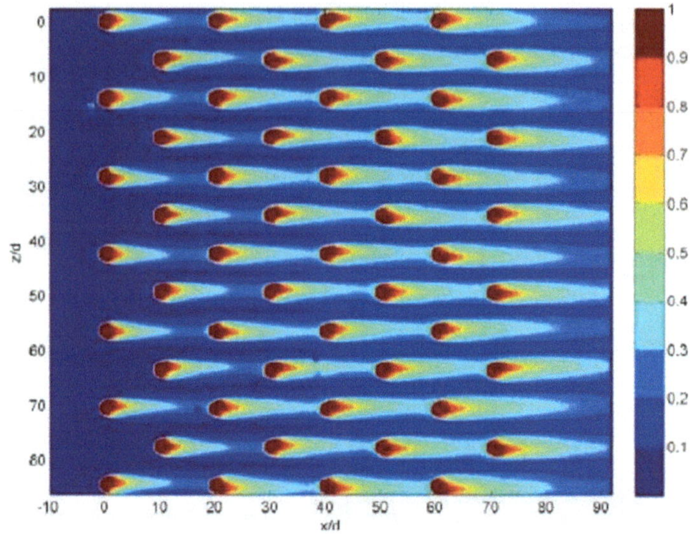

Fig. 15.3 Details of the film cooling

(thermal image) of that object. Many infrared cameras also include a visible light camera that automatically captures a standard digital image with each pull of the trigger. By blending these images it is easier to correlate problem areas in infrared image with the actual equipment. IR-Fusion technology combines a visible light image with an infrared thermal image with pixel-for-pixel alignment. The intensity of the visible light image and the infrared image van be varied to more clearly see the problem in the infrared image or locate it within the visible light image.

15.2 Temperature Measurement Using LC

The most accurate method of temperature measurement is using liquid crystal (LC). In this section we present three cases, where different test objects are instrumented with liquid crystals. The first case is a simple flat plate, the second case is a plate with concave curvature surface, and the third case is a test object that creates positive, zero- and negative pressure gradients.

The liquid crystal technique used in this investigation has been used by several researchers. Simonich and Mofat [262] and Hippensteele et al. [263] applied the liquid crystal technique to a turbine blade. They developed a convenient method of using liquid crystals with a composite heater sheet to determine the local heat transfer coefficient distribution on both sides of a turbine blade that was subject to a steady inlet flow condition. The original heater sheet consisted of a plastic with a carbon impregnated coating which gave a heat transfer coefficient error of ±6.2

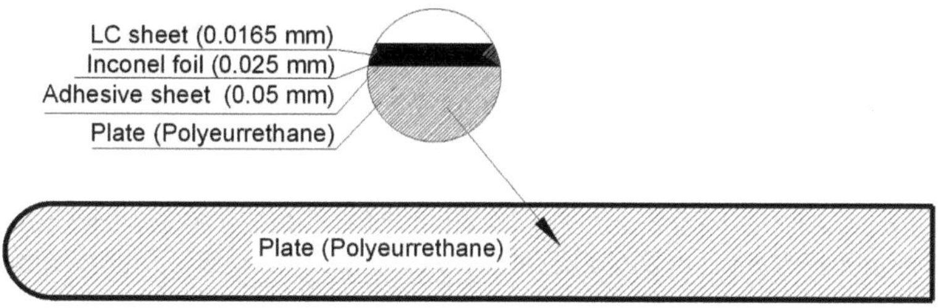

Fig. 15.4 Details of liquid crystal attachment to the concave side of a curved plate

For the measurement of temperature distribution along the surface of a test object, the temperature of the LC-sheet relative to the temperature of the incoming flow must increase. This is facilitated by supplying power to the Inconel sheet, which is tacwelded to two copper plates on both sides of the plate as shown in Fig. 15.4. The power required for heating the plate is supplied by a DC power supply. The electric current passing through the copper plates and thus to the Inconel sheet is measured with a multi-meter connected across a shunt resistor. Two separate multi-meters are used to measure the voltages in the circuit. One multi-meter is connected across the output leads of the power supply to measure the voltage output, and the other is connected across the plate terminals. These two measurements are used to calculate the power losses in the cable resistance.

15.2.1 Working Principle of Liquid Crystals

Each LC-sheet is designed to handle certain maximum temperature. Once the LC is exposed to a specific temperature, its color changes. The LC-sheet shows a broad spectrum of colors. Among the many color spectra (green, yellow, blue, red etc.), there is one distinguished band. In the example we present here the distinguished band is the yellow band that represents a specific temperature, for which the band is calibrated. The yellow band of the liquid crystal is used to record the data. The location of the yellow band is controlled through the power supply to the plate, and the voltage and the current readings across the plate terminals are recorded for different locations of the yellow band on the plate.

15.2.2 Calibration of Liquid Crystals

At the beginning of each test, the yellow band of the liquid crystal is calibrated for temperature. The yellow band is used for measurements because it occurs over the narrowest temperature band and has good uniformity. Calibration consists of adjusting the power from the power supply until the yellow band is directly on top of the surface flush mounted thermocouple. The thermocouples as calibrators can be arranged on both sides of the plate for simultaneous calibration of the liquid crystal temperature. Attention must be paid not to install thermocouples close to the side walls of the test section, thus not to be exposed to secondary flow. Once steady state has been reached, the temperature is recorded. This calibration is performed under no flow conditions and repeated several times. The yellow band temperature was found to be approximately 47.5 °C for each calibration performed.

15.2.3 How to Measure Surface Temperature with LC

Starting from the leading edge,the power is adjusted such that the position of the yellow band is identical with the target location on the test object, in this case the plate leading edge. The power is then changed to move the yellow band away from the leading edge in a definite increment ΔS in longitudinal direction. To identify the longitudinal location, a grid should be drawn on one or both sides of the plate with an $\Delta S = 10\,\text{mm}$ or smaller as shown in the following sections. It must be noted that, in order to achieve steady state operation, certain time has to elapse before moving to the next point. This process is continued until the full longitudinal length of the plate for both surfaces is completely mapped. This process is the same for all cases here and all wake passing frequencies for periodic unsteady inlet flow conditions. The collected data are reduced based upon a heat flux analysis which entails determining all the energy losses on a flux basis and subtracting them from the heat flux of the Inconel foil. The analysis of the collected data is used for calculation of the heat transfer coefficient and the related dimensionless quantities such as Nusselt number. In this context, it should be noted that all the energy losses on a flux basis are subtracted from the heat flux of the Inconel foil. This leaves the convective portion of the energy equation, and from this, the heat transfer coefficient can be obtained with the following equation:

$$h = \frac{Q_{conv}}{\left(T_{yel} - T_{\infty}\right)} \tag{15.2}$$

with Q_{conv} as the convective portion of the energy equation per unit area and is defined by

$$Q_{conv} = Q_{foil} - Q_{rad} - Q_{cond} \tag{15.3}$$

with Q_{foil} as the heat flux of the Inconel foil and Q_{rad} the radiation heat flux emitting from the surface of the liquid crystals. These two quantities are given by

$$Q_{foil} = \frac{V\,I}{A_{foil}} \tag{15.4}$$

and the radiation heat flux is calculated by Max Plank equation:

$$Q_{rad} = \varepsilon\sigma\left(T_{yl}^4 - T_\infty^4\right) \tag{15.5}$$

with the emissivity of the liquid chrystal $\varepsilon = 0.85$ and σ as the Stephan-Bolzmann constant. The amount of radiation reflected back from the top wall may be less than than two percent of the input power and as a result it can neglected. The third loss, the conduction loss, is found to be negligible based upon a two-dimensional finite difference nodal analysis.

15.2.4 Case Studies with LC-Measurement

In the following, we present several cases dealing with temperature measurements on flat plate, concave and concave side of a side of a curved plate. In case of the latter, we also present the measurement results in terms of Nusslet number and heat transfer coefficient.

15.2.5 Flat Plate LC Cover

To perform the temperature measurement with the objective to determine the heat transfer coefficient (or Nusselt number), the test object must be prepared first. As the test object, we use a flat plate, where its surface will be covered by an Inconel heater foil and LC-sheet as shown in Fig. 15.4. It is specially manufactured from a *non-conductive material* with a thin Inconel heater foil attached to its surface via a two-sided temperature resistant adhesive tape. On top of this foil a sheet of liquid crystal is attached for temperature measurements. It is also possible to spray LC on the test object surface; in this case one should pay attention that the LC-layer has a uniform thickness. This is the simplest case that allows measuring the heat transfer coefficient along the surface of a plate.

15.2.6 Curved Plate Exposed to a Periodic Unsteady Flow

We already discussed this case, but in the context of heat transfer measurement with LC, for the sake of completeness, a brief recapitulation is given in this section, Fig. 15.5. As seen, the curved plate is covered by a grid that indicated the position of the yellow line. The yellow arrow shows the direction of yellow line movement.

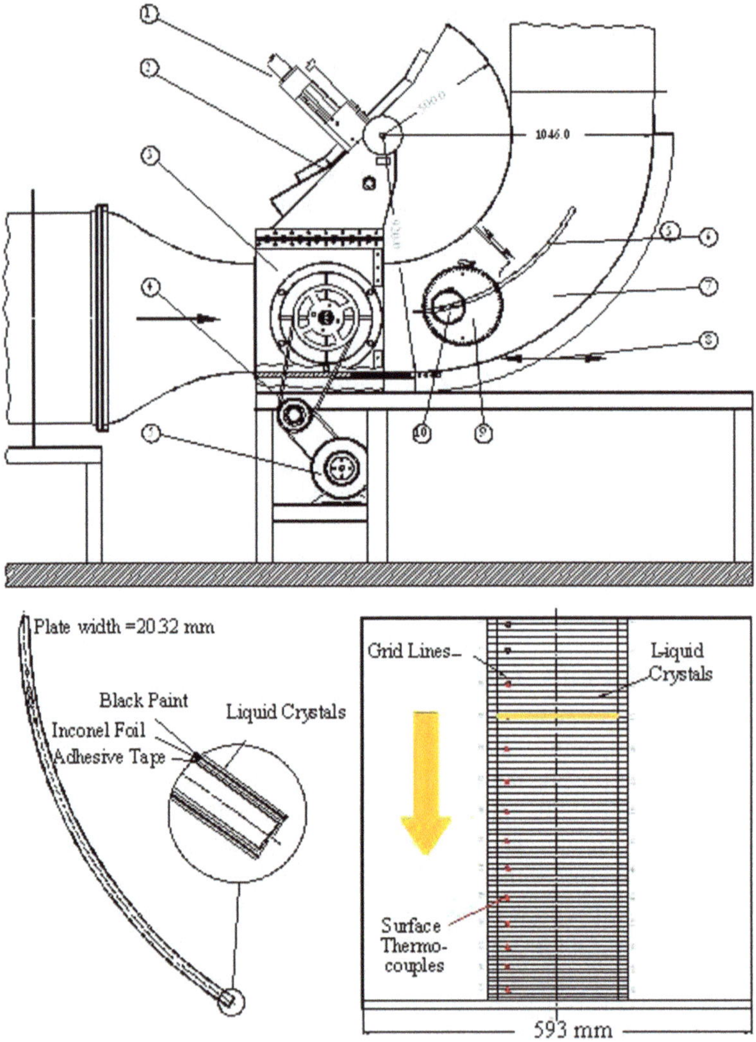

Fig. 15.5 Details of liquid crystal attachment to the concave side of a curved plate

The curved plate Fig. 15.5 (bottom), is installed in the test section of the facility shown in Fig. 15.5 (top). The plate shows the yellow line at an arbitrary position defined by the voltage and current of the power supply. Using the heat transfer coefficient Eq. (15.2), we obtain the Nusselt number

$$Nu_x = \frac{h_x x}{k} \tag{15.6}$$

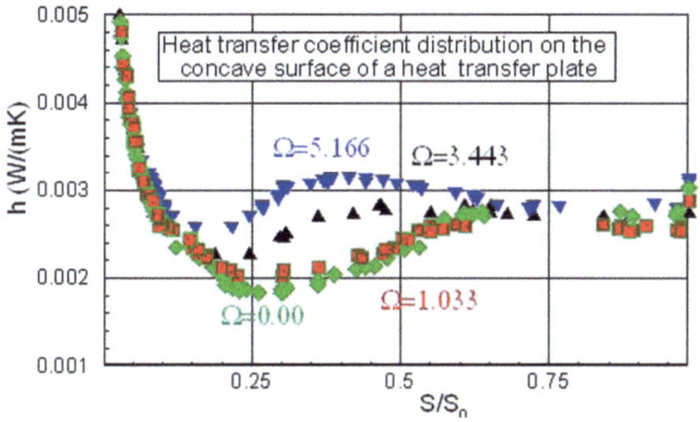

Fig. 15.6 Heat Transfer coefficient along the concave side of a curved plate

with h_x as the local heat transfer coefficient at the longitudinal location x and k the thermal conductivity of the working medium.

15.2.7 Curved Plate Concave Side Heat Transfer

The local heat transfer coefficient distribution on concave surface is shown in Fig. 15.6. The experiments were performed at wake reduced frequencies of $\Omega =$ 0.000, 1.033, 1.725, 3.443, and 5.166 with $\Omega = 0.00$ as the steady inlet flow condition. The figure shows the effect of the wakes on hear transfer coefficient. As seen she start and end of transition shifts towards the leading edge as the wake passing increases. The beginning of transition occurs for $\Omega = 0.000, 1.033, 1.725, 3.443,$ and5.166 around $s/s0 = 0.35, 0.27, 0.26, 0.21,$ and 0.17, respectively. For the aforementioned values of Ω, the end of transition occurs around $s/s_0 = 0.63, 0.60, 0.58,$ 0.38, and 0.34, respectively. The elevated average heat transfer coefficient occurs for higher values of Ω mainly due to a higher freestream turbulence level generated by stronger wake mixing and turbulent activities inside the boundary layer.

In full accord with the aerodynamic picture of the boundary layer transition discussed in previous Chapters, the heat transfer is predominantly dictated by freestream turbulence.

15.2.8 Curved Plate Convex Side Heat Transfer

For the convex side, Fig. 15.7, the shape of the Nusselt number distribution remains constant for all values of Ω (positive and negative, i.e., counterclockwise and clockwise rotation of the wake generator). The reason for this is a separation bubble that occurs just downstream of the leading edge of the curved plate. The flow separation on the convex side of the plate was caused by an incidence angle of $\beta = -8°$ relative to the convex surface. This incident caused a separation bubble to form at

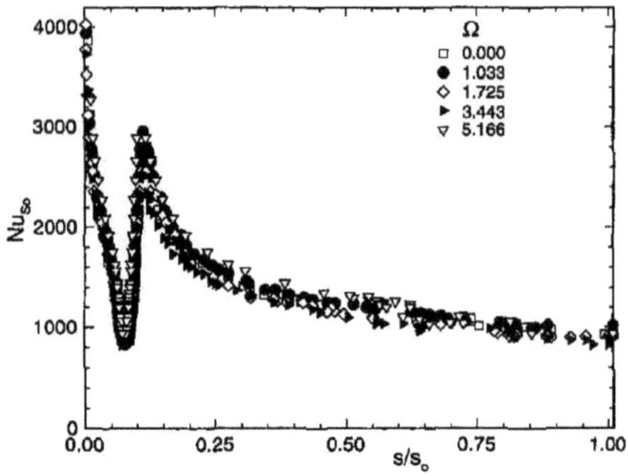

Fig. 15.7 Nusselt number distribution along the convex side of curve plate

$s/s_0 = 0.075$. Once the flow finally reattaches at $s/s_0 = 0.125$, it is fully turbulent which is evident by the increased heat transfer after reattachment.

Figure 15.7 exhibits the distribution of the Nusselt number vs. the dimensionless parameter s/s_0 on the convex surface of the curved plate. The distribution of the Nusselt number is presented for five different reduced frequencies Ω. As Fig. 15.7 shows. The boundary layer starts with the laminar portion and continues to arrive at the beginning of the transition at about $s/s_0 \approx 0.075$ and reaches its end at $s/s_0 \approx 0.125$

15.2.9 Turbine Blade

This case represents a more complicated case but yet an interesting case that has a far reaching practical application. The turbine blade with the Inconel foil and LC-sheet is shown in Fig. 15.8 with the details shown in Fig. 15.8 (left) and Inconel- and LC-sheet attachment Fig. 15.8 (right).

The heat transfer coefficient measurements along the convex (or suction side SS) and the concave side (or pressure side PS)were extensively treated in Chap. 12. Here we present just a representative case, Fig. 15.9 where the distribution of the heat transfer coefficient is presented for three different Re-number with the highest h-distribution corresponding to highest Re-number in this test.

The abscissa $s/s_0 > 0$ represents the heat transfer coefficient on convex surface (suction surface), while $s/s_0 < 0$ represent the concave surface (pressure surface) heat transfer coefficient. Highest heat transfer coefficient is found at the stagnation point with $s/s_0 = 0$

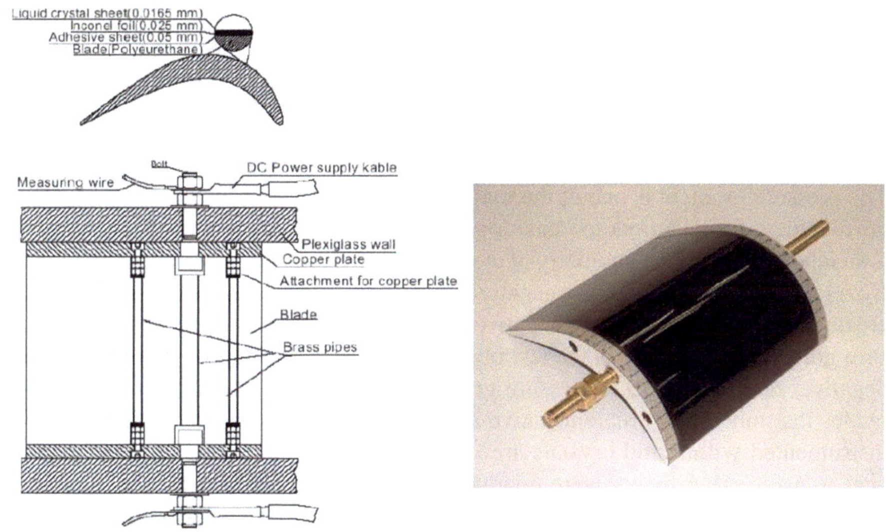

Fig. 15.8 Details of liquid crystal attachment to a turbine blade

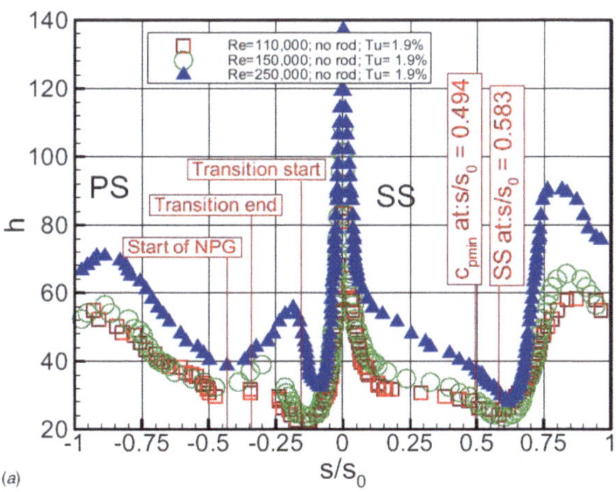

Fig. 15.9 Heat Transfer coefficient distribution along the suction and pressure surface on a turbine blade

For this case study, we utilize the cascade facility described in Chap. 11, where the heat transfer investigations were performed on the test object, namely a typical highly loaded turbine blade installed in aircraft engines. Extensive experimental investigations were carried out and reported in a series of publications. In what follows, we present the quint essential of that investigations with the objective to provide the reader with deeper insight into the advanced heat transfer.

15.3 Boundary Layer Parameters

The parameters defining the aerodynamic behavior of a boundary layer were discussed in Chap. 11. In this section we provide a detailed insight into the heat transfer behavior of a test object that is exposed to different flow parameters such as:
The pressure gradient is one of the major parameters that affects the boundary layer development from laminar to transitional to turbulent and separation. This parameter is an inherent part of the geometry of the test object. Examples of such test objects are: Nozzles and diffusors creating negative, positive pressure gradients. Turbine blades create on their section surface negative, while on the pressure surface positive pressure gradients. Therefor, as the test object, we choose one that inherently produces negative, zero and positive pressure gradients. To this end, we resort to the turbine blades that inherently create negative and positive pressure gradient. These blades, instrumented with liquid crystals are installed into the test section, where detailed heat transfer experiments were conducted. Details of heat transfer measurements using the liquid crystal technique are presented in an extensive literature review in [262].

The turbine blade prepared for heat transfer investigations is shown in Fig. 15.10. It is specially manufactured from a nonconductive material with a thin Inconel heater foil attached to its surface via a two-sided temperature resistant adhesive tape as shown in Fig. 15.10. On top of this tape, a sheet of liquid crystal is attached for temperature measurements. The power required for heating the blade is supplied by a direct current (DC) power supply system, which controls the positioning of the yellow band. Data are collected an reduced based upon a constant heat flux analysis. They determine all energy losses and subtract them from the heat flux of the Inconel foil.

The heat transfer experiment presented in this section investigates the individual and combined effects of steady and periodic unsteady wake flows and freestream turbulence intensity Tu on heat transfer behavior of the boundary layer including the laminar, the transitional, the turbulent and the flow separation.

Figure 15.11 shows a composite picture that included the development of the flow and heat transfer quantities from the leading edge to trailing edge of the test object. It includes (a) the pressure distribution, (b) the velocity contour, (c) the fluctuation contour, and (d) a representative time averaged heat transfer coefficient (HTC)distribution for for a periodic unsteady wake flow at $\Omega = 3.18$(d). Furthermore, the diagrams in Fig. 15.11a–d deliver a coherent picture of the separation bubble. It depicts four distinct intervals that mark different events along the suction surface. An initially strong negative pressure gradient starting from the leading edge preserves the stable laminar boundary layer until the pressure minimum at $S/S_0 = 0.494$ has been reached. The laminar boundary layer characterized by the lack of lateral turbulence fluctuations, is not capable of transferring mass, momentum, and energy to the boundary layer, resulting in a steep drop of HTC from leading edge to $S/S_0 = 0.494$, where the pressure gradient changes the sign. The HTC drops further at a larger slope and assumes a minimum at the start of the separation bubble

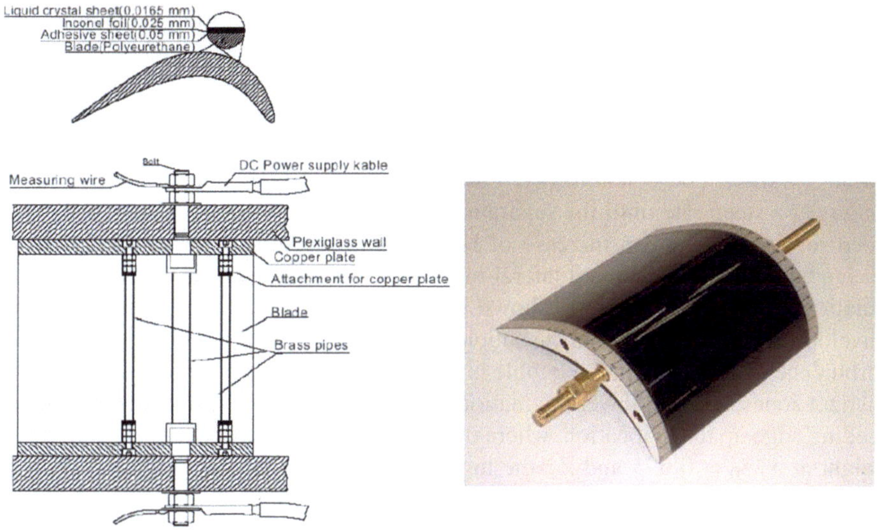

Fig. 15.10 Details of heat transfer blade with LC-Instrumentation

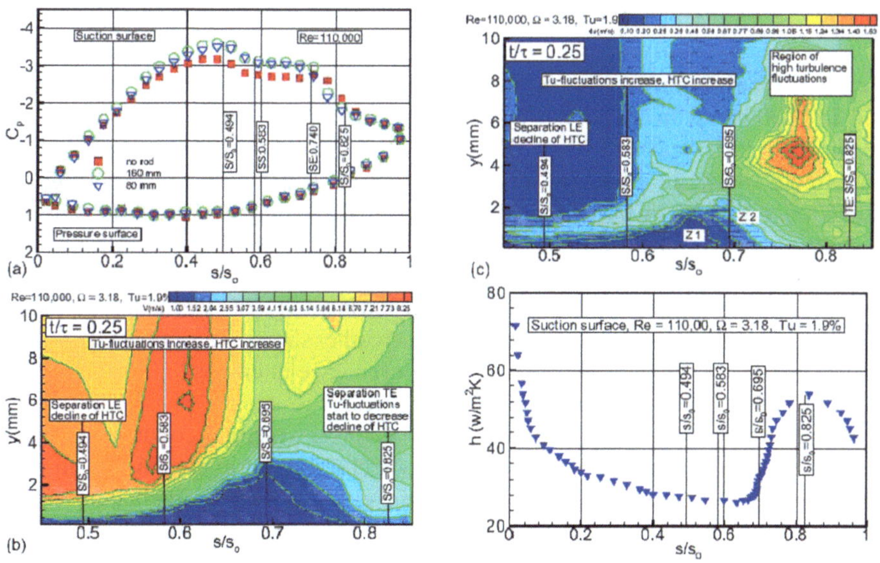

Fig. 15.11 Composite picture of **a** pressure distribution, **b** inception of separation bubble, **c** impact of the wake on separation bubble and **d** time averaged heat transfer coefficient

$S/S_0 = 0.583$. Passing through the pressure minimum, the initially stable laminar (or nonturbulent) boundary layer encounters a change in pressure gradient from negative to positive, causing it to become unstable and to separate at $S/S_0 = 0.583$. This point marks the leading edge of the separation bubble Fig. 15.11b, c. From this point on, the turbulence activities outside the bubble continuously increase, causing the heat transfer coefficient to rise. Further increase of HTC beyond $S/S_0 = 0.583$ occurs at a steep rate until the separation bubble trailing edge at $S/S_0 = 0.825$ has been reached. The steep increase of HTC within the separation bubble is due to an increased longitudinal and lateral turbulence fluctuation caused by the flow circulation within the bubble. As shown in Fig. 15.11c, the extent of low turbulence envelope is much smaller than the bubble size itself, Fig. 15.11b. This implies that the turbulence activity within the bubble is nonuniform and can be subdivided into two distinct zones, Z1 with lower fluctuation activities that occupies the bubble from the leading edge up to the location, where the bubble lateral extent reaches its maximum height at $S/S_0 = 0.695$ and Z2 the higher fluctuation zone beyond the maximum height, Fig. 15.11c.

15.3.1 Parameter Variation at Steady Inlet Flow Condition

Heat transfer experiments were carried out at Reynolds numbers of 110,000, 150,000, and 250,000 based on the suction surface length and the cascade exit velocity while keeping the turbulence intensity TU constant constant as depicted in Fig. 15.12. This figure shows the heat transfer distribution on suction side (SS) and pressure side (PS) with $S/S_0 = 0$ as the stagnation point. These diagrams exhibit the impact of the geometry on pressure distribution and thus the pressure gradient. They also reveal the significant influence of the Re-number and TU on heat transfer coefficient. Note that in each of these diagrams the turbulence intensity TU was kept constant. A comparison of the these two diagrams with each other shows the significant increase of heat transfer coefficient caused by higher turbulence intensity TU.

In presenting the heat transfer results, we prefer to use the plain heat transfer coefficient rather than the Nusselt number, which uses thermal conductivity and a constant characteristic length, such as the blade chord, to form the Nusselt number. Three sets of plots are generated to extract the effect of each individual parameter on heat transfer coefficients. Each figure includes the pressure as well as the suction surface heat transfer coefficient h as a function of dimensionless surface length s/s_0. While according to the equation of motion the boundary layer behavior is completely decoupled from thermal boundary layer behavior, the latter is through the equation of energy directly coupled with the boundary layer aerodynamics. Thus, a detailed description of heat transfer behavior is directly coupled with the aerodynamic results.

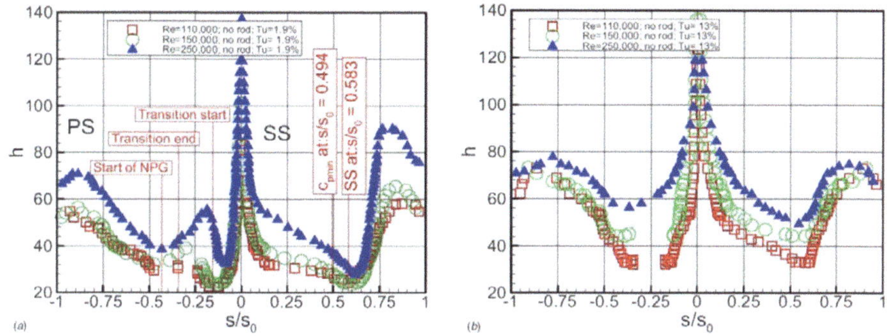

Fig. 15.12 Impact of the Reynolds number on heat transfer coefficient: Diagram **a** has the turbulence intensity of 1.9%, while diagram **b** has a turbulence intensity of 13% for steady inlet flow condition

15.3.2 Parameter Variation at Unsteady Inlet Flow Condition

In this section we are using a periodic unsteady wake flow generated by a system of rods attached to the two timing belts as discussed in Chap. 11. The parameters are: (1) Re-number, (2) reduced frequency Ω and (3) the turbulence intensity TU Periodic Unsteady Inlet Flow Condition, Variation of Re Number at Constant Tu.

In Fig. 15.13, the periodic unsteady inlet flow conditions was generated for reduced frequencies $\Omega = 1.59$ that is created by a rod spacing $SR = 160$ mm.

Keeping the reduced frequency and the turbulence intensity constant, heat transfer measurements were carried out for Re = 110,000, 150,000, and 250,000. HTC distribution along the suction and the pressure surfaces is shown in Fig. 15.13 for $\Omega = 1.59$ and a urbulence intnsity of $TU = 1.9\%$. The HTC distributions in Fig. 15.13 exhibit a systematic change for all three Re cases but a substantial increase for Re = 250,000. This is due to the fact that of the three Reynolds cases, the one with Re = 250,000 was able to influence the laminar portion of the boundary layer and, thus, the heat HTC behavior. Comparing the periodic unsteady wake flow case in Fig. 15.13 with the steady flow case shown in Fig. 15.12, only a marginal change in HTC can be observed. This is due to the fact that wakes generated by the translating rods with an $\Omega = 1.59$ are far apart from each other. Consequently, the turbulence activities of their vortical cores are not mutually interacting and, therefore, they are unable to substantially affect the total turbulence picture of the flow leading to almost the same HTC picture.

Keeping the reduced frequency $\Omega = 1.59$, $SR = 160$ mm and increasing the turbulence intensity to $Tu = 3\%$ in Fig. 15.13 we observe only a minor increase in HTC for the suction surface compared to in Fig. 15.12 the pressure surface, however, a major increase in HTC is clearly visible, where the transition length almost completely disappeared. In this case, it seems that the wake unsteadiness is about to submerge in the stochastic high frequency freestream turbulence generated by Grid TG1. Further increasing the turbulence intensity by subsequently attaching Grids

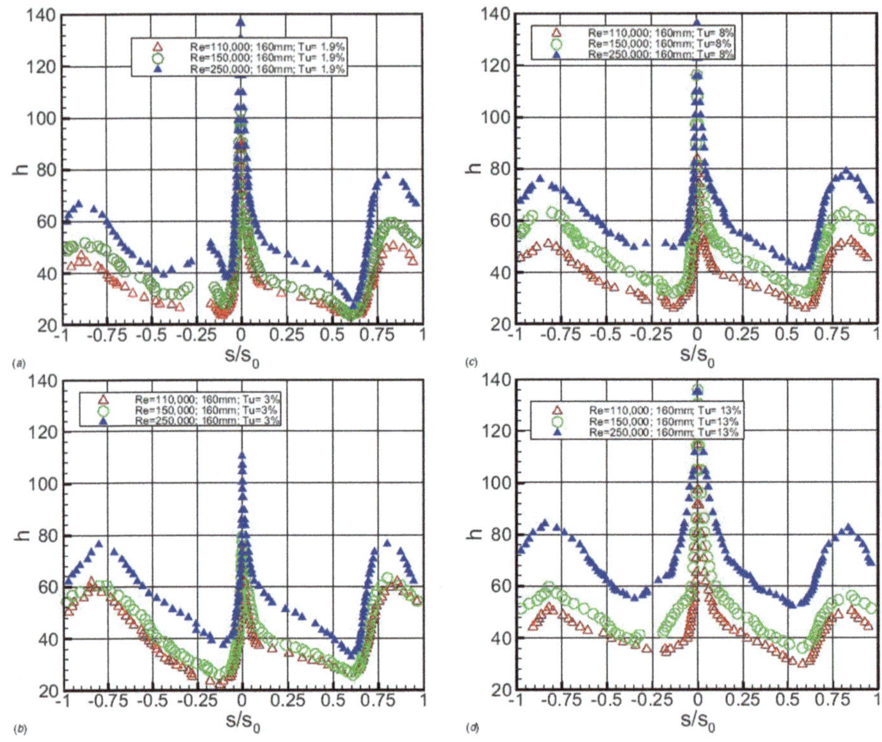

Fig. 15.13 Impact of the Reynolds number on heat transfer coefficient: Diagram **a** has the turbulence intensity of 1.9%, while diagram **b** has a turbulence intensity of 13% for steady inlet flow condition

TG2 and TG3 rid specifications are listed in Tables 1 and 2 has not brought a substantial change compared to the case shown in Fig. 15.13b. Results are presented in Figs. 15.13c, d, where a systematic shift of HTC toward slightly higher values is shown for Re=110,000 and 150,000 and substantially higher values for Re=250,000.

15.4 Temperature Measurement in Rotating Frame

In Section 15.3 we used liquid crystal method to measure surface temperature on a turbine blade in a stationary frame. This method is simple to use and delivers accurate results for measuring the surface temperature of test objects installed in a stationary test section, where the test object is neither moving nor rotating. However for the following reasons it is not appropriate for application in a rotating frame:

(1) It requires installation of heaters to position the yellow line. However, it is not possible to position the yellow bands while the test object is rotating,

(2) Attaching a heater- and a liquid crystal sheet changes the geometry of test object,
(3) The attachment of the above sheets to a rotating test object, because of centrifugal force causing both sheets to detach from the surface,
(4) It cannot be used for measuring the film cooling heat transfer because of surface holes.

An alternative technique is using the pressure/temperature sensitive paints (PSP/TSP) presented in this section. The measurement of surface temperature using Pressure/Temperature Sensitive Paint (PSP/TSP)is distinctively different from the methods discussed above. The temperature measurement capability of this method is far superior to the methods mentioned previously, however the data acquisition system is more complex than the two previously discussed methods. Considering this circumstance, the author decided to introduce the pressure/temperature sensitive paints technique for measuring the blade surface temperature in a high speed rotating turbine at Texas A&M Turbomachinery Performance and Flow Research Laboratory (TPFL). It is worth noting that the theoretical basis of this topic is neither taught in graduate experimental heat transfer classes nor mentioned in existing heat transfer textbooks.

In order to provide the reader with the basic knowledge of this measurement technique, we first present the a brief description of this technique and for more details refer to this chapter.

The Pressure Sensitive Paint (PSP) consists of photo luminescent molecules held together by a binding compound as shown in Fig. 15.14. The luminous particles in the PSP emit light when excited, with the emitted light intensity being inversely proportional to the partial pressure of oxygen in the surroundings. The emitted light intensity can be recorded using a charged-coupled Device (CCD), CCD-camera Fig. 15.15, and the corresponding oxygen partial pressures can be obtained by calibrating the emitted intensity against the partial pressure of oxygen. For details of PSP/TSP-measurement, data acquisition, calibration and data analysis we refer to paper published by Schobeiri and his co-workers in the bibliography at the end of this Chapter.

The image intensity obtained from PSP by the camera during data acquisition is normalized with a reference image intensity taken under no-flow conditions. Background noise in the optical setup is removed by subtracting the image intensities with the image intensity obtained under no-flow conditions without excitation. The resulting intensity ratio can be converted to pressure ratio using the previously determined calibration curve and can be expressed as:

$$\frac{I_{ref} - I_{blk}}{I - I_{blk}} = \left[\frac{(P_{O_2})_{air}}{(P_{O_2})_{ref}}\right) = f(P_{ratio}) \tag{15.7}$$

Where I denotes the intensity obtained for each pixel and $f(P_{ratio})$ is the relation between intensity ratio and pressure ratio obtained after calibrating the PSP. Further details in using PSP for pressure measurements are given among others in McLachlan and Bell.

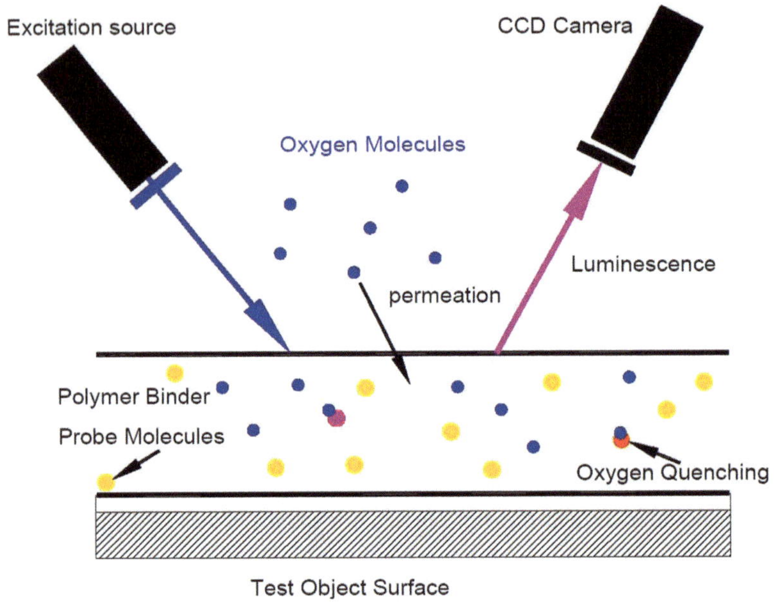

Fig. 15.14 Working principle of using PSP-measurement technique

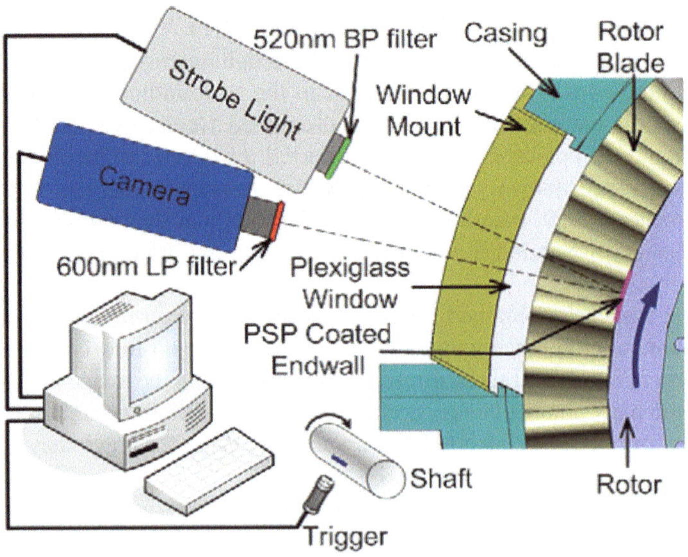

Fig. 15.15 Optical setup for PSP data acquisition

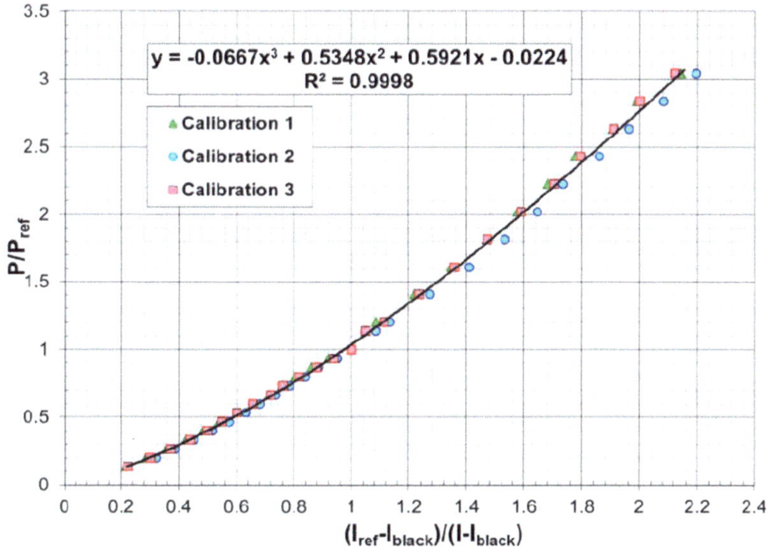

Fig. 15.16 Sample PSP-calibration curve

Similar to any other temperature measurement method, the results obtained by using PSP must be calibrated. The calibration for PSP is performed using a vacuum chamber, where its pressure is from 0 to 2 atm with corresponding emitted intensity that is recorded for each pressure setting.

The calibration curve, Fig. 15.16, is a typical calibration curve for PSP. Generally, the calibration is sensitive to temperature with higher temperatures resulting in lower emitted light intensities. To obtain film cooling effectiveness, air and nitrogen were used alternately as coolant. Nitrogen which has approximately the same molecular weight as the air displaces the oxygen molecules on the surface causing a change in the emitted light intensity from PSP. By noting the difference in emitted light intensity and subsequently the partial pressures between the air and nitrogen injection cases, the film cooling effectiveness can be determined using the following equation:

$$\eta = \frac{C_{air} - C_{mix}}{C_{N_2} - C_{air}} = \frac{\left(P_{O_2}\right)_{air} - \left(P_{O_2}\right)_{mix}}{\left(P_{O_2}\right)_{air}} \tag{15.8}$$

Where C_{air}, C_{mix} and C_{N2} are the oxygen concentrations of mainstream air, air/ nitrogen mixture and nitrogen on the test surface respectively and are directly proportional to the partial pressure of oxygen. The definition of adiabatic film cooling effectiveness is:

$$\eta = \frac{T_f - T_m}{T_C - T_m} \tag{15.9}$$

with T_c, T_f and T_m as coolant temperature, local film temperature and mainstream temperature. In the following we present three case studies dealing with measuring film cooling effectiveness of turbine blades operating at high rpm up to 3000. The results show the capability of measuring the film cooling effectiveness.

15.5 Case Studies, Heat Transfer in Rotating Frame

Because of its engineering relevance, we present three representative cases that deal with film cooling heat transfer in rotating frame. This three cases include measurement of film cooling heat transfer on:

(1) Turbine blade tip rotating at three different rpm,
(2) Turbine leading edge film cooling at three different blowing ratios,
(3) Turbine end wall film cooling at one rpm and six blowing ratios.

The Case Studies reported in this and other chapters of this book are the results of several years of experimental research. In this context it should be noted that establishing an experimental facility, instrumentation, data acquisition and analysis is a very time consuming and expensive undertaking. In case of aerodynamic investigations, the RANS based Computational Fluid Dynamics (CFD) codes can be used that deliver reasonable results. Particularly, it can be utilized for systematic parameter studies. In case of temperature calculation, however, because of coupling the Navier-Sokes equations with energy equation and the lack of properly modeling the impact of the velocity fluctuation components on temperature calculation, errors ranging from 50% to over 100% are observed. It should be noted that, prior to undertaking the experiments that are presented here as Case Studies, extensive CFD-investigation were carried out. Comparing the results from CFD with the experiment revealed the same type of errors mentioned above.

15.5.1 Rotating Heat Transfer, Case I

Figure 12.13 shows the film cooling effectiveness measured on tip of turbine blades with different geometries. Two tip geometries are presented. The first tip geometry of Fig. 12.13a shows the film cooling effectiveness contour for a plane tip with tip film cooling holes.

The second tip geometry belongs to a blade with squealer tip with film cooling holes shown in Fig. 12.13b. Film cooling heat transfer investigation for both blade tips were conducted at 3000.0, 2550.0, and 2000.0 rpm. The experiments were performed at $TPFL$.

The purpose of this experiment was to verify the validity of CFD-results. Particularly to determine whether the number of the holes, the locations of the cooling holes

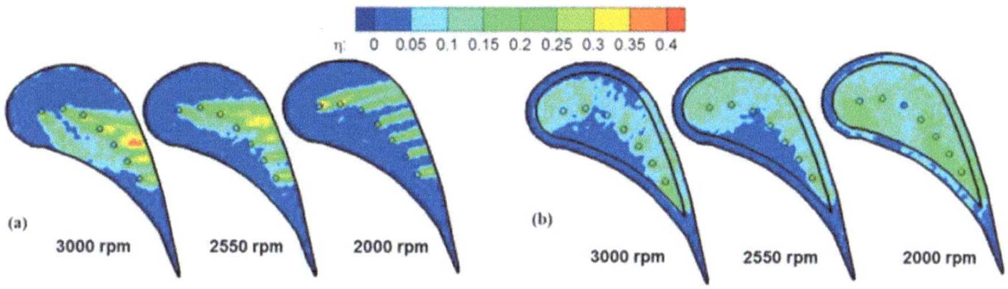

Fig. 15.17 Turbine blade geometries with: **a** plane tip and **b** squealer tip

and the amount of injected cooling air would establish a full coverage of film for the surface to be cooled. As seen from Fig. 15.17, only the tip of the blade with squealer rotating at 2000 rpm delivers a full coverage. As a result, the location of the film cooling holes, their diameters, their distributions and the amount of injected coolant must be redesigned. Comparing the results from CFD with the experiment revealed the same type of errors mentioned above.

15.5.2 Rotating Heat Transfer, Case II

This case shows the film cooling effectiveness distribution on the surface of a turbine blade that is running at 2400 rpm. Three rows of film cooling holes, each having five holes are incorporated in and around the stagnation line. A borescope attached to a traversing system takes data from hub to tip and after analysis using Eqs. (15.1)–(15.3), the effectiveness picture is created as shown in Fig. 15.19. Once the image is obtained, the data has to be further analyzed to obtain quantitative results as the following diagram, Fig. 15.18 shows.

The significance of film cooling experiments for an engineering application is to achieve a uniform surface cooling through the cooling air injection. Looking at Fig. 12.13a it becomes obvious that the film did not cover the entire surface of the turbine blade tip. This information compels the designer to modify the position of the cooling holes. The situation looks quite different, if the blade tip has a squealer geometry. In this case the film is trapped within the squealer, particularly at lower rpm.

The purpose of this experiment was to verify the sufficient film coverage around the leading edge of the validity of the CFD-prediction. The spectrum of the cooling effectiveness is shown in Fig. 15.18 with the minimum $\eta = 0.0$ and the maximum $\eta = 0.6$.

As seen in Fig. 15.19, traversing of the leading edge area from hub to tip of the blade is accomplished by a borescope attached to an Argon Laser system that serves as the exitation source. The emitted light from PSP transmitted via a camera to data acquisition system for further processing, which after data analysis delivers

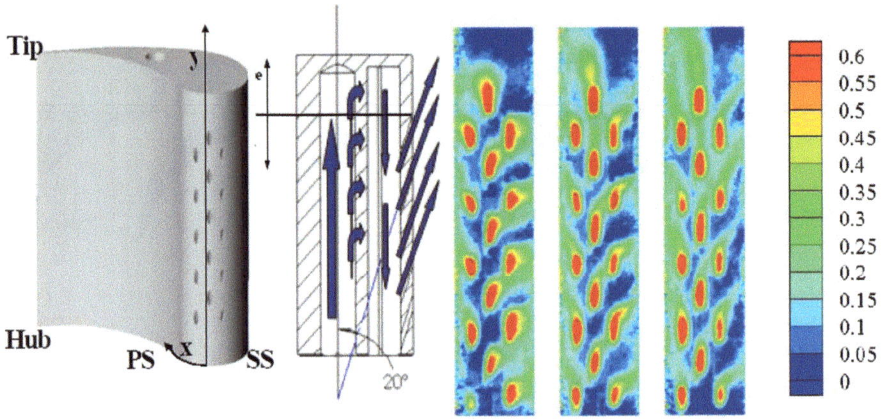

Fig. 15.18 Turbine blade with leading edge film cooling holes (left), ejection of coolant via internal channels through the cooling holes (middle). The contour plot of film cooling effectiveness for three blowing ratios $M = 0.5,\ 1.0\ 2.0$

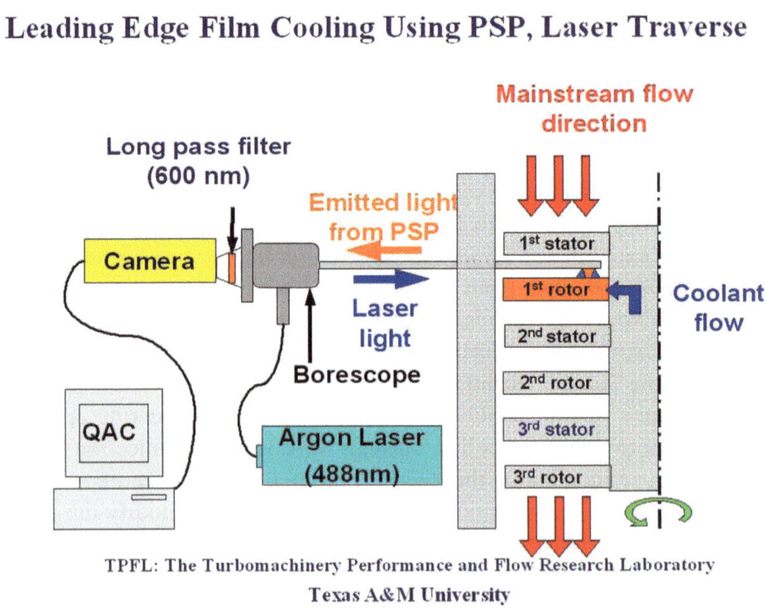

Fig. 15.19 Data acquisition setup and the control system for moving the borescope

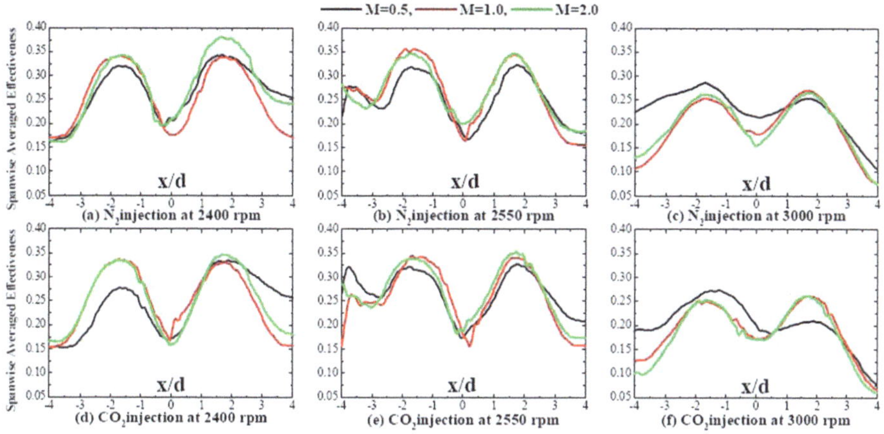

Fig. 15.20 Spanwise averaged quantitative results from the PSP measurement

the contour plots Fig. 15.18. The three columns of film cooling pictures belong to the blowing ratios $M = 0.5,\ 1.0,\ 2.0$ (from the left to right). As seen, the leading edge of the turbine blade is not completely covered. Here again, this information compels the designer to redesign the blade and place more holes at the leading edge.

Finally we have extracted the quantitative results from the PSP measurement and plotted in Fig. 15.20. As seen, the spanwise averaged effectiveness is plotted against the ratio x/D with x as the stream-wise distance from leading edge stagnation and D the hole diameter. The quasi periodic pattern with $\eta_{\max} \cong 0.35$ and $\eta_{\min} \cong 0.15$ shows a significant fluctuation that is not acceptable. With increasing the number of holes and reducing the spacing, a better coverage can be achieved.

15.5.3 Rotating Heat Transfer, Case III

Figure 15.21 shows film-cooling effectiveness on the rotor platform for downstream film cooling holes for the reference speed, 3000 rpm, and all mass flow ratios $M = \frac{\rho_c V_c}{\rho_\infty V_\infty}$. As expected film-cooling effectiveness is maximum near the coolant hole exit. As we proceed downstream of the holes, effectiveness magnitude diminishes as the coolant mixes with the mainstream flow. Peak effectiveness values occur as we approach $M_{holes} = 1$ and are approximately equal to $M_{holes} = 0.7$ exactly where the coolant ejects out of the holes. Effectiveness values and film distribution begin to decrease below and above $M_{holes} = 0.75$ and $M_{holes} = 1.25$, respectively. M_{holes} in this range provides good film-cooling protection on the platform covering most of the downstream passage surface. The contribution of each hole toward effective film coverage also varies depending on its location on the platform surface. Since coolant density is assumed to be the same as that of the mainstream, M_{holes} is dependent

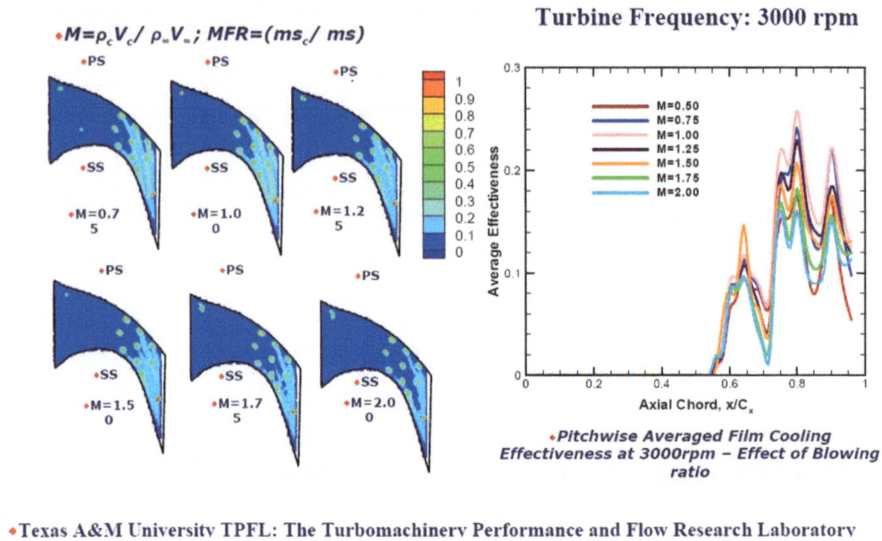

Fig. 15.21 Details of the cooling effectiveness on the blade platform

only on the exit velocity of the coolant gas. For $M_{holes} = 1.0$, the velocity of the coolant ejecting out of the individual holes is approximately the same as that of the mainstream relative velocity at the first stage rotor exit. As the coolant flow velocity approaches the mainstream relative velocity, it appears that the ejected coolant has just the right momentum to adhere to the platform surface, displacing the mainstream boundary layer and minimizing the effects of the secondary flows.

This allows the coolant to provide better film coverage and higher effectiveness magnitudes as minimal coolant is dissipated into the mainstream flow before providing any protection. At $M_{holes} < 0.75$ the coolant quantity for film cooling is small and is incapable of providing any effective protection on the platform surface. The lower momentum prevents the coolant from penetrating the boundary layer on the platform surface, hindering the development of an effective thermal barrier. The low momentum coolant tends to get carried away by the higher momentum mainstream flow decreasing the effectiveness. On the contrary, the ejected coolant for $M_{holes} > 0.75$ possesses larger momentum and has a tendency to lift-off as it leaves the coolant holes.

Appendix A
Tensor Operations in Orthogonal Curvilinear Coordinate Systems

A.1 Change of Coordinate System

The vector and tensor operations we have discussed in the foregoing chapters were performed solely in rectangular coordinate system. It should be pointed out that we were dealing with quantities such as velocity, acceleration, and pressure gradient that are independent of any coordinate system within a certain frame of reference. In this connection it is necessary to distinguish between a coordinate system and a frame of reference. The following example should clarify this distinction. In an absolute frame of reference, the flow velocity vector may be described by the rectangular Cartesian coordinate x_i:

$$\boldsymbol{V} = \boldsymbol{V}(x_1, x_2, x_3) = \boldsymbol{V}(\boldsymbol{X}). \tag{A.1}$$

It may also be described by a cylindrical coordinate system, which is a non-Cartesian coordinate system:

$$\boldsymbol{V} = \boldsymbol{V}(x, r, \theta) \tag{A.2}$$

or generally by any other non-Cartesian or curvilinear coordinate ξ_i that describes the flow channel geometry:

$$\boldsymbol{V} = \boldsymbol{V}(\xi_1, \xi_2, \xi_3). \tag{A.3}$$

By changing the coordinate system, the flow velocity vector will not change. It remains invariant under any transformation of coordinates. This is true for any other quantities such as acceleration, force, pressure or temperature gradient. The concept of invariance, however, is generally no longer valid if we change the frame of reference. For example, if the flow particles leave the absolute frame of reference and enter the relative frame of reference, for example a moving or rotating frame,

M. T. Schobeiri, *Advanced Fluid Mechanics and Heat Transfer for Engineers and Scientists*, https://doi.org/10.1007/978-3-030-72925-7

its velocity will experience a change. In this Chapter, we will pursue the concept of quantity invariance and discuss the fundamentals that are needed for coordinate transformation.

A.2 Co- and Contravariant Base Vectors, Metric Coefficients

As we saw in the previous chapter, a vector quantity is described in Cartesian coordinate system x_i by its components:

$$\boldsymbol{V} = \boldsymbol{e}_i V_i = \boldsymbol{e}_1 V_1 + \boldsymbol{e}_2 V_2 + \boldsymbol{e}_3 V_3 \tag{A.4}$$

with e_i as orthonormal unit vectors (Fig. A.1 left). The same vector transformed into the curvilinear coordinate system ξ_k (Fig. A.1 right) is represented by:

$$\boldsymbol{V} = \boldsymbol{g}_k V^k = \boldsymbol{g}_1 V^1 + \boldsymbol{g}_2 V^2 + \boldsymbol{g}_3 V^3 \tag{A.5}$$

where $\boldsymbol{g}_k$ are the base vectors and $\boldsymbol{V}^k$ the components of $\boldsymbol{V}$ with respect to the base $\boldsymbol{g}_k$ in a curvilinear coordinate system. For curvilinear coordinate system, we place the indices diagonally for summing convenience. Unlike the Cartesian base vectors $\boldsymbol{e}_i$, that are orthonormal vectors (of unit length and mutually orthogonal), the base vectors $\boldsymbol{g}_k$ do not have unit lengths. The base vectors $\boldsymbol{g}_k$ represent the rate of change of the position vector $\boldsymbol{x}$ with respect to the curvilinear coordinates ξ_i.

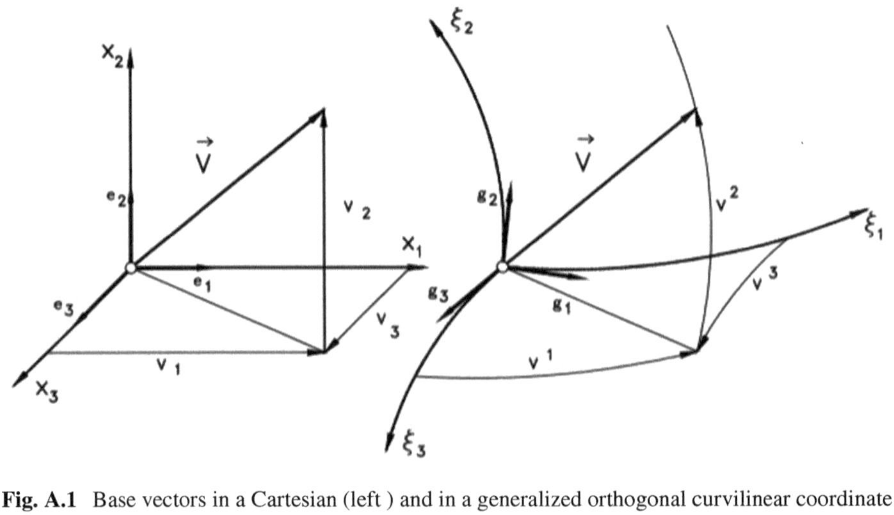

Fig. A.1 Base vectors in a Cartesian (left) and in a generalized orthogonal curvilinear coordinate system (right)

$$\boldsymbol{g}_k = \frac{\partial \boldsymbol{x}}{\partial \xi_k} = \frac{\partial(\boldsymbol{e}_i x_i)}{\partial \xi_k}. \tag{A.6}$$

Since in a Cartesian coordinate system the unit vectors $\boldsymbol{e}_i$, are not functions of the coordinates x_i, Eq. (A.6) can be written as:

$$\boldsymbol{g}_k = \boldsymbol{e}_i \frac{\partial x_i}{\partial \xi_k}. \tag{A.7}$$

Similarly, the *reciprocal base* vector $\boldsymbol{g}^k$ defined as:

$$\boldsymbol{g}^j = \boldsymbol{e}_m \frac{\partial \xi_j}{\partial x_m}. \tag{A.8}$$

As shown in Fig. A.2, the covariant base vectors $\boldsymbol{g}_2$, $\boldsymbol{g}_2$, and $\boldsymbol{g}_3$ are tangent vectors to the mutually orthogonal curvilinear coordinates ξ_1, ξ_2, and ξ_3. The reciprocal base vectors $\boldsymbol{g}a^1$, $\boldsymbol{g}^2$, $\boldsymbol{g}^3$, however, are orthogonal to the planes described by $\boldsymbol{g}_2$ and $\boldsymbol{g}_3$, $\boldsymbol{g}_3$ and $\boldsymbol{g}_1$, and $\boldsymbol{g}_1$ and $\boldsymbol{g}_2$, respectively. These base vectors are interrelated by:

$$\boldsymbol{g}_k \cdot \boldsymbol{g}^j = \boldsymbol{e}_i \cdot \boldsymbol{e}_m \frac{\partial x_i}{\partial \xi_k} \frac{\partial \xi_j}{\partial x_m} = \delta_{im} \frac{\partial x_i}{\partial \xi_k} \frac{\partial \xi_j}{\partial x_m} = \frac{\partial \xi_j}{\partial \xi_k} \equiv \delta_k^j \tag{A.9}$$

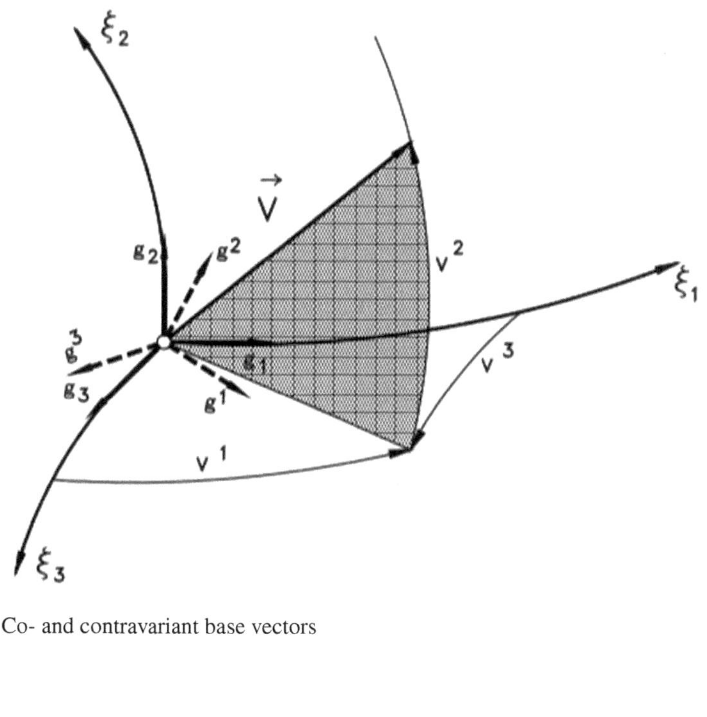

Fig. A.2 Co- and contravariant base vectors

where $\boldsymbol{g}_k$ and $\boldsymbol{g}^j$ are referred to as the covariant and contravariant base vectors, respectively. The new Kronecker delta δ_k^j from Eq. (A.9) has the values:

$$\boldsymbol{g}_k \cdot \boldsymbol{g}^j = \delta_k^j, \delta_k^j = 1 \text{ for } k = j, \delta_k^j = 0 \text{ for } k \neq j.$$

The vector $\boldsymbol{V}$ written relative to its contravariant base is:

$$\boldsymbol{V} = \boldsymbol{g}^k V_k = \boldsymbol{g}^1 V_1 + \boldsymbol{g}^2 V_2 + \boldsymbol{g}^3 V_3. \tag{A.10}$$

Similarly, the components V_k and V^k are called the covariant and contravariant components, respectively. The scalar product of covariant respectively contravariant base vectors results in the covariant and covariant metric coefficients:

$$g_{ij} = \boldsymbol{g}_i \cdot \boldsymbol{g}_j, g^{ij} = \boldsymbol{g}^i \cdot \boldsymbol{g}^j. \tag{A.11}$$

The mixed metric coefficient is defined as

$$g_i^j = \boldsymbol{g}_i \cdot \boldsymbol{g}^j. \tag{A.12}$$

The covariant base vectors can be expressed in terms of the contravariant base vectors. First we assume that:

$$\begin{aligned}
\boldsymbol{g}^1 &= A^{11}\boldsymbol{g}_1 + A^{12}\boldsymbol{g}_2 + A^{13}\boldsymbol{g}_3 \\
\boldsymbol{g}^2 &= A^{21}\boldsymbol{g}_1 + A^{22}\boldsymbol{g}_2 + A^{23}\boldsymbol{g}_3 \\
\boldsymbol{g}^3 &= A^{31}\boldsymbol{g}_1 + A^{32}\boldsymbol{g}_2 + A^{33}\boldsymbol{g}_3.
\end{aligned} \tag{A.13}$$

Generally the contravariant base vector can be written as

$$\boldsymbol{g}^i = A^{ij}\boldsymbol{g}_j. \tag{A.14}$$

To find a direct relation between the base vectors, first the coefficient matrix A^{ij} must be determined. To do so, we multiply Eq. (A.14) with $\boldsymbol{g}^k$ scalarly:

$$\boldsymbol{g}^i \cdot \boldsymbol{g}^k = A^{ij}\boldsymbol{g}_j \cdot \boldsymbol{g}^k. \tag{A.15}$$

This leads to $g^{ik} = A^{ij}\delta_j^k$. The right hand side is different from zero only if $j = k$. That means:

$$g^{ik} = A^{ik}. \tag{A.16}$$

Introducing Eqs. (A.16) into (A.14) results in a relation that expresses the contravariant base vectors in terms of covariant base vectors:

$$\boldsymbol{g}^i = g^{ij}\boldsymbol{g}_j. \tag{A.17}$$

The covariant base vector can also be expressed in terms of contravariant base vectors in a similar way:

$$\boldsymbol{g}_k = g_{kl}\boldsymbol{g}^l. \tag{A.18}$$

Multiply Eq. (A.18) with (A.17) establishes a relationship between the covariant and contravariant metric coefficients:

$$\boldsymbol{g}^i \cdot \boldsymbol{g}_k = g^{ij} g_{kl} \boldsymbol{g}_j \cdot \boldsymbol{g}^l, \text{ and } \delta^i_k = g^{ij} g_{kl} \delta^1_j. \tag{A.19}$$

Applying the Kronecker delta on the right hand side results in:

$$g^{ij} g_{kj} = \delta^i_k. \tag{A.20}$$

A.3 Physical Components of a Vector

As mentioned previously, the base vectors $\boldsymbol{g}_i$ or $\boldsymbol{g}^j$ are not unit vectors. Consequently the co- and contravariant vector components V_j or V^1 do not reflect the physical components of vector $\boldsymbol{V}$. To obtain the physical components, first the corresponding unit vectors must be found. They can be obtained from:

$$\boldsymbol{g}^*_i = \frac{\boldsymbol{g}_i}{|\boldsymbol{g}_i|} = \frac{\boldsymbol{g}_i}{\sqrt{\boldsymbol{g}_i \cdot \boldsymbol{g}_i}} = \frac{\boldsymbol{g}_i}{\sqrt{g_{(ii)}}}. \tag{A.21}$$

Similarly, the contravariant unit vectors are:

$$\boldsymbol{g}^{*i} = \frac{\boldsymbol{g}^i}{|\boldsymbol{g}^i|} = \frac{\boldsymbol{g}^i}{\sqrt{\boldsymbol{g}^i \cdot \boldsymbol{g}^i}} = \frac{\boldsymbol{g}^i}{\sqrt{g^{(ii)}}} \tag{A.22}$$

where $\boldsymbol{g}^*_i$, represents the unit base vector, $|\boldsymbol{g}^i|$ the absolute value of the base vector. The expression (ii) denotes that no summing is carried out, whenever the indices are enclosed within parentheses. The vector can now be expressed in terms of its unit base vectors and the corresponding physical components:

$$\boldsymbol{V} = \boldsymbol{g}_i V^i = \boldsymbol{g}^*_i V^{*i} = \frac{\boldsymbol{g}_i}{\sqrt{g_{(ii)}}} V^{*i}. \tag{A.23}$$

Thus the covariant and contravariant physical components can be easily obtained from:

$$V^*_i = \sqrt{g^{(ii)}} V_i, \; V^{*i} = \sqrt{g_{(ii)}} V^i. \tag{A.24}$$

A.4 Derivatives of the Base Vectors, Christoffel Symbols

In a curvilinear coordinate system, the base vectors are generally functions of the coordinates itself. This fact must be considered while differentiating the base vectors. Consider the derivative:

$$\boldsymbol{g}_{i,j} \equiv \frac{\partial \boldsymbol{g}_i}{\partial \xi_j} = \frac{\partial}{\partial \xi_j}\left(e_k \frac{\partial x_k}{\partial \xi_i}\right) = e_k \frac{\partial^2 x_k}{\partial \xi_j \partial \xi_i}. \tag{A.25}$$

Similar to Eq. (A.7), the unit vector $\boldsymbol{e}_k$ can be written:

$$\boldsymbol{e}_k = \boldsymbol{g}_n \frac{\partial \xi_n}{\partial x_k}. \tag{A.26}$$

Introducing Eq. (A.26) into (A.25) yields:

$$\boldsymbol{g}_{i,j} = \frac{\partial^2 x_k}{\partial \xi_j \partial \xi_i} \frac{\partial \xi_n}{\partial x_k} \boldsymbol{g}_n \equiv \Gamma^n_{ij} \boldsymbol{g}_n = \Gamma_{ijn} \boldsymbol{g}^n \tag{A.27}$$

with Γ_{ijn}, and Γ^n_{ij} as the Christoffel symbol of first and second kind, respectively with the definition:

$$\Gamma_{ijn} = \frac{\partial^2 x_k}{\partial \xi_i \partial \xi_j} \frac{\partial \xi_k}{\partial x_n}, \Gamma^n_{ij} = \frac{\partial^2 x_k}{\partial \xi_j \partial \xi_i} \frac{\partial \xi_n}{\partial x_k}. \tag{A.28}$$

From (A.28) follows that the Christoffel symbols of the second kind is related the first kind by:

$$\Gamma^k_{ij} = \Gamma_{ijm} g^{mk}. \tag{A.29}$$

Since the Christoffel symbols convertible by using the metric coefficients, for the sake of simplicity, in what follows, we use the second kind. The derivative of contravariant base vector is:

$$\boldsymbol{g}^j_i \equiv \frac{\partial \boldsymbol{g}^j}{\partial \xi_i} = -\Gamma^j_{ik} \boldsymbol{g}^k. \tag{A.30}$$

The Christoffel symbols are then obtained by expanding Eq. (A.28):

$$\Gamma^k_{ml} = \Gamma^k_{lm} = \frac{1}{2} g^{kn} (g_{mn,l} + g_{nl,m} - g_{lm,n}) \tag{A.31}$$

$$\Gamma^k_{ml} = \Gamma^k_{lm} = \frac{1}{2} g^{(kk)} (g_{mk,l} + g_{kl,m} - g_{lm,k}). \tag{A.32}$$

In Eq. (A.31), the Christoffel symbols are symmetric in their lower indices. Furthermore, the fact that the only non-zero elements of the metric coefficients are the diagonal elements allowed the modification of the first equation in (A.31) to arrive at (A.32). Again, note that a repeated index in parentheses in an expression such as $g^{(kk)}$ does not subject to summation.

A.5 Spatial Derivatives in Curvilinear Coordinate System

The differential operator ∇, Nabla, is in curvilinear coordinate system defined as:

$$\nabla = g^i \frac{\partial}{\partial \xi_i}. \tag{A.33}$$

A.5.1 Application of ∇ to Tensor Functions

In this chapter, the operator ∇ will be applied to different arguments such as zeroth, first and second order tensors. If the argument is a zeroth order tensor which is a scalar quantity such as pressure or temperature, the results of the operation is the gradient of the scalar field which is a vector quantity:

$$\nabla p = \boldsymbol{g}^i \frac{\partial p}{\partial \xi_i} \equiv \boldsymbol{g}^i p_i. \tag{A.34}$$

The abbreviation "i" refers to the derivative of the argument, in this case p, with respect to the coordinate ξ_i. If the argument is a first order tensor such as a velocity vector, the order of the resulting tensor depends on the operation character between the operator ∇ and the argument. For divergence and curl of a vector using the chain rule, the differentiations are :

$$\nabla \cdot \boldsymbol{V} = \left(\boldsymbol{g}^i \frac{\partial}{\partial \xi_i}\right) \cdot (\boldsymbol{g}_j V^j) = \boldsymbol{g}^i \cdot \left(\frac{\partial \boldsymbol{g}_j}{\partial \xi_i} V^j + \frac{\partial \boldsymbol{V}^j}{\partial \xi_i} g_j\right) \tag{A.35}$$

$$\nabla \times \boldsymbol{V} = \left(\boldsymbol{g}^i \frac{\partial}{\partial \xi_i}\right) \times (\boldsymbol{g}^j V_j) = \mathbf{g^i} \times \mathbf{g^j V_{j,i}} + \boldsymbol{g}^\mathbf{i} \times \mathbf{g_i^j V_j}. \tag{A.36}$$

Implementing the Christoffel symbol, the results of the above operations are the divergence and the curl of the vector $\boldsymbol{V}$. It should be noticed that a scalar operation leads to a contraction of the order of tensor on which the operator is acting. The scalar operation in (A.35) leads to:

$$\nabla \cdot \boldsymbol{V} = V_i^i + V^J \Gamma_{ij}^i. \tag{A.37}$$

The vector operation yields the rotation or curl of a vector field as:

$$\nabla \times \boldsymbol{V} = \boldsymbol{g}^i \times \boldsymbol{g}^j (V_{j,i} - \Gamma^k_{ij} V_k) = \frac{1}{\sqrt{g}} \epsilon^{ijk} \boldsymbol{g}_k (V_{j,i} - \Gamma^k_{ij} V_k) \tag{A.38}$$

with ϵ^{ijk} as the permutation symbol that functions similar to the one for Cartesian coordinate system and $\sqrt{g} = \sqrt{|g_{ij}|}$.

The gradient of a first order tensor such as the velocity vector **V** is a second order tensor. Its index notation in a curvilinear coordinate system is:

$$\nabla \boldsymbol{V} = \boldsymbol{g}^i \boldsymbol{g}_j (V^j_i + V^k \Gamma^j_{ik}). \tag{A.39}$$

A scalar operation that involves ∇ and a second order tensor, such as the stress tensor $\boldsymbol{\Pi}$ or deformation tensor $\boldsymbol{D}$, results in a first order tensor which is a vector:

$$\nabla \cdot \Pi = \left(\boldsymbol{g}^m \frac{\partial}{\partial \xi_m} \right) \cdot (\boldsymbol{g}_i \boldsymbol{g}_j \pi^{ij}) = \boldsymbol{g}^m \cdot (\boldsymbol{g}_k \boldsymbol{g}_l)(\pi^{kl}_m + \pi^{nl} \Gamma^k_{nm} + \pi^{kn} \Gamma^1_{nm}). \tag{A.40}$$

The right hand side of (A.40) is reduced to:

$$\nabla \cdot \Pi = \boldsymbol{g}_j (\pi^{mj}_m + \pi^{nj} \Gamma^m_{nm} + \pi^{mn} \Gamma^j_{mn}). \tag{A.41}$$

By calculating the shear forces using the Navier-Stokes equation, the second derivative, the Laplace operator Δ, is needed:

$$\Delta = \nabla \cdot \nabla = \nabla^2 = \left(\boldsymbol{g}^i \frac{\partial}{\partial \xi_i} \right) \cdot \left(\mathbf{g}^{\mathbf{j}} \frac{\partial}{\partial \xi_{\mathbf{j}}} \right). \tag{A.42}$$

This operator applied to the velocity vector yields:

$$\Delta \boldsymbol{V} = \boldsymbol{g}_m g^{ik} [V^m_{ik} + V^n_i \Gamma^m_{nk} + V^n_k \Gamma^m_{ni} - V^m_j \Gamma^j_{ik} + V^p (\Gamma^n_{pi} \Gamma^m_{nk} - \Gamma^j_{ik} \Gamma^m_{pj} + \Gamma^m_{pi,k})]. \tag{A.43}$$

A.6 Application Example 1: Inviscid Incompressible Flow Motion

As the first application example, the equation of motion for an inviscid incompressible and steady low is transformed into a cylindrical coordinate system, where it is decomposed in its three components r, θ, z. The coordinate invariant version of the equation is written as:

$$\boldsymbol{V} \cdot \nabla \boldsymbol{V} = -\frac{1}{\rho} \nabla p. \tag{A.44}$$

The transformation and decomposition procedure is shown in the following steps.

A.6.1 Equation of Motion in Curvilinear Coordinate Systems

The second order tensor on the left hand side can be obtained using Eq. (A.38):

$$\nabla \boldsymbol{V} = \boldsymbol{g}^i \boldsymbol{g}_j (V_i^j + V^k \Gamma_{ik}^j). \tag{A.45}$$

The scalar multiplication with the velocity vector $\boldsymbol{V}$ leads to:

$$\boldsymbol{V} \cdot \nabla \boldsymbol{V} = \boldsymbol{g}_m \boldsymbol{V}^m \cdot \boldsymbol{g}^i \boldsymbol{g}_j (V_i^j + V^k \Gamma_{ik}^j). \tag{A.46}$$

Introducing the mixed Kronecker delta:

$$\boldsymbol{V} \cdot \nabla \boldsymbol{V} = \delta_m^i \boldsymbol{g}_j V^m (V_i^j + V^k \Gamma_{ik}^j). \tag{A.47}$$

For an orthogonal curvilinear coordinate system the mixed Kronecker delta is:

$$\begin{aligned} \delta_m^i &= 1 \text{ for } i = m \\ \delta_m^i &= 0 \text{ for } i \neq m. \end{aligned} \tag{A.48}$$

Taking this into account, Eq. (A.47) yields:

$$\boldsymbol{V} \cdot \nabla \boldsymbol{V} = \boldsymbol{g}_j \boldsymbol{V}^i (V_i^j + V^k \Gamma_{ik}^j). \tag{A.49}$$

Rearranging the indices

$$\boldsymbol{V} \cdot \nabla \boldsymbol{V} = \boldsymbol{g}_i (V^j V_j^i + V^j V^k \Gamma_{kj}^i). \tag{A.50}$$

The pressure gradient on the right hand side of Eq. (A.43) is calculated form Eq. (A.32):

$$\nabla p = \boldsymbol{g}^i \frac{\partial p}{\partial \xi_i} = \boldsymbol{g}^i p_i. \tag{A.51}$$

Replacing the contravariant base vector with the covariant one using Eq. (A.51) leads to:

$$\nabla p = \boldsymbol{g}^i \frac{\partial p}{\partial \xi_i} = \boldsymbol{g}_i g^{ji} p_j. \tag{A.52}$$

Incorporating Eqs (A.50) and (A.52) into Eq. (A.43) yields:

$$\boldsymbol{g}_i (V^j V_j^i + V^j V^k \Gamma_{kj}^i) = -\frac{1}{\rho} \boldsymbol{g}_i g^{ji} p_j. \tag{A.53}$$

In i-direction, the equation of motion is:

$$V^j V^i_j + V^j V^k \Gamma^i_{kj} = -\frac{1}{\rho} g^{ji} p, j. \tag{A.54}$$

A.6.2 Special Case: Cylindrical Coordinate System

To transfer Eq. (A.40) in any arbitrary curvilinear coordinate system, first the coordinate system must be specified. The cylinder coordinate system is related to the Cartesian coordinate system is given by:

$$x_1 = r\cos\Theta, x_2 = r\sin\Theta, x_3 = z. \tag{A.55}$$

The curvilinear coordinate system is represented by:

$$\xi_1 = r, \xi_2 = \Theta, \xi_3 = z. \tag{A.56}$$

A.6.3 Base Vectors, Metric Coefficients

The base vectors are calculated from Eq. (A.7).

$$\boldsymbol{g}_k = \boldsymbol{e}_i \frac{\partial x_i}{\partial \xi_k}. \tag{A.57}$$

Equation (A.57) decomposed in its components yields:

$$\begin{aligned}
\boldsymbol{g}_1 &= \boldsymbol{e}_1 \frac{\partial x_1}{\partial \xi_1} + e_2 \frac{\partial x_2}{\partial \xi_1} + e_3 \frac{\partial x_3}{\partial \xi_1} \\
\boldsymbol{g}_2 &= e_1 \frac{\partial x_1}{\partial \xi_2} + e_2 \frac{\partial x_2}{\partial \xi_2} + e_3 \frac{\partial x_3}{\partial \xi_2} \\
\boldsymbol{g}_3 &= e_1 \frac{\partial x_1}{\partial \xi_3} + e_2 \frac{\partial x_2}{\partial \xi_3} + e_3 \frac{\partial x_3}{\partial \xi_3}.
\end{aligned} \tag{A.58}$$

The differentiation of the Cartesian coordinates yields:

$$\begin{aligned}
\boldsymbol{g}_1 &= e_1 \cos\theta + e_2 \sin\theta \\
\boldsymbol{g}_2 &= -e_1 r \sin\theta + e_2 r \cos\theta \\
\boldsymbol{g}_3 &= e_3.
\end{aligned} \tag{A.59}$$

The co- and contravariant metric coefficients are:

$$(g_{ij}) = \begin{pmatrix} 1 \ 0 \ 0 \\ 0 \ r^2 \ 00 \ 0 \ 1 \end{pmatrix}, (g^{ij}) = \begin{pmatrix} 1 & 0 & 0 \\ 0 & 1/r^2 & 0 \\ 0 & 0 & 1 \end{pmatrix}. \tag{A.60}$$

The contravariant base vectors are obtained from:

$$\begin{aligned} \boldsymbol{g}^i &= g^{ij}\boldsymbol{g}_j \\ \boldsymbol{g}^1 &= g^{11}\boldsymbol{g}_1 + g^{12}\boldsymbol{g}_2 + g^{13}\boldsymbol{g}_3 \\ \boldsymbol{g}^2 &= g^{21}\boldsymbol{g}_1 + g^{22}\boldsymbol{g}_2 + g^{23}\boldsymbol{g}_3 \\ \boldsymbol{g}^3 &= g^{31}\boldsymbol{g}_1 + g^{32}\boldsymbol{g}_2 + g^{33}\boldsymbol{g}_3. \end{aligned} \tag{A.61}$$

Since the mixed metric coefficient are zero, Eq. (A.61) reduces to:

$$\boldsymbol{g}^1 = g^{11}\boldsymbol{g}_1, \boldsymbol{g}^2 = g^{22}\boldsymbol{g}_2, \boldsymbol{g}^3 = g^{33}\boldsymbol{g}_3. \tag{A.62}$$

A.6.4 Christoffel Symbols

The Christoffel symbols are calculated from Eq. (A.30)

$$\Gamma^k_{ml} = \Gamma^k_{lm} = \frac{1}{2} g^{(kk)}(g_{mk,l} + g_{kl,m} - g_{lm,k}). \tag{A.63}$$

To follow the calculation procedure, one zero- element and one non-zero element are calculated:

$$\begin{aligned} \Gamma^1_{11} &= \frac{1}{2} g^{11}\left(\frac{\partial g_{11}}{\partial \xi_1} + \frac{\partial g_{11}}{\partial \xi_1} - \frac{\partial g_{11}}{\partial \xi_1}\right) = 0 \\ \Gamma^1_{22} &= \frac{1}{2} g^{11}\left(\frac{\partial g_{21}}{\partial \xi_2} + \frac{\partial g_{12}}{\partial \xi_2} - \frac{\partial g_{22}}{\partial \xi_1}\right) = -r. \end{aligned} \tag{A.64}$$

All other elements are calculated similarly. They are shown in the following matrices:

$$(\Gamma^1_{lm}) = \begin{pmatrix} 0 & 0 & 0 \\ 0 & -r & 0 \\ 0 & 0 & 0 \end{pmatrix}, (\Gamma^2_{lm}) = \begin{pmatrix} 0 & 1/r & 0 \\ 1/r & 0 & 0 \\ 0 & 0 & 0 \end{pmatrix}, (\Gamma^3_{lm}) = \begin{pmatrix} 0 & 0 & 0 \\ 0 & 0 & 0 \\ 0 & 0 & 0 \end{pmatrix}. \tag{A.65}$$

Introducing the non-zero Christoffel symbols into Eq. (A.54), the components in g_l, g_2, and g_3 directions are:

$$V^1 V^1_1 + V^2 V^1_2 + V^3 V^1_3 + \Gamma^1_{22} V^2 V^2 = -\frac{1}{\rho} g^{11} p_1 \tag{A.66}$$

$$V^1 V_1^2 + V^2 V_2^2 + V^3 V_3^2 + 2\Gamma_{21}^2 V^2 V^1 = -\frac{1}{\rho} g^{22} p_2 \tag{A.67}$$

$$V^1 V_1^3 + V^2 V_2^3 + V^3 V_3^3 = -\frac{1}{\rho} g^{33} p_3. \tag{A.68}$$

A.6.5 Introduction of Physical Components

The physical components can be calculated from Eqs. (A.21) and (A.24):

$$\begin{aligned} &V_i^* = \sqrt{g^{(ii)}} V_i,\ V^{*i} = \sqrt{g_{(ii)}} V^i \\ &V^{*1} = \sqrt{g_{(11)}} V_1,\ V^{*2} = \sqrt{g_{(22)}} V^2;\ V^{*3} = \sqrt{g_{33}} V^3 \\ &V^{*1} = \sqrt{1} V^1;\ V^{*2} = \sqrt{r^2} V^2;\ V^{*3} = \sqrt{1} V^3. \end{aligned} \tag{A.69}$$

The V^i-components expressed in terms of V^{*i} are:

$$V^1 = V^{*1};\ V^2 = \frac{1}{r} V^{*2};\ V^3 = V^{*3}. \tag{A.70}$$

Introducing Eq. (A.70) into Eqs. (A.66)–(A.68) results in:

$$V^{*1} V_1^{*1} + \frac{V^{*2}}{r} V_2^{*1} + V^{*3} V_3^{*1} + \Gamma_{22}^1 \frac{V^{*2} V^{*2}}{r^2} = -\frac{1}{\rho} g^{11} p_{,1} \tag{A.71}$$

$$V^{*1} \frac{V_{,1}^{*2}}{r} - V^{*1} \frac{V^{*2}}{r^2} + \frac{V^{*2}}{r^2} V_{,2}^{*2} + V^{*3} \frac{V_{,3}^{*2}}{r} + \frac{2}{r} \Gamma_{21}^2 V^{*2} V^{*1} = -\frac{1}{\rho} g^{22} p_{,2} \tag{A.72}$$

$$V^{*1} V_{,1}^{*3} + \frac{V^{*2}}{r} V_{,2}^{*3} + V^{*3} V_{,3}^{*3} = -\frac{1}{\rho} g^{33} p_{,2}. \tag{A.73}$$

According to the definition:

$$\xi_1 = r;\ \xi_2 = \Theta;\ \xi_3 = z \tag{A.74}$$

the physical components of the velocity vectors are:

$$V^{*1} = V_r;\ V^{*2} = V_\Theta;\ V^{*3} = V_z \tag{A.75}$$

and insert these relations into Eqs. (A.71)–(A.73), the resulting components in r, Θ, and z directions are:

$$
\begin{aligned}
V_r\frac{\partial V_r}{\partial r}+\frac{V_\Theta}{r}\frac{\partial V_r}{\partial \Theta}+V_z\frac{\partial V_r}{\partial z}-\frac{V_\Theta^2}{r}&=-\frac{1}{\rho}\frac{\partial p}{\partial r}\\
V_r\frac{\partial V_\Theta}{\partial r}+\frac{V_\Theta}{r}\frac{\partial V_\Theta}{\partial \Theta}+V_z\frac{\partial V_\Theta}{\partial z}+\frac{V_r V_\Theta}{r}&=-\frac{1}{\rho}\frac{\partial p}{r\partial \Theta}\\
V_r\frac{\partial V_z}{\partial r}+\frac{V_\Theta}{r}\frac{\partial V_z}{\partial \Theta}+V_z\frac{\partial V_z}{\partial z}&=-\frac{1}{\rho}\frac{\partial p}{\partial z}.
\end{aligned}
\tag{A.76}
$$

A.7 Application Example 2: Viscous Flow Motion

As the second application example, the Navier-Stokes equation of motion for a viscous incompressible flow is transferred into a cylindrical coordinate system, where it is decomposed in its three components r, θ, z. The coordinate invariant version of the equation is written as:

$$
\boldsymbol{V}\cdot\nabla\boldsymbol{V}=-\frac{1}{\rho}\nabla p+\nu\nabla^2\boldsymbol{V}. \tag{A.77}
$$

The second term on the right hand side of Eq. (A.77) exhibits the shear stress force. It was treated in Sect. A.5, Eq. (A.42) and is the only term that has been added to the equation of motion for inviscid flow, Eq. (A.44).

A.7.1 Equation of Motion in Curvilinear Coordinate Systems

The transformation and decomposition procedure is similar to the example in Section A.6. Therefore, a step by step derivation is not necessary.

$$
\begin{aligned}
\boldsymbol{g}_i(V^jV^i_{,j}+V^jV^k\Gamma^i_{kj})=-\frac{1}{\rho}\boldsymbol{g}_ig^{ji}p_{,j}+\nu\boldsymbol{g}_m[&V^m_{,ik}+\\
&V^n_{,i}\Gamma^m_{nk}+V^n_{,k}\Gamma^m_{ni}-V^m_{,j}\Gamma^j_{ik}+\\
&V^p(\Gamma^n_{pi}\Gamma^m_{nk}-\Gamma^j_{ik}\Gamma^m_{pj}+\Gamma^m_{pi,k})]g^{ik}.
\end{aligned}
\tag{A.78}
$$

A.7.2 Special Case: Cylindrical Coordinate System

Using the Christoffel symbols from Sect. A.6.4 and the physical components from Sect. A.6.5, and inserting the corresponding relations these relations into Eq. (A.78), the resulting components in r, Θ, and z directions are:

$$V_r \frac{\partial V_r}{\partial r} + \frac{V_\Theta}{r} \frac{\partial V_r}{\partial \theta} + V_z \frac{\partial V_r}{\partial z} - V_{\frac{\Theta^2}{r}} = -\frac{1}{\rho} \frac{\partial p}{\partial r} +$$
$$\nu \left(\frac{\partial^2 V_r}{\partial r^2} + \frac{1}{r^2} \frac{\partial^2 V_r}{\partial \Theta^2} + \frac{\partial^2 V_r}{\partial z^2} - 2\frac{\partial V_\Theta}{r^2 \partial \Theta} + \frac{\partial V_r}{r \partial r} - \frac{V_r}{r^2} \right) \tag{A.79}$$

$$V_r \frac{\partial V_\Theta}{\partial r} + \frac{V_\Theta}{r} \frac{\partial V_\Theta}{\partial \Theta} + V_z \frac{\partial V_\Theta}{\partial z} + \frac{V_r V_\Theta}{r} = -\frac{1}{\rho} \frac{\partial p}{r \partial \Theta}$$
$$+ \nu \left(\frac{\partial^2 V_\Theta}{\partial r^2} + \frac{1}{r^2} \frac{\partial^2 V_\Theta}{\partial \Theta^2} + \frac{\partial^2 V_\Theta}{\partial z^2} + \frac{2}{r^2} \frac{\partial V_r}{r^2 \partial \Theta} + \frac{1}{r} \frac{\partial V_\Theta}{\partial r} - \frac{V_\Theta}{r^2} \right)$$
$$V_r \frac{\partial V_z}{\partial r} + \frac{V_\Theta}{r} \frac{\partial V_z}{\partial z} + V_z \frac{\partial V_z}{\partial z} = -\frac{1}{\rho} \frac{\partial p}{\partial z}$$
$$+ \nu \left[\frac{\partial^2 V_z}{\partial r^2} + \frac{\partial^2 V_z}{r^2 \partial \Theta^2} + \frac{\partial^2 V_z}{\partial z^2} + \frac{1}{r} \frac{\partial V_z}{\partial r} \right]. \tag{A.80}$$

Appendix B
Physical Properties of Dry Air

Table B.1 Enthalpy h, specific heat at constant pressure c_p, entropy s, viscosity μ and thermal conductivity κ as a function of temperature T pressure $p = 1$ bar

T (C)	h (kJ/kg)	c_p (kJ/kg K)	s (kJ/kg K)	μ (kg/ms)10^6	κ (J/ms K)10^3
0.000	0.010	1.003	6.774	12.794	24.210
10.000	10.043	1.003	6.811	17.744	24.893
20.000	20.080	1.004	6.845	18.190	25.571
30.000	30.121	1.004	6.879	18,632	26.243
40.000	40.167	1.005	6.912	19.069	26.910
50.000	50.219	1.005	6.943	19.503	27.572
60.000	60.277	1.006	6.974	19.933	28.229
70.000	70.343	1.007	7.004	20.359	28.880
80.000	80.417	1.008	7.033	20.781	29.527
90.000	90.500	1.009	7.061	21.199	30.169
100.000	100.593	1.010	7.088	21.613	30.806
110.000	110.697	1.011	7.115	22.024	31.439
120.000	120.812	1.012	7.141	22.431	32.067
130.000	130.940	1.013	7.166	22.834	32.690
140.000	141.080	1.015	7.191	23.234	33.309
150.000	151.235	1.016	7.216	23.630	33.924
160.000	161.404	1.018	7.239	24.023	34.534
170.000	171.588	1.019	7.263	24.412	35.140
180.000	181.788	1.021	7.285	24.798	35.742
190.000	192.004	1.022	7.308	25.180	36.340
200.000	202.238	1.024	7.329	25.559	36.934
210.000	212.489	1.026	7.351	25.935	37.524
220.000	222.759	1.028	7.372	26.308	38.110
230.000	233.047	1.030	7.393	26.677	38.692
240.000	243.355	1.032	7.413	27.043	39.271
250.000	253.683	1.034	7.433	27.407	39.846
260.000	264.032	1.036	7.452	27.767	40.417
270.000	274.401	1.038	7.472	28.124	40.985
280.000	284.791	1.040	7.491	28.478	41.549
290.000	295.203	1.042	7.509	28.829	2.1104
300.000	305.637	1.044	7.528	29.177	42.667

M. T. Schobeiri, *Advanced Fluid Mechanics and Heat Transfer for Engineers and Scientists*, https://doi.org/10.1007/978-3-030-72925-7

Table B.2 Enthalpy h, specific heat at constant pressure c_p, entropy s, viscosity μ and thermal conductivity κ as a function of temperature T pressure $p = 1$ bar

T (C)	h (kJ/kg)	C_p (kJ/kg K)	s (kJ/kg K)	μ (kg/ms)10^6	κ (J/ms K)10^3
300.000	305.637	1.044	7.528	29.177	42.667
310.000	316.093	1.047	7.546	29.523	43.221
320.000	326.572	1.049	7.564	29.865	43.772
330.000	337.074	1.051	7.581	30.205	44.320
340.000	347.598	1.054	7.598	30.542	44.865
350.000	358.146	1.056	7.615	30.877	45.406
360.000	368.718	1.058	7.632	31.209	45.945
370.000	379.313	1.061	7.649	31.538	46.481
380.000	389.932	1.063	7.665	31.864	47.013
390.000	400.575	1.065	7.681	32.188	47.543
400.000	411.242	1.068	7.697	32.510	48.070
410.000	421.933	1.070	7.713	32.829	48.595
420.000	432.648	1.073	7.729	33.145	49.116
430.000	443.388	1.075	7.744	33.459	49.635
440.000	454.151	1.078	7.759	33.771	50.151
450.000	464.939	1.080	7.774	34.081	50.665
460.000	475.751	1.082	7.789	34.388	51.177
470.000	486.587	1.085	7.804	34.693	51.685
480.000	497.448	1.087	7.818	34.995	52.192
490.000	508.332	1.090	7.833	35.296	52.696
500.000	519.240	1.092	7.847	35.594	53.197
510.000	530.172	1.094	7.861	35.890	53.697
520.000	541.128	1.097	7.875	36.184	54.194
530.000	552.107	1.099	7.889	36.476	54.688
540.000	563.110	1.101	7.902	36.766	55.181
550.000	574.135	1.104	7.916	37.054	55.671
560.000	585.184	1.106	7.929	37.340	56.160
570.000	596.256	1.108	7.942	37.624	56.646
580.000	607.351	1.111	7.955	37.907	57.130
590.000	618.468	1.113	7.968	38.187	57.612
600.000	629.607	1.115	7.981	38.465	58.092
610.000	640.769	1.117	7.994	38.742	58.570
620.000	651.952	1.119	8.006	39.017	59.046
630.000	663.157	1.122	8.019	39.290	59.521
640.000	674.384	1.124	8.031	39.561	59.993
650.000	685.631	1.126	8.044	39.831	60.464
660.000	696.900	1.128	8.056	40.099	60.932
670.000	708.190	1.130	8.068	40.365	61.339
680.000	719.500	1.132	8.080	40.630	61.864
690.000	730.830	1.134	8.091	40.893	62.327

Table B.3 Enthalpy h, specific heat at constant pressure c_p, entropy s, viscosity μ and thermal conductivity κ as a function of temperature T pressure $p = 1$ bar

T (C)	h (kJ/kg)	C_p (kJ/kg K)	s (kJ/kg K)	μ (kg/ms)10^6	κ (J/ms K)10^3
710.000	753.550	1.138	8.115	41.415	63.249
720.000	764.940	1.140	8.126	41.673	63.707
730.000	776.349	1.142	8.138	41.930	64.163
740.000	787.777	1.144	8.149	42.186	64.618
750.000	799.223	1.146	8.160	42.440	65.071
760.000	810.689	1.147	8.172	42.692	65.522
770.000	822.172	1.149	8.183	42.944	65.972
780.000	833.674	1.151	8.194	43.193	66.420
790.000	845.193	1.153	8.204	43.442	66.866
800.000	856.730	1.155	8.215	43.689	67.311
810.000	868.284	1.156	8.226	43.935	67.754
820.000	879.855	1.158	8.237	44.180	68.196
830.000	891.443	1.160	8.247	44.423	68.636
840.000	903.047	1.162	8.258	44.665	69.075
850.000	914.669	1.163	8.268	44.906	69.511
860.000	926.306	1.165	8.278	45.146	69.947
870.000	937.959	1.166	8.289	45.384	70.381
880.000	949.627	1.168	8.299	45.621	70.813
890.000	961.311	1.169	8.309	45.857	71.243
900.000	973.011	1.171	8.319	46.093	71.672
910.000	984.725	1.172	8.329	46.326	72.100
920.000	996.454	1.174	8.339	46.559	72.526
930.000	1.008.198	1.175	8.348	46.791	72.950
940.000	1.019.956	1.177	8.358	47.022	73.373
950.000	1.031.728	1.178	8.368	47.251	73.794
960.000	1.043.515	1.179	8.377	47.480	74.213
970.000	1.055.315	1.181	8.387	47.708	74.631
980.000	1.067.129	1.182	8.396	47.934	75.047
990.000	1.078.956	1.183	8.406	48.160	75.462
1.000.000	1.090.796	1.185	8.415	48.385	75.875
1.010.000	1.102.650	1.186	8.424	48.609	76.286
1.020.000	1.114.516	1.187	8.434	48.832	76.696
1.030.000	1.126.395	1.189	8.443	49.054	77.104
1.040.000	1.138.287	1.190	8.452	49.275	77.511
1.050.000	1.150.191	1.191	8.461	49.495	77.915
1.060.000	1.162.108	1.192	8.470	49.714	78.318
1.070.000	.174.0361	1.193	8.479	49.932	78.719
1.080.000	1.185.977	1.195	8.488	50.150	79.119
1.090.000	1.197.929	1.196	8.496	50.367	79.516
1.100.000	1.209.893	1.197	8.505	50.583	79.912

Table B.4 Enthalpy h, specific heat at constant pressure c_p, entropy s, viscosity μ and thermal conductivity κ as a function of temperature T pressure $p = 1$ bar

T (C)	h (kJ/kg)	C_p (kJ/kg K)	s (kJ/kg K)	μ (kg/ms)10^6	κ (J/ms K)10^3
0.000	0.0100	1.003	6.12	17.294	24.210
10.000	10.043	1.003	6.349	17.744	24.893
20.000	20.080	1.004	6.383	18.190	25.571
30.000	30.121	1.004	6.417	18.632	26.243
40.000	40.167	1.005	6.450	19.069	26.910
50.000	50.219	1.005	5.481	19.503	27.572
60.000	60.277	1.006	6.512	19.933	28.229
70.000	70.343	1.007	6.542	20.359	28.880
80.000	80.417	1.008	6.571	20.781	29.527
90.000	90.500	1.009	6.599	21.199	30.169
100.000	100.593	1.010	6.626	21.613	30.806
110.000	110.697	1.011	6.653	22.024	31.439
120.000	120.812	1.012	6.679	22.431	32.067
130.000	130.940	1.013	6.704	22.834	32.690
140.000	141.080	1.015	6.729	23.234	33.309
150.000	151.235	1.016	6.754	23.630	33.924
160.000	161.404	1.018	6.777	24.023	34.534
170.000	171.588	1.019	6.801	24.412	35.140
180.000	181.788	1.021	6.823	24.798	35.742
190.000	192.004	1.022	6.846	25.180	36.340
200.000	202.238	1.024	6.868	25.559	36.934
210.000	212.489	1.026	6.889	25.935	37.524
220.000	222.759	1.028	6.910	26.308	38.110
230.000	233.048	1.030	6.931	26.677	38.692
240.000	243.356	1.032	6.951	27.043	39.271
250.000	253.684	1.034	6.971	27.407	39.846
260.000	264.032	1.036	6.990	27.767	40.417
270.000	274.401	1.038	7.010	28.124	40.985
280.000	284.791	1.040	7.029	28.478	41.549
290.000	295.203	1.042	7.047	28.829	42.110
300.000	305.637	1.044	7.066	29.177	42.667

Table B.5 Enthalpy h, specific heat at constant pressure c_p, entropy s, viscosity μ and thermal conductivity κ as a function of temperature T pressure $p = 1$ bar

T (C)	h (kJ/kg)	C_p (kJ/kg K)	s (kJ/kg K)	μ (kg/ms)10^6	κ (J/ms K)10^3
310.000	316.093	1.047	7.084	29.523	43.221
320.000	326.572	1.049	7.102	29.865	43.772
330.000	337.074	1.051	7.119	30.205	44.320
340.000	347.598	1.054	7.136	30.542	44.865
350.000	358.146	1.056	7.154	30.877	45.406
360.000	368.718	1.058	7.170	31.209	45.945
370.000	379.313	1.061	7.187	31.538	46.481
380.000	389.932	1.063	7.203	31.864	47.013
390.000	400.575	1.065	7.220	32.188	47.543
400.000	411.242	1.068	7.235	32.510	48.070
410.000	421.933	1.070	7.251	32.829	48.595
420.000	432.648	1.073	7.267	33.145	49.116
430.000	443.388	1.075	7.282	33.459	49.635
440.000	454.151	1.078	7.297	33.771	50.151
450.000	464.939	1.080	7.312	34.081	50.665
460.000	475.751	1.082	7.327	34.388	51.177
470.000	486.587	1.085	7.342	34.693	51.685
480.000	497.448	1.087	7.356	34.995	52.192
490.000	508.332	1.090	7.371	35.296	52.696
500.000	519.240	1.092	7.385	35.594	53.197
510.000	530.172	1.094	7.399	35.890	53.697
520.000	541.128	1.097	7.413	36.184	54.194
530.000	552.107	1.099	7.427	36.476	54.688
540.000	563.110	1.101	7.440	36.766	55.181
550.000	574.135	1.104	7.454	37.054	55.671
560.000	585.184	1.106	7.467	37.340	56.160
570.000	596.256	1.108	7.480	37.624	56.646
580.000	607.351	1.111	7.493	37.907	57.130
590.000	618.468	1.113	7.506	38.187	57.612
600.000	629.607	1.115	7.519	38.465	58.092
610.000	640.769	1.117	7.532	38.742	58.570
620.000	51.9526	1.119	7.545	39.017	59.046
630.000	663.157	1.122	7.557	39.290	59.521
640.000	674.384	1.124	7.569	39.561	59.993
650.000	685.631	1.126	7.582	39.831	60.464
660.000	696.900	1.128	7.594	40.099	60.932
670.000	708.190	1.130	7.606	40.365	61.399
680.000	719.500	1.132	7.618	40.630	61.864
690.000	730.830	1.134	7.630	40.893	62.327

Table B.6 Enthalpy h, specific heat at constant pressure c_p, entropy s, viscosity μ and thermal conductivity κ as a function of temperature T pressure $p = 1$ bar

T (C)	h (kJ/kg)	C_p (kJ/kg K)	s (kJ/kg K)	μ (kg/ms)10^6	κ (J/ms K)10^3
710.000	753.550	1.138	7.653	41.415	63.249
720.000	764.940	1.140	7.664	41.673	63.707
730.000	776.349	1.142	7.676	41.930	64.163
740.000	787.777	1.144	7.687	42.186	64.618
750.000	799.223	1.146	7.698	42.440	65.071
760.000	810.689	1.147	7.710	42.692	65.522
770.000	822.172	1.149	7.721	42.944	65.972
780.000	833.674	1.151	7.732	43.193	66.420
790.000	845.193	1.153	7.743	43.442	66.866
800.000	856.730	1.155	7.753	43.689	67.311
810.000	868.284	1.156	7.764	43.935	67.754
820.000	879.855	1.158	7.775	44.180	68.196
830.000	891.443	1.160	7.785	4.4234	68.636
840.000	903.047	1.161	7.796	44.665	69.075
850.000	914.669	1.163	7.806	44.906	69.511
860.000	926.306	1.165	7.816	45.146	69.947
870.000	37.9599	1.166	7.827	45.384	70.381
880.000	949.627	1.168	7.837	45.621	70.813
890.000	961.311	1.169	7.847	45.857	71.243
900.000	973.011	1.171	7.857	46.093	71.672
910.000	984.725	1.172	7.867	46.326	72.100
920.000	996.454	1.174	7.877	46.559	72.526
930.000	1.008.198	1.175	7.887	46.791	72.950
940.000	1.019.956	1.177	7.896	47.022	73.373
950.000	1.031.728	1.178	7.906	47.251	73.794
960.000	1.043.515	1.179	7.916	47.480	74.213
970.000	1.055.315	1.181	7.925	47.708	74.631
980.000	1.067.129	1.182	7.934	47.934	75.047
990.000	1.078.956	1.183	7.944	48.160	75.462
1.000.000	1.090.796	1.185	7.953	48.385	75.875
1.010.000	1.102.650	1.186	7.963	48.609	76.286
1.020.000	1.114.516	1.187	7.972	48.832	76.696
1.030.000	1.126.395	1.189	7.981	49.054	77.104
1.040.000	1.138.287	1.190	7.990	49.275	77.511
1.050.000	1.150.191	1.191	7.999	49.495	77.915
1.060.000	1.162.108	1.192	8.008	49.714	78.318
1.070.000	1.174.036	1.193	8.017	49.932	78.719
1.080.000	1.185.977	1.195	8.026	50.150	79.119
1.090.000	1.197.929	1.196	8.035	50.367	79.516
1.100.000	1.209.893	1.197	8.043	50.583	79.912

Table B.7 Enthalpy h, specific heat at constant pressure c_p, entropy s, viscosity μ and thermal conductivity κ as a function of temperature T pressure $p = 1$ bar

0.000	0.010	1.003	6.114	17.294	24.210
10.000	10.043	1.003	6.150	17.744	24.893
20.000	20.080	1.004	6.184	18.190	25.571
30.000	30.121	1.004	6.218	18.632	26.243
40.000	40.167	1.005	6.251	19.069	26.910
50.000	50.219	1.005	6.282	19.503	27.572
60.000	60.277	1.006	6.313	19.933	28.229
70.000	70.343	1.007	6.343	20.359	28.880
80.000	80.417	1.008	6.372	20.781	29.527
90.000	90.500	1.009	6.400	21.199	30.169
100.000	100.593	1.010	6.427	21.613	30.806
110.000	110.697	1.011	6.454	22.024	31.439
120.000	120.812	1.012	6.480	22.431	32.067
130.000	130.940	1.013	6.506	22.834	32.690
140.000	141.080	1.015	6.530	23.234	33.309
150.000	151.235	1.016	6.555	23.630	33.924
160.000	161.404	1.018	6.578	24.023	34.534
170.000	171.588	1.019	6.602	24.412	35.140
180.000	181.788	1.021	6.624	24.798	35.742
190.000	192.004	1.022	6.647	25.180	36.340
200.000	202.238	1.024	6.669	5.5592	36.934
210.000	212.489	1.026	6.690	25.935	37.524
220.000	222.759	1.028	6.711	26.308	38.110
230.000	233.047	1.030	6.732	26.677	38.692
240.000	243.355	1.032	6.752	27.043	39.271
250.000	253.683	1.034	6.772	27.407	39.846
260.000	264.032	1.036	6.792	27.767	40.417
270.000	274.401	1.038	6.811	28.124	40.985
280.000	284.791	1.040	6.830	28.478	41.549
290.000	295.203	1.042	6.848	28.829	42.110
300.000	305.637	1.044	6.867	29.177	42.667

Table B.8 Enthalpy h, specific heat at constant pressure c_p, entropy s, viscosity μ and thermal conductivity κ as a function of temperature T pressure $p = 1$ bar

T (C)	h (kJ/kg)	C_p (kJ/kg K)	s (kJ/kg K)	μ (kg/ms)10^6	κ (J/ms K)10^3
310.000	316.093	1.047	6.885	29.523	43.221
320.000	326.572	1.049	6.903	29.865	43.772
330.000	337.074	1.051	6.920	30.205	44.320
340.000	347.598	1.054	6.938	30.542	44.865
350.000	358.146	1.056	6.955	30.877	45.406
360.000	368.718	1.058	6.971	31.209	45.945
370.000	379.313	1.061	6.988	31.538	46.481
380.000	389.932	1.063	7.004	31.864	47.013
390.000	400.575	1.065	7.021	32.188	47.543
400.000	411.242	1.068	7.037	32.510	48.070
410.000	421.933	1.070	7.052	32.829	48.595
420.000	432.648	1.073	7.068	33.145	49.116
430.000	443.388	1.075	7.083	33.459	49.635
440.000	454.151	1.078	7.098	33.771	50.151
450.000	464.939	1.080	7.113	34.081	50.665
460.000	475.751	1.082	7.128	34.388	51.177
470.000	486.587	1.085	7.143	34.693	51.685
480.000	497.448	1.087	7.158	34.995	52.192
490.000	508.332	1.090	7.172	35.296	52.696
500.000	519.240	1.092	7.186	35.594	53.197
510.000	530.172	1.094	7.200	35.890	53.697
520.000	541.128	1.097	7.214	36.184	54.194
530.000	552.107	1.099	7.228	36.476	54.688
540.000	563.110	1.101	7.241	36.766	55.181
550.000	574.135	1.104	7.255	37.054	55.671
560.000	585.184	1.106	7.268	37.340	56.160
570.000	596.256	1.108	7.281	37.624	56.646
580.000	607.351	1.111	7.295	37.907	57.130
590.000	618.468	1.113	7.307	38.187	57.612
600.000	629.607	1.115	7.320	38.465	58.092
610.000	640.769	1.117	7.333	38.742	58.570
620.000	51.9526	1.119	7.346	39.017	59.046
630.000	663.157	1.122	7.358	39.290	59.521
640.000	674.384	1.124	7.370	39.561	59.993
650.000	685.631	1.126	7.383	39.831	60.464
660.000	696.900	1.128	7.395	40.099	60.932
670.000	708.190	1.130	7.407	40.365	61.399
680.000	719.500	1.132	7.419	40.630	61.864
690.000	730.830	1.134	7.431	40.893	62.327
700.000	742.180	1.136	7.442	41.155	62.789

Table B.9 Enthalpy h, specific heat at constant pressure c_p, entropy s, viscosity μ and thermal conductivity κ as a function of temperature T pressure $p = 1$ bar

T (C)	h (kJ/kg)	C_p (kJ/kg K)	s (kJ/kg K)	μ (kg/ms)10^6	κ (J/ms K)10^3
710.000	753.550	1.138	7.454	41.415	63.249
720.000	764.940	1.140	7.465	41.673	63.707
730.000	776.349	1.142	7.477	41.930	64.163
740.000	787.777	1.144	7.488	42.186	64.618
750.000	799.223	1.146	7.499	42.440	65.071
760.000	810.689	1.147	7.511	42.692	65.522
770.000	822.172	1.149	7.522	42.944	65.972
780.000	833.674	1.151	7.533	43.193	66.420
790.000	845.193	1.153	7.544	43.442	66.866
800.000	856.730	1.155	7.554	43.689	67.311
810.000	868.284	1.156	7.565	43.935	67.754
820.000	879.855	1.158	7.576	44.180	68.196
830.000	891.443	1.160	7.586	4.4234	68.636
840.000	903.047	1.161	7.597	44.665	69.075
850.000	914.669	1.163	7.607	44.906	69.511
860.000	926.306	1.165	7.617	45.146	69.947
870.000	37.9599	1.166	7.628	45.384	70.381
880.000	949.627	1.168	7.638	45.621	70.813
890.000	961.311	1.169	7.648	45.857	71.243
900.000	973.011	1.171	7.658	46.093	71.672
910.000	984.725	1.172	7.668	46.326	72.100
920.000	996.454	1.174	7.678	46.559	72.526
930.000	1.008.198	1.175	7.688	46.791	72.950
940.000	1.019.956	1.177	7.697	47.022	73.373
950.000	1.031.728	1.178	7.707	47.251	73.794
960.000	1.043.515	1.179	7.717	47.480	74.213
970.000	1.055.315	1.181	7.726	47.708	74.631
980.000	1.067.129	1.182	7.736	47.934	75.047
990.000	1.078.956	1.183	7.745	48.160	75.462
1.000.000	1.090.796	1.185	7.754	48.385	75.875
1.010.000	1.102.650	1.186	7.764	48.609	76.286
1.020.000	1.114.516	1.187	7.773	48.832	76.696
1.030.000	1.126.395	1.189	7.782	49.054	77.104
1.040.000	1.138.287	1.190	7.791	49.275	77.511
1.050.000	1.150.191	1.191	7.800	49.495	77.915
1.060.000	1.162.108	1.192	7.809	49.714	78.318
1.070.000	1.174.036	1.193	7.818	49.932	78.719
1.080.000	1.185.977	1.195	7.827	50.150	79.119
1.090.000	1.197.929	1.196	7.836	50.367	79.516
1.100.000	1.209.893	1.197	7.844	50.583	79.912

References

1. Navier, C. L. M. H. (1823). Mémoire sur les lois du mouvement des fluides. *Memoires de l'Academie Royale des Sciences, 6*, 389–416.
2. Stokes, G. G. (1851). On the effect of internal friction of fluids on the motion of pendulums. *Transactions of the Cambridge Philosophical Society, 9*(II), 8–106.
3. Prandtl, L. (1904). Über Flüßigkeitsbewegung bei sehr kleiner Reibung,Verh. 3. Intern. Math. Kongr., Heidelberg 1904, S. 484ß491, Nachdruck: Ges. Abh. S. 575ß584ö Berlin, Göttingen, Heidelberg, Springer 1961.
4. von Kármán, Th. (1921). Über laminare und turbulente Reibung. *Zeitschrift für angewandte Mathematik und Mechanik, 1*(1921), 233–252.
5. Reynolds, O. (1883). An experimental investigation of the circumstances which determine whether the motion of water shall be direct Sinuous and of the law or resistance in parallel channels. *Philosophical Transactions of the Royal Society, 174*, 935–982.
6. Eric, F. R. (1960). *Rheology-theory and practice* (Vol. 3). Academic Press.
7. Navier, C. L. M. H. (1823). Mémoire sur les lois du mouvement des fluides. *Memoires de l'Academie Royale des Sciences, 6*, 389–416.
8. Stokes, G. G. (1851). On the effect of internal friction of fluids on the motion of pendulums. *Transactions of the Cambridge Philosophical Society, 9*(II), 8–106.
9. Prandtl, L. (1904). Über Flüßigkeitsbewegung bei sehr kleiner Reibung, Verh. 3. Intern. Math. Kongr., Heidelberg 1904, S. 484ß491, Nachdruck: Ges. Abh. S. 575ß584ö Berlin, Göttingen, Heidelberg, Springer 1961.
10. von Kármán, Th. (1921). Über laminare und turbulente Reibung. *Zeitschrift für angewandte Mathematik und Mechanik, 1*(1921), 233–252.
11. Reynolds, O. (1883). An experimental investigation of the circumstances which determine whether the motion of water shall be direct Sinuous and of the law or resistance in parallel channels. *Philosophical Transactions of the Royal Society, 174*, 935–982.
12. Eric, F. R. (1960). *Rheology-theory and practice* (Vol. 3.). Academic Press.
13. Aris, R. (1962). *Vectors, tensors, and the basic equations of fluid mechanics*. Dover Publication.
14. Spurk, J. (1997). *Fluid mechanics*. Springer.
15. White, F. M. (1974). *Viscous fluid flow*. McGraw-Hill.
16. Truesdell, C., & Noll, W. (1965). *Handbuch der Physik*. Springer.
17. Truesdell, C. (1952). *Archive for Rational Mechanics and Analysis, 1*, 125 (1952).

M. T. Schobeiri, *Advanced Fluid Mechanics and Heat Transfer for Engineers and Scientists*, https://doi.org/10.1007/978-3-030-72925-7

18. Ferziger, H. J., & Peric, M. (1996). *Computational methods for fluid dynamics* (3rd ed.). Springer
19. Vavra, M. H. (1960). *Aerothermodynamics and flow in turbomachines*. Wiley.
20. Traupel, W. (1977). *Thermische Turbomaschinen, Bd.I*. Springer.
21. Horlock, J. H. (1966). *Axial flow compressors*. Butterworth.
22. Horlock, J. H. (1966). *Axial flow turbine London*. Butterworth.
23. Schobeiri, M. T. (2005). *Turbomachinery flow physics and dynamic performance*. Springer, 3-54022368-1
24. Schobeiri, M. T. (1990). Optimum design of a low pressure double inflow radial steam turbine for open-cycle ocean thermal energy conversion. *Journal of Turbomachinery, 112*, 71–77.
25. Schobeiri, M. T. (1990). Thermo-fluid dynamic design study of single- and double-inflow radial and single-stage axial steam turbines for open-cycle ocean thermal energy conversion, Net power producing experiment facility. *ASME Transaction, Journal of Energy Resources, 112*, 41–50.
26. Prandtl, L. (1904). Über Flüssigkeitsbewegung bei sehr kleiner Reibung, 3. Internat. Math. Kongr. Heidelberg, 104 also in Prandtl gesammelte Abhandlungen. Springer.
27. Prandtl, L. (1961). Über den Reibungswiderstand strömender Luft, Ergebnisse der Aerodynamischen Versuchsanstalt Göttingen. In *Gesammelte Abhandlung*. Springer.
28. Prandtl, L. (1922). Über die Entstehung von Wirbeln in idealen Flüssigkeiten, mit Anwendung auf die Tragflügeltheorie und andere Aufgaben. Vortr. Geb. Hydro-u. Aerodyn., Innsbruck (1922), 18–33. Nachdruck: Ges. Abhandlungen, S. 696–713.
29. Betz, A. (1964). *Konforme Abbildung* (2nd ed.). Springer.
30. Spurk, J. H. (1997). *Fluid mechanics*. Springer. ISBN 3-540-61651-9.
31. Prandtl, L., & Tietjens, O. G. (1934). *Applied hydro-and aeromechanics*. Dover Publications Inc.
32. Wylie, R. (1975). *Advanced engineering mathematics* (4th ed.). McGraw-Hill Kogakusha Ltd. 0-07-072180-7
33. Koppenfels, W., & Stallmann, F. (1959). *Praxis der konformen Abbildung*. Springer.
34. Vavra, M. H. (1960). *Aero-thermodynamics and flow in turbomachines*. Wiley.
35. Thompson, W., & Kelvin, L. (1869). On Vortex Motion. *Philosophical Transactions of the Royal Society of Edinburgh, 25*(1868), 217–260.
36. Helmholtz, H. (1858). Über die Integrale der hydrodynamischen Gleichungen, welche den Wirbelbewegungen entsprechen. *Zeitschrift der reinen und der angewandten Mathematik, 55*(1858), 25–55.
37. Kotschin, N. J., Kiebel, L. A., Rose, N. W. (1954). *Theoretische Hydrodynamik, Bd. I*. Akademie-Verlag.
38. Schlichting, H., & Trockenbrodt, E. (1969). *Aerodynamik des Flugzeuges* (2nd ed.). Springer.
39. Schlichting, H. (1979). *Boundary layer theory* (7th ed.). McGraw-Hill.
40. Jeffery, G. B. (1915). The two-dimensional steady motion of a viscous fluid. *The London, Edinburgh, and Dublin Philosophical Magazine and Journal of Science, 29*, 455–465.
41. Hamel, G. (1916). Spiralförmige Bewegung zäher Flüssigkeiten. *JahresBericht der deutschen Mathematikervereinigung, 25*, 34–60.
42. Schobeiri, M. T. (1980). Geschwindigkeit- und Temperaturvereitlungen in Hamelscher Spiralströmung, Zeitschrift für angewandte Mathematik und Mechanik. *ZAMM, 60*, 195.
43. Schobeiri, M. T. (1990). The influence of curvature and pressure gradient on the flow and velocity distribution. *International Journal of Mechanical Sciences, 32*(10), 851–861.
44. Schobeiri, M. T. (1976). Näherungslösung der Navier-Stokes Differential gleichungen für eine zweidimensionale stationäre Laminarströmung konstanter Viskosität in konvexen und konkaven Diffusoren und Düsen, Zeitschrift für angewandte Mathematik und Physik. *ZAMP, 27*, 9.
45. Milsaps, K., & Pohlhausen, K. (1953). Thermal distribution in Jeffery-Hamel flow. *Journal of Aeronautical Sciences, JAS, 20*, 187.
46. Spurk, J. H. (1997). *Fluid mechanics*. Springer. ISBN 3-540-61651-9.

47. Oseen, C. W. (1910). Über die Stokes'sche Formel und über eine verwandte Aufgabe in Hydrodynamik. In *Arkiv for Matematik, Astronomi Och Fysik* (vol. 6, No. 29).
48. Reynolds, O. (1883). On the experimental investigation of the circumstances which determine whether the motion of water shall be direct of sinous, and the law of resistance in parallel channels. *Philosophical Transactions of the Royal Society, 174*, 935–982.
49. Schlichting, H. (1950). Über die Entstehung der Turbulenzentstehung. *Forschung im Ingenieurwesen, 16*, 65–78.
50. Schlichting, H. (1959). Entstehung der Turbulenz. In *Handbuch der Physik, 8*(1), 351–450. (Springer-Verlag Edition, McGraw-Hill Book Company).
51. Schlichting, H. (1979). *Boundary layer theory* (7th ed.). McGrawHill Book Company.
52. Schubauer, G. B., & Klebanof, P. S. (1955). *Contributions on the mechanics of boundary layer transition*. NACA TN 3489 (1955) and NACA Rep. 1289 (1956).
53. McCormick, M. E. (1968). An analysis of the formation of turbulent patches in the transition. Boundary Layer. *Journal of Applied Mechanics, 35*, 216.
54. White, F. M. (1974). *Viscose fluid flow*. McGraw-Hill.
55. Mayle, R. E. (1991). The role of Laminar-turbulent transition in gas turbine engines. *Journal of Turbomachinery, 113*, 509–537.
56. Squire, S. B. (1933). On the stability of three-dimensional distribution viscous fluid between parallel walls. *Proceedings of the Royal Society, A, 142*, 621–628.
57. Orr, W. M. F. (1907). The stability of the steady motions of a perfect liquid and of a viscous liquid. Part I: A perfect liquid; Part II: A viscous liquid. *Proceedings of the Royal Irish Academy, 27*, 9–68; 69–138.
58. Sommerfeld, A. (1908). Ein Beitrag zur hydrodynamischen Erklärung der turbulenten Flüssigkeitsbewegungen. Atti, del 4. Congr. Internat. Di Mat. III, pp. 116–124, Roma.
59. Orszag, S. A. (1971). Accurate solution of the Orr-Sommerfeld stability equation. *Journal of Fluid Mechanics, 50*(part 4), 680–703.
60. Morkovin, M. V. (1969). On the many faces of transition. In C. S. Wells (Ed.), *Viscous drag reduction* (pp. 1–31). Plenum Press.
61. Emmons, H. W. (1951). The laminar-turbulent transition in boundary layer part I. *Journal of the Aeronautical Sciences, 18*(7), 490–498.
62. Schubauer, G. B., & Klebanoff, P. S. (1955). NACA Report 1289.
63. Dhawan, S., & Narasimha, R. (1958). Some properties of boundary layer flow during the transition from laminar to turbulent motion. *Journal of Fluid Mechanics, 3*, 418–436.
64. Chakka, P., & Schobeiri, M. T. (1997). *Modeling of unsteady boundary layer transition on a curved plate under periodic unsteady flow condition*. ASME Paper No. 97-GT-399, presented at the IGTI, International Gas Turbine Congress, in Orlando ASME.
65. Schobeiri, M. T., & Chakka, P. (2002). Prediction of turbine blade heat transfer and aerodynamics using unsteady boundary layer transition model. *International Journal of Heat and Mass Transfer, 45*, 815–829.
66. Herbs, R. (1980). *Entwicklung von Grenzschichten bei instationärer Zuströmung*, Dissertation D17. Technische Universität Darmstadt.
67. Pfeil, H., Herbst, R., & Schroder, T. (1982). *Investigation of laminar-turbulent transition of boundary layers disturbed by wakes*. ASME Turbo Expo Pap. 82-GT-124.
68. Schobeiri, M. T., & Radke R. (1994). *Effect of periodic unsteady wake flow and pressure gradient on boundary layer transition along a the concave surface of a curved plate*. ASME-paper 94-GT-327, presented at the International Gas Turbine and Aero-Engine Congress and Exposition in Hague, Netherlands, June 18–16, 1994.
69. Halstead, D. E., Wisler, D. C., Okiishi, T. H., Walker, G. J., Hodson, H. P., & Shin, H.-W. (1997). Boundary layer development in axial compressors and turbines: Part 3 of 4. *ASME Transactions, Journal of Turbomachinery, 119*, 225–237.
70. Schobeiri, M. T., Jose, J., & Pappu, K. (1996). Development of two dimensional wakes within curved channel: Theoretical framework and experimental investigations. *ASME Journal of Turbomachinery, 118*, 506–518.

71. Schobeiri, M. T., Read, K., & Lewalle, J. (2003). Effect of unsteady wake passing frequency on boundary layer transition, experimental investigation and wavelet analysis. *Journal of Fluids Engineering, 125*, 251–266.
72. Chavary, R., & Tutu, N. K. (1978). Intermittency and preferential transport of heat in a round jet. *Journal of Fluid Mechanics, 88*, 133–160.
73. Taylor, G. I. (1921). Diffusion by continuous movements. *The Proceedings of the London Mathematical Society, 2*(20), 196–211.
74. von Kármán, Th. (1937). On the statistical theory of turbulence. *Proceedings of the National Academy of Sciences of the United States of America, 4*, 137.
75. Hinze, J. O. (1975). *Turbulence* (2nd ed.). McGraw Hill Book company.
76. Rotta, J. C. (1972). *Turbulente Strömungen*. Teubner-Verlag Stuttgart.
77. Richardson, L. F. (1922). *Weather prediction by numerical process*. Cambridge University Press.
78. Kolmogorov, A. N. (1941). Local structure of turbulence in incompressible viscous fluid for very large Reynolds number. *Doklady Akademia Nauk, SSSR, 30*, 299–303.
79. Grant, H., Stewart, R. W., & Moilliet, A. (1962). Turbulence spectra from a tidal channel. *Journal of Fluid Mechanics, 12*, 241.
80. Onsager, L. (1945). *Physical Review, 68*, 286.
81. Weizsäcker, C. F. (1948). *Zeitschrift Physik, 124*, 628, also Proceedings of the Royal Society, London, *195A*, 402.
82. Bradshaw, P., & Perot, J. B. (1993). A note on turbulent energy dissipation in the viscous wall region. *Physics of Fluids A: Fluid Dynamics, 5*, 3305.
83. Launder, B. E., Reece, G. I., & Rodi, W. (1975). Progress in the development of reynolds-stress turbulent closure. *Journal of Fluid Mechanics, 68*, 537–566.
84. Launder, B. E., & Spalding, D. B. (1974). The numerical computation of turbulent flows. *Computer Method in Applied mechanics and Engineering, 3*, 269–289.
85. Boussinesq, J. (1887). *Memoires Presentes par Divers Savants a l'Academie des Sciences* (vol. 23, p. 46).
86. Prandtl, L. (1925). Über die ausgebildete Turbulenz. *ZAMM, 5*, 136–139.
87. Schlichting, H. (1979). *Boundary layer theory* (7th ed.). McGraw-Hill Book Company.
88. Wilcox, D. (1993). *Turbulence modeling for CFD* (p. 91011). DCW Industries Inc.
89. Müller, T. (1991). *Untersuchungen von Geschwindigkeitsprofilen und deren Entwicklung in Strömungsrichtung in zweidimensionalen transitionalen Grenzschichten anhand eines Geschwindigkeitsmodells*, Dissertation. Technische Hochschule Darmstadt, Germany D 17.
90. Van Driest, E. R. (1951). Turbulent boundary layer in compressible fluids. *Journal of Aeronautical Sciences, 18*(145–160), 216.
91. Kays, W. M., & Moffat, R. J. (1975). The behavior of transpired turbulent boundary layers. In *Studies in Convection, Vol. 1: Theory, Measurement and Application*. Academic Press.
92. Smith, A. M. O., & Cebeci, T. (1967). *Numerical solution of the turbulent boundary layer equations*. Douglas aircraft division report DAC 33735.
93. Klebanoff, P. S. (1954). *Characteristics of turbulence in boundary layer with zero pressure gradient*. NACA TN 3178.
94. Baldwin, B. S., & Lomax, H. (1978). *Thin layer approximation and algebraic model for separated turbulent flows*. AIAA Paper 78-257.
95. Kolmogorov, A. M. (1942). *Equations of Turbulent motion of an incompressible fluid*. Akad. Nauk SSR, Seria Fiz. VI, No. 1-2.
96. Prandtl, L. (1945). Über ein neues Formelsystem für die ausgebildete Turbulenz, Nachrichten der Akademie der Wissenschaften. *Göttingen. Mathematisch-Physikalische Klasse, 1945*, 6.
97. Launder, B. E., & Spalding, D. B. (1972). *Mathematical models of turbulence*. Academic Press.
98. Chou, P. Y. (1945). On the velocity correlations and the solution of the equation of Turbulent fluctuations. *Quarterly of Applied Mathematics, 3*, 38.
99. Jones, W. P., & Launder, B. E. (1972). The prediction of laminarization with a two-equation model of turbulence. *International Journal of Heat and Mass Transfer, 15*, 301–314.

100. Menter, F. R. (1993). *Zonal two-equation k-T turbulence models for aerodynamic flows*. AIAA Technical Paper, 93-2906.
101. Rodi, W., & Scheurer, G. (1986). Scrutinizing the k-g model under adverse pressure gradient conditions. *The Journal of Fluids Engineering, 108*(1986), 174–179.
102. Menter, F. R. (1992). Influence of freestream values on k-T turbulence model predictions. *AIAA Journal, 30*(No. 6).
103. Wicox, D. (2008). *Private communications*. DCW Industries Inc.
104. Menter, F. R. (1993). *Zonal two equation k-g turbulence models for aerodynamic flows*. AIAA Paper 93-2906.
105. Menter, F. R. (1994). Two-equation Eddy-viscosity turbulence models for engineering applications. *AIAA Journal, 32*, 269–289.
106. Menter, F. R., Kuntz, M., & Langtry, R. (2003). Ten years of experience with the SST turbulence model. In K. Hanjalic, Y. Nagano, & M. Tummers (Eds.), *Turbulence, heat and mass transfer* (Vol. 4, pp. 625–632). Begell House Inc.
107. Menter, F. (2008). CFX, Germany, Private communications.
108. Durst, F., Pereira, J. C. F., & Tropea, C. (1993). The plane Symmetric sudden-expansion flow at low Reynolds numbers. *Journal of Fluid Mechanics, 248*, 567–581.
109. Schobeiri, M. T. (2012). *Turbomachinery flow physics and dynamic performance* (2nd ed.). Springer. 978-3642-24674-6
110. Schobeiri, M. T., Gilarranz, J., & Johansen, E. (1999). Final report on: Efficiency, performance, and interstage flow field measurement of Siemens Westinghouse HP-Turbine Blade Series 9600 and 5600.
111. Schobeiri, M. T., Gillaranz, J. L., & Johansen, E. S. (2000). Aerodynamic and performance studies of a three stage high pressure research turbine with 3-D blades, design point and off-design experimental investigations. In: *Proceedings of ASME Turbo Expo 2000, 2000-GT-484*.
112. Schobeiri, M. T., Abdelfatta, S., & Chibli, H. (2011). Investigating the cause of CFD-deficiencies in accurately predicting the efficiency and performance of high pressure turbines: a combined experimental and numerical study, submitted for publication. The document along with the entire experimental and numerical data are also available at Turbomchinery Performance and Flow Research Laboratory, Texas A&M University.
113. ANSYS INC. (2009). *ANSYS-CFX release documentation* (12.0 ed.).
114. Denton, J. (1992). The calculation of three-dimensional viscous flow through multistage turbomachines. *Journal of Turbomachinery, 114*.
115. Schobeiri, M. T., & Pappu, K. (1999). Optimization of trailing edge ejection mixing losses downstream of cooled turbine blades: a theoretical and experimental study. *ASME Journal of Fluids Engineering, 121*, 118125.
116. Schobeiri, M. T., & Chakka, P. (2002). Prediction of turbine blade heat transfer and aerodynamics using unsteady boundary layer transition model. *International Journal of Heat and Mass Transfer, 45*, 815–829.
117. Schobeiri, M. T., & Radke, R. E. (1994). *Effects of periodic unsteady wake flow and pressure gradient on boundary layer transition along the concave surface of a curved plate*. ASME Paper 94-GT-327, presented at the International Gas Turbine and Aero-Engine Congress and Exposition, Hague, Netherlands, June 13–16, 1994.
118. Schobeiri, M. T., Read, K., & Lewalle, J. (2003). Effect of unsteady wake passing frequency on boundary layer transition, experimental investigation and wavelet analysis. *Journal of Fluids Engineering, 125*, 251–266 (a combined two-part paper, this paper received the ASME-2004 FED-Best Paper Award).
119. Wright, L., & Schobeiri, M. T. (1999). The effect of periodic unsteady flow on boundary layer and heat transfer on a curved surface. *ASME Transactions, Journal of Heat Transfer, 120*, 22–33.
120. Schobeiri, M. T., Öztürk, B., & Ashpis, D. (2003). On the physics of the flow separation along a low pressure turbine blade under unsteady flow conditions. ASME 2003-GT-38917, presented at International Gas Turbine and Aero-Engine Congress and Exposition, Atlanta, Georgia,

June 16–19, 2003, also published in ASME Transactions. *Journal of Fluid Engineering, 127*, 503–513.
121. Schobeiri, M. T., & Öztürk, B. (2004). Experimental study of the effect of the periodic unsteady wake flow on boundary layer development, separation, and re-attachment along the surface of a low pressure turbine blade. ASME 2004-GT-53929, presented at International Gas Turbine and AeroEngine Congress and Exposition, Vienna, Austria, June 14–17. (2004). also published in the ASME Transactions. *Journal of Turbomachinery, 126*(4), 663–676.
122. Schobeiri, M. T., Öztürk, B., & Ashpis, D. (2005). *Effect of reynolds number and periodic unsteady wake flow condition on boundary layer development, separation, and re-attachment along the suction surface of a low pressure turbine blade*. ASME Paper GT2005-68600.
123. Schobeiri, M. T., Öztürk, B., & Ashpis, D. (2005). *Intermittent behavior of the separated boundary layer along the suction surface of a low pressure turbine blade under periodic unsteady flow conditions*. ASME Paper GT2005-68603.
124. Öztürk, B., & Schobeiri, M. T. (2006). *Effect of turbulence intensity and periodic unsteady wake flow condition on boundary layer development, separation, and re-attachment over the separation bubble along the suction surface of a low pressure turbine blade*. ASME, GT2006-91293.
125. Traupel, W. (1977). *Thermische turbomaschinen* (3rd ed.). Springer, ISBN3-540-0739-4.
126. Dzung, L. S. (1971). Konsistente Mittewerte in der Theorie der Turbomaschinen für kompressible Medien. *BBC-Mitteilung, 58*, 485–492.
127. Reichardt, H. (1950). Gesetzmäßigkeiten der freien Turbulenz. VDI-ForschHeft 414.2. Auflage Düsseldorf, VDI-Verlag.
128. Prandtl, L. (1942). Bemerkung zur Theorie der freien Turbulenz. *Zeitschrift für angewandte Mathematik und Mechanik (ZAMM), 22*(5), 241–254.
129. Eifler, J. (1975). *Zur Frage Der Freien Turbulenten Strömungen, Insbesondere Hinter Ruhenden Und Bewegten Zylindern*, Dissertation D17. Technische Hochschule Darmstadt.
130. Schobeiri, M. (2012). *Turbomachinery flow physics and dynamic performance* (2nd ed.). Springer. ISBN 978-3-642-24675-3.
131. Schobeiri, M. T., John, J., & Pappu, K. (1996). Development of two-dimensional wakes within curved channels: theoretical framework and experimental investigation. *Journal of Turbomachinery, 118*, 506.
132. Schobeiri, M. T., Pappu, K., & John, J. (1995). Theoretical and experimental study of development of two-dimensional steady and unsteady wakes within curved channels. *ASME Transactions, Journal of Fluid Engineering, 117*, 593–598.
133. Schobeiri, M. T. (2008). Influence of curvature and pressure gradient on turbulent wake development in curved channels. *Transactions of the ASME, Journal of Fluid Engineering, 139*, 1–1.
134. Raj, R., & Lakshminarayana, B. (1973). Characteristics of the wake behind a cascade of airfoils.
135. Wattendorf, F. I. (1935). A study of the effect of curvature on fully-developed turbulent flow. *Proceedings of the Royal Society, London, 148*, 565.
136. Eskinazi, S., & Yeh, H. (1956). An investigation on fully developed turbulent flows in a curved channel. *Journal of the Aeronautical Sciences, 23*, 23.
137. Prandtl, L. (1904). Über Flüßigkeitsbewegung bei sehr kleiner Reibung, Verh. 3. Intern. Math. Kongr., Heidelberg 1904, S. 484–491, Nachdruck: Ges. Abh. S. 575–584; Springer.
138. Blasius, H. (1908). Grenzschichten in Flüssigkeiten mit kleiner Reibung. *Zeitschrift für Angewandte Mathematik und Physik, 56*, 1–37.
139. Falkner, V. M., & Skan, S. W. (1930). Some approximate solutions of the boundary layer equations. *Philosophical Magazine, 12*, 865–896 (ARC RM 1314).
140. Hartree, D. (1937). On an equation occurring in Falkner and Skan's approximate treatment of the equation of the boundary layer. *Mathematical Proceedings of the Cambridge Philosophical Society, 33*(Part II), 223–239.
141. Schlichting, H. (1979). *Boundary layer theory* (7th ed.). McGraw-Hill Book Company.
142. Spurk, J. (1997). *Fluid mechanics*. Springer.

143. Pohlhausen, K. (1921). Zur näherungsweisen Lösung der Differentialgleichungen der laminaren Reibungsschicht. *ZAMM, 1*, 252–268.
144. Von Kármán, Th. (1921). Über laminare und turbulente Reibung. *ZAMM, 1*, 233–253.
145. Ludwieg, H., & Tillman, W. (1949). *Untersuchungen über die Wandschubspannung in turbulenten Reibungsschichten*. Ingenieur Archiv 17, 288-299, Summary and translation in NACA-TM-12185 (1950).
146. Coles, D. E. (1956). The law of the wake in turbulent boundary layer. *Journal of Fluid Mechanics, 1*, 191–226.
147. White, F. M. (1974). *Viscose fluid flow*. McGraw-Hill.
148. Lehman, K. (1979). *Untersuchungen turbulenter Grenzschichte*. Dissertation, D17, Technische Hochschule Darmstadt.
149. Rotta, J. C. (1972). *Turbulente Strömungen*. Teubner-Verlag Stuttgart.
150. Truckenbrodt, E. (1980). *Fluidmechanik* (Vol. 2). Springer.
151. Pfeil, H., & Sticksel, W. H. (1981). Influence of the pressure gradient on the law of the wall. *AIAA Journal, 20*(3), 342–346.
152. Pfeil, H., & Amberg, T. (1989). Differing development of the velocity profiles of three-dimensional turbulent boundary layers. *AIAA Journal, 27*, 1456–9.
153. Prandtl, L. (1927). Über den Reibungswiderstand strömender Luft," Ergebnisse, AVA, Göttingen, 3. Liefg. Seite 1-5, 4. Liefg. Seite 18-29.
154. Prandtl, L. (1932). Zur turbulenten Strömung in Rohren, und längs Platten. AVA, Göttingen, vierte Serie.
155. Nikuradse, J. (1932). *Gesetzmäßigkeit der turbulenten Strömung in glatten Rohren* (p. 356). Forsch. Arb. Ing. -Wesen.
156. Launder, B. E., & Spalding, D. B. (1972). *Mathematical models of turbulence*. Academic Press.
157. Reichardt, H. (1950). *Gesetzmäßigkeiten der freien Turbulenz*. VDI-Forsch. - Heft 414. 2. Auflage Düsseldorf, VDI-Verlag.
158. Laufer, J. (1952). *The structure of turbulence in fully developed pipe flow*. NACA Report 1174.
159. Anderson, P. S., Kays, W. M., & Moffat, R. J. (1972). *The turbulent boundary layer on a porous plate: an experimental study of the fluid mechanics for adverse free stream pressure gradients*. Report No.HMT15, Department of Mechanical Engineering, Stanford University, CA.
160. Wieghardt, K. (1944). Über die turbulente Strömungen in Rohr und längs einer Platte. *Zeitschrift für angewandte Mathematik und Mechanik, ZAMM, 24*, 294.
161. Sticksel, W. J. (1984). *Theoretische und experimentelle Untersuchungen turbulenter Grenzschichtprofile*. Dissertation D 17, Fachbereich Maschinenbau: Technische Hochschule Darmstadt.
162. Coles, D. E., & Hirst, E. A. (1968). *Computation of turbulent boundary layers, AFSOR-IFP—Stanford Conference* (Vol. II). Thermoscience Division, Stanford University.
163. Coles, D. E. (1956). Law of the wake in turbulent boundary layer. *Journal of Fluid Mechanics, 1*(2), 191–226.
164. Clauser, F. H. (1954). Turbulent boundary layers in adverse pressure gradients. *Journal of Aeronautical Sciences, 21*, 91–108.
165. Spurk, J. H. (1997). *Fluid mechanics*. Springer. ISBN 3-540-61651-9
166. Mellor, G. I., & Gibson, D. M. (1966). Equilibrium turbulent boundary layers. *Journal Fluid Mechanics, 24*, 225–256.
167. Prandtl, L., & Schlichting, H. (1934). Werft, Reederei, Hafen 15 1-4. Nachdruck: Gesammelte Abhandlung, Seite (pp. 649–662). Springer
168. Falkner, V. M. (1943). The resistance of a smooth flat plate with turbulent boundary layer. *Aircraft Engineering, 15*, 65–69.
169. Blasius, H. (1913). Ähnlichkeitsgesetz bei Reibungsvorgängen in Flüssigkeiten. Forschung. Arb. Ing-Wes, Nr. 134, Berlin.
170. Crawford, M. E., & Kays, W. M. (1976). NASA CR-2742.

171. Cebeci, T., & Bradshaw, P. (1974). *Physical and computational aspects of convective heat transfer*. Springer.
172. Kays, W. M., Crawford, M. E. (1980). *Konvective heat and mass transfer*. McGraw-Hill Series in Mechanical Engineering.
173. Schobeiri, M. T., & Chakka, P. (2002). Prediction of turbine blade heat transfer and aerodynamics using unsteady boundary layer transition model. *International Journal of Heat and Mass Transfer, 45*(2002), 815–829.
174. Chakka, P., & Schobeiri, M. T. (1999). Modeling of unsteady boundary layer transition on a curved plate under periodic unsteady flow condition: aerodynamic and heat transfer investigations. *ASME Transactions, Journal of Turbo Machinery, 121*, 8897.
175. Schobeiri, M. T., Read, K., & Lewalle, J. (2003). Effect of unsteady wake passing frequency on boundary layer transition, experimental investigation and wavelet analysis. *ASME Transactions, Journal of Fluids Engineering, 125*, 251–266 (a combined two-part paper).
176. Schobeiri, M. T. (2008). Influence of curvature and pressure gradient on turbulent wake development in curved channels. *ASME Transactions, Journal of Fluids Engineering*. 130.
177. Schobeiri, M. T., John, J., & Pappu, K. (1996). Development of two-dimensional wakes within curved channels, theoretical framework and experimental investigation, an honor paper. ASME Transactions. *Journal of Turbomachinery, 118*, 506–518.
178. John, J., & Schobeiri, M. T. (1996). Development of two-dimensional wakes in a curved channel at positive pressure gradient. *ASME Transactions, Journal of Fluid Engineering, 118*, 292–299.
179. Eifler, J. (1975). *Zur Frage der Freien Turbulenten Strömungen, Insbesondere Hinter Ruhenden und Bewegten Zylindern, dissertation*. Technische Hochschule Darmstadt.
180. Schobeiri, M. T., Öztürk, B., & Ashpis, D. (2007). Effect of reynolds number and periodic unsteady wake flow condition on boundary layer, development, separation, and intermittency behavior along the suction surface of a low pressure turbine blade. *Transactions of the ASME, 92*, 129.
181. Durst, F., Zanoun, E.-S., & Pashstrapanska, M. (2001). In-situ calibration of hot wires close to highly heat-conducting walls. *Experiments in Fluids, 31*, 103–110.
182. Bruun, H. H. (1995). *Hot-wire anemometry*. Oxford University Press.
183. Hippensteele, S. A., Russell, L. M., & Stepka, S. (1981). Evaluation of a method for heat transfer measurements and thermal visualization using a composite of a heater element and liquid crystals. *ASME Journal of Heat Transfer, 105*, 184–189.
184. Wright, L., & Schobeiri, M. T. (1999). The effect of periodic unsteady flow on boundary layer and heat transfer on a curved surface. ASME Transactions. *Journal of Heat Transfer, 120*, 2233.
185. Schobeiri, M. T. (2006). Advances in unsteady aerodynamics and boundary layer transition, W.IT. *Book, United Kingdom: Flow Phenomena in Nature, 2*, 573–605.
186. Schobeiri, M. T., & Chakka, P. (2002). Prediction of turbine blade heat transfer and aerodynamics using unsteady boundary layer transition model. *International Journal of Heat and Mass Transfer, 45*, 815–829.
187. Wilcox, D. (1993). Wilcox, D. (1993). *Turbulence modeling for CFD* (p. 91011). DCW Industries Inc.
188. Schobeiri, M. T., & Öztürk, B. (2004). Experimental study of the effect of the periodic unsteady wake flow on boundary layer development, separation, and re-attachment along the surface of a low pressure turbine blade. In ASME 2004-GT-53929, presented at International Gas Turbine and Aero-Engine Congress and Exposition, Vienna, Austria, June 14–17. (2004). also published in the ASME Transactions. *Journal of Turbomachinery, 126*(4), 663–676.
189. Schobeiri, M. T., Öztürk, B., & Ashpis, D. (2005). *Effect of reynolds number and periodic unsteady wake flow condition on boundary layer development, separation, and re-attachment along the suction surface of a low pressure turbine blade*. ASME Paper GT2005-68600.
190. Öztüurk, B., & Schobeiri, M. T. (2007). Effect of turbulence intensity and periodic unsteady wake flow condition on boundary layer development separation and re-attachment along the suction surface of a low pressure turbine blade. *ASME Transaction, Journal of Fluid Engineering, 129*, 747–763. (a major paper).

191. Kaszeta, R. W., & Simon, T. W. (2002). *Experimental investigation of transition to turbulence as affected by passing wakes*. NASA/CR—2002-212104.
192. Schobeiri, M.T., Öztürk, B., & Ashpis, D. (2007). Effect of reynolds number and periodic unsteady wake flow condition on boundary layer development, separation, and intermittency behavior along the suction surface of a low pressure turbine blade. A combined two part paper: Part I: Boundary Layer Development, Part II: Intermittency Behavior. *ASME-Transactions, Journal of Turbomachinery, 129/107*, 92–107 (a major paper).
193. Roberts, S. K., & Yaras, M. I. (2003). *Effects of periodic-unsteadiness, free stream turbulence and flow reynolds number on separation-bubble transition*. ASME GT-2003-38262.
194. Herbst, R. (1980). *Entwicklung von Strömungsgrenzschichten bei instationärer Zuströmung in Turbomaschinen*, Dissertation D-17. Technische Hochschule Darmstadt.
195. Pfeil, H., Herbst, R., & Schröder, T. (1983). Investigation of the laminar turbulent transition of boundary layers disturbed by wakes. *ASME Journal of Engineering for Power, 105*, 130–137.
196. Schobeiri, M. T., & Radke, R. (1994). *Effects of periodic unsteady wake flow and pressure gradient on boundary layer transition along the concave surface of a curved plate*. ASME Paper No. 94-GT-327.
197. Halstead, D. E., Wisler, D. C., Okiishi, T. H., Walker, G. J., Hodson, H. P., & Shin, H.-W. (1997). Boundary layer development in axial compressors and turbines: Part 3 of 4. *ASME Journal of Turbomachinery, 119*, 225–237.
198. Schobeiri, M. T., Öztürk, B., & Ashpis, D. (2005). On the physics of the flow separation along a low pressure turbine blade under unsteady flow conditions. ASME 2003-GT-38917, presented at International Gas Turbine and Aero-Engine Congress and Exposition, Atlanta, Georgia, June 16–19, 2003, also published in ASME Transactions. *Journal of Fluid Engineering, 127*, 503–513.
199. Schobeiri, M. T., & Öztürk, B. (2004). Experimental study of the effect of the periodic unsteady wake flow on boundary layer development, separation, and re-attachment along the surface of a low pressure turbine blade. ASME 2004-GT-53929, presented at International Gas Turbine and Aero-Engine Congress and Exposition, Vienna, Austria, June 14–17. (2004). also published in the ASME Transactions. *Journal of Turbomachinery, 126*(4), 663–676.
200. Langtry, R. B., & Menter, F. R. (2005). *Transition modeling for general CFD applications in aeronautics*. AIAA Paper, 2005-522.
201. Emmons, H. W. (1951). The laminar-turbulent transition in boundary layer-part I. *Aerospace Science and Technology, 18*, 490–498.
202. Dhawan, S., & Narasimha, R. (1958). Some properties of boundary layer flow during the transition from laminar to turbulent motion. *Journal of Fluid Mechanics, 3*, 418–436.
203. Abu-Ghannam, B. J., & Shaw, R. (1980). Natural transition of boundary layers–the effects of turbulence, pressure gradient and flow history. *Journal of Mechanical Engineering Science, 22*, 213–228.
204. Gostelow, J. P., & Blunden, A. R. (1989). Investigations of boundary layer transition in an adverse pressure gradient. *ASME Journal of Turbomachinery, 111*, 366–375.
205. Dullenkopf, K., & Mayle, R. E. (1994). ASME Paper No. 94-GT-174.
206. Gostelow, J. P., Melwani, N., & Walker, G. J. (1995). *Effects of streamwise pressure gradient on turbulent spot development*. ASME Paper No. 95-GT-303.
207. Walker, G. J. (1989). Modeling of transitional flow in laminar separation bubbles. In *9th International Symposium on Air Breathing Engines*, pp. 539–548.
208. Hodson, H. P. (1990). Modeling unsteady transition and its effects on profile loss. *Journal of Turbomachinery, 112*, 691–701.
209. Paxson, D. E., & Mayle, R. E. (1991). Laminar boundary layer interaction with an unsteady passing wake. *Journal of Turbomachinery, 113*, 419–427.
210. Orth, U. (1992). *Unsteady boundary-layer transition in flow periodically disturbed by wakes*. ASME Paper No. 92-GT-283.
211. Hedley, B. T., & Keffer, F. J. (1974). Turbulent/non-turbulent decisions in an intermittent flow. *Journal of Fluid Mechanics, 64*, 625–644.

212. Kovasznay, L. S. G., Kibens, V., & Blackwelder, R. F. (1970). *Journal of Fluid Mechanics, 41*, 283.
213. Antonia, R. A., & Bradshaw, P. (1971). *Imp. College Aero. Rep.*, No. 71-04.
214. Bradshaw, P., & Murlis, J. (1973). *Imp. College Aero. Tech. Note.*, No. 73-108.
215. Mayle, R. E. (1991). The role of laminar-turbulent transition in gas turbine engines. *Journal of Turbomachinery, 113*, 509–537.
216. Launder, B. E., & Spalding, D. B. (1972). *Mathematical models of turbulence*. Academic Press.
217. Crawford, M. E., & Kays, W. M. (1976). *STAN5 (TEXSTAN version)—A program for numerical computation of two dimensional internal and external boundary layer flow*. NASA CR-2742.
218. Schmidt, R. C., & Patankar, S. V. (1991). Simulating boundary layer transition with low-Reynolds-number $k - \epsilon$ turbulence models: part I-An evaluation of prediction characteristics; Part II–An approach to improving the predictions. *Journal of Turbomachinery, 113*, 10–26.
219. Schobeiri, M. T., Pappu, K., & Wright, L. (1995). *Experimental study of the unsteady boundary layer behavior on a turbine cascade*. ASME Paper No. 95-GT-435.
220. Schobeiri, M. T., McFarland, E., & Yeh, F. (1991). *Aerodynamic and heat transfer investigations on a high reynolds number turbine cascade*. NASA TM-103260.
221. John, J., & Schobeiri, M. T. (1993). A simple and accurate method of calibrating X-Probes. *ASME Journal of Fluids Engineering, 115*, 148–152.
222. Schobeiri, M. T., Read, K., & Lewalle, J. (1995). *Effect of unsteady wake passing frequency on boundary layer transition: Experimental investigation and wavelet analysis*. ASME Paper No. 95-GT-437.
223. Schobeiri, M. T., John, J., & Pappu, K. (1996). Development of two dimensional wakes within channel: theoretical framework and experimental investigations. *ASME Journal of Turbomachinery, 118*, 506–518.
224. Wright, L., & Schobeiri, M. T. (1999). The effect of periodic unsteady flow on boundary layer and heat transfer on a surface. *ASME Trans. J. Heat Transfer, 120*, 22–33.
225. Patankar, S. V., & Spalding, D. B. (1970). *Heat and mass transfer in boundary layers* (2nd ed.). International Textbook Company Ltd.
226. Kays, W. M., & Krawford, M. E. (1966). *Convective heat and mass transfers*. McGraw-Hill Company.
227. Hippensteele, S. A., Russell, L. M., & Stepka, S. (1981). Evaluation of A method for heat transfer measurements and thermal visualization using a composite of a heater element and liquid crystals. *ASME Journal of Heat Transfer, 105*, 184–189.
228. Han, J.-C., Zhang, L., & Ou, S. (1993). Influence of unsteady wake on heat transfer coefficient from a gas turbine blade. *ASME Journal of Heat Transfer, 115*, 904–911.
229. Spurk, J. (1997). *Fluid mechanics*. Springer.
230. Prandtl, L., Oswatisch, K., & Wiegarhd, K. (1984). *Führer durch die Strömung-lehre* (8th ed.). Vieweg Verlag.
231. Shapiro, A. H. (1954). *The dynamics and thermodynamics of compressible fluid flow* (Vol. I). Ronald Press Company.
232. Schobeiri, M. T. (1998). A new shock loss model for transonic and supersonic axial compressors with curved blades. *AIAA, Journal of Propulsion and Power, 14*(4), 470–478.
233. Moeckel, J. D. (1942). *Approximate method for predicting form and location of E-detached shock waves*. NACA TN 1921.
234. Liniger W., & Willoughby R. (1970). Efficient integration methods for stiff systems of ordinary differential equations. *SIAM. Numerical Analysis, 7*(1).
235. Bruun, H. H. (1995). *Hot-wire anemometry*. Oxford University Press.
236. Betchov, R. (1948). L'inflluence de la conduction thermique sue les anemometres a fil chaud. *Proceedings of the Koninklijke Nederlandse Akademie van Wetenschappen, 51*, 721–730.
237. King, C. F. (1914). On the convection of heat from small cylinders in a stream of fluid: Determination of the convection constants of small platinum wires with application to hot-wire anemometry. *Philosophical Transactions of the Royal Society, A214*, 374–432.

238. Kramer, H. (1946). Heat transfer from spheres to flowing media. *Physica, 12*, 61–80.
239. Hinze, J. O. (1959). *Turbulence*. McGraw-Hill.
240. Champagne, F. H., Sleicher, C. A., & Wehrmann, O. H. (1967). Turbulence Measurements with Inclined Hot-wires, Part 1: Heat transfer experiments with inclined hot-wire. *Journal of Fluid Mechanics, 28*, 153–175.
241. Webster, C. A. G. (1962). A note on the sensitivity to yaw of a hot-wire anemometer. *Journal of Fluid Mechanics, 13*, 307–312.
242. Jorgensen, F. E. (1971). Directional sensitivity of wire and fibre film probes. *DISA Information, 11*, 31–37.
243. Bradshaw, P. (1971). *An introduction to turbulence and its measurement*. Pergamon Press.
244. Bruun, H. H., Nabhani, N., Al-Kayiem, H. H., Fardad, A. A., Khan, M. A., & Hogarth, E. (1990). Calibration and analysis of X hot-wire probe signals. *Measurements of Science Technology, 1*, 782–785.
245. Lekakis, I. C., Adrian, R. J., & Jones, B. G. (1989). Measurement of velocity vectors with orthogonal and non-orthogonal triple-sensor probes. *Experiments in Fluids, 7*, 228–240.
246. Schröder, T. (1985). *Entwicklung des instationären Nachlauf's hinter quer zur Strömungsrichtung bewegten Zylindern und dessen Einfluß auf das Umschlagverhalten von ebenen Grenzschichten stromabwärts angeordneter Versuchskörper*. Dissertation D-17, Technische Hochschule Darmstadt.
247. Willmarth, W. W., & Bogar, T. J. (1977). Survey and new measurements of turbulent structure near the wall. *Physics of Fluids, 20*, S9–S91.
248. Johnson, F. D., & Eckelmann, H. (1984). A variable angle method of calibration for X-probes applied to wall-bounded turbulent shear flow. *Experiments in Fluids, 2*, 121–1130.
249. Leuptow, R. M., Breuer, K. S., & Haritonidis, J. H. (1988). Computer-aided calibration of X-probes using a look-up table. *Experiments in Fluids, 6*, 115–118.
250. Browne, L. W. B., Antonia, R. A., & Chua, L. P. (1989). Calibration of X-probes for turbulent flow measurements. *Experiments in Fluids, 7*, 201–208.
251. John, J., & Schobeiri, T. (1997). A simple and accurate method of calibrating X-probes. *Transaction of the ASME, Journal of Fluid Engineering, 115*, 148–152.
252. Yavuzkurt, S. (1984). A guide to uncertainty analysis of hot-wire data. *Transaction of the ASME, Journal of Fluids Engineering, 106*, 181186.
253. Moffat, R. J. (1982). Contributions to the theory of single-sample uncertainty analysis. Transaction of the ASME. *Journal of Fluids Engineering, 106*, 181–186.
254. Kupferschmied, P., Koppel, P., Gizzi, W., Roduner, C., & Gyarmath, G. (2000). Time-resolved flow measurements with fast-response aerodynamic probes in turbomachines. *Measurement Science and Technology, 11*(2000), 1036–1054.
255. Treaster, A. L., & Yocum, A. M. (1979). The calibration and application of five-hole probes. *ISA Transactions, 18*(3), 23–34.
256. Ostowari, C., & Wentz, W. H. (1983). Modified calibration techniques of a five-hole probe for high flow angles. *Experiments in Fluids, 1*(3), 166–168.
257. Reichert, B. A., & Wendt, B. J. (1994). *A new algorithm for five-hole probe calibration, data reduction, and uncertainty analysis*. NASA, Lewis Research Center, Internal Fluid Mechanics Division.
258. Wendt, B., & Reichert, A. B. (1993). A new algorithm for five-hole probe calibration and data reduction and its application to a rake-type probe. *Journal of Fluid Measurement and Instrumentation, 161*, 29–35.
259. Town, J., Camci, C., & Camci, C. (2011). Sub-miniature five-hole probe calibration using a time efficient pitch and yaw mechanism and accuracy improvements. In *ASME paper, GT2011-46391, Proceedings of ASME Turbo Expo Turbine Technical Conference, IGTI 2011*, June 6–10, 2011, Vancouver, Canada.
260. Pappu, K. (1997). *Theoretical and experimental study on periodic unsteady wake development, unsteady boundary layer development, and trailing edge ejection mixing loss optimization in a linear turbine cascade*. Dissertation, Texas A&M University.

261. Rubner, K., & Bohn, D. (1972). Verfahren für die Auswertung der Messergebnisse von Fünflochsonden. *Zeitschrift Flugwissenschaften, 20.*
262. Simonich, J. C., & Moffat, R. J. (1984). Liquid crystal visualization of surface heat transfer on a concavely curved turbulent boundary layer. *ASME Journal of Engineering for Gas Turbines and Power, 106*, 619–627.
263. Hippensteele, S. A., Russell, L. M., & Stepka, F. S. (1983). Evaluation of a method for heat transfer measurements and thermal visualization using a composite of a heater element and liquid crystals. *ASME Journal of Heat Transfer, 105*, 184–189.
264. You, S. M., Simon, T., & Kim, J. (1989). Free-stream turbulence effects on convex-curved turbulent boundary layers. *ASME Journal of Heat Transfer, III*, 66–76.
265. Russell, L. M., Hippensteele, S. A., & Poinsatte, P. E. (1993). *Measurements and computational analysis of heat transfer and flow in a simulated turbine blade internal cooling passage*. NASA TM-106189.
266. https://www.fluke.com/en-us/products/thermal-cameras.
267. McLachlan, B., & Bell, J. (1995). Pressure-sensitive paint in aerodynamic testing. *Experimental Thermal and Fluid Science, 10*, 470–485.
268. Suryanarayanan, A., Mhetras, S. P., & Schobeiri, M. T. (2009). Film-cooling effectiveness on a rotating blade platform. *Journal of Turbomachinery, 131*(1), 011014.
269. kire khar A., Mhetras, S. P., Schobeiri, M. T. . (2009). Film-cooling effectiveness on a rotating blade platform. *Journal of Turbomachinery, 131*(1), 011014.
270. Aris, R. (1962). *Vector, tensors and the basic equations of fluid mechanics*. Prentice-Hall. Inc.
271. Brand, L. (1947). *Vector and tensor analysis*. Wiley.
272. Klingbeil, E. (1966). *Tensorrechnung für Ingenieure*. Bibliographisches Institut.
273. Lagally, M. (1944). *Vorlesung über Vektorrechnung, dritte Auflage. Akademische Verlagsgeselschaft*. Leipzig University
274. Vavra, M. H. (1960). *Aero-thermodynamics and flow in turbomachines*. Wiley.
275. Spurk, -Ing. H. (1997). *Fluid Mechanics Problems and Solutions (FMP&S).*

Index

M. T. Schobeiri, *Advanced Fluid Mechanics and Heat Transfer for Engineers and Scientists*, https://doi.org/10.1007/978-3-030-72925-7

MIX
Papier aus verantwortungsvollen Quellen
Paper from responsible sources
FSC® C105338

If you have any concerns about our products,
you can contact us on
ProductSafety@springernature.com

In case Publisher is established outside the EU,
the EU authorized representative is:
Springer Nature Customer Service Center GmbH
Europaplatz 3, 69115 Heidelberg, Germany

Printed by Libri Plureos GmbH
in Hamburg, Germany